CW01376514

Non-destructive evaluation (NDE) of polymer matrix composites

Related titles:

Advanced fibre-reinforced polymer composites in civil engineering
(ISBN 978-0-85709-418-6)

Nanotechnology in eco-efficient construction
(ISBN 978-0-85709-544-2)

Handbook of seismic risk analysis
(ISBN 978-0-85709-268-7)

Details of these books and a complete list of titles from Woodhead Publishing can be obtained by:

- visiting our web site at www.woodheadpublishing.com
- contacting Customer Services (e-mail: sales@woodheadpublishing.com; fax: +44 (0) 1223 832819; tel.: +44 (0) 1223 499140 ext. 130; address: Woodhead Publishing Limited, 80 High Street, Sawston, Cambridge CB22 3HJ, UK)
- contacting our US office (e-mail: usmarketing@woodheadpublishing.com; tel.: (215) 928 9112; address: Woodhead Publishing, 1518 Walnut Street, Suite 1100, Philadelphia, PA 19102-3406, USA)

If you would like e-versions of our content, please visit our online platform: www.woodheadpublishingonline.com. Please recommend it to your librarian so that everyone in your institution can benefit from the wealth of content on the site.

We are always happy to receive suggestions for new books from potential editors. To enquire about contributing to our Composites Science and Engineering series, please send your name, contact address and details of the topic/s you are interested in to gwen.jones@woodheadpublishing.com. We look forward to hearing from you.

The team responsible for publishing this book:
Commissioning Editor: Francis Dodds
Publications Coordinator: Adam Davies
Project Editor: Sarah Lynch
Editorial and Production Manager: Mary Campbell
Production Editor: Adam Hooper
Cover Designer: Terry Callanan

© Woodhead Publishing Limited, 2013

Woodhead Publishing Series in Composites Science and Engineering:
Number 43

Non-destructive evaluation (NDE) of polymer matrix composites

Techniques and applications

Edited by
Vistasp M. Karbhari

WP
WOODHEAD
PUBLISHING

Oxford   Cambridge   Philadelphia   New Delhi

© Woodhead Publishing Limited, 2013

Published by Woodhead Publishing Limited,
80 High Street, Sawston, Cambridge CB22 3HJ, UK
www.woodheadpublishing.com
www.woodheadpublishingonline.com

Woodhead Publishing, 1518 Walnut Street, Suite 1100, Philadelphia, PA 19102-3406, USA

Woodhead Publishing India Private Limited, 303 Vardaan House, 7/28 Ansari Road, Daryaganj, New Delhi – 110002, India
www.woodheadpublishingindia.com

First published 2013, Woodhead Publishing Limited
© Woodhead Publishing Limited, 2013; Chapter 13 © the Commonwealth of Australia, 2013. The publisher has made every effort to ensure that permission for copyright material has been obtained by authors wishing to use such material. The authors and the publisher will be glad to hear from any copyright holder it has not been possible to contact.
The authors have asserted their moral rights.

This book contains information obtained from authentic and highly regarded sources. Reprinted material is quoted with permission, and sources are indicated. Reasonable efforts have been made to publish reliable data and information, but the authors and the publishers cannot assume responsibility for the validity of all materials. Neither the authors nor the publishers, nor anyone else associated with this publication, shall be liable for any loss, damage or liability directly or indirectly caused or alleged to be caused by this book.

Neither this book nor any part may be reproduced or transmitted in any form or by any means, electronic or mechanical, including photocopying, microfilming and recording, or by any information storage or retrieval system, without permission in writing from Woodhead Publishing Limited.

The consent of Woodhead Publishing Limited does not extend to copying for general distribution, for promotion, for creating new works, or for resale. Specific permission must be obtained in writing from Woodhead Publishing Limited for such copying.

Trademark notice: Product or corporate names may be trademarks or registered trademarks, and are used only for identification and explanation, without intent to infringe.

British Library Cataloguing in Publication Data
A catalogue record for this book is available from the British Library.

Library of Congress Control Number: 2013936444

ISBN 978-0-85709-344-8 (print)
ISBN 978-0-85709-355-4 (online)
ISSN 2052-5281 Woodhead Publishing Series in Composites Science and Engineering (print)
ISSN 2052-529X Woodhead Publishing Series in Composites Science and Engineering (online)

The publisher's policy is to use permanent paper from mills that operate a sustainable forestry policy, and which has been manufactured from pulp which is processed using acid-free and elemental chlorine-free practices. Furthermore, the publisher ensures that the text paper and cover board used have met acceptable environmental accreditation standards.

Typeset by Toppan Best-Set Premedia Limited, Hong Kong
Printed by Lightning Source

Contents

	Contributor contact details	xv
	Woodhead Publishing Series in Composites Science and Engineering	xxi

Part I Non-destructive evaluation (NDE) and non-destructive testing (NDT) techniques — 1

1 Introduction: the future of non-destructive evaluation (NDE) and structural health monitoring (SHM) — 3
V. M. KARBHARI, University of Texas at Arlington, USA

1.1	Introduction	3
1.2	Non-destructive evaluation (NDE) and structural health monitoring (SHM)	4
1.3	Conclusion and future trends	10
1.4	References	11

2 Non-destructive evaluation (NDE) of composites: acoustic emission (AE) — 12
J. Q. HUANG, The Boeing Company, USA

2.1	Introduction	12
2.2	Fundamentals of acoustic emission (AE)	13
2.3	Acoustic emission (AE) testing	23
2.4	Comparisons	27
2.5	Future trends	28
2.6	Sources of further information and advice	30
2.7	References	32

3 Non-destructive evaluation (NDE) of composites: eddy current techniques — 33
H. HEUER, Dresden University of Technology and Fraunhofer Institute for Non-Destructive Testing, Germany and M. H. SCHULZE and N. MEYENDORF, Fraunhofer Institute for Non-Destructive Testing, Germany

3.1	Introduction	33
3.2	Eddy current testing: principles and technologies	35
3.3	High-frequency eddy current imaging of carbon fiber materials and carbon fiber reinforced polymer (CFRPs) composites	43
3.4	Analytical methods for data processing	48
3.5	Conclusion	54
3.6	References	54

4	**Non-destructive evaluation (NDE) of composites: introduction to shearography**	**56**
	D. Francis, Cranfield University, UK	
4.1	Introduction	56
4.2	The theoretical principles of shearography	57
4.3	The practical application of shearography	62
4.4	Shearography for non-destructive evaluation (NDE) of composite materials	70
4.5	Comparing shearography with other techniques	75
4.6	Future trends	76
4.7	Sources of further information and advice	79
4.8	References	80

5	**Non-destructive evaluation (NDE) of composites: digital shearography**	**84**
	Y. Y. Hung and L. X. Yang, Oakland University, USA and Y. H. Huang, Singapore-MIT Alliance for Research and Technology (SMART) Centre, Singapore	
5.1	Introduction	84
5.2	Principles of digital shearography	85
5.3	The practical application of digital shearography	98
5.4	Using digital shearography to test composites	101
5.5	Conclusion	109
5.6	Acknowledgment	110
5.7	References and further reading	110

6	**Non-destructive evaluation (NDE) of composites: dielectric techniques for testing partially or non-conducting composite materials**	**116**
	R. A. Pethrick, University of Strathclyde, UK	
6.1	Introduction	116
6.2	Low-frequency dielectric measurement of partially conductive and insulating composite materials	118

6.3	Low-frequency dielectric cure monitoring	120
6.4	Low-frequency dielectric measurement of water ingress into composite structures	124
6.5	High-frequency measurements of dielectric properties	127
6.6	Conclusion	132
6.7	Acknowledgements	133
6.8	References	134
7	**Non-destructive evaluation (NDE) of composites: using ultrasound to monitor the curing of composites** W. STARK, BAM Federal Institute for Materials Research and Testing, Germany and W. BOHMEYER, Sensor and Laser Technique, Germany	136
7.1	Introduction	136
7.2	Types of thermosets used in composites	137
7.3	Methods for monitoring composites	139
7.4	Monitoring the degree of curing and the mechanical properties of composites	143
7.5	Online process monitoring using ultrasound	145
7.6	Using ultrasonic online process monitoring in practice: monitoring curing	154
7.7	Using ultrasonic online process monitoring in practice: automotive engineering	160
7.8	References	178

Part II Non-destructive evaluation (NDE) techniques for adhesively bonded applications — 183

8	**Non-destructive evaluation (NDE) of composites: dielectric methods for testing adhesive bonds in composites** R. A. PETHRICK, University of Strathclyde, UK	185
8.1	Introduction	185
8.2	The use of dielectric testing in cure monitoring	188
8.3	The use of dielectric testing to check bond integrity	193
8.4	The use of dielectric testing to assess ageing of bonded joints	200
8.5	Conclusion	216
8.6	Acknowledgements	217
8.7	References	217

9	**Non-destructive evaluation (NDE) of aerospace composites: methods for testing adhesively bonded composites**	**220**
	B. EHRHART and B. VALESKE, Fraunhofer Institute for Non-Destructive Test Methods (IZFP), Germany and C. BOCKENHEIMER, Airbus Operations GmbH, Germany	
9.1	Introduction	220
9.2	Adhesive bonding in the aerospace industry	221
9.3	The role of non-destructive testing (NDT) in testing adhesive bonds	222
9.4	Non-destructive testing (NDT) methods	227
9.5	Challenges in non-destructive testing (NDT) of adhesive bonds	233
9.6	Conclusion	235
9.7	Sources of further information and advice	236
9.8	References	236
10	**Non-destructive evaluation (NDE) of composites: assessing debonding in sandwich panels using guided waves**	**238**
	S. MUSTAPHA, University of Sydney, Australia and National ICT Australia (NICTA), Australia and L. YE, University of Sydney, Australia	
10.1	Introduction	238
10.2	Processing of wave signals	240
10.3	Numerical simulation of wave propagation	246
10.4	Debonding detection and assessment in sandwich beams	252
10.5	Debonding detection in sandwich panels using time reversal	259
10.6	Conclusion and future trends	271
10.7	References	274
11	**Non-destructive evaluation (NDE) of composites: detecting delamination defects using mechanical impedance, ultrasonic and infrared thermographic techniques**	**279**
	B. S. WONG, Nanyang Technological University, Singapore	
11.1	Introduction	279
11.2	Using mechanical impedance: disbonding in aluminium honeycomb structures	281
11.3	Using ultrasonic 'C' scanning: carbon fibre-reinforced (CFR) composites	292
11.4	Using infrared thermography	295
11.5	Conclusion: comparing different techniques	301
11.6	References	305

Part III Non-destructive evaluation (NDE) techniques in aerospace applications   307

12 Non-destructive evaluation (NDE) of aerospace composites: application of infrared (IR) thermography   309
O. LEY and V. GODINEZ, Mistras Group Inc., USA

12.1	Introduction: thermography as a non-destructive evaluation (NDE) technique	309
12.2	Heat propagation in dynamic thermography	315
12.3	Thermography in aerospace composites	321
12.4	Conclusion	332
12.5	References and further reading	332

13 Non-destructive evaluation (NDE) of aerospace composites: flaw characterisation   335
N. RAJIC, Defence Science and Technology Organisation, Australia

13.1	Introduction	335
13.2	Fundamentals of heat diffusion	337
13.3	Non-destructive evaluation (NDE) of delaminations and planar inclusions	346
13.4	Non-destructive evaluation (NDE) of impact damage	354
13.5	Non-destructive evaluation (NDE) of porosity	356
13.6	Experimental demonstration	359
13.7	Future trends	363
13.8	References	364

14 Non-destructive evaluation (NDE) of aerospace composites: detecting impact damage   367
C. MEOLA and G. M. CARLOMAGNO, University of Naples Federico II, Italy

14.1	Introduction	367
14.2	Effectiveness of infrared thermography	369
14.3	On-line monitoring	372
14.4	Non-destructive evaluation (NDE) of different composite materials	384
14.5	Conclusion and future trends	392
14.6	Acknowledgements	393
14.7	References	393

15 Non-destructive evaluation (NDE) of aerospace composites: ultrasonic techniques   397
D. K. HSU, Iowa State University, USA

15.1	Introduction	397
15.2	Inspection of aerospace composites	398
15.3	Ultrasonic inspection methods for aerospace composites	402
15.4	Ultrasonic inspection of solid laminates	406
15.5	Ultrasonic inspection of sandwich structures	415
15.6	Ultrasonic non-destructive testing (NDT) instruments for aerospace composites	419
15.7	Conclusion	420
15.8	References	420

16 Non-destructive evaluation (NDE) of aerospace composites: acoustic microscopy — 423
B. R. TITTMANN, C. MIYASAKA and M. GUERS,
The Pennsylvania State University, USA, H. KASANO,
Takushoku University, Japan and H. MORITA,
Ishikawajima – Harima Heavy Industries Co. Ltd, Japan

16.1	Introduction	423
16.2	Case study: damage analysis using scanned image microscopy	424
16.3	Case study: damage analysis using acoustic microscopy	431
16.4	Future trends: using embedded ultrasonic sensors for structural health monitoring of aerospace materials	440
16.5	Conclusion	444
16.6	References	446

17 Non-destructive evaluation (NDE) of aerospace composites: structural health monitoring of aerospace structures using guided wave ultrasonics — 449
M. VEIDT and C. K. LIEW, The University of Queensland, Australia

17.1	Introduction	449
17.2	Structural health monitoring (SHM) transducer systems	451
17.3	Guided wave (GW) structural health monitoring (SHM) systems for composite structures	455
17.4	Conclusion	466
17.5	References	468

Part IV Non-destructive evaluation (NDE) techniques in civil and marine applications — 481

18 Non-destructive evaluation (NDE) of composites: techniques for civil structures — 483
U. B. HALABE, West Virginia University, USA

18.1	Introduction	483
18.2	Infrared thermography	484
18.3	Ground penetrating radar (GPR)	501
18.4	Digital tap testing	507
18.5	Issues and challenges in using non-destructive evaluation (NDE) techniques	508
18.6	Future trends	510
18.7	References	512

19	**Non-destructive evaluation (NDE) of composites: application of thermography for defect detection in rehabilitated structures**	**515**
	A. SHIRAZI, Simpson Gumpertz & Heger, USA and V. M. KARBHARI, University of Texas at Arlington, USA	
19.1	Introduction	515
19.2	Principles of infrared (IR) thermography	517
19.3	Using infrared (IR) thermography in practice: application to a bridge deck assembly	518
19.4	Data collection methodology	522
19.5	Assessing results	526
19.6	Conclusion	539
19.7	References	540

20	**Non-destructive evaluation (NDE) of composites: using shearography to detect bond defects**	**542**
	F. TAILLADE, M. QUIERTANT and K. BENZARTI, Institut Français des Sciences et Technologies des Transports, de l'Aménagement et des Réseaux (IFSTTAR), France, C. AUBAGNAC, CETE de Lyon, France and E. MOSER, Dantec Dynamics, Germany	
20.1	Introduction	542
20.2	Shearography	544
20.3	The role of shearography in detecting defects	546
20.4	Field inspection of a fiber-reinforced polymer (FRP)-strengthened bridge: a case study	552
20.5	Conclusion	554
20.6	References	554

21	**Non-destructive evaluation (NDE) of composites: use of acoustic emission (AE) techniques**	**557**
	W. CHOI, North Carolina A&T State University, USA and H-D. YUN, Chungnam National University, Korea	
21.1	Introduction	557

21.2	Testing acoustic techniques	559
21.3	Challenges in using acoustic emission	566
21.4	Conclusion	572
21.5	References	572

22 Non-destructive evaluation (NDE) of composites: microwave techniques — 574

M. Q. FENG, Columbia University, USA and G. ROQUETA and L. JOFRE, Universitat Politècnina de Catalunya (UPC), Spain

22.1	Introduction	574
22.2	Electromagnetic (EM) properties of materials	577
22.3	Sensing architectures	589
22.4	Microwave surface imaging of fiber-reinforced polymer reinforced concrete (FRP RC) structures	595
22.5	Microwave sub-surface imaging of fiber-reinforced polymer reinforced concrete (FRP RC) structures	602
22.6	Future trends	613
22.7	Sources of further information and advice	614
22.8	References and further reading	614

23 Non-destructive evaluation (NDE) of composites: using fiber optic sensors — 617

Y. DONG, New Mexico Institute of Mining and Technology, USA

23.1	Introduction	617
23.2	Fiber optic sensing technologies	619
23.3	Fiber optic sensors (FOSs) integrated with fiber-reinforced polymer (FRP) reinforcements	624
23.4	Fiber optic sensors (FOSs) monitoring fiber-reinforced polymer (FRP) concrete interfacial bond behavior	625
23.5	Field applications of fiber optic sensors (FOSs) to fiber-reinforced polymer (FRP) rehabilitated structures	627
23.6	Future trends	628
23.7	References	629

24 Non-destructive evaluation (NDE) of composites: infrared (IR) thermography of wind turbine blades — 634

N. P. AVDELIDIS, Université Laval, Canada and T-H. GAN, Brunel University, UK

24.1	Introduction	634
24.2	Wind turbines	635
24.3	Infrared thermography (IRT)	636

24.4	Signal processing techniques	639
24.5	Quality assurance and structural evaluation of glass fibre reinforced polymer (GFRP) wind turbine blades	644
24.6	Infrared thermography (IRT) standards	645
24.7	Conclusion	645
24.8	Acknowledgements	647
24.9	References	647
25	**Non-destructive evaluation (NDE) of composites for marine structures: detecting flaws using infrared thermography (IRT)**	**649**
	A. SURATKAR, A. Y. SAJJADI and K. MITRA, Florida Institute of Technology, USA	
25.1	Introduction	649
25.2	Infrared thermography (IRT)	650
25.3	Case study: non-destructive evaluation (NDE) of defects in a boat hull	651
25.4	Assessing the effectiveness of infrared thermography (IRT)	654
25.5	Conclusion	665
25.6	References	666
	Index	*669*

Contributor contact details

(* = main contact)

Editor and Chapter 1

Vistasp M. Karbhari
University of Texas at Arlington
321 Davis Hall, Box 19125
Arlington, TX 76019-0125
USA

E-mail: vkarbhari@uta.edu

Chapter 2

Jerry Q. Huang
The Boeing Company
2201 Seal Beach Boulevard
Seal Beachm, CA 90740
USA

E-mail: jerry.q.huang@boeing.com

Chapter 3

Professor Dr Henning Heuer*
Dresden University of Technology
 and Fraunhofer Institute for
 Non-Destructive Testing (IZFP)
Maria-Reiche Str. 2
01109 Dresden
Germany

E-mail: henning.heuer@izfp-d.
 fraunhofer.de

Martin H. Schulze and Professor
 Norbert Meyendorf
Fraunhofer Institute for
 Non-Destructive Testing (IZFP)
Dresden
Germany

Chapter 4

Dr D. Francis
Department of Engineering
 Photonics
School of Engineering
Cranfield University
Bedfordshire MK43 0AL
UK

E-mail: daniel.francis@cranfield.
 ac.uk

Chapter 5

Professor Emeritus Y. Y. Hung*
and Professor L. X. Yang
Department of Mechanical
  Engineering
Oakland University
Rochester
Michigan
USA

E-mail: hung@oakland.edu; yang2@oakland.edu

Y. H. Huang
MIT 3D Optical Systems Group
Singapore-MIT Alliance for
  Research and Technology
  (SMART) Centre
Singapore 117543
Singapore

E-mail: howard.huang@smart.mit.edu

Chapters 6 and 8

Dr Richard A. Pethrick
WestCHEM
Department of Pure and Applied
  Chemistry
University of Strathclyde
Thomas Graham Building
295 Cathedral Street
Glasgow G1 1Xl
UK

E-mail: r.a.pethrick@strath.ac.uk

Chapter 7

Dr Wolfgang Stark*
BAM Federal Institute for
  Materials Research and Testing
Unter den Eichen 87
D-12205 Berlin
Germany

E-mail: wolfgang.stark@bam.de

Dr Werner Bohmeyer
Sensor and Laser Technique
Schulstr. 15
D-15366 Neuenhagen
Germany

E-mail: service@pyrosensor.de

Chapter 9

Bastien Ehrhart* and Professor
  Bernd Valeske
Fraunhofer Institute for Non-
  Destructive Test Methods (IZFP)
Campus E3.1
66123 Saarbruecken
Germany

E-mail: bastien.ehrhart@izfp.fraunhofer.de; bernd.valeske@izfp.fraunhofer.de

Dr Clemens Bockenheimer
Airbus Operations GmbH
Airbus Allee 1
28199 Bremen
Germany

E-mail: clemens.bockenheimer@airbus.com

Chapter 10

Samir Mustapha* and Lin Ye
Laboratory of Smart Materials and
 Structures (LSMS)
Centre for Advanced Materials
 Technology (CAMT)
School of Aerospace, Mechanical
 and Mechatronic Engineering
Building J07, Maze Crescent
Darlington Campus
University of Sydney
NSW 2006
Australia

E-mail: samir.mustapha@sydney.
 edu.au; lin.ye@sydney.edu.au

Chapter 11

Dr B. Stephen Wong
School of Mechanical and
 Aerospace Engineering
Nanyang Technological University
Singapore 639798
Singapore

E-mail: mbwong@ntu.edu.sg

Chapter 12

Obdulia Ley* and Velery Godinez
Mistras Group Inc.
195 Clarksville Rd
Princeton Junction,
NJ 08550
USA

E-mail: obdulia.ley@mistrasgroup.
 com

Chapter 13

Dr N. Rajic
Defence Science and Technology
 Organisation
506 Lorimer Street
Fishermans Bend
Victoria 3207
Australia

E-mail: nik.rajic@dsto.defence.gov.au

Chapter 14

Dr Carosena Meola* and Professor
 Giovanni Maria Carlomagno
Department of Aerospace
 Engineering
University of Naples *Federico II*
Via Claudio 21
80125 Naples
Italy

E-mail: carmeola@unina.it;
 carmagno@unina.it

Chapter 15

Professor D. K. Hsu (retired)
Center for Nondestructive
 Evaluation
Iowa State University
1915 Scholl Road
Ames, IA 50011
USA

E-mail: dkhsu@iastate.edu

Chapter 16

B. R. Tittmann*, C. Miyasaka and
 M. Guers
The Pennsylvania State University
University Park, PA 16802
USA

E-mail: brt4@psu.edu

H. Kasano
Takushoku University
Tokyo 193-0985
Japan

H. Morita
Ishikawajima-Harima Heavy
 Industries Co. Ltd
Yokohama
Japan

Chapter 17

Associate Professor Martin Veidt*
 and Dr Chin Kian Liew
School of Mechanical and Mining
 Engineering
The University of Queensland
Brisbane
Queensland 4072
Australia

E-mail: m.veidt@uq.edu.au

Chapter 18

Professor Udaya B. Halabe
Department of Civil and
 Environmental Engineering
Constructed Facilities Center
Benjamin M. Statler College of
 Engineering and Mineral
 Resources
West Virginia University
395 Evansdale Drive
Morgantown, WV 26506-6103
USA

E-mail: udaya.halabe@mail.wvu.
 edu

Chapter 19

Ali Shirazi
Simpson Gumpertz & Heger
4000 MacArthur Boulevard, Suite
 710
Newport Beach, CA 92660
USA

Vistasp M. Karbhari*
University of Texas at Arlington
321 Davis Hall, Box 19125
Arlington, TX 76019-0125
USA

E-mail: vkarbhari@uta.edu

Chapter 20

Frédéric Taillade*, Marc Quiertant
 and Karim Benzarti
Université Paris-Est
Institut Français des Sciences et
 Technologies des Transports, de
 l'Aménagement et des
 Réseaux
Boulevard Newton
Champs sur Marne
F-77447 Marne la Vallée Cedex 2
France

E-mail: frederic.taillade@ifsttar.fr

Christophe Aubagnac
Departement Laboratoire d'Autun
CETE de Lyon
BP 141
F-71404 Autun cedex
France

Eberhard Moser
Dantec Dynamics
D-89077 Ulm
Germany

Chapter 21

Dr Wonchang Choi*
North Carolina A&T State
  University
North Carolina
USA

E-mail: wchoi@ncat.edu

Dr Hyun-Do Yun
Chungnam National University
Korea

Chapter 22

Dr Maria Q. Feng*
Department of Civil Engineering
  and Engineering Mechanics
SMaRT Laboratory
Columbia University
New York, NY 10069
USA

E-mail: mfeng@columbia.edu

Gemma Roqueta
Department Teoria del Senyal i
  Comunicacions
Campus Nord-UPC
Building D3-114
Universitat Politècnina de
  Catalunya (UPC)
Jordi Girona 1-3
08034 Barcelona
Spain

E-mail: gemma.roqueta@gmail.com

Lluis Jofre
Signal Theory and Communications
  Department (TSC)
Universitat Politècnica de
  Catalunya (UPC)
Campus Nord-UPC
Building D3
Jordi Girona 1-3
08034 Barcelona
Spain

E-mail: jofre@tsc.upc.edu

Chapter 23

Yongtao Dong
Department of Civil and
  Environmental Engineering
New Mexico Institute of Mining
  and Technology
801 Leroy Place
Socorro, NM 87801
USA

E-mail: ydong@nmt.edu

Chapter 24

Professor Nico P. Avdelidis
Université Laval
Quebec
Canada

E-mail: nico.avdel@gmail.com

Professor Tat-Hean Gan*
Brunel University
Uxbridge
UK

E-mail: tat-hean.gan@brunel.ac.uk

Chapter 25

Ashish Suratkar*, Dr Amir Yousef
   Sajjadi and Professor Kunal
   Mitra
Department of Mechanical
   Engineering
Florida Institute of Technology
Melbourne, FL 32901
USA

E-mail: asuratkar2009@my.fit.edu;
   asajjadi@my.fit.edu; kmitra@fit.
   edu

Woodhead Publishing Series in Composites Science and Engineering

1 **Thermoplastic aromatic polymer composites**
   F. N. Cogswell
2 **Design and manufacture of composite structures**
   G. C. Eckold
3 **Handbook of polymer composites for engineers**
   Edited by L. C. Hollaway
4 **Optimisation of composite structures design**
   A. Miravete
5 **Short-fibre polymer composites**
   Edited by S. K. De and J. R. White
6 **Flow-induced alignment in composite materials**
   Edited by T. D. Papthanasiou and D. C. Guell
7 **Thermoset resins for composites**
   Compiled by Technolex
8 **Microstructural characterisation of fibre-reinforced composites**
   Edited by J. Summerscales
9 **Composite materials**
   F. L. Matthews and R. D. Rawlings
10 **3-D textile reinforcements in composite materials**
   Edited by A. Miravete
11 **Pultrusion for engineers**
   Edited by T. Starr
12 **Impact behaviour of fibre-reinforced composite materials and structures**
   Edited by S. R. Reid and G. Zhou
13 **Finite element modelling of composite materials and structures**
   F. L. Matthews, G. A. O. Davies, D. Hitchings and C. Soutis
14 **Mechanical testing of advanced fibre composites**
   Edited by G. M. Hodgkinson
15 **Integrated design and manufacture using fibre-reinforced polymeric composites**
   Edited by M. J. Owen and I. A. Jones

16 **Fatigue in composites**
 Edited by B. Harris
17 **Green composites**
 Edited by C. Baillie
18 **Multi-scale modelling of composite material systems**
 Edited by C. Soutis and P. W. R. Beaumont
19 **Lightweight ballistic composites**
 Edited by A. Bhatnagar
20 **Polymer nanocomposites**
 Y-W. Mai and Z-Z. Yu
21 **Properties and performance of natural-fibre composites**
 Edited by K. Pickering
22 **Ageing of composites**
 Edited by R. Martin
23 **Tribology of natural fiber polymer composites**
 N. Chand and M. Fahim
24 **Wood-polymer composites**
 Edited by K. O. Niska and M. Sain
25 **Delamination behaviour of composites**
 Edited by S. Sridharan
26 **Science and engineering of short fibre reinforced polymer composites**
 S-Y. Fu, B. Lauke and Y-M. Mai
27 **Failure analysis and fractography of polymer composites**
 E. S. Greenhalgh
28 **Management, recycling and reuse of waste composites**
 Edited by V. Goodship
29 **Materials, design and manufacturing for lightweight vehicles**
 Edited by P. K. Mallick
30 **Fatigue life prediction of composites and composite structures**
 Edited by A. P. Vassilopoulos
31 **Physical properties and applications of polymer nanocomposites**
 Edited by S. C. Tjong and Y-W. Mai
32 **Creep and fatigue in polymer matrix composites**
 Edited by R. M. Guedes
33 **Interface engineering of natural fibre composites for maximum performance**
 Edited by N. E. Zafeiropoulos
34 **Polymer-carbon nanotube composites**
 Edited by T. McNally and P. Pötschke
35 **Non-crimp fabric composites: Manufacturing, properties and applications**
 Edited by S. V. Lomov

© Woodhead Publishing Limited, 2013

36 **Composite reinforcements for optimum performance**
   Edited by P. Boisse
37 **Polymer matrix composites and technology**
   R. Wang, S. Zeng and Y. Zeng
38 **Composite joints and connections**
   Edited by P. Camanho and L. Tong
39 **Machining technology for composite materials**
   Edited by H. Hocheng
40 **Failure mechanisms in polymer matrix composites**
   Edited by P. Robinson, E. S. Greenhalgh and S. Pinho
41 **Advances in polymer nanocomposites: Types and applications**
   Edited by F. Gao
42 **Manufacturing techniques for polymer matrix composites (PMCs)**
   Edited by S. Advani and K-T. Hsiao
43 **Non-destructive evaluation (NDE) of polymer matrix composites: Techniques and applications**
   Edited by V. M. Karbhari
44 **Environmentally friendly polymer nanocomposites: Types, processing and properties**
   S. S. Ray
45 **Advances in ceramic matrix composites**
   Edited by J. Low
46 **Ceramic nanocomposites**
   Edited by R. Banerjee and I. Manna

Part I
Non-destructive evaluation (NDE) and non-destructive testing (NDT) techniques

1
Introduction: the future of non-destructive evaluation (NDE) and structural health monitoring (SHM)

V. M. KARBHARI, University of Texas at Arlington, USA

DOI: 10.1533/9780857093554.1.3

Abstract: As the use of fiber reinforced polymer matrix composites increases there is an increasing need for ensuring both a high level of quality control and that products and systems respond as required over extended periods of time. Thus, while there is a continuing need for methods of non-destructive evaluation (NDE), there is also a need to go further than just detection to the assessment of the effect of the defect on performance and life. It is important to keep in mind that the field is developing extremely rapidly and methods of testing and evaluation are increasingly being integrated with complex models and simulations along with technological advances that make it possible to combine specific methods with the means of assessment to enable rapid structural health monitoring (SHM) in the field.

Key words: non-destructive evaluation (NDE), structural health monitoring (SHM), defects, service life, capacity, durability, damage tolerance, design, performance.

1.1 Introduction

As the use of fiber reinforced polymer matrix composites increases in applications ranging from aerospace and military field to those in civil infrastructure, the need for ensuring a high level of quality control is also increased. Simultaneously, there is a need to ensure that products and systems respond as required over extended periods of time. Thus, while there is a continuing need for methods of non-destructive evaluation (NDE), especially those capable of rapidly detecting even smaller sized defects, there is also a need to go further than just detection to the assessment of the effect of the defect on performance and life. This extension of NDE which includes the steps of diagnosis and prognosis is often known as structural health monitoring (SHM). While the focus of this book is on methods of NDE, it is important to keep in mind that the field is developing extremely rapidly and methods of testing and evaluation are being increasingly being integrated with complex models and simulations along with

Some elements of this chapter are adapted from *Structural health monitoring of civil infrastructure systems* published by Woodhead Publishing Limited.

technological advances that make it possible to combine specific methods with the means of assessment to enable rapid SHM in the field. In many cases the SHM of a component, structure, or system can be conducted autonomously, or at least, without removing that item from service.

1.2 Non-destructive evaluation (NDE) and structural health monitoring (SHM)

Thompson (2009) attempted to provide differentiation between NDE and SHM by stating that NDE refers to 'those situations in which the relative positions of a measurement system and the component under inspection can be manipulated with respect to one another,' whereas SHM refers to 'those situations in which fixed sensors are placed on a structure, either in-situ or ex-situ.' While this does provide a means of differentiation at a simplistic level the use of the word 'sensor' has to be taken in the broadest context. One could also consider that NDE focuses on time-based, off-line, monitoring using passive systems that provide disparate and partial information about the condition of a material, component, or structure, facilitating time-based maintenance and an asset management approach that is time-consuming and could necessitate downtime. SHM, on the other hand, focuses on condition-based, on-line monitoring using combinations of active and passive systems that provide a comprehensive assessment of the 'health' of the material, component, or structure, facilitating proactive maintenance that results in 'intelligent' asset management.

Traditional NDE technologies are based on the detection of damage or defects (often caused during the manufacturing process or intrinsic in the raw constituent materials) which in the case of composites can range from air bubbles, voids, and blisters, fiber/fabric misalignment and/or wrinkling, fiber failure, matrix crazing and cracking, resin-rich or poor sites, layer separation and delaminations, as well as to aspects such as bond failures, crushing of cores and core shear. In addition aspects such as impact damage (both externally visible and internal), moisture entrapment and other aspects that could result in longer-term deterioration need to be pin-pointed and their extent determined. There are a range of NDE techniques used with composites and some of the most commonly used techniques can be classified as (a) visual inspection, (b) tap testing, (c) dye-penetrant testing, (d) stress-coating testing, (e) ultrasonic inspection, (f) acoustic emission techniques, (g) dielectric techniques, (h) X-ray radiography, (i) thermography, (j) eddy-current techniques, (k) microwave techniques, (l) Shearography, and (m) electron probe imaging. Each of these has its specific field of applicability although there is a level of overlap as related to defect size and accuracy of detection. Although a number of these techniques are 'passive', some are dependent on frequencies of excitation and a range is shown in Fig. 1.1.

1.1 Frequency spectrum for selected NDE techniques. UV = ultraviolet, ESR = election spin resonance, NMR = nuclear magnetic resonance. (After Summerscales, n.d.)

Irrespective of the NDE technique the key element is the detection of the smallest possible defect or damage sufficiently in advance of an event that causes failure. This essentially leads to the development of a 'critical sized flaw' representing a threshold beyond which 'safe use' cannot be guaranteed. Often this indicates a level beyond which defect growth (such as in the case of a crack) is rather rapid. This can be used in the development of a damage tolerance methodology as depicted in Fig. 1.2.

While concepts of damage tolerance have been used successfully and are increasingly effective, with the development of better and more refined and accurate methods of NDE the challenge is not just that of detecting damage. Rather, it is the use of sensing technologies in conjunction with materials and structural models (and simulations) to facilitate the inference of future behavior/response in terms of performance characteristics such as remaining capacity and life. The migration from NDE to SHM is thus a key aspect in the development of prognosis of the 'health' of the unit under consideration.

While interest in SHM has increased as related to both research and implementation, its basis and motivation can be traced to the very earliest endeavors of humankind to conceptualize, construct, worry about deterioration, and then attempt to repair (or otherwise prolong the life) of a structure. This is largely in response to the fact that over time all structures deteriorate and it is essential that the owner/operator has a good idea as to the extent of deterioration and its effect on remaining service-life and capacity, and has sufficient information to make a well-informed decision regarding optimality of repair. Thus it represents an attempt at deriving knowledge about the actual condition of a structure, or system, with the aim of not just knowing that its performance may have deteriorated, but rather to be able to assess remaining performance levels and life. This ability will,

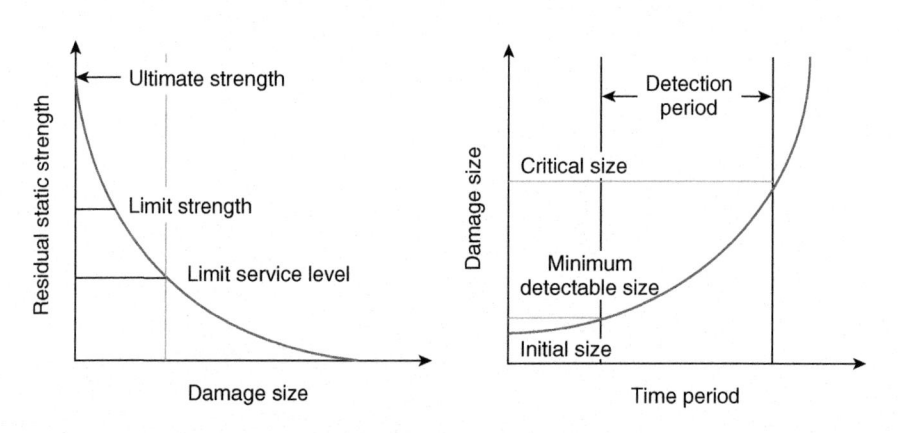

1.2 Concept of damage tolerance.

at some point in the near future, enable those associated with the operation of civil infrastructure systems to handle both the growing inventory of deteriorating and deficient systems and the need for the development of design methods that inherently prescribe risk to a system based on usage and hence differentiate between systems based on frequency of use and type of operating environment. Further, such a system would enable decisions related to resource allocation to be made on a real-time basis rather than years ahead, thereby allowing for maintenance plans to be based on the actual state of a structure and need rather than a time-based schedule. This would allow for real-time resource allocation and so enable a more optimal approach to maintenance and replacement of structural inventory.

In its various forms, over the years, SHM has been represented as the process of conventional inspection, inspection through a combination of data acquisition and damage assessment, and more recently as the embodiment of an approach enabling a combination of non-destructive testing and structural characterization to detect changes in structural response. It has also often been considered as a complementary technology to systems identification and non-destructive damage detection methods. A decade ago Housner *et al*. (1997) defined it as 'the use of *in-situ*, non-destructive sensing and analysis of structural characteristics, including the structural response, for detecting changes that may indicate damage or degradation.' While this definition provides a basis for the development of a good data management system in that it enables the collection of incremental indicators of change, it falls short of the goal of considering the effect of deterioration on performance and thence on the estimation of remaining service life. These management systems thus focus on processing collected data, but are unable to measure or evaluate the rate of structural deterioration, and more importantly from an owner's perspective, are unable to predict remaining service life and level of available functionality (e.g. the load levels that the structure can be subjected to within the pre-determined level of reliability and safety). In essence a true system should be capable of determining and evaluating the serviceability of the structure, the reliability of the structure, and the remaining functionality of the structure in terms of durability. This functionality has an analogy to the health management system used for humans wherein the patient undergoes a sequence of periodic physical examination, preventive intervention, surgery, and recovery (Aktan *et al*., 2000). Thus one would expect that a health monitoring system not only provides an indication of 'illness' but also enabled an assessment of its cause and extent, as well as the effect of that illness.

Intrinsically, owners and operators of equipment and systems ranging from aircraft to bridges and pipelines need the knowledge of the integrity and reliability of the network, structural system, and/or components in real time such that they can not only evaluate the state of the structure but also assess when preventive actions need to be taken. This would allow them to take timely decisions on whether functionality has been impaired to a point as a result of an event (or series of events) where the structure had to be removed from operation or shut down to prevent accidents, or whether it could remain in use with a pre-specified level of reliability. Thus, what is needed is an efficient method to collect data from a structure in-service and process the data to evaluate key performance measures such as serviceability, reliability and durability. In the context of civil structures, the definition by Housner *et al.* (1997) is modified and structural health monitoring is defined as 'the use of *in-situ*, non-destructive sensing and analysis of structural characteristics, including the structural response, for the purpose of estimating the severity of damage/deterioration and evaluating the consequences thereof on the structure in terms of response, capacity, and service-life' (Karbhari, 2005). Essentially, a SHM system then must have the ability to collect, validate, and make accessible operational data on the basis of which decisions related to service-life management can be made. While this task may not have been possible a decade ago, the recent progress in (a) sensor technology, (b) methods of damage identification and characterization, (c) computationally efficient methods of analysis, and (d) data communication, analysis and interrogation, has made it possible for one to consider SHM as a tool not just for operation and maintenance but also for the eventual development of a true reliability- and risk-based methodology for design. The advancements in forms of energy harvesting and storage, in addition to the integration of types of distributed sensor networks have also enabled rapid progress in this area, moving the concept from research to field implementation.

SHM intrinsically includes the four operations of acquisition, validation, analysis, prognosis, and management and thus a SHM system inherently consists of the five aspects of (1) sensors and sensing technology, (2) diagnostic signal generation, (3) signal transmission and processing, (4) event identification and interpretation, and (5) integration into an operative system for systems life management. It must be emphasized that although a number of NDE techniques are incorporated in the overall methodology of SHM of structures, there is a distinct difference between NDE and SHM. NDE is dependent on the measurement of specific characteristics and provides an assessment of state at a single point in time without necessarily enabling the assessment of the effect or extent of deterioration, whereas SHM requires the diagnosis and prognosis (interpretation) of events and sequences of events with respect to parameters such as capacity and remaining service-life. It is this aspect that also differentiates SHM from mere monitoring of use (usage monitoring). Usage monitoring is now fairly common and consists of the acquisition of data from a system related to response to external and internal excitations. SHM, in contrast, also includes the interrogation of this data to quantify the change in state of the system and thence the prognosis of aspects such as capacity and remaining service-life. The former is hence a necessary, but not sufficient, part of the latter. Also, the concept of monitoring prescribes that it be an ongoing, preferably autonomous, process rather than one that is used at preset intervals of time through human intervention. Thus SHM is essentially the basis for condition-based, rather than time-based, monitoring and the system should be integrated with the use of real-time data on aging and degradation into the assessment of structural integrity and reliability.

Unfortunately, typical systems do not use an integrated approach to the design, implementation, and operation of the SHM system, resulting in the benefits of the system often not being realized. Too often, a disproportionate emphasis is placed on the collection of data rather than on the management of this data and the use of decision-making tools that would support the ultimate aim of using the collected data to effect better management of the infrastructure system. Typically, systems collect data on a continuous or periodic basis and transmit the data to a common point. The data is then compared with results from a numerical model that simulates the original structure. The weakness is that most systems do not attempt to update the model to reflect aging and deterioration, or changes made through routine maintenance or even rehabilitation. Thus while it is easy to note that a change has taken place and even that a predetermined performance threshold has been reached, the approach does not enable prediction of future response, nor the identification of 'hot spots' that may need to be further monitored. While some systems incorporate a systems identification or non-destructive damage evaluation algorithm to rapidly process the data, these are the exception, not the rule. Even here, there is a gap between the management of this data, and its use towards the ultimate goals of estimating capacity and service-life (Sikorsky and Karbhari, 2003).

In the recent past SHM systems have often been used to showcase a specific sensor technology/NDE method or to provide a means for collecting data for ongoing research into the effect of external factors, rather than as a means of diagnosis and prognosis of the structural system under consideration. While the enhancement of systems safety (through use of the SHM as a mechanism of warning of failure) is desirable, SHM systems by themselves cannot ensure a higher level of safety, or even a better method of maintenance. SHM systems by themselves also cannot ensure a decrease in the level of maintenance, or even an increase in the periods between maintenance. If appropriately designed, however, they can reduce the amount of unnecessary inspections and ensure that deterioration/degradation is tracked such that the owner/operator has consistent and updated estimates of deterioration (quantity and general location), capacity, and remaining service life. One method of doing that is through the appropriate establishment of performance thresholds with each of these being compared with current response through use of data gathered from the sensor network and integrated into an appropriately designed structural response simulation. Periods of inspection are then set based on the rate with which performance degrades between these thresholds. Obviously the models have to be updated each time to ensure that the simulations mimic the field response. This also enables the prediction of potential 'hot spots' in the structure providing a means of identifying where, and when, maintenance is needed. The integration of analysis tools that account for variability due to variation in materials, and the time- and environment-based deterioration thereof, would also be essential if a true assessment is to be made.

Despite significant challenges, tremendous advances have been made in the actual development and implementation of SHM systems, capable of serving as true tools for health monitoring, i.e. not just being able to state that the 'patient' is sick, but rather of being able to pinpoint the location and reason, as well as the effect of the incapacity. These are essentially based on the ability to acquire data, transmit it, interrogate it, and then make decisions based on the cumulative sets of data stored in the data base. Thus, in effect, the SHM system is essentially a decision system that is fronted by sensors and backed by a knowledge base. The transition from maintenance that is largely unscheduled (i.e. as a result of an event that causes a visible change) to a condition-based mechanism to one that is predicted based on the use of a SHM system also has tremendous efficiencies in terms of cost. Current maintenance measures often result in significant downtime for the structural system which leads to losses such as through delays in traffic and in delivery of goods and services, which would also be dramatically reduced through migration to a SHM system-based maintenance philosophy.

1.3 Conclusion and future trends

In considering the future of NDE and SHM it is important that one not only considers its effect in terms of gains from the monitoring, diagnosis and prognosis of state, but that one also considers the future potential for use of the data and prognosis for the development of new methods of design that would not only be more structurally efficient, but which could intrinsically benefit from a more comprehensive understanding of reliability and risk associated with individual systems. Classically structures are designed by considering the relationship between the *capacities* of the components of the structure, and of the assembled system, to the *demand* anticipated for that structure. While the methodology has served fairly well for centuries it does have some major disadvantages.

It is conceivable that, in the near future, the use of an appropriately designed SHM system based on the use of appropriate NDE would enable further understanding of response through data analysis and interrogation, which would lead to better and more refined methods of design. In addition the use of real-time data enables immediate updating of risk and resource allocation, thereby linking design, construction and service-life maintenance together and enabling the development of a new paradigm for future development of design codes and specifications – one not based on use of inordinately high factors of safety due to uncertainty, but one based on a continuous assessment and monitoring of risk and health. Design would thus be predicated on optimizing structural efficiency with the integration of reliability based on local specifics rather than through global use of large factors of safety. This would re-envision design and maintenance.

1.4 References

Aktan, A.E., Catbas, F.N., Grimmelsman, K.A. and Tsikos, C.J. 'Issues in Infrastructure Health Monitoring for Management,' *ASCE Journal of Engineering Mechanics*, **126**[7], pp. 711–724, 2000.

Housner, G.W., Bergman, L.A., Caughey, T.K., Chassiakos, A.G., Claus, R.O., Masri, S.F., Skelton, R.E., Soong, T.T., Spencer, B.F. and Yao, J.T.P. 'Structural Control: Past, Present, and Future' *ASCE Journal of Engineering Mechanics*, **123**[9], pp. 897–971, 1997.

Karbhari, V.M. 'Health Monitoring, Damage Prognosis and Service-Life Prediction – Issues Related to Implementation,' in *Sensing Issues in Civil Structural Health Monitoring*, Ed. F. Ansari, Springer, pp. 301–310, 2005.

Sikorsky, C. and Karbhari, V.M. 'An Assessment of Structural Health Monitoring capabilities to Load Rate Bridges,' *Proceedings of the First International Conference on Structural Health Monitoring and Intelligent Infrastructure*, A.A. Balkema Publishers, pp. 977–985, 2003.

Summerscales, J. 'MATS 324 Notes, Composites Design and Manufacture,' University of Plymouth, n.d.

Thompson, R.B. 'NDE Simulations: Critical Tools in the Integration of NDE and SHM,' *Proceedings of SPIE*, Vol. 7294, 729402, 2009.

2
Non-destructive evaluation (NDE) of composites: acoustic emission (AE)

J. Q. HUANG, The Boeing Company, USA

DOI: 10.1533/9780857093554.1.12

Abstract: This chapter introduces the acoustic emission (AE) technique, which is a powerful and efficient non-destructive evaluation (NDE) method for polymer matrix composites (PMCs). AE fundamentals are discussed, with particular attention to AE techniques for PMCs, which are generally inhomogeneous and anisotropic. The typical AE testing process is described, and compared with other NDE methods. The chapter concludes with a short commentary on likely future trends and some suggested sources of further information.

Key words: acoustic emission (AE), crack initiation and growth, damage detection and location, incipient failure, impact damage, non-destructive evaluation (NDE), polymer matrix composites, real-time and continuous monitoring, structural damage, structural health monitoring.

2.1 Introduction

Acoustic emission (AE) has been studied and used over the last six decades, and the science behind acoustic emissions from materials under stress is therefore well established. The technology has been applied to materials research, non-destructive evaluation (NDE), and structural health monitoring. AE was first employed as an NDE method for polymer matrix composites (PMCs) in the early 1980s, and its use has since increased alongside the expanding applications of PMCs, which in the last decade have come to be used for primary structures such as airframes in commercial and military airplanes. This chapter briefly introduces the AE technique used in the NDE of PMCs.

The fundamentals of AE are described, including sources of AE in PMCs, various types of noise, wave propagation, and AE signal detection. The AE testing process for the NDE of PMCs is outlined, with details of test plan preparation, AE system set-up and calibration, conducting tests and data analysis. A comparison of the AE technique with other NDE methods for PMCs is provided, where advantages and limitations are discussed. Future trends and sources of further information are covered in the final sections.

Because numerous papers have already been written on AE for the NDE of homogeneous and isotropic materials such as metals, this chapter focuses on AE for the NDE of PMCs, which are typically inhomogeneous and anisotropic. More general information on AE and techniques can be found in Section 2.6.

2.2 Fundamentals of acoustic emission (AE)

2.2.1 AE

AE refers to a transient elastic wave generated by rapid energy release from a local source within a material under stress. The sources of acoustic emission are the mechanisms of fracture or deformation phenomena within materials, such as crack initiation and growth, micro-structural separation, and movement between material phases. The energy of acoustic emission is generally proportional to the strain energy released from the newly created crack surfaces. AE waves propagate according to the sound velocity of the material, and can be detected and located with a properly arranged AE sensor array.

For an intact PMC material, AEs can start at a stress level as low as 20% of the ultimate strength. If there is no structural damage, AE measurements, such as the number of AE hits and cumulative energy, usually accelerate steadily as load increasing before macro-damages in the composite material occur.

An important phenomenon of AE was discovered by Josef Kaiser in 1950, and is called the Kaiser effect. It describes an irreversible phenomenon, stating that materials emit only under unprecedented stress. This effect holds true for most homogenous materials such as metals. The Kaiser effect fails most noticeably in situations where time-dependent mechanisms control the deformation,[1] and in composite materials, in which frictional AE from damaged regions is one of the AE mechanisms. However, the principle of the Kaiser effect is very useful in the detection of damage and defects in PMCs with a designed loading pattern including the holding periods and reloading runs.

As an extension of the Kaiser effect, the Felicity effect was introduced to address the breakdown of the Kaiser effect, by stating that emission is often observed at loads lower than the previous maximum, especially when the material is in poor condition or close to failure. The Felicity ratio is a quantitative measure, which is defined as the ratio of the stress at which the Felicity effect occurred to the previous maximum stress applied. Like the Kaiser principle, the Felicity ratio can be used as a valuable tool in detecting damage in PMCs. Some successful examples of using the Felicity ratio as a diagnostic tool in the testing of fiberglass vessels and storage tanks can be found in references 1 and 2.

In principle, the acoustic emission detection technique is similar to that of earthquake detection. Waves are generated by a rapid release of energy from the source within the material or medium under stress, rather than from external excitation as in ultrasonic testing, so that the event can be detected and located as it occurs. For this reason, the terms 'stress wave emission' and 'micro-seismic activity' have been used in AE literature. The fundamentals of AE theory and techniques generally apply to PMCs, but special considerations concerning the natures of inhomogeneous materials and anisotropic properties have to be accounted for in data analysis and interpretation.

As a non-destructive testing method, AE examination has demonstrated capabilities in material research, structural health monitoring (SHM), detecting leaks, corrosion, and various incipient failures in in-service structures or equipment. The AE technique offers some unique and advantageous capabilities, such as:

- real-time and continuous monitoring of the state of a material or structure under stresses or environmental attack;
- detection of growing defects or damage during its service-life;
- unattended and remote monitoring;
- global monitoring of a large structure by a sparse sensor network;
- detection of damage that is inaccessible to the more traditional NDE methods; and
- detection of incipient failures long before the final failure, which can prevent catastrophic failures.

Because of these unique capabilities, AE examination has found more and more applications in materials research, SHM, and in-line inspection for quality control of PMCs.

2.2.2 Acoustic emission sources in PMCs

Typical AE sources are related to the common damage mechanisms found in PMCs, including matrix crazing and cracking, delamination, fiber breakage, and splitting along fibers. AE signals generated by debond at the interface in a bonded structure, and impact by a foreign object to the composite, are two major AE sources in real-world applications of polymer matrix composites. Debond is a failure at the interface of the two bonded parts. Impact damage is one of the most problematic types of damage to PMCs due to their weaker strength in the direction normal to the fiber plane, and the fact the threat of impact damage throughout the life of PMCs is ever-present. Fortunately, debond and impact events are easily detected by the AE technique.

It should be pointed out that the failure of a PMC is a process of progressive damage that involves micro and macro-structural damages. It usually starts with micro-cracks in the polymer matrix and may propagate in both the transverse direction and along the fiber. At higher stress levels, certain types of macro-damage can develop, such as delamination between plies due to transverse stress. As stress increases, fiber can be fractured in tension or shear depending on competing stresses governed by composite lay-up (ply orientations) and the state of local stresses. The presence of defects and stress concentration in a PMC structure with a non-flat geometry will affect the damage locations and failure process. The complicated failure process is reflected in AE detection of different types of signals, in terms of signal features. Though there may be one dominant type of AE signals from matrix crazing and cracking in the early stages of a PMC failure process, the types of AE signals become mixed because they come from mixed AE sources, and the number of detected AE signals increases exponentially once other damage mechanisms start to contribute towards the final stage of the failure process.

2.2.3 Unwanted AE sources: types of noise

Extraneous noise may pose challenges to AE testing because various types of noise can be recorded as part of AE data. Severe noise condition can be either a show stopper for a proposed AE test or a problem in the post-analysis in extracting the useful information from the noise-dominant data. For this reason, extraneous noise must be controlled, if not completely eliminated, to a level at which an AE test can be conducted without significant difficulties in data analysis and interpretation.

To effectively control extraneous noise, we need to understand the sources of noise. The noise affecting AE testing can generally be categorized into three types:

- mechanical noise such as hitting, fretting, and frictional rubbing that act directly on structure surfaces or to the body through a load joint;
- thermally induced noise due to temperature effects on the piezoelectric element and thermal stresses in sensors, cables, and connectors, produced by thermal shock, rapid temperature change, or overheating; and
- electrical magnetic interference (EMI) including cross-talking between adjacent cables and the 'antenna' effect due to poor shielding or damage in the cables or a bad connection.

In most cases, noise can be eliminated or controlled to an acceptable level. Some commonly used countermeasures for controlling extraneous noise include adjusting gain and threshold for data acquisition, selecting proper filters in preamplifiers, using guard sensors, parameter-based software filters, and quick fixes of the test conditions that cause the noise problem. In the real world, however, there are cases in which extraneous noise cannot be controlled by quick and simple means. In these cases, AE techniques can be used only if true information can extracted from the data with help from effective software filters and data analysis techniques.

There are often questions about background when AE testing is being considered for certain applications. It is worth pointing out that the 'background noise' often referred to is in the audible range and usually not a big concern, unless it falls into one of the three categories mentioned above. Audible noise acts on the structure under AE testing through a big mismatch of acoustic impedance at the interface between air and the structure surface. A good example of AE applications in an extremely noisy environment was a successful flight experiment that used AE for structural health monitoring of the composite liquid hydrogen (LH2) tank of a space vehicle. Successful detection of acoustic emission from the composite structure was demonstrated even with the extremely high level of noise from the rocket engine and LH2 thermal effects.[3]

2.2.4 AE wave basics

Acoustic emission wave propagation in PMCs obeys the same basic principles of ultrasonic waves traveling in homogeneous and isotropic materials, i.e.:

- *Reflection* – occurs at the interface of two dissimilar materials, at angles equal to the angles of incidence.
- *Refraction* – the bending of a wave as it passes from one medium to another. If sound energy is partially transmitted beyond the interface, the transmitted wave may be refracted or bent, depending on the relative velocities of the respective media, and partially converted to a mode of propagation different from that of the incident wave.
- *Diffraction* – the bending of waves around corners (point reflector, edge reflector and apertures).
- *Dispersion* – velocity changes with frequency.
- *Scattering* – the reflection of a wave into many directions.
- *Attenuation* – sound wave decrease in intensity as they travel away from their source, due to geometrical spreading, scattering and absorption.
- *Impedance* – determines the amount of reflection and transmission of a wave when the wave is passing from one medium to another.

The boundary effects on wave propagation at and across the interface of tow mediums are illustrated in Fig. 2.1.

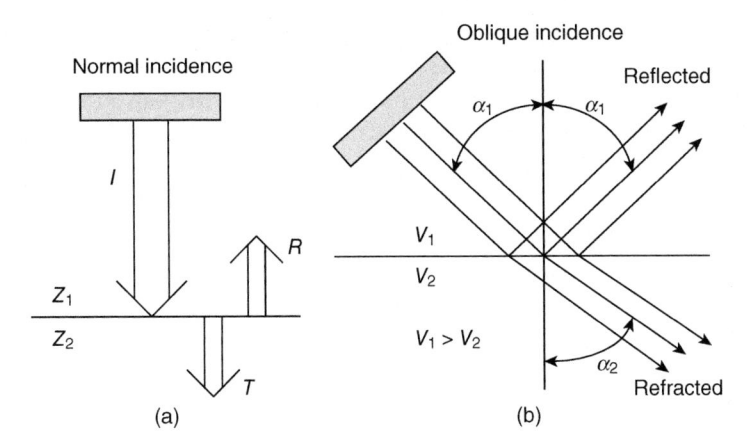

2.1 Illustration of wave propagation at and across the interface of two mediums: (a) in normal incidence and (b) in oblique incidence.

The wave reflection and refraction obey Snell's law, *which is given* as:

$$\frac{\sin\alpha_1}{\sin\alpha_2} = \frac{V_1}{V_2} \qquad [2.1]$$

where V_1 and V_2 are the wave velocity in medium 1 and medium 2, and impedance, Z, is given by:

$$Z = \rho * V \qquad [2.2]$$

where ρ is the density and V is the sound velocity of the medium. Coefficient of energy reflection for normal incidence, R:

$$R = \frac{E_R}{E_I} = \frac{(Z_2 - Z_1)^2}{(Z_2 + Z_1)^2} \qquad [2.3]$$

Coefficient of energy transmission for normal incidence, T:

$$T = E_T/E_I = 1 - R, \quad T + R = 1, \quad E_T + E_R = E_I \qquad [2.4]$$

Although these wave propagation characteristics have not been addressed much in AE literature and practice guidance, understanding them will be very helpful in assessing the suitability of AE techniques for certain NDE applications, especially for bonded structures, design test specimens for noise reduction, and selecting coupling media for better wave transmission from structure to sensor. When facing questions on the applicability of AE technique in a high intensity noise environment, explaining the big mismatch in impedance ($Z = V\rho$) from air to the composite could be crucial to the decision-making process.

Some characteristics of wave propagation in PMCs are different from those in homogeneous and isotropic materials, such as velocity and attenuation. PMCs are usually inhomogeneous and anisotropic. In an isotropic material, the sound velocity is determined by the elastic moduli and material density, while the sound velocity in PMCs is a function of direction, material properties and geometry. Since most PMCs are geometrically plate-like laminates, there are numerous detailed studies on Lamb waves (elastic waves propagate in solid plates) in composite laminates and some of these focus on AE applications.[4–7]

Understanding how AE waves propagate in plate-like composites can be very useful in AE data analysis and interpretation. The plate-like composites are typically made of fiber reinforced lay-ups consisting of laminas or woven fiber fabrics, and have a plate shape, but are not necessary flat. An important feature of the plate-like composites to AE applications is that they can be treated as a two-dimensional (2D) plate for the purposes of wave propagation analysis, and the 2D location algorithm can be used to determine the location of the AE source origin. The major results from plate wave studies are provided as follows.

For the symmetric and orthotropic laminated plate, the motions in x, y, z directions are given by:

$$A_{11}\frac{\partial^2 u}{\partial x^2} + A_{66}\frac{\partial^2 u}{\partial y^2} + (A_{12} + A_{66})\frac{\partial^2 v}{\partial x \partial y} = \rho\frac{\partial^2 u}{\partial t^2} \quad [2.5]$$

$$(A_{12} + A_{66})\frac{\partial^2 u}{\partial x \partial y} + A_{66}\frac{\partial^2 v}{\partial x^2} + A_{22}\frac{\partial^2 v}{\partial y^2} = \rho\frac{\partial^2 v}{\partial t^2} \quad [2.6]$$

$$D_{11}\frac{\partial^4 w}{\partial x^4} + 2(D_{12} + 2D_{66})\frac{\partial^4 w}{\partial x^2 \partial y^2} + D_{22}\frac{\partial^4 w}{\partial y^4} = -\rho\frac{\partial^2 w}{\partial t^2} \quad [2.7]$$

where A_{ij} and D_{ij} are in-plane stiffnesses and bending stiffnesses, respectively, defined by

$$(A_{ij}, D_{ij}) = \int_{-\frac{h}{2}}^{\frac{h}{2}} Q_{ij}^{(k)}(1, z^2)\,dz \quad [2.8]$$

and ρ' is the integral density of the plate, defined by

$$\rho' = \int_{-\frac{h}{2}}^{\frac{h}{2}} \rho^{(k)}\,dz \quad [2.9]$$

$Q_{ij}^{(k)}$ and $\rho^{(k)}$ are the reduced stiffnesses and the density of the kth layer of the laminate, respectively.

Considering plane waves propagation in a plate, let the displacements be:

$$u = A_1 e^{i(\mathbf{k}\cdot\mathbf{x}-\omega t)}, \quad v = A_2 e^{i(\mathbf{k}\cdot\mathbf{x}-\omega t)}, \quad w = A_3 e^{i(\mathbf{k}\cdot\mathbf{x}-\omega t)} \quad [2.10]$$

where A_1, A_2 and A_3 are amplitudes of the displacements, μ, v and ω, respectively, and $\mathbf{k}$ and $\mathbf{x}$ are the wave number and the position vectors in the $x - y$ plane.

When transverse wavelengths of plate waves are quite large compared with the plate thickness so that the plate reduces to the thin-plate, solving the equations yields the results of interest, which will be helpful for analyzing wave signals of acoustic emission from plate-like PMCs.

By solving the above equations simultaneously, it can be shown that a plate wave consists of two wave modes; extensional and flexural, which travel at different velocities and frequencies.

It can be shown, to the lowest order, $|u_z / u_x| = 0(kH)$ for an extension wave so that the predominant particle motion in the plate is in the direction of wave propagation, and $|u_x / u_z| = 0(kH)$ for a flexural wave so that the predominant particle motion in the plate is perpendicular to the direction of wave propagation.

For an extensional wave propagating in the x-direction, substituting [2.10] into [2.5] with $\mathbf{k}\mathbf{x} = k_x x$ and $\omega = kc_e$, the resulting dispersion relation is given by

$$c_e = \sqrt{A_{11}/\rho h} \quad [2.11]$$

Similarly, from [2.6], the velocity for an extensional wave propagating in the y-direction is given by

$$c_e = \sqrt{A_{22}/\rho h} \quad [2.12]$$

For a flexural wave propagating in the x-direction, substituting the third equation of [2.10] into [2.7] with $\omega = kc_f$, we obtain

$$c_f = (D_{11}/\rho h)^{1/4} \omega^{1/2} \quad [2.13]$$

and for a flexural wave propagating in the y-direction,

$$c_f = (D_{22}/\rho h)^{1/4} \omega^{1/2} \quad [2.14]$$

where ρ is the plate or overall density, i.e. $\rho = \rho'/h$, and h is the plate thickness.

These results indicate, similar to homogeneous and isotropic plate, that the extensional wave is non-dispersive and its velocity is determined by the plate properties only, and the flexural wave is dispersive and its velocity

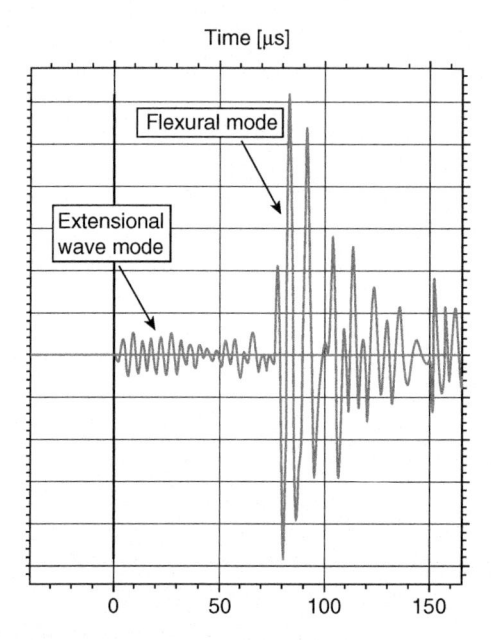

20   Non-destructive evaluation (NDE) of polymer matrix composites

2.2 The extensional and flexural wave modes can be clearly identified in this detected AE signal from a drop-weight impact on a composite panel.

varies with the square root of frequency. These results are valid only for plates satisfying the 'thin plate' conditions used to derive the results, i.e. the plate thickness ≪ wave length. To help visualize the extensional and flexural wave modes, Fig. 2.2 shows an AE signal detected after a drop-weight impact on a carbon fiber reinforced panel, which clearly shows two distinguishable wave modes contained in the AE signal. A special case of the symmetric and orthotropic laminated plates is the quasi-isotropic laminate, which has the same in-plane stiffness and bending stiffness in all directions, and thus the same velocities in all directions.

In general, sound waves travel fastest in the fiber direction and slowest in the transverse direction. Attenuation in PMCs is exhibited in a similar manner to sound velocity. Higher attenuation is observed in PMCs than in metals. When choosing AE sensors, preamps, determining sensor spacing and setting data acquisition parameters, these characteristics should be considered.

2.2.5   AE detection

AE techniques can be used to detect damage and damage growth in PMCs. The region of the damage can also be located by AE techniques such as triangulation and zonal methods.

AE sensing

AE detection and source location can be accomplished using a proper AE sensor array and location algorithms. AE sensors, a type of piezoelectric device that converts mechanical forces to electrical signals, must be attached directly to the structure with good acoustic coupling between the sensor and structural surface. The acoustic coupling can be achieved by applying a suitable gel-type material, like ultrasonic gel, or adhesive at the interface. For easy removal after testing, the sensors can be attached to the structure or specimen with gel at the interface and secured in place with an attachment device or simply with adhesive tapes or clamps. For permanent sensor installation, an adhesive such as a selected epoxy can be used to bond the sensors to the structure and provide the necessary acoustic coupling, strength, and durability.

AE testing is normally performed in the ultrasonic range. The chosen lower frequency bound is usually between 20 and 100 kHz. AE signals from PMCs decay rapidly at high frequency and hardly show any significant contents over 1000 kHz.

When selecting AE sensors for PMC applications, the following should be considered and compared for specific applications:

- sensitivity in the frequency range of interest;
- sensor specifications such as dimensions, weight, configurations (integral or non-integral cable and preamp, connector location, single-ended or differential), and operating frequency and temperature ranges.

Whether resonant or broadband (or wideband) sensors should be used depends on the needs of the application. Usually, a resonant sensor provides higher sensitivity while a broadband sensor provides a relatively flat sensitivity over a broad frequency range. It should be pointed out that resonant sensors still have a range of frequency, but vary in sensitivity over the range with the highest peak at the resonant frequency.

The preamplifier is the first amplifier following the sensor, and has the important function of converting the sensor impedance to one suitable for driving the long signal cables and additional electronic components or units. A preamplifier usually contains filters that reject undesirable signals. Depending on noise type, high-pass, low-pass, and bandpass filters can be used. High-pass filters eliminate low-frequency mechanical noise such as vibration in structures and background noise. Low-pass filters are used to limit EMI and radio frequency (RF) signals.

A bandpass filter provides a single transmission band extending from lower cut-off frequency to upper cutoff frequency, and is therefore used when investigating a specific frequency range.

AE measurements

A local material change giving rise to an acoustic emission is generally called an event. One AE event may be detected by multiple sensors and multiple times at one sensor if the reflected signal exceeds the threshold. A signal detected by the system is called a hit. The AE system records the hit counts and feature data of each hit. The events are determined by AE system software. There are two types of acoustic emissions as shown in Fig. 2.3:

- burst emission, which is the discrete signal related to an individual emission event; and
- continuous emission, which features a sustained signal level caused by rapidly occurring AE events.

An acoustic emission is detected as a transient wave signal. Each of the detected AE signals can be recorded as two types of data; parameter data and waveform data.

AE parameter data is obtained by measuring waveform features of an individual AE hit, or by measuring AE activity over a period of time. The measurable parameters from a waveform include amplitude, energy, count or ring-down count, rise time, and duration. As shown in Fig. 2.4, the

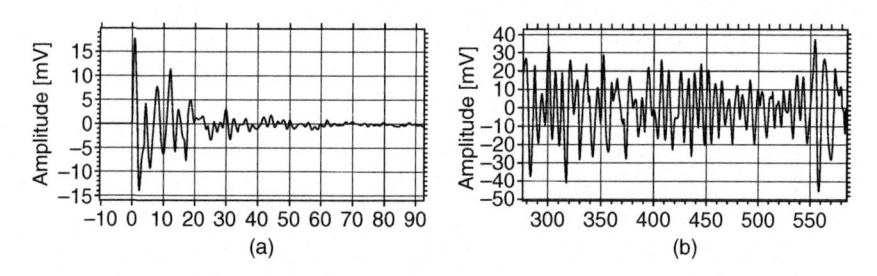

2.3 (a) Burst emission; (b) continuous emission.

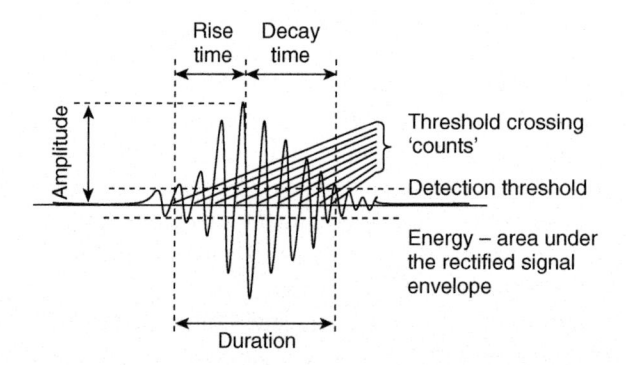

2.4 Illustration of an AE waveform and measureable parameters.

amplitude is the peak value of the AE signal; the count or ring-down count is the number of times the AE signal exceeds the detection threshold; the rise time is the time between AE signal start and the peak amplitude; the decay time is the time between the peak amplitude and the last threshold crossing; the duration is the time between the AE signal start and end. Some other parameter data can be obtained by measuring AE activity over a period time, such as RMS (root mean square), AE hits, and rates.

2.3 Acoustic emission (AE) testing

An AE test consists of three basic steps: planning, system set-up and calibration, and conducting the test. Each of these steps is briefly described in the following sections.

2.3.1 Test plan

An AE test plan must be prepared using detailed information and procedures according to the requirements for the test. The test plan must include, but need not be limited to, the following basic information:

- test requirements;
- number of sensors to be used, sensor locations and installation method;
- how and where to set up the AE system, preamps, and cables;
- data acquisition parameter settings;
- parametric data (load, strain, temperature, etc.) input and acquisition requirements;
- analysis and report approaches;
- safety requirements (which are especially important for composite loading tests).

2.3.2 System set-up and calibration

The AE system should be set up according to the test plan. Figure 2.5 is a schematic illustration of an AE system set-up for a specimen test. As shown, the specimen is under load. An AE sensor is attached to the specimen with either adhesive bonding or gel couplant at the interface between the sensor and specimen. The sensor is then connected to the preamp input end, and the output from preamp is connected to the AE data acquisition unit. The acquired data is digitized and then fed into the AE computer for data storage, analysis, and display.

Some parametric data, such as load, strain, and temperature, may be needed in AE data analysis and interpretation and can be recorded by the AE system through its parametric channels. Thus, the parametric and AE

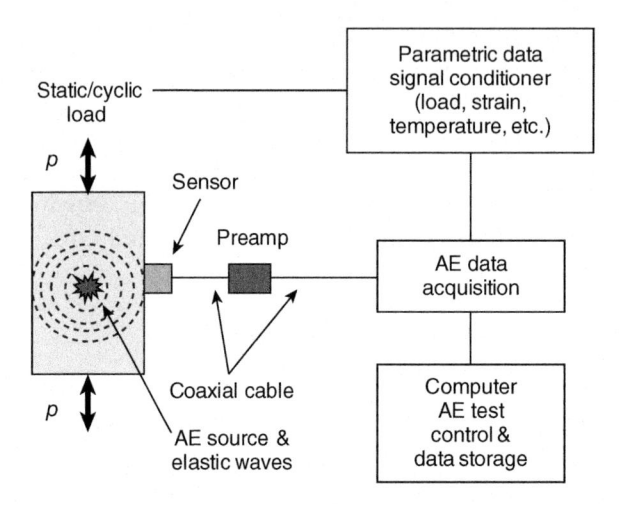

24   Non-destructive evaluation (NDE) of polymer matrix composites

2.5 Schematic illustration of AE system set-up for a specimen loading test.

data are synchronized in the AE system as they are recorded, which makes the AE and parametric data correlation much easier for real-time monitoring or in post-analysis.

What is not shown in Fig. 2.5 is the setting for data acquisition, which is an important part of AE testing. The AE data acquisition setting includes settings for the gain, detection threshold, sampling rate, times for defining an AE event, and parameters for source location, amongst others. These settings should be determined based on wave propagation characteristics for the current material and structure, and the level of noise. Improper settings can affect AE system performance. For example, the selection of values for the threshold and gain can affect AE detection, noise rejection, and the dynamic range (the difference, in decibels, between the overload level and the minimum signal level) of the AE system. Raising the threshold above the noise level is often an effective way to reject noise, but, if it reduces the dynamic range significantly, lowering the gain in preamp should also be considered.

After sensor installation or before an AE test, a calibration of sensors must be conducted.[8] This serves two purposes: (1) checking the system set-up including the system performance at the current settings and the quality of sensor installation, and (2) establishing a set of reference data generated by repeatable, simulated AE sources.

Calibration can be carried out using the pencil-lead break test, called the Hsu–Nielsen source,[9] or using the auto-pulsing function provided by the AE system, before and after the AE test. The Hsu–Nielsen source is created using a mechanical pencil to break a 2H 0.5 or 0.3 mm diameter pencil lead approximately 2.5 mm (0.1 inches) from the tip with a 30 degree angle between the lead and the specimen surface to generate a simulated AE burst signal. Alternatively, the calibration can be conducted using the AE system-equipped pulsing function in which one of the installed AE sensors acts as a pulser to send a pulse signal into the structure or specimen and other sensors as receivers. This process repeats among the installed AE sensors until every sensor has functioned as a pulser, which will generate a table or list of the calibration results of all installed sensors. The wave velocities can also be obtained from the calibration data. If necessary, this calibration process can be performed after an AE test or periodically during a long period of AE monitoring.

2.3.3 Conducting the test

The actual AE test is relatively simple. The test is conducted by starting AE data acquisition following the test procedures defined in the test plan. In most cases, the AE data is displayed in real time during the test for situation awareness or for monitoring purpose.

When AE technique is used for structural health monitoring or as part of *in-situ* NDE, the real-time monitoring has a responsibility in addition to the data collection. The real-time monitoring may be required to compare AE data with pre-determined criteria, according to the test requirements for status or conditions of the test article under AE monitoring. In AE tests that require real-time monitoring, the AE data is usually used in decision-making, for instance, 'go or no-go' for a higher loading during a load test or 'pass or no-pass' at the end of an in-line inspection by AE technique.

It is good practice to record or take notes of all necessary information about the test article, load and test conditions and anything that happened during the test, which may help in data analysis and interpretation.

2.3.4 Data analysis

The purpose of AE data analysis is to extract useful information from the raw data, to interpret the data correctly, and to relate it to the integrity of the structure based on the AE findings and the evaluation criteria. For the purposes of NDE, the AE data analysis is normally performed in three steps: (1) process data, (2) interpret data, and (3) evaluate structural integrity of PMCs.

Data processing may involve using various filtering techniques to remove unwanted noise or certain parts of the data for clarity, and converting time domain data into frequency or other domains for data feature extraction. Data processing and interpretation may be carried out at the same time, because the data must be assessed for significance to the expected structural response to the applied loading and test conditions. The evaluation process involves assessment of structural integrity based on the information extracted from AE data against criteria that was set for certain AE parameters and their statistical distributions.

As mentioned earlier, there are two sets of data that can be obtained from an AE test, AE parameter data and waveform data, and so there are two possible approaches to AE data analysis: AE parameter-based analysis and waveform-based analysis. In the analysis of AE data for NDE of PMCs, the first step is usually analyzing the parameter data because it takes less effort and time, while waveform data is only used when necessary. By examining the waveform features, more information about an individual AE hit may be obtained, which is helpful in identifying the source of the hit and more accurately determining the source location. Figure 2.6 shows a typical waveform and its power spectrum of an EMI signal identified in an AE test. By examining the features exhibited in the waveform, frequency spectrum, and its parameter values such as amplitude, energy, rise time duration, proper adjustment of the data acquisition settings and use of front-end filters can effectively eliminate this type of noise.

It is worth pointing out that waveform-based analysis is very useful in AE source characterization and location, but time consuming unless the process is automated. Some advanced AE data analysis methods such as modal AE,[4–7] pattern recognition[7,10,11] and neural networks have been used in AE studies of PMCs.

Depending on the NDE requirements for a specific AE test, data analysis may need to be performed at one of the following stages: (1) real-time analysis, (2) post-test analysis, and (3) both real-time and post-test analyses. The real-time AE data analysis is used when it is necessary to know the structural response in terms of AE data to the current test conditions. In real-time analysis, AE data is processed, plotted and displayed in the

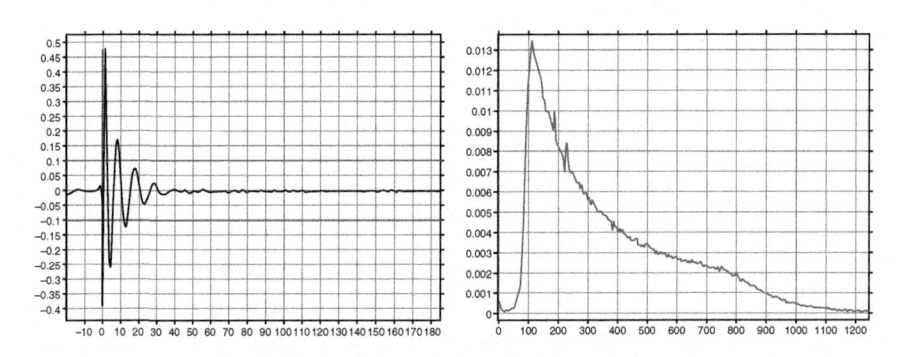

2.6 An EMI signal identified from waveform data analysis: left, waveform; right, spectrum.

pre-designed graphical layouts during the test. Real-time analysis is often used in structural tests for the purposes of qualification, acceptance, and proof load, where AE technique is used as a dynamic NDE method.

Post-test AE data analysis is performed if real-time monitoring is not required. Post-test analysis is often used in research into PMC materials and periodical reviews of data in long-term in-service monitoring applications.

In some cases, both real-time and post-test analyses are required. This may be necessary when real-time analysis is inconclusive, or if further analysis is needed for more detailed information about damage or an anomaly detected during the test, such as the type of damage, when and where the damage occurred, or to confirm that no damage occurred during the test. For example, the AE technique can be used to study damage tolerance of a PMC structure with certain built-in crack arrest mechanisms, in which the post-analysis was to correlate AE results with the damage process involving crack initiation, arrest, propagation, and the crack path.

AE source location can be performed in several ways, including the linear, zonal, and triangulation methods, depending on the structural geometry and accuracy requirement. For a linear structure, the linear location method is sufficient. For 2D geometry, which is not necessarily flat, the zonal and triangulation methods are suitable solutions. The former gives a zone location based on what channels detected the AE event, and the latter gives a point location based on the times arrived at three or more sensors. For a 3D location issue in a thick composite structure or section, a minimum of four sensors is required to detect the AE event. The triangulation method can be used in a 3D coordinate system to calculate the source location.

2.4 Comparisons

The AE technique has particular advantages and limitations for use in NDE of polymer matrix composites. The AE technique offers some unique capabilities such as:

- real-time detection;
- continuous monitoring;
- rapid and global inspection;
- suitable for remote and unattended NDE operations;
- can be used at any stage of a composite structure lifespan from manufacturing to service.

The AE technique has proven to be a very cost effective and time-saving NDE method for PMC structures when it is used as an *in-situ* NDE inspection method during manufacturing and the consequent proof load testing, or to recertify an in-service composite structural to its design capability.[12]

In addition, AE has also exhibited some superior capabilities in detecting certain types of damage:

- Because the AE method detects the damage as it occurs, it is more effective in detecting damages that cause no apparent deformation, such as delamination or cracks along the fiber direction in thick composites or tapered composite sections. Similarly, AE can be effective in detecting the so-called 'kissing bond' (a bond with no strength, or which has failed, but is perfectly closed) and weak bond conditions. These flaws or damage conditions are difficult to detect with other NDE methods.
- AE sensing is more sensitive than other NDE methods such as ultrasonic and radio graphic testing,[1] thus, the AE technique can be used to detect micro damage, crack initiation, and incipient failure.

Like all NDE methods, the AE technique has certain limitations:

- AE testing must be conducted under load (except for impact damage detection).
- It is not suitable for determining the size of damage or flaw.
- Data analysis and interpretation may be difficult in some cases.
- Data interpretation mainly relies on AE experts, other than a process that is developed from NDE standards made of representative materials and structural features.
- Noise, attenuation and anisotropic properties of PMCs may pose challenges for AE testing and data analysis.

2.5 Future trends

Looking back over the last six decades, during which the AE technique has developed into a powerful NDE method, it can be seen that two elements have driven the maturation of the AE technique: technology development and application development. These two elements interact to push and pull development forward. Therefore, future trends in AE can be predicted by examining the trends of these two development elements.

The development of AE technology addresses challenges from applications that 'pull' for more capabilities from AE hardware and software. AE manufacturers are trying to meet the capability demand and to keep up the pace at which new technologies become available. However, perhaps because of limited market for AE equipment, the gap between demand and the AE system capability remains wide. Hardware development has to absorb new technologies fast.

The development of AE applications is concerned with the use of AE techniques for specific applications. It helps improve understanding of the physics associated with the AE from various failure mechanisms of composite and anisotropic wave propagation characteristics. It also drives the development of AE-based NDE methods including techniques and processes for conducting NDE of composites. Application development has been mainly carried out by domain experts who understand collectively AE, the NDE process, and composite materials.

PMCs possess a number of superior properties such as strength to weight ratio, fatigue, corrosion, and chemical resistance, and significant part count reduction. As a result, polymer matrix composites have found more and more applications in the aerospace, defense, automobile, marine, chemical and oil industries.[13] One example is the breakthrough in usage of high-performance PMCs in primary structures such as the wings and fuselage of large civil aircraft such as the Boeing 787 and Airbus 350. Owing to the rapid increase in industrial use of PMCs, demands to develop effective and reliable NDE methods to meet challenges in PMC inspection are high. AE techniques are being used and developed more than ever before due to the unique capabilities of AE for NDE of PMCs. AE for PMC applications is experiencing a transition from more fundamental studies aimed at explaining the physics associated with acoustic emission from composites to more practical applications. It is reasonable to expect that the future trends for AE applications will involve a shift from the lab-centric AE applications to more 'real-world' applications such as manufacturing NDE, structural tests, in-service inspection, and SHM.

Based on the increasing demand for using AE's unique capabilities such as real-time, continuous and remote monitoring while under loading or during transportation, it is possible to suggest the following future trends:

- The incorporation of new technologies such as wireless, microelectromechanical systems (MEMS), and remote operation into AE systems to meet the high demand for more capability and flexibility for the various application circumstances.
- The integration of new materials such as nano-materials into sensors for improved properties in sensitivity, stability, and durability in extreme environments.
- The miniaturization and weight reduction of AE hardware including sensors, cables, preamps and data acquisition systems.
- The incorporation of new capabilities into AE software suites, including more powerful digital filters for handling extraneous noise in real-world applications, artificial intelligent tools for data processing and analysis including waveform analyses, and user-friendly interfaces for incorporating the customer algorithms for data processing and automated processes.

Future developments for AE application development for NDE of PMCs are expected in the following areas:

- Improved understanding of the meaning of data related to various AE mechanisms in PMCs.
- Standardization of AE processes for NDE applications.
- Quantitative or measurable AE criteria.
- Modeling and simulation of AE behavior and wave propagation in various composites systems.
- Integration of AE test results with structural models and structural analysis tools.

2.6 Sources of further information and advice

There are many sources of information on AE in the form of books, journals, conference papers, and technical reports. Since the Internet has become an important part of our lives, a large amount of information on AE technology is now available online. Searching for key words such as 'acoustic emission in composites' through Google yielded over one million results in 2012, and the number is growing. Because more and more polymer matrix composites are being used, especially in the aerospace and automobile industries, some conferences have expanded to include NDE and SHM in composites. As a result, conference materials including research papers, presentations, and conference proceedings, have become a growing source of information on AE. It is impossible to list all of the sources of information, but some from each of the categories mentioned are listed below.

Introductory literature

Some introductory literature on AE can be found at the Acoustic Emission Working Group website, http://www.aewg.org/, under acoustic emission literature.

Information provided by manufacturers

More introductory articles on acoustic emission techniques and information on training courses, certification and equipment can be found on the websites of AE equipment manufacturers:

- Physical Acoustics Corporation, a member of MISTRAS Group Inc., www.pacndt.com
- Vallen Systeme GmbH, the Acoustic Emission Company, www.vallen.de

Journals on AE research and applications

- *Journal of Acoustic Emission,* Acoustic Emission Group, Editor Kanji Ono.

- *NDT & E International,* Elsevier Ltd, Editor-in-Chief: D. E. Chimenti
- *Research in Nondestructive Evaluation*, American Society for Nondestructive Testing, Editor-in-Chief: John C. Duke, Jr.

Standards/practice guides/definitions/terminology

- *Non-destructive Testing Handbook*, Third Edition: Volume 6, Acoustic Emission Testing. American Society for Non-destructive Testing, Technical Editors: R. Miller and E. Hill, 2005.
- ASNT Publications: http://www.asnt.org/, American Society for Nondestructive Testing
- ASTM Publications: http://www.astm.org/, ASTM International
- ANSI Publications: http://www.ansi.org/, American National Standards Institute

Websites

- Acoustic Emission Working Group, www.aewg.org
- European Working Group on Acoustic Emission, www.ewgae.eu
- The Latin American Acoustic Emission Group (GLEA), www.cnea.gov.ar/cac/endye/glea/
- Physical Acoustics Corporation, a member of MISTRAS Group Inc., www.pacndt.com
- Vallen Systeme GmbH, The Acoustic Emission Company, www.vallen.de

Conferences

- Acoustic Emission Working Group (AEWG) meeting in the United States
- European Conference on Acoustic Emission Testing (EWGAE)
- International AE Symposium (IAES)
- International Conference or World Meeting (ICAE)
- International Conference of AE from Reinforced Composite or Composite Materials (AERC/AECM)
- Structural Health Monitoring Workshop, Stanford, California.
- SPIE SHM and NDE

Others

- The Acoustic Emission Collection of the Grainger Engineering Library at the University of Illinois, USA at Urbana-Champaign numbers over 4500 items. http://shiva.grainger.uiuc.edu/ae/opent1.asp

- Compilations of Author Index files of various AE conferences and meetings can be found at the AEWG website, http://www.aewg.org/

2.7 References

1. *Nondestructive Testing Handbook*, Third Edition, Volume Six, *Acoustic Emission Testing*, American Society for Nondestructive Testing, 2005, Editors R. K. Miller and E. v. K. Hill.
2. Fowler, T. J. and R. S. Scarpellini, 'Acoustic Emission Testing of RFP Equipment.' *Chemical Engineering*, New York, NY, McGraw-Hill, October and November, 1980.
3. Huang, Q. and G. Nissen, 'Structural Health Monitoring of DC-XA LH2 Tank Using Acoustic Emission', presentation and paper at the International Workshop on Structural Health Monitoring, Stanford University, Stanford, California, September, 1997.
4. Prosser, W. H. and M. R. Gorman (1994), 'Plate Mode Velocities in Graphite/Epoxy Plates,' *Journal of the Acoustical Society of America*, **96**(2), Pt. 1, 902–907.
5. Gorman, M. R. (1991), 'Plate Wave Acoustic Emission,' *Journal of the Acoustical Society of America*, **90**(1), 358–364.
6. Gorman, M. R. and S. M. Ziola (1991), 'Plate Waves Produced by Transverse Matrix Cracking,' *Ultrasonics*, **29**, 245–251.
7. Huang, Q. 'Characterization of Acoustic Emission from Failure Processes of Carbon Fiber-Epoxy Composites by Pattern Recognition', Ph.D. dissertation, UCLA, 1993.
8. ASTM E2661 / E2661M - 10 Standard, Practice for Acoustic Emission Examination of Plate-like and Flat Panel Composite Structures Used in Aerospace Applications, ASTM International, West Conshohocken, PA, 2010.
9. ASTM E976 - 10 Standard, Guide for Determining the Reproducibility of Acoustic Emission Sensor Response, ASTM International, West Conshohocken, PA, 2010.
10. Huang, Q. and K. Ono, 'Application of Advanced AE Source Characterization Method in Airframe Industry', ASNT Fall Conference, Long Beach, California, 1993.
11. Ono, K. and Q. Huang, 'Pattern Recognition Analysis of Acoustic Emission Signals', *Progress in Acoustic Emission VII*, eds. T. Kishi et al., the Japanese Society for NDI, 1994, pp. 69–78, the 12th International Acoustic Emission Symposium.
12. Goggin, P., J. Huang, E. White, and E. Haugse, 'Challenges for SHM Transition to Future Aerospace Systems' (keynote speech and paper), *Proceedings of Structural Health Monitoring*, ed. F.-K. Chang, the 4th International Workshop on Structural Health Monitoring, Stanford University, California, 2003.
13. Trends in advanced composite usage observed during the review of 2010 JEC Awards applications, *JEC Composites*, 2010.

3
Non-destructive evaluation (NDE) of composites: eddy current techniques

H. HEUER, Dresden University of Technology and Fraunhofer Institute for Non-Destructive Testing, Germany and M. H. SCHULZE and N. MEYENDORF, Fraunhofer Institute for Non-Destructive Testing, Germany

DOI: 10.1533/9780857093554.1.33

Abstract: Industrial mass-producing carbon fiber reinforced polymer (CFRP) processes (e.g. resin transfer molding) require non-destructive testing (NDT) methods that can be applied to dry textile multilayer materials and to final components as well. Questions to be solved are fiber orientations, gaps, local defects, etc. By analyzing the electrical properties of carbon fiber materials – the fiber distribution and the dielectric properties can be inspected non-destructively. Based on high-frequency eddy current techniques, structural and hidden defects such as missing carbon fiber bundles, lanes, fringes and angle deviations for hidden layers can also be detected. Carbon fiber based materials show a low electrical conductivity, which is sufficient to measure deviations in the material by using eddy current techniques. Eddy current methods show a high potential for inline integration due to the absence of couplings, e.g., compared to ultrasonic.

Key words: non-destructive testing (NDT) on dry carbon fiber materials, carbon fiber reinforced polymer (CFRP) textural analyses, high frequency eddy current imaging.

3.1 Introduction

Along the value added chain of carbon fiber reinforced polymer (CFRP) products many different physical testing methods are successfully applied. However, looking in more detail at the process chain, there is a gap where standard non-destructive evaluation (NDE) methods cannot be applied. For automated mass production facilities based on textile production processes it is important to acquire quality parameters before the resin infiltration step. From this follows an increasing demand for NDE methods that can be applied inline to dry multilayered carbon textiles such as multiaxial non-woven materials which typically have three to five layers.

Hidden defects that arise early in the production step may have far-reaching consequences with a high risk of damage progression and even increased costs as products have to be discarded after machining. The final component tests are usually performed with ultrasonic, X-ray or thermograph methods close to the final product stage and without the possibility of repair.

With exact knowledge of the semi-finished product quality, process parameters can be adjusted in time to reduce rejects resulting from defects. If missing or misaligned fiber bundles are detected inline, the machine can be stopped and readjusted, resulting in less material wastage. In addition, subsequent process steps can be controlled by utilizing the incoming product quality data, e.g. gaps between fiber bundles will influence the behavior during polymer or resin infiltration, so by knowing the gap's size and density, the infiltration parameters can by adapted. Also, for non-crimp carbon fabric production, the properties of the raw material may be aligned to the cut out process for later component assembly in order to increase yield.

The eddy current technique can measure these values by using variations in the electric and dielectric properties of the materials. The method works without contact, and no coupling or X-ray protection is needed, and it is well suited for automation. Figure 3.1 shows an eddy current image of a dry 3 axial material that illustrate the principle capability of testing carbon materials by eddy current methods.

3.1 Eddy current image of dry 3 axial (45°/0°/–45°) carbon fiber non-woven fabric (30 × 30 cm).

Typical material properties that can be detected are the orientation of different layers, missing bundles or gaps, overlaps, and fringes. Also local defects such as metal insertions can be detected. High frequency eddy current (HF EC) can be applied non-destructively to the textile and to the final consolidated product.

3.2 Eddy current testing: principles and technologies

Eddy current technology is a well-established non-destructive method for the characterization of surfaces or material incontinuities by analyzing conductivity and permeability variations. A primary magnetic field is generated when alternating current is applied to an induction coil. Eddy currents are generated in a conductive specimen when the coil is placed near that specimen (Fig. 3.2). The eddy current flow in the specimen generates a secondary magnetic field opposed to the primary field. The total field change is determined by conductivity differences in the sample, which cause an impedance change measured in the pick-up coil. If the material properties are changed, e.g. due to a deviation of current paths resulting from cracks or insertion in the sample, the secondary field changes and causes an impedance shift in the pick-up coil. The measured values from the pick-up coil are evaluated on the complex impedance plane.

An important parameter for eddy current measurement is the frequency of the excited alternating current. Due to the skin effect, the depth of eddy current excitation decreases with increasing frequency. The point where the eddy current density has decreased to 1/e, or approximately 37% of the surface density, is called the standard depth of penetration (δ or 1δ) and is used as a criterion for ideal measurement [1]. The standard penetration depth of the eddy current into a material is affected by the frequency of the excitation current, the electrical conductivity, and magnetic permeability of the sample, as given by:

$$\delta = \sqrt{2/\omega\sigma\mu} = 1/\sqrt{\pi f \mu_0 \mu_r \sigma} \qquad [3.1]$$

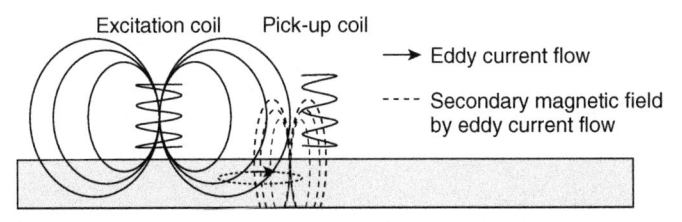

3.2 Schematic diagram of probe and specimen configuration for eddy current testing.

where ω is the angular frequency, σ is the electrical conductivity of the specimen, f is the frequency, μ is the permeability, μ_0 is the absolute permeability, and μ_r is the relative permeability of the specimen.

The magnetic permeability is regarded as 1 in diamagnetic substances such as carbon with a 2π bond, as it has four paired electrons in bonding orbital. The standard depth of penetration also depends on the sensor diameter [2]. The eddy current density normalized to the density at the surface follows the ratio R_s/δ where R_s is the coil radius and δ the standard penetration depth [3].

Although the eddy current penetrates deeper than one standard depth of penetration, its density decreases rapidly with depth. Figure 3.3 shows the correlation between frequency and penetration. At two standard depths of penetration (2δ) the eddy current density decreases to $1/e^2$, or 13.5% of the surface density. At three standard depths (3δ), the eddy current density decreases to $1/e^3$, or 5% of the surface density.

Furthermore, the density of the eddy current is influenced by the frequency itself. The modified Faraday's law for a coil or wire explains the relation between magnetic flux and time domain, namely:

$$V_L = -N \cdot d\phi/dt \qquad [3.2]$$

where V_L is the induced voltage, N is number of turns in the coil, and $d\phi/dt$ is the rate of change of magnetic flux in webers per second. Equation (3.2) simply states that the induced voltage is proportional to the rate of change of the magnetic flux. In other words, since a frequency is an inverse

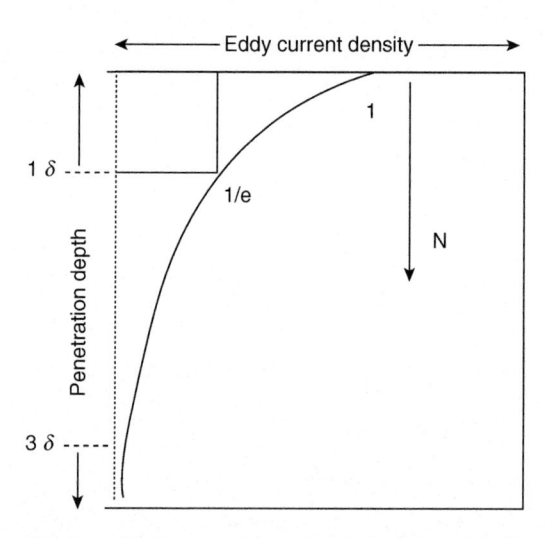

3.3 Standard penetration depth and density of eddy current.

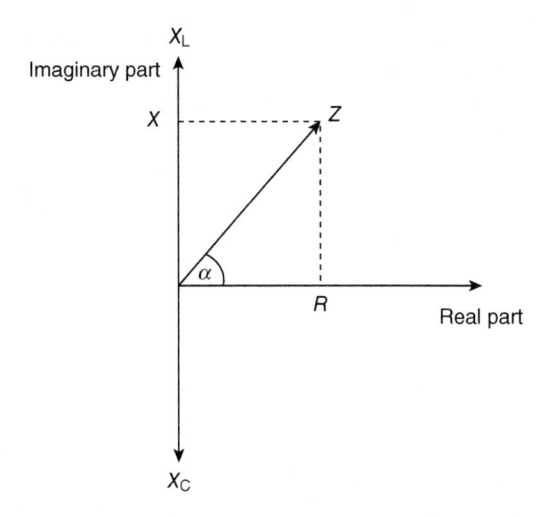

3.4 Complex impedance plane. Changes in inductors (*L*) or capacitors (*C*) are shown on the imaginary axis and changes in the electrical conductivity (*R*) are shown on the real axis.

time, when the frequency of the flux increases the eddy current signal, which can equally be regarded as induced voltage (V_L), also increases. Higher frequencies therefore represent a good option through which the sensitivity of the eddy current method may be increased for low conductive materials such as carbon textiles and CFRPs due to the extended trade-off between penetration depths vs. signal amplitude.

The electrical impedance Z (Fig. 3.4) measured on the pick-up coil is represented as a complex quantity Z where the real part is the resistance R and the imaginary part is the frequency-dependent reactance X, composed of inductive reactance X_L and capacitive reactance X_C. The eddy current probe coils (induction coils) are mainly inductors with minor capacitor effects. Since the impedance strongly depends on frequency, parasitic effects coming from different probe-to-surface distances (so-called lift-off) can be compensated for by using different excitation frequencies.

3.2.1 Eddy current testing of carbon fiber materials

Classical eddy current testing (ECT) is applied on isotropic conductive materials such as aluminum, copper, steel or titanium. Unidirectional single layered carbon fiber material has a conductivity of $\sigma = 5*10^6$ S/m in longitudinal and $\sigma = 1*10^3$ S/m in the lateral direction. Through variation of fiber orientation, stacking sequence, fiber and matrix combinations, quality and permittivity of the resin, fiber density and coating polymers around the fiber rovings, these values differ. The packaging density is the ratio of fiber and polymer. In a raw carbon textile material the permittivity is characterized by the fiber coating and the surrounding air. In a CFRP the air is substituted by resin polymers. The complex permittivity generates a displacement current in the material.

Mook and Lang have discussed the basic principles of eddy current testing on CFRPs and have demonstrated its application [4] [5]. Figure 3.5 shows the main effects of applying an alternating magnetic field to strongly anisotropic CFRP material. The eddy current propagation in one plane in the material is influenced by three main parameters:

1. ***Fiber/volume ratio*:** The fiber/volume ratio determines the amount of conductive carbon fibers in a volume section. The ratio is normally defined by the structural design of the CFRP and depends on the number of filaments per fiber bundle and the number of fiber bundles themselves. The microscopic filaments with a diameter of a few µm are combined into a fiber bundle with some thousand filaments (e.g. noted as 5k, 10k, 50k bundle). The amount of conductive material and its packaging density defines the average electrical conductivity of the material which needs to taken into account for general measurement parameterization (frequency, coil diameter).

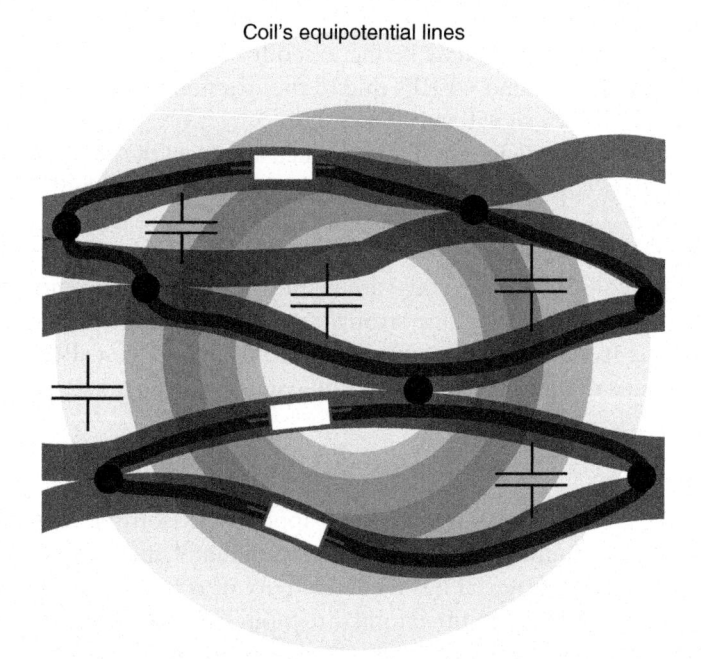

3.5 Electric and dielectric behaviour of carbon fiber matrix by Lange/Mook (Mook *et al.* [5]).

2. ***Electrical connection between fiber bundles:*** Depending on the chemical and structural conditions of the interfaces of filaments, the fiber bundles, and the packaging density due to mechanical pressing, the electrical contact between neighboring bundles can vary. Identical materials can provide different degrees of eddy current propagation due to the quality of internal electrical connections during compression. This effect identifies differences in handling or mechanical processing.
3. ***Capacitive effect due to properties of the carbon surrounding dielectric:*** In addition to the electrical connection of fiber bundles, the dielectric properties of the material also influence the complex impedance. In the case of a textile material, air acts as the surrounding dielectric material, in the case of a CFRP after consolidation, the dielectric material is the resin or polymer. Embedded water also influences the signal. The Cole–Cole plot is often used to describe the frequency-dependent behavior of dielectric relaxation in polymers. Every dielectric material has a frequency-dependent dipole orientation and ionic conduction resulting in characteristic cutoff frequencies in the Cole–Cole plot. With the usage of high-frequency EC this phenomenon can be measured. The dependence of frequency on the normalized absolute eddy current signal is shown in Fig. 3.6.

The experiment shows that for typical one or three axial fabrics with air as the dielectric, the optimal frequency is in the range between 1 and 50 MHz.

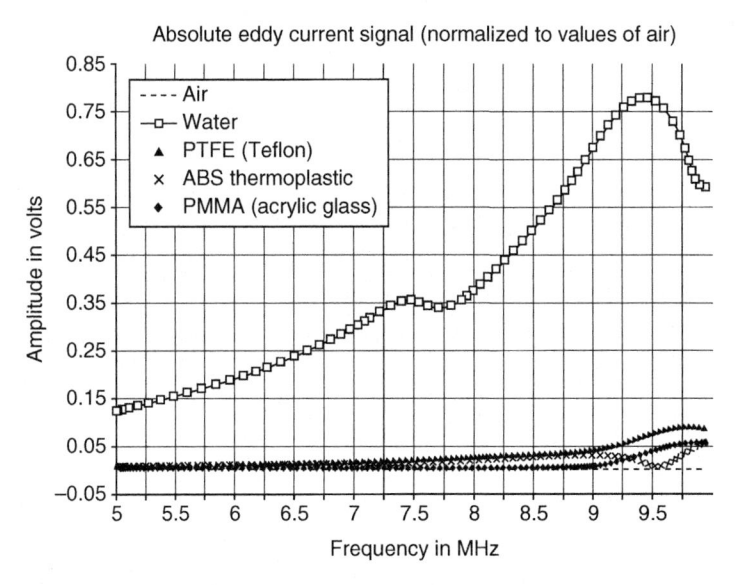

3.6 Frequency properties of different dielectric materials by usage of ECT.

The poorer the conductive matrix is, the higher the measuring frequency must be set.

3.2.2 Device technology for high eddy current imaging

Eddy current instruments are available in many configurations by different commercial suppliers. Typically, the instruments are designed for manual handling of a singe coil sensor or array probe with wheel tracker. The parameterization of such standard devices is optimized for NDE tasks such as crack detection or material identification for metallic specimens.

Also, mechanical manipulators like x-y-z axle scanners or robot-based manipulators are in commercial use to scan an eddy current probe over a three-dimensional surface, e.g. rotor blades. Due to the high electrical conductivity and limited skin depth (see Section 3.2) of metallic specimens, the frequency range of standard instruments start from the low kHz range up to single digit MHz range. For low conductive carbon-based materials, the frequency range should be extended up to 50 MHz, and in some special cases up to 80 MHz is required. This frequency range was usually observed only in laboratory-based equipment [6] [7]. Initiated by the increasing demand for high-frequency eddy current measurements, instruments operating in the range of 100 KHz up to 100 MHz were developed and are now commercially available. The results shown in the following sections were acquired with EddyCus® instruments by Fraunhofer IZFP in Dresden, Germany. Combined with a precision X-Y-Z manipulator, high-resolution EC images can be acquired. The picture in Fig. 3.7 shows an experimental

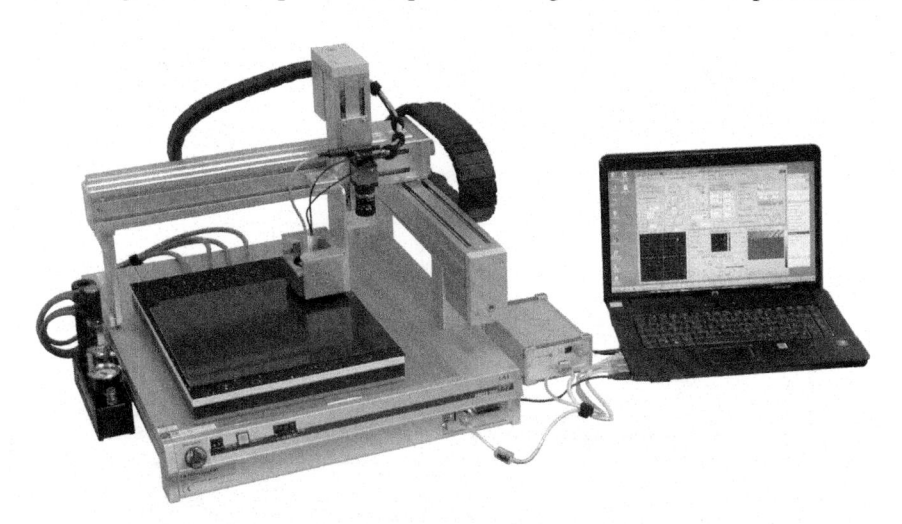

3.7 Setup of the HF EC imaging system for laboratory use. The box between scanner and laptop contains the EC instruments and control unit.

setup for laboratory use. The sensors glide lightly and nearly pressure-free over the surface. Surface roughness can almost be balanced against lift-off effects. The standard system can capture a maximum surface area of 300 mm by 300 mm with a maximum speed of 300 mm/sec at a sampling rate of 3,000 samples per second. In addition, the EddyCus® software provides sequential multi-frequency data acquisition with up to four frequencies. The instrument allows the acquisition of complex eddy current signals in amplitude and phase shift in the complex impedance plane as a time plot by C-scan.

To perform eddy current measurements in a frequency range above 1 MHz, mechanical vibrations (lift-off variations) and electromagnetic disturbance become issues requiring control. Also, electrical conductive dust in carbon-contaminated environments becomes a real issue. To solve this practical problem, a more robust set-up can be used that is shielded against dust and is mostly electromagnetic compatibility (EMC) safe as shown in Fig. 3.8. The technical parameters of the instrument used for the data acquisition is given in Table 3.1.

The eddy current sensor needs to be optimized for carbon materials. Different sensor types, such as absolute, differential and compensated sensors, can be used for single sensor probing. The single sensor configuration is a directional sensor with a high spatial resolution due to its high-focusing

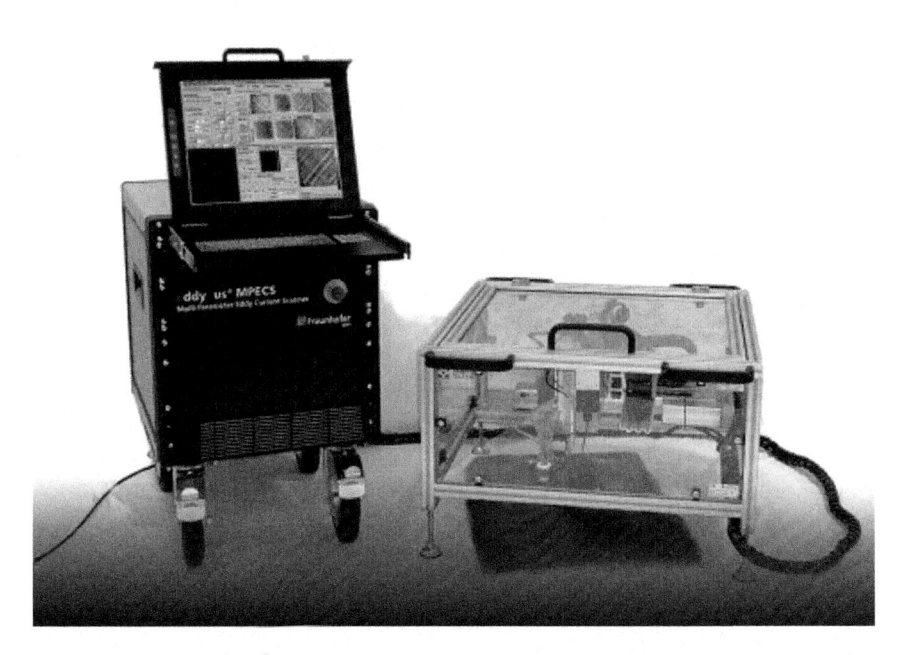

3.8 EddyCus® HF EC Scanner for operation in non-laboratory environment.

42   Non-destructive evaluation (NDE) of polymer matrix composites

Table 3.1 Technical specification of HF EC Scanner EddyCus®

Description	Value
Maximum scan area	300 mm * 400 mm
Scan resolution of X direction	25 μm to 1 mm
Scan resolution of Y direction	25 μm to 10 mm
Scan speed	1 mm/s to 300 mm/s
Frequency range	100 kHz to 100 MHz
Discrete frequencies	4 at a time
Sensor type	Absolute, half-transmission single sensor

3.9 (a, b) Sensor mounting kit, (c) anisotropic single sensor.

point-spread function (PSF) [8] [9]. The applied sensor, Fig. 3.9, is of absolute half-transmission anisotropic type with a lateral pitch of 3.5 mm and about 20 turns, which can be revolved in various measurement angles.

The description of the lateral hardware resolution of such a sensor is difficult as its focusing point looks like the density function of a normal probability curve $N(0,1)$. This means that even the smallest conductivity and permeability variations such as cracks can cause significant signal changes.

3.10 Scalable 16 sensor demonstrator line array.

Therefore, material discontinuities cannot be reliably measured in real-world dimensions and the description of the actual sensor resolution is complex. In fact, anisotropic sensors [10] have a variation in resolution due to different measuring angles of the sensor and the use of frequencies in the range from 1 MHz up to 50 MHz. Rotation of the sensor and static measurements with sensor-specific sample angles of the textile material or CFRP improve the separation of different layers enormously.

This sensor principle was utilized to develop a sensor line optimized for carbon materials' testing by eddy current with 16 individual sensor pairs to increase the measuring cycle speed by parallel measurement (Fig. 3.10). However, the array concept is viable only for absolute planar objects, which are unfortunately rarely encountered in real applications.

3.3 High-frequency eddy current imaging of carbon fiber materials and carbon fiber reinforced polymer materials (CFRPs) composites

3.3.1 Method

As described in Section 3.1, the effect to be measured is the variation of the complex impedance of the pick-up coil during movement over the surface of a conductive material where an eddy current flow has been excited. The measured impedance consists of a real and an imaginary part that can be visualized in a Nyquist plot (Fig. 3.11). In accordance with

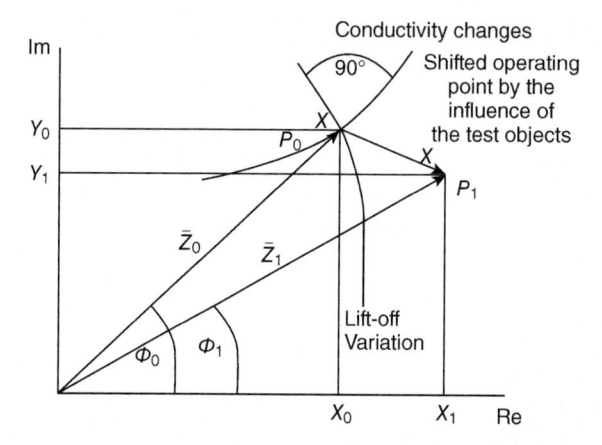

3.11 Nyquist plot visualization of complex impedance change during probe movement.

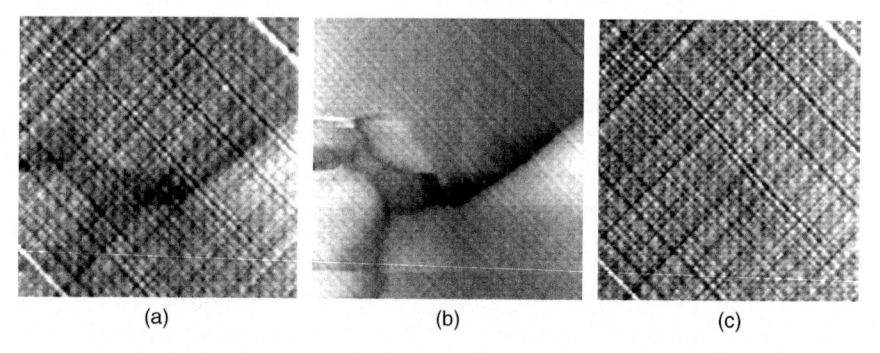

3.12 Data visualization: (a) the gray coloured real part; (b) the imaginary part; (c) the real part after rotation of the Z vector by 115°.

changes of electrical and dialectical properties and lift-off variations, the operating point P moves from starting point P_0 to the new operating point P_1, indicated by changes of the imaginary part (Y), the real part (X) or the magnitude (Z), and the phase angle (ϕ) of each measuring point.

The actual impedance value needs to be related with the x–y information of the manipulator to create an eddy current image. The example in Fig. 3.12 shows how the information of imaginary and real values differs. In image (a) the information from the carbon fiber structure clearly dominates. The orientation of the bundles is apparent as a function of gap size, e.g. in the top right-hand corner. By using the projection of the working point P to the imaginary axle (90° phase rotated in comparison to (a)) it can be found that additional information are visible. The black shadow in image

(b) comes from dielectric effects such as topographical effects, or different dielectric behavior of the specimen (e.g. by water insertion).

By performing a vector rotation, by rotating the phase angle, the image can be optimized. Figure 3.12(c) shows the vector rotated image of the real part with an applied rotation of 115°. It is observed that the capacity influences have more or less vanished and only the fiber orientation is visible. This method can be applied after data acquisition or in real time when the scanning device is running. All eddy current C-scans shown in this chapter are the result of vector rotation from the acquired complex values. The projected vector length is placed into a 2D matrix denoted EC image. With this technique it is possible to separate and maximize defects or material parameters and/or eliminate parasitic effects such as impedance changes of the instrument cables. It is possible that different layer depths may be recognizable with this technique. The zero compensation is carried out either in air or in a homogeneous and verified error-free area on the sample.

3.3.2 Testing raw multiaxial materials without resin

Properties in multidirectional materials during textile manufacturing processes such as missing fiber bundles, lanes, fringes or angle errors may have far-reaching consequences and a high risk of damage progression of the post-processed CFRP. The images of Fig. 3.13 are acquired with a scanning step width of $1.5*1.5$ mm^2, except sample #5, where a miniaturized single coil probes was used. The EC images were acquired at scanning speeds of about 60 mm/s with frequencies ranging from 2 M to 50 MHz.

To show the functionality of high-resolution eddy current imaging, a microscopic eddy current image for sample #5 was captured. The size of the square scan area is $50*50$mm^2 at a scan step width of 78 125 μm in both directions. An image with 640*640 pixels results. The main material properties (dark line from lower left to upper right) are shown very clearly. Also the non- or slightly conductive weaving treads are visible. The distance between the two is about 3.2 mm.

3.3.3 Testing non-woven laminated materials

Beside the application of HF EC imaging to raw textile materials without resin this method can also be applied to final consolidated laminates (Fig. 3.14). Additionally therefore to the case of raw materials, some added effects related to the resin infiltration can be inspected.

Two main kinds of material properties in laminated CFRPs can be observed. First, process faults such as lanes, missing bundles or undulations can be inspected and second, the non-visible delamination between the

46   Non-destructive evaluation (NDE) of polymer matrix composites

Sample picture (1st layer)	Picture (3rd layer)	EC image (front)	EC image (rear)
#1	Missing bundles / Tapered lane	Defect in 2nd layer	

3-layer RCF with small tapered lanes in the 1st and 3rd layer, four missing fiber bundles in the 3rd and one horizontal in the 2nd layer – sample size [mm]: 210*295

#3	Lanes / Missing bundles		Small wave

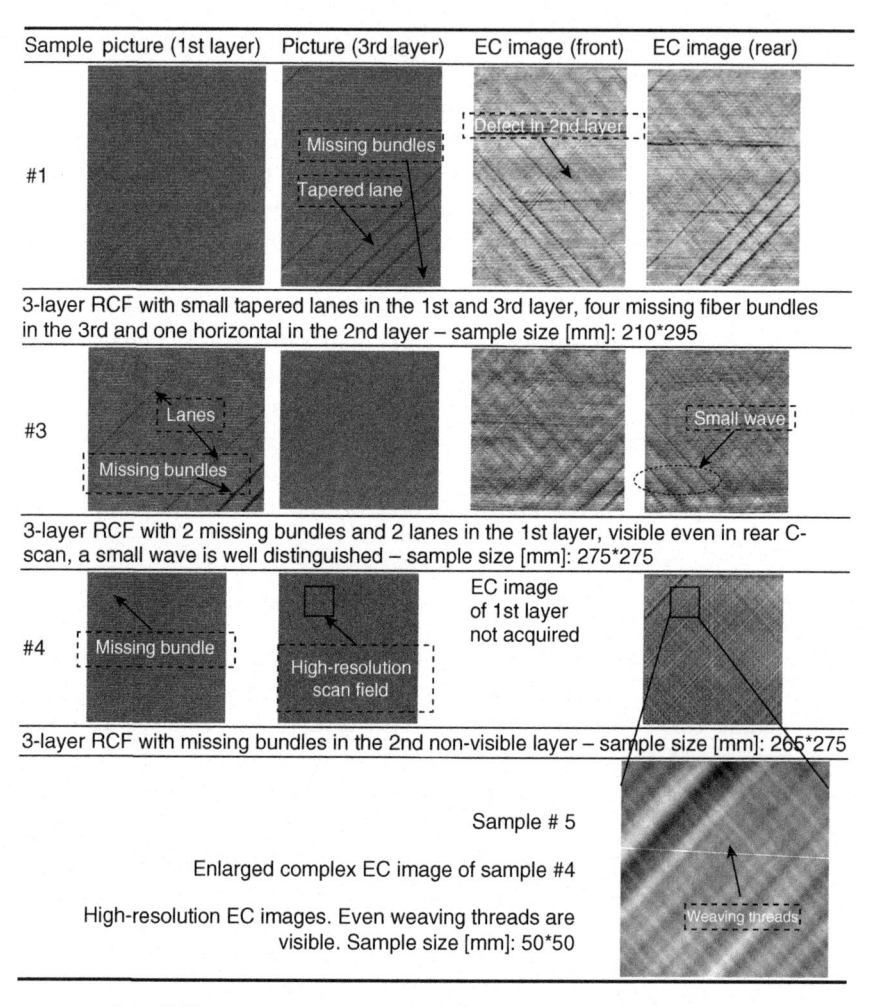

3-layer RCF with 2 missing bundles and 2 lanes in the 1st layer, visible even in rear C-scan, a small wave is well distinguished – sample size [mm]: 275*275

#4	Missing bundle	EC image of 1st layer not acquired / High-resolution scan field	

3-layer RCF with missing bundles in the 2nd non-visible layer – sample size [mm]: 265*275

Sample # 5

Enlarged complex EC image of sample #4

High-resolution EC images. Even weaving threads are visible. Sample size [mm]: 50*50

Weaving threads

3.13 RCF samples #1, #3 and #4, #5.

epoxy resin and the fiber layers, which may be caused by an abnormally high in-plane or orthogonal pressure of the CFRP, is observable. Discontinuities can be located in different layers of the dry carbon material and also in the laminated material. The eddy current amplitude and phase images represent integral information about material characteristics. An adapted set-up of measuring frequencies allows penetration of the sample to various depths of the CFRP layers. To verify that the depth is high enough for inspection of the entire volume, a tin plate (metal reference) was attached under sample #7. By increasing the measurement frequency, the visibility

Eddy current techniques 47

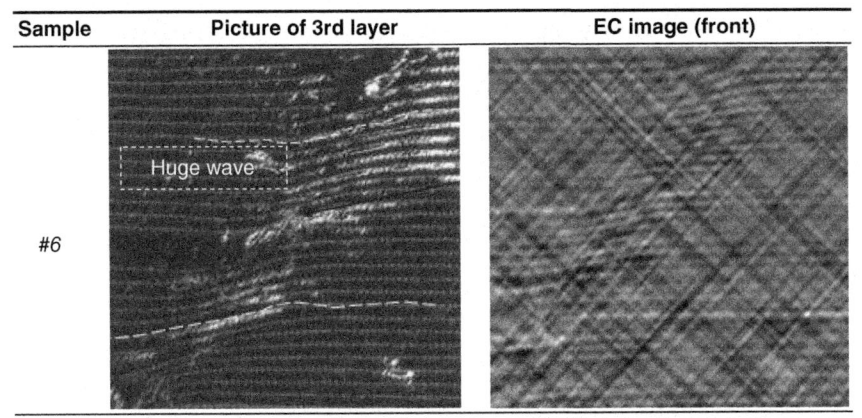

Sample	Picture of 3rd layer	EC image (front)
#6	Huge wave	

A huge wave in the 3rd layer was formed in this 5-layer CFRP during the production process. The defect which is a cause of material fatigue is clearly visible in EC. Sample size (l*w*h) [mm]: 200*200*3

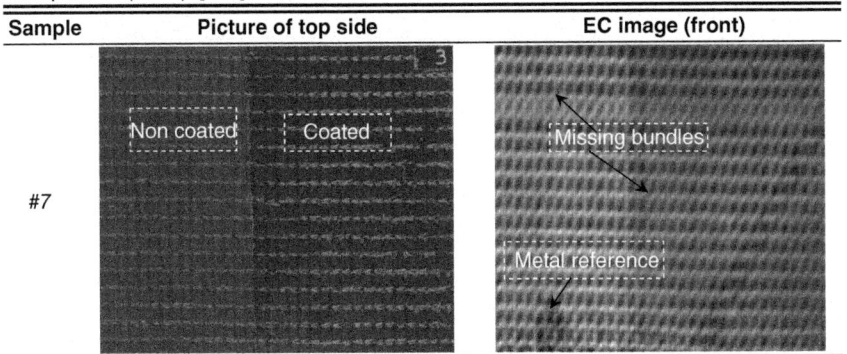

Sample	Picture of top side	EC image (front)
#7	Non coated / Coated	Missing bundles / Metal reference

2 bundles in the 2nd (45°) and 3rd (0°) layer of this 5 layered CFRP were removed. Picture of production process was unavailable. Both defects are visible in EC image. By horizontal high pass filtering the 45° defect stands out. By vertical filtering the corresponding effect will be present. Sample size (l*w*h) [mm]: 200*180*3

3.14 CFRP samples #6 and #7.

of this plate in the image decreases due to the skin effect. Penetration depth also depends on packaging and epoxy density, number of rovings and layers, and the number of different angle orientations.

The second specified patterns of defects are delamination effects between the carbon fiber and the epoxy resin layer, and cracks of the fiber plate extending through different layers. For damage initialized with a delamination, the epoxy loses its adhesive properties to the carbon fiber and, ultimately, the fiber will be destroyed. This is the reason for the high priority of delamination detection.

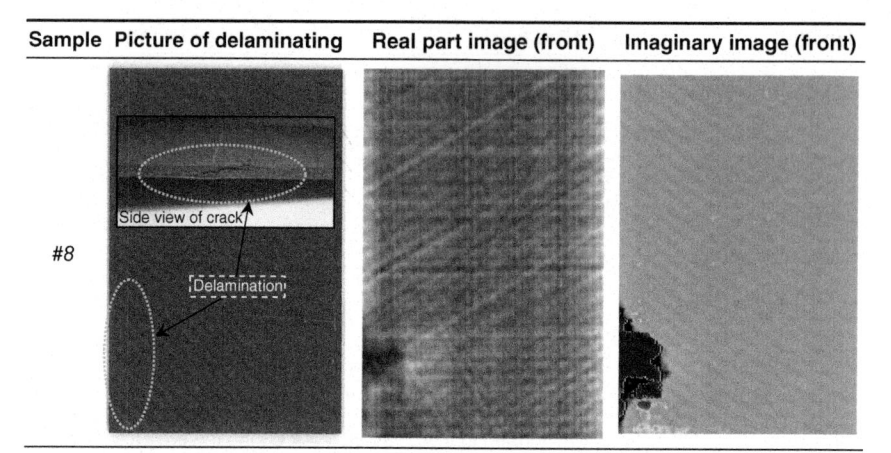

48 Non-destructive evaluation (NDE) of polymer matrix composites

Sample	Picture of delaminating	Real part image (front)	Imaginary image (front)
#8	Side view of crack / Delamination		

Delamination in a 12-layered CFRP combined with some inner material cracks. Signal optimization due to the use of phase image analysis.
Sample size (*l*w*h*) [mm]: 100*150*3

3.15 Delaminating in CFRP.

In sample #8 of Fig. 3.15, too high in-plane compressive forces were inducted into the 12 layer CFRP with the angle settings of 2 × 90°/03°/120°/30°/150°/60°. The first six layers are visible in the complex C-scan image. If there were defects, as in samples #6 or #7, they would become visible. In the real image of Fig. 3.15 the delamination is clearly visible due mainly to the interrupted cross-connection of neighborhood fibers. In the imaginary image, dielectric influences due to the created air insertions are responsible for the contrast.

The sensor is not optimized for the required frequencies suitable to detect deeper defects. In other words, the chaotic conductivity arrangements which are higher than the lift-off signal resulting from lower measuring frequencies make it very difficult to inspect these deeper areas. This explains why these layers and the real volume extension of the delamination might not be complete. Larger sensors with lower resolution but higher penetration depths at higher frequencies are the solution.

3.4 Analytical methods for data processing

Algorithms for image processing of NDE data are discussed in several chapters of this book and can also be applied to eddy current images. In the following section the focus is on the textural analysis of multilayer non-woven fabrics, a major field of application of HF EC.

3.4.1 Data processing by 2D fast Fourier transformation (2D FFT)

By two-dimensional fast Fourier transformation (2D-FFT) processing the spatial image with an intensity value for each pixel, the image can be transformed into the frequency domain. 2D-FFT produces two output pictures, a real and an imaginary image for the spatial image input. In real-world images, which have a natural distribution, it is usually assumed that the real part represents image intensity and the imaginary part is 0. The technique is often used to reduce image noise or blur by filtering distinguished frequencies.

For images with textural information the imaginary part of the FFT is not 0. For the image $f(x,y)$, its 2D FFT $F(u,v)$ is represented by using the formula below, where u, v are the coordinates of the transformed image:

$$F(u,v) = \frac{1}{MN} \sum_{x=0}^{M} \sum_{y=0}^{N} f(x,y) * e^{-j2\pi\left(\frac{u*x}{M}\right)+\left(\frac{v*y}{N}\right)} \qquad [3.3]$$

with: $j = \sqrt{(-1)}$.

The inverse transformation can be obtained by using the formula:

$$f(x,y) = \sum_{u=0}^{M} \sum_{v=0}^{N} F(u,v) * e^{j2\pi\left(\frac{u*x}{M}\right)+\left(\frac{v*y}{N}\right)} \qquad [3.4]$$

Real and imaginary parts of the complex 2D-FFT output vectors are visualized with its absolute values:

$$ABS[F(u,v)] = \sqrt{(REAL)^2 + (IMAGINARY)^2} \qquad [3.5]$$

Some simple examples of textural images and their specific 2D-FFT are shown in Fig. 3.16. As you can see the image frequencies increases from image center to the edges. Rotation of the source image results in a rotation of the frequency maxima in the 2D-FFT plot. The higher the values of different orientated image frequencies, the more local the maxima in the frequency domain. The examples given in Fig. 3.16 are used to explain the potential of 2D-FFT for textural analyses of non-woven structures.

In relation to the texture of CFRPs, the image frequency correlates with gap size and fiber bundle size whereas the rotation of the frequency maxima represents layer orientation. If a structure contains a discrete number of different gap sizes between fiber bundles, a discrete number of image frequencies will occur (3rd example). Due to the simple interpretable frequency signal, 2D-FFT is a probate method for determination of textural information of non-woven fabrics. Figure 3.17 (b) shows a 2D-FFT for a

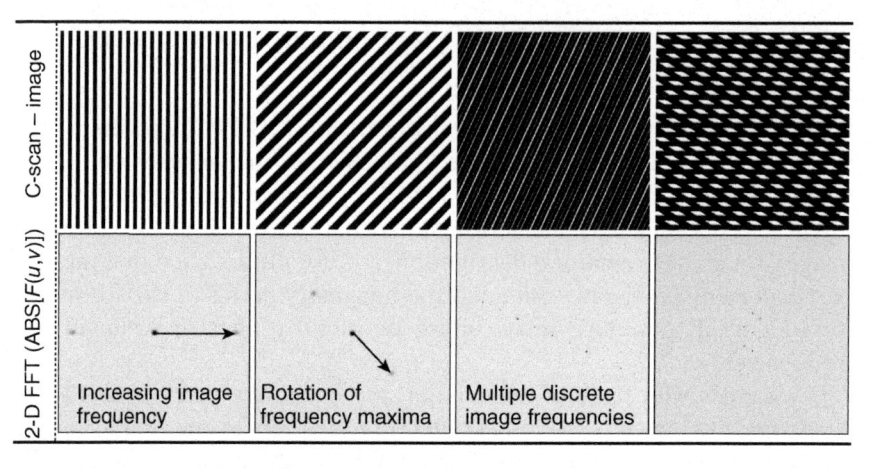

3.16 Basic principles of 2D fast Fourier transformation (images and their corresponding 2D-FFT).

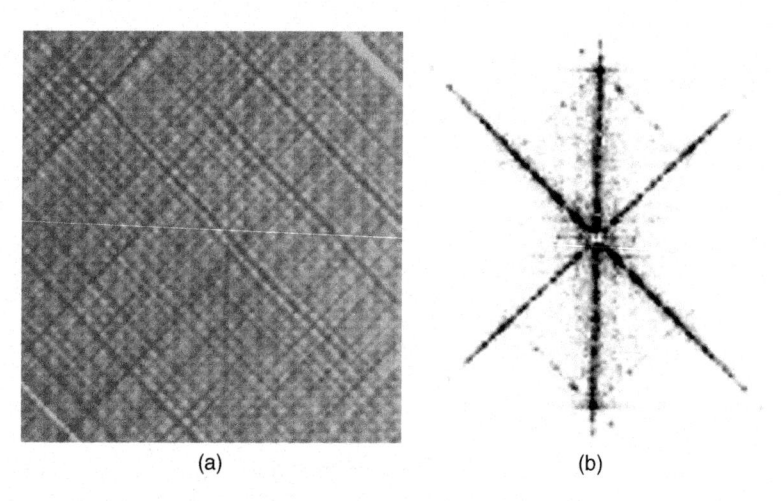

3.17 (a) HF EC image of a 3 axial non-woven fabric (30 × 30 cm^2) and (b) the corresponding 2D-FFT.

real HF EC image of a 3 axial non-woven fabric. Clearly visible are the three characteristic lines indicating the 45°, 0° and −45° layers.

As described in Fig. 3.16; 3rd example, multiple discrete points will arise if multiple discrete fiber bundle gaps exist in the original image. With HF EC images of real materials, often a broad distribution of different gap sizes exist, each of which causes lines in the 2D-FFT.

3.4.2 Image processing by phase-independent filter types

In Fig. 3.18 the effects of using a low- and a high-pass filter are shown. High-frequency image parts become invisible by performing a low-pass filtering. The use of a phase-independent high-pass filter on its own is not a good way to characterize fiber orientations. For textural analysis the relevant image frequencies needs to be marked by using special filter-like squares, triangles or polygons. Several examples are described in Fig. 3.19. All images were processed with ImageJ (http://imagej.nih.gov/ij/index.html). ImageJ and its Java source code are freely available in the public domain.

3.4.3 Textural analysis

To extract textural information such as missing bundles, lanes, misalignments of fiber bundles or of whole layers from the eddy current images, an automated classification process is needed. In the examples shown in Fig. 3.19, different orientated filters were applied. As a first step, the layer to be investigated has to be selected in the 2D-FFT spectra. This can be done by

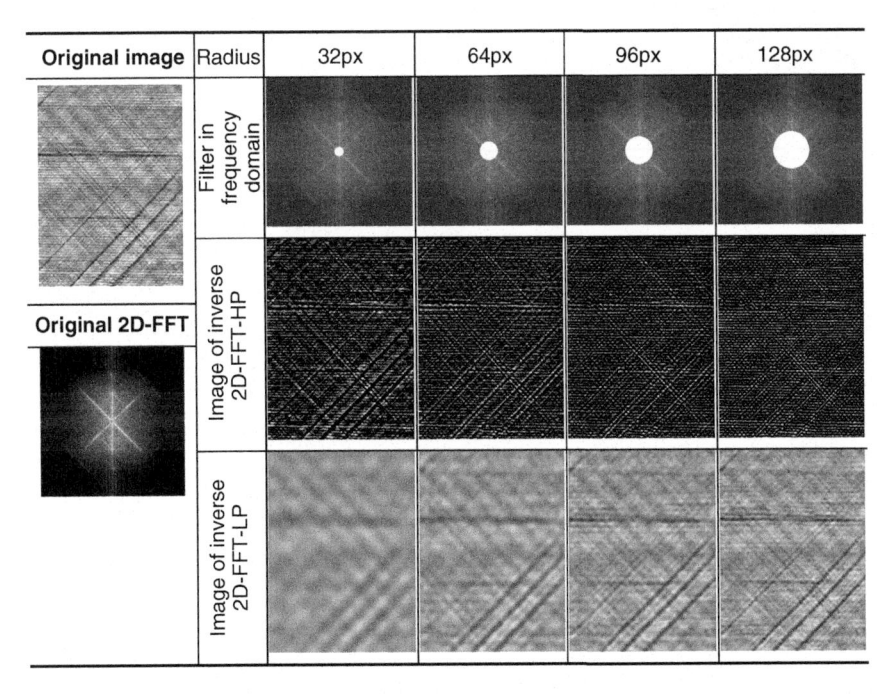

3.18 EC image on triaxial (45°, 0, −45°) non-woven fabrics (20 × 30 cm²) with high- and low-pass filter by using radial windowing in frequency domain.

Filtered frequency domain	Inverse 2D-FFT	Threshold	Classification	Regions of interest with measured area values
				# Area 1 50 2 37 3 26 4 24 5 32 6 45 7 310 8 1116 9 75 10 33 11 417 12 39 13 17 14 28
				# Area 1 152 2 8 3 9 4 14 5 13 6 1372 7 11 8 541 9 1429 10 5 11 1272 12 9 13 17
				# Area 1 2 2 43 3 35 4 52 5 20 6 30 7 318 8 1 9 16 10 104 11 60

3.19 Results and textural analyses of 2D-FFT processed images.

suppressing the frequency from the layers that should not be observed by using filters, requiring a priori knowledge of the layer design. The first column of Fig. 3.19 shows the 2D-FFT with filter settings for each observed layer. As a second step, an inverse 2D-FFT needs to be performed to reconstruct the artificial single layer image. The second column of Fig. 3.19 shows the inverse 2D-FFT after filtering. The remaining image contains mostly information from the selected layer. By defining a threshold value, critical image areas can be distinguished from non-critical areas. In this case the threshold minima were set to maximum entropy 25 [11]. The classification itself takes place by labeling each region of interest (ROI) and counting the pixel area.

Another parameter to be measured is the layer orientation. The amplitude maxima in the frequency domain can be used for precise angle determination for each layer. In Fig. 3.20 the whole procedure for orientation analysis is shown. One hidden layer in the specimen was displaced by 94.5°

Eddy current techniques 53

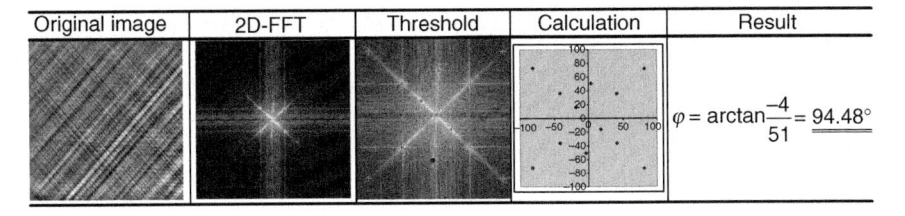

Original image	2D-FFT	Threshold	Calculation	Result
				$\varphi = \arctan\dfrac{-4}{51} = 94.48°$

3.20 Measurement of fiber orientation.

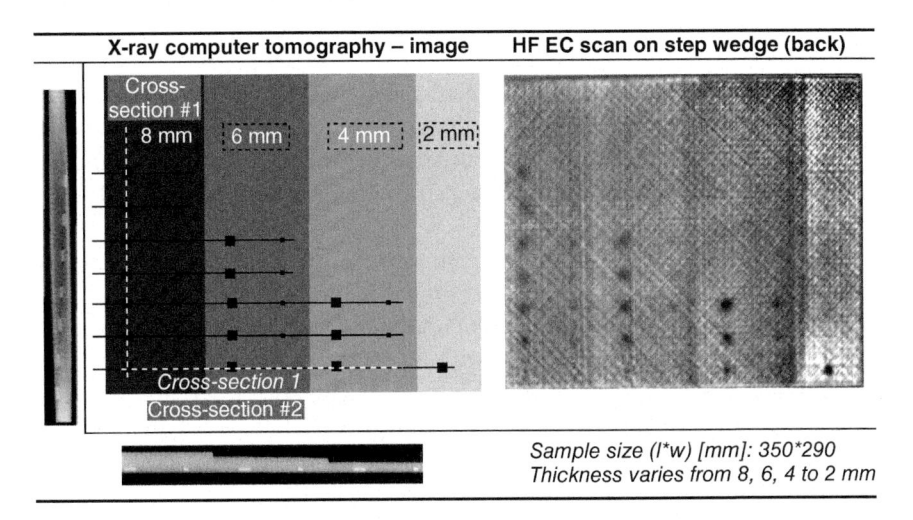

3.21 Characterization of penetration depth (sample # 9).

$-90° = 4.5°$. This example illustrates that the layer orientation can be measured with an accuracy of better than 1°.

3.4.4 Depth resolution in consolidated components

As discussed, the penetration depth strongly relates to the probe, and used frequency to the material configuration. Figure 3.21 shows an experimental set-up which illustrates the penetration depth for a given material. A wedge-like plate with stepwise enhanced wall thickness was produced. For the basis substrate, a 2 mm plate was used, laminated with a smaller second 2 mm plate, etc. Inside the 2 mm plate, artificial defects were inserted at different depths. The X-ray CT images visualize the position and depth of the artificial defects. A reference specimen with four different thicknesses and defects inserted at different thicknesses was used. The step wedge was scanned from the back. The laminated test structures of cross-section #1 and #2 are visible over the entire depth range in the EC image. The defect amplitude for cross-section #1 decreased because of the increasing test volume that needs to be penetrated. The same effect was visible in cross-section #2. In this case a penetration depth of at least 8 mm was achieved.

3.5 Conclusion

In the decision matrix of non-destructive testing methods for CFRPs at the early product stages, the eddy current method can provide unique information when compared with other NDE methods. CFRPs have a low electrical conductivity which makes them suitable for the application of quality testing using eddy currents. Over recent years, high-resolution eddy current imaging technology has been substantially developed. Industrially applied systems such as the EddyCus® system are now available. HF EC imaging is a proven technology for inspection of raw carbon fiber fabrics and final, consolidated, CFRP.

Eddy current-based methods are interesting owing to the simplicity of machinery integration. Unlike ultrasonic-based methods, no coupling is required. Because of this non-contact application, eddy current sensors can be used directly in the process flow without affecting the material properties and quality. Textural analyses and fault testing can performed at high resolution to document and guarantee the quality of RCF material and finished components. Knowledge of the product quality very early in the value adding chain helps to increase the yield of the production process, and the safety of the final product.

3.6 References

1. NDT Resource Center, Depth of Penetration & Current Density, 'http://www.ndt-ed.org/EducationResources/CommunityCollege/EddyCurrents/Physics/depthcurrentdensity.htm'.
2. D.J. Hagemaier, 'Eddy-current standard depth of penetration', *Materials Evaluation*, Vol. 43, No. 11, pp. 1438–1441, 1985.
3. Z. Molttl, 'The quantitative relation between true and standard depth of penetration of air-cored probe coils in eddy current testing', *NDT International*, Vol. 23, No. 1, pp. 11–18, 1990.
4. R. Lange, G. Mook, 'Structural analysis of CFRP using eddy current methods', *NDT & E International*, Vol. 27, Issue 5, pp. 241–248, 1994.
5. G. Mook, R. Lange, O. Koeser, 'Non-destructive characterisation of carbon-fibre-reinforced plastics by means of eddy-currents', *Composites Science and Technology*, Vol. 61, Issue 6, pp. 865–873, 2001.
6. B.A. Abu-Nabah, P.B. Nagy, 'High-frequency eddy current conductivity spectroscopy for residual stress profiling in surface-treated nickel-base superalloys', *NDT & E International*, Vol. 40, Issue 5, pp. 405–418, 2007.

7. H. Heuer, S. Hillmann, M. Klein, N. Meyendorf, 'Sub surface material characterization using high frequency eddy current spectroscopy', *MRS Proceedings*, Vol. 1195, Issue 1, 2009.
8. G. Mook, F. Michel, J. Simonin, Electromagnetic imaging using probe arrays, 17th World Conference on Nondestructive Testing, 25–28 Oct 2008, Shanghai, China.
9. A. Yashan, W. Bisle, T. Meier, 'Inspection of hidden defects in metal–metal joints of aircraft structures using eddy current technique with GMR sensor array'. In: 9th European Conference on NDT – ECNDT Berlin, 25–29 Sep 2006, Paper Tu.4.4.4.
10. Y. Wuliang, P.J. Withers, U. Sharma, A.J. Peyton, 'Noncontact characterisation of carbon-fiber-reinforced plastics using multifrequency eddy current sensors', *IEEE Transactions on, Instrumentation and Measurement*, Vol. 58, pp. 738–743, 2009.
11. J.N. Kapur, P.K. Sahoo, A.C.K. Wong, 'A new method for gray-level picture thresholding using the entropy of the histogram', *Graphical Models and Image Processing*, Vol. 29, No. 3, pp. 273–285, 1985.

4
Non-destructive evaluation (NDE) of composites: introduction to shearography

D. FRANCIS, Cranfield University, UK

DOI: 10.1533/9780857093554.1.56

Abstract: Shearography is an optical technique that relies on interference of laser speckle patterns to visualise variations in surface strain. As it provides full-field measurements and is particularly robust against external vibrations for an interferometric system, it is well suited to industrial non-destructive testing. This chapter discusses the principle of operation of shearography, explaining the process of fringe formation and interpretation and the use of phase analysis to provide quantitative data and improve fringe contrast and measurement sensitivity. The application of shearography for non-destructive evaluation of composite materials is then considered and is compared with other well-established techniques.

Key words: shearography, speckle interferometry, non-destructive evaluation (NDE), phase analysis, surface strain measurement.

4.1 Introduction

Speckle shearing interferometry, or shearography (Hung, 1982), is an interferometric optical measurement technique that utilises the laser speckle effect. It is derived from a similar technique, electronic speckle pattern interferometry (ESPI) (Sharp, 1989), which is related to holography. These techniques are appealing for use in non-destructive testing owing to their non-contact nature and ability to provide a full-field measurement. The basic principle of speckle interferometry involves the comparison of speckle patterns recorded before and after a load or displacement is applied to the component under test. Correlation of the speckle patterns produces a fringe pattern, with the fringes representing a contour map of the measurand. Further processing is required to determine the phase of the correlation fringe pattern and this is usually done to improve the fringe contrast and to provide quantitative data. Shearography offers a number of advantages over ESPI for industrial testing. Unlike ESPI, which directly measures the displacement of the surface of the object under investigation, shearography measures the spatial derivative of displacement and is therefore directly sensitive to surface strain. In addition, shearography is remarkably resilient to environmental disturbances for an interferometric technique and is therefore better suited to operation in industrial environments where it has become well established (Hung, 1997). It is commonly used for the detection of defects in composite materials (Hung, 1996) where the presence of flaws is visualised as strain anomalies in the region of the defect which become apparent when the component is placed under an applied load.

In this chapter the principle of operation of shearography is explained. The concept of laser speckle is introduced and the theory of speckle interferometric fringe formation is presented in Section 4.2. Experimental configurations and the image processing requirements for temporal phase analysis are discussed in Section 4.3. Section 4.4 describes shearography as a tool for non-destructive evaluation (NDE) of composite materials and Section 4.5 compares it with other well-established techniques. A brief discussion on future trends and sources of further information follow in Sections 4.6 and 4.7.

4.2 The theoretical principles of shearography

This section introduces laser speckle and then explains the principles of fringe formation and fringe pattern interpretation.

4.2.1 Laser speckle

Shearography is reliant on the laser speckle effect, which occurs whenever light with sufficient temporal and spatial coherence is scattered from an optically rough surface whose topographical features are greater than the wavelength of the light. When this occurs, a complex granular pattern is produced which is the result of the constructive and destructive interference of light scattered from different points on the surface. Laser speckle is termed differently depending on whether or not it is imaged. In the non-imaged case it is known as objective speckle (Fig. 4.1a) and the average speckle size $\langle \sigma_O \rangle$ is given by (Cloud, 2007)

$$\langle \sigma_O \rangle = \frac{\lambda D}{A} \tag{4.1}$$

where A is the diameter of the illuminated area, D is the distance between the object surface and the detector and λ is the wavelength of the laser light. When the pattern is imaged it is referred to as subjective speckle (Fig. 4.1b) and the average speckle size is dependent on the parameters of the imaging system (Cloud, 2007)

$$\langle \sigma_S \rangle = \lambda \frac{F(1+M)}{a} \tag{4.2}$$

where F/a is the F-number of the lens (focal length over aperture) and M is the magnification. In shearography it is subjective speckle that is relevant

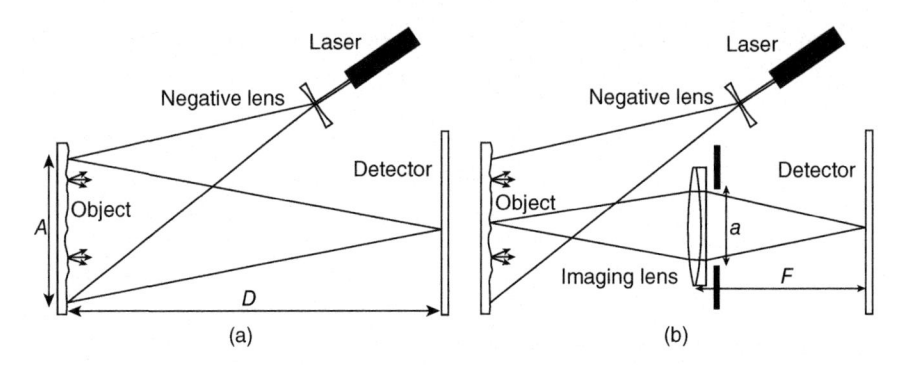

4.1 The formation of objective speckle (a) and subjective speckle (b).

because measurements are made by imaging the surface of the component under investigation. The F-number of the imaging lens can be selected to match the speckle size with the resolution of the camera used.

4.2.2 Coherent addition of speckle patterns

Interferometric laser speckle techniques, such as shearography, rely on the interference of laser speckle with another light field. In shearography this is an identical but spatially shifted (sheared) speckle pattern, whereas in ESPI this tends to be a reference beam with a smooth wavefront. The resulting light field is referred to as an interferometric speckle pattern, the intensity of which can be expressed as

$$I_R(x, y) = a_1^2(x, y) + a_2^2(x, y) + 2a_1(x, y)a_2(x, y)\cos[\Delta\psi(x, y)] \quad [4.3]$$

at each point in the image plane (x, y). Here a_1 and a_2 are the amplitudes of the light distributions that follow the two different paths from the object through the optical system to the image plane. $\Delta\psi$ is the phase difference of light traversing the two paths and varies randomly in space due to the speckle effect. This intensity distribution is recorded as a digital image and is referred to as the reference frame. Applying a load to the object under test causes its surface to deform, resulting in a change in the phase difference between the light in the two paths through the interferometer. The intensity of the speckle pattern after deformation can be expressed as

$$I_S(x, y) = a_1^2(x, y) + a_2^2(x, y) + 2a_1(x, y)\cos[\Delta\psi(x, y) + \Delta\phi(x, y)] \quad [4.4]$$

where $\Delta\phi$ is the change in phase due to the deformation of the object. This phase change is observed as a change in the intensity of the individual speckles, which cycles from dark to light to dark again as the phase change varies from 0 to 2π. The image recorded after deformation is referred to as the signal frame. The reference and signal frames appear as speckle patterns with no obvious qualitative difference between them. To visualise the surface deformation the two images are correlated, typically by subtracting the intensity values at each pixel location (x, y). This is easily achieved digitally with the image being rectified afterwards to take into consideration any negative values. The resulting intensity distribution can be written as

$$I = I_S - I_R = 2a_1 a_2 [\cos(\Delta \psi) - \cos(\Delta \psi + \phi)] \qquad [4.5]$$

where the (x, y) dependencies have been dropped for clarity. This can be rearranged using a trigonometrical identity to give

$$I = 4a_1 a_2 \sin\left(\Delta \psi - \frac{\phi}{2}\right) \sin\left(\frac{\phi}{2}\right) \qquad [4.6]$$

The first term $4a_1 a_2$ represents the background intensity and the second term $\sin(\Delta \psi - \phi/2)$ corresponds to the random speckle intensity. The third term $\sin(\phi/2)$ is an intensity modulation which maximises the function whenever $\phi = (2m+1)\pi$ and minimises it whenever $\phi = 2m\pi$, where m is an integer. The resulting image is a fringe pattern with the fringes representing contours of equal phase difference. This image is referred to as a correlation fringe pattern. Figure 4.2 shows the typical correlation fringe patterns observed using ESPI and shearography for an out-of-plane displacement of a couple of micrometres. The graphs show the approximate variation in displacement and displacement gradient with position for a line horizontally across the image. The fringes were obtained from a point out-of-plane displacement of a flat plate clamped around its perimeter. The field of view corresponds to a region of approximately 10 cm^2 on the object's surface.

4.2.3 Interpretation of the fringe pattern

In shearography, interferometric speckle is created through the coherent addition of two laterally displaced speckle patterns. This lateral shift is known as the image shear and is typically a few millimetres, or about 10–20% of the image size. The component that provides the image shear is known as the shearing device and typical examples include a prism in front of one half of the imaging lens (Hung et al., 1978), a Michelson interferometer (Steinchen et al., 1996) or a diffractive optical element (DOE) (Mihaylova et al., 2004). The shearing device causes light scattered from each point on the object's surface to follow two different paths to the image plane. Therefore, light arriving at each pixel in the camera's detector array is made up of light scattered from two points on the surface separated by the shear distance.

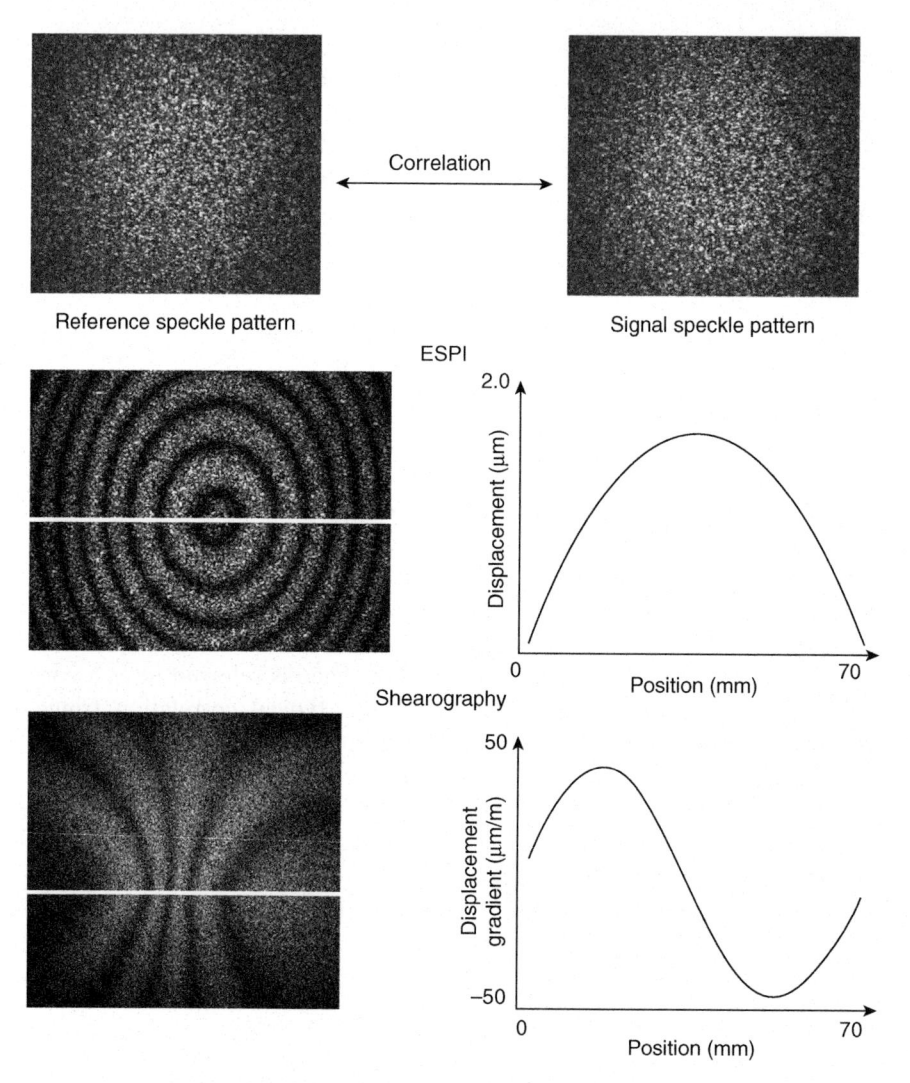

4.2 Correlation of interferometric speckle patterns recorded before (reference frame) and after (signal frame) deformation of the object results in a fringe pattern. The fringes correspond to surface displacement for ESPI and displacement gradient for shearography. The plots show the approximate variation in displacement and displacement gradient for the central row of the two fringe patterns (highlighted by the white horizontal lines).

Consider a point P on the surface of the object positioned at the coordinate location (x, y, z), as shown in Fig. 4.3. Light from the source S is scattered from P and follows one of the two paths through the shearing device before arriving at the detector D at pixel location (x_D, y_D, z_D). Light

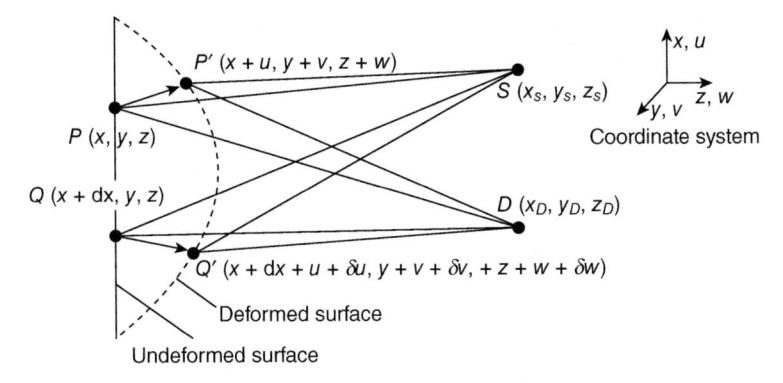

4.3 The optical paths from the point source *S* to two surface points *P* and *Q* separated by the shear amount d*x* and then to a pixel in the detector array *D*. After deformation the surface points are displaced to *P′* and *Q′*. The coordinate system indicates that *u*, *v* and *w* are the displacement components in the *x*, *y* and *z* directions.

scattered from the neighbouring point Q, separated from P by the shearing amount dx, is located at $(x+dx, y, z)$ and follows the other path through the shearing device to the pixel D. After a load is applied to the object, the surface deforms and P is shifted to P' at $(x+u, y+v, z+w)$, where (u, v, w) is the displacement vector at P. The point Q meanwhile is shifted to Q' at $(x+dx+u+\delta u, y+v+\delta v, z+w+\delta w)$.

The change in the path length due to the deformation for the point P is given by

$$\Delta L_P = (SP' + P'D) - (SP + PD) \quad [4.7]$$

where

$$SP = \left[(x-x_S)^2 + (y-y_S)^2 + (z-z_S)^2\right]^{1/2}$$
$$PD = \left[(x-x_D)^2 + (y-y_D)^2 + (z-z_D)^2\right]^{1/2}$$
$$SP' = \left[(x+u-x_S)^2 + (y+v-y_S)^2 + (z+w-z_S)^2\right]^{1/2}$$
$$P'D = \left[(x+u-x_D)^2 + (y+v-y_D)^2 + (z+w-z_D)^2\right]^{1/2}$$

[4.8]

This can be simplified with the aid of a binomial expansion (Hung and Liang, 1979) to give

$$\Delta L_P = Au + Bv + Cw \quad [4.9]$$

where A, B and C are geometrical factors dependent on the observation and illumination positions. The point Q can be analysed similarly resulting in

$$\Delta L_Q = A(u+\delta u) + B(v+\delta v) + C(w+\delta w) \quad [4.10]$$

The difference in path length between Q and P is therefore

$$\Delta L_Q - \Delta L_P = A\delta u + B\delta v + C\delta w \qquad [4.11]$$

The factors A, B and C are equivalent to the x, y and z components of the sensitivity vector which is defined as the bisector of the observation vector and the illumination vector and is expressed as

$$\mathbf{k} = \hat{o} - \hat{i} \qquad [4.12]$$

If the shear distance dx is small compared to the observation and illumination distances, the difference in displacement between P and Q approximates to the displacement derivative. The optical phase difference of light scattered from the two points can then be expressed as

$$\Delta\phi = \frac{2\pi}{\lambda}\left(k_x \frac{\partial u}{\partial x} + k_y \frac{\partial v}{\partial x} + k_z \frac{\partial w}{\partial x}\right)\mathrm{d}x \qquad [4.13]$$

where k_x, k_y and k_z are the x, y and z components of k.

The strain component that shearography is sensitive to is therefore dependent on the observation and illumination directions of the optical configuration. Typically, a shearography measurement would consist of a combination of the two in-plane and one out-of-plane strain components. However, the configuration can be adjusted to isolate one or more of the strain components as will be discussed in Section 4.3. Equation [4.13] assumes shear in the x-direction. An equivalent expression can be derived with shear in the y-direction. This means that there are a total of six orthogonal strain components that can be measured with shearography, sufficient to fully characterise surface strain.

4.3 The practical application of shearography

This section describes practical shearography configurations and methods of isolating specific strain components. Commonly used image processing routines for improving measurements are also discussed.

4.3.1 Out-of-plane sensitive shearography

A typical optical configuration of a shearography system, based on a Michelson interferometer, is shown in Fig. 4.4. Light from the laser is expanded to illuminate a region of interest on the surface of the test object, forming a speckle pattern. Scattered light arriving at the beamsplitter is split with half of the energy transmitted to the reference mirror and half reflected to the shearing mirror. The shearing mirror is oriented at a slight angle with

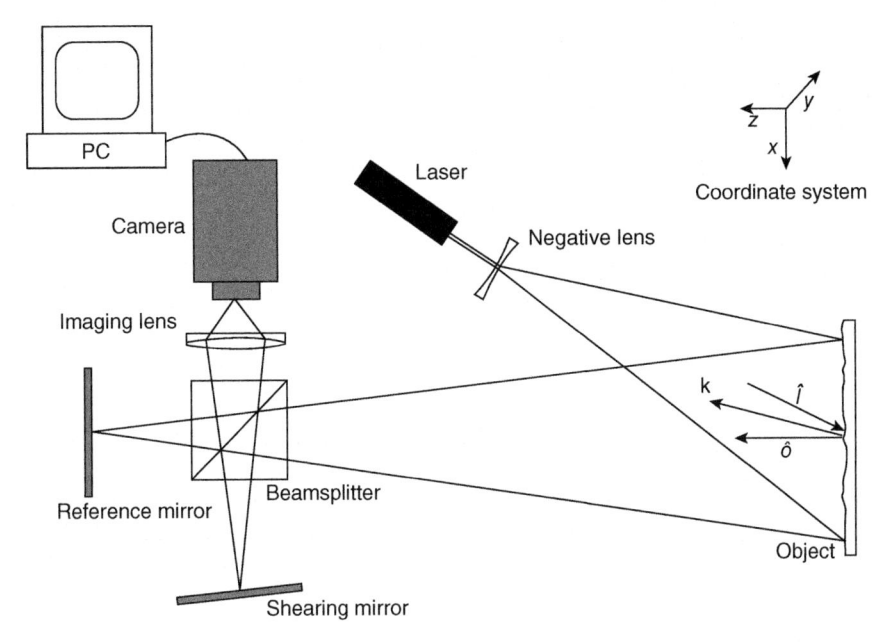

4.4 A conventional shearography arrangement based on a Michelson interferometer. The strain component the instrument is sensitive to is determined by the sensitivity vector **k**, which is predominantly in the out-of-plane (*z*) direction.

respect to the optical axis of the interferometer, which results in an image that is laterally shifted relative to the one from the reference mirror. Light reflected from the two mirrors recombines at the beamsplitter, forming an interferometric speckle pattern which is imaged by the camera.

A shearography measurement is sensitive to a strain component that is defined by the sensitivity vector **k**. This component contains contributions from the three orthogonal strain components as stated in equation [4.13]. To isolate just the out-of-plane component, the observation and illumination directions need to be collinear. When this is the case, the *x* and *y* components of the sensitivity vector become zero and the phase becomes

$$\Delta\phi = \frac{4\pi}{\lambda}\frac{\partial w}{\partial x}\mathrm{d}x \qquad [4.14]$$

However due to practical difficulties, achieving collinear illumination and observation can be problematic. Another solution to completely isolate the out-of-plane component is to use two illumination (or observation) directions as discussed in Section 4.3.2. However, in a configuration like that shown in Fig. 4.4, where the angle between the illumination and observation directions is small, the contribution from the in-plane components is small and the system is largely sensitive to the out-of-plane component. Generally this is considered to be sufficient for out-of-plane strain measurement.

Although the Michelson interferometer is more complex than a simple prism-based shearing device and is less light efficient (half the light is lost on the second pass through the beamsplitter) it offers a number of other practical advantages. The shearing mirror offers a convenient method of adjusting the shearing amount, and hence the sensitivity, and provides a simple way to switch between x and y shear directions. The reference mirror can be equipped with a piezoelectric transducer (PZT) which allows the phase within the interferometer to be adjusted. This allows the temporal phase stepping (Creath, 1985) technique to be implemented. Temporal phase stepping is an important technique for determining the interferometric phase $\Delta\phi$ of the fringe pattern, which is required for quantitative analysis and also results in much higher contrast images than the correlation fringe patterns obtained simply by subtraction (such as shown in Fig. 4.2). This is discussed in more detail in Section 4.3.3.

4.3.2 Multiple channel shearography

A measurement channel is a term used to define one illumination and one observation direction. A single measurement channel system is sensitive to one component of strain defined by the sensitivity vector as discussed in Section 4.3.1. In order to isolate one of the in-plane strain components, a system with two measurement channels can be used. The experimental arrangement of a typical system to do this is shown in Fig. 4.5. The arrangement consists of a pair of illumination beams oriented symmetrically about the observation axis. The object is illuminated sequentially by the first channel (channel with $\mathbf{k}_1$ in Fig. 4.5) and then by the second channel (channel with $\mathbf{k}_2$ in Fig. 4.5) (Steinchen et al., 1999). Images are recorded from the two channels before and after deformation and the phase for each is calculated. The phase of the fringe pattern obtained from the first channel is given by (Hung and Wang, 1996)

$$\Delta\phi_1 = \frac{2\pi}{\lambda}\left[(1+\cos\theta)\frac{\partial w}{\partial x} + \sin\theta\frac{\partial u}{\partial x}\right]dx \qquad [4.15]$$

where θ is the angle between the observation and illumination vectors. Similarly, the phase of the fringe pattern from the second channel is

$$\Delta\phi_2 = \frac{2\pi}{\lambda}\left[(1+\cos\theta)\frac{\partial w}{\partial x} - \sin\theta\frac{\partial u}{\partial x}\right]dx \qquad [4.16]$$

Introduction to shearography 65

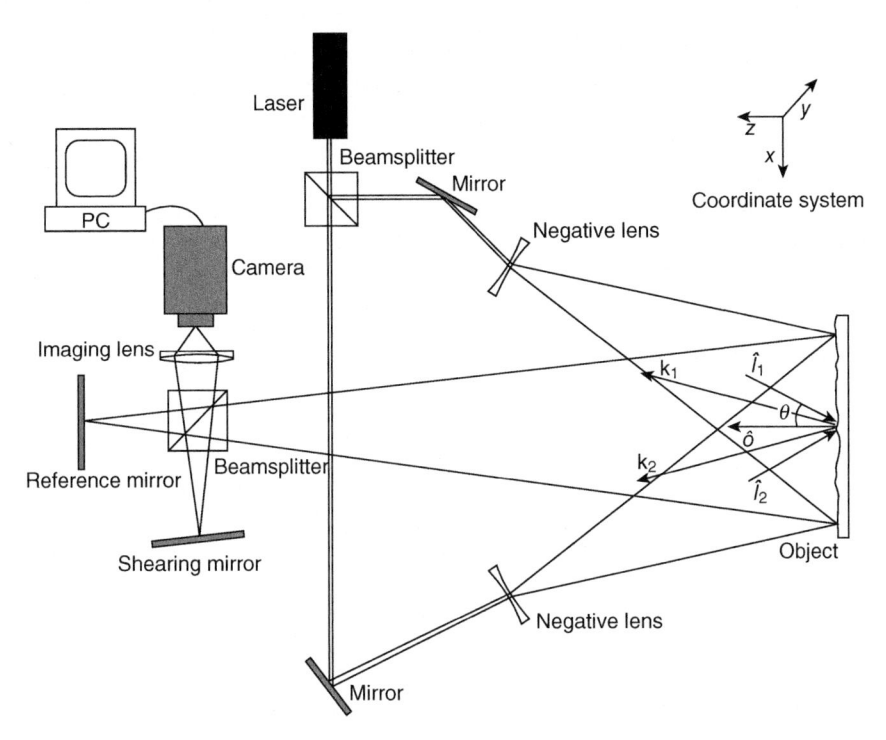

4.5 A dual-beam shearography system used for measuring in-plane strain. The two beams are aligned symmetrically and offer two sensitivity vectors, **k**$_1$ and **k**$_2$. Measurements from the two channels can be combined to provide the out-of-plane strain ($\partial w/\partial x$) or the in-plane strain ($\partial v/\partial x$).

The out-of-plane strain can be isolated by summing the two phase measurements to give

$$\Delta\phi_1 + \Delta\phi_2 = \frac{4\pi}{\lambda}\left[(1+\cos\theta)\frac{\partial w}{\partial x}\right]dx \qquad [4.17]$$

The in-plane strain can be isolated by subtracting the two phase measurements to give

$$\Delta\phi_1 - \Delta\phi_2 = \frac{4\pi\sin\theta}{\lambda}\frac{\partial u}{\partial x}dx \qquad [4.18]$$

In this example, the $\partial u/\partial x$ component is isolated because the illumination beams are both located in the *x–z* plane. Aligning them in the *y–z* plane instead results in the $\partial v/\partial x$ component being isolated. This configuration only allows one in-plane and the out-of-plane components to be isolated.

To obtain all three components available from a single shear direction with a single measurement, a system with three illumination directions can be used (Aebischer and Waldner, 1997). If the illumination positions are placed at three corners of a square centred on the observation axis and collimating optics are used, the in-plane components can be isolated by subtracting measurements from channels on the same side of the square and the out-of-plane component can be obtained by summing the measurements from the channels diagonally opposite, as illustrated in Fig. 4.6. In other cases the orthogonal strain components can be obtained using the coordinate transformation (James and Tatam, 1999)

$$\begin{pmatrix} \frac{\partial u}{\partial x} \\ \frac{\partial v}{\partial x} \\ \frac{\partial w}{\partial x} \end{pmatrix} = \frac{2\pi}{\lambda} \begin{pmatrix} k_{x1} & k_{y1} & k_{z1} \\ k_{x2} & k_{y2} & k_{z2} \\ k_{x3} & k_{y3} & k_{z3} \end{pmatrix}^{-1} \begin{pmatrix} \Delta\phi_1 \\ \Delta\phi_2 \\ \Delta\phi_3 \end{pmatrix} dx^{-1} \qquad [4.19]$$

where the numerical subscripts refer to measurement channels 1–3 respectively.

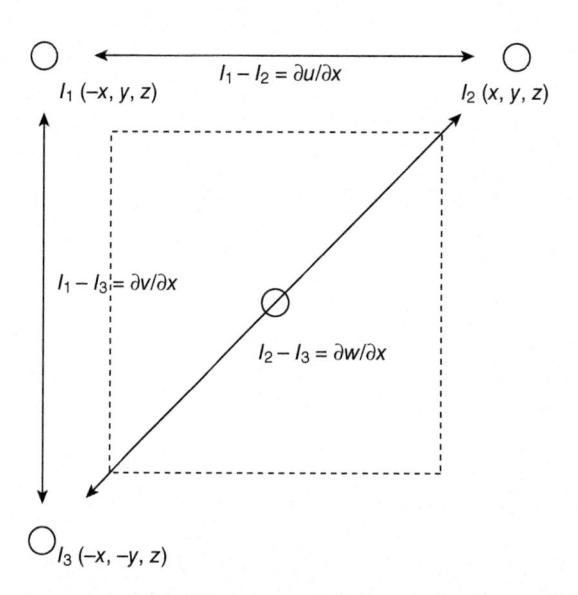

4.6 The locations of illumination positions l_1–l_3 at three corners of a square. Combination of measurements from different channels isolates a different strain component.

4.3.3 Temporal phase evaluation

Temporal phase stepping is a technique that is often used in shearography to obtain quantitative measurements of surface strain variations. It is also regularly used in qualitative investigations due to the fact that it allows images of much higher contrast than typical correlation fringe patterns (such as the one shown in the bottom left of Fig. 4.2) to be obtained. Also, since the phase is determined at each pixel, rather than just at the fringe maxima and minima, much greater sensitivity is achieved (Steinchen et al., 1998). Temporal phase stepping involves the recording of a series of interferograms with a known phase difference between them and combining them using a phase stepping algorithm. The resulting image is known as a wrapped phase map, in which the calculated phase values are bound between $-\pi$ and $+\pi$. Using a Michelson shearing interferometer, phase stepping can be implemented easily by mounting the reference mirror on a PZT (piezo-electric transducer) which can be controlled using a DAC (digital-to-analogue converter).

The intensity I at each pixel (x, y) in a speckle interferogram can be expressed as

$$I(x, y) = I_0(x, y)\{1 + \gamma_0(x, y)\cos[\phi(x, y)]\} \quad [4.20]$$

where I_0 is the distribution of background light, γ_0 is the intensity modulation and ϕ is the phase. Since there are three unknowns in equation [4.20], at least three phase stepped interferograms are required to compute the phase. The three-step algorithm is given by (Creath, 1993)

$$\phi = \tan^{-1}\left(\sqrt{3}\frac{I_3 - I_1}{2I_2 - I_1 - I_3}\right) \quad [4.21]$$

where $I_1 - I_3$ are the intensity values at each pixel for the three phase-stepped interferograms, which each have a relative phase step of $2\pi/3$. The (x, y) dependencies have been dropped for clarity. When implemented, phase stepping algorithms tend to use the extended arctan function (known as atan2 in a numerous programming languages) which results in phase measurements that are wrapped modulo 2π. More sophisticated algorithms that require more than three phase stepped interferograms exist and are more resilient to uncertainties due to miscalibration of the phase shifting device. A particularly popular one is the five-step algorithm (Hariharan et al., 1987) in which the images have a $\pi/2$ phase step between them.

$$\phi = \tan^{-1}\left(\frac{2(I_2 - I_4)}{2I_3 - I_5 - I_1}\right) \quad [4.22]$$

68   Non-destructive evaluation (NDE) of polymer matrix composites

The procedure followed to obtain a wrapped phase map is illustrated in Fig. 4.7, presented here using synthesised data. Before loading, a series of five phase-stepped interferometric speckle patterns is recorded. The images are then combined on a pixel-wise basis using equation [4.22], resulting in the phase distribution corresponding to the reference state. After loading, the process is repeated and the signal phase distribution is obtained. The reference and signal phases are then subtracted using (Steinchen *et al.*, 1998)

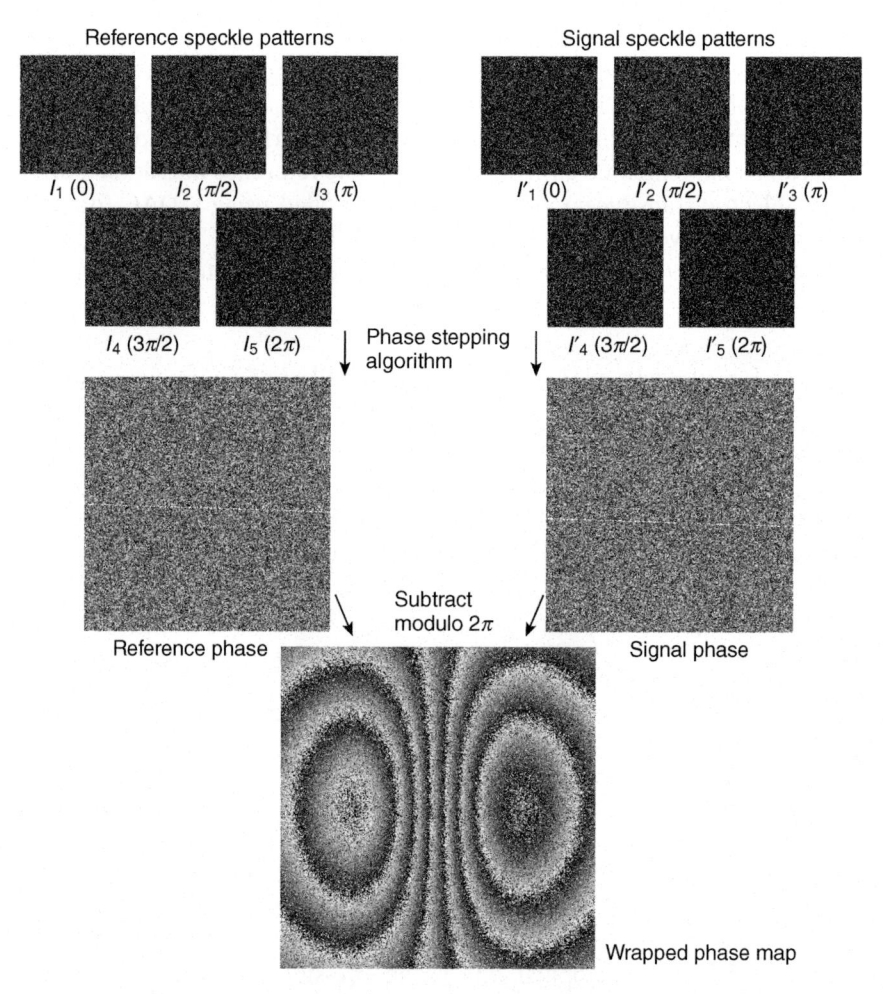

4.7 The procedure followed in temporal phase stepping for shearography: phase stepped interferograms (I_1–I_5, added phase values are shown in brackets) are combined with a phase stepping algorithm before and after deformation. The signal and reference phases are subtracted to produce the wrapped phase map.

Introduction to shearography 69

$$\Delta\phi = \begin{cases} \phi_S - \phi_R & \text{for } \phi_S \geq \phi_R \\ \phi_S - \phi_R + 2\pi & \text{for } \phi_S < \phi_R \end{cases} \quad [4.23]$$

to obtain the wrapped phase map, where black points represent $-\pi$ and white points represent $+\pi$.

4.3.4 Processing phase maps

Wrapped phase maps obtained using interferometric speckle techniques are often corrupted with high-frequency noise, typically due to speckle decorrelation (Aebischer and Waldner, 1999). Image processing using a low-pass averaging filter is often used to reduce the noise content. A low-pass filter kernel typically takes the form

$$K = \frac{1}{9}\begin{bmatrix} 1 & 1 & 1 \\ 1 & 1 & 1 \\ 1 & 1 & 1 \end{bmatrix} \quad [4.24]$$

A two-dimensional convolution applied between the image and the filter kernel has the result of smoothing out features in the image. The operation is equivalent to placing the kernel at each pixel (x, y) and replacing the central value with the average of all values within the 3×3 neighbourhood. Using a larger filter kernel, e.g. 5×5 or 7×7 enhances the smoothing effect. The 1/9 in equation [4.24] ensures that the total average value of the output image is the same as the input image. Simply applying the filter to the wrapped phase map would result in blurring of the desired phase fringe discontinuities. Instead, the sine and cosine of the wrapped phase are calculated and filtered, as shown in Fig. 4.8. Recalculating the phase map with an arctan function produces a wrapped phase map with reduced high-frequency noise content that is of much higher contrast. The phase map can be further improved by repeating the sine-cosine filtering process, ensuring an intermediate phase recalculation after each filter pass (Aebischer and Waldner, 1999).

The wrapped phase map can be further processed to remove the fringe discontinuities resulting in a continuous, quantitative measure of the phase, which is proportional to the displacement derivative in the direction of the sensitivity vector. This process is known as phase unwrapping and is illustrated in Fig. 4.9. In simple form, the process involves adding 2π to pixel values whenever the difference between adjacent pixels is greater than π and subtracting 2π whenever the difference exceeds $-\pi$. This simple procedure is successful only if the wrapped phase map is particularly free from noise and there are no breaks in the fringe discontinuities. In practice this is not usually the case and so more sophisticated unwrapping algorithms

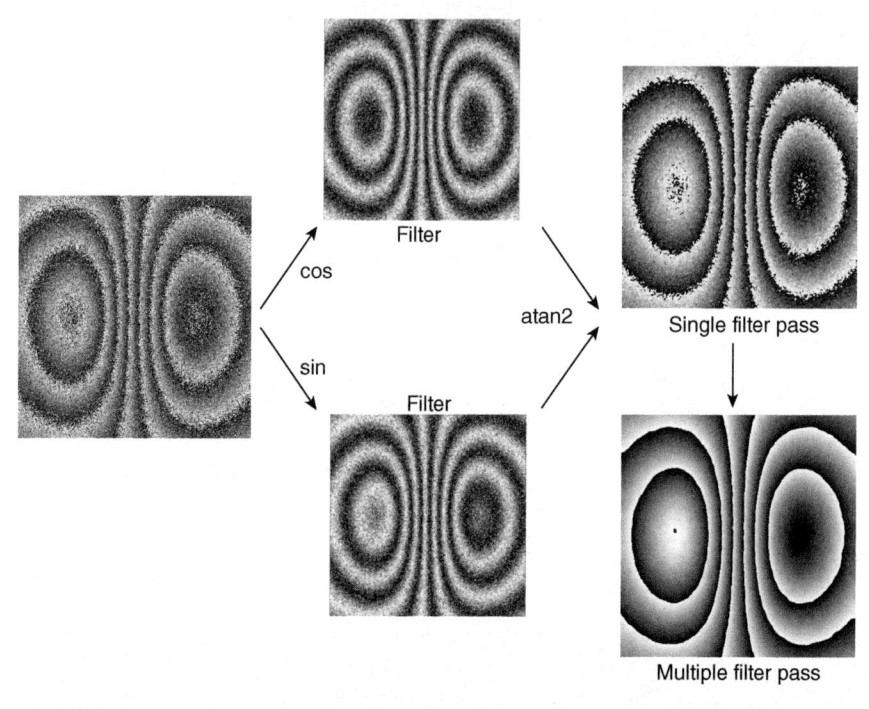

4.8 To reduce noise in the wrapped phase map, the sine and cosine are calculated and a low-pass smoothing filter is applied. The filtered images are recombined with an arctan function, resulting in a well-filtered phase map, which can be improved further by repeated filtering.

are used. A popular algorithm is the branch cut algorithm (Goldstein *et al.*, 1988) which is relatively straightforward to implement and therefore comparatively fast, though it is not as robust as some of the more intensive algorithms. The book by Ghiglia and Pritt (1998) details a number of phase unwrapping algorithms and provides C++ source code for each of them. In the example shown in Fig. 4.9, the phase is unwrapped relative to an arbitrary point in the phase map. In practical shearography, some prior knowledge of the experimental conditions is required to know which point is at zero phase. The rest of the phase should then be unwrapped relative to this pixel.

4.4 Shearography for non-destructive evaluation (NDE) of composite materials

Shearography has developed into a very important tool for non-destructive testing due to its ability to provide full-field, non-contact measurement. It

Introduction to shearography

Wrapped phase

Unwrapped phase

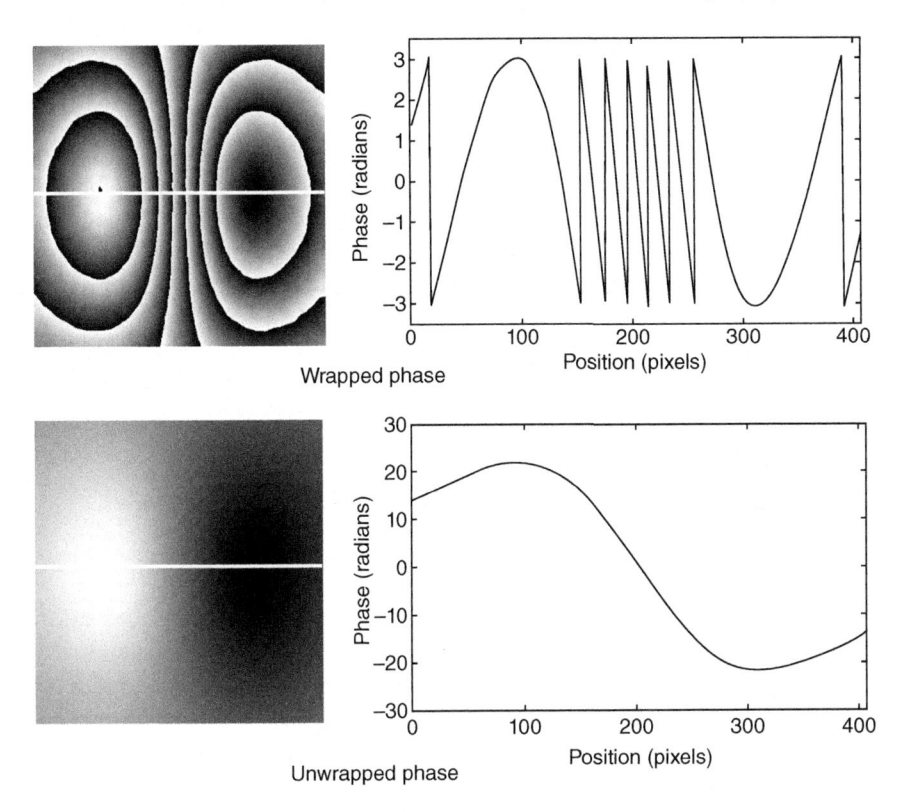

4.9 The process of phase unwrapping: the multiply filtered wrapped phase map from Fig. 4.8 is shown in the top left with a plot of phase vs. position for the central row (highlighted by the white horizontal line through the phase map) shown in the top right. After the fringe discontinuities are removed by the unwrapping process, a continuous phase distribution is obtained as shown in the phase map in the bottom left.

is particularly well suited to industrial testing compared with holography and ESPI because of its short coherence length requirements, and therefore resilience to environmental disturbances, and its direct sensitivity to strain. Shearography has particular application in sub-surface defect detection and identification in composite materials. Defects such as delaminations or cracks are a common problem in composite material manufacture and service and non-destructive testing is required to locate and characterise them. The surface of a component in the region of a delamination responds differently to defect-free regions when an external load is applied. Loading therefore results in surface strain anomalies in the defected regions which can be observed using shearography as anomalies in the fringe pattern. The magnitude of the delamination can be estimated from the fringe pattern and this information can be used to assess whether or not the defect compromises the structural integrity of the component. Since non-destructive testing of composite materials using shearography usually requires only qualitative investigation, phase analysis is not essential. However, wrapped phase fringe patterns tend to be much higher contrast and offer much better sensitivity. This improvement in sensitivity can mean the difference between a defect being detected or missed completely (Steinchen et al., 1998). The ability to unwrap the phase map offers a semi-quantitative measurement of the magnitude of a particular defect by indicating the surface strain distribution in the direction of the sensitivity vector. The rest of this section discusses some of the loading techniques that are used in NDE and compares shearography with other well-established NDE techniques.

4.4.1 Loading techniques

Static loading

When a test component is loaded statically deformation is induced after recording the reference images and ceases before the signal frames are recorded. The phase measured is directly proportional to the strain induced between the two sets of recordings. The method of loading should be chosen carefully to minimise the influence of rigid body motion during the process which can cause speckle decorrelation (Hung et al., 2007). For composite material testing, vacuum, pressure or thermal loading is often used. Vacuum loading involves placing the component under test within a vacuum cell. Images are recorded before inducing the vacuum after which a negative pressure differential is produced that deforms the surface. After loading, a second set of images is recorded and combined with the first set to generate the fringe pattern. The pressure required to deform the surface enough to induce a strain that may be detected using shearography is generally low (10–100 Pa for example). This is dependent on the surface elasticity of the component and the magnitude and depth of the defect (Taillade et al., 2011). Vacuum loading is popular in industrial testing due to its ease of operation. Some commercial shearography systems, such as the Steinbichler ISIS mobile 3000 (Steinbichler, 2012), incorporate a vacuum loading cell within the shearography instrument itself. Internal pressure loading is a convenient method for testing hollow components, such as tubes and pipes. Stress is induced by pumping air or water into the component. Yang et al. (1995) describe testing of a fibre reinforced plastic tube using shearography with internal pressure loading.

Another common method of stressing is through the application of a thermal load, for instance using infra-red lamps (Groves et al., 2009). Heat causes the material to expand, deforming the surface. This is a quasi-static

Introduction to shearography 73

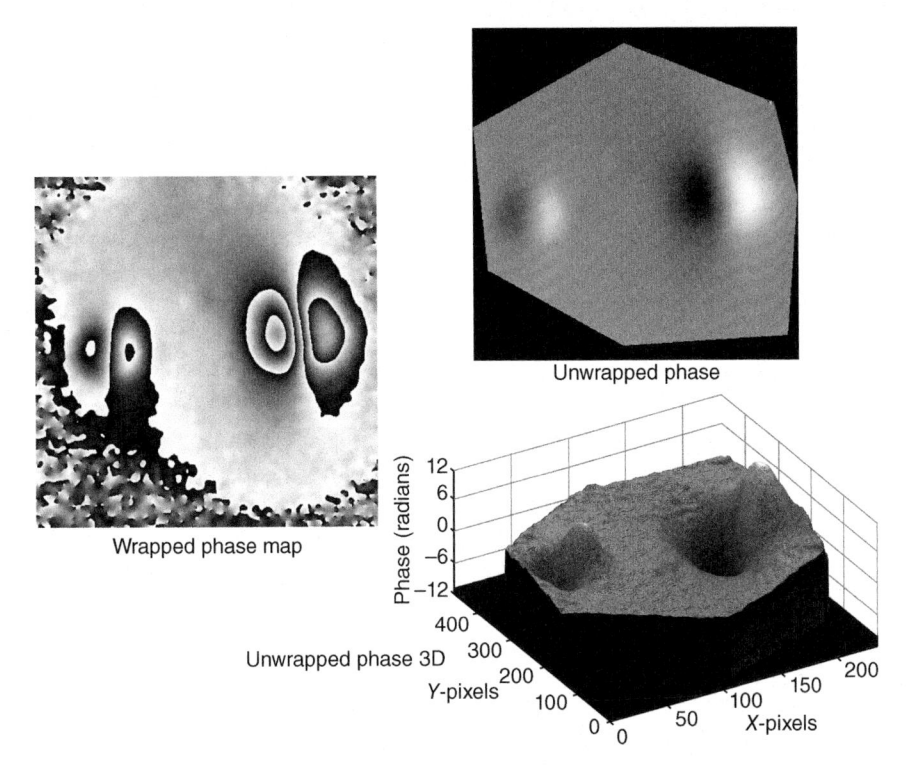

4.10 An example of non-destructive testing of a composite panel using shearography: The panel was loaded thermally and revealed two defects of different magnitude.

loading technique in that the thermally induced deformation diminishes gradually over time as the object cools. It is often appropriate to wait for a short period (a minute or so) before recording phase stepped images after thermal stressing to ensure that the temperature gradient is reduced. Figure 4.10 shows an example of shearography testing of a composite panel that possesses two defects. The component was thermally loaded and phase-stepped images were captured as the object cooled. Unwrapping the resulting filtered wrapped phase map and plotting the resulting phase distribution in three dimensions clearly shows the difference in the magnitude of the defects.

Vibration

Another method of stressing a component is dynamic loading by inducing a resonant vibration using for example, a piezoelectric shaker or an acoustic wave (Hung *et al.*, 2000a). Unlike static loading which can only be used for closed boundary flaws, dynamic loading has the advantage over static loading in that it can be used to detect open boundary delaminations (Hung et al., 2000b). If the frequency of the vibration that is applied is harmonic and significantly higher than the camera frame rate (typically kHz compared with frame rates of tens to hundreds of Hz) then time-averaged measurements can be performed using continuous wave illumination. Correlation of consecutive frames containing time-averaged speckle in the presence of harmonic vibration results in an intensity distribution given by (Valera and Jones, 1995)

$$I_{ave}(x, y) \propto J_0^2 \left[\frac{4\pi}{\lambda} \frac{\partial A(x, y)}{\partial x} dx \right] \qquad [4.25]$$

where A is the amplitude of the vibration and J_0 is the zero order Bessel function. The maximum intensity in the fringe pattern corresponds to points where the derivative of the amplitude equals zero and fringe brightness drops off rapidly with increasing fringe order. The brightest Bessel fringe corresponds to points of maximum vibrational amplitude, in contrast to ESPI where they correspond to the vibrational nodes. The speckle pattern intensity averaged over many vibrational cycles in a single frame does not change significantly from frame to frame. Therefore when correlating images by subtraction, a π phase shift applied between frames can be introduced to maximise fringe contrast. With time-averaged shearography, sequential subtraction is used where the signal frame from one fringe pattern is used as the reference frame for the next and so on, unlike in static shearography where acquired signal frames are subtracted from a single reference frame. Figure 4.11a shows a typical time-averaged shearography fringe pattern taken from a surface vibrating at 440 Hz with images captured at a frame rate of a few hertz. The distribution of the J_0^2 zero order Bessel function is shown in Fig. 4.11b. The brightness of the Bessel function is approximately six times that of first order fringe, as can be seen from Fig. 4.11b. The difference in intensity between these two fringes appears less in the experimental fringe pattern due to some contrast enhancements done in Photoshop to improve the presentation.

The use of stroboscopic illumination offers a number of additional benefits, in particular the resulting fringe patterns correspond to a cosinusoidal distribution and therefore do not suffer the intensity drop that affects the Bessel fringes obtained with continuous wave illumination. In addition, phase stepping can be implemented by synchronising the repetition rate of the illumination with the applied vibration (Steinchen et al., 1997). A recent publication (De Angelis et al., 2012) reports on a vibration-based technique that can be used quantify the depth and magnitude of defects within composites. This was validated using a range of prefabricated delaminations in fibre-reinforced composite test plates. A piezoelectric shaker was used to

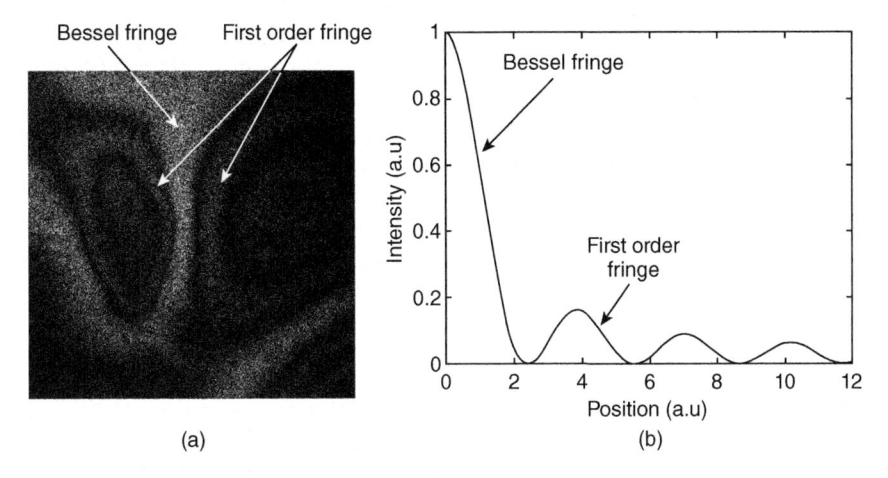

4.11 A typical time-averaged shearography fringe pattern, taken from a surface vibrating at 440 Hz (a). The fringe contrast has been enhanced in Photoshop for presentation purposes. The intensity distribution of the J_0^2 zero order Bessel function (b) clearly shows the drop in intensity with increasing fringe order.

induce vibrations, the frequency of which was tuned until the resonance modes were detected. The resonance frequency was found to increase with delamination depth and decrease with delamination size.

4.5 Comparing shearography with other techniques

There are numerous experimental techniques for industrial non-destructive testing that offer different benefits depending on the application. This section discusses two approaches that have been previously compared with shearography in the literature. In ultrasonic testing (Ryžek *et al.*, 2006; Garnier *et al.*, 2011), a high-frequency sound wave (typically 1–100 MHz) emitted by a transducer is transmitted through the component under test. When the sound wave arrives at an interface within the component, a proportion of the sonic energy is reflected and detected by the sensing element of the transducer. This results in a signal that can be viewed on an oscilloscope. Defects in the component are observed as peaks in the detector signal that are in addition to peaks associated with the front and rear faces of the component. The position of the peaks in the trace provides an indication of the depth of the defect and the shape of the peak suggests information on the type and magnitude of the defect. Ultrasonic testing is essentially a point-by-point technique and two-dimensional data is built up by measuring at adjacent points sequentially (ultrasonic C-scan). Measurements can therefore be quite time consuming for even moderately large objects. Hung and Ho (2005) illustrate this with the testing of composite panel of area 30 cm² which took one second to test using shearography but ten minutes with an ultrasonic C-scan. With ultrasonic testing there is no need to stress the component, unlike in shearography, which avoids introducing additional wear during the testing procedure. However, fluid coupling is required because high-frequency sound waves do not travel through air.

Thermography is an infrared imaging technique that has frequently been compared with shearography for non-destructive testing (Hung *et al.*, 2007, 2009; Taillade *et al.*, 2011; Garnier *et al.*, 2011). Like shearography, thermography is a non-contact measurement technique that provides a full-field measurement. The component under test is heated by a relatively long exposure to an infrared lamp (~ minutes) and imaged by an infrared camera. Subsurface defects are observed as hotspots in the thermal image. Because of the complexity of the thermal conduction processes in materials containing defects, such measurements can only be considered qualitative. Quantitative measurements can be achieved with pulsed thermography where components are subjected to short thermal pulses (~ ms) with high peak powers. Likewise, thermal images are recorded with an infra-red camera as the component cools. Plotting the logarithm of the thermal variance against the log of the cooling time of a single pixel results in a linear relationship for defect-free regions; however, a deviation from linearity is observed in the presence of defects. The time at which the deviation from linearity occurs indicates the depth of the defect. Since the process depends on the thermal response of the component rather than the mechanical response in shearography, it is not susceptible to rigid body motions and can therefore be used in a wider range of environments. However variations in the surface thermal emissivity of some components may result in the detection of false positives. The comparison between these NDE techniques is summarised in Table 4.1.

4.6 Future trends

Shearography has become a well-established technique for qualitative NDE and numerous commercial systems are on the market. Current research is aimed at the development of fully quantitative systems and the challenges that these systems present. Fully quantitative shearography requires the use of at least three measurement channels. Measurements from the three channels with two orthogonal shear directions can be combined using a matrix transformation (equation [4.19]) to yield all six surface strain components. The data from the separate channels can be multiplexed temporally (James and Tatam, 1999) or by wavelength (Kästle *et al.*, 1999). Wavelength multiplexing can reduce the overall recording time but requires additional wavelength selective components, resulting in a more complex

Table 4.1 Comparison of shearography, thermography and ultrasound

	Shearography	Thermography	Ultrasound
Loading	Mechanical, thermal or vibrational	Thermal	None
Measurand	Strain	Thermal emissivity	Acoustic vibration
Surface contact	Non-contact	Non-contact	Contact
Requirements	Diffuse scattering surface	Infra-red lamps and cameras	Fluid coupling
Advantage	Rapid, full-field measurements	Rapid, full-field measurements	Direct depth information
Disadvantage	Depth information not directly available with static loading	Detection of false positives due to variations in surface thermal emissivity	Time-consuming two-dimensional measurements with C-scan
Limitation	Rigid body motion	Materials' thermal properties	Spatial resolution limited by transducer

system. To make quantitative measurements from non-planar test objects a correction is required to account for the object shape. Groves *et al.* (2004) demonstrated a method of determining object slope, the derivative of shape, using shearography. The method involves translating the diverging lens of the illumination beam between exposures in a direction perpendicular to the illumination axis. A quantitative shearography system incorporating this technique has been used to investigate pipe welds (Groves *et al.*, 2003). Strain measurements from the system compared well with strain gauge and FBG (fibre Bragg grating) measurements (Groves *et al.*, 2007). Goto and Groves (2010) investigated the influence of errors in quantitative shearography using finite element models. Charrett *et al.* (2011) showed that an improvement in accuracy of 33% could be achieved through the introduction of a fourth measurement channel.

Shearography instrumentation has also been developed for the investigation of dynamic events such as rotations, transient vibrations and shocks. In cases where the motion of the object is non-harmonic, the time averaged techniques discussed in Section 4.4.1 cannot be used. The solution involves using pulsed lasers to effectively freeze the motion of the object. Measurements are made by correlation of images recorded from subsequent laser pulses and show the surface strain variation occurring in the time between the two pulses. However, because of the timescales involved, temporal phase stepping cannot be used.

Alternative phase analysis techniques that are suitable for dynamic measurements include spatial phase stepping and the spatial carrier technique. Spatial phase stepping involves recording phase stepped images simultaneously with different cameras (Kujawinska, 1993) or different regions of the same camera (García et al., 1999). Diffractive optics are used to separate the images and impose the required phase shifts. The spatial carrier technique involves the introduction of a carrier frequency either by illuminating the object with sequential pulses of the output from the laser source and ensuring that each pulse has a different divergence (Fernández et al., 2000) or by using a Mach–Zehnder shearing interferometer (Pedrini et al., 1996). Fourier transform techniques are then used to recover the phase information (Takeda et al., 1982). The principle of the Mach–Zehnder shearing interferometer is illustrated in Fig. 4.12. The two mirrors are aligned such

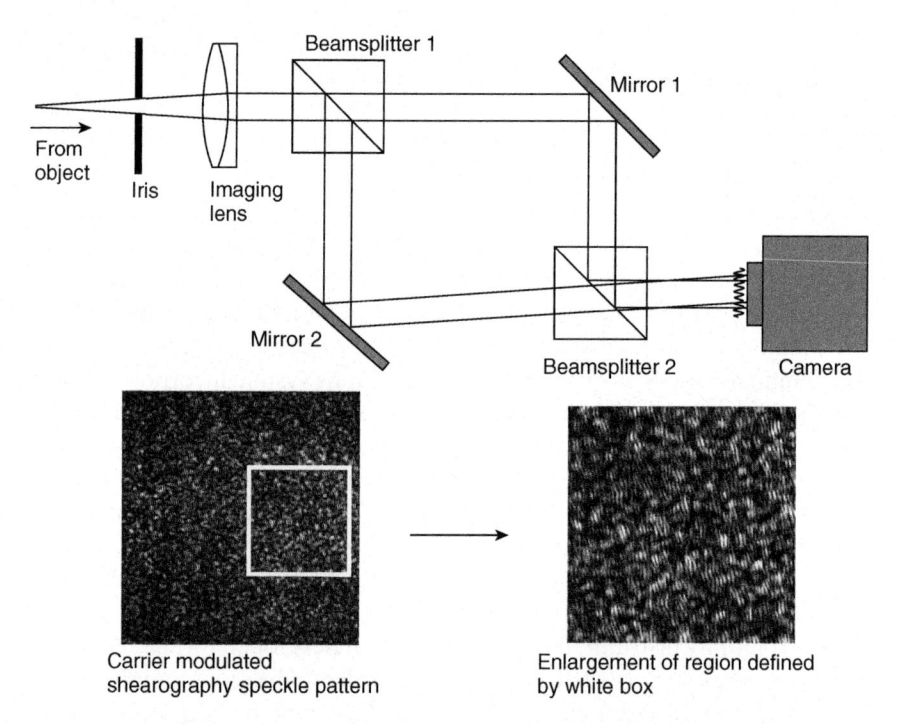

4.12 The principle of the Mach–Zehnder shearing interferometer: the mirrors are aligned to provide a tilt between the wavefronts, which produces a carrier fringe pattern if the speckle size is large enough. The fringes can be seen by viewing a magnified region of the speckle pattern.

that a tilt is introduced between the interfering wavefronts producing a carrier frequency. The speckle size is increased using the iris at the input of the interferometer to a size sufficient to support the carrier fringes (> six pixels). The lower part of the figure shows a typical carrier modulated shearography speckle pattern. The enlargement clearly shows the high-frequency carrier fringes.

Applications include the investigation of the propagation of shock waves through composite materials (Santos *et al.*, 2004) and analysis of the propagation of Lamb waves through fibre-reinforced composites (Focke *et al.*, 2008). Francis *et al.* (2008) developed a quantitative shearography system for dynamic measurements using fibre-optic imaging bundles. The bundles were used to transport images with different observation directions, and therefore different sensitivity vectors, to a CCD camera where they were spatially multiplexed. Phase analysis was performed using the spatial carrier technique in a Mach–Zehnder shearing interferometer.

4.7 Sources of further information and advice

The key textbook in the field is *Digital shearography* by Steinchen and Yang (2003). The book details the complete theoretical background of shearography before moving on to describe specific applications such as non-destructive testing and strain measurement. Other more general textbooks on speckle interferometry include *Digital speckle interferometry and related techniques* edited by Rastogi (2001) and *Holographic and speckle interferometry* by Jones and Wykes (1989). A good text on temporal and spatial phase analysis and related techniques is *Interferogram analysis, digital fringe pattern measurement techniques* edited by Robinson and Reid (1993).

In addition to these books there are a number of review papers that provide useful technical background to shearography. The author's own review paper (Francis *et al.*, 2010) covers shearography technology and applications with a focus on more recent technological developments, particularly multi-channel, quantitative shearography. The review of Chen (2001) summarises the state of the art at the time with a focus on dynamic measurements, both transient and harmonic. Hung and Ho's review (2005) focuses more on industrial applications, in particular non-destructive testing and residual stress measurement. Later reviews by Hung *et al.* review and compare shearography and thermography for adhesive bond evaluation (2007) and for non-destructive testing (2009).

A number of companies develop commercial shearography systems aimed at the industrial non-destructive testing market. Dantec Dynamics, a leading supplier of laser-based measurements systems, offer a range of shearography systems for non-destructive testing in automotive, aerospace and wind turbine industries (Dantec, 2012). Steinbichler Optotechnik offer a mobile shearography system, ISIS, and Intact, a system specifically for tyre testing (Steinbichler, 2012). The non-destructive testing products developed by Optonor (Optonor and Trandheim, 2012) in Trondheim have more of a focus on vibrational excitation. Their SNT 4050 shearography system incorporates a mechanical shaker and a function generator which provides vibrational frequencies up to 5 kHz. Laser Technology Inc. (2012) in the USA provides a range of shearography systems for aerospace, automotive and marine industries.

4.8 References

Aebischer H A and Waldner S (1997), 'Strain distributions made visible with image-shearing speckle pattern interferometry', *Opt. Laser. Eng.*, **26**, 407–20.

Aebischer H A and Waldner S (1999), 'A simple and effective method for filtering speckle interferometric phase fringe patterns', *Opt. Commun.*, **162**, 205–10.

Charrett T O H, Francis D and Tatam R P (2011), 'Quantitative shearography: error reduction by using more than three measurement channels', *Appl. Opt.*, **50**:2, 134–46.

Chen F (2001), 'Digital shearography: state of the art and some applications', *J. Electron. Imaging*, **10**:1, 240–51.

Cloud G (2007), 'Optical methods in experimental mechanics, part 27: speckle size estimates', *Exp. Techniques*, May/June, 19–22.

Creath K (1985), 'Phase-shifting speckle interferometry', *Appl. Opt.*, **24**:18, 3053–8.

Creath K (1993), 'Temporal phase measurement methods', in Robinson D W and Reid G T, *Interferogram analysis, digital fringe pattern measurement techniques*, Bristol, IOP Publishing, 94–140.

Dantec Dynamics GmbH, Ulm, DE (2012), http://www.dantecdynamics.com/Default.aspx?ID=665 [accessed 09/02/12]

De Angelis G, Meo M, Almond D P, Pickering S G and Angioni S L (2012), 'A new technique to detect defect size and depth in composite structures using digital shearography and unconstrained optimization, *NDT&E Int.*, **45**, 91–6.

Fernández A, Doval Á F, Kaufmann G H, Dávila A, Blanco-García J, Pérez-López C and Fernández J L (2000), 'Measurement of transient out-of-plane displacement gradients in plates using double-pulsed subtraction TV shearography', *Opt. Eng.*, **39**:8, 2106–13.

Focke O, Hildebrand A, Von Kopylow C and Calomfirescu M (2008), 'Inspection of Lamb waves in carbon fiber composites using shearographic interferometry', *Proc. SPIE*, 6934, 693403-1–12.

Francis D, James S W and Tatam R P (2008), 'Surface strain measurement of rotating objects using pulsed laser shearography with coherent fibre-optic imaging bundles', *Meas. Sci. Technol.*, **19**, 105301–13.

Francis D, Tatam R P and Groves R M (2010), 'Shearography technology and applications: a review', *Meas. Sci. Technol.*, **21**, 102001–29.

García B B, Moore A J, Pérez-López C, Wang L and Tschudi T (1999), 'Spatial phase-stepped interferometry using a holographic optical element', *Opt. Eng.*, **38**:12, 2069–74.

Garnier C, Pastor M-L, Eyma F and Lorrain B (2011), 'The detection of aeronautical defects in situ on composite structures using non destructive testing', *Comp. Struct.*, **93**, 1328–36.

Ghiglia D C and Pritt M D (1998), *Two dimensional phase unwrapping: theory, algorithms and software*, Chichester, Wiley.

Goldstein R M, Zebker H A and Werner C L (1988), 'Satellite radar interferometry: two-dimensional phase unwrapping', *Radio Sci.*, **23**, 713–20.

Goto D T and Groves R M (2010), 'Error analysis of 3D shearography using finite element modelling', *Proc. SPIE*, **7718**, 771816-1-12.

Groves R M, James S W and Tatam R P (2003), 'Pipe weld investigation using shearography', *Strain*, **39**, 101–5.

Groves R M, James S W and Tatam R P (2004), 'Shape and slope measurement by source displacement in shearography', *Opt. Laser. Eng.*, **41**, 621–34.

Groves R M, Chehura E, Li W, Staines S E, James S W and Tatam R P (2007), 'Surface strain measurement: a comparison of speckle shearing interferometry and optical fibre Bragg gratings with resistance foil strain gauges', *Meas. Sci. Technol.*, **18**, 1175–84.

Groves R M, Pradarutti B, Kouloumpi E, Osten W and Notni G (2009), '2D and 3D non-destructive evaluation of a wooden panel painting using shearography and terahertz imaging', *NDT&E Int.*, **42**, 543–9.

Hariharan P, Oreb B F and Eiju T (1987), 'Digital phase shifting interferometry: a simple error compensating algorithm', *Appl. Opt.*, **26**, 2504–6.

Hung Y Y (1982), 'Shearography: a new optical method for strain measurement and non-destructive evaluation', *Opt. Eng.*, **21**, 391–5.

Hung Y Y (1996), 'Shearography for non-destructive evaluation of composite structures', *Opt. Laser. Eng.*, **24**, 161–82.

Hung Y Y (1997), 'Digital shearography versus TV-holography for non-destructive evaluation', *Opt. Laser. Eng.*, **26**, 421–36.

Hung Y Y and Ho H P (2005), 'Shearography: an optical measurement technique and applications', *Mat. Sci. Eng. R*, **49**, 61–87.

Hung Y Y and Liang C Y (1979), 'Image shearing camera for direct measurement of surface strains', *Appl. Opt.*, **18**:7, 1046–51.

Hung Y Y and Wang J Q (1996), 'Dual-beam phase shift shearography for measurement of in-plane strains, *Opt. Laser. Eng.*, **24**, 403–13.

Hung Y Y, Turner J L, Tafralian M, Hovanesian J D and Taylor C E (1978), 'Optical method for measuring contour slopes of an object', *Appl. Opt.*, **21**, 391–5.

Hung Y Y, Luo W D, Lin L and Shang H M (2000a), 'NDT of joined surfaces using digital time-integrated shearography with multiple-frequency sweep', *Opt. Laser. Eng.*, **33**, 369–82.

Hung Y Y, Luo W D, Lin L and Shang H M (2000b), 'Evaluating the soundness of bonding using shearography', *Comp. Struct.*, **50**, 353–62.

Hung Y Y, Chen Y S, Ng S P, Shepard S P, Hou Y and Lhota J R (2007), 'Review and comparison of shearography and pulsed thermography for adhesive bond evaluation', *Opt. Eng.*, **46**:5, 051007-1-16.

Hung Y Y, Chen Y S, Ng S P, Liu L, Huang Y H, Luk B L, Ip R W L, Wu C M L and Chung P S (2009), 'Review and comparison of shearography and active thermography for nondestructive evaluation', *Mat. Sci. Eng. R*, **64**, 73–112.

James S W and Tatam R P (1999), 'Time-division-multiplexed 3D shearography', *Proc. SPIE.*, **3744**, 394–403.

Jones R and Wykes K (1989), *Holographic and speckle interferometry*, Cambridge, Cambridge University Press.

Kästle R, Hack E and Sennhauser U (1999), 'Multiwavelength shearography for quantitative measurements of two-dimensional strain distributions', *Appl. Opt.*, **38**:1, 96–100.

Kujawinska M (1993), 'Spatial phase measurement methods', in Robinson D W and Reid G T, *Interferogram analysis, digital fringe pattern measurement techniques*, Bristol, IOP Publishing, 94–140.

Laser Technology Inc, Norristown PA, USA (2012), http://www.laserndt.com/products/home.htm [accessed 09/02/12].

Mihaylova E, Naydenova I, Martin S and Toal V (2004), 'Electronic speckle pattern shearing interferometer with a photopolymer holographic grating', *Appl. Opt.*, **43**:12, 2439–42.

Optonor A S, Trandheim, N O (2012), http://www.optonor.no/Contact.aspx [accessed 09/02/12].

Pedrini G, Zou Y-L and Tiziani H J (1996), 'Quantitative evaluation of digital shearing interferogram using the spatial carrier method', *Pure Appl. Opt.*, **5**, 313–21.

Rastogi P K (2001), *Digital speckle interferometry and related techniques*, Chichester, Wiley.

Robinson D W and Reid G T (1993), *Interferogram analysis, digital fringe pattern measurement techniques*, Bristol, IOP Publishing.

Ružek R, Lahonka R and Jironč J (2006), 'Ultrasonic C-Scan and shearography NDI techniques evaluation of impacts defects identification', *NDT&E Int.*, **39**, 132–42.

Santos F, Vaz M and Monteiro J (2004), 'A new set-up for pulsed digital shearography applied to defect detection in composite structures', *Opt. Laser. Eng.*, **42**, 131–40.

Sharp B (1989), 'Electronic speckle pattern interferometry (ESPI)', *Opt. Laser. Eng.*, **21**, 241–55.

Steinbichler Optotechnik GmbH, Neubeuern, DE (2012), http://www.steinbichler.de/en/main/shearography—ndt.htm [accessed 18/01/12].

Steinchen W and Yang L X (2003) *Digital Shearography: Theory and application of digital speckle pattern shearing interferometry*, Bellingham, SPIE Press.

Steinchen W, Yang L X and Schuth M (1996), 'TV-shearography for measuring 3D-strains', *Strain*, May, 49–58.

Steinchen W, Yang L X and Kupfer G (1997), 'Digital shearography for non-destructive testing and vibration analysis', *Exp. Techniques*, July/August, 20–3.

Steinchen W, Yang L, Kupfer G and Mäckel (1998), 'Non-destructive testing of aerospace composite materials using digital shearography', *Proc. Instn. Mech. Engrs.*, **212**:G, 21–30.

Steinchen W, Kupfer G, Mäckel P and Vössing F (1999), 'Determination of strain distribution by means of digital shearography', *Measurement*, **26**, 79–90.

Takeda M, Ina H and Kobayashi S (1982), 'Fourier-transform method of fringe-pattern analysis for computer-based topography and interferometry', *J. Opt. Soc. Am.*, **72**:1, 156–9.

Taillade F, Quiertant M, Benzati K and Aubangac C (2011), 'Shearography and pulsed stimulated infrared thermography applied to a nondestructive evaluation of FRP strengthening systems bonded on concrete structures', *Constr. Build. Mater.*, **25**, 568–74.

Valera J D R and Jones J D C (1995), 'Vibration analysis by modulated time-averaged speckle shearing interferometry', *Meas. Sci. Technol.*, **6**, 965–70.

Yang L X, Steinchen W, Schuth M and Kupfer G (1995), 'Precision measurement and nondestructive testing by means of digital phase shifting speckle pattern and speckle pattern shearing interferometry', *Measurement*, **16**, 149–60.

5
Non-destructive evaluation (NDE) of composites: digital shearography

Y. Y. HUNG and L. X. YANG, Oakland University, USA and Y. H. HUANG, Singapore-MIT Alliance for Research and Technology (SMART) Centre, Singapore

DOI: 10.1533/9780857093554.1.84

Abstract: The increasing usage of composite materials has called for more effective techniques for inspecting the integrity of composite structures, as composite materials generally have higher likelihood of having material imperfections. This chapter reviews shearography and its applications in non-destructive testing. Shearography is an interferometric technique for full-field, non-contacting measurement of surface deformation (displacement or strain). It was developed to overcome several limitations of holography by eliminating the reference beam. Consequently, it is less sensitive to environmental disturbances and is a practical tool that can be used in field/factory settings. In non-destructive testing, shearography reveals defects in an object by identifying defect-induced deformation anomalies. Shearography has already received considerable industrial acceptance, in particular, for nondestructive testing of tires and aerospace structures. Other applications of shearography include strain measurement, material characterization, residual stress evaluation, leak detection, hermetic seal evaluation, vibration studies and 3D shape measurement. This chapter focuses on its application in non-destructive testing with an emphasis on composite materials.

Key words: digital shearography, non-destructive testing (NDT), strain measurement, composite materials.

5.1 Introduction

The high-strength/weight advantage of composite materials has received increasing popularity not only in the aerospace industry but also in other sectors, particularly the automotive industry striving to enhance fuel efficiency by reducing vehicle weight. Composite materials, however, have a higher likelihood of having defects due to the material complexity. Therefore, non-destructive testing (NDT) plays an important role in ensuring the integrity of composite structures. Shearography, because of its numerous advantages, has emerged to be a practical tool for inspecting composite materials.

Shearography is an interferometric technique developed to address several limitations of holography. Its significant advantages include (1) not requiring a reference light beam, thus leading to simple optical set-ups, reduced requirements for coherence length of the laser, and vibration isolation during measurement; and (2) direct measurement of surface strains (first-order derivatives of surface displacements). These distinct advantages have rendered shearography as a practical measurement tool and it has already gained wide industrial acceptance for NDT. For instance, the rubber industry routinely uses shearography for evaluating tires, and the aerospace industry has adopted it for NDT of aircraft structures, in particular, composite structures (Ibrahim *et al.*, 2004; Lobanov *et al.*, 2009; Ruzek & Bechal, 2009). Other applications of shearography include: measurement of strains (Steinchen *et al.*, 1998b), material properties (Lee *et al.*, 2004a, 2008), residual stresses (Hung *et al.*, 1988, 1997, 1999), 3D shapes, vibrations (Steinchen *et al.*, 1997; Yang *et al.*, 1998), as well as leakage detection (Hung & Shi, 1998).

Three versions of shearography are in existence using (1) photographic recording; (2) thermoplastic recording; and (3) digital recording. The photographic version (Hung, 1982; Hung *et al.*, 1994) uses photographic emulsion as the recording media and hence can produce very high-resolution quality images. However, it has lost its popularity due to the slow and costly photographic process. The thermoplastic version (Hung and Hovanesian, 1990) was developed to address the limitations of photographic recording. It uses a reusable thermoplastic plate which is instantly processed after recording. This bypasses the inconvenient wet photographic process, but the cost of the thermoplastic plates has deterred industrial users. Furthermore, the production of thermoplastic plates of consistent quality also presents a problem. The full capability of shearography was not uncovered until the introduction of digital shearography (Hung, 1989a), which uses video sensors as the recording media and digital processing technology for data acquisition and analysis. This chapter details the principles of digital shearography and its applications in nondestructive testing with an emphasis on composite materials.

5.2 Principles of digital shearography

5.2.1 Description of the technique

A schematic diagram of shearography is shown in Fig. 5.1. The test object is illuminated with an expanded laser point source and imaged by a video camera connected to a microcomputer for recording and processing. An image-shearing device is placed in front of the camera lens. Two practical image-shearing devices will be described in a later section. The image-shearing device brings two non-parallel rays of light scattered from two different object points to interfere with each other. Since the angle of the two interfering rays is very small, the spatial frequency of the interference

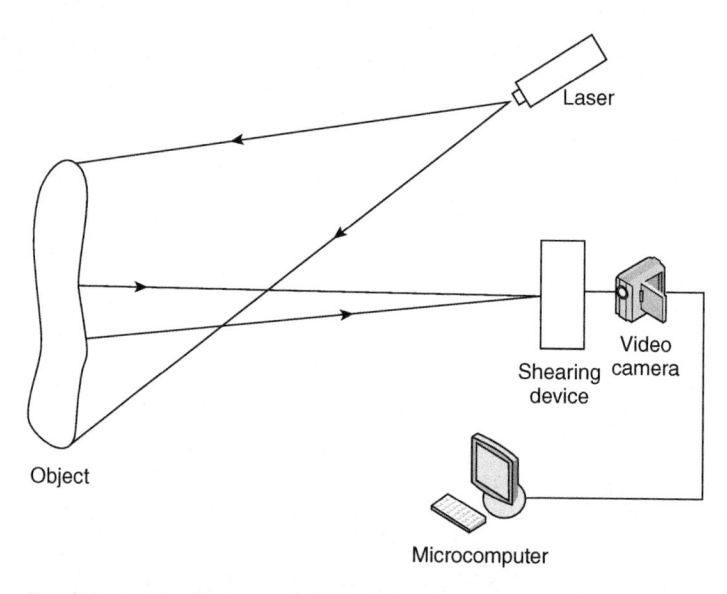

5.1 Schematic diagram of digital shearography. The two types of shearing devices will be presented in a later section.

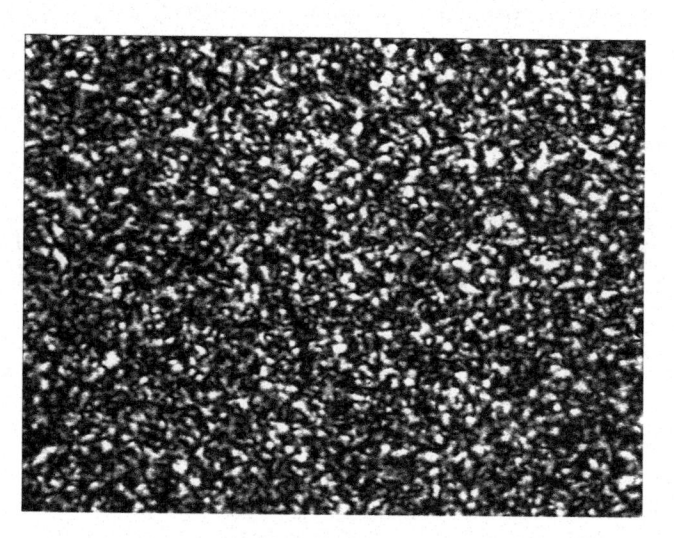

5.2 A typical speckle pattern (random interference pattern).

patterns is so low that can be resolved and recorded by video image sensor of low resolution capability.

A typical digitized speckle image is illustrated in Fig. 5.2. As the object surface is diffusely reflective, the interference of the scattered rays yields a random pattern known as speckle pattern. When the object is slightly deformed, the speckle pattern will be slightly changed. Practice of digital shearography involves digitizing sequentially two speckle patterns of the test object, one before and another after deformation, into the microcomputer. As will be shown in Section 5.2.3, the difference of the two speckle patterns enables reconstruction of an interpretable fringe pattern that depicts surface deformation (surface displacement or displacement-derivatives) The use of a frame grabber with on-board processing capability further allows real-time image acquisition, subtraction and reconstruction of fringe patterns at video rate. Figure 5.3 is an example of the fringe pattern. The fringe lines are loci of the deflection derivative of a plate (the plate was fixed along the boundaries and deflected by a point-load at the center) with respect to x (Fig. 5.3a) and y (Fig. 5.3b). Should an object have a subsurface defect, the defect will produce a deformation anomaly which is translated into an anomaly in the fringe pattern that allows the defect to be revealed. Figure 5.4 is a fringe pattern depicting deformation of a

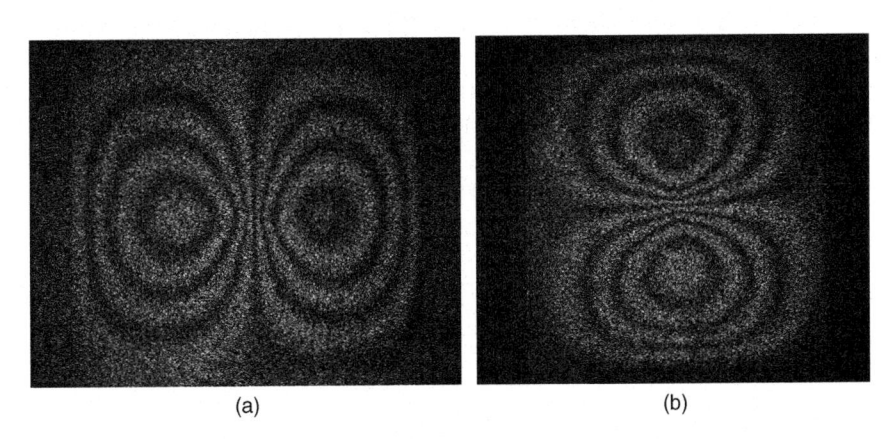

5.3 Fringe patterns depicting the deflection derivatives of a rectangular plate clamped along its boundaries and subjected to uniform pressure: (a) $\partial w/\partial x$; (b) $\partial w/\partial y$.

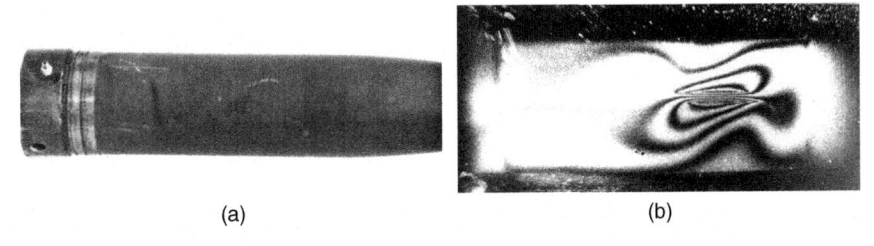

5.4 Fringe anomaly (right) reveals a crack in a steel pressure vessel (left).

pressure vessel. A subsurface crack is clearly revealed by the fringe anomaly. This is the principle of NDT using shearography.

5.2.2 Shearing devices

While there are several shearing techniques, only the two most practical are presented here.

Doubly refractive crystal

Double refraction or birefringence (Hung, 1989b) is the decomposition of a ray of light into two rays when it passes through certain optically anisotropic materials, such as crystals of calcite or boron nitride. This character also occurs in certain plastics and liquid crystals. For convenience, a commercially available optical device known as Wollaston prism may be used. A Wollaston prism consists of two orthogonal calcite prisms, cemented together on their base (traditionally with Canada balsam) to form two right-angle triangle prisms with perpendicular optic axes. It separates incoming randomly polarized or unpolarized rays into two orthogonal linearly polarized outgoing rays with an angular separation.

As illustrated in Fig. 5.5(a), a ray passing through a Wollaston prism is split into two angularly separated rays. Consequently, when imaging through

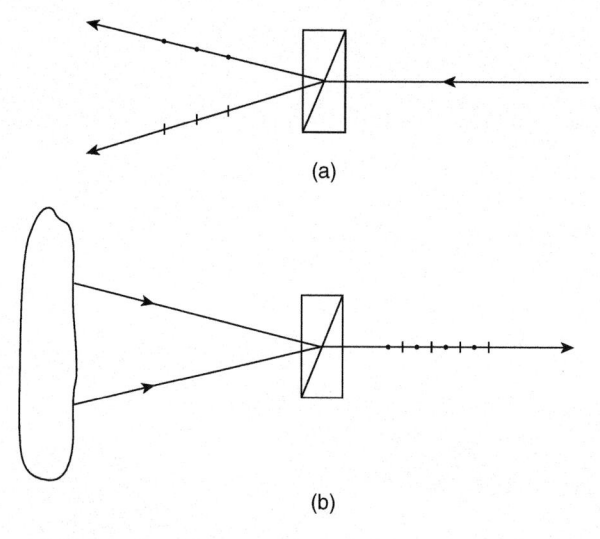

5.5 Illustration of a Wollaston prism as a shearing device. (a) One light beam is split into two angularly separated and orthogonally polarized beams. (b) Conversely, two non-parallel rays scattered from two different points on the object are combined and become collinear.

the prism, two sheared-images are detected. Conversely, two rays scattered from the diffused object surface are combined and become collinear. This is illustrated in Fig. 5.5(b). Scattered rays from a diffused surface are not polarized. However, each ray passing through the prism is split into two orthogonally polarized rays. Because of the orthogonal polarization they will not interfere with each other. To enable interference, a polarizer with its polarization axis oriented at an azimuth of 45° is required. As the illuminated object surface is generally optically rough, interference of the two sheared-images produces a speckle pattern embedded in the shearographic image. As explained above, the difference of two shearographic images before and after deformation will produce a fringe pattern depicting the surface deformation.

Michelson shearing interferometer

A second practical shearing device is a modified Michelson interferometer (Hung, 1974; Leendertz and Butters, 1973). A schematic diagram of the device is shown in Fig. 5.6. By tilting mirror 2 of the Michelson interferometer by a very small angle, the measuring system can brings the laser rays

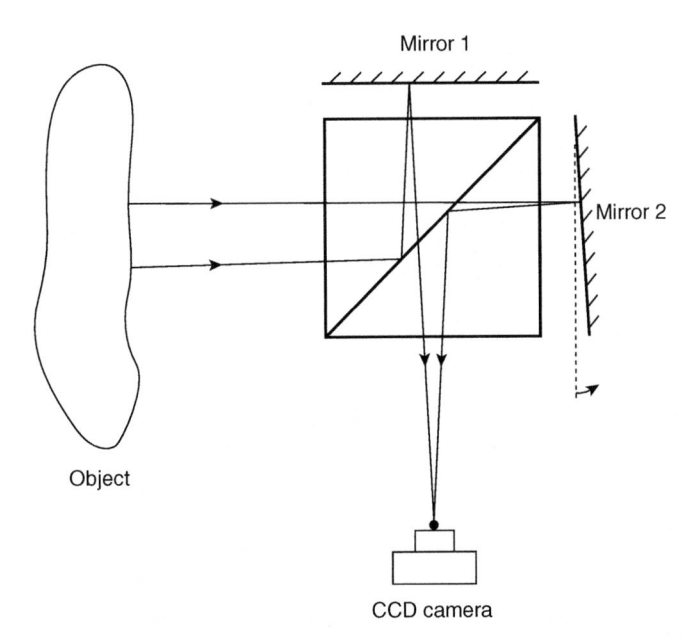

5.6 Illustration of a Michelson shearing interferometer as a shearing device.

from two points on the object surface into one point on the image plane of the CCD camera. These laser rays interfere with each other, producing a speckle pattern, the intensity difference of the speckle patters before and after deformation generate a visible fringe pattern, i.e. a so-called shearogram. The advantage of using the modified Michelson interferometer as a shearing device is the ease of changing shearing amount and the direction. In addition the phase shift technique may be introduced, as described in Section 5.2.5 (Steinchen et al., 1995; Yang et al., 1995).

5.2.3 Principles of fringe formation

A shearographic speckle image may be mathematically represented as follows:

$$\mathbf{I} = \mathbf{I}_o (1 + \mu \cos \phi) \quad [5.1]$$

where $\mathbf{I}$ is the intensity distribution of the speckle pattern received at the image plane of the camera; $\mathbf{I}_o$ is the intensity of the laterally sheared images (which is the background of the speckle pattern); μ is the amplitude of modulation of the speckle pattern; and ϕ is a random phase angle. After the object is deformed, the intensity distribution of the speckle pattern is changed slightly to $\mathbf{I}'$, which is described by the following equation:

$$\mathbf{I}' = \mathbf{I}_o [1 + \mu \cos(\phi + \Delta)] \quad [5.2]$$

where Δ denotes the phase change due to surface deformation.

Note that both Eqs. (5.1) and (5.2) contain the random phase angle ϕ; therefore no readily interpretable information can be deduced by looking at the speckle images. However, computing the difference of the intensities of the two speckle patterns (Equations (5.1) and (5.2)) yields a visible fringe pattern described by the following equation:

$$\mathbf{I}_d = 2\mathbf{I}_o \left[\mu \sin\left(\phi + \frac{\Delta}{2}\right) \sin\left(\frac{\Delta}{2}\right) \right] \quad [5.3]$$

where $\mathbf{I}_d$ is pixel-by-pixel intensity difference. Eq. (5.3) shows the formation of a fringe pattern depicting the fringe-phase Δ, in which the dark fringes correspond to $\Delta = 2n\pi$, with $n = 0,1,2,3,...$ being the fringe orders, and the bright fringes correspond to half fringe orders. Figure 5.3(a) shows a typical shearographic fringe pattern for a fully clamped rectangular plate under uniform lateral load when image-shearing is along the x-direction, and Fig. 5.3(b) shows the corresponding fringe pattern when image-shearing is along the y-direction. Note that the absolute value of $\mathbf{I}_d$ is used in the display of the fringe patterns, as an image cannot have negative value.

5.2.4 Fringe interpretation

The phase change Δ is induced by the change in the relative optical path length of light scattered from two neighboring object points, $P(x, y, z)$ and $P(x+\Delta x, y, z)$, considering that the direction of image-shearing is parallel with the x-axis and the amount of shearing is Δx. Referring to Fig. 5.7, it may be shown that Δ is related to the relative displacement $(\Delta u, \Delta v, \Delta w)$ of the two neighboring points as follows:

$$\Delta = \frac{2\pi}{\lambda}(A\delta u + B\delta v + C\delta w) \qquad [5.4]$$

where (u, v, w) and $(u+\Delta u, v+\Delta v, w+\Delta w)$ are, respectively, the displacement vectors of $P(x, y, z)$ and $P(x+\Delta x, y, z)$, λ is the wavelength of the laser, and A, B, C are sensitivity factors that are related to the positions of the illumination point $S(x_s, y_s, z_s)$ and the camera $O(x_0, y_0, z_0)$ in the following manner.

$$\begin{aligned} A &= (x-x_0)/R_o + (x-x_s)/R_s \\ B &= (y-y_0)/R_o + (y-y_s)/R_s \\ C &= (z-z_0)/R_o + (z-z_s)/R_s \end{aligned} \qquad [5.5]$$

with $R_o^2 = x_0^2 + y_0^2 + z_0^2$ and $R_s^2 = x_s^2 + y_s^2 + z_s^2$.

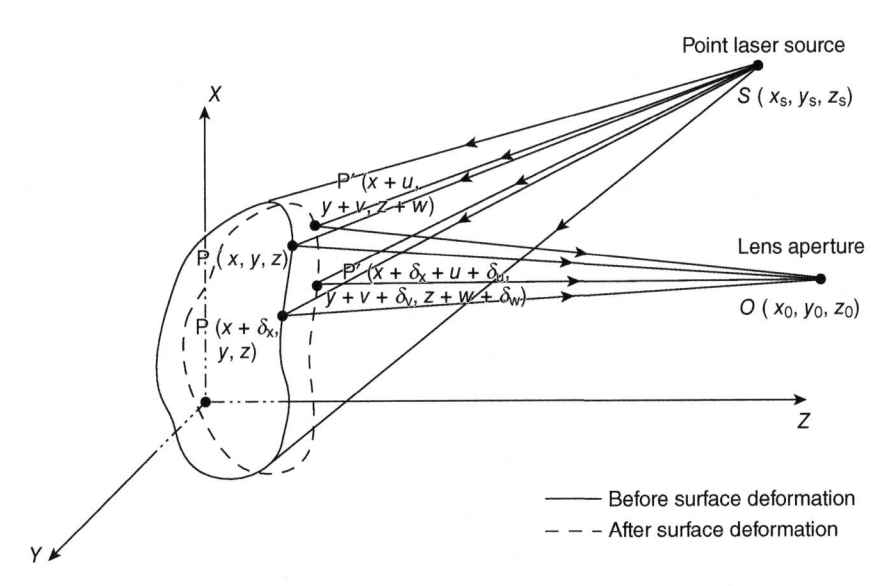

5.7 Diagram for derivation of optical path length change due to surface deformation.

The equation $z = z(x, y)$ describes the object surface and is not an independent variable when surface points are considered. Equation (5.4) may be rewritten in the following form:

$$\Delta = \left(\frac{2\pi}{\lambda}\right)\left[A\frac{\delta u}{\delta x} + B\frac{\delta v}{\delta x} + C\frac{\delta w}{\delta x}\right]\delta x \qquad [5.6]$$

Slight image-shearing gives rise to a small value of δx. The relative displacement terms in Equation (5.6) may subsequently be treated as the partial derivatives of u, v and w with respect to the direction of image-shearing x. Thus, Equation (5.6) becomes

$$\Delta = \left(\frac{2\pi}{\lambda}\right)\left[A\frac{\partial u}{\partial x} + B\frac{\partial v}{\partial x} + C\frac{\partial w}{\partial x}\right]\delta x \qquad [5.7]$$

Changing the shearing in another direction by rotating the image-shearing device or tilting the mirror of the Michelson interferometer in another direction will result in fringe patterns depicting displacement derivatives with respect to the current direction of image-shearing. Hence, if image-shearing is along the y-axis by an amount δy, Equation (5.7) is modified to the following:

$$\Delta = \left(\frac{2\pi}{\lambda}\right)\left[A\frac{\partial u}{\partial y} + B\frac{\partial v}{\partial y} + C\frac{\partial w}{\partial y}\right]\delta y \qquad [5.8]$$

The fringe patterns in Fig. 5.3(a) and (b) for a laterally deflected plate therefore represent, respectively, $\partial w/\partial x$ and $\partial w/\partial y$. It is possible to use a multiple image-shearing camera for simultaneous recording of displacement derivatives with respect to x and y (Hung & Durelli, 1979; Long, 1996; Steinchen & Yang, 2003, Chapter 2).

In NDT applications, quantitative precision measurement of surface deformation is not necessary. If the geometrical dimensions of the object are small compared to the distances of illumination and recording, the sensitivity factors may be approximated as follows:

$$\begin{aligned} A &= \sin\alpha \\ B &= \sin\beta \\ C &= (1+\cos\gamma) \end{aligned} \qquad [5.9]$$

where α and β are the angles of illumination made, respectively, with the yz-plane and the xz-plane; γ is the angle between the illumination beam and the z-axis, which is here designated as the imaging direction. To determine a complete state of surface deformation, each component of the displacement derivatives $\left[\frac{\partial u}{\partial x}, \frac{\partial v}{\partial x}, \frac{\partial w}{\partial x}\right]$ and $\left[\frac{\partial u}{\partial y}, \frac{\partial v}{\partial y}, \frac{\partial w}{\partial y}\right]$ must be measured. However, as shown in Eqs. (5.7) and (5.8), the phase Δ measures a combination of the displacement derivative components. Therefore it is necessary to take three measurements with three different sensitivity factors (A, B, C) to generate three equations of Eq. (5.7) and three equations of Eq. (5.8) for the various displacement derivative components to be separated. The process for measurement and separation of the displacement derivative components is quite complicated, which will not be elaborated here. Readers are directed to Steinchen & Yang, (2003, Chapter 7) for further details.

In most NDT applications, generally only the out-of-plane information, i.e. $\partial w/\partial x$ and $\partial w/\partial y$, is required. This information can be simply acquired by arranging the optical layout so that each of α and β is approximately equal to zero. In this situation, Eqs. (5.7) and (5.8) are reduced to:

$$\Delta = \left(\frac{2\pi}{\lambda}\right)\left(C\frac{\partial w}{\partial x}\right)\delta x \qquad [5.10]$$

$$\Delta = \left(\frac{2\pi}{\lambda}\right)\left(C\frac{\partial w}{\partial y}\right)\delta y \qquad [5.11]$$

Hence, the measurement is sensitive to only $\partial w/\partial x$ and $\partial w/\partial y$, the derivatives of out-of-plane displacement. One may use a multiple-image shearing camera (Yang et al., 2004) to allow simultaneous measurement of the two displacement derivative components. It is also possible to measurement surface displacement instead of displacement derivative using a large shear scheme with a reference object. The principles can be found in Hung et al. (2003).

5.2.5 Fringe phase determination

Traditional methods of shearographic analysis rely on the reconstruction of visible fringes and the correct identification of fringe orders (Equation (5.3)). Therefore, only at locations where the dark and bright fringes appear can the fringe orders be correctly determined; elsewhere the fringe orders are generally estimated using linear interpolation. The following describes an automated process in which the phase-change Δ is directly determined at every digitized point without having to rely on fringe reconstruction and fringe order identification. It will be seen that this has constituted a significant advantage of digital shearography.

Perusal of Eqs. (5.1) and (5.2) shows there are four unknowns, namely, I_o, μ, ϕ and Δ. By superimposing known phases to Eqs. (5.1) and (5.2), additional equations are generated to solve for these unknowns. This is achieved by phase-shift techniques (to be described later for each shearing device). The phase-shift techniques allow a known phase to be introduced to generate the additional equations. Although there are several phase-shift algorithms (Creath, 1990), the four-frame algorithm is described here. Digitizing four speckle patterns of Eq. (5.1) with a phase-shift increment of $\pi/2$ generates the following four equations:

$$\begin{aligned}
\mathbf{I}_1 &= \mathbf{I}_o[1+\mu\cos(\phi+0)] \\
\mathbf{I}_2 &= \mathbf{I}_o[1+\mu\cos(\phi+\pi/2)] \\
\mathbf{I}_3 &= \mathbf{I}_o[1+\mu\cos(\phi+\pi)] \\
\mathbf{I}_4 &= \mathbf{I}_o[1+\mu\cos(\phi+3\pi/2)]
\end{aligned}$$
[5.12]

The random phase ϕ is thus determined using the following equation.

$$\phi = \arctan(\mathbf{I}_4 - \mathbf{I}_2)/(\mathbf{I}_1 - \mathbf{I}_3)$$
[5.13]

The process is repeated for the deformed speckle pattern of Eq. (5.2), yielding

$$\phi' = \phi + \Delta = \arctan(\mathbf{I}_4' - \mathbf{I}_2')/(\mathbf{I}_1' - \mathbf{I}_3')$$
[5.14]

The computed phases ϕ and ϕ' are wrapped into the range of $-\pi$ and $+\pi$, the phase-change distribution Δ of deformation is subsequently determined by subtracting ϕ from ϕ' according to the following equation (Steinchen & Yang, 2003, Chapter 4):

$$\Delta = \begin{cases} \phi' - \phi & \text{for } \phi' \geq \phi \\ \phi' - \phi + 2\pi & \text{for } \phi' < \phi \end{cases}$$
[5.15]

The addition of 2π if $\phi' < \phi$ keeps the phase-change distribution Δ in a range from 0 to 2π rather than from -2π to 2π. Different unwrapping algorithms have been proposed to unwrap the phase change distribution Δ (Liu & Yang, 2007; Macy 1983). As an example, Fig. 5.8(a) and (b) show the wrapped and unwrapped phase distributions for a rectangular plate clamped along its boundary deformed by point load applied transverse at the center. The phase-shift technique for phase determination in digital shearography depends on the type of shearing devices used. For the two shearing devices presented in this article, the associated phase-shift techniques are described below.

Phase-shifter for Wollaston prism shearing device

For Wollaston shearing device, phase-shifting is achieved by employing a phase-shifter of a variable phase retarder. The variable wave retarder is a transmissive element made of liquid crystal. It has two orthogonally polarized axes (referred to as a slow axis and a fast axis) of transmission.

Digital shearography 95

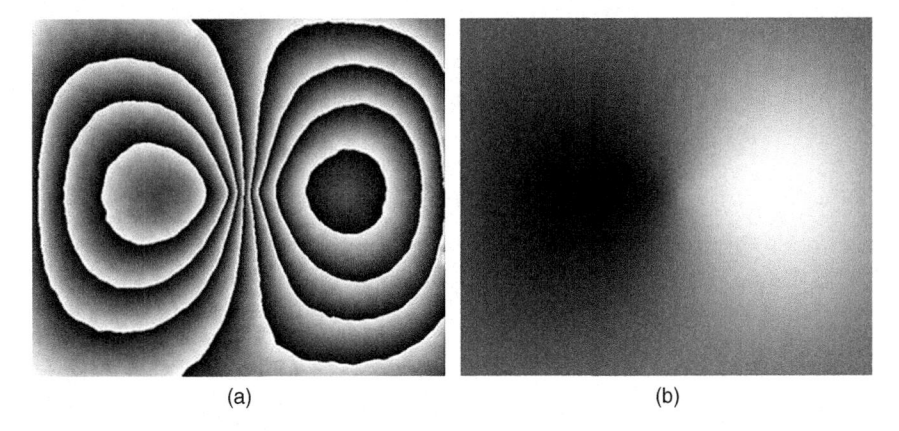

(a) (b)

5.8 The phase distributions of w/x for a rectangular plate clamped along its boundary deformed by point load applied transversely at the center: (a) wrapped phase; (b) unwrapped phase.

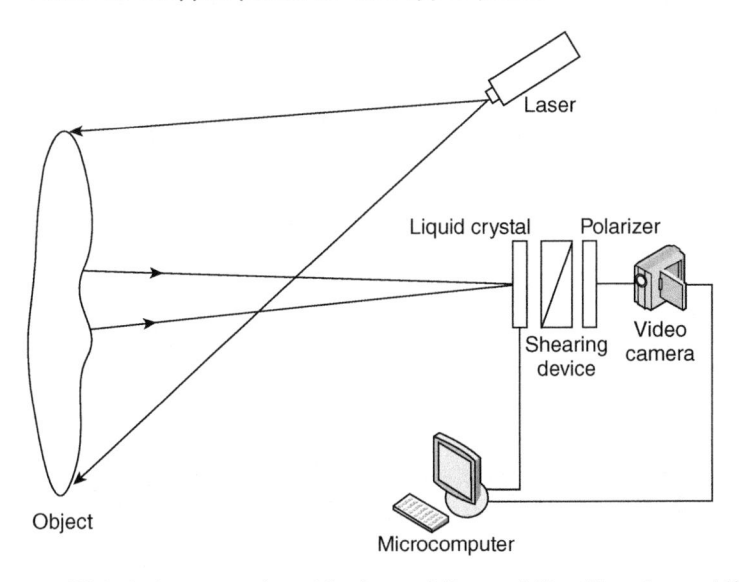

5.9 Digital shearography with phase-shift capability. The phase-shift device is a computer-controlled variable wave-retarder made of liquid crystal associated with a Wollaston prism.

The light transmitted by the two axes (slow axis and fast axis) incurs a different degree of retardation. The relative retardation of the light transmitted by the fast axis and the slow axis can be varied by controlling the voltage applied to the liquid crystal. This device is commercially available (suggested source: Medowlark Optics).

Figure 5.9 shows an optical layout for using the phase-shifter with a Wollaston prism shearing device. The wave retarder is placed in front of the shearing device, and at the same time, its slow and fast axes are aligned with the two polarization axes of Wollaston prism. This allows manipulation of the phase difference between two sheared wavefronts transmitted by the Wollaston prism.

Phase-shifter for Michelson shearing interferometer

Figure 5.10 illustrates digital shearography with phase-shift capability using a Michelson shearing interferometer. The phase is shifted by controlling the translation of mirror 2 driven by a piezoelectric crystal transducer (PZT). Theoretically, the 90-, 180- and 270-deg phase shifts can be obtained by moving the PZT driven mirror to small displacements of $\lambda/8$, $\lambda/4$ and $3\lambda/8$, respectively. If a relationship between voltage and displacement of a PZT is accurately known, a calibration for moving the given displacements can be done easily by changing the voltages. If a relationship between voltage and displacement is not accurately known, an auto- and self-calibration method is needed. The details of the principles can be found in Yang & Thorsten (2008) and Wu *et al.* (2013).

5.2.6 Fringe enhancement and multiplication technique

Shearographic fringe patterns obtained by direct digital subtraction of Eq. (5.3) are generally adequate for NDT applications. However, the fringe

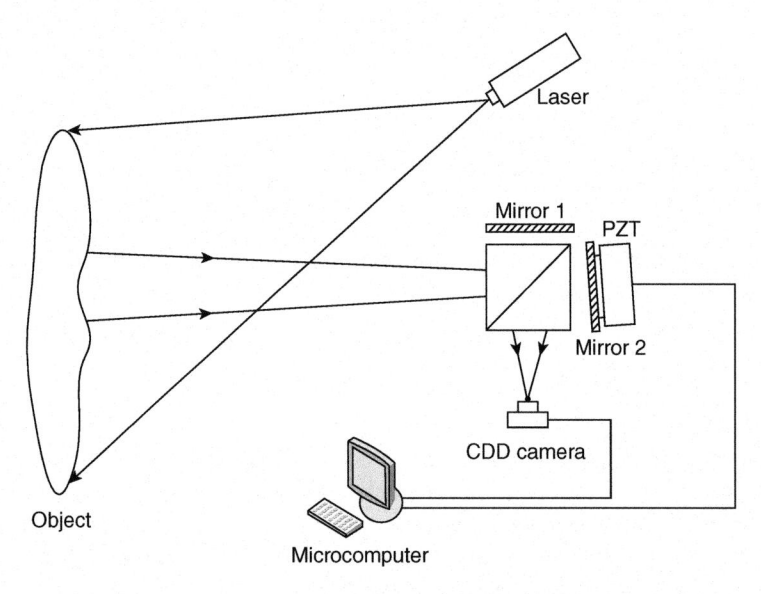

5.10 Digital shearography with phase-shift capability. The phase-shift device associated with a Michelson shearing interferometer. Phase is shifted by controlling the translation of the mirror using a PZT.

patterns obtained are of poor quality due to speckle noise. This problem can be overcome by obtaining the fringe phase using the phase-shift technique described in the above section. After determining Δ, a fringe pattern called the phase map of the shearogram can be obtained. For an 8 bit hardware resolution, there are 256 gray levels within one fringe (from black to white). The multiplication of fringe numbers can be achieved by a modulo operation:

$$G'(x, y) = [G(x, y) * M] \bm{mod} 256 \qquad [5.16]$$

where $G'(x, y)$ is the gray level in the new phase map after multiplication, $G(x, y)$ the gray level of the original phase map. M is the multiplication number.

Mod means Modulo operation. $[G(x, y) * M]$ **modulo** 256 (abbreviated as $[G(x, y) * M]$ **mod** 256) can be thought of as the remainder, on division of $[G(x, y) * M]$ by 256. For instance, $G(x, y) = 200$ and $M = 3$, $[G(x, y) * M) = 600$, the expression '600 mod 256' would evaluate to 88 because 600 divided by 256 leaves a remainder of 88. After this operation, the fringe number will be increased by a factor of M.

Should the hardware resolution is 10 bit, there are 1024 gray levels within one fringe, Eq. (5.16) becomes:

$$G'(x, y) = [G(x, y) * M] \bm{mod} 1024 \qquad [5.17]$$

Figure 5.11a presents a smoothed phase map of the shearogram revealing three delaminations in a composite honeycomb sample. Figure 5.11b shows the multiplied fringe pattern of Fig. 5.11a with M 6. Obviously, the fringe multiplication technique can be used to enhance flaw visibility in NDT application.

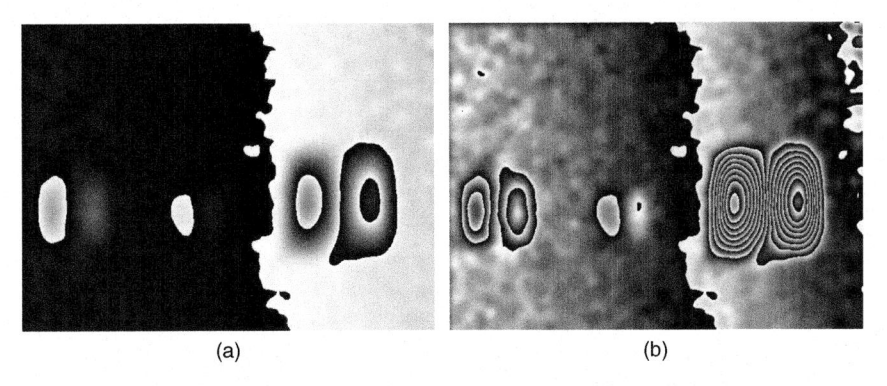

(a) (b)

5.11 (a) Smoothed phase map of shearogram revealing three delaminations in a composite honeycomb sample; (b) multiplied fringe pattern with *M* number of 3.

5.3 The practical application of digital shearography

5.3.1 How does shearography detect flaws?

When a test object containing a flaw is stressed, a strain concentration at the vicinity of the defect is induced. If the flaw is not too deeply embedded within the object, the induced strain concentration would cause anomalies in the surface strain distribution. Subsequently, these anomalies are translated into fringe anomalies if two speckle patterns of the object, taken one before and another after loading, are compared. Thus, shearography reveals flaws, both surface and internal, through identification of anomalies in the fringe pattern provided the flaw is not too remote from the surface.

Figure 5.12 shows a comparison between holographic and shearographic NDT. Assuming that a specimen has a delamination, a sufficient loading, e.g. by a thermal or a vacuum loading, will cause buckling on the sample surface. Holography detects a flaw by looking for displacement anomaly induced by the defect and the anomaly area looks like a circular or elliptical fringe pattern (depending on the defect styles). Shearography detects the flaw by looking for an anomaly of deformation derivative induced by the defect and the anomaly area looks like a butterfly pattern. To demonstrate

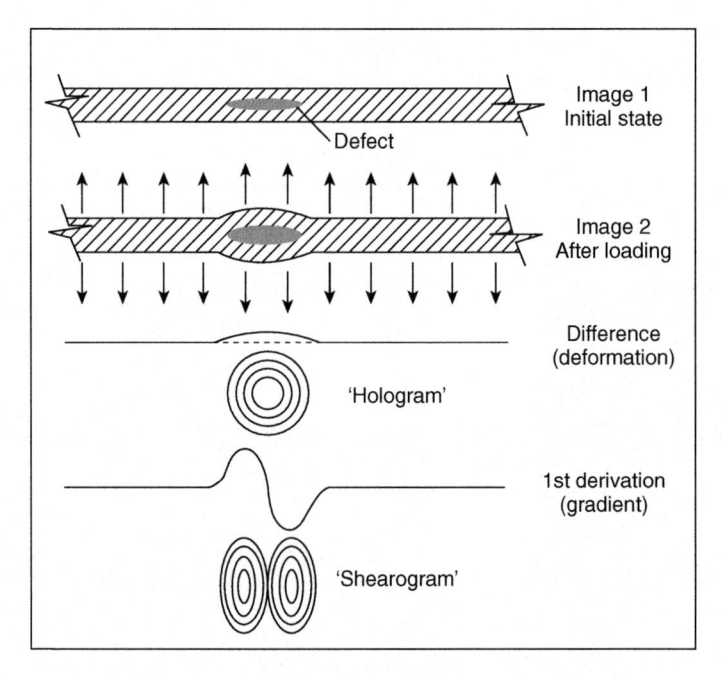

5.12 Interpretation and comparison of the results between holographic and shearographic inspection.

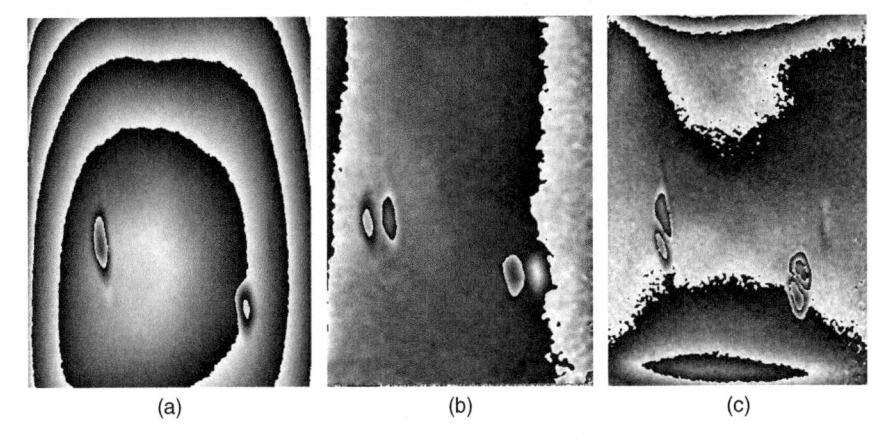

5.13 (a) Measuring result of holography, anomalous elliptical fringe patterns appear at the debond positions. (b) and (c) Measuring results of shearography with *x* and *y* shearing directions, respectively, anomalous butterfly fringe patterns appear at the debond positions and they are orientated in the shearing direction.

the concept, a laminated can with two debonds was tested by both holography and shearography using a thermal loading. Figure 5.13a shows the holographic measuring results in which elliptical anomalous fringes appear at the positions of debonds. Figure 5.13b and c shows the measuring results of digital shearography with x and y shearing directions, respectively. Butterfly patterns appear at the positions of the debonds and the orientations of the butterflies depend on the x and y shearing direction.

For practical applications, holography has a high requirement for environmental conditions such as rigid body motion. Whereas digital shearography measures the derivatives of deformation, it is not sensitive to rigid body motion and is better suited for industrial applications.

5.3.2 Shearography versus ultrasound

Shearography has found wide applications in NDT and it has already received industrial acceptance as a useful NDT tool, particularly for composite structures such as tires and honeycomb structures. It is particularly effective in revealing delaminations or debonds. Figure 5.14 shows a comparison of test result obtained using digital shearography and C-scan ultrasonic testing on a composite sample. Both techniques readily reveal an edge pullout and a delamination. With digital shearography, the flaws are revealed in one second, whilst the ultrasonic technique requires point-by-point scanning along the test surface and, at the same time, proper fluid coupling between the transducer and the test surface must be ensured (Choi *et al.*,

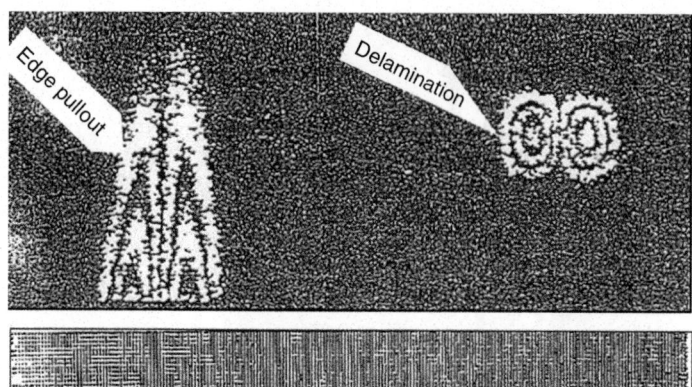

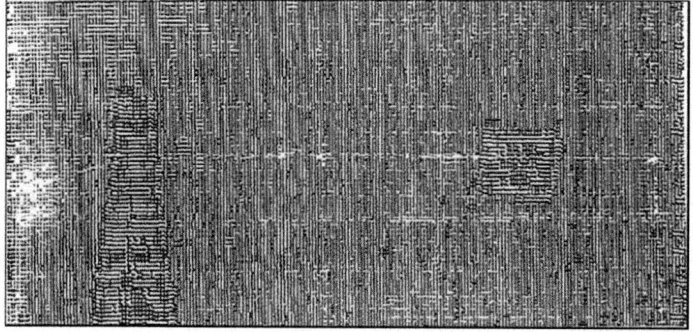

5.14 Comparison of shearography with C-scan ultrasound. An edge pullout and a delamination in a composite panel are revealed by both techniques. Time required: 10 minutes for ultrasound and 1 second for shearography. Fluid coupling is needed in ultrasonic testing.

2008). Note that the ultrasound image is of low quality, as the result was obtained in production environments using a fast C-scan process. Also this is the original ultrasound image without digital enhancement. A limitation of shearography, however, is the need to apply suitable stress increments to the test object during inspection, since the underlying principle of this technique is based upon changes in the state of the test surface.

5.3.3 Methods of stressing

Application of shearography in NDT has therefore essentially become the development of a practical means of stressing the object that would readily reveal flaws. Ideally, the stress increment should be similar to the service stresses so that flaws that are critical and detrimental to the service life of the object would be revealed, and cosmetic flaws that do not undermine the structural integrity of the test object can be ignored. This would minimize unnecessary rejects during inspection. In this regard, shearography has an advantage over ultrasonic testing, as the latter detects flaws by identifying in-homogeneities in the object and does not provide direct information on the criticality of the flaws. Exact duplication of the actual stress increment for shearographic testing, however, is generally difficult. Therefore, various practical means of stressing the object must be developed. In developing these methods, an important precaution to be taken is restricting rigid body motion of the object during stressing, as excessive rigid body motion would cause speckles de-correlation, resulting in degradation of fringe quality (Petersen, 1991; Yura et al., 1993). This limitation can be somewhat overcome by a compensation technique (Hung & Wang, 1997).

The various stressing methods that are in current use and yet do not cause intolerable rigid body motion of the test object include the use of pressure, vacuum, thermal, electrical induction, and acoustical and mechanical vibration excitations. The use of microwave that excites water molecules is particularly effective for detecting the presence of moisture in plastics and non-metallic composites, but safety precautions must be taken seriously.

Digital shearography has already received considerable acceptable for NDT, in particular for composite materials. While many industrial reports on the subject are not available in the public domain, the authors have collected some relevant published papers for readers' convenience (Abou-Khousa et al., 2007; Akhter et al., 2009; Ambu et al., 2006; Amenabar et al., 2011; Balageas et al., 2000; Bosia et al., 2002; Cawley, 1994; Corigliano et al., 2004; Dragan & Swiderski, 2010; Fomitchov et al., 1997; Francis et al., 2010; Garnier et al., 2011; Gryzagoridis et al., 1995; Hatta et al., 2005; Hofmann et al. 2008; Huang et al., 2007; Hung, 1998, 1999; Hung et al., 2000, 2009; Lai et al., 2009; Lee et al., 2004b; Liu et al., 1996; Lobanov et al., 2007; Nokes & Cloud, 1993; Okafor et al., 2001; Pang & Wu, 1996; Qin et al., 1997; Restivo et al., 2008; Richardsona et al., 1998; Ruzeke Behal, 2009; Shang et al., 1995; Steinchen et al., 1998a; Toh et al., 1990; Wang et al., 1998, 2008; Wu et al., 2011; Yang & Ettemeyer, 2003; Yap et al., 2004; Zhang & Richardson, 2007). Presented below are some examples of early work of the authors.

5.4 Using digital shearography to test composites

5.4.1 Detection of debonds in laminated composites

Shearography is particularly effective in detecting debonds in laminated composite materials. Illustrated here is a multi-layer composite sample with known debonds at the interface of various layers. The stressing method used is partial vacuum. The result is shown in Fig. 5.15. When partial vacuum is applied, the debond behaves as a plate subjected to uniform pressure. The fringe lines are loci of the derivative of the plate deflection and they appear in the form of a butterfly. The size of a debond is approximately equal to the size of the butterfly. The depth of a debond can be estimated from the

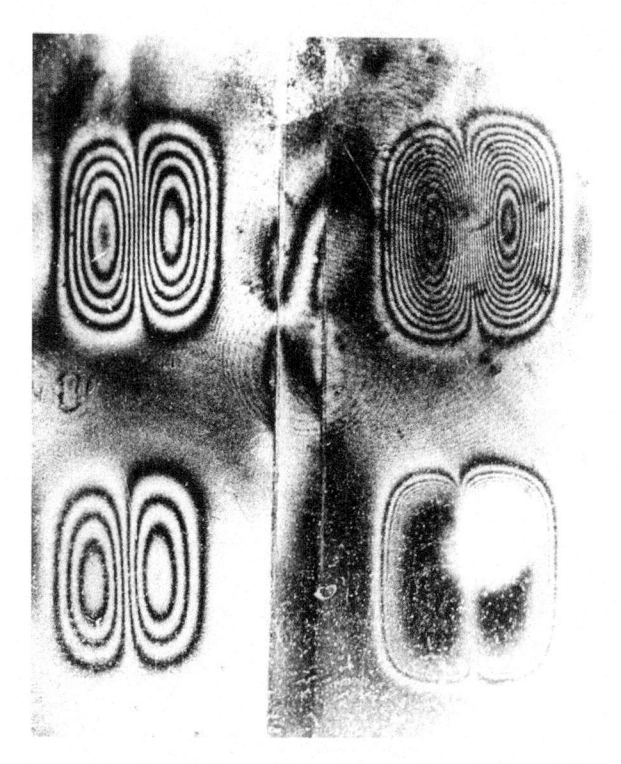

5.15 Fringe patterns of debonds at the various depths (3, 6, 9 and 12 mm) of a multi-layer composite material sample. Note that the one closest to the surface has the highest fringe density and vice versa.

fringe density. For debonds of the same size, the one closer to the surface has a higher fringe density and vice versa. This is illustrated in Fig. 5.15, which shows four debonds of approximately the same size but located at depths of 3, 6, 9 and 12 mm from the surface. This can be explained by the plate theory which predicts that the deflection derivative is inversely proportional to the flexural rigidity of the plate, and the flexural rigidity is proportional to the cube of the plate thickness. The model considers an equivalent plate of thickness equal to the depth of the debond measured from the surface, and the shape and size of the equivalent plate follows that of the debond. For equal size debonds, the density of the fringes is inversely proportional to the cube of the depth.

5.4.2 Pneumatic tire inspection

Shearography has been widely adopted by the rubber industry for inspecting debonds and other imperfections in tires. The technique was specified

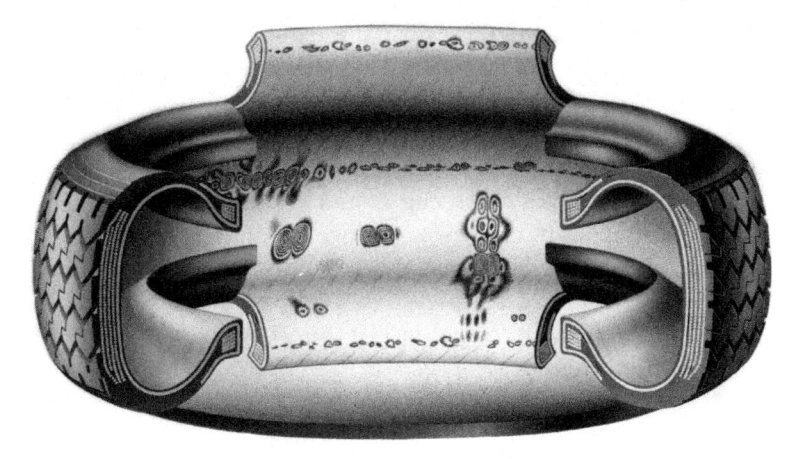

5.16 Simulated fringe patterns revealing several debonds in a truck tire using partial vacuum stressing.

by the US Federal Aviation Administration (FAA) for inspecting aircraft tires. Tires are complex laminated composite structures and vacuum stressing is well suited for testing tires. The test tire without mounting wheel is placed inside the vacuum chamber in which the pressure can be reduced. During testing, the tire interior is illuminated with laser light and imaged by a shearographic camera; two shearographic images are sequentially digitized with the chamber pressure slightly reduced (by around 50 mm of Hg). Computing the difference of the two images produces a fringe pattern allowing tire imperfections to be identified. Figure 5.16 shows a simulated fringe pattern revealing delaminations in a truck tire.

5.4.3 Skin–core debonds in honeycomb materials

Shearography is very effective for detecting skin–core debonds in honeycomb structures. The most popular loading methods for inspecting such kinds of flaws are vacuum and thermal stressing (Yang, 2006). The vacuum stressing is equivalent to a uniform tensile force applied to the sample surface which pulls the surface outward. Thus a delamination causes the surface directly above the flaw to bulge out slightly, which can be detected by digital shearography. Methods using vacuum loading can be divided into two categories, i.e. entire vacuum stressing and partial vacuum stressing. In the entire vacuum stressing method, the whole sample is placed in a vacuum chamber and the surface under test is observed through a transparent window. The shearographic sensor is usually set in the outside of the chamber. If the object surface to be tested is unable to be observed through a transparent window, e.g. NDT of a tire, the sensor should be put in the inside of the chamber. In this case, the vacuum chamber should have a relatively large volume. For larger objects such as the wing of an aircraft, enclosing the object in a vacuum chamber is impractical. Therefore, partial vacuum stressing is utilized. In this test, the compact digital shearographic sensor is installed within a suction cup. The plastic ring is applied around all peripheries of the suction cup mouth so that the suction cup is sealed against the test surface when applying the initial vacuum. The portable digital shearographic system with a partial vacuum loading is currently commercially available, such as from Dantec-Dynamic GmbH, Germany, Steinbichler GmbH, Germany, and LTI Inc. USA.

Figure 5.17a and b show a portable system from Dantec-Dynamic and a measuring result using this system for inspecting a honeycomb panel with two delaminations, respectively. The mouth of the suction cup, i.e. the measuring area, is about 0.09 m^2 (1 square foot). During measurements, two phase distributions are calculated at the initial vacuum and an additional vacuum and then a digital shearogram can be obtained by a digital subtraction. The measurement can be finished less than half a minute, mainly depending on the time for applying the vacuum. The system is well suited for the NDT of panels with a flat surface or curved surfaces without a significant change in curvature. Should the surface be too dark or too shiny, a white spray (easily removable) should be applied to improve the intensity of light reflected.

Another useful loading system for inspecting skin–core debonds in honeycomb materials is thermal stressing. In this type of testing, the object under test is radiated with heat. The temperature gradient developed

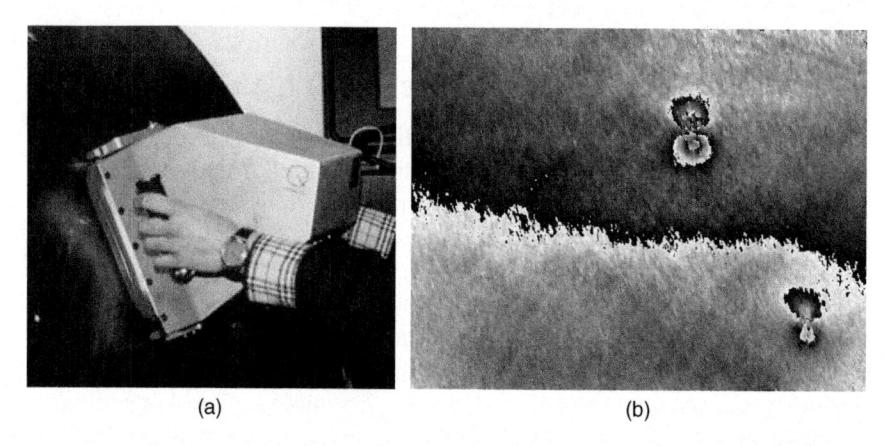

5.17 (a) A commercially available portable digital shearographic NDT system with a partial vacuum loading (from Dantec-Dynamic); (b) a measuring result using this system for a honeycomb panel with two delaminations.

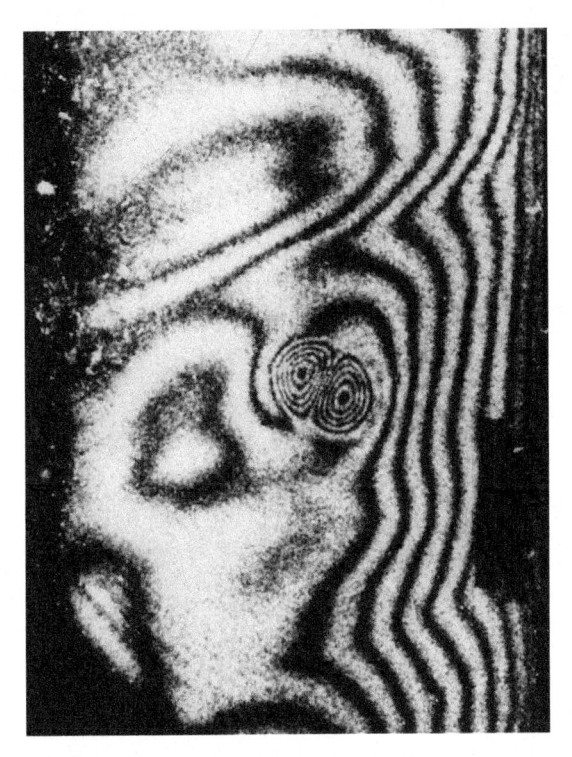

5.18 Revelation of a debond in a filament-winded composite pressure vessel. The stressing method is internal pressurization.

induces thermal strains in the object. Should a delamination or a debond develop in the object, heat will make the delamination or the debond expand causing the material above the defect to bulge and generate a strain anomaly on the surface. The defect can be found easily by digital shearography. A simple way to transport a thermal load into the material is to use a heat gun or a high-power (infrared) light. Figure 5.18 shows a measuring result in a honeycomb structure using thermal stressing.

5.4.4 NDT of pressure vessels

Loading with internal pressure is ideal for pressure vessels, pipes and those structures that can be pressurized (Hung, 1989a, 1996). Since internal pressurization represents the actual stressing for pressure vessels and pipes, the magnitude of the internal loading can be applied as big as the maximum of the actual loading. Figure 5.19 shows a shearographic measuring result revealing a delamination in a composite cylinder. The stressing method is internal pressurization. Figure 5.20 shows the shearographic test results for

106 Non-destructive evaluation (NDE) of polymer matrix composites

5.19 Fringe pattern revealing a delamination in a honeycomb panel of an aircraft. The stressing method is partial vacuum.

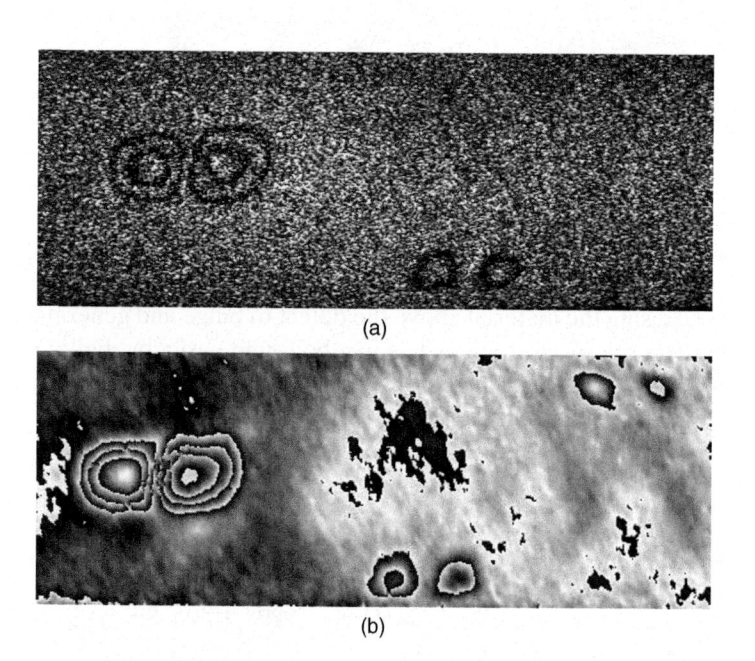

(a)

(b)

5.20 Shearographic inspection of a glass fiber-reinforced plastic tube subjected to a 0.6 MPa internal pressure: (a) two delaminations were found in the intensity subtraction shearogram; (b) three defects have been found in the phase map of shearogram, and the image displayed is the original phase map multiplied by a value of 2.

© Woodhead Publishing Limited, 2013

a glass fiber-reinforced plastic tube subjected to an internal pressure. Two delaminations were found by a 0.6 MPa internal pressure in the intensity subtraction shearogram, whereas three delaminations have been found in that the phase map of shearogram at the same loading, which demonstrates that the phase map has higher measuring sensitivity than the intensity subtraction image.

5.4.5 Crack detection

For inspecting crack in composites, a thermal or pulsed thermal stressing can be used. Figures 5.21 and 5.22 demonstrate shearographic NDT of cracks in composite materials using thermal and pulsed thermal stressing, respectively.

5.4.6 Evaluation of bonding integrity of composite assemblies

Dynamic loading includes harmonic and non-harmonic excitation. Harmonic excitation can usually be generated in a sample by attaching a piezoelectric crystal in combination with an amplifier or the use of a shaker. The purpose of doing a shearographic NDT with harmonic excitation is to excite the debonds/flaws into a controlled resonant state. A problem can arise from a mix of the vibration induced by the nature frequency of the sample

5.21 Shearogram revealing micro-cracks in a glass fiber-reinforced plastic plate under thermal stressing.

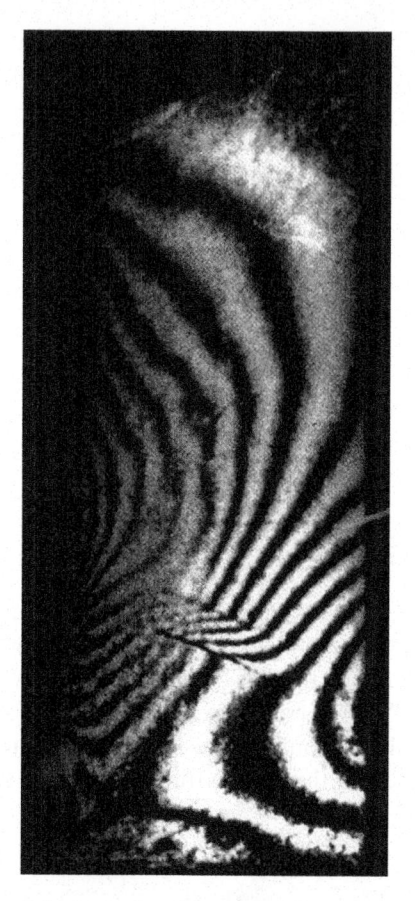

5.22 Detection of a crack on a composite turbine blade using pulsed thermal stressing.

itself with the vibration of the flaws, since this makes the identification of flaw location difficult. However, a non-harmonic and acoustic excitation can reduce such effect and has achieved great success for NDT of debonds and voids in composite materials (Hung, 1997). It should, however, be emphasized that NDT by a dynamic load is more complicated than by a static or quasi-static load, such as a vacuum stress, a thermal loading and an internal pressure. Thus, experience using this method is required for obtaining good results.

This is a very useful and practical technique for inspecting bonding integrity of composite structures. It uses a broadband acoustic excitation with an image peak-intensity recording algorithm. During testing, the test object is excited by a sound wave of varying frequencies within a preselected band. When the frequency of excitation coincides with the natural frequency of

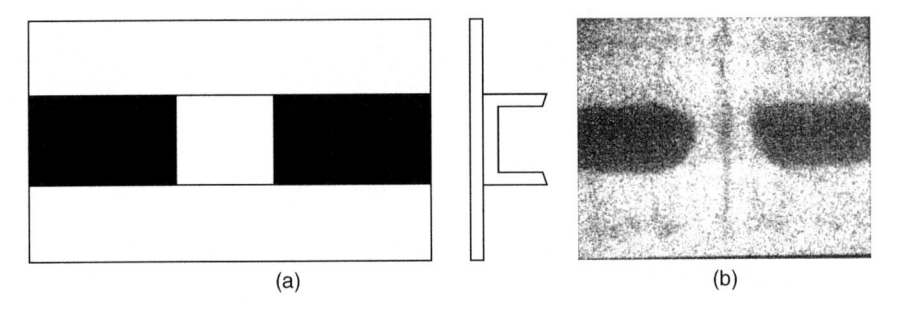

5.23 Shearographic inspection of adhesive-bonded composite assemblies using the multi-frequency vibration excitation: (a) A composite plate was bonded to U-shape channel; (b) the white patch around the center revealing a non-bonded area. The dark areas (on both sides of the non-bonded area) are indicative of good bonding.

the debonded area, the area is in resonance and hence vibrates with a relatively larger deformation. The deformation (displacement-derivatives) of the area is captured with a peak image-intensity storage algorithm. In the peak image-intensity acquisition operation, the current image-difference is continuously compared pixel-by-pixel with the preceding values stored in the computer memory – the higher pixel intensity of the two will be stored in computer memory. Thus, debonded areas, whose deformation rate is higher than that of the non-defective areas, manifest as bright patches and the bonded area will appear dark. An example of such a result is shown in Fig. 5.23.

5.5 Conclusion

A review of shearography is given and its application for non-destructive testing, particularly for composites, has been illustrated. It is an optical method which enjoys the advantages of being a full-field method that does not require contacting or scanning. It also allows the size location of the detected flaws to be more readily assessed. A major difference of shearography from other NDT techniques is the mechanism of detecting defects. It reveals defects by looking for defect-induced deformation anomalies (or strain concentrations). The requirement to deform the test object is a limitation for objects that cannot be conveniently stressed to produce the required deformation. However, if the test object can be deformed in a mode similar to the actual loading in service, shearography can be used to reveal critical detects and ignore cosmetic defects. Shearography is applicable to virtually any material, provided the surface of the test object is diffusely reflective. For shiny surfaces, additional surface preparation is needed to make the surface diffusely reflective. This is normally achieved by simply spraying the surface with a non-glossy white paint or removable white coating.

5.6 Acknowledgment

The authors sincerely appreciate the help from Ran Wu, Jianfei Sun, Xin Xie, Xu Chen, and Nan Xu, who are all graduate students of Optical Laboratory at Oakland University for drawing the principal images and preparing some experimental results.

5.7 References and further reading

Abou-Khousa M A, Ryley A, Kharkovsky S, Zoughi R, Daniels D, Kreitinger N and Steffes G (2007), 'Comparison of x-ray, millimeter wave, shearography and through-transmission ultrasonic methods for inspection of honeycomb composites', *Review of Progress in Quantitative Nondestructive Evaluation*, Vols 26A and 26B, Aip Conference Proceedings 894, 999–1006.

Akhter N, Jung H C, Chang H S and Kim K S (2009), 'Location of delamination in laminated composite plates by pulsed laser holography', *Optics and Lasers in Engineering*, **47**(5), 584–588.

Ambu R, Aymerich F, Ginesu F and Priolo P (2006), 'Assessment of NDT interferometric techniques for impact damage detection in composite laminates', *Composites Science and Technology*, **66**(2), 199–205.

Amenabar I, Mendikute A, Lopez-Arraiza A, Lizaranzu M, and Aurrekoetxea J (2011), 'Comparison and analysis of non-destructive testing techniques suitable for delamination inspection in wind turbine blades', *Composites Part B – Engineering*, **42**(5), 1298–1305.

Balageas D, Bourasseau S, Dupont M, Bocherens E, Dewynter-marty V and Ferdinand P (2000), 'Comparison between non-destructive evaluation techniques and integrated fiber optic health monitoring systems for composite sandwich structures', *Journal of Intelligent Material Systems and Structures*, **11**(6), 426–437.

Bosia F, Botsis J, Facchini M and Giaccari P (2002), 'Deformation characteristics of composite laminates – Part 1: speckle interferometry and embedded Bragg grating sensor measurements', *Composites Science and Technology*, **62**(1), 41–54.

Cawley P (1994), 'The rapid nondestructive inspection of large composite structures', *Composites*, **25**(5), 351–357.

Choi M, Park J, Kim W and Kang K (2008), 'Inspection of impact damage in honeycomb composite by ESPI, thermography and ultrasonic testing', *International Journal of Modern Physics B*, **22**(9–11), 1033–1038.

Corigliano A, Papa E and Pavan A (2004), 'Study of the mechanical behaviour of a macroscopic glass-polyester composite by ESPI method and numerical simulations', *Composites Science and Technology*, **64**(12), 1829–1841.

Creath K (1990), 'Phase-measurement techniques for nondestructive testing', *Proceedings of SEM Conference on Hologram Interferometry and Speckle Metrology*, Baltimore, Maryland, November 5–8, 473–478.

Dragan K and Swiderski W (2010), 'Studying efficiency of NDE techniques applied to composite materials in aerospace applications', *ACTA Physica Polonica A*, **117**(5), 878–883.

Fomitchov P, Wang L S and Krishnaswamy S (1997), 'Advanced image-processing techniques for automatic nondestructive evaluation of adhesively-bonded structures using speckle interferometry', *Journal of Nondestructive Evaluation*, **16**(4), 215–227.

Francis D, Tatam R P and Groves R M (2010), 'Shearography technology and applications: a review', *Measurement Science & Technology*, **21**(10), 102001.

Garnier C, Pastor M L, Eyma F and Lorrain B (2011), 'The detection of aeronautical defects in situ on composite structures using non destructive testing', *Composite Structure*, **93**(5), 1328–1336.

Gryzagoridis J, Findeis D and Schneider D R (1995), 'The impact of optical NDE methods in vessel fracture protection', *International Journal of Pressure Vessels and Piping*, 61(2–3), 457-469.

Hatta H, Aly-Hassan M S, Hatsukade Y, Wakayama S, Suemasu H. and Kasai N (2005), 'Damage detection of C/C composites using ESPI and SQUID techniques', *Composites Science And Technology*, **65**(7–8), 1098–1106.

Hofmann D, Pandarese G, Revel G M, Tomasini E P and Pezzoni R (2008), 'Optimization of the excitation and measurement procedures in nondestructive testing using shearography', *Review of Scientific Instruments*, **79**(11), 115105.

Huang S J, Lin H L and Liu H W (2007), 'Electronic speckle pattern interferometry applied to the displacement measurement of sandwich plates with two "fully potted" inserts', *Composite Structures*, **79**(2), 157–162.

Hung Y Y (1974), 'A speckle-shearing interferometer: a tool for measuring derivatives of surface displacement', *Optical Communications*, **11**(2), 132–135.

Hung Y Y (1982), 'Shearography: a new optical method for strain measurement and nondestructive testing', *Optical Engineering*, **21**(3), 391–395.

Hung Y Y (1989a), 'Shearography: a novel and practical approach to nondestructive testing', *Journal of Nondestructive Evaluation*, **8**(2), 55–68.

Hung Y Y (1989b), 'Apparatus and method for electronic analysis of test object', US Patent 4,887,899.

Hung Y Y (1996), 'Shearography for non-destructive evaluation of composite structures', *Optics and Lasers in Engineering*, **24**(2–3), 161–168.

Hung Y Y (1997), 'Automated nondestructive shearographic inspection of debonds in composites using multi-frequency vibrational stressing', *Proceedings of 1997 SEM Spring Conference*, Bellevue, Washington, June 2–4, 98–99.

Hung Y Y (1998), 'Computerized shearography and its application for nondestructive evaluation of composites', Chapter 17, *Manual on Experimental Methods of Mechanical Testing of Composites*, edited by C.H. Jenkins, published by Society for Experimental Mechanics.

Hung Y Y (1999), 'Applications of digital shearography for testing of composite structures', *Composites Part B – Engineering*, **30**(7), 765–773.

Hung Y Y and Durelli A J (1979), 'Simultaneous measurement of three displacement-derivatives using a multiple image shearing interferometric camera', *Journal of Strain Analysis*, **14**(3), 81–88.

Hung Y Y and Hovanesian J D (1990), 'Fast detection of residual stresses in an industrial environment by thermoplastic-based shearography', *Proceedings of the 1990 SEM Spring Conference on Experimental Mechanics*, Albuquerque, New Mexico, June 4–6, 769–775.

Hung Y Y and Shi D H (1998), 'Technique for rapid inspection of hermetic seals of microelectronic packages using shearography', *Optical Engineering*, **37**(5), 1406–1409.

Hung Y Y and Wang J Q (1997), 'Technique for compensating excessive rigid body motion in NDT of large structures using shearography', *Optics and Lasers in Engineering*, 26(2–3), 249–258.

Hung Y Y, Long K W, Hovanesian J D and Hathaway R (1988), 'Fast detection of residual stresses by shearography', *Proc. SPIE Industrial Laser Interferometry*, **955**, 26–35.

Hung Y Y, Tang S and Hovanesian J D (1994), 'Real-time shearography for measuring time-dependent displacement derivatives', *Experimental Mechanics*, **34**(1), 89–92.

Hung Y Y, Long K W and Wang J W (1997), 'Measurement of residual stress by phase shift shearography', *Optics and Lasers in Engineering*, **27**(1), 61–73.

Hung Y Y, Shang H M, Lin L, Zhang Y B and Wen X Y (1999), 'Evaluation of residual stresses in plastics and composites by shearography', SAE Paper #1999-01-1254, 1999 SAE International Congress and Exposition, Detroit, Michigan, March 1–4.

Hung Y Y, Luo W D, Lin L and Shang H M (2000), 'Evaluating the soundness of bonding using shearography', *Composite Structures*, **50**(4), 353–362.

Hung Y Y, Shang H M and Yang L X (2003), 'Unified approach for holography and shearography in surface deformation measurement and nondestructive testing', *Optical Engineering*, 42(5), 1197–1207.

Hung Y Y, Chen Y S, Ng S P, Liu L, Huang Y H, Luk B L, Ip R W L, Wu C M L and Chung P S (2009), 'Review and comparison of shearography and active thermography for nondestructive evaluation', *Material Science & Engineering R – Reports*, **64**(5–6), 73–112.

Ibrahim J S, Petzing J N and Tyrer J R (2004), 'Deformation analysis of aircraft wheels using a speckle shearing interferometer', *Proceedings of the Institution of Mechanical Engineers Part G – Journal of Aerospace Engineering*, **218**(G4), 287–295.

Lai W L, Kou S C, Poon C S, Tsang W F, Ng S P and Hung Y Y (2009), 'Characterization of flaws embedded in externally bonded CFRP on concrete beams by infrared thermography and shearography', *Journal of Nondestructive Evaluation*, **28**(1), 27–35.

Lee J R, Molimard J, Vautrin A and Surrel Y (2004a), 'Digital phase-shifting grating shearography for experimental analysis of fabric composites under tension', *Composites Part A – Applied Science and Manufacturing*, **35**(7–8), 849–859.

Lee J R, Molimard J, Vautrin A and Surrel Y (2004b), 'Application of grating shearography and speckle shearography to mechanical analysis of composite material', *Composites Part A – Applied Science and Manufacturing*, **35**(7–8), 965–976.

Lee J R, Yoon D L, Kim J S and Alain V (2008), 'Investigation of shear distance in Michelson interferometer-based shearography for mechanical characterization', *Measurement Science & Technology*, **19**(11), 115303.

Leendertz J A and Butters J N (1973), 'An image shearing speckle pattern interferometer for measuring bending moments', *Journal of Physics: E, Scientific Instrumentation*, **6**(11), 1107–1110.

Liu S and Yang L X (2007), 'Phase unwrapping in a regional sense – fringe estimation, region segmentation, and sequence expansion', *Optical Engineering*, **46**(5), 05012-1 – 05012-9.

Liu W, Tan Y and Shang H M (1996), 'Singlemode optical fiber electronic speckle pattern interferometry', *Optics and Lasers in Engineering*, **25**(2–3), 103–109.

Lobanov L M, Pivtorak L V, Kyyanets and Savyts'ka O M (2007), 'Nondestructive testing of composite and metallic pipes by the method of electronic shearography', *Materials Science*, **43**(4), 522–527.

Lobanov L M, Bychkov S A, Pivtorak V A, Derecha V Y, Kuders'kyi V O, Savyts'ka O M and Kyyanets I V (2009), 'On-line monitoring of the quality of elements of aircraft structures by the method of electron shearography', *Materials Science*, **45**(3), 366–371.

Long K W (1996), '3-Beam phase shift shearography for simultaneous measurement of in-plane and out-of-plane displacements and its applications to residual stress measurements', PhD Dissertation, Oakland University.

Macy W W (1983), 'Two-dimensional fringe pattern analysis', *Applied Optics*, **22**(23), 3898–3901.

Nokes J P and Cloud G L (1993), 'The application of interferometric techniques to the nondestructive inspection of fiber-reinforced materials', *Experimental Mechanics*, **33**(4), 314–319.

Okafor A C, Otieno A W, Dutta A and Rao V S (2001), 'Detection and characterization of high-velocity impact damage in advanced composite plates using multi-sensing techniques', *Composite Structures*, **54**(2–3), 289–297.

Pang L Y and Wu X P (1996), 'Nondestructive testing by ESPI and quasi-phase shift gradient technique', *Optics and Lasers in Engineering*, **25**(2–3), 93–101.

Petersen M O (1991), 'Decorrelation and fringe visibility: on the limiting behaviour of various electronic speckle pattern correlation interferometers', *Journal of the Optical Society of America A*, **8**(7), 1082–1089.

Qin S, Shang H M and Giam C L (1997), 'Holographic assessment of edge delamination on foam-adhesive bonded joints', *Composites Part B – Engineering*, **28**(3), 177–183.

Restivo G, Isaicu G A and Cloud G L (2008), 'Low-cost non-destructive inspection by simplified digital speckle interferometry', *Journal of Nondestructive Evaluation*, **27**(4), 135–142.

Richardsona M O W, Zhang Z Y, Wisheart M, Tyrer J R and Petzing J (1998), 'ESPI non-destructive testing of GRP composite materials containing impact damage', *Composites Part A – Applied Science and Manufacturing*, **29**(7), 721–729.

Ruzek R and Behal J (2009), 'Certification programme of airframe primary structure composite part with environmental simulation', *International Journal of Fatigue*, **31**(6), 1073–1080.

Ruzek R, Lohonka R and Jironc J (2006), 'Ultrasonic C-scan and shearography NDI techniques evaluation of impact defects identification', *NDT & E International*, **39**(2), 132–142.

Shang H M, Lim E M and Lim K B (1991a), 'Computer-aided assessment of debonds in laminates from shearographic fringe patterns', *Journal of Materials Processing Technology*, **25**(1), 55–67.

Shang H M, Toh S L, Chau F S, Shim V P W and Tay C J (1991b), 'Locating and sizing Disbonds in glassfibre-reinforced plastic plates using shearography',

Transactions ASME, Journal of Engineering Materials and Technology, **113**(1), 99–103.

Shang H M, Tham L M and Chau F S (1995), 'Shearographic and holographic assessment of defective laminates with bond-lines of different elasticities', *Transactions ASME, Journal of Engineering Materials and Technology*, **117**(3), 322–329.

Steinchen W and Yang L X (2003), *Digital Shearography – Theory and application of digital speckle pattern shearing interferometry*, Belingham, Washington, SPIE Press.

Steinchen W, Yang L X, Schuth M and Kupfer G (1995), 'Application of shearography to quality assurance', *Journal of Materials Processing Technology*, **52**(1), 141–150.

Steinchen W, Yang L X and Kupfer G (1997), 'Digital shearography for nondestructive testing and vibration analysis', *Experimental Techniques*, July/August, 20–23.

Steinchen W, Yang L X, Kupfer G and Mackel P (1998a), 'Non-destructive testing of aerospace composite materials using digital shearography', *Proceedings of the Institution of Mechanical Engineerings Part G – Journal of Aerospace Engineering*, **212**(G1), 21–30.

Steinchen W, Yang L X, Kupfer G, Maechel P and Voessing F (1998b), 'Strain analysis by means of digital shearography: potential, limitations and demonstration', *Journal of Strain Analysis*, **332**, 171–182.

Toh S L, Chau F S, Shim V P W, Tay C J and Shang H M (1990), 'Application of shearography in nondestructive testing of composite plates', *Journal of Materials Processing Technology*, **23**(3), 267–275.

Wang W C, Day C H, Hwang C H and Chiou T B (1998), 'Nondestructive evaluation of composite materials by ESPI', *Research in Nondestructive Evaluation*, **10**(1), 1–15.

Wang W C, Su C W and Liu P W (2008), 'Full-field non-destructive analysis of composite plates', *Composites Part A – Applied Science and Manufacturing*, **39**(8), 1302–1310.

Wu S J, He X Y and Yang L X (2011), 'Enlarging the angle of view in Michelson interferometer-based shearography by embedding a 4f system', *Applied Optics*, **50**(21), 3789–3794.

Wu S J, Zhu L Q, Feng Q B and Yang L X (2013), 'Digital shearography with *in situ* phase shift calibration based on optical method', Accepted by *Optics and Lasers in Engineering*.

Yang L X (2006), 'Recent developments of digital shearography for NDT', *Material Evaluation*, **64**(7), 703–709.

Yang L X and Ettemeyer A (2003), 'Strain measurement by 3D-electronic speckle pattern interferometry: potentials, limitation and applications', *Optical Engineering*, **42**(5), 1257–1266.

Yang L X and Thorsten S (2008), 'Digital speckle interferometry in engineering', Chapter 22, *New Directions in Holography and Speckle*, edited by Caulfield H J and Vikram C S, American Scientific Publishers, Stevenson Ranch, California.

Yang L X, Steinchen W, Schuth M and Kupfer G (1995), 'Precision measurement and nondestructive testing by means of digital phase shifting speckle pattern and speckle pattern shearing interferometry', *Measurement*, **16**(3), 149–160.

Yang L X, Steinchen W, Kupfer G, Maeckel P and Voessing F (1998), 'Vibration analysis by digital shearography', *Optics and Laser in Engineering*, **30**, 199–212.

Yang L X, Chen F, Steinchen W and Hung Y Y (2004), 'Digital shearography for nondestructive testing: potentials, limitations and applications', *Journal of Holography and Speckle*, **1**(2), 69–79.

Yap A U J, Tan A C S and Quan C (2004), 'Non-destructive characterization of resin-based filling materials using electronic speckle pattern interferometry', *Dental Materials* **20**(4), 377–382.

Yura H T, Hanson S G and Grum T P (1993), 'Speckle statistics and interferometric decorrelation effects in complex ABCD optical systems', *Journal of the Optical Society of America A*, **10**(2), 316–323.

Zhang Z Y and Richardson M O W (2007), 'Low velocity impact induced damage evaluation and its effect on the residual flexural properties of pultruded GRP composites', *Composite Structures*, **81**(2), 195–201.

6
Non-destructive evaluation (NDE) of composites: dielectric techniques for testing partially or non-conducting composite materials

R. A. PETHRICK, University of Strathclyde, UK

DOI: 10.1533/9780857093554.1.116

Abstract: This chapter describes the use of high- and low-frequency methods for the non-destructive characterization of insulating or low-conductivity composite structures. The low-frequency measurements allow study of the progress of cure through the use of interdigitated electrodes buried in the laminate structure. Water ingress into composites can also be mapped by measurements in the frequency range 1 Hz to 1 kHz. The use of large format probes has made it possible to explore the water distribution in ship hulls and similar structures which have been exposed to water for prolonged periods of time. Measurements at microwave frequencies allow defects to be detected within the composite and determine the location of water- and air-filled voids. At very high frequencies the measurements are sensitive to the stand-off distance and in order to achieve the highest resolution it is necessary to use computer models to assist with the analysis of data. With the use of appropriate vector analysis and phase shifting it is possible to achieve mapping of defects which are difficult to detect by other methods.

Key words: composites, millimetre microwave techniques, interdigitated electrodes, low-frequency dielectric methods, cure, water absorption.

6.1 Introduction

Non-destructive testing (NDT) is used for the examination of safety critical structures and includes inspection of aircraft, offshore marine structures, boat hulls, transport vehicles – train carriages, Formula 1 racing cars, wind turbines, chemical plant and distribution lines, etc. Composites are designed to combine stiffness, strength and weight at an appropriate cost. The typical polymer composite will usually be based on a thermosetting resin, but can occasionally be constructed using a high-temperature thermoplastic polymer. Reinforcement will be achieved using glass, carbon, Kevlar, natural or metal fibres. The resultant composite material has the advantages of high specific strength and stiffness, tailored directional properties, excellent fatigue resistance, dimensional stability and environmental resistance, together with ease of fabrication.

However, composites are susceptible to impact damage, moisture pick-up and lightning strikes, have a relatively high cost, do not yield plastically in regions of high stress concentration and are subject to random property variation due to the nature of the manufacturing process. Advanced composites are used in airframe construction for floor beams, doors, aerodynamic fairings, control surfaces, rudders, elevators and ailerons, owing to their low weight and high stiffness. Wind turbine blades are increasingly being constructed from composites and have similar design requirements to those used for rudders in aircraft construction. Fibre reinforced polymer composite laminates may suffer delamination as a result of faults introduced during manufacture or as a consequence of environmental exposure. The detection of delaminations has challenged the NDT community for many years (Jones *et al.*, 1988; Sansalone, 1997; Zoughi and Bakhtiari, 1990; Zoughi *et al.*, 2002).

Infrared thermography provides a very useful method for the detection of cracks and defects which lie close to the surface (Jones *et al.*, 1998); however, when studying composite materials with low thermal conductivity, depth resolution appears to be a problem. Ultrasonic methods often suffer from high-attenuation problems in composites and X-rays fail to find defects because of lack of contrast. In general, ultrasonic methods have the disadvantage of requiring the transducer to be in physical contact with the surface or strongly coupled through a fluid layer. In many situations, it is desirable to have physical separation between probe and the article being examined. Dielectric methods are applicable to the study of these systems, sensing changes in impedance and therefore being particularly sensitive to the presence of air- or water-filled void structures and also do not requiring physical contact with the article.

In Chapter 8 of this book the focus is placed primarily on metal bonded structures. Some carbon fibre composites are also sufficiently conductive to allow examination of internal adhesive bonds using the methods discussed for metals. The focus of Chapter 8 is also on the non-destructive analysis of adhesively bonded structures. Because the substrates being bonded together were intrinsically electrically conducting, the adhesive bond forms a capacitor at low frequency and a waveguide at high frequency. If, however, the substrates are non-conducting then accessing the adhesive bond becomes a totally different problem. Dielectric methods can be applied to non-conducting materials, but the interpretation of the observed changes to the electric fields used becomes more complex.

In general, a dielectric material placed in an electric field will induce distortion of that field. Measurement of the changes in the electric and magnetic components of an electric field allow characterization of the dielectric properties of the media. In a conventional dielectric measurement, the material being examined will be placed between two conducting electrodes and the capacitance and conductance of the material determined. Measurement of these quantities as a function of frequency allows calculation of the frequency dependence of real and imaginary parts of the complex dielectric permittivity. Normally, it is not possible to sandwich a composite between two electrodes and often access to the structure is restricted to one side. As a consequence, measurement of the dielectric properties has to be obtained from an analysis of the distortion of the electric field when it is brought into proximity with the dielectric.

6.2 Low-frequency dielectric measurement of partially conductive and insulating composite materials

Measurements on partially conducting and insulating composites require use of precisely defined electric fields and often the use of models to assist with the analysis of the data. As with the study of adhesive bonds, the approach taken depends on the frequency used in the observation, which in turn influences the type of quality of information which can be obtained. In Chapter 8 it will be shown that dielectric measurements at frequencies of the order of 10–100 Hz can be used to follow dipole and conduction processes which change with the state of cure of the resin. The usual approach, which is employed when the composite is non-conducting, is to sense the dielectric properties using interdigitated electrodes which are buried in the laminate. The structure of an interdigitated electrode is shown in Fig. 6.1.

The size and shape of the electrode can be changed to address the requirements of the particular NDT problem (Maloney and Smith, 1992). Small interdigitated electrodes structures containing about 20 digits and with overall dimensions of 5 × 3 mm^2 are available commercially and can be used as disposable items for monitoring cure in autoclaves. Larger arrays can be created for mapping large surfaces areas. In the conventional electrode, alternate digits are at high and low potentials. The penetration of the electric field depends on the separation and spacing of the digits but will be typically of the order of 2–5 mm using digits of width 80 µm and spacing 120 µm. An increased penetration can be achieved if instead of neighbouring digits being at high and low potentials, an intermediate potential is applied on the second electrode and the third is at the high potential Fig. 6.2a.

The field created by this configuration is thrown further into the sample and penetrations of the order of 10 mm or greater can be achieved. The electrode shown in Fig. 6.1 was designed to operate at a frequency of 1 kHz and allowed examination of a composite structure which was 30 mm thick.

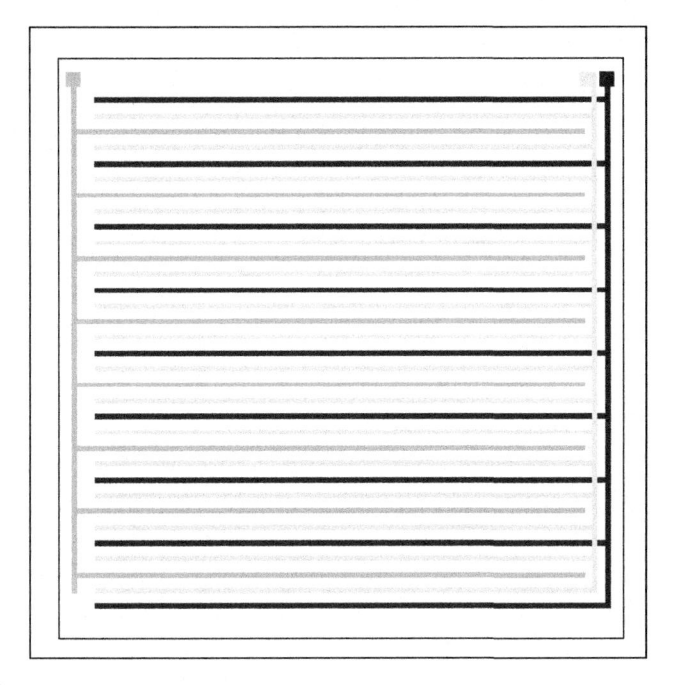

6.1 Schematic of an interdigitated electrode structure for use in low-frequency dielectric examination of composite structures.

Suitable design of the electrode structure allows a maximum sensitivity to be achieved for a particular frequency of operation and the electrodes are easily created by etching commercially available copper coated polymide film. To increase the detection sensitivity, the electrode structure forms one arm of a Wheatstone bridge circuit Fig. 6.3. The other elements of the bridge circuit can be adjusted to remove the resistance and capacitance contributions from the cables and leads connecting the electrode to the circuit and increase the sensitivity of the measurement of the change rather than having to measure the total values of the components.

The relationship between the input voltage (V_1) and the output voltage (V_0) and the resistance (R_m) and capacitance (C_m) of the composite are:

$$R_m = \frac{|Z_m|^2}{\text{Re}|Z_m|} \quad C_m = \frac{\text{Im}|Z_m|}{\omega|Z_m|^2} \quad [6.1]$$

where Z_m is the impedance of the composite of equivalent resistance R_m and capacitance C_m. The ratio of the voltages is:

$$\frac{V_0}{V_1} = \frac{Z_m}{Z_m + Z_1} - \frac{1}{2} \quad [6.2]$$

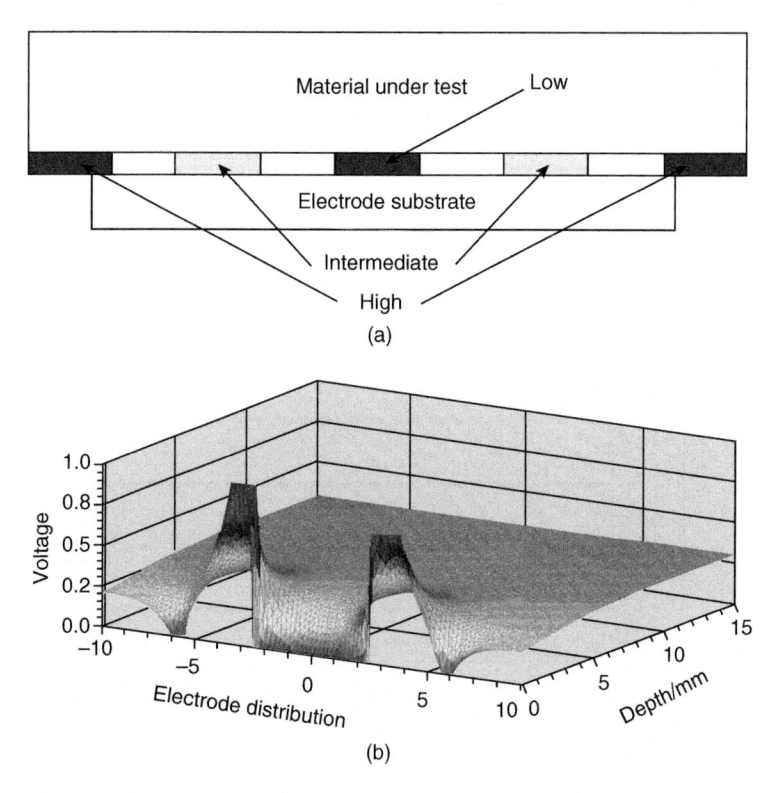

6.2 (a) Voltage distribution for an electrode operating in non-standard mode and (6), field distribution from electrode structure with an intermediate applied potential.

where Z_1 is the impedance of R_1 and C_1. The equivalent resistance and capacitance of the composite are obtained by substituting Z_m from equation (6.2) into equation (6.1). A further modification of the above configuration which adds greater sensitivity is to measure the current via an electrometer rather than voltage balance. A number of circuits have been described but they all are similar in principle (Hayward *et al.*, 1984; Maistros and Partridge, 1998).

6.3 Low-frequency dielectric cure monitoring

As indicated in Chapter 8, changes in the dielectric properties of a thermosetting resin can be directly related to the degree of cure in that material. However, interdigitated electrode structures impose limitations with regards the frequency range which can be scanned and in practice it is usual for these measurements to be performed at a single frequency rather than over

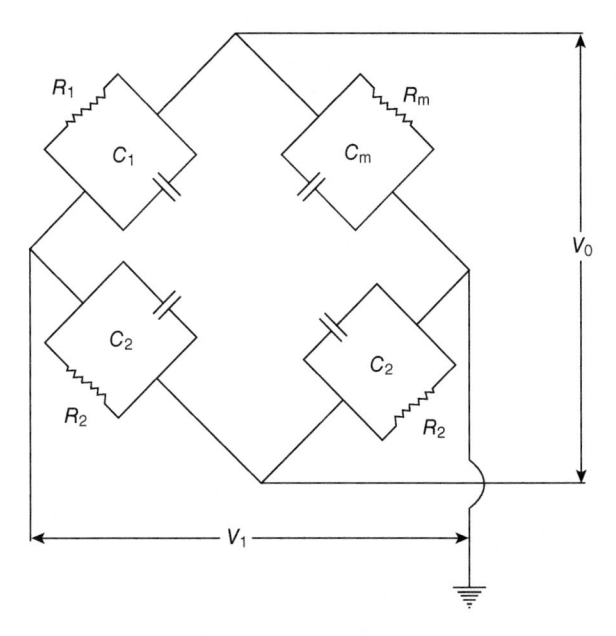

6.3 Wheatstone bridge configuration for low-frequency dielectric measurements.

a broad frequency range. The constraints of the digitations of the electrode limit the depth which is being sensed, so that cure monitoring may often be carried out on thick laminated sections using probes which are placed at different depths through out the composite structure. Heat transfer to and from the mould and the effects of changes in the thermal conductivity as cure progresses may all have a profound effect on the extent to which cure is achieved at different parts of the same structure. The types of change observed are illustrated in Fig. 6.4.

Because of limitations imposed by the electrodes, the range of frequencies available may be little more than a decade or two. As a consequence the changes in the conductivity contribution may not be as clearly resolved as when the broader frequency measurements are performed. However, examination of the changes in the dielectric permittivity and dielectric loss with time allow both the conduction and dipolar contributions to be identified from the dielectric data at low frequency in ε' (Fig. 6.4a) and ε'' (Fig. 6.4b). The initial dielectric loss is very large and is not shown in Fig. 6.4b at low frequency. As cure proceeds, loss rapidly reduces and drops to a value of ε'' of 0.1 after ~7000 seconds; this marks the point at which gelation of the resin is occurring. The peak in the dielectric loss occurring after ~10 000 seconds and the drop in the dielectric permittivity ε' are associated with the gel being transformed into a rigid glass and the peak observed is associated

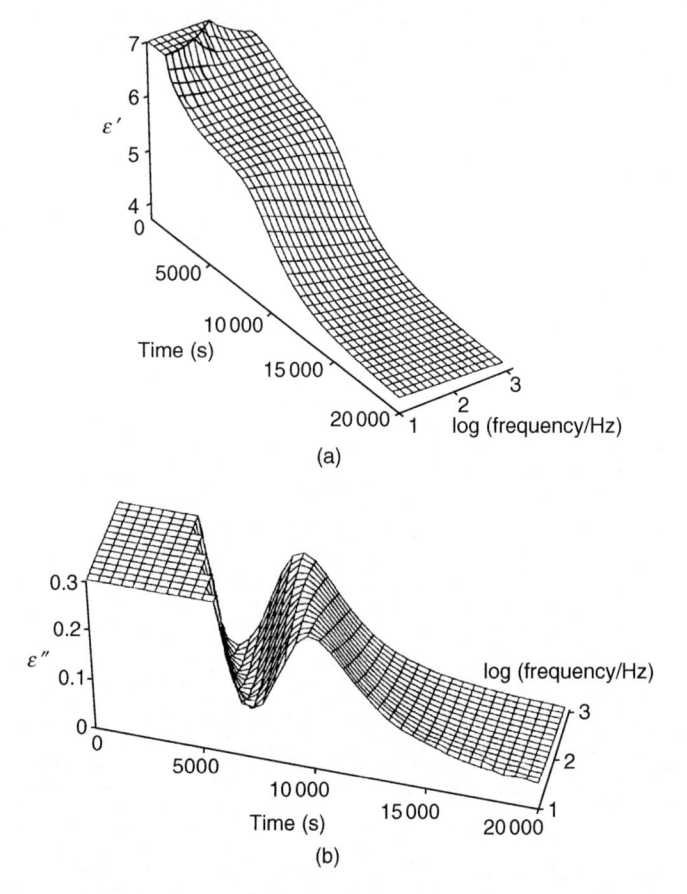

6.4 Variation of the dielectric permittivity ε' (a) and dielectric loss ε'' (b) for the cure of a typical thermoset resin.

with dipolar relaxation of the matrix. The gel point and vitrification points are very important in determining the state of cure. It has been shown by Kranbuehl (1990) that if $\log(\varepsilon'' \times \omega)$ is plotted against cure time, the initial curves follow a common trace and only separate during the vitrification stage Fig. 6.5.

Initially, the laminate ply will be a semi-solid and will not wet the probe; a low electric signal is obtained. As the temperature is raised, the electrode is wetted and the signal increases [A]. The initial dielectric loss is dominated by ionic conductivity and when the values of ε'' are multiplied by ω, a constant value is obtained. In the region between [A] and [B] the resin is in a viscous liquid state and the ionic conductivity will be fairly constant. A blip in the trace marks the increase in the viscosity. Just beyond [B] is the

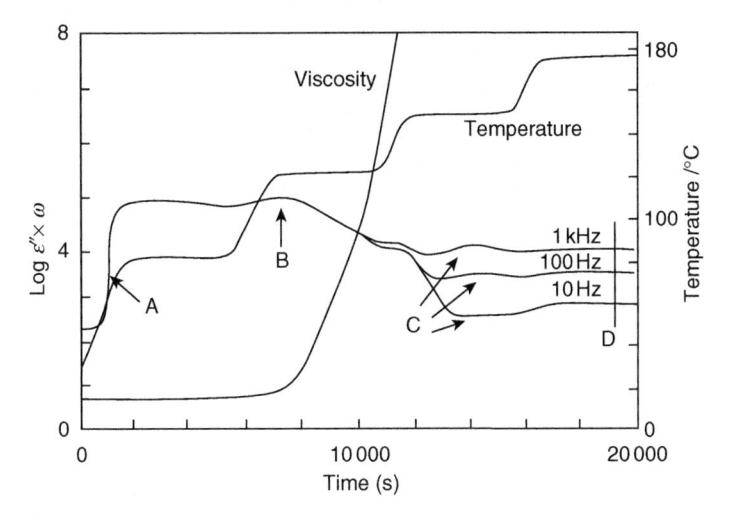

6.5 Schematic of the variation of the dielectric loss ε'' multiplied by the frequency measured at three frequencies, 10 Hz, 100 Hz and 1 KHz, the variation of the temperature using in the cure process and the viscosity as a function of cure time.

gelation point and the composite is transformed from a rubber to a glass. The glass will, however, contain a large number of unreacted monomer species which will plasticize the material and allow dipole rotation to occur once the temperature is raised [C]. Further reaction occurs and the dipole process moves to lower frequency as indicated by the small blip in the curve at about 15 000 seconds. At [D] the composite is fully cured and dipole motion will have ceased. This form of display is useful to map the changes in viscosity but is not so easily used to follow the cessation of the dipole motion which marks the glass transition process.

This approach has been used to determine the different rates of cure in thick laminate stacks and show that the surface is curing faster than the centre of a 192 ply laminate stack (Kranbuehl, 1990). A number of workers have established correlations between the changes in the dielectric characteristics of the curing laminate and other physical changes, such as the reaction exotherm as monitored by differential scanning calorimetry or infrared monitoring of the consumption of the monomer during the cure process (Kazilas and Partridge, 2005; Kim and Gille, 1996; Maistros and Partridge, 1998). Remote monitoring of the state of sure of a resin is now a well-established NDT technique and allows optimization of cure of composite structures being fabricated in autoclaves, etc. (Kazilas and Partridge, 2005).

6.4 Low-frequency dielectric measurement of water ingress into composite structures

Water ingress into composites can produce significant damage as a result of delamination leading to loss of mechanical strength. Figure 6.6 shows the type of damage found on a composite structure which has been exposed to moisture and high temperature for a prolonged period of time. The section of trunking shown in Fig. 6.6 had contained steam at a pressure of several bar and was observed to show blistering and delamination when inspected during a routine maintenance close-down of the plant. The blisters are a consequence of explosive decompression: the moisture penetrates the composite and subsequently segregates in regions between the glass laminates. When the pressure is released, blisters are formed within the laminate structure. The blisters will eventually compromise the structural integrity and lead to delamination. The low-frequency NDT method can be used to follow water ingress into composite structures Fig. 6.7.

A sample of glass reinforced plastic (GRP) was exposed at 70 °C to an environment of 100% humidity for a period of 208 days. The dielectric characteristics of the composite initially consists of a dielectric permittivity which falls between the frequencies of 10^{-1} Hz and 10^5 Hz, with a dielectric loss feature located below 10 Hz and above 10^4 Hz. The lower frequency loss feature is characteristic of the interfacial polarization observed with

6.6 A section of composite high-pressure trunking which exhibits decompression blisters.

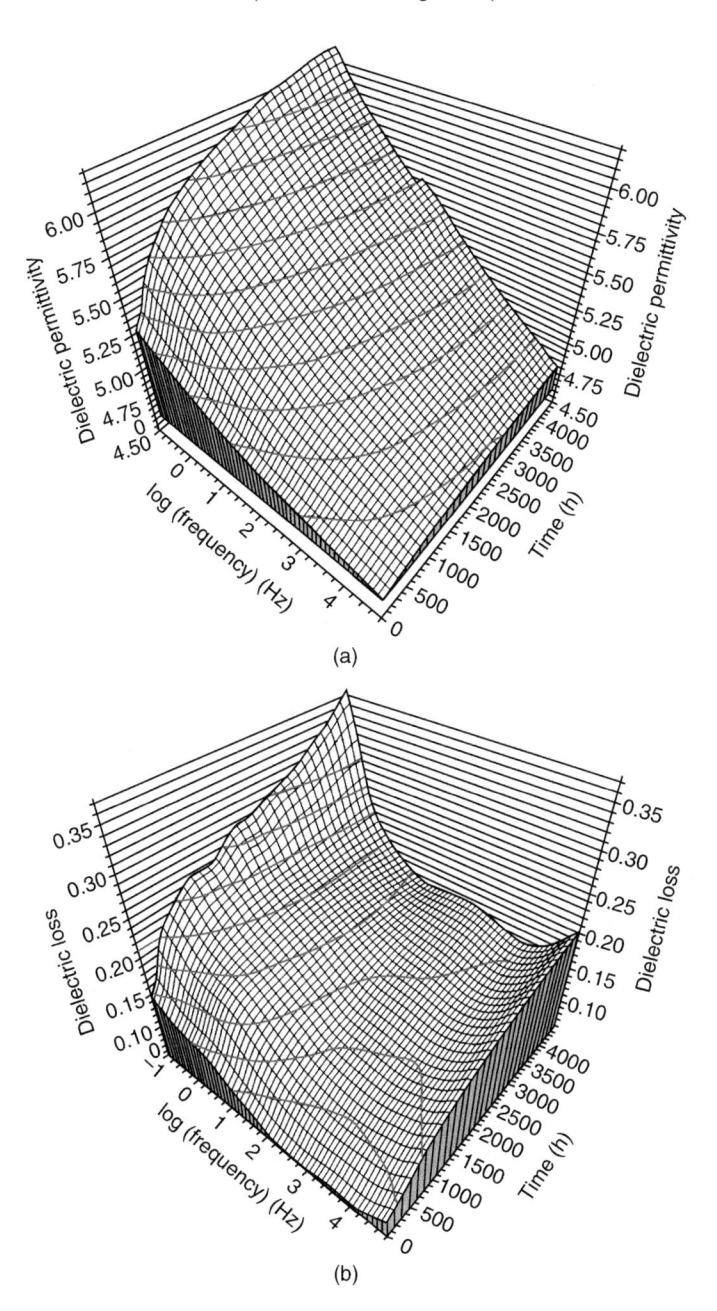

Dielectric techniques for testing composite materials   125

6.7 Variation of dielectric permittivity (a) and dielectric loss (b) for a glass reinforced plastic (GRP) composite material exposed at 70 °C.

heterogeneous materials and reflects the trapping at internal interfaces of charge carriers migrating between the glass fibres of the matrix. The higher-frequency process, located at ~10^4 Hz, is the local dipole rotation of the vinyl ester used in the resin and is a librational motion of the ester group (Pethrick et al., 2010).

The absorption of water by the composite leads to significant increases in the dielectric permittivity with time (Fig. 6.7a) and corresponding increases in the dielectric loss with time (Fig. 6.7b). Close inspection of the changes in the dielectric loss indicates that there is a marked increase at ~10^{-1} Hz up to 1500 hours, followed by a second step increase, which develops into a well-defined peak. As the water content increases so the mobility of the ionic impurities increases and there is a corresponding increase in the contribution from interfacial polarization. The increase in the loss peak at ~10^5 Hz corresponds to the interaction of molecular water with the ester dipoles. The overall changes in the dielectric permittivity parallel those in the dielectric loss and are shown in Fig. 6.7b. A plot of the change in the dielectric permittivity measured at 10 Hz against the square root of time shows a linear variation which is typical of a diffusion controlled process (Fig. 6.8).

The change in the permittivity is significantly smaller that that observed with more polar resins, such as amine cured epoxy formulations. The vinyl polyester resin contains a large amount of styrene and is intrinsically non-polar and hydrophobic. The observed change in permittivity would correspond to a water uptake of ~ 0.18 sw/w%. This level of water does not produce significant plasticization of the resin, but can, if it is segregated into inter-laminate regions, induce blistering and cracking as shown in Fig. 6.6.

Using a 25 cm^2 electrode array it was possible to survey a boat hull which had been in use for 30 years. The dielectric measurements indicate that the

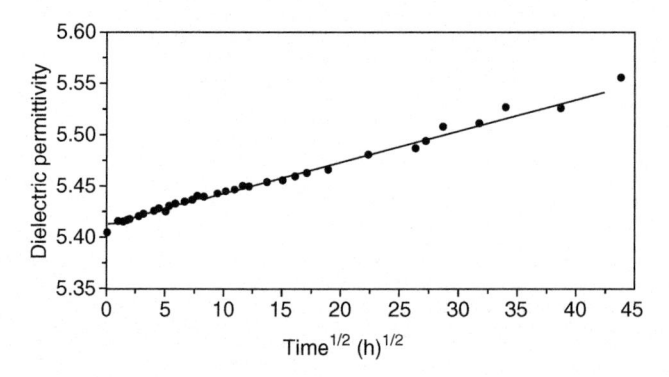

6.8 Variation of the dielectric permittivity measured at 10 Hz plotted against the square root of time.

moisture level in the composite was of the order of 2%. Variations in the level of water adsorbed were detected above and below the water line and fore and aft in the region of the engine compartment. A comparison was made of the extent of water between composites exposed to deionized and salt water (Boinard *et al.* 1999, 2000). It was found that salt water is less aggressive than deionized water in terms of attack on the GRP composite structures.

Low-frequency measurements can provide very useful information on the nature of the state of cure of composite materials and the extent to which a composite may have adsorbed water, but provide limited information in terms of the location of defects within the laminate structure. As discussed in Chapter 8, use of high-frequency techniques allows visualization of the internal structure and location of the size and shape of defects.

6.5 High-frequency measurements of dielectric properties

At frequencies above 10 MHz guided wave propagation becomes appropriate. There are two different approaches which can be used for the study of microwave propagation in composite media: the near-field and far-field approaches.

6.5.1 Near-field approach

The near-field approach involves scanning the substrate with a probe which is constructed so as to focus a propagating electric wave into a small region. Probes have different shape and form but might be a rectangular shaped block of low-loss dielectric material sharpened to a pyramidal tip, which is partially metallized and terminated by a micron sized plane facet. In the typical configuration the entire energy of the incident wave is concentrated as a very strong near-field and the electrical characteristics are sensitive to fluctuation in the dielectric properties close to the tip. The impedance of the waveguide is measured using a network analyser ideally operating over the frequency range 10 MHz to 30 GHz. Moving the probe across the surface allows the generation of an impedance map and high-resolution images can be created. Gray and Zoughi (1997) have demonstrated that the detection sensitivity can be greatly enhanced by adding an additional back gap; however, this is not always practical when access is only possible from one side. It may also be difficult to achieve a constant stand-off distance, i.e. the distance between the antenna and object surface, making on-site evaluation sometimes difficult. The near field technique has, however, been shown to be sensitive to disbonds in multilayer laminates. The high detection capability associated with this technique stems from the fact that near-field microwave signals are sensitive to minute variations in the dielectric properties and geometry of the medium in which they propagate. A method for the quantitative estimation of the disband thickness has been proposed in terms of a maximum-likelihood (ML) disbond thickness in which values obtained using different frequencies are compared and showed a good correlation with actual measurements. (Abou-Khousa and Zoughi, 2007).

Near-field measurements have also been undertaken on carbon fibre reinforced plastic (CFRP) employing dual-polarized reflectometer for detecting disbands. The system has the advantage that it is capable of overcoming the problem of definition of the standoff distance on the measurements. It can simultaneously generate three images of a defect: two at orthogonal polarizations and one after the influence of standoff distance variations is removed using the information provided by the two images. The dual-polarized reflectometer operates at X-band (8.2–12.4 GHz) and consists of a computer-controlled 2D scanning mechanism, two microwave circuits incorporated into a dual-polarized square waveguide probe, and a conditioning circuit (real-time signal processing section) that automatically removes the standoff distance variation (Kharkovsky et al., 2008).

The signal is applied to the sample using an open ended square waveguide probe which simultaneously supports two orthogonally polarized dominant TE10 mode microwave signals. The two signals irradiate and receive reflected signals from the same area of the CFRP. In designing the probe, it was important to achieve isolation between the two signals but allow combination of the information from the two signals to allow the automatic and real-time standoff distance variation removal from the perpendicularly polarized data/image. Isolation was achieved by operating two orthogonally polarized ports operating at 9.6 GHz for parallel polarization and 11.6 GHz for perpendicular polarization. A standoff distance of 7–11 mm was found to give the best results with a distance of 8 mm being optimal. A number of samples of dimensions $380 \times 520 \times 80$ mm^3 were examined which contained an internal disbond created by the introduction of a thin sheet of foam. In this study the CFRP was mounted on a concrete slab. Defects left from the manufacturing process caused significant surface roughness, which led to significant variation in the value of the standoff distance. Using a 2D scanning system an area of 140×140 mm^2 was scanned and an image produced by recording the raster output voltages from the two orthogonally polarized ports, normalizing each data matrix with respect to its highest voltage value and producing a grey scale image. The disbonded region is clearly visible in the image for perpendicular polarization, demonstrating the utility of the method for the detection of disband and demonstrated in-field on a bridge which contained bent members.

6.5.2 Far-field approach

Although it is easier to perform the measurements in the far-field approach, there may be difficulties in interpretation of edge reflection due to the small aperture size (Ganchev et al., 1995a). This problem can be overcome by employing a focusing dielectric lens. In addition, the phase variation with distance in free space or inside of a dielectric layer is a linear function of the electrical thickness of the layer, making examination of thin laminates difficult. Generally the far-field (plane wave) approach is preferred as the modelling of fields is relatively straight forward compared with the mathematically cumbersome near-field approach. General characteristics of plane waves can be found in various electromagnetic textbooks (Chipman, 1968; Glazier and Lamont, 1958; Pozar, 2005; Ramo et al., 1995) and will not be discussed here. Consider plane wave propagation through a laminate containing glass fibers orientated at 0° and 90° (Fig. 6.9).

The relationships between the fields between two adjacent interfaces can be expressed by:

$$\begin{bmatrix} E_0 \\ H_0 \end{bmatrix} = \begin{bmatrix} \cosh \gamma_{1z} l & z_{z1} \sinh \gamma_{1z} l \\ z_{z1}^{-1} \sinh \gamma_{1z} l & \cosh \gamma_{1z} l \end{bmatrix} \cdot \begin{bmatrix} E_1 \\ H_1 \end{bmatrix} \quad \text{and} \quad \begin{bmatrix} E_0 \\ H_0 \end{bmatrix} = [M_1] \begin{bmatrix} E_2 \\ H_2 \end{bmatrix} \qquad [6.3]$$

where E and H are respectively the electric and magnetic fields and the subscripts '0' and '2' refer to the fields at the first and second interface and '1' represents the properties of the medium between the two interfaces. The thickness of the laminate is 'l' and the intrinsic impedance of the material parallel to 'l' is z_z. The parameter γ_z is a complex quantity which is a combination of the attenuation – α_z and the wavenumber – β_z and is given by:

$$\gamma_z = \alpha_z + j \cdot \beta_z \quad \text{and} \quad j = \sqrt{-1} \qquad [6.4]$$

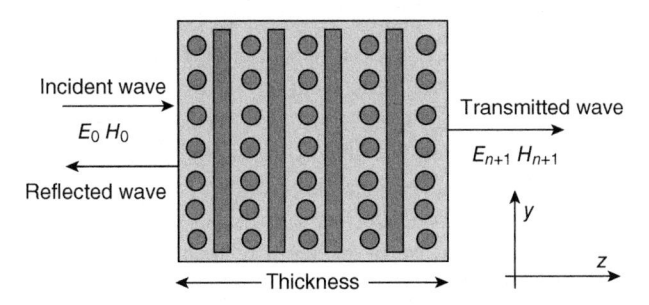

6.9 Schematic of plane wave propagation through a multilayered laminate.

The effective permittivity (ε_{eff}) of a laminate at an angle θ between the fibre direction and the x-axis is:

$$\varepsilon_{\text{eff}} = \varepsilon_1 \cos^2 \theta + \varepsilon_2 \sin^2 \theta \qquad [6.5]$$

where ε_1 and ε_2 are respectively the permittivity of the ply when the direction of the fibres is parallel to the x-axis and parallel to the y-axis.

For a material in which there are a series of laminate layers, as in Fig. 6.9. The extensions of equation (6.3) are:

$$\begin{bmatrix} E_0 \\ H_0 \end{bmatrix} = [M_1][M_2]\ldots[M_n]\begin{bmatrix} E_{n+1} \\ H_{n+1} \end{bmatrix} = \begin{bmatrix} m_{T11} & m_{T12} \\ m_{T21} & m_{T22} \end{bmatrix}\begin{bmatrix} E_{n+1} \\ H_{n+1} \end{bmatrix} = [M_T]\begin{bmatrix} E_{n=1} \\ H_{n+1} \end{bmatrix} \qquad [6.6]$$

The reflection coefficient (ρ) and transmission coefficient (τ) have the form:

$$\rho = \frac{E_{0,-}}{E_{0,+}} = \frac{m_{T11} \cdot z_{n+1} + m_{T12} - z_0(m_{T21} \cdot z_{n+1} + m_{T22})}{m_{T11} \cdot z_{n+1} + m_{T12} + z_0(m_{T21} \cdot z_{n+1} + m_{T22})} \qquad [6.7]$$

$$\tau = \frac{E_{n+1,+}}{E_{0,+}} = \frac{2 \cdot z_{n+1}}{m_{T11} \cdot z_{n+1} + m_{T12} + z_0(m_{T21} \cdot z_{n+1} + m_{T22})} \qquad [6.8]$$

where z_{n+1} is the intrinsic impedance of the transmitted field, z_0 is the intrinsic impedance of the incident field and $E_{i,+/-}$ are the integration constant for fields positive and negative to the travelling wave (Seo et al., 2006; Teo et al., 2006).

This model can be used to simulate the response for a glass fibre laminate structure and it is possible to introduce the effects of delamination, which may include a combination of matrix cracks, fibre breakage or pull-out. Carbon fibre or graphite laminates cannot be modelled because the high conductivity of the carbon fibres almost completely attenuates the microwaves. Application of the above model to a laminate structure indicates that for 48 ply laminate with each laminate of thickness 0.125 mm and assuming that the medium is loss-less, peaks and troughs in the reflection and transmission coefficients arise as a consequence of constructive and destructive interference. The location of the peaks and troughs are defined in terms of wavelength which are defined by the total thickness of the laminate.

When the laminate thickness is equal to multiples of the quarter wavelength, i.e. 16.2 and 27 GHz, the reflection coefficient will be maximized and when the laminate thickness is equal to multiples of half wavelength, i.e. at 10.8 and 21.6 GHz, there will be no reflection at all. The introduction of losses into the media changes the amplitude of the peaks and troughs but their location is found to stay fairly constant. Air gaps between laminates have a significant effect on the standing wave pattern and hence on the coefficients of ρ and τ. The minimum air gap which appears to be identifiable is ~0.25 mm and for easy recognition the air gap has to be 0.4 mm, which is effectively a delamination. On a frequency scan, a number of peaks can be identified which coincide with those predicted by the above simulation. Detailed analysis indicates that a number of reflections arise as a consequence of the waveguide used to illuminate the surface and the distance between the waveguide and the sample being studies. Some of these peaks which are internal reflections of the waveguide do not change whereas other which arise from reflections involving the surface do.

Studies of disbands at varying locations through a laminate have shown that the sensitivity varies with distance from the surface. Under optimal conditions, disbonds with a thickness of 0.03 mm can be detected, but the ability to detect such a defect depends on its depth. The standoff distance is very important and for operation at 5 GHz, a separation between the waveguide and the sample of 10 mm appeared to be optimal (Ganchev *et al.*, 1995a, 1995b). The form of the waveguide used to create the probe is critical and an analysis of the complexities which arise when elliptic dielectric resonator antennae are used indicates that a complex field can arise which will itself add additional features to the analysis (Tadjalli *et al.*, 2006).

Examination of laminate structures of different resin compositions has been successfully undertaken. An example of such a structure would be a GRP laminate which has been over-coated with a pure resin gel coat. Such structures are often fabricated by building up successive layers one at a time. This method of fabrication introduces the possibility that one layer may not effectively wet out the previous layer and leads to the possibility of subsequent delamination in use. Neural network analysis of such structures is concerned with the detection of delamination or regions which may contain air voids. In terms of impedance, the problem which must be addressed is the separation of variation of the dielectric characteristics from those of the thickness of the layers making up the overall composite structure. This problem requires the use of an accurate forward electromagnetic propagation model. In order to achieve this level of resolution it is necessary to include the influence of higher-order modes in analysis of the wave propagation. Validation of the model was achieved by comparison of the results with measured values of the layer thickness and dielectric properties. The model uses an iterative inverse method and a good correspondence between experiment and theory (Table 6.1). The test structure was created using Teflon layers sandwiched between thicker rubber layers. The structure contained two thin Teflon layers and the fit of the model to the real dimensions was very good (Ghasr *et al.*, 2009).

Provided the conductivity is not very high it is possible to examine materials containing carbon fibres or carbon black. The dielectric permittivity

Table 6.1 Data for the Teflon films in a rubber–Teflon sandwich structure

Layer	Relative complex permittivity	Thickness (mm)	Estimated thickness (mm)
Rubber	4.80-j0.17	3.175	
Teflon	2.00-j6E-4	0.381	0.385
Rubber	5.31-j0.22	6.35	
Teflon	2.00-j6E-4	0.508	0.518
Rubber	4.80-j0.17	3.175	

can be shown to be directly related to the carbon black loading and can be modelled in term of its ability to form chains within the material (Daly *et al.*, 1986; Ganchev *et al.*, 1994). Studies have also been undertaken of the dielectric properties obtained from an analysis of the frequency dependence of the impedance data. (Lee and Springer, 1984; Qaddoumi *et al.*, 1996). Although this initial study was carried out on a filled waveguide, similar information can be obtained from the end of line measurements. The changes in the dielectric properties as a consequence of cure are discernible even at high frequency for cured and partially cured resin using the open-ended rectangular waveguide. As in the analysis of the composites, the standoff distance is an important parameter in achieving optimum sensitivity for the measurements. The impedance data obtained by scanning the surface in a raster mode allowed microwave images to be created using data obtained at 10 and 24 GHz.

A particularly interesting study involved the use of microwave NDT for the analysis of space shuttle external fuel tank insulating foam (Kharkovsky *et al.*, 2006). Following the space shuttle Columbia's catastrophic accident there has been an increasing requirement to have a robust method for the life cycle NDT of the external fuel tank spray on foam insulation (SOFI). Using both near- and far-field millimetre wave NDT techniques images of anomalies in foam panels were observed. The study used continuous-wave reflectometers at single frequencies of 33.5, 70 or 100 GHz representing a relatively wide range of millimetre wave spectrum (K-band (26.5–40 GHz) to W-band (75–110 GHz)). Simple image processing techniques improved the images and allowed observation of relatively smaller anomalies. This demonstrates the utility of the microwave approach.

6.6 Conclusion

The utility of dielectric NDT testing lies in terms of its ability both to produce high-definition images of defects buried within composite structures and to indicate whether a material has absorbed significant amounts of moisture or contains air-filled voids. The wider use of the method relies on the ability to perform the measurements easily and the cost of the equipment being sufficiently inexpensive to allow wide applications of the method. Some of these barriers have restricted the use of the method for specialist applications but clearly the advantages which it presents will make its use more widely accepted in the future.

As has been noted, despite the significant improvements in analysis which have taken place over the last ten years, it has been the portability and expense of the techniques which have to some extent limited their application. This problem has been addressed to some extent with a novel approach which uses a rapidly rotary scanning portable Q band device. The raster scanning method, although relatively simple in design and construction, is inherently slow. In order to obtain the highest resolution images, synthetic aperture focusing is used and this necessitates use of a vector network analyser, which itself is bulky, not easily mounted on a scanning platform and expensive moreover, the system is not readily adapted for *in situ* measurements.

To address these problems a device has been proposed which consists of a rotating scanning element operating at 35–45 GHz and is constructed to be portable (Ghasr *et al.*, 2011). The rotary scanner system produces a coherent (amplitude and phase) and accurate data suitable for synthetic aperture imaging and the 10 GHz bandwidth allows the generation of relatively high-resolution millimetre wave holographical images. The system is capable of scanning a circular area of ~120 cm diameter, at a step size of 5 mm, in about 15 min. It is, however, very clear that microwave techniques are capable of adding additional NDT information on the state of cure and occurrence of defects in composite which is not otherwise available and despite the difficulties of operation the potential of the methods has been successfully demonstrated.

With the wider use of composites in wind turbines, aircraft wings and related large structures there will be an increasing demand for NDT methods which are capable of providing the maintenance engineer with early warning of possible failure and the fabrication engineer with a greater level of confidence in the quality of the product produced. Whilst the methods described appear expensive and sophisticated, their wider acceptance will allow cheaper products to be created which could be easily used by skilled operatives.

6.7   Acknowledgements

The author wishes to acknowledge the help of his various co-workers over the years and in particular Dr David Hayward who carried out many of the high-frequency measurements and designed the low-frequency probe used in the assessment of the GRP hull.

6.8 References

Abou-Khousa M., Zoughi R., (2007) Disbond thickness evaluation employing multiple-frequency near-field microwave measurements, *IEEE Trans Instrumentation Measurement*, **56** (4) 1107–1113.

Boinard P., Boinard E., Pethrick R.A., Banks W.M., Crane R.L., (1999) Dielectric spectroscopy as a non-destructive technique to assess water sorption in composite materials, *Sci Eng Composite Mater*, **8** (4) 175–179.

Boinard E., Pethrick R.A., Dalzel-Job J., Macfarlane C.J., (2000) Influence of resin chemistry on water uptake and environmental ageing in glass fibre reinforced composites-polyester and vinyl ester laminates, *J Mater Sci*, **35** (8) 1931–1937.

Chipman R.A., (1968) *Transmission Lines*, Schaum's Outline Series, McGraw-Hill Book Company, New York, London.

Daly J.H., Hayward, D., Pethrick R.A., (1986) Studies of dielectric-constant measurements and tribo-electric charging of pigmented polymer systems, *J Phys D Appli Phys*, **19** (5) 885–896.

Ganchev S.I., Bhattacharyya J., Bakhtiari S., Qaddoumi N., Brandenburg D., Zoughi R., (1994) Microwave diagnosis of rubber compounds, *IEEE Trans Microwave Theory Techniques*, **42** (1) 18–24.

Ganchev S.I., Qaddoumi N., Bakhtiari S., Zoughi R., (1995a) Calibration and measurement of dielectric properties of finite thickness composite sheets with open-ended coaxial sensors, *IEEE Trans Instrumentation Measurement*, **44** (6) 1023–1029.

Ganchev S.I., Qaddoumi N., Ranu E., Zoughi R., (1995b) Microwave detection optimization of disbond in layered dielectrics with varying thickness, *IEEE Trans Instrumentation Measurement*, **44** (2) 326–328.

Ghasr M.T., Simms D., Zoughi R., (2009) Multimodal solution for a waveguide radiating into multilayered structures – dielectric property and thickness evaluation, *IEEE Trans Instrumentation Measurement*, **58** (5) 1505–1513.

Ghasr M.T., Pommerenke D., Case J.T., McClanahan A., Aflaki-Beni A., Abou-Khousa M., Kharkovsky S., Guinn K., De Paulis F., Zoughi R., (2011) Rapid rotary scanner and portable coherent wideband Q-band transceiver for high-resolution millimeter-wave imaging applications, *IEEE Trans Instrumentation Measurement*, **60** (1) 186–197.

Glazier E.V.D., Lamont H.R.L., (1958) *Transmission and Propagation*, Her Majesty's Stationary Office, London.

Gray S., Zoughi R., (1997) Dielectric sheet thickness variation and disband detection in multilayered composites using an extremely sensitive microwave approach, *Mater Eval*, **55** (1) 42–48.

Hayward D., Mahboubian Jones G.M.B., Pethrick, RA, (1984), Low-frequency dielectric measurements (10^{-4} to 6×10^4 Hz) – a new computer-controlled method, *J Phys E Sci Instruments*, **17** (8) 683–690.

Jones T., Polansky D., Berger H., (1988) Radiation inspection methods for composites, *NDT Int* **21** (4) 277–282.

Kazilas M.C., Partridge I.K., (2005) Exploring equivalence of information from dielectric and calorimetric measurements of thermoset cure – a model for the relationship between curing temperature, degree of cure and electrical impedance, *Polymer*, **46** 5868–5878.

Kharkovsky S., Case J.T., Abou-Khousa M.A., Zoughi R., Hepburn F.L., (2006) Millimeter-wave detection of localized anomalies in the space shuttle external fuel tank insulating foam, *IEEE Trans Instrumentation Measurement*, **55** (4) 1250–1257.

Kharkovsky S., Ryley A.C., Stephen V., Zoughi R., (2008) Dual-polarized near-field microwave reflectometer for noninvasive inspection of carbon fiber reinforced polymer-strengthened structures, *IEEE Trans Instrumentation Measurement*, **57** (1) 168–175.

Kim J.S., Gille D., (1996) Analysis of dielectric sensors for the cure monitoring of resin matrix composite materials, *Sensors Actuators B* **30** 159–164.

Kranbuehl D., (1990) *FDMS Sensing for Continuous On-line in-situ Monitoring and Control of Cure, Flow and Cure of Polymer Measurement and Control*, RAPRA Technology Limited, Shawbury, UK.

Lee W., Springer G., (1984) Interaction of electromagnetic radiation with organic matrix composites, *J Compos Mater* **18** 357–386.

Maistros G.M., Partridge I.K., (1998) Monitoring autoclave cure in commercial carbon fibre/epoxy composites, *Composites Part B* **29B** 245–250.

Maloney J., Smith G., (1992) The efficient modelling of thin material sheets in the finite-difference time-domain (FDTD) method, *IEEE Trans Antennas Propagation* **40** (3) 323–330.

Pethrick R.A., Amornsakchai T., North A.M., (2010) *Introduction to Molecular Motion in Polymers*, CRC Press, Taylor Francis, Boca Raton, FL.

Pozar D., (2005) *Microwave Engineering*, 3rd ed, Wiley, New York, Chichester.

Qaddoumi N., Ganchev S., Zoughi R., (1996) Microwave diagnosis of low-density fiberglass composites with resin binder, *Res Nondestr Eval*, **8** 177–188.

Ramo S., Whinnery J., Van Duzer T., (1995) *Fields and Waves in Communication Electronics*, 3rd ed., Wiley, New York, Chichester.

Sansalone M., (1997) Impact-echo: the complete story, *ACI Structural J* **94** (6) 777–786.

Seo I., Chin W., Lee D., (2006) Characterization of electromagnetic properties of polymeric composite materials with free space method, *Composite Structures*, **66** 533–542.

Tadjalli S., Sebak A., Denidni T., (2006) Resonance frequencies and far field patterns of elliptical dielectric resonator antenna: analytical approach, *Prog Electromagnetics Rese, PIERS*, **64**, 81–98.

Teo Y-H., Wang X., Chiu W-K., (2006) Simulations of microwave propagation in delaminated unidirectional glass–epoxy laminate, *Composite Structures*, **75** 422–427.

Zoughi R, Bakhtiari S., (1990) Microwave non destructive detection and evaluation of disbonding and delamination in layered-dielectric-slab, *IEEE Trans Instrumentation Measurement*, **39** (6) 1059–1663.

Zoughi R., Lai J., Munoz K., (2002) A brief review of microwave testing of stratified composite structures: a comparison between plane wave and near field approaches, *Mater Eval*, **60** (2) 171–177.

7
Non-destructive evaluation (NDE) of composites: using ultrasound to monitor the curing of composites

W. STARK, BAM Federal Institute for Materials Research and Testing, Germany and W. BOHMEYER, Sensor and Laser Technique, Germany

DOI: 10.1533/9780857093554.1.136

Abstract: Typical composite materials for lightweight construction consist of a fibre (glass or carbon) reinforced thermosetting matrix. The impregnation of the fabric is made with liquid resin – usually epoxy resin or unsaturated polyester resin are used. The main processing methods are resin transfer moulding (RTM) and prepreg technology often used in combination with autoclave curing. After shaping, the resin must be cured. The final properties of the produced parts depend on the curing degree and the final glass transition temperature. Analytical methods and the use of ultrasound for online cure monitoring in the manufacturing process are introduced. Examples for the application of ultrasonic cure monitoring for the production of composite parts for traffic systems are also covered.

Key words: curing behaviour, curing degree, glass transition, rheology, differential scanning calorimetry (DSC), dynamic mechanical analysis (DMA), sound velocity, sound damping, resin transfer moulding (RTM), prepreg.

7.1 Introduction

For all means of transportation there are requirements to increase energy efficiency. This implies both improvements to engine design, and a reduction in weight whilst maintaining strength, such has been achieved in the air craft industry through the use of carbon fibre composites. A carbon fibre fraction of more than 50% by weight is the aim for the Boeing 787 Dreamliner and the Airbus A350 XWB. In the automotive industry, weight reduction is particularly important for electric cars where the heaviness of the batteries must be offset by light weight components, especially the car body.

Materials which fulfil the demands of high strength in combination with low weight include the composites. Typical for a composite is the combination of fibre enhancement – often carbon or glass fibre – and a polymer matrix. This matrix can be built up of thermoplastic or thermoset material. At present, thermoset materials are considered optimal as they offer more scope for building complex parts. Thermosets are characterized by mixtures of a reactive resin and a hardener. They are liquids with low viscosity at room temperature or slightly higher and can infiltrate a fibre structure (fabric) easily. The reactivity of the resin and hardener gives rise to the possibility for a change of state from a liquid to a stiff solid by the chemical curing reaction, which is a crosslinking process. The reaction is started mainly by higher temperatures, but the chemistry can be set to run at room temperature also.

The most widely used thermoset matrix for cured composites are epoxy and unsaturated polyester resins. Polyurethane resins are under development for shorter cycle times. Resin and hardener are mixed in a prescribed ratio by the processor and then infiltrated into the roving. One such method is the resin transfer moulding (RTM) technique. Here the fabric is placed in a closed mould and the freshly mixed resin–hardener mixture is pumped into the mould, which has been preheated to curing temperature. After a period of time, sufficient to reach the required degree of curing, the mould is opened and the part removed.

An alternative process starts with a so-called prepreg material, where the prepreg manufacturer impregnates the fabric with the resin–hardener mixture. The reactive prepreg must be stored at typically −18 °C to avoid precuring before processing. Prepregs are shipped as coil. With spreading machines many layers are placed one upon the other until the default thickness is reached. This final thickness can be up to few centimetres. Such prepreg stacks are wrapped by a vacuum-tight foil. They are cured for many hours in autoclaves with pressures up to 10 bar and high curing temperatures. These parts can have dimensions as large as 5×20 m^2. The final properties of composite parts are strongly dependent on the degree of curing reached in the manufacturing process. In the aerospace industry a curing degree close to 100% is required and this must be checked and documented during the production process.

7.2 Types of thermosets used in composites

Thermosetting materials can reach their final properties by a technical crosslinking reaction. Before crosslinking they are liquids or low melting point materials and after, they form solid, stiff, materials. The most popular systems for composite manufacturing are epoxy resins. Epoxy resins are characterized by an oxygen atom connected with two adjacent carbon atoms in a chemical compound. This ring with oxygen, also named the oxirane ring, builds an approximate equilateral triangle. It is highly strained and therefore epoxy resins are highly reactive. Two oxirane rings at either end of the basic molecule are necessary for a meshwork formation. Crosslinking can be carried out by hardeners such as anhydrides, amines or phenols. Figure 7.1 illustrates the formula for the widely used epoxy bisphenol A diglycidyl ether (produced from a reaction between epichlorohydrin and bisphenol-A) with a diamine hardener.

In the crosslinking reaction the oxirane ring is opened and an OH-group is formed with a hydrogen atom from the NH-group of the amine. The amine molecule is joined up with the free valence of the CH_2-group. Through an additional reaction of the second part of the diamine, a chemical bridge between adjacent epoxy molecules is formed – see Fig. 7.2. Using polyfunctional amines three-dimensional networks result. Such a heavily crosslinked polymer is rigid and strong. The crosslinking process is known as curing. Depending on the kind of resin, hardener and activators, curing can proceed

7.1 Epoxy bisphenol A diglycidyl ether and diamine hardener.

7.2 Illustration of the crosslinked network structure of bisphenol A diglycidyl ether and diamine hardener.

at room temperature, though many formulations need high temperatures for the reaction.

When using epoxy, knowledge of the curing process is of great importance. Final properties, in particular the glass–rubber transition temperature (T_g), are determined by the chemical structure and the curing conditions. Also, optimization of the process, especially knowledge of the time for gelation and vitrification, is important. At gelation the molecular weight has increased so much that the viscosity becomes infinite and the resin can no longer flow and fill the mould. At vitrification the reaction comes to a standstill because molecular mobility reduction reduces contact between the reactants. Other classes of resins with similar use as epoxy are unsaturated polyester (UP) and polyurethane (PU). A specific class of UP based compounds are the bulk moulding compound (BMC) from UP with short glass fibres and sheet moulding compound (SMC) with glass fibre matting.

Moulding compounds are a special subclass of thermosetting materials. The well-known phenolic resin based Bakelite has been used for more than 100 years. Beside phenolic resin with hexamethylenetetramine hardener (PF), melamine-formaldehyde (MF) and urea- formaldehyde (UF) have found widespread application. Beyond the thermosets, other crosslinked polymers include elastomers (rubber). As distinguished from thermosetting materials, the number of crosslinking bridges is rather low.

7.3 Methods for monitoring composites

7.3.1 Rheology

One methodology for the measurement of rheological characteristics involves enclosing the material in the gap between two plates or cones, and to rotate or oscillate one of the plates, as illustrated in Fig. 7.3. For curing materials, only the oscillating modus is of interest. Measured parameters are the torque, the amplitude and the phase between both. The complex shear modulus may also be calculated from the torque and deformation angle, and is formed by the storage modulus G' and loss modulus G''. From both, an effective viscosity ($|\eta^*|$) can be a calculated using the Cox-Merz rule [1]. Figure 7.3 shows the crosslinking reaction of epoxy resin registered by an oscillating rheometer.

A sharp rise of viscosity was observed after approximately 2 min. This rise of viscosity, comprising more than three orders of magnitude, is an indication of gelation. G' and G'' approach each other and are equal at about four minutes. This characteristic point is often defined as the gel point [2]. After 10 min, a maximum in loss modulus G'' appears. This is a typical indication of a rubber to glass transition [3–5], as a consequence of which, the compound solidifies. In Fig. 7.4 an example of basic investigations

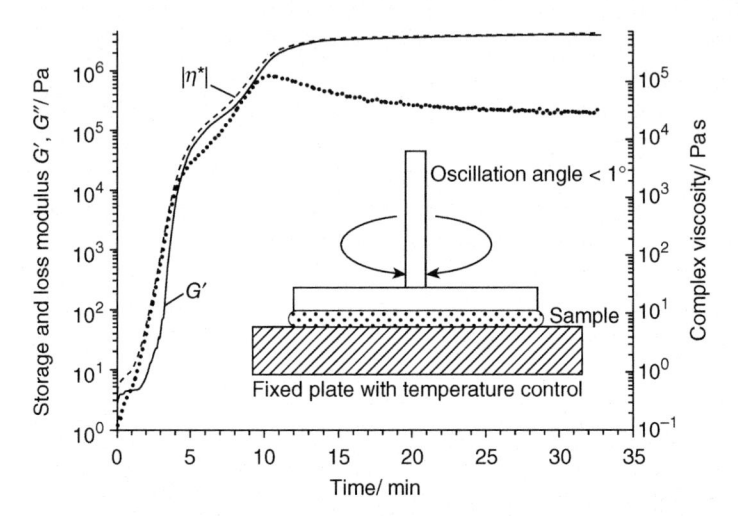

7.3 Storage and loss modulus and calculated complex viscosity vs. curing time from rheological measurements, isothermal epoxy resin curing at 100 °C, logarithmic y-axis. Insert: measuring principle of a rheometer.

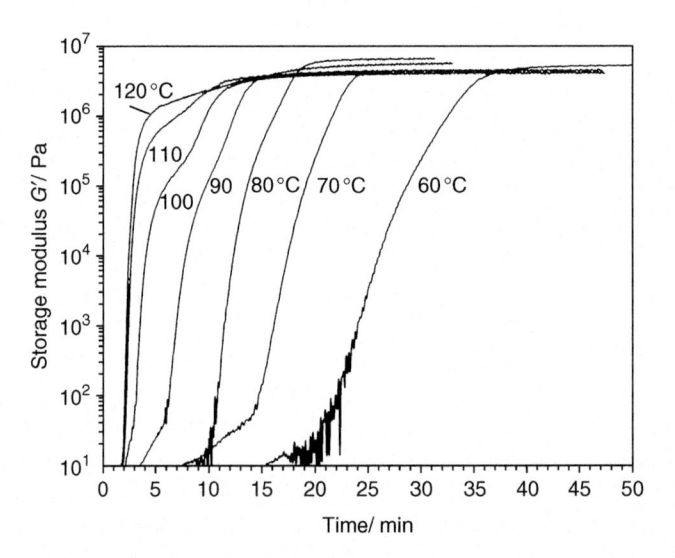

7.4 Storage modulus vs. curing time from rheological measurements, epoxy resin curing, influence of process temperature, logarithmic y-axis.

on the influence of process temperature on epoxy resin curing is summarized.

7.3.2 Differential scanning calorimetry (DSC)

Differential scanning calorimetry (DSC) is used to detect the heat flow rate between two crucibles with high accuracy, as illustrated in Fig. 7.5. One of the crucibles contains the material to investigate, the other is usually empty. The temperature of the chamber can be either constant or on a heating/cooling ramp realizing a specific temperature change. For polymers, characteristic physical conversions such as melting, crystallization, evaporation and glass transition lead to a heat flow between crucibles. The crosslinking reaction is exothermic and can therefore be detected well with DSC.

The DSC curve for a reactive epoxy is given in Fig. 7.5. The crosslinking reaction provides an exothermic peak with a maximum at 140 °C. The reaction enthalpy measured in J/g can be calculated from time integral of that peak. DSC is a popular method for determining the degree of curing α:

$$\alpha_{DSC} = 1 - \frac{\Delta H_r}{\Delta H_t}$$

where ΔH_r is the remaining reaction enthalpy, and ΔH_t the total reaction enthalpy.

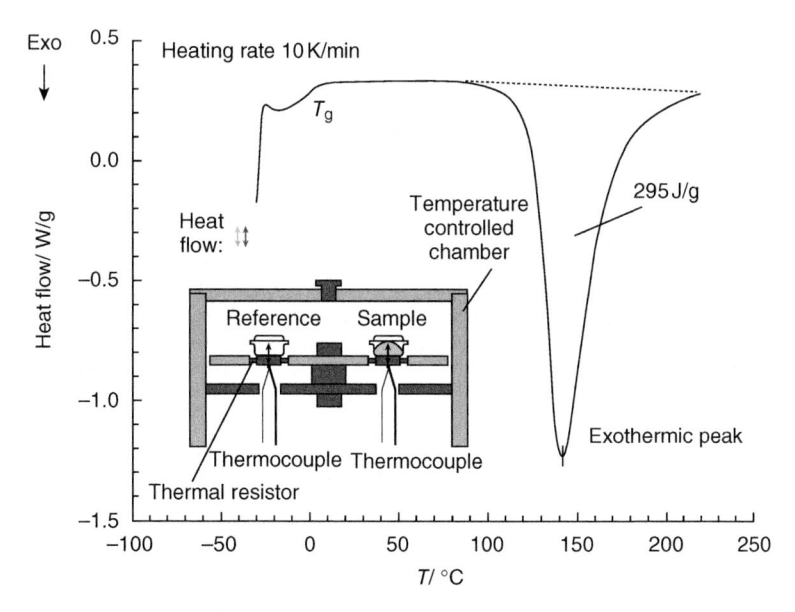

7.5 DSC heat flow vs. temperature, epoxy resin, starting from the frozen state to a temperature of full crosslinking. Insert: measuring principle of DSC.

7.3.3 Dynamic mechanical analysis (DMA)

With dynamic mechanical analysis (DMA), a sample is excited by a sinusoidal mechanical deformation. The amplitude of the force and the deformation, and the phase relation between them are measured. The storage modulus E' and the loss modulus E'' may then be calculated. Figure 7.6 illustrates the storage and loss modulus as function of temperature for a sample prepared from a cured epoxy.

The storage modulus (E') varies across the temperature range from about 80 to 150 °C, where a decrease of approximately one order of magnitude is observed. This change is accompanied by a peak in the loss modulus. For higher frequencies, the transition shifts to higher temperatures. This behaviour illustrates the glass–rubber transition of the cross-linked resin. The glass temperature T_g can be defined as the peak of the loss modulus, or the half height of the step in storage modulus [5]. Only at full curing can the final glass transition temperature ($T_{g\infty}$) be reached. Therefore, DMA analysis is a reliable method to determine whether curing was well done.

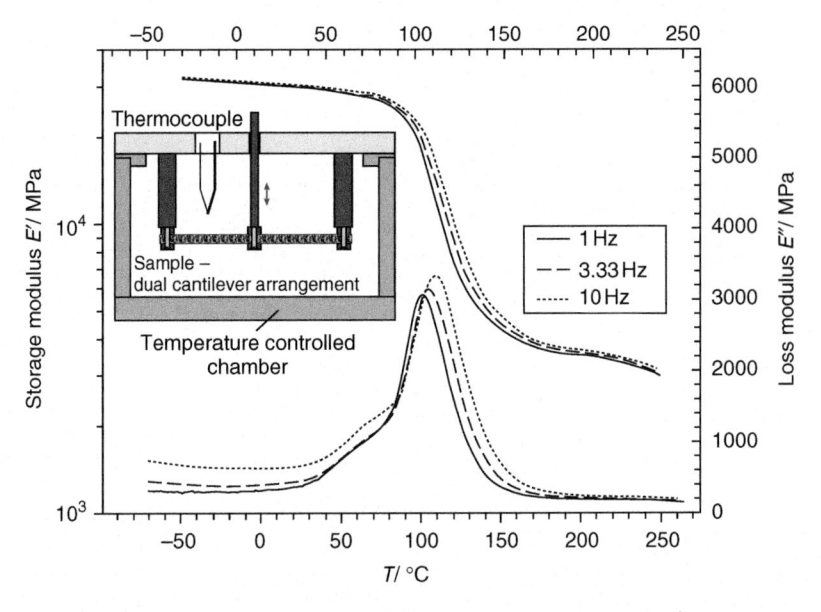

7.6 DMA, storage and loss modulus vs. temperature, sample epoxy resin, heating rate 1 K/min (E' – logarithmic axis). Insert: measuring principle of DMA.

7.4 Monitoring the degree of curing and the mechanical properties of composites

Curing increases the glass transition temperature and determines the final mechanical properties. Figure 7.7 shows the change of T_g resulting from stepwise curing. A typical error during thermoset curing is an insufficient curing time. In the case of phenolic resin, insufficient curing and consequently too low a glass transition temperature leads to bladders after demoulding, as shown in Fig. 7.8, which illustrates a defective PF part.

Incomplete crosslinking can be detected via analytic methods. For the case of a broken carbon composite bicycle fork, too minor a degree of curing was detected by DMA, as shown in Fig. 7.9. In the first measurement of the failed component a T_g of 111 °C was detected. By increasing the temperature to 250 °C in the DMA run, the component was able to postcure. In the second measurement this sample showed a notable increase of T_g from 111 °C to 141 °C, illustrating imperfect curing. In practice the benefits of reaching a very high curing often is underestimated. Figure 7.10 illustrates that for a UF-moulding compound, the impact strength rises steadily with curing time [6].

Wolfahrt *et al.* [7] changed the degree curing of epoxy prepregs from 68 to 100% and found that the fracture toughness increased with degree of curing, remaining nearly constant after 90%. A distinct correlation between degree of curing and crack propagation speed for glass fibre epoxy composite was reported by Trappe *et al.* [8]. A change of degree of curing from

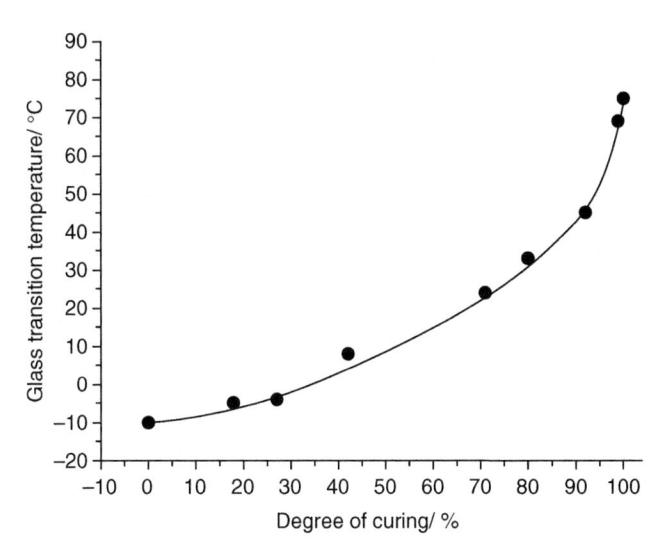

7.7 Epoxy resin, glass transition temperature vs. degree of curing, DSC.

7.8 PF-part (Bakelite) taken from the mould in an incomplete state of crosslinking. Such manufacturing defects can easily be avoided by utilization of online process monitoring.

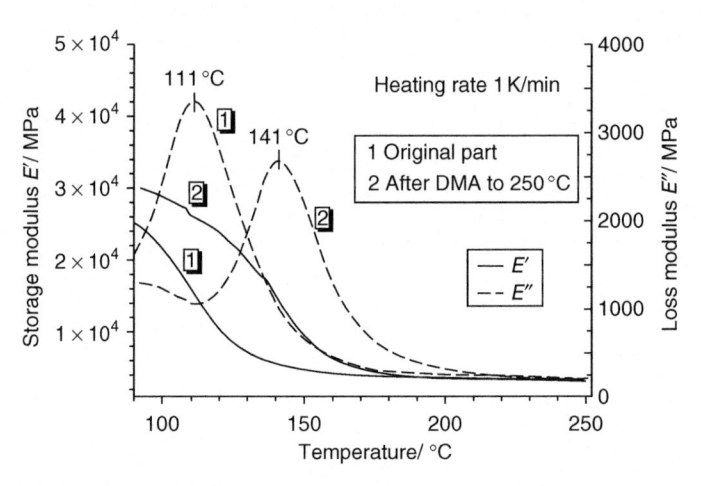

7.9 Incorrect bicycle part, storage and loss modulus vs. temperature, detection of wrong glass transition temperature by DMA.

92% to 96% produced a decrease of crack propagation speed by a factor of 10. For dental composites a correlation between degree of curing, fracture toughness [9] and wear depth [10] was reported.

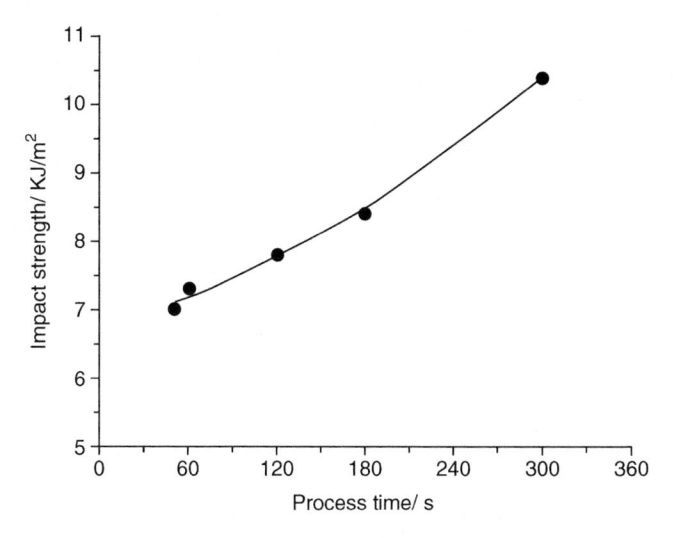

7.10 Impact strength vs. process time, parts from UF-moulding compound.

7.5 Online process monitoring using ultrasound

In the 1940s, the first successful experiments were completed using ultrasonics as a diagnostic tool. In the following years the enormous potential of ultrasonics for non-destructive testing was realized (see e.g. Krautkraemer and Krautkraemer [11]), including the study of the curing process [12]. Fundamental work was undertaken by Alig and coworkers [13–19], and Challis *et al.* [20–24]. A good overview on methods for cure monitoring including ultrasound is given in Lodeiro and Mulligan [25].

With ultrasonic cure monitoring, the sound velocity of a longitudinal wave v_s^{long} in a medium and the damping of this wave α^{long} are connected with the longitudinal storage modulus L' and the longitudinal loss modulus L'' as follows:

$$L' = (v_s^{long})^2 \cdot \rho \quad \text{and} \quad L'' = 2(v_s^{long})^3 \cdot \rho \cdot \alpha^{long} / \omega \qquad [7.1]$$

where ρ is the density, ω the angular frequency and α the damping.

L' and L'' can be summarized as the complex longitudinal modulus L^*:

$$L^* = L' + iL''$$

The longitudinal wave modulus is connected with the compression modulus K^* and the shear modulus G^* and finally also with the modulus of elasticity E^* and the shear modulus G^*:

$$L^* = K^* + \frac{4}{3}G^* = \frac{4G^* - E^*}{3 - \frac{E^*}{G^*}}$$ [7.2]

The crosslinking process causes a characteristic change to the modulus of elasticity and shear modulus. Hence also a characteristic change of the longitudinal wave modulus exists and can be used for characterization of the crosslinking.

7.5.1 Working principles and use of transducers

Figure 7.11 illustrates the principle behind ultrasonic measurement, and Fig. 7.12 shows the propagation of a longitudinal wave. The working principle of the ultrasonic sensors rests upon the electrical excitation of a piezoelectric ceramic. Its orientation is chosen such that longitudinal waves are emitted by thickness oscillations. A thickness change of a piezoelectric material then generates a voltage, so the same sensor can be used as transmitter or receiver. Figure 7.13 shows such a transducer and typical dimensions are given in Fig. 7.14. The relatively small sensors can be included in a mould or a test set-up easily.

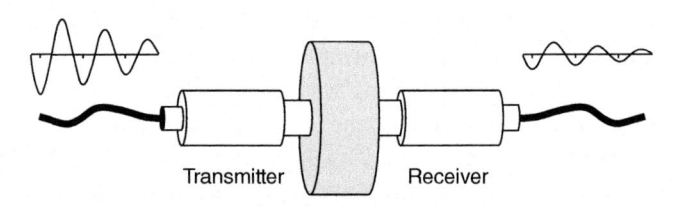

7.11 Scheme of an ultrasonic measuring arrangement – through transmission.

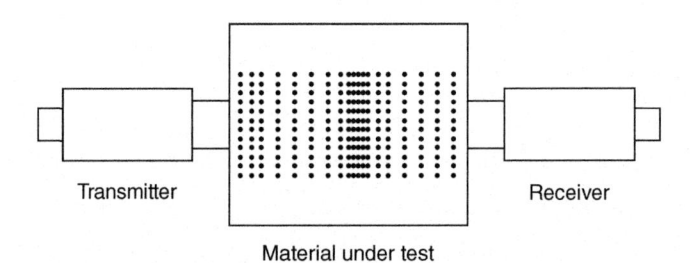

7.12 Illustration of the propagation of longitudinal waves (compression waves) between the sensors.

Using ultrasound to monitor the curing of composites   147

7.13 Photo of an ultrasonic transmitter (product of GE Company).

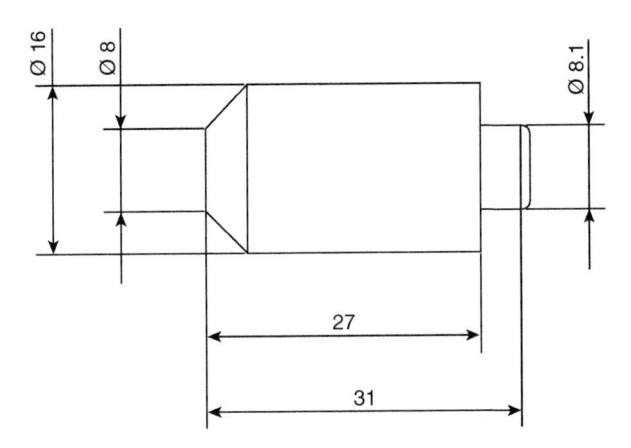

7.14 Technical drawing of an ultrasonic sensor.

7.5.2 Parameters to be measured

To determine the sonic parameters in the media under test two parameters are needed: the sound velocity and the damping. To determine the sound velocity, the thickness of the material and the pulse transit time must be measured. Many methods to determine the thickness of the material between fixed or variable plates are known, but most of them suffer from temperature-dependent errors. It is sometimes possible to increase the material thickness to reduce the influence of this inaccuracy. Our compression mould was completed with a dial gauge with USB data output. It works successfully for a probe thickness of between 1 and 5 mm. For a given mould cavity, the thickness of the cavity can be measured or is known from construction.

The easiest method to determine the travelling time is to measure the difference between a reference signal, where the transmitter is in direct

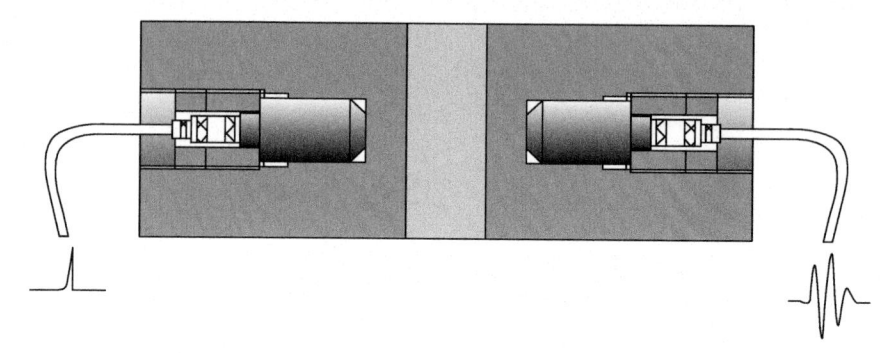

7.15 Typical measuring set-up for hidden ultrasonic sensors in the wall of the mould.

contact to the receiver without a test material, and the signal that has passed through the material under test. In this way the travelling times in both sensors (and in the wall in front of the sensors) need not be known. A correct determination of the damping (sound attenuation) in the material is a complex task. The damping is influenced by different parameters: acoustic impedance mismatching between sensor and material and its change during curing, scattering of the incident wave, and voids in the material [26]. It is therefore helpful to normalize the measured amplitude by the amplitude of the reference signal.

In principle it is possible to use through transmission (two sensors in a line) or to measure the echo from the opposite wall using one sensor only. The second method demands a short excitation pulse for a clear separation between excitation and reflection. This is always difficult for a sample thickness in the order of millimetres. Furthermore, reflections in the sensor indicate a more complicated data evaluation.

The sensors can be fixed in a hidden position or they are arranged in the surface of the mould having direct contact to the material under test. The second method delivers higher amplitudes but normally it is used in the lab where an imprint on the sample is accepted.

The widely used arrangement of hidden transmitter and receiver is shown in Fig. 7.15.

7.5.3 Excitation of the transmitter

There are different possibilities for the electrical excitation of a piezoelectric transducer in the kHz or MHz region (see e.g. [11]). From non-destructive testing (NDT) applications it is known that a continuous excitation leads to interferences between the incoming and reflected signals arising from each boundary layer so that interpretation of the received signal is complex.

To avoid such problems, adequate sensors and a short excitation pulse may be used. For such a mode of excitation, the influence of interferences from different boundaries may be minimized. Two methods for excitation are well known: one of these uses a single pulse having a rise or fall time in the 10–100 ns regions and amplitude in the range between 10 and 500 V. This pulse can be generated by switching on (or switching off) a static voltage at the sensor or by an external pulser. In such a case the free oscillation of the sensor is mainly determined by its resonance frequency. An additional damping of the piezo ceramic element reduces the duration of the oscillation. However, with such an excitation a narrow banded signal will not be generated, it will feature a frequency spectrum. The appearance of multiple frequencies can cause problems because there is an influence of the frequency on the sound velocity (dispersion) and on attenuation.

Another possibility is burst excitation using a group of positive and/or negative rectangular pulses forming the excitation signal or an excitation using a free programmable pulse generator. The effective frequency of excitation is determined by the duration of each pulse and can be varied. For example a central frequency of 4 MHz is generated by rectangular pulses with alternating polarity having the same duration of 125 ns. Changing the duration of each pulse shifts the central frequency. Using this method for excitation, the frequency of the transducer signal can be varied within the limits given by the datasheet of the sensor. For instance, a transducer with a central frequency of 4 MHz can be stimulated to transmit signals between 3 and 6 MHz. The frequency spectrum broadens when the number of burst pulses decreases, as may be illustrated by a Fourier transformation.

Such an excitation offers the possibility to apply a frequency where the damping in the material under test is more favourable. Figure 7.16 shows the output signal of a transmitter for different excitation pulses. The damping of the excitation pulse in the test material is a critical parameter, as will be shown later. It depends on the amount of carbon or glass fibres as well as on the concentration of gas bubbles in the material and on a strong loss process during the softening of the material. Damping increases with material thickness. In nearly all practical applications the loss of intensity caused by these effects is the most critical influence. There are limits for increasing the excitation power of the transmitter (durability) and consequently the detection system has to be optimized to reach a good signal to noise ratio. Another way to reduce the influence of losses in the material is to reduce the excitation frequency. It is well known that the damping increases as frequency rises. A reduction of frequency leads to larger transmitter diameters and reduces the time resolution of the system. An optimum has to be found and typical frequencies range between 500 kHz and 5 MHz.

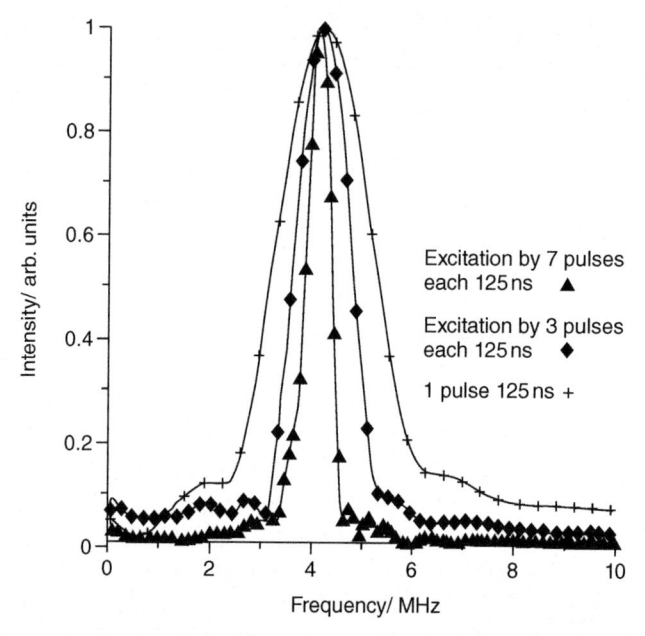

7.16 Output signal for different numbers of excitation pulses. One burst pulse consists of a rectangular pulse having duration time of 125 ns, 3 burst pulses means a group of three pulses with alternating polarity each pulse 125 ns (1/250 ns = 4 MHz).

An additional effect appearing in composite materials which has an influence on the received signal is the reflection of the ultrasonic wave at internal interfaces in the material (transition from resin to fibre) or at internal gas bubbles [14, 27]. These reflections generate signals with an additional delay and consequently constitute interference. The influence of such interference increases with pulse duration.

7.5.4 Signal detection

The 'efficiency' of ultrasonic transducers is low. An excitation of the transmitter with more than 100 V leads to an output signal at the receiver in direct contact to the transmitter of about 1 V and up to 10 V. For losses of 50 to 60 dB in the material, and additional losses through mismatching, an output signal as low as 10 µV can be expected. It is challenging to measure such small signals in the MHz region. First of all an excellent broad band pre- and post-amplifier system combined with band filters is necessary. Band filters are an effective way to improve the signal to noise ratio. Furthermore, an averaging of the measured signals is helpful in increasing the signal to noise ratio. For repetition rates of 1000 groups of excitation pulses per second, averaging using 500 groups delivers two datasets for the calculation of sound velocity and damping per second what is enough for all current applications. For very fast processes the data acquisition rate can be increased. Every dataset includes process time, pulse shape, pulse transit time, attenuation, mould and/or resin temperature, and thickness and is saved automatically.

The amplifier should have an automatic gain control because the input signal may vary by four to five orders of magnitude. Afterwards an analogue to digital converter (ADC) with at least eight bit resolution and a digitalization rate of 100 MHz is necessary. Using such components, and adequate software, the time resolution is in the order of one nanosecond.

7.5.5 Signal evaluation

Different software methods are used to determine the transit time. The simplest consists in the direct determination of the transit time by a comparison between reference and measured signal using maxima or minima or zero-crossing. More sophisticated methods are based on correlation functions or Fourier transformation. The problems arising from data analysis for composite materials, together with discussion of a more sophisticated data evaluation, are discussed in McHugh [26], and Hiramatsu et al. [28].

7.5.6 Testing crosslinking

Two main parameters of the crosslinking process are of interest: the transit time and the attenuation of the sound wave in the material. The first parameter is often used directly to give an information about the crosslinking process. The velocity of sound may be interpreted such that a strong increase at the beginning and a stationary value at the end of the process also contains valuable information about the crosslinking process. A typical dataset for a prepreg curing is depicted in Fig. 7.17, which shows the sound velocity and the attenuation of the ultrasonic signal together with the linearly increasing temperature. The attenuation in the material shows two distinct regions: a very strong attenuation (up to 75 dB) during the softening of the material (60 to 120 min) and a clear reduction of the damping during crosslinking (130 to 200 min) which is similar to the increase of sound velocity here.

7.5.7 Compression mould for laboratory testing

A small mould developed for use in the laboratory is shown in Fig. 7.18. It is fixed in a press and is heated from the back. The sensors are arranged in

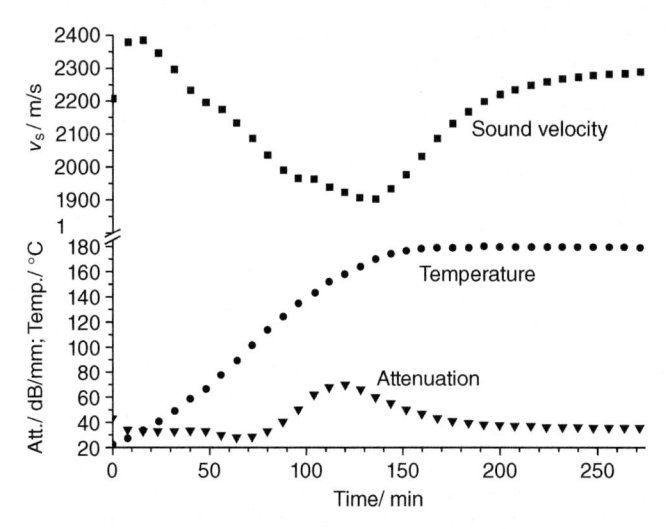

7.17 Sound velocity, damping and mould temperature vs. process time, curing process for a T-ramp of 1 K/min up to 180 °C followed by a constant temperature. The strong attenuation during softening (up to 75 dB/mm is shown). The velocity increase at the end of the process is a clear indication of the crosslinking process.

7.18 Opened measuring cell. The arrow depicts on the sensor in the lower part.

both parts in such a way that they are on a line. The aim of this tool is to realize an incoming goods control, and to optimize manufacturing conditions, especially when new basic materials with unknown curing parameters have to be qualified for the production process.

Using ultrasound to monitor the curing of composites   153

In Fig. 7.19 a scheme of a complete measuring station is depicted, consisting of the mould with two sensors and heaters, a distance measuring system and the excitation and detection unit controlled by a computer. In Figure 7.20 is a photograph of the main components (pulse generation, pulse detection and a 16 channel multiplexer). The system is controlled by a PC and was prepared for application in industry.

The ultrasonic measuring system was developed in cooperation with BAM (www.bam.de), ISK-Iserlohn (www.isk-iserlohn.de) and SLT (www.pyrosensor.de) and is named US-plus® [29–32]. The specific configuration is characterized by a single channel set up or up to 16 channels for flow front detection and homogeneity investigations characterized by:

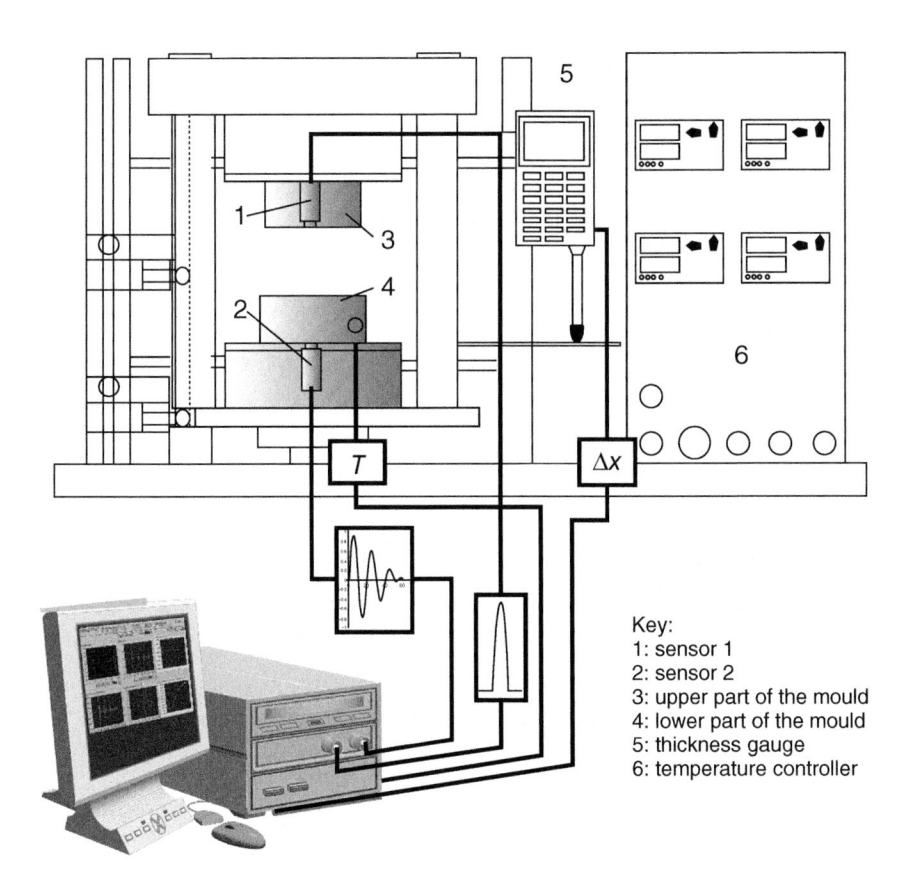

7.19 Schematic drawing of the complete computer controlled measuring system consisting of temperature controlled mould, distance measurement gauge, pulse generation and data acquisition unit.

154 Non-destructive evaluation (NDE) of polymer matrix composites

7.20 Main item of the US-plus® system designed for use in the production process. A PC is needed during use.

- automatic detection of the Us-signals;
- online information about the curing process with automatic data recording and graphical readout;
- comparison of curing curves for analysis of reproducibility, comparison between materials and influence of change of process parameters;
- comparison with master curves and automatic alarm when predefined limit values are exceeded.

This computer controlled system is designed for use in industry for online logging of the curing process and for flow front detection. It has some input ports for data from the mould and can send signals to stop the curing process. All measured data are archived automatically. A comparison with a master curve is possible. Different online graphs are available.

In Fig. 7.21 a laboratory device based on the US-plus® system is shown. It includes a small pneumatic press containing sensors for one or some channels. The temperature of the mould is computer controlled. Heating ramps or an isothermal mode of operation are possible, also in combination, up to 180 °C. The main application is the curing diagnostic of different materials in the lab and the check of incoming goods for different industrial application, e.g. injection moulding.

7.6 Using ultrasonic online process monitoring in practice: monitoring curing

The curing process leads to a characteristic increase of storage modulus as measured with a rheometer. The sound velocity is a function of the storage modulus (see eq. 7.1). The idea of ultrasound online cure monitoring is to bring the measurement of storage modulus into the mould with the help of small, temperature-stable ultrasound sensors. In addition to the storage modulus, a distinct increase of sound velocity is expected. The following

7.21 Complete laboratory device for the registration of the curing process, based on the computer controlled ultrasound system US-plus®.

examples show ultrasound results for curing of epoxy, unsaturated polyester, phenolic moulding compound and elastomer vulcanization. The data can be attained either by small laboratory tools (typical for incoming goods control) or from production tools.

7.6.1 Testing curing of phenolic resins

The PF-compound type used came from former Bakelite GmbH, Letmathe. It was possible to obtain a special mixture without the hexamethylenetramine hardener. The results for 3 mm thick plates are presented in Fig. 7.22. In the compression moulding process the change of sound velocity was recorded online. The mould was preheated to 160 °C. The room temperature compound was filled in and the mould closed. For the moulding compound without hardener the softening and melting lead to a typical continuous reduction of sound velocity. The melting process is finished after about 70 s. The moulding compound with hardener initially shows the same decrease

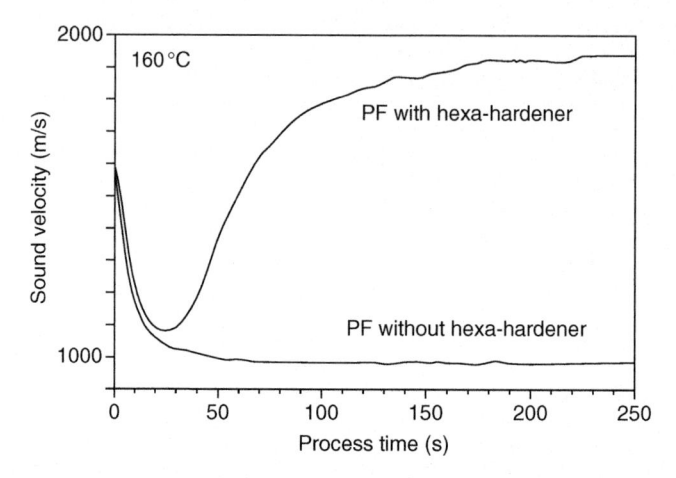

7.22 Sound velocity vs. process time, compression moulding of phenolic moulding compound for the commercial mixture and a mixture without hardener.

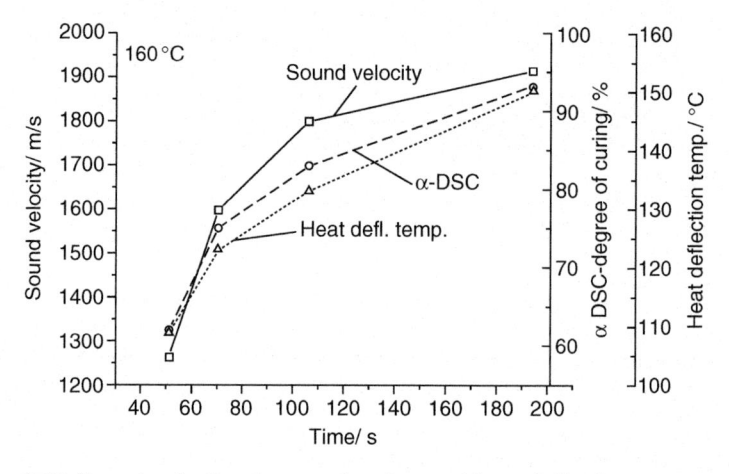

7.23 Sound velocity, degree of curing and heat deflection temperature vs. process time, compression moulding of phenolic moulding compound.

by softening but from circa 25 s the curing causes an increase in sound velocity.

To investigate the correlation between actual sound velocity, state of curing, and mechanical properties at four definite times (51, 70, 106, 194 s) the compression process was interrupted and the formed plate quickly removed. From the plate, samples for DSC were cut, the heat deflection temperature (HDT) (DIN EN ISO 75) was determined and a bending test to destruction was performed. In Fig. 7.23 beside the sound velocity for

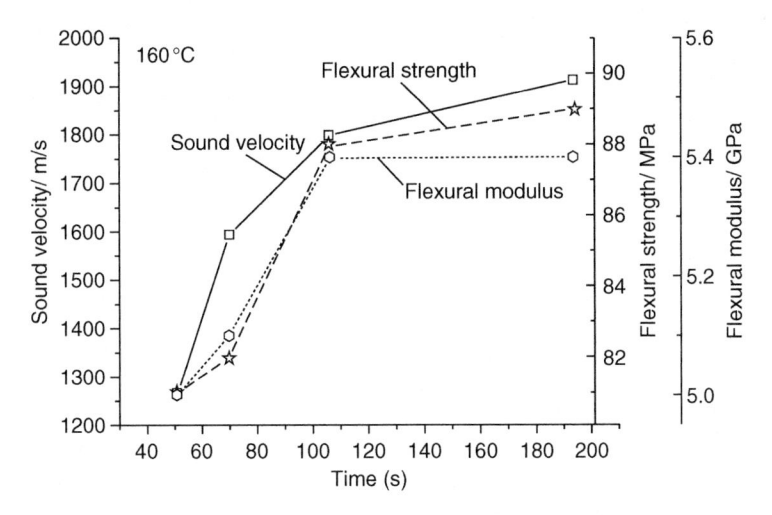

7.24 Sound velocity, flexural strength and flexural modulus vs. process time, compression moulding of phenolic moulding compound.

these four termination times the degree of curing (α-DSC) and the HDT are summarized. For more clarity the results for a room temperature bending test – the flexural modulus from the rise of the stress–strain curve – and the flexural strength are given in Fig. 7.24. The increase in sound velocity and improvement of HDT and of mechanical properties correlate well with the rise of the degree of curing. Therefore the progress of sound velocity rise gives a direct indication of the degree of curing and the final properties of the produced parts.

7.6.2 Testing curing of liquid epoxy and polyester resins

A measuring cell for liquids was developed to investigate casting resins. The lower part of the mould was formed as a cavity, the upper part as a stamp. Both parts were encased with heating elements and thermocouples for temperature control were incorporated. The mould could be completely disassembled to remove the cured block. The sound velocity during isothermal curing for two kinds of epoxy resin and an unsaturated polyester resin are given. The epoxy resins were selected to have different glass transition temperatures when fully cured ($T_{g\infty}$). Epoxy resins with low T_g (here 120 °C) are used in numerous technical applications from the transport sector through to high-voltage insulation. The results for isothermal curing at temperatures from 70 to 140 °C are illustrated in Fig. 7.25.

The higher the temperature the faster the marked increase of sound velocity manifests. This illustrates the Arrhenius behaviour of reaction

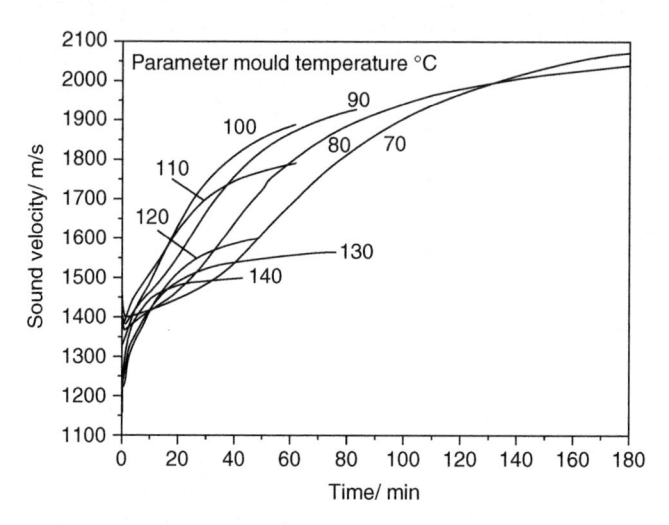

7.25 Sound velocity vs. process time at different temperatures, epoxy resin (low T_g type) curing.

kinetics where an increase of temperature by 10 K increases the reaction speed by a factor of about two. The lessening of the end value of the sound velocity with increasing temperature is a result of the general temperature dependence of the modulus of a fully cured resin. The greater reduction between 110 and 120 °C is a result of passing through $T_{g\infty}$. At 130 and 140 °C the final state is rubber-like; at 110 and 120 °C it is transient.

The epoxy resin from the second example is a high end product for aircraft manufacturing. Its maximal glass transition temperature $T_{g\infty}$ is in the range of 200 °C. In the manufacturing process it is typically used in carbon fibre prepreg lay technology. Curing is carried out in autoclaves at 180 °C for many hours. The results for temperatures from 140 to 180 °C are illustrated in Fig. 7.26. The influence of curing temperature is illustrated very clearly. With rising temperatures the sound velocity at the beginning is reduced, the time for increase and fall of the curve is also reduced. For 170 and 180 °C a diminution of the final value appears. This is indicative of the proximity of the processing temperature to T_g. The times found in rheology experiments for gelation and vitrification are given in the description.

In the next example, shown in Fig. 7.27, the curing of unsaturated polyester resin for manufacturing of wind blades is described. Curing was performed at 60, 70 and 80 °C. The object was to observe the effect of a substantial heat release by the exothermic crosslinking reaction so that an additional thermocouple was incorporated into the resin. Initially, the resin warms to the mould temperature and in this phase the sound velocity decreases by reduction of resin viscosity. At all manufacturing temperatures

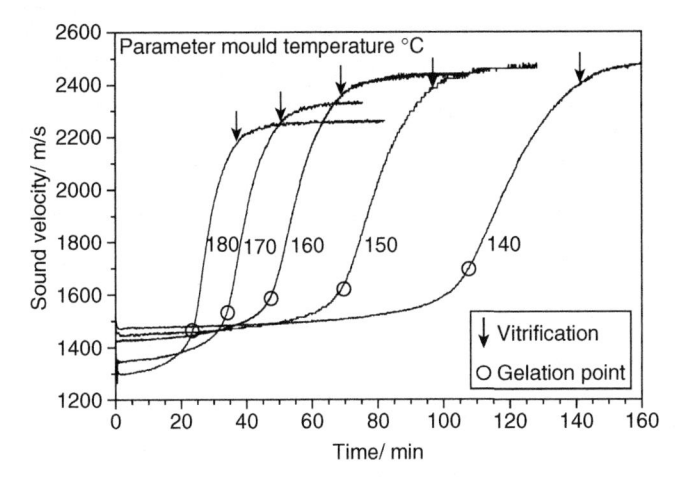

Using ultrasound to monitor the curing of composites   159

7.26 Sound velocity vs. process time at different temperatures, epoxy resin (high T_g type), additional the times for gelation and vitrification from rheology are marked.

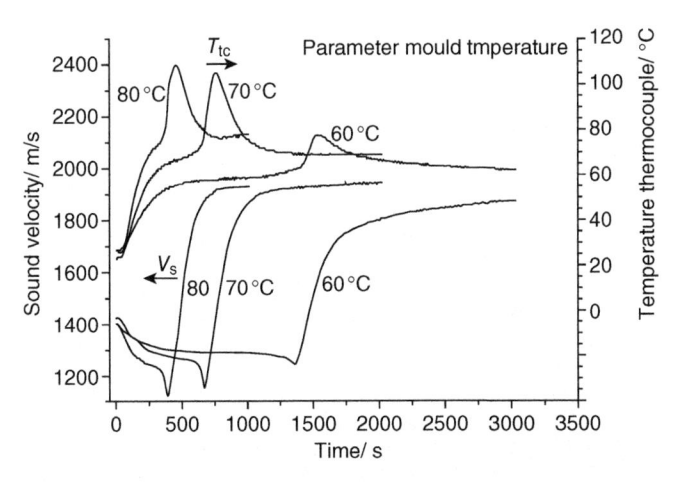

7.27 Sound velocity and thermocouple resin temperature vs. process time at three mould temperatures, unsaturated polyester resin.

a temperature rise appears when the reaction starts. At 70 and 80 °C this rise is significant. The sound velocity shows that in the early state of the reaction, this reduction of viscosity is substantially reduced by this effect before it increases. Such information cannot be gleaned through rheological measurements because the heat flow to the measuring plates is too high, or through monitoring of the component manufacturing process as it is impossible to incorporate thermocouples into the resin. Ultrasound curves yield this important additional information on strong temperature rise. For flow front detection it is necessary to connect a number of sensors from the tool so the measuring unit was extended by a multiplexer set-up with up to 8 or 16 measuring lines.

7.7 Using ultrasonic online process monitoring in practice: automotive engineering

7.7.1 Belt pulleys

In recent times belt pulleys in the power-train have changed from transformed sheet steel to injection moulded mineral filled phenolic resin [33, 34]. An example is shown in Fig. 7.28. For this, a special injection-compression moulding process, with mould temperature of around 160 °C, was developed [34]. In the process, an incompletely closed mould (preset gap thickness 1 ... 10 mm) is filled by an injection moulding machine before it is closed by high pressure. This causes a very homogeneous mould filling and high compaction. The closed mould is held for sufficient time to reach a high degree of curing.

During development of the technology, the ultrasound measuring system was used to gain information about the process [35]. Two ultrasound sensors were incorporated into the mould, of which Fig. 7.29 gives a schematic view.

7.28 Belt pulley made from phenolic moulding compound. Sensor position is marked.

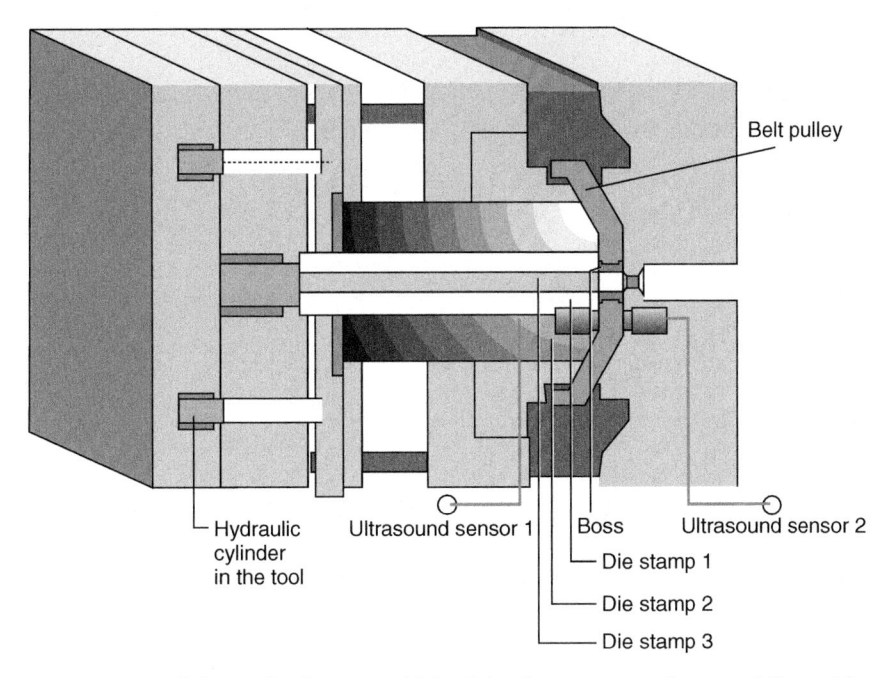

7.29 Schematic view, mould for injection compression moulding with incorporated ultrasonic measuring line.

Typical records of the change of ultrasound velocity during the manufacturing process of the phenolic moulding compound are given in Fig. 7.30 [36]. During the injection phase, the first sensor signal arises from the filling of the gap between the sensors. The temperature increase of the compound leads to a lowering of the viscosity of the melt, which is accompanied by a reduction in sound velocity. After about 12 s the crosslinking process dominates. The main part of the reaction takes 30 to 40 s. After that the reaction speed is reduced but does not fully come to an end during the 75 s cycle time. A sloped rise of sound velocity is clear evidence of a reaction slowed down by vitrification. The shift of glass transition temperature by crosslinking brings the glass transition temperature close to the mould temperature. The material solidifies, the viscosity is dramatically reduced, and contact between the reacting agents, resin and hardener, is also thus reduced. This process is to be expected in this case as the $T_{g\infty}$ of the phenolic moulding compound is >250 °C and the mould temperature is 160 °C, i.e. the latter is considerably below this final glass transition temperature. The reproducibility of the ultrasound curve shape is very good. It is easy to see deviations in the manufacturing process, for example in mould temperature or compound composition, very early.

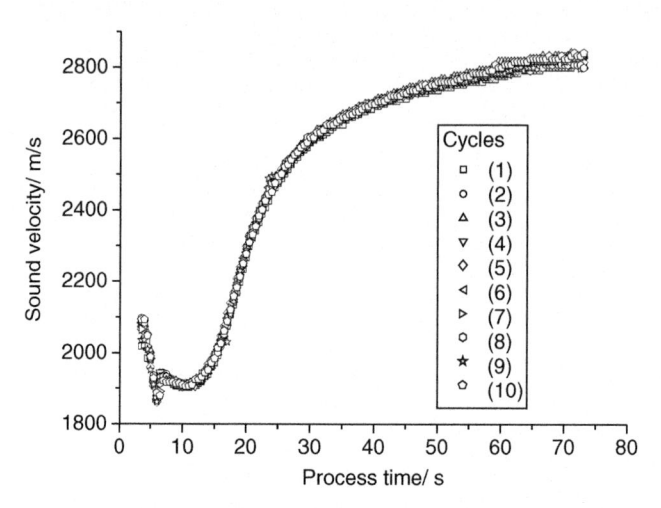

7.30 Sound velocity vs. process time, injection moulding process of phenolic moulding compound, test of reproducibility.

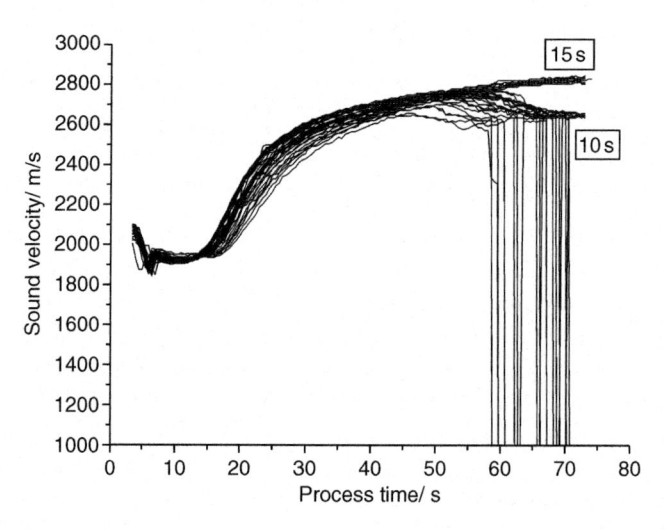

7.31 Sound velocity vs. process time, injection moulding process of phenolic moulding compound, influence of switching time from injection pressure to hold pressure.

In Fig. 7.30 a superposition of 10 cycles is illustrated. The aim here is to illustrate the reproducibility of the process. The curves demonstrate the high precision of the measuring and manufacturing technology. During injection moulding, for specific process adjustments, shrinkage of the resin due to crosslinking emerged. An example is given in Fig. 7.31 which shows data relating to a change in the time the post-stress was held, from 10 s to 15 s, during the phase of intensive curing.

Reducing the time of post-pressure by 5 s lowered the sound velocity, apparent in the flat part of the curve. In some cases the signal was interrupted before the process ended. Reduced sound velocity, or the signal interruption, are indications of the influence of shrinkage. Obviously at first only the pressure in the compound is reduced without a detachment from the mould wall.

Beside optimization and permanent documentation of the manufacturing process online, process monitoring offers the profitable control of incoming compound. This can thus be moved from the laboratory into the production process. Deviations from stored master curves of the sound velocity can be detected immediately by the measuring system and trigger an alarm. An unexpected additional result was the correlation between the final value of sound velocity and the test of belt pulley strength, as shown in Fig. 7.32 [35].

The higher the sound velocity the better the strength values. A sound velocity less than about 90% of the maximum value is an indication of a mistake in the technological process. Through the ultrasound measuring system, control of the final sound velocity is possible for every produced part and parts so that those not reaching the predefined sound velocity level can be automatically rejected.

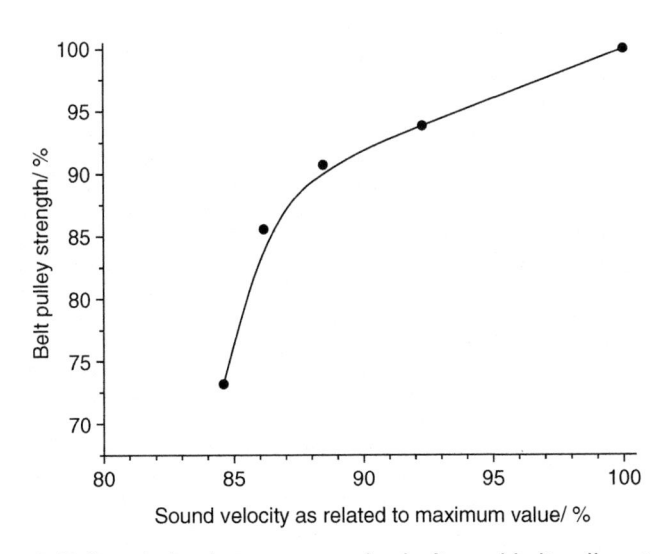

7.32 Correlation between sound velocity and belt pulley strength.

7.7.2 Headlights

Modern automotive headlights with clear glass panels realize the form of the light output by concave mirrors. A technology has been developed to produce such mirrors is injection moulding of BMC. BMC is a composite material of UP resin and short glass fibres [37–39].

An example of a BMC head light is given in Fig. 7.33. To improve the production stability, and product quality, the ultrasound measuring system was installed in a BMC test compression mould, as shown in Fig. 7.34 [40]. The mould was developed by ISK [41].

Initial tests of the method found differences in curing behaviour between two BMC suppliers – named material A and B – as shown in Fig. 7.35. By varying the temperature between 140 and 160 °C, an overlapping region of similar reaction speed could be found for both materials. On the basis of this insight, a quick shift between both suppliers seemed possible without voluminous tests in the production line. Process optimization was also a possible task. For every temperature an optimal process time could be determined and finally a cycle time reduction could be initialized. The possibilities for reduction of curing time seemed to be up to 20%.

7.7.3 Lorry cabins

RTM is a current method for composite manufacturing [42–46]. The ultrasound measuring system was applied to the process for fabrication of large-volume cabin parts for lorries. The aim was to detect the resin flow front

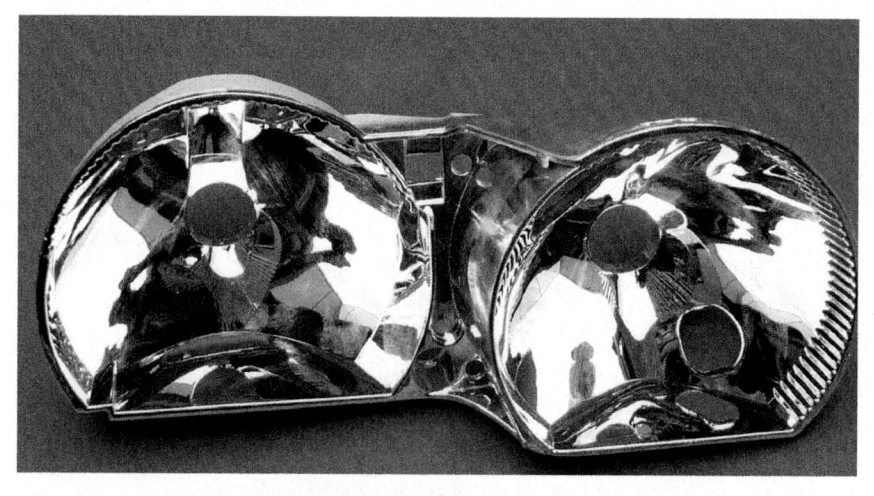

7.33 Headlight from BMC produced by injection moulding process, product of Automotive Lighting, Reutlingen.

7.34 Compression mould with incorporated ultrasound sensors for incoming goods control. Photo: ISK Iserlohn.

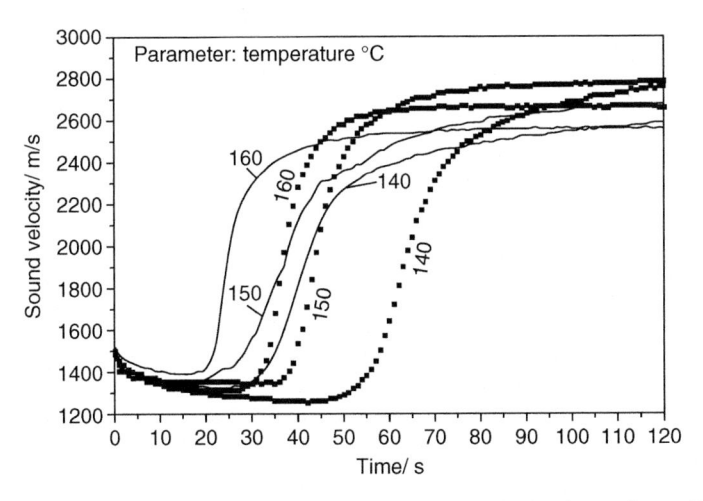

7.35 Sound velocity vs. process time, curing behaviour of two BMC materials (material A, solid curves; material B, dotted curves) at three typical manufacturing temperatures, 140, 150 and 160 °C.

propagation and the uniformity of the curing process in a large-volume tool. An RTM mould is shown schematically in Fig. 7.36. The bifid mould establishes a gap. The gap is filled with inlayed fibre mat. A medium filling state is optimal. The two components – resin and hardener – are pumped to a specific mixing ratio into a preheated mixing chamber and from there, via a runner, direct into the mould. The reactive resin spreads in the gap and

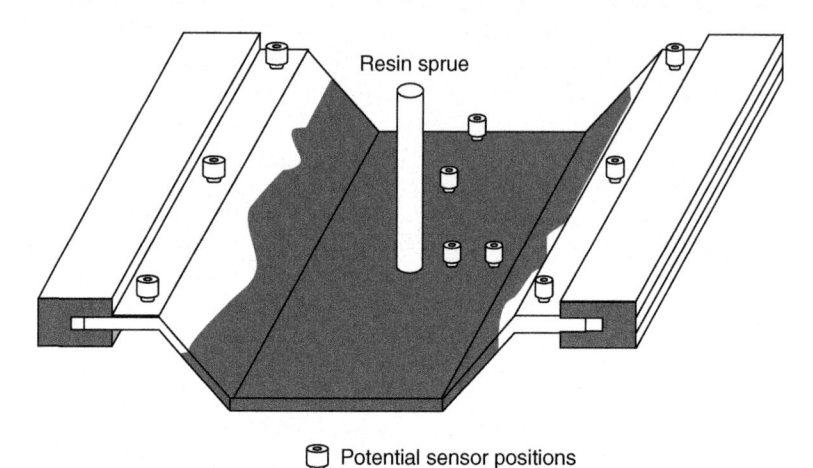

7.36 Schematic view of an RTM mould with partial filling and possible sensor positions.

finally reaches the rising gates. Here the displaced air passes off together with the resin. For flow front detection and cure monitoring an arrangement of ultrasound sensors can be realized at critical positions.

In a research project, together with IKV Aachen (http://www.ikv-aachen.de/en/) and partners from industry, namely Bakelite (today Momentive http://ww2.momentive.com/home.aspx), Fritzmeier (http://www.fritzmeier.de/dt/Sitemap/map_home.html) and BAM, the ultrasound method was adapted to the RTM process for production of lorry cabin roofs, as shown in Fig. 7.37.

In the RTM mould five ultrasound measuring lines were installed. Figure 7.38 shows a photograph of the upper part of the mould and marks for the sensor positions. The sensors were inserted from the back in blind holes behind the wall of the mould, as shown in Fig. 7.39.

Typical measurement results are given for the first 4 min in Fig. 7.40, which shows the mould filling. As expected from the assembly of the ultrasound measuring lines, the resin reaches line 1 first since it is closest to the sprue, 2 and 3 are reached at the same time, and 4 and 5 later but not at the same time as expected from the geometry. Figure 7.41 shows the signal progress for the complete manufacturing process. After having reached the measuring line, the resin viscosity decreases due to warming, as indicated by the decrease of sound velocity. The resin is injected with a temperature of 80°C and the mould temperature is 160°C, so the resin temperature increases substantially by the time it has reached the measuring line. This is accompanied by a reduction of sound velocity caused by the viscosity

Using ultrasound to monitor the curing of composites 167

7.37 Photograph of the RTM part for lorry cabin produced by Fritzmeier GmbH.

7.38 Upper part of the RTM mould with signs showing the position of the sensors: 1 in vicinity to the inlet nozzle (arrow), 2, 3, 4 and 5 close to the outlet risers.

168   Non-destructive evaluation (NDE) of polymer matrix composites

7.39 Insertion of the sensors into the mould from the back.

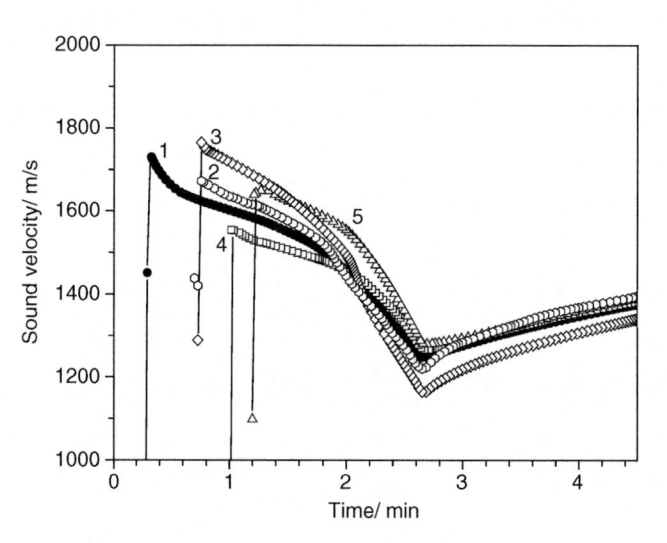

7.40 Sound velocity vs. process time, flow front detection by signal onset at five measuring lines.

decrease of the resin. After about 2.5 min the temperature has reached a value such that the curing reaction begins to dominate.

The curing process at all five sensor positions is rather similar. Characteristic for all ultrasonic curves is an apparent two step process. After a period of deceleration, starting from about 8 min, the sound velocity change increases again. A possible explanation is the exothermic character of the

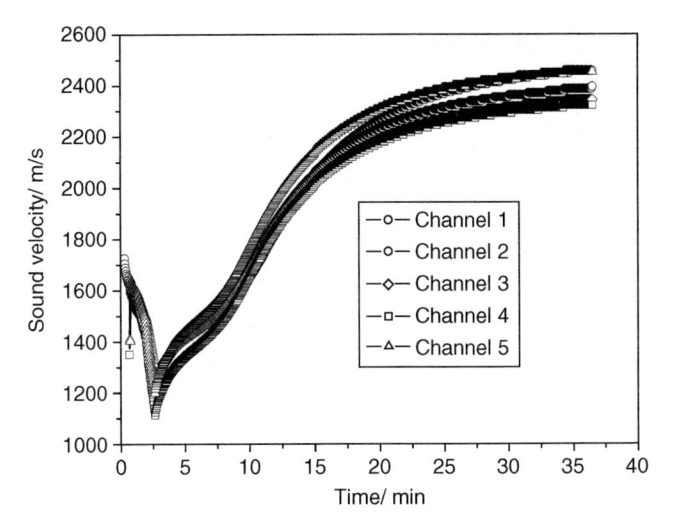

7.41 Sound velocity vs. process time for the complete RTM process.

crosslinking reaction. In thick components, the thermal conduction is low, so an increase of temperature may lead to an acceleration of reaction speed. The final continuous small rise of the sound velocity is an indication for a slowdown of reaction speed due to vitrification. The sound velocity data for the five measuring lines show that the curing run is rather even. That means that the temperature distribution in the mould is nearly perfect. Differences in the end value can be explained by different glass fibre contents of the composite in the measuring line and inexactnesses in the determination of the gap thickness. The RTM process with ultrasound process monitoring is described in more detail in Schmachtenberg and Schulte zur Heide [44] and Schmachtenberg *et al.* [47].

7.7.4 High-pressure vessels for hydrogen storage

Fuel cell drive systems need a low weight tank for hydrogen. For gas storage 700 bar pressure vessels are under development. An example of a test in the BAM climate test chamber is shown in Fig. 7.42 [48]. An inliner, for example a polyethylene flask, is encapsulated by an epoxy impregnated carbon fibre prepreg wrapping. The prepreg is cured at high temperatures, for example in an autoclave. The maximum curing temperature is limited by the inner polyethylene flask. To give the construction engineers a method for process optimization and to define standards for quality control of the production process, a measuring cell for investigation of prepreg curing was developed by SLT and is shown in Fig. 7.43.

7.42 CFK-hydrogen cylinders under test in a temperature-controlled high-pressure test apparatus in the BAM. Photo: BAM Federal Institute for Materials Research and Testing.

7.43 Mini-compression gadget with two heat regulated plates and incorporated ultrasound sensors for process monitoring in laboratory scale.

Using ultrasound to monitor the curing of composites 171

The test gadget is based on a small compression mould consisting of a fixed upper plate and a lower plate moved by a pneumatic punch. Both plates have heating elements and separate sensors for controlling the temperature. By careful design and use of a thermocouple and a control circuit per plate, a homogenous temperature distribution is realized. The diameter of the plates is 50 mm. Additionally to isothermal measurements, temperature courses, for example ramps, can be programmed. The complete manufacturing process can be simulated.

In Fig. 7.44 the change of sound velocity and DMA storage modulus E' of a prepreg with a temperature programme is shown. The temperature programme was adapted to the manufacturing process: first linear heating with 1 K/min to 110 °C and then a hold at this temperature for long period of time. Initially, the reduction of viscosity with rising temperature leads to a decrease of the E' and v_s curves before the crosslinking reaction sets in at 110 °C, leading to a distinct rise. Ultrasound and DMA results correspond well. The higher the frequency, the steeper the rise of the curves at curing. This is a consequence of the frequency dependence of the glass transition temperature. For higher frequencies, the impact of vitrification occurs earlier and is more clearly seen.

At room temperature, where the resin is liquid, the prepreg has the higher sound velocity as caused by the high amount of stiff carbon fibre fillers. The changes with increasing temperature are very similar for both kinds of samples though the prepreg always shows the higher value of sound velocity. The critical temperatures are very similar, also at cooling, the knee in the curves appears at a similar temperature.

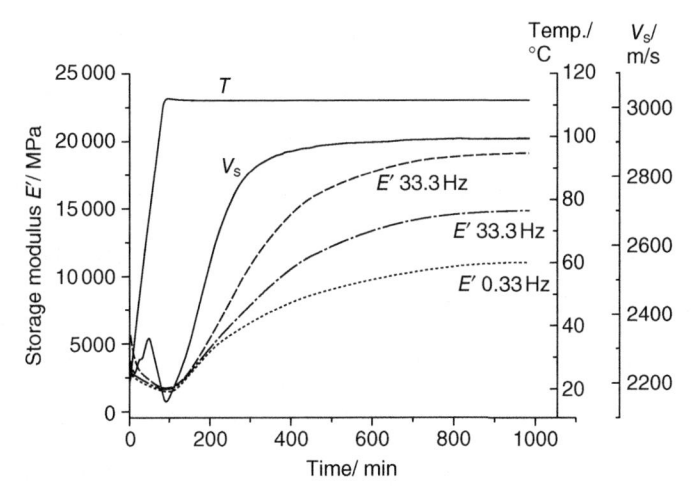

7.44 Storage modulus, sound velocity and temperature vs. process time, epoxy prepreg, linear heating and isothermal curing at 110 °C.

A linear heating experiment with a 1 K/min ramp for prepreg is shown in Fig. 7.45.

In the heating run, the softening of the resin with increasing temperature is at first dominant. Near 105 °C the crosslinking begins to dominate, causing an increase in the modulus and thus in the sound velocity. With depletion of the reactants the increase of sound velocity comes to an end (about 130 °C) and from here it decreases continuously. This is an indication that the glass transition temperature of the fully cured material was exceeded. An additional support for this interpretation is the identical temperature dependence in the cooling run, where both curves are virtually the same. Below about 100 °C the slope changes and the temperature dependence is reduced. This seems to be caused by passing the rubber to glass transition.

To prove this more in detail with a fully cured prepreg sample, DMA for five frequencies were measured. The result is given in Fig. 7.46. At circa 100 °C, at the onset temperature, a stepped decrease of E' which is typical for a transition from the glass to the rubber state appears. The frequency dependence shows the characteristic behaviour. A similar experiment with the fully cured prepreg was carried out with the ultrasound measuring system, as shown in Fig. 7.47. A change of sound velocity temperature function at about 115 °C is seen. This decrease is also indicative of passing the glass-rubber transition. The higher temperature for ultrasound (4 MHz) in comparison to DMA is due to the frequency dependence of T_g.

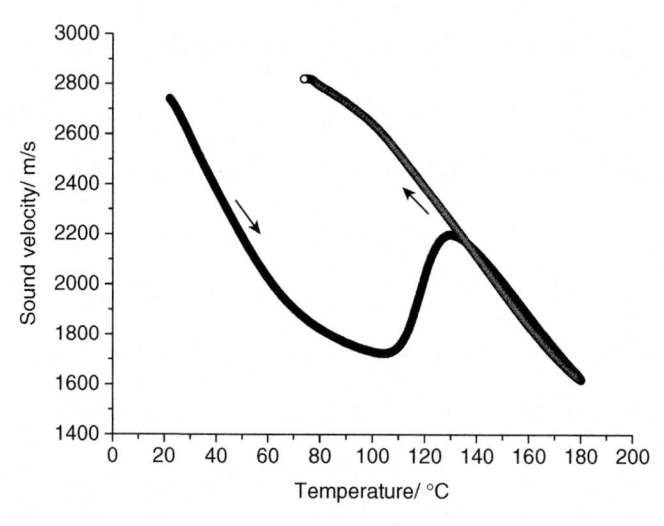

7.45 Sound velocity vs. temperature at linear heating and cooling of prepreg, temperature rate 1 K/min.

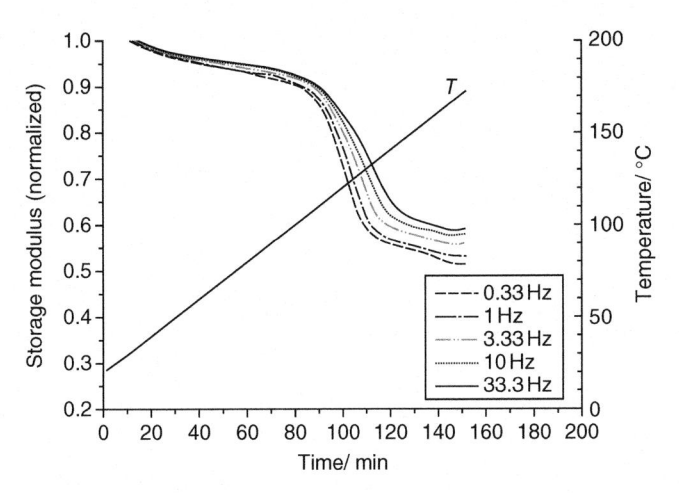

7.46 Storage modulus *E'* of fully cured epoxy prepreg vs. time, sample from fig. 7.44, heating rate 1 K/min.

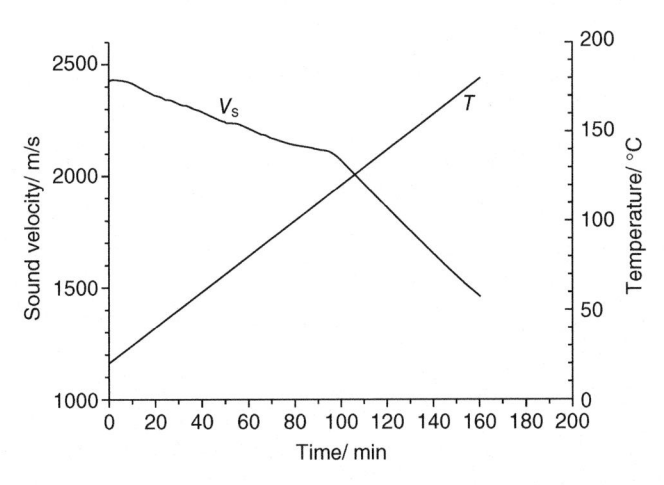

7.47 Sound velocity of fully cured epoxy resin vs. time, sample from Fig. 7.44 at linear temperature rise, heating rate 1 K/min.

A similar linear heating experiment with a 1 K/min ramp was carried out on the basic resin of the prepreg material to compare the behaviour of the prepreg and the pure epoxy resin crosslinking. The results are given in Fig. 7.48. At room temperature, where the resin is liquid, the prepreg has the higher sound velocity caused by the high amount of stiff carbon fibre fillers. The changes with increasing temperature are very similar for both samples, although the prepreg shows the higher sound velocity. The critical

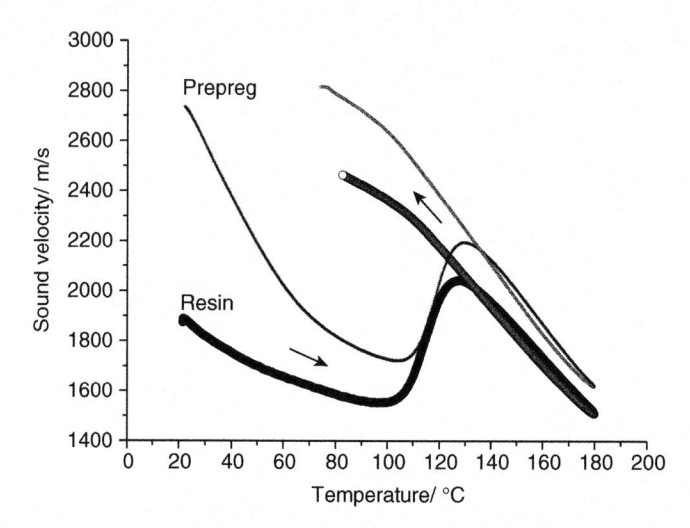

7.48 Change of sound velocity at linear heating and cooling of prepreg and merge resin, temperature rate 1 K/min.

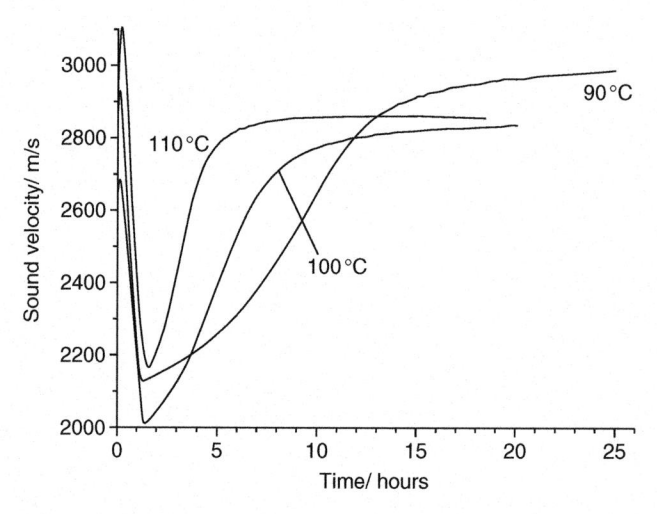

7.49 Change of sound velocity by prepreg crosslinking for three temperatures.

temperatures are very similar; also at cooling, the knee in the curves appears at a similar temperature.

To simulate the manufacturing process for three temperatures the isothermal curing process was recorded as shown in Fig. 7.49. The mould was preheated to the selected temperature. The room temperature stored prepreg was deposited on the lower plate and the mould closed. The initial sharp decrease of sound velocity is a consequence of viscosity reduction

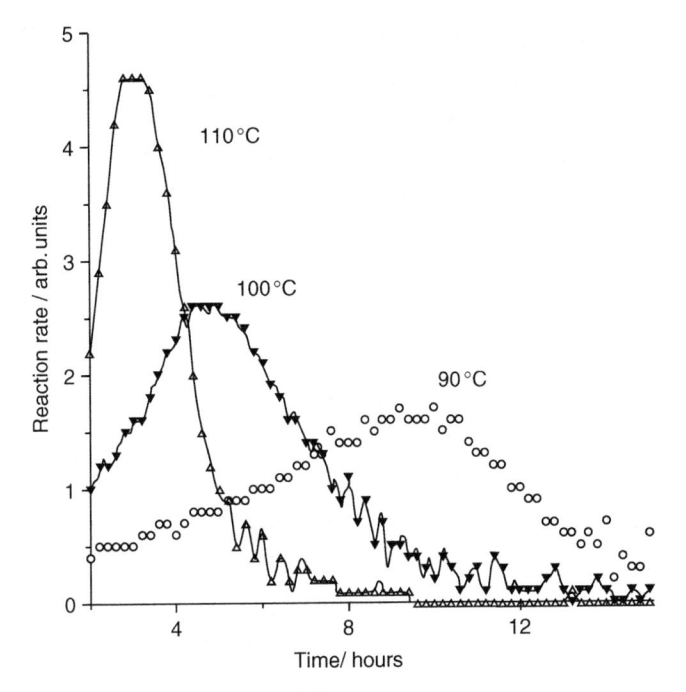

[Graph: Reaction rate vs. Time/hours showing curves for 110°C, 100°C, and 90°C]

7.50 Reaction rate vs. process time concerning results in Fig. 7.49, prepreg crosslinking for three temperatures.

due to heating of the resin. The higher the temperature the earlier the curing process manifests. At 110°C a plateau is reached. This is indicative of full curing. At 90°C the appearance of vitrification reduces the reaction speed and causes a long-lasting rise.

The first derivative, i.e. the reaction rate, for the sound velocity curves is shown in Fig. 7.50. The first deviation was also calculated for the same data. This is an expression of the reaction rate. The picture is very clear: the reaction rate and the time for the reaction maximum are distinct functions of the temperature. For another kind of prepreg (quick curing resin) the reproducibility of the ultrasound method has been tested. The results are shown for 120°C mould temperature in Fig. 7.51 for seven measurements. The distinctions in the curves at the beginning are due to a different thickness change of the sample during the softening period. The curing process observed above about 10 min is very similar and the end of the crosslinking process does not differ remarkably.

7.7.5 Autoclave curing

After initial investigations with the small laboratory mould, the manufacturing of prepreg parts in an autoclave were performed. A photograph of

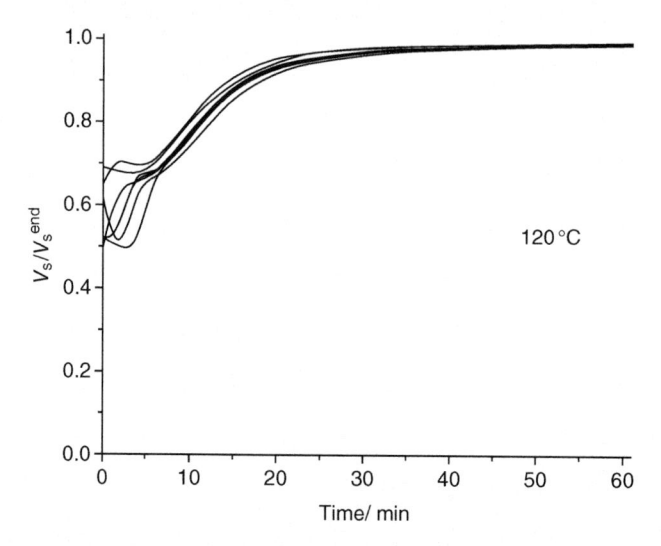

7.51 Comparison of the change of sound velocity of glass fiber prepreg material – normalization on the end valu.

7.52 Ultrasound measuring assembly in front of an autoclave for prepreg curing. Photo from J. Döring, BAM.

the measuring assembly is given in Fig. 7.52. The sensors were incorporated into the mould inside the autoclave. The coaxial measuring cables were connected with coaxial feed through in a measuring flange – seen in the picture on the left side close to the door. The results were analogous to those of the laboratory mould.

7.7.6 Vulcanization of tyre rubber

After investigating ultrasonic cure monitoring of thermosetting reactive materials, experimental scope was extended to include the vulcanization of rubber [49–53]. Here an example for a tyre mixture is given. During automobile tyre production, every new mixed rubber compound is controlled by a measurement of the vulcanization process for product control. The standard instruments in the tyre rubber industry are specially adapted plate–plate rheometers also known as a curemeter or MDR (moving die rheometer). The rubber compound is placed between two heated plates which are pressed together until the sample is formed. The lower plate is fixed; the upper plate oscillates with a small angle. The torque to hold the oscillation angle constant is recorded as function of time. The main information from such measurements are the time to start the reaction, known as the scorch time, at two or five percent increase of torque and the real part of the complex torque. The critical times to reach 10, 50 and 90% of the total change of torque by vulcanisation are defined by such measurements. Details are provided in a standard [54].

To demonstrate the capabilities of the ultrasound cure monitoring method it is convenient to compare it with the well-established curemeter measurements. In Fig. 7.53 the real part of the torque from the curemeter S' is depicted together with the sound velocity at isothermal vulcanisation of a tyre rubber mixture at 160 °C. The data for the first 200 s are separately shown in Fig. 7.54. During the initial phase, both measurements give comparable results. The ultrasound data could also be used for determination

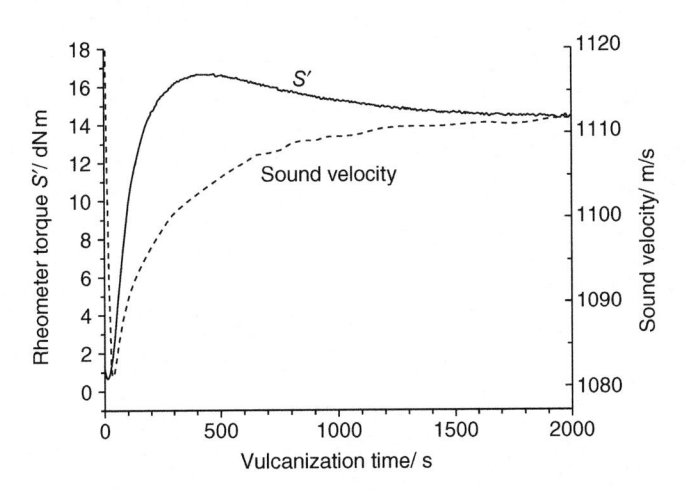

7.53 Real part of torque S' from curemeter and sound velocity process time, vulcanisation of a rubber compound at 160 °C.

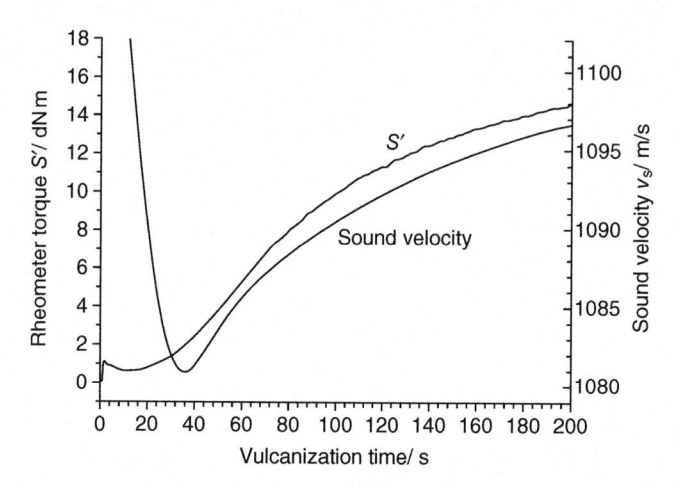

7.54 Real part of torque S' from curemeter and sound velocity process time, vulcanisation of a rubber compound at 160 °C, illustration of the begin of reaction in Fig. 7.52.

of a scorch time. The small difference between both methods can be understood by differences in the time for warming up of the compound. The curemeter curve in Fig. 7.52 shows a maximum in torque S' and a reduction over a longer period. This phenomenon is typical for sulfur crosslinked rubber and named reversion and is due to the thermal degradation of sulfur bonds.

Surprisingly, reversion is not detected by ultrasound. A possible cause of this difference is the difference in the amplitude of mechanical excitation. We have found a pronounced influence of the angle of deflection in the curemeter such that the higher the angle the higher the reversion.

Summarizing the rubber tests it can be stated that ultrasound can be used in a manner similar to a curemeter as a vulcanization monitoring system which can be applied in the mould. In addition, it gives direct information on reaction processes not resulting from reversion.

7.8 References

1. Mezger, T.G., *The Rheology Handbook: For users of rotational and oscillatory rheometers:* 2nd revised edition. 2006, Hannover: Vincentz Network GmbH & Co. KG.
2. Winter, H.H., Can the gel point of a crosslinking polymer be detected by the G'–G'' Crossover? *Polymer Engineering and Science*, 1987. **27**(22): p. 1698–1702.
3. Ehrenstein, G.W., G. Riedel, and P. Trawiel, *Praxis der thermischen Analyse von Kuststoffen*. 2003, München: Carl Hanser Verlag.

4. Sperling, L.H., *Introduction to physical polymer science*, 4th edition. 2006, New Jersey: John Wiley & Sons, Inc.
5. Ehrenstein, G.W., G. Riedel, and P. Trawiel, *Thermal Analysis of Plastics Theory and Practice*. 2004, München: Carl Hanser.
6. Jaunich, M. and W. Stark, Influence of curing on properties of urea resin. *Materials Testing-Materials and Components Technology and Application*, 2009. **51**(11–12): p. 828–834.
7. Wolfahrt, M., G. Pilz, and R.W. Lang, *Einfluss des Aushärtegrades auf wesentliche Werkstoffeigenschaften eines Epoxidharz-Matrixwerkstoffes*. Internet publication. http://www.dgm.de/download/tg/706/706_78.pdf.
8. Trappe, V., S. Guenzel, and M. Jaunich, Correlation between crack propagation rate and cure process of epoxy resins. *Polymer Testing*, 2013. **31**(5): p. 654–659.
9. Ferracane, J.L., H.X. Berge, and J.R. Condon, In vitro aging of dental composites in water – effect of degree of conversion, filler volume, and filler/matrix coupling. *Journal of Biomedical Materials Research*, 1998. **42**(3): p. 465–472.
10. Ferracane, J.L., et al., Wear and marginal breakdown of composites with various degrees of cure. *Journal of Dental Research*, 1997. **76**(8): p. 1508–1516.
11. Krautkraemer, J. and H. Krautkraemer, *Ultrasonic Testing of Materials*. 1990, Berlin: Springer.
12. Rogers, G.M., The structure of epoxy resins using NMR and GPC techniques, *Journal of Applied Polymer Science*, 1953. **16**: p. 1972.
13. Alig, I., et al., Ultrasound for tracking epoxy-resin curing. *Zeitschrift für Chemie*, 1987. **27**(5): p. 195.
14. Alig, I. and D. Lellinger, Frequency-dependence of ultrasonic velocity and attenuation in 2-phase composite systems with spherical scatterers. *Journal of Applied Physics*, 1992. **72**(12): p. 5565–5570.
15. Alig, I., et al., Dynamic light-scattering and ultrasonic investigations during the cure reaction of an epoxy-resin. *Journal of Applied Polymer Science*, 1992. **44**(5): p. 829–835.
16. Alig, I., et al., Monitoring of film formation, curing and ageing of coatings by an ultrasonic reflection method. *Progress in Organic Coatings*, 2007. **58**(2–3): p. 200–208.
17. Alig, I., et al., Polymerization and network formation of UV curable materials monitored by hyphenated real-time ultrasound reflectometry and near-infrared spectroscopy (RT US/NIRS). *Progress in Organic Coatings*, 2006. **55**(2): p. 88–96.
18. Alig, I. and S. Tadjbakhsch, Film formation and crystallization kinetics of polychloroprene studied by an ultrasonic shear wave reflection method. *Journal of Polymer Science Part B-Polymer Physics*, 1998. **36**(16): p. 2949–2959.
19. Alig, I., et al., Ultrasonic and DSC investigations of the curing process of diandiglycidether in dependence on the curing temperature. *Acta Polymerica*, 1989. **40**(9): p. 590–595.
20. Challis, R.E., et al., Models of ultrasonic bulk wave propagation in thermosets with spatially varying viscoelastic properties, in *Review of Progress in Quantitative Nondestructive Evaluation*, Vols 24A and 24B, D. O. Thompson and D.E. Chimenti, Editors. 2005, Melville: American Institute of Physics, p. 1258–1265.
21. Challis, R.E., et al., Viscoelasticity of thin adhesive layers as a function of cure and service temperature measured by a novel technique. *Journal of Applied Polymer Science*, 1992. **44**(1): p. 65–81.

22. Challis, R.E., et al., Ultrasonic measurements related to evolution of structure in curing epoxy resins. *Plastics Rubber and Composites*, 2000. **29**(3): p. 109–118.
23. Challis, R.E., et al., Scattering of ultrasonic compression waves by particulate filler in a cured epoxy continuum. *Journal of the Acoustical Society of America*, 1998. **103**(3): p. 1413–1420.
24. Challis, R.E., et al., Following network formation in an epoxy/amine system by ultrasound, dielectric, and nuclear magnetic resonance measurements: A comparative study. *Journal of Applied Polymer Science*, 2003. **88**(7): p. 1665–1675.
25. Lodeiro, M.J. and D.R. Mulligan, *Cure Monitoring Techniques for Polymer Composites, Adhesives and Coatings*. Measurement Good Practice Guide No 75, Engineering and Process Control Division National Physical Laboratory, 2005. Teddington, Middlesex.
26. McHugh, J., *Ultrasound Technique for the Dynamic Mechanical Analysis (DMA) of Polymers*. Dissertation 2007. Technische Universität Berlin, Fakultät III – Prozesswissenschaften.
27. Cents, A.H.G., et al., Measuring bubble, drop and particle sizes in multiphase systems with ultrasound. *Aiche Journal*, 2004. **50**(11): p. 2750–2762.
28. Hiramatsu, N., S. Taki, and K. Matsushige, Ultrasonic spectroscopy for teflon glass bead composite system. *Japanese Journal of Applied Physics Part 1 – Regular Papers Short Notes & Review Papers*, 1989. **28**: p. 33–35.
29. Doering, J., et al., *Contribution to Ultrasound Cure Control for Composite Manufacturing*. NDT.NET/ http://www.ndt.net/article/wcndt00/papers/idn482/idn482.htm, 2000. Proceeding 15th World Conference for Non Destructive Testing WCNDT, Roma, 2000.
30. Doering, J., W. Stark, and G. Splitt, *On-line Process Monitoring of Thermosets by Ultrasonic Methods*. NDT.NET, 1998. **3**(11): p. Proceedings of the European Conference on Non Destructive Testing, Copenhagen, 1998.
31. Kuerten, C., J. Doering, and W. Stark, *US-Plus – Cure monitoring for thermosets*. http://www.isk-iserlohn.de/us-plus.htm.
32. Bohmeyer, W., *US-plus measuring system*. http://www.pyrosensor.de/usplus.html.
33. Bittmann, E., Thermosets. *Kunststoffe-Plast Europe*, 2005. **95**(10): p. 168–172.
34. Berthold, J., Processing of thermosetting resins: pioneering spirit needed for survival of the species. *Kunststoffe-Plast Europe*, 2003. **93**(10): p. 101.
35. Hoster, B., et al., *Online-Aushärtungskontrolle bei der Verarbeitung von vernetzenden Kunststoffen mit Ultraschall*, Annual Conference German Society of Non-Destructive Testing, Salzburg, 17–19 May 2004, 2004.
36. Straet, T., et al., Ultrasonic measurements on thermoset moulding compounds. *Kunststoffe-Plast Europe*, 2005. **95**(7): p. 73–77.
37. Kuels, N., In drei Schritten zum Erfolg. *Kunststoffe*, **2008**(7): p. 60–61.
38. Moritzer, E., SMC und BMC, Neue Anwendungen – neue Märkte. *Kunststoffe*, **2006**(11): p. 112–115.
39. Sommer, M., SMC and BMC: New applications – new markets. Kunststoffe-Plast Europe, 2001. **91**(10): p. 226–227.
40. Doring, J., et al., Ultrasound process control yields mechanical parameters of thermosetting plastics. *Materialprufung*, 2007. **49**(5): p. 238–242.
41. ISK, *Iserlohner Kunstoff-Technologie GmbH*. http://www.isk-iserlohn.de/ueber_uns.htm.

42. Fries, E., J. Renkl, and S. Schmidhuber, Das schwarze Gold des Leichtbaus. *Kunststoffe* **2011**(9): p. 52–56.
43. Bittmann, E., Automobilindustrie belebt das Geschäft: Duroplaste. *Kunststoffe* **2006**(3): p. 76–82.
44. Schmachtenberg, E. and J. Schulte zur Heide, Online process control. *Kunststoffe-Plast Europe*, 2003. **93**(12): p. 82.
45. Merz, P., RTM parts with a Class A surface – emission-free processing method for unsaturated, liquid polyester resins. *Kunststoffe-Plast Europe*, 2000. **90**(5): p. 108.
46. Merz, P., Glass-fibre-reinforced polymers (GRP). *Kunststoffe-Plast Europe*, 2001. **91**(10): p. 354–354.
47. Schmachtenberg, E., J. Schulte zur Heide, and J. Topker, Application of ultrasonics for the process control of resin transfer moulding (RTM). *Polymer Testing*, 2005. **24**(3): p. 330–338.
48. Mair, G., http://www.bam.de/de/kompetenzen/fachabteilungen/abteilung_3/fg32/fg32_ag4.htm.
49. Stark, W., J. Doering, and J. Kelm, *Test apparatus for a sample of a rubber mixture, to establish its vulcanizing behavior, uses ultrasonic sensors to register ultrasonic signals through the material for accurate readings without rubber damage*, Patent DE 101 38 790 A1, 2001.
50. Stark, W., J. Doering, and J. Kelm, *Monitoring vulcanization comprises treating vulcanization mixture with ultrasound, determining propagation with high accuracy and amplitude frequently and storing and displaying them as function of time*, Patent DE 101 38 791 A1, 2001.
51. Jaunich, M. and W. Stark, Monitoring the vulcanization of rubber with ultrasound: influence of material thickness and temperature. *Polymer Testing*, 2009. **28**(8): p. 901–906.
52. Jaunich, M., W. Stark, and B. Hoster, Monitoring the vulcanization of elastomers: comparison of curemeter and ultrasonic online control. *Polymer Testing*, 2009. **28**(1): p. 84–88.
53. Hoster, B., M. Jaunich, and W. Stark, *Monitoring of the vulcanisation process by ultrasound during injection moulding*. NDT.NET/ http://www.ndt.net/article/ndtnet/2009/hoster.pdf, 2009 (9).
54. ISO6502, *Rubber – Guide to the use of curemeters*. International Standard.

Part II
Non-destructive evaluation (NDE) techniques for adhesively bonded applications

8
Non-destructive evaluation (NDE) of composites: dielectric methods for testing adhesive bonds in composites

R. A. PETHRICK, University of Strathclyde, UK

DOI: 10.1533/9780857093554.2.185

Abstract: This chapter considers the non-destructive testing (NDT) of conducting adhesively bonded structures. The application of the dielectric NDT technique is illustrated by application to metal and composite–adhesive bonded joints. This approach requires that the substrates, metal or composite, should be electrically conducting. The chapter is divided into two parts; the use of low-frequency broad-band methods for the characterisation of cure of the resin matrix and monitoring water absorption and high-frequency time domain measurements for monitoring ageing and location of defects. Dielectric NDT studies of the cure processes allow identification of the points at which gelation and vitrification of the resin matrix occur. These points are critical for the optimisation of the resin cure process in the fabrication of composite material. Water possesses a large dipole and its ingress into the resin matrix produces large changes in the dielectric properties of the matrix. Dielectric NDT can be used to study water absorption and diffusion in the matrix. High-frequency time domain measurements allow the location of defects through analysis of the wave propagation within the adhesive bond. Application of the NDT method for the location of defects is illustrated in the context of both metal and composite bonded structures.

Key words: adhesive bonded structures, dielectric non-destructive testing, water absorption, carbon fibre reinforced composites, cure.

8.1 Introduction

Adhesive bonding as an alternative to riveting or welding is a well-established technique and has been used in aircraft manufacture for well over 50 years (Bishopp, 1997, 2005). Over the last 20 years there has been increased use of polymer composites and many structures may contain both metal and composites either as separate entities or bonded together. Norman de Bruyne first demonstrated bonding in 1935, but it was not until 1942 that aluminium alloy lap joints were prepared with breaking stresses of over 2000 p.s.i. (~13.8 MPa). The adhesive used for these first demonstrations was called 'Redux', a phenolic vinyl adhesives, created by research at

Duxford. This adhesive package was used in the construction of the 'Comet' in 1949 which was the first extensively adhesive bonded airframe. Adhesives provide the ability to join dissimilar materials to give light weight, but strong and stiff structures. Whilst initially adhesives were used for metal–metal bonding they also play a very important role in bonding composite structures and creation of metal–composite structures; such as the honey comb and the new composite material GLARE, which is based on a aluminium–glass fibre multilayered laminate construction (Pethrick, 2010).

The Redux adhesive was difficult to use because it required both pressure and heat to produce the bond. Epoxy adhesives do not give off gases when they cure and are the preferred systems for many applications. The original Redux bonds, however, survived for over 40 years without significant loss of strength and demonstrated the potential for the use of polymeric adhesives to create engineering structures. Epoxy resin used for adhesives and matrix resins consist of an epoxy component based usually on the diglycidylether of Bisphenol A (DGEBA), which is either reacted with itself, with an amine or acid anhydride (Fig. 8.1).

The process of conversion of the mobile low molecular weight starting materials into a highly crosslinked high-modulus solid is referred to as the *cure* process. The common feature of all polymerisation processes is the conversion of a small molecule into a long chain. If the growing entity has more than two functionalities, a branched network structure will be formed. The monomer will usually be a free-flowing liquid which is converted into a rigid solid by the polymerisation process. This basic chemistry is common to all the resins used for the creation of the matrix in composites and structural adhesives. This cure process occurs without the evolution of volatiles and the final product should be a dense matrix structure. Such reactions may proceed either at room or elevated temperatures. The curing agents for the two-part amine reaction are commonly polyamine-amides, aliphatic polyamines (e.g. diethylenetriamine or $N,N,$-diethyl amino-propylamine). When higher temperatures are used, less reactive aromatic amine

8.1 Schematic of various types of epoxy resin polymerisation.

Dielectric methods for testing adhesive bonds in composites

hardeners (e.g. diamino diphenyl sulphone) or dicyandiamide are used to create a single component formulation which at room temperature is uncreative but becomes reactive on heating. The epoxy addition polymerisation reaction involves intermolecular reactions of the epoxy group and is driven by certain aromatic amines, BF_3-amine complexes or imidazole complexes and occurs at elevated temperatures. The catalyst is termed latent in that it only initiates the reaction once the resin has been heated to high temperature.

Most epoxy resins are naturally brittle and need to be toughened. There are two commonly used approaches to toughening epoxy resins; the addition of a rubber, usually carboxyl-terminated butadiene-acrylonitrile (CTBN) or a thermoplastic such as polyethersulphone (Bucknall and Partridge, 1983; Bucknall et al., 1994; Bucknall, 2007; MacKinnon et al., 1992, 1993, 1994, 1995; Pethrick et al., 1996). The advent of nano particle technology has created the possibility of further enhancement of physical properties using dispersions of nano-silica (Hsieh et al., 2011).

In the context of non-destructive testing (NDT), a good adhesive or composite matrix requires that the polymerisation processes summarised in Fig. 8.1 are effectively carried out and that the majority of the reactive functionalities are converted into a network polymer structure. Cure of the resin is usually carried out in a vacuum bag or in an autoclave which makes inspection of the process difficult. The skilled operative will know intuitively when this process has been completed and will not over-cook the resin. If the resin is subjected to a high temperature for longer than is necessary, the network which is initially formed can become degraded and the network structure which gives it strength will be destroyed. In industry, the process monitoring the cure of an adhesive usually involves the observation of a change in colour of the resin from a honey yellow to a light pink. If the colour is allowed to become blood red this implies that the bond has been over-cured and will lose strength. In the case of composites, cure is not normally allowed to proceed to completion and the final product should retain a honey yellow colour. Excessive darkening of the surface of a composite may indicate that the heat generated during the cure process is beginning to degrade the polymer which has been formed (Maxwell et al., 1981; Maxwell and Pethrick, 1983). In production, it is desirable to be able to remotely monitor the cure process and dielectric monitoring has been shown to have a role in this context.

Coupon monitoring is often used in manufacturing and allows observation of the progress of the chemical processes through the use of techniques such as infrared and Raman spectroscopy and differential scanning calorimetry; however, these methods are not ideal for the monitoring of large or complex structures. The thicker the structure being formed the higher the probability that the article may exotherm and degradation may follow.

Dielectric NDT has a role in both monitoring the cure of the resin and in the case of adhesive bonding checking the integrity of the joint formed. Unfortunately, adhesive joints and composite structures can exhibit ageing characteristics when subjected to hostile environments. Water is a very aggressive medium, inducing corrosion of metals, modifying the physical properties of the resin and causing delamination in composites. Dielectric NDT provides the possibility of assessing whether a joint structure is undergoing ageing and the extent to which the processes have progressed. In principle, the method allows assessment of the likely hood that the joint will fail and hence provides a useful health monitor for old adhesive bonded structures.

The role of dielectric NDT can be considered under two headings:

- Assessment of the state of cure of the adhesive joint and check whether a good joint has been formed – dielectric NDT in manufacture.
- Assessment as to whether a joint has undergone significant ageing and to what degree it may have lost strength – dielectric NDT for structural health monitoring.

These roles are discussed in the following sections. Under the first heading, methods divide into those used for cure monitoring and those used to check the integrity of the joint or composite structure.

8.2 The use of dielectric testing in cure monitoring

The basic underlying principle behind dielectric cure monitoring is that as the chemical reaction (Fig. 8.1) proceeds, the molecular weight of the entities present increase and their ability to move will decrease. Measurement of the frequency dependence of the dielectric permittivity – $\varepsilon'(\omega)$ and loss $\varepsilon''(\omega)$ show a series of features which are characteristic of the mobility of the molecules forming the resin (Fig. 8.2).

At low frequency, a large peak is observed which is associated with the motion of charged species through the uncured resin and is indicative of the local viscosity of the resin. Initially at frequencies above those used in the measurements, individual epoxy resin dipoles undergo relatively free rotation and lead to the observation of a apparent plateau in the dielectric permittivity $\varepsilon'(\omega)$ and a loss feature which is essentially lost in the more dominant conduction process. As cure proceeds, which is indicated as a movement of the traces towards the front of the graph (Fig. 8.2), the large peak decreases and essentially disappears into the plateau of the constant dielectric permittivity $\varepsilon'(\omega)$ (Fig. 8.2a). The drop in the peak reflects the increase in the viscosity of the resin as cure proceeds. There is a corresponding drop in $\varepsilon''(\omega)$, the characteristic $1/\omega$ slope of the conduction developing into a well-defined peak associated with dipole relaxation. This latter feature

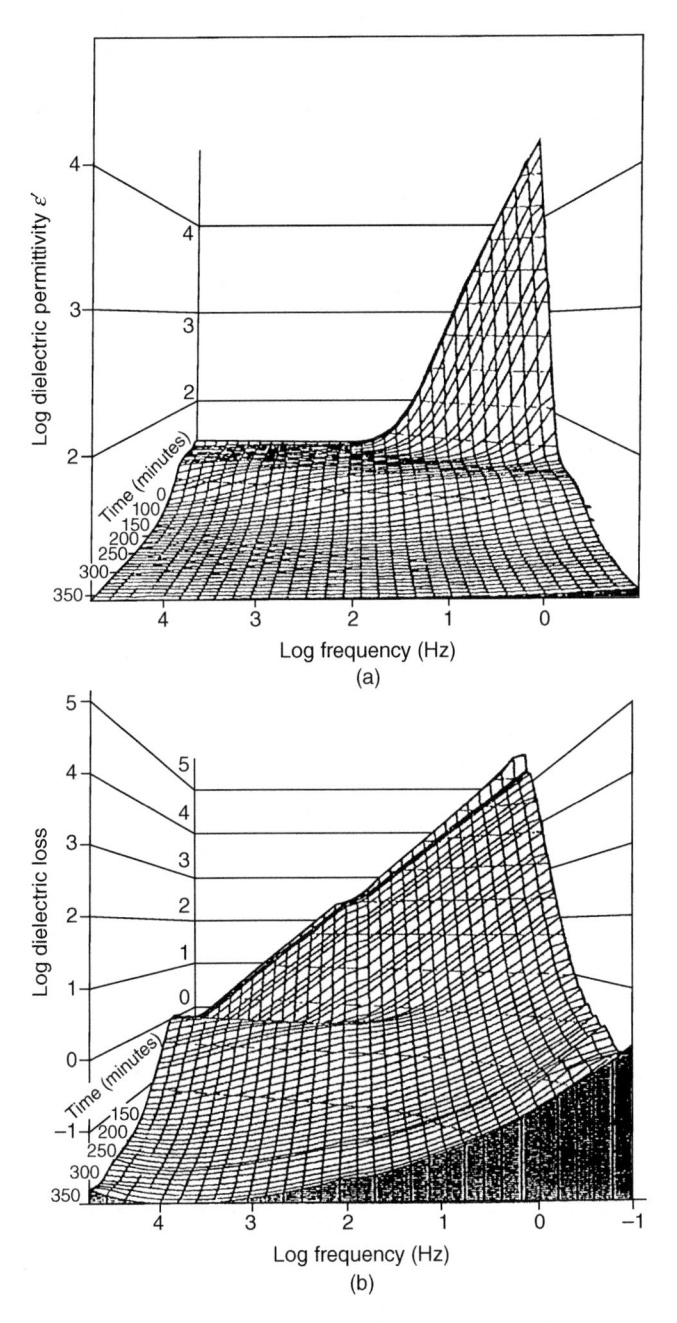

8.2 Plots of the dielectric permittivity ε' (a) and dielectric loss ε'' (b) for a typical epoxy cure system. Note that the frequency decreases from left to right and the time from front to back.

moves to lower frequency and eventually disappears, as indicated by the contour lines (Fig. 8.2b). As the molecular weight of the resin increases, so the rate at which dipole rotation occurs is slowed down and eventually when the material passes through its glass to rubber transition (T_g) dipole rotation ceases to occur (Pethrick et al., 2010).

The slope of the variation of $\varepsilon''(\omega)$ with frequency of the conductivity is a measure of the motion of ionic species through the matrix and contributes to the dielectric loss through a conductivity term. Most resins will contain a low level of ionic impurities which are left behind from the synthesis method; for instance, when epoxy resins are produced by reaction with epichlorohydrine and chlorine is liberated during the synthesis and usually removed by precipitation. However, unless specifically removed, low levels of chlorine ions are usually left in the resin. The mobility of these ions is a direct reflection of the microscopic viscosity and is reduced as the resin cures. As a consequence of the increasing viscosity there is a large decease in the $\varepsilon''(\omega)$ as cure proceeds. The increase in viscosity of the resin is a sensitive probe of the progress of the cure reactions and allows the dielectric method to be used to remotely follow the cure reaction when adhesive joints or composite structures are being fabricated. In the former, the metal or carbon fibre substrates can act as electrodes and give direct access to the resin being cured.

The frequency dependence of the dielectric loss $\varepsilon''(\omega)$ can be described in terms of the following equation:

$$\varepsilon''(\omega) = \frac{(\varepsilon_0 - \varepsilon_\infty)\omega\tau}{(1+\omega^2\tau^2)} + \frac{\sigma}{\omega} \qquad [8.1]$$

where ε_0 and ε_∞ are respectively the low- and high-frequency limiting values of the dielectric permittivity and are directly related to the strength of the dipole relaxation, ω is the frequency of measurement (radial) and τ is the relaxation time of the dipole motion. The first term in equation (8.1) describes the frequency dependence of a simple dipole relaxation process, whilst the second term is the contribution to the loss arsing from mobile ion conduction and is reflected in the conductivity σ. Using the above equation, the time dependence of the conductivity and the movement of the dipole processes as a function of the extent of cure can be calculated (Fig. 8.3).

The decrease in the conductivity as cure proceeds is shown in Fig. 8.3a. As the temperature used in the cure process is increased so that rate increases at which the conductivity decreases increases. As the cure time increases so the dipole relaxation time will slow down and move through the frequency window (Fig. 8.3b). The initial state of the material is soft and rubbery and this corresponds to the relaxation frequency lying above 10^6 Hz. As cure proceeds, so the dipole relaxation process slows and eventually the

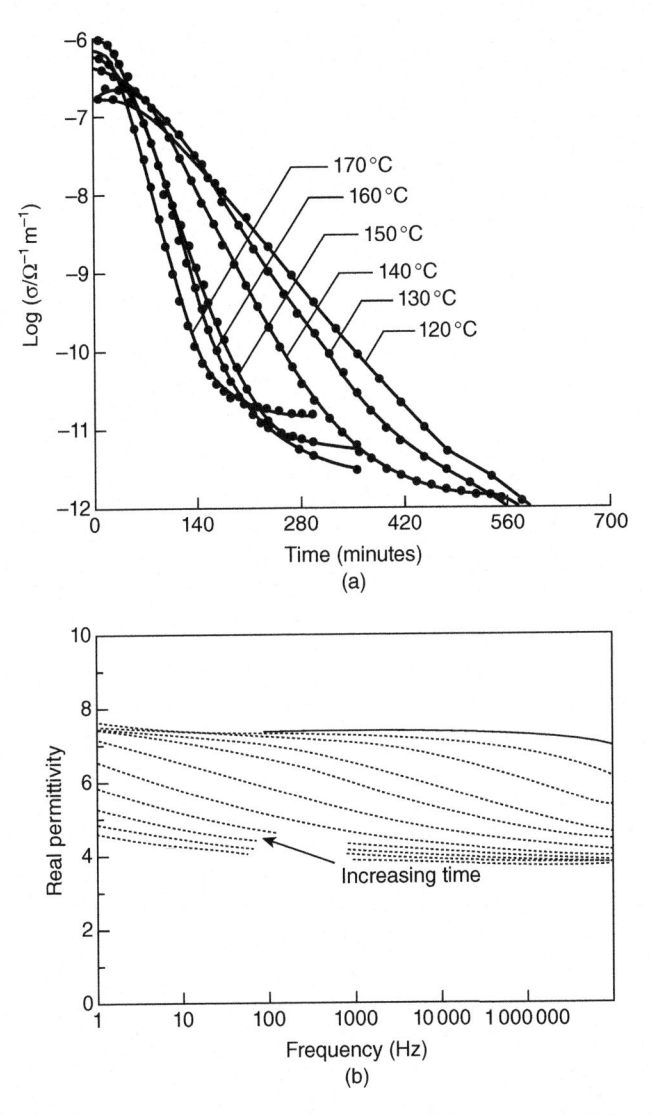

8.3 Variation of the conductivity (a) and the location of the dipole relaxation (b) with cure time.

relaxation frequency will be well below 1 Hz. The dipole relaxation process has dropped below the observation window and the material is now a glass. Vitrification marks the point at which the material is capable of taking a mechanical load. The point at which the conductivity has dropped to a low level corresponds with the point at which gelation has occurred in the system. Gelation and vitrification are the two critical points which allow optimisation of the cure process. In order to understand how temperature

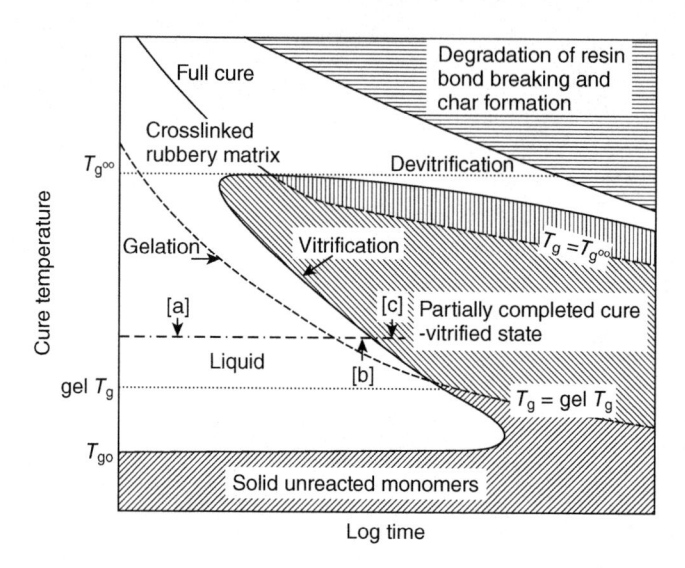

8.4 Time temperature transformation diagram for a typical cure system.

influences the rate at which gelation and vitrification occurs it is useful to use time temperature transformation plots (Fig. 8.4).

Figure 8.4 summarises the critical phase changes which occur during a cure process. If we consider the resin being cured isothermally (the hashed horizontal line in Fig. 8.4), it starts as a low-viscosity liquid. As the reaction proceeds, the molecular weight of the polymer molecules produced will slowly increase and there will be a small increase in the viscosity designated by the almost horizontal line marked [a]. In terms of the plot in Fig. 8.4, the first significant point which is observed is the formation of a three-dimensional loosely crosslinked network; the *gelation* point. The gelation point is designated [b] on the line. This is a critical point in cure as beyond this point the resin cannot be made to change shape and the structure which it has formed will be retained throughout the subsequent cure process. The material is at this point essentially a rubber. As further reaction occurs the resin will become more crosslinked and will undergo shrinkage to a greater or lesser extent. The polymer chains in the resin, however, are still relatively mobile and can undergo rotation. It is this motion which is detected as the dipole relaxation. As further crosslinking takes place, the dipoles will become increasing restricted in their motion and the rate at which they can rotate will be slowed down. Ultimately, crosslinking will be sufficiently extensive to freeze the motion of the dipoles and the material has been transformed from a rubber to a glass. This occurs at point [c], the vitrification point. The material may not be completely reacted at this point and it may require the system to be heated to a higher temperature to complete the reaction and develop the full strength required for a particular application. The value of the glass transition temperature – T_g, reflects the quanity of uncrosslinked chains in the material and will have its maximum value T_g^∞, only when all the reactive functions have been consumed and the matrix is completely crosslinked.

Dielectric monitoring can follow the processes occurring inside an auto clave or in a vacuum bagged system where direct observation is difficult. The equipment used often is operated at a single frequency or over a limited frequency window. Measurements of the conductivity and capacitance map well onto the changes which are presented in Fig. 8.4. The conductivity can be directly related to the change in viscosity and hence identification of the *gelation*, whereas the dipole relaxation indicates chain mobility and allows determination of the *vitrification*. The changes in the conductivity (Fig. 8.3a), indicate the steady increase in the viscosity as cure proceeds. The measured conductivity contains a small contribution from the presence of the dipole relaxation; however, this contribution is initially several orders of magnitude smaller than the contribution from ionic mobility. The changes shown in Fig. 8.3 correspond to the variation of the slope of the low-frequency process with time (equation (8.1)). In order to see the ion mobility, it is necessary to take measurements at low frequency and typically measurements are made at 10 Hz or below. The change in the dielectric permittivity is directly related to the capacitance measurements and this falls initially as the ion conduction disappears and then falls a second time as the dipole motion is frozen out of the system. It is possible using relatively simple automated measuring system to identify the critical points in the curves and both the gelation and vitrification points. More detailed discussions of the application of the dielectric method and the interpretation of the data are presented elsewhere (Garden *et al.*, 2007; Kazilas and Partridge, 2004; Maistros and Partridge, 1998; Pethrick 1998; Skordos and Partridge, 2004).

8.3 The use of dielectric testing to check bond integrity

The dielectric NDT approach described in this chapter focuses on configurations where the bond forms a dielectric layer between two conducting surfaces. The bond may be formed between two conducting layers, metal to metal, metal to carbon fibre composite or between two carbon fibre composite layers. The essential feature in these structures is that the substrate is assumed to have a high electrical conductivity. Metals naturally have a high electrical conductivity; however, the typical cured carbon fibre reinforced plastic (CFRP) is sufficiently conductive to allow the internal bonded layer to be examined by the same method (Boinard *et al.*, 2002).

Whilst being able to monitor the cure of an adhesive bond is also desirable it is also important to confirm that a good bond has been formed. If we are bonding two conducting substrates together, then the insulating resin layer creates the ideal capacitor or alternatively a wave/guide. Using broad band dielectric measurements (10^{-3} to 10^{10} Hz) it is possible to not only determine whether a good bond has been formed but also to explore whether the bond is fully cured and, if aged, whether significant corrosion in the case of a metal or disbonding in the case of CFRP, has occurred at the adhesive substrate interface.

8.3.1 Visualisation of joint structure

For the structural visualisation of the joint it is necessary to perform measurements at frequency greater than 1 MHz and typically measurements may require to be carried out at up to 20 GHz or higher. The approach adopted is to launch an electrical pulse created from a network analyser into the waveguide and observe either or both the transmitted or reflected signals. A schematic of the measurement system is shown in Fig. 8.5.

The measurements are performed using a commercial network analyser which allows exploration of the reflection coefficient in the time domain and also measurement of the impedance as a function of frequency, usually displayed in the form of a Smith's chart. The detailed of the apparatus used are presented elsewhere (Crane *et al.*, 2006; Li *et al.*, 1997). The pulse is generated by the network analyser and linked to a probe via a semi-flexible

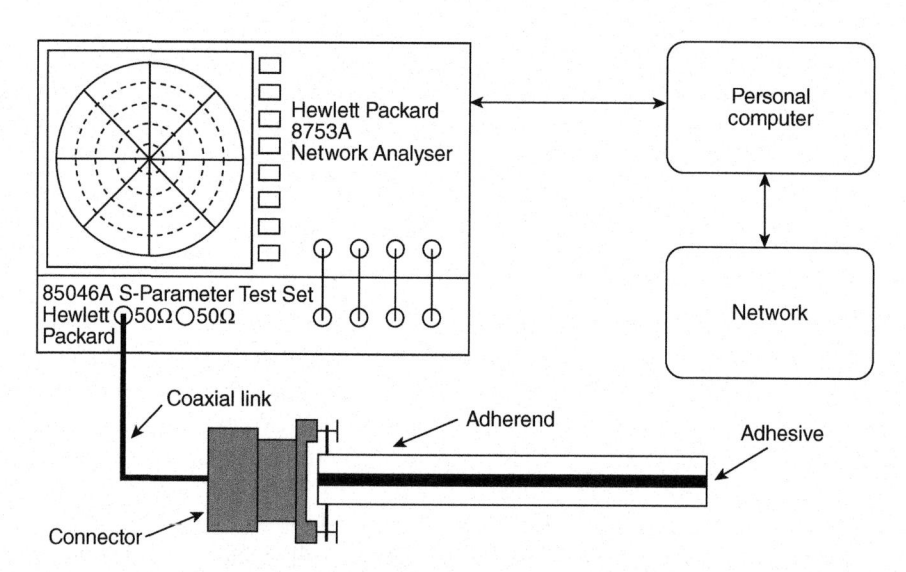

8.5 Schematic of the high-frequency dielectric measurement system.

50Ω coaxial cable. The probe contains two point contacts which are spring loaded to achieve reproducible contact with the joint. Different designs have been developed to accommodate different joint structures. One of the contacts carries the signal and the other is earthed. This connector to the joint represents a mismatch and will reflect the pulse and this form the initial marker for the measurements. The pulse created by the equipment depends on the network analyser used and for an instrument capable of operating to 20 GHz, the pulse width is a few pico-seconds in width. The network analyser is initially calibrated by connection short circuits, open circuits and a 50Ω connector to the line. Examples of the type of signals which are observed are presented in Fig. 8.6.

For an ideal joint (Fig. 8.6a), the first negative reflection arises from the connection between the measuring probe and the front face of the joint. The first positive peak as associated with the signal which has been reflected from the end of the joint. The transit time is directly proportional to the length of the joint and the velocity of the pulse is dictated by the dielectric permittivity of the resin making the bond. The third, fourth and, fifth peaks are associated with pulses which have made multiple transits of the bond and their amplitude decreases because of the loss in the bond and the impedance mismatch between the probe and the bond. The low-level noise between the peaks is a consequence of the interaction of the pulse with the probe.

Adhesive bonding is used extensively in aircraft manufacture for bonding stringers to the fuselage and combining composite subsection in wings, etc. Whether they are top hat or T stringers depends on where they are located on the fuselage surface; however, both are likely to be shaped as that the adhesive cross-section changes along the length of the joint. If the cross-section changes the pulse will see a change in electrical impedance and be partly reflected and partly transmitted. The trace obtained for a joint in which there is a single step is shown in Fig. 8.6b. The location of the change in impedance is marked by the second negative peak. The solid black line represents the trace obtained when the signal travels from left to right and the dotted trace when it is propagating from right to left. To further illustrate the effect of change in cross-section a joint with three steps is presented in Fig. 8.6c. The amplitude of the peaks increases which the reduction in the cross-section, which reflects increased electrical impedance at this point in the joint.

The location of the peak provides information on the distance along the joint and the amplitude the cross-section at that point in the line. Fig. 8.6d shows a joint in which there is a neck located half way down the joint. As the pulse propagates down the joint it first experiences a high impedance step and then a step down to the original low impedance. The beginning of the step is marked by the negatively going peak and the end of the step by

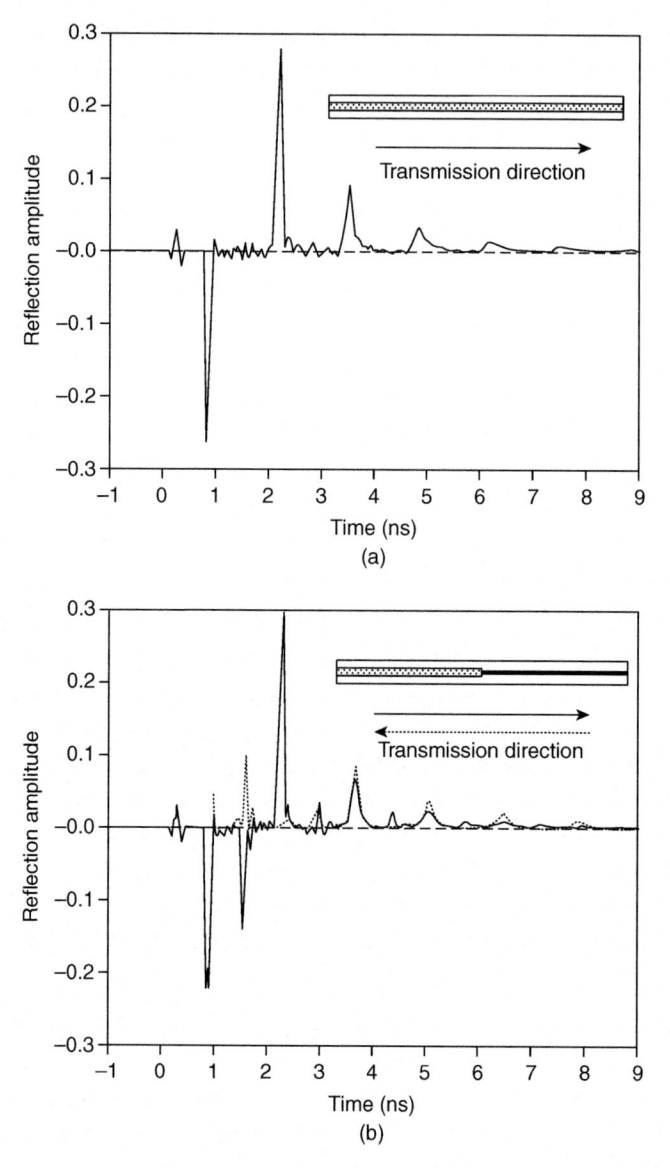

196 Non-destructive evaluation (NDE) of polymer matrix composites

8.6 Reflection traces for time domain reflectivity measurements on various metal joint structures: (a) ideal joint, (b) joint with single cross-section change, (c) joint with three steps in cross-section, (d) joint with cross-section reduction at mid-point.

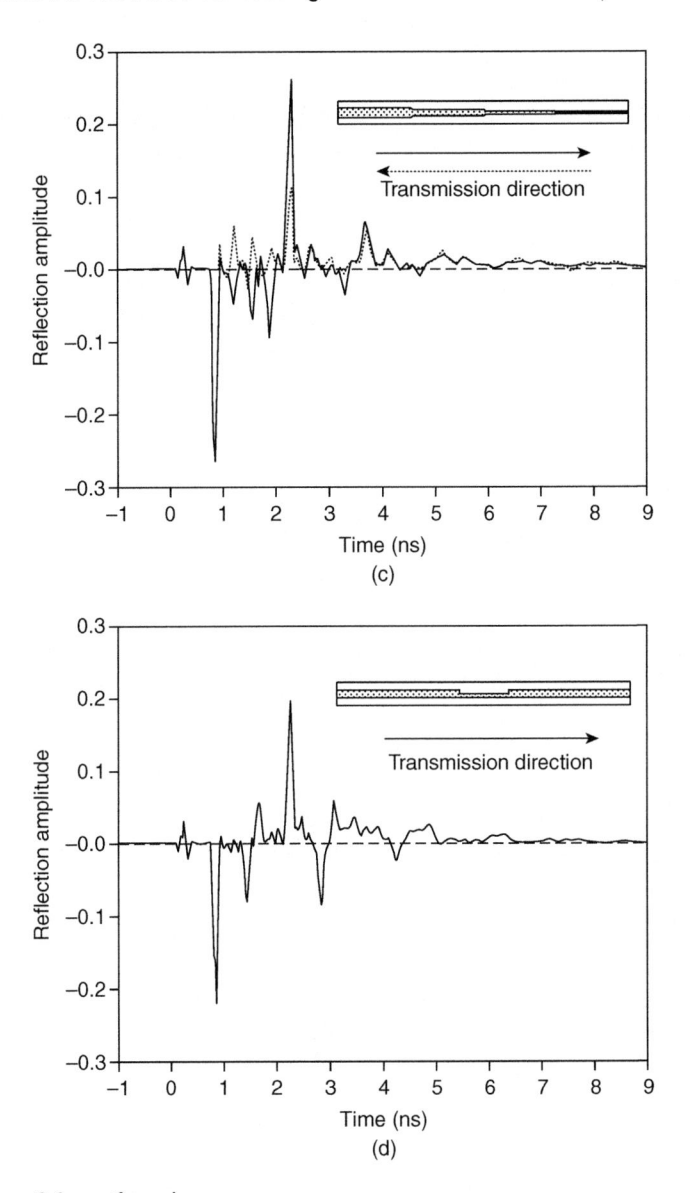

8.6 continued

a positive going peak, the total length of the joint is marked by the large positive going peak. The subsequent peak pattern reflects the effects of the various reflections interacting in a constructive and destructive manner. Using the peaks in the reflection amplitude it is possible to identify the distance between the initial peak and the various features identified. The typical error in the distances measured from the time traces is less than 1.5% when operating at 20 GHz. This study was carried out with metal substrates but the same effects are observed with CFRP bonds, except that the losses associated with the low conductivity of the substrate broaden the peaks and attenuate the higher-order reflections.

From an analysis of the Smith's chart (Fig. 8.7a), it is possible to map the impedance variation as a function of frequency. This data can be used to compute the high-frequency variation of the dielectric permittivity and dielectric loss. Above a frequency of ~100 MHz, the effects of the physical structure of the wave guide joint will influence the impedance data and a series of peaks are observed associated with resonances which correspond to the reflections observed in the time domain. Certain of the observed peaks arise from the phase shifts and reflections created in the connector between the 50 Ω line and the joint being examined. By combining the time and frequency domain data it is possible to solve the so-called *inverse problem*. Taking the time domain data and a mathematical model, the distribution of impedance changes can be used to create a visualisation of the impedance distance profile and hence the internal structure of the bond. The mathematical analysis used is summarised elsewhere (MacKay *et al.*, 2005, 2007). For an accurate representation to be produced it is necessary

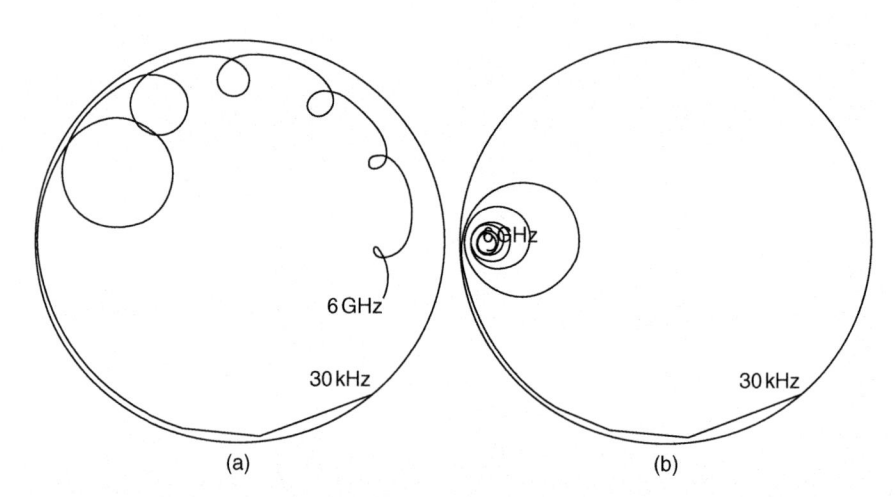

8.7 Typical Smith's chart for adhesive bonded structure (a), and revised plot when correction has been made for the connector using inverse modelling theory (b).

to include in the model the frequency dependence of the dielectric permittivity and dielectric loss. This information is directly obtained from the frequency-dependent observations.

The principal problem which emerges when comparing the simulation with the real dimensions of the structure is the effect of the impedance changes which occur in the connector. This problem can be solved by adding an adjustable phase shift to the impedance data, so as to constrain the final point on the impedance curve to lie on the horizontal, i.e. to constrain the wave to end up with a phase shift which is a multiple of 90° (Fig. 8.7b). Using the software generated by this approach, it is possible to map the impedance changes down the joint and abstract the high-frequency dielectric properties of the resin. The adjustable parameters in the model are the dielectric permittivity, dielectric loss, bond line cross-section and the conductivity of the interface between the adhesive and adherent. In most media, the dielectric permittivity will vary with frequency and is simply the model using the form:

$$\varepsilon'(\omega) = \frac{(\varepsilon_0 - \varepsilon_\infty)}{(1 + \omega^2 \tau^2)} + \varepsilon_\infty \qquad [8.2]$$

where $\varepsilon'(\omega)$ is the permittivity at the effective frequency ω of measurement, ε_0 and ε_∞ are the low- and high-frequency limiting values of the dielectric permittivity and τ is the characteristic frequency for the dipole relaxation process. The data on the magnitude of the parameter can be obtained from the analysis of the frequency dependence of the impedance.

The cross-section of the bond line in the joint may vary down its length and this variation is calculated by integrating the signal response as a function of the phase shift. The other adjustable parameter is the conductivity of the interface, which is influenced by the extent to which the propagating signal penetrates the conducting substrate and adds a phase shift to the propagating signal. For conducting media the penetration will be relatively small and the surface conductivity is dominated by the interface.

In a joint the appropriate mode of propagation is the transverse electro magnetic (TEM) guided wave and the voltage reflection coefficient at the input end of a joint is given by:

$$\rho = \frac{Z_{in} - Z_{co}}{Z_{in} + Z_{co}} \qquad [8.3]$$

where Z_{in} and Z_{co} are respectively the input and characteristic impedance of the joint. The input impedance can also be expressed in terms of the characteristics of the line:

$$Z_{in} = Z_o \cdot \frac{Z_L + Z_o \cdot \tanh(\gamma.l)}{Z_o + Z_L \cdot \tanh(\gamma.l)} \qquad [8.4]$$

where Z_L, Z_o, γ and l are respectively the termination and characteristic impedance of the joint, the propagation factor and the length of the joint. The parameters Z_o and γ can be defined into terms of the electrical characteristics of the equivalent circuit:

$$Z_0 = \sqrt{\frac{R+j\omega L}{G+j\omega C}} \quad \text{and} \quad \gamma = \sqrt{(R+j\omega L)(G+j\omega L)} \qquad [8.5]$$

where R, L, G and C are the resistance, inductance, conductivity and capacitance per unit length, respectively, ω is the angular frequency and $j^2 = -1$. The expressions for the above parameters can be derived for electromagnetic field theory and have the following form:

$$R = \frac{2}{b}\sqrt{\frac{\omega\mu_c}{2\sigma_c}} \quad L = L_i + L_e \quad G = \sigma_d \cdot \frac{b}{a}$$

$$C = \varepsilon_0 \cdot \varepsilon^* \cdot \frac{b}{a} \quad L_i = \frac{2}{\omega b}\sqrt{\frac{\omega\mu_c}{2\sigma_c}} \quad L_e = \mu_0 \cdot \mu_d \cdot \frac{a}{b} \qquad [8.6]$$

where σ_c and μ_c are the conductivity and permittivity of the conductor, L_i and L_e are the conductor internal and external inductance per unit length, a, b, σ_d, μ_d and ε^* are thickness width, conductivity, permeability and complex permittivity of the dielectric, respectively, and μ_0 and ε_0 are the permeability and permittivity of free space. Although there are a number of parameters in these equations, the main variables are a and b, the height and width of the joint. Using these equations it is possible to simulate the reflectively distance profiles and also the frequency dependence of the reflection coefficient (MacKay et al., 2005). Iterative solution using both sets of data yields realistic values for the joint cross-section and the frequency dependence of the dielectric properties at high frequency. A very important parameter which emerges from the analysis is the conductivity of the interface. Comparison of experiment and theory using a frequency of 3 GHz in the excitation pulse is shown in Fig. 8.8.

At the lower frequency the peaks are broader as a consequence of the greater losses in the material but the shape and the location of the peaks have a good fit with the theory. A more detailed discussion of the approach used and the method of calculation is presented elsewhere (Crane et al., 2006; MacKay et al., 2005, 2007). The model based on the above approach has been successfully used to study pristine metal and composite CFRP joints and also explore the changes which occur to joints when they are aged.

8.4 The use of dielectric testing to assess ageing of bonded joints

In many aerospace applications the bonds which are created are expected to last the life of the aircraft. In the initial design, 30 years was assumed to

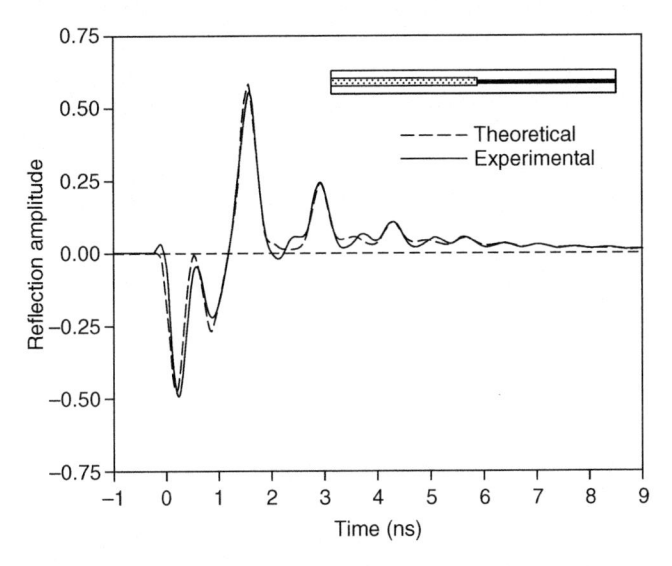

8.8 Comparison of theoretical simulation and experiment for reflection coefficient measurements made at 3 GHz.

be a reasonable life expectancy for a civil aircraft; however, there are a significant number of Boeing 747s and similar aircraft that are still flying but were built well over 35 years ago. Many of the current civil aircraft contain significant amounts of composite structures in wings, rudders and ailerons as well as in flooring, bulkheads, etc. The composites tend to be either glass reinforced plastic (GRP) as in GLARE, which is a glass reinforced aluminium layered composite structure, novolac cured honeycomb bonded onto aluminium and carbon fibre reinforced box structures in wing sections. It is therefore appropriate to consider the ageing of metal composite bonded structures as well as pure composite bonded structures.

8.4.1 Ageing of metal–metal joints: the case of aluminium

It is desirable to be able to check whether bonds have or about to lose a significant amount of strength. In this context, dielectric NDT has the potential of playing an important role. Adhesive bonds are created by carefully etching and anodising the aluminium prior to bonding. The oxide layer which is created will be typically about 200 nm think and have a high degree of pore structure which enhances the lock and key contribution to the interfacial strength. The surface will usually be primed using a very low-viscosity epoxy resin which fills the pores and creates the necessary bridge to the film adhesive. The cured bond structure will be able to achieve significant load-bearing characteristics and has allowed both significant weight saving to be incorporated into aircraft design and the creation of a profile which could not be created by other means. An aircraft in its service life will be subjected to a series of stresses and also to different environments. In the case of composites, water ingress can detect kissing bonds and lead to dewetting and ultimately delamination. The principal reason for loss in bond strength is the ingress of moisture into the bond line. The changes observed with metal and adhesive bonds are very similar; the dielectric data, however, for metal bonds is clearer because the effects of attenuation in the CFRP which diminishes the signal strength are not present.

Testing a good adhesive bond

A study was undertaken of the changes which occur to the dielectric characteristics when they are exposed to a warm humid environment (Li *et al.*, 1997). To accelerate the ageing process these test were carried out at 70 °C and therefore reflect changes which will at ambient temperature occur over many years. The changes in the dielectric characteristics for a well-bonded structure are presented in Fig. 8.9. For a well-cured joint there is very little change in the dielectric permittivity over the frequency range from 1 MHz to 1 GHz. The dipole relaxation in this system has been frozen out and

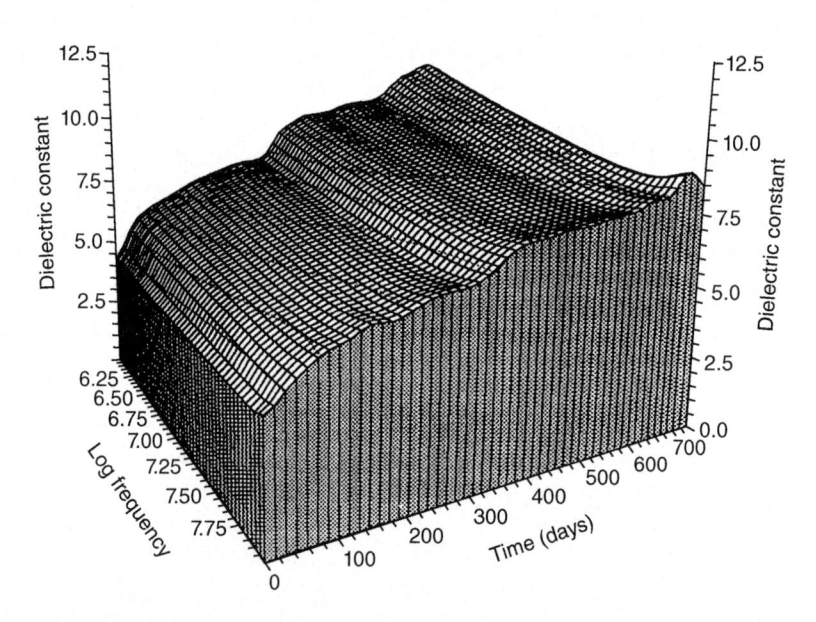

8.9 Variation with time and frequency of the dielectric permittivity as a function of time of exposure to water at 70 °C for a good aluminium–epoxy–aluminium joint structure.

pendant dipole motions occur at lower frequencies than those used in this study. As the ageing progresses there is initially a fairly rapid increase in the permittivity which parallels the uptake of moisture into the resin system. Water has a dielectric permittivity of ~80 and hence it is possible easily to observe the uptake moisture with a resolution of better than 0.1%. The water entering the resin and occupies any free space that may exist. Water can also bind itself to the polymer molecules via the pendant hydroxyl groups which are created during the cure process (Fig. 8.1). The water is a very good plasticiser for epoxy resins and after 300 days will have lower the glass–rubber transition from a value of ~120 °C in the cure adhesive to ~70 °C. When the resin is cured there will be stresses left in the matrix when it becomes plasticised by the ingress of moisture, which will allow it to relax, and there are consequent dimensional changes in the thickness and width of the adhesive. The shrinkage of the adhesive appears as an apparent increase in the permittivity, if the dimensions in the calculation are held constant at their initial values.

Subsequent ageing leads to small increases in the permittivity values which are a reflection of the slow and continuous increase in the amount of water which is taken into the bond line. However, it will at this point be reaching a plateau value. The time domain reflection (TDR) coefficient plots obtained for this joint structure are shown in Fig. 8.10.

The pulse pattern is similar to that shown in Fig. 8.6 except that the measurements were performed at 3 GHz rather than at 20 GHz and hence the peaks are broader and the resolution is slightly lower. The initial trace at the bottom of Fig. 8.10 shows the expected pulse train, deceasing in amplitude and reflecting a good joint structure. As water enters the joint so the dielectric loss increases and the amplitude of the higher order peaks decreases quite rapidly. The location of the first peak moves along the time axis consistent with the increase in dielectric permittivity as the water is taken into the resin matrix. Whilst the amplitude of the peak is slightly decreased the general pattern of the noise between the initial reflection from the connector and the first reflected pulse stays essentially constant, indicating that the joint structure has not changed significantly. Subsequent studies have shown that for a *good* bond it is possible to recover ~80% of the original strength on subsequent dehydration of the adhesive layer.

Testing a joints with poor interfaces

A study was carried out of a joint in which the oxide growth was not carefully controlled (Fig. 8.11) (Li et al., 1997). The initial traces observed over the frequency range 10 MHz to 1 GHz looks initially the same as for the good joint, the dielectric permittivity increasing regularly with time reflecting the ingress of moisture into the bond. There is a change in the

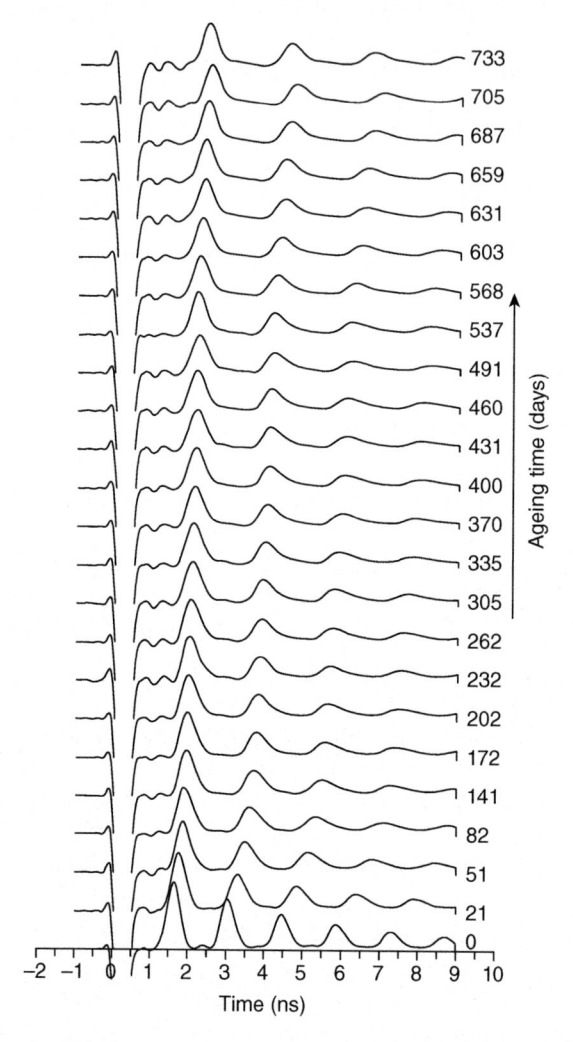

8.10 Variation of the TDR traces as a function of time of exposure to water at 70 °C for a good aluminium–epoxy–aluminium joint structure.

permittivity after ~300 days which reflects the stress relaxation of the resin. However, this dimensional change in the joint is marked by a significant increase in the permittivity and the development of marked frequency dependence. The frequency dependence in the permittivity is a result of water interacting with the surface oxide creating hydroxyl groups which will create a dipole feature around 1 MHz (Li *et al.*, 1997). The observed undulations in the dielectric permittivity are a consequence of conversion of oxide to hydroxide – corrosion of the interface region. The permittivity values after ~450 days show a decrease which has to be interpreted as being

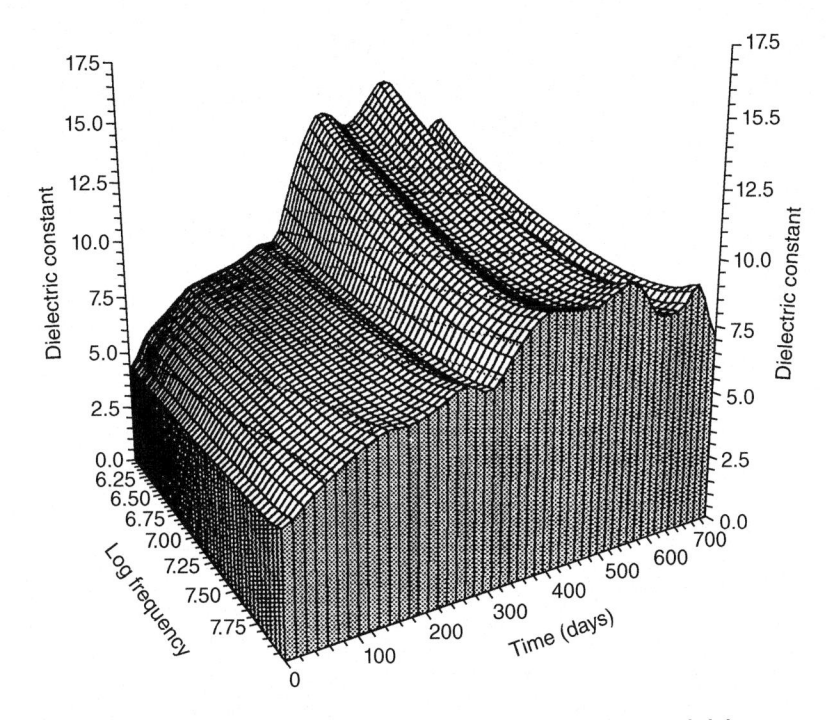

8.11 Variation with time and frequency of the dielectric permittivity as a function of time of exposure to water at 70 °C for a poorly bonded aluminium–epoxy–aluminium joint structure.

associated with a drop in the dielectric characteristics as a consequence of air being introduced into the joint structure!

The TDR traces for this join were also measured and are shown in Fig. 8.12. The initial traces look very similar to those of the good joint. However, inspection of the traces after 100 days indicates that the falling edge of the reflection peak is broadening and masking the modulations which were initially observed after this peak. After 305 days, a second peak is observed to be building into the trace at about 2.5 ns. The traces are ceasing to be horizontal, indicating that there is significant dispersion of the signal before it is reflected. At about 500 days the reflection peak appears to move to the right, which is consistent with a drop in the permittivity and this is associated with air being introduced into the joint. The pulse pattern after the first reflection has become highly irregular as a consequence of the joint structure becoming physically disrupted. At 700 days, the joint is rapidly losing its strength and the bond line is becoming visibly disrupted. The TDR traces are therefore giving a visual indication that regions of disbonding are being created within the joint. Whereas the good bond was apparently still in good condition, the poor bond had visibly deteriorated.

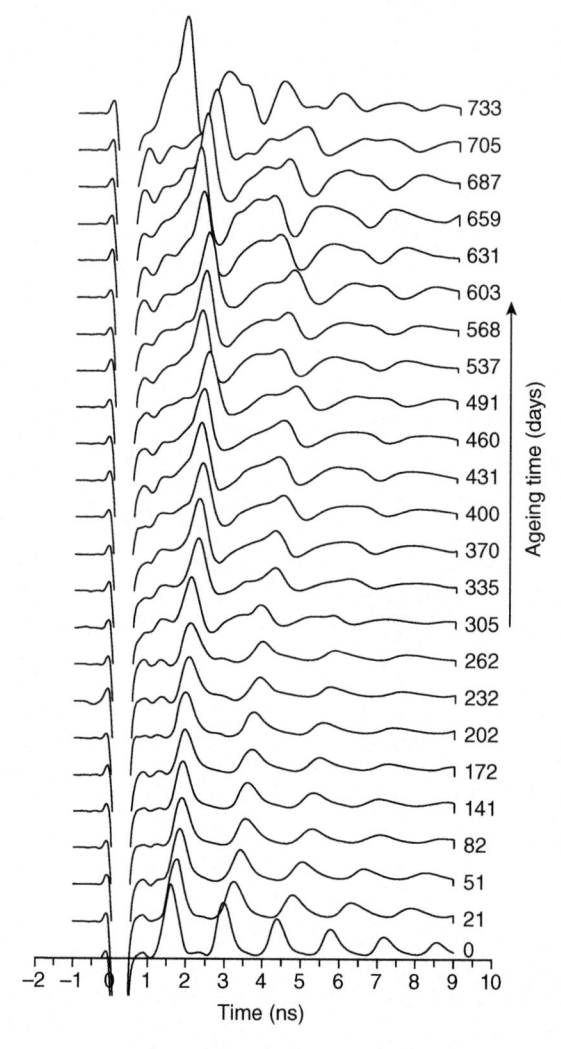

8.12 Variation of the TDR traces as a function of time of exposure to water at 70 °C for a good aluminium–epoxy–aluminium joint structure.

Using the modelling software it is possible to determine the way in which the surface resistivity, bond line thickness, and dielectric permittivity vary as a function of ageing time (Fig. 8.13) (Armstrong *et al.*, 2004). For the first ~350 days there is little change in the thickness of the joint; however, at about 400 days there is a significant increase in the thickness associated with the now plasticised resin undergoing stress relaxation. The permittivity increase at 3 MHz has increased by ~1, which corresponds to a water increase of ~1.25%. Some of the water will be relaxing at lower frequency and the total water absorbed will probably be ~3–4%. This amount of water

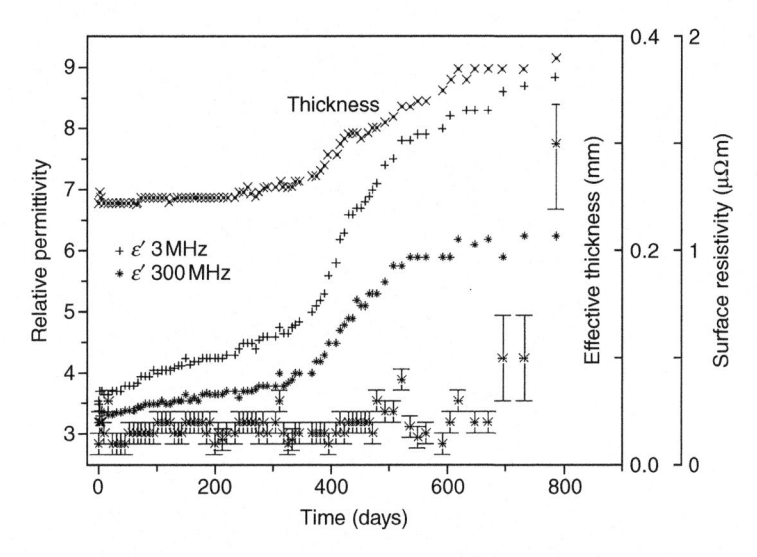

8.13 Variation with ageing time of the surface resistivity, bond line thickness and dielectric permittivity vary as a function of ageing for an aluminium–epoxy–aluminium joint aged at 70 °C.

is sufficient to lower the T_g of the matrix from ~110 °C to about 70 °C, allowing the stress relaxation to occur. The rise in the permittivity at 300 MHz is indicative of water being located in voids and relaxing at a frequency of 12 GHz, above the frequency of observation. The data indicates that the stress relaxation allows voids to be created in the resin and these are being filled with water. After 600 days the 300 MHz permittivity appears to plateau, indicating that the growth of water filled voids is complete. The increase in the permittivity at 3 MHz is consistent with the oxide growth and its associated relaxation at 3 MHz. As would be expected there is a corresponding change in the surface resistivity as the oxide is converted into hydroxide and this occurs after ~600 days' exposure. The dielectric NDT of these joints allows a detailed picture to be created of the various stages of the ageing process and this correlates well with the changes in the measured mechanical strength (Armstrong *et al.*, 2004). The mechanical strength of a badly deteriorated joint can have dropped to below 20% of its original strength after exposure and the strength is not recovered on dehydration. The extent to which the strength is lost depends critically on the stability of the interface and the nature of the changes which are induced on hydration.

8.4.2 Ageing of metal–composite joints

Use of composites in aircraft construction introduces the problem of bonding materials with dissimilar thermal expansion coefficient together.

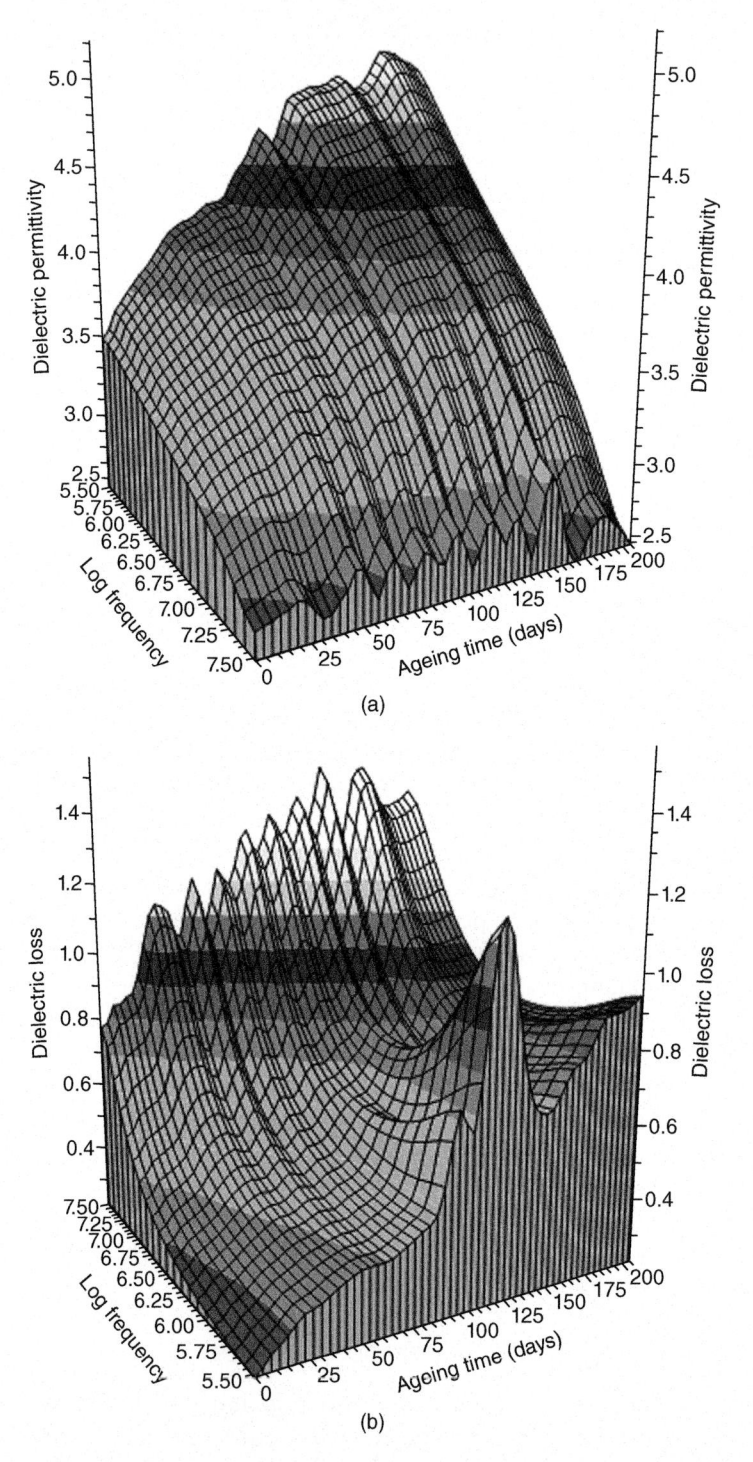

8.14 Dielectric permittivity ε' (a) and dielectric loss ε'' (b) for a CFRP–epoxy–aluminium joint aged at 70 °C as a function of ageing time.

Typically this requires the use of special designed fixtures which allow relative movement of the components. However with careful design and the use of the correct adhesives it is possible to achieve good bonds between CFRP composite panels and aluminium sheets (Halliday et al., 1999). The dielectric permittivity ε' and dielectric loss ε'' (Fig. 8.14) exhibit very similar characteristics to those observed with aluminium–aluminium joints.

The dielectric permittivity ε' shows the initial steady increase over the first 50 days, followed by the step increase as the adhesive stress relaxes and then modulation as voids are created. The dielectric loss ε'' exhibits the expected increase at low frequency associated with the water entering the matrix but the peak at high frequency which is observed at around 125 days is the water filling the voids in the resin. The water is plasticising the resin and after 160 days the voids collapse and the high-frequency peak is decreased. The adhesive used to construct these bonds is a lower-temperature adhesive and less moisture tolerant that that used with the aluminium–epoxy–aluminium joints. The adhesive bond is retaining its integrity as indicated by the time domain traces (Fig. 8.15). The TDR traces are limited to the first period and subsequent peaks are so attenuated as not to be observed. However, it is very clear that there are not significant changes in the traces over the period of study and the adhesive bond strength has only decreased by 25% from its original value. The joints obtained were surprisingly durable.

8.4.3 Ageing of composite–composite joints

Carbon fibres are sufficiently electrically conductive to allow a fully cured CFRP laminate to look like a conductive substrate. A study has been undertaken looking at the effects of fibre alignment on the dielectric characteristics of CFRP–epoxy–CFRP bonds (Boinard et al., 2002). The study examined five different lay-ups in which the sequence was varied and most importantly examined the direction of the fibre which interfaces with the adhesive. It was found that provided there was sufficient contact between the fibres in the various layers, the TDR response was little affected by the internal orientation of the fibres. However the directions of the fibres next to the adhesive were found to have a significant effect (Fig. 8.16). If the 90° layers are sandwiched between the 0° layers, no effects in the reflection time are noticeable; however, if the 90° layer is at the interface then there is shifted relative to the location of the 0° peak.

Effect of water uptake on the adherent dielectric properties

Dielectric spectroscopy is ideally suited to monitoring water uptake by the composite. A [90$_2$/0/90$_2$/0/90] CFRP laminates were immersed in 60 °C

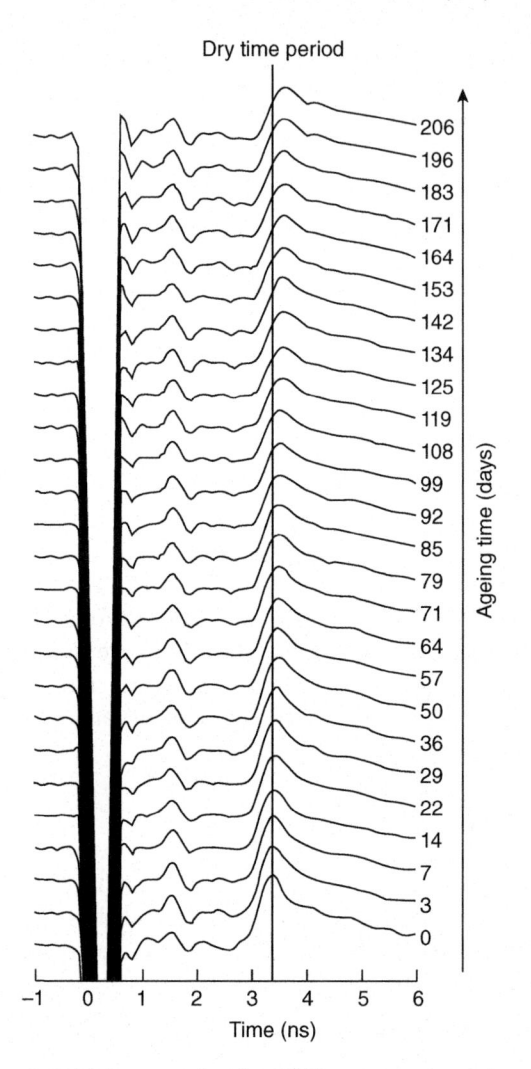

8.15 TDR traces for the CFRP–epoxy–aluminium joint aged at 70 °C as a function of time.

water bath for up to 5500 hours. The desorption process was investigated by heating the sample an air-circulated oven at 60 °C. The mass variation, change in electrical volume resistivity, dielectric and TDR measurements were performed, and the resistivity and mass uptake plots are shown in Fig. 8.17.

The laminate absorbs ~2% of water over the 5500 hours of exposure. As the water goes into the polymeric material, the resistivity initially increases as a consequence of the water swelling the matrix and reducing the contact between the conducting carbon fibres. Removal of the water causes the

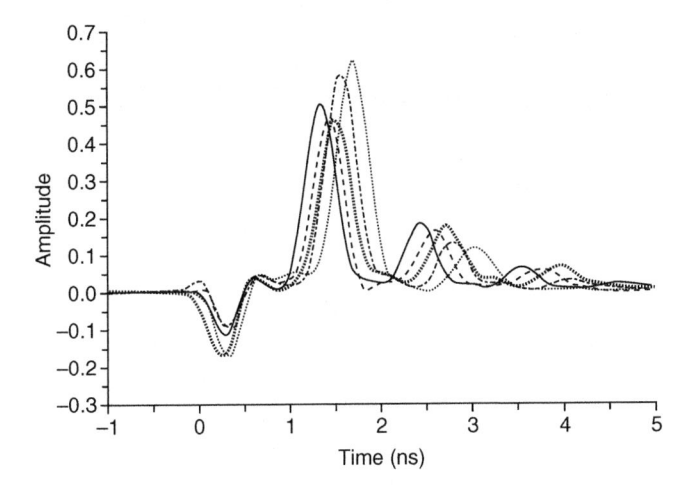

8.16 Dielectric TDR of different carbon fibre lay-up plate joints, last lamina at the interface being either 0° or 90°.

contacts between fibres to be recovered and the resistivity to fall. The decrease in resistivity is faster on desorption than its loss on adsorption; the surface layer losing moisture faster than the bulk and restoring an electrically conducting path through the material. During the swelling of the matrix, some interlayer delamination can occur and this will lead to a non-recovery of the resistivity during the drying cycle as was observed by ultrasonic C-scan of the laminates. TDR studies of pulse propagation within the laminate indicate that whilst small changes in the shape of the pulse could be detected at the time of flight, it stayed constant during the water uptake and desorption.

Effect of adhesive thickness on dielectric measurements

If the adhesive layer becomes very thin, <0.1 mm, then the possibility of carbon fibre bridging occurs and as a consequence the two laminate layers being bonded are short circuited and the signature of the bond is not detected. However, if the bond layer is too thick, >50 mm, then the dielectric damps the signal propagation and the bond appears like an open circuit. The optimum response is observed at about 1 to 5 mm, which is close to a typical adhesive layer used in practical fabrications (Boinard, 2002).

Carbon fibre bonded joints

Carbon fibre composites have sufficient electrical conductivity to be able to form a capacitor or wave guide structure analogous to that which allowed

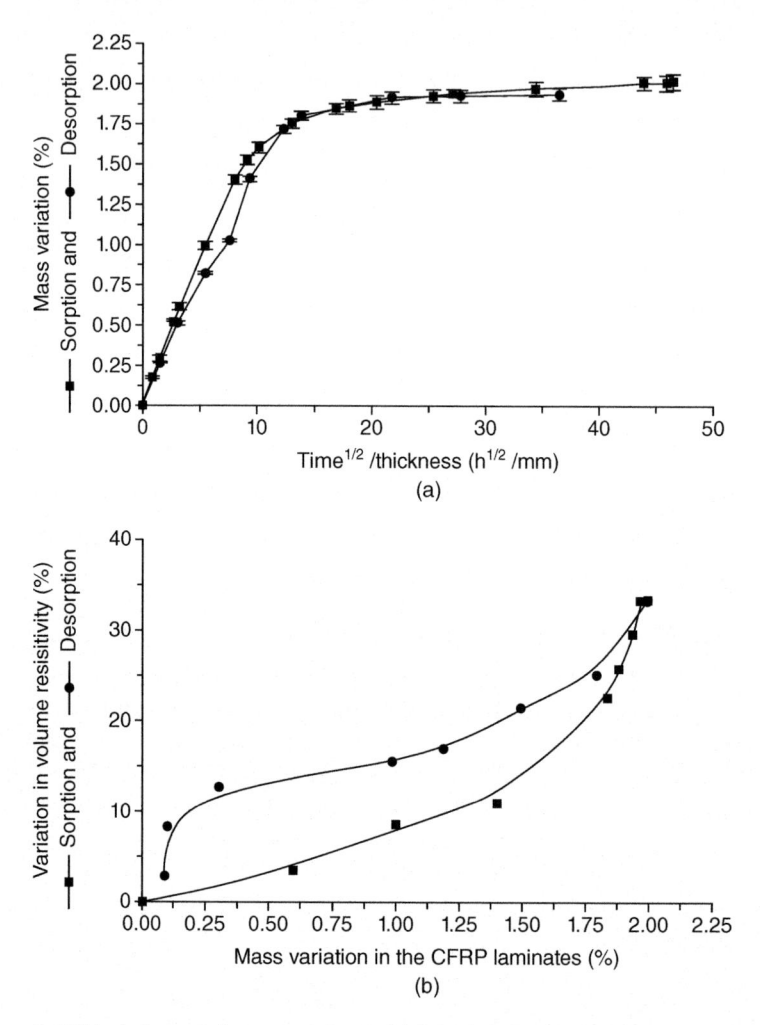

8.17 Variation of the mass change (a) and resistivity (b) during the water absorption and sorption processes plotted against the mass variation in the CFRP laminate for a study carried out at 60 °C.

study of the adhesive bond in aluminium joints. Studies of adhesively bonded carbon fiber reinforced plastic (CFRP/AF 163-2K) joints aged at 75 °C and exposed to cryogenic temperatures periodically showed a good correlation between the water uptake and changes in dielectric properties. Absorption, desorption and then re-absorption behaviour of the joints was studied and both the adhesive and adherend layers were found to adsorb significant amounts of water (Fig. 8.18).

Initially, the majority of water uptake occurs in the CFRP layers, with the adhesive taking much longer to reach high levels of water content. During

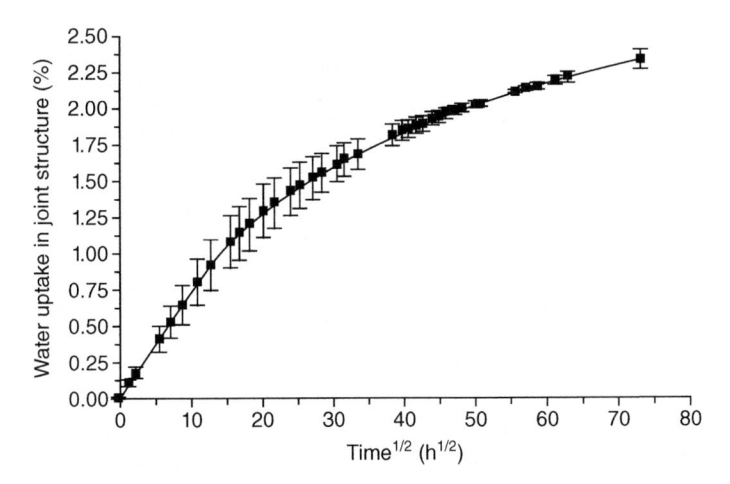

8.18 Water absorption against square root of time for CFRP–epoxy bonded joints exposed to moisture at 60 °C.

the sorption process, the composite joints do not reach an equilibrium value but steadily increase with exposure time. The non-equilibrium phenomenon can be attributed to the effects of plasticisation processes, microcracking and crazes formation and matrix/reinforcement interface disbondings. The changes in the dielectric permittivity ε' and dielectric loss ε'' for the joint as a function of exposure time are shown in Fig. 8.19.

The changes in permittivity and loss are very similar to those observed with the metal jointed structures. The creation of the large peak at low frequency is associated with a combination of plasticisation of the matrix and the increased freedom of motion of a series of charge carriers in the system leads to the development of space charge in addition to conduction effects. The water entering the matrix assists the plasticisation and hence stress relaxation, which eventually leads to an increase in the void content. This initially is seen as a drop in the permittivity as time proceeds. In Fig. 8.19a the peaks are associated with the creation of the space charge effects and then the subsequent reduction in the growth of the permittivity when air-filled voids are created (Boinard, 2002; Boinard et al., 2005).

The TDR traces of the joint as a function of ageing time are shown in Fig. 8.20a. The original trace is the furthest a way and the vertical line indicates the location of the original peak. As the exposure time is increased so the peak shifts along the time axis in proportional to the increased permittivity of the adhesive. The structure of the TDR traces does not change significantly with time, which reflects the stability of the joint and the fact that it is maintaining its strength despite being exposed to high levels of moisture. The variation of the permittivity as a function of time is shown in

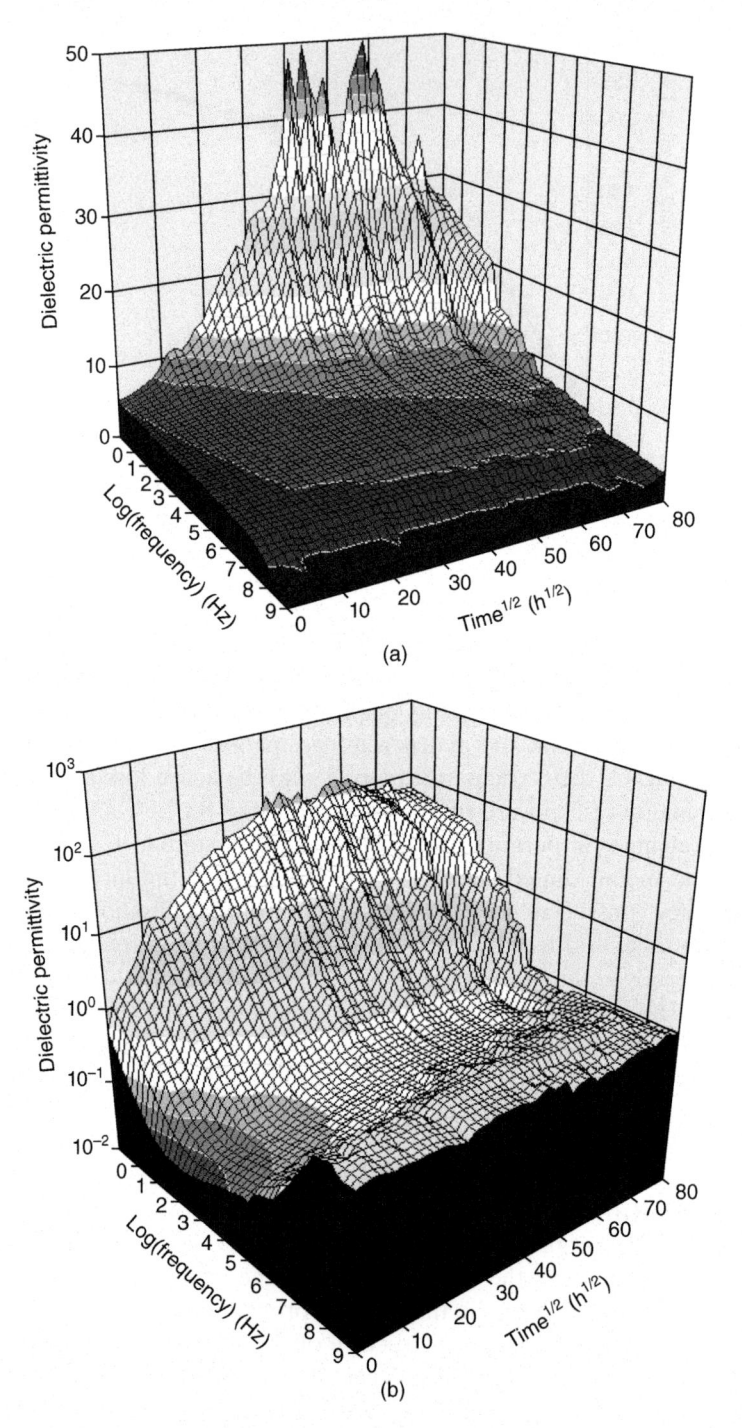

8.19 Changes in the dielectric permittivity ε' and dielectric loss ε'' for the joint as a function of exposure time of a CFRP–epoxy laminate structure.

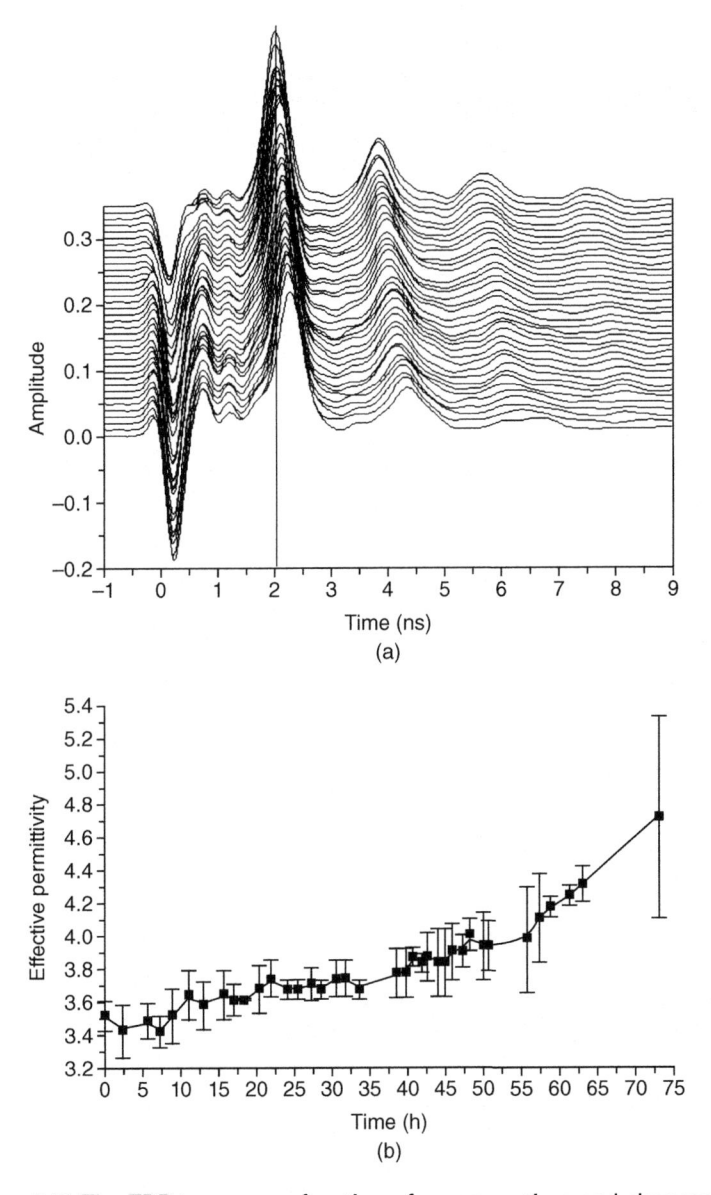

8.20 The TDR traces as a function of exposure time and changes in the dielectric permittivity ε' for a CFRP–epoxy laminate structure exposure to moisture at 60 °C.

Fig. 20b. The initial increase in the permittivity is slow, indicating the fact that the laminate will also be absorbing moisture and hence the amount in the bond will be proportionately lower than would have been expected from the measurement of the total water uptake. The increase in the later stages of exposure is consistent with the increase in the void content and water filling voids as well as being located in the resin. The initial dielectric and mechanical characteristics are to a large extent recovered when the joints are dried.

The study explored the periodically exposing of joints to cryogenic temperatures. Procedure was to simulate the effects of flight were temperature of the order of $-50\,°C$ may be experienced. Initial water uptake was greatest in the cryogenically frozen samples; however, over time the greater leaching and desorption suffered by frozen joints resulted in a drop in weight relative to non-frozen samples. Dielectric and mechanical tests on adhesively bonded CFRP joints aged in a series of solvents; deicing fluids (ethylene glycol), aviation fuel, moisture, brine and paint strippers and solvents (dichloromethane and butanone) have been undertaken (Banks *et al.*, 2004, Boinard, 2002, Boinard *et al.*, 2005). The ingress of the solvent leads to swelling of the joints and plasticisation of the adhesive with consequential changes in the mechanical properties. Drying of the joints indicated that in certain cases it was possible for the mechanical properties of the joints to be recovered to almost their original values. In other cases the recovery was limited and reflected changes occurring within the adhesive layer.

8.5 Conclusion

The dielectric neural network technique can characterise the cure processes in thermoset resin systems and allow identification of both the gelation and vitrification points. Being able to study the variation of the critical points as a function of temperature allows assessment of the cure process in structures, such as autoclaves where visual access may be difficult. Using microwave capability it is possible to map the impedance variations within a cured joint and locate voids and kissing bonds. Comparison of the time domain traces for pristine and aged joints allows regions where defects are being generated to be identified. Use of modelling software allows determination of the resistivity of the interface and this parameter is a sensitive indicator of the onset of corrosion of the oxide layer in materials such as aluminium. The principles developed for the study of aluminium–epoxy bonded structures are equally applicable to the study of bonding in CFRP laminate structures, the principal difference being that because of the lower electrical conductivity of the laminate and the potentially higher dielectric losses the observation of the higher-order peaks is rarely possible. However in favourable situations it is possible to quantitatively determine the moisture content and the defect structure of the bond line within the laminate structure. The dielectric method has significant potential application for the study of the state of joints in ageing aircraft and estimation of the bond strength remaining in bonded structures.

8.6   Acknowledgements

The author wishes to acknowledge the support of EPSRC, AFSOR, DTI, Civil Aviation Authority and BAe Systems for support of the project in various ways. The help of various coworkers is gratefully acknowledged and in particular Dr David Hayward who assisted in the supervision of several of the PhD students involved in the project.

8.7   References

Armstrong G.S., Banks W.M., Pethrick R.A., Crane R.L., and Hayward D., (2004) Dielectric and mechanical studies of the durability of adhesively bonded aluminium structures subjected to temperature cycling. Part 2: examination of the failure process and effects of drying, *Proceedings of the Institution of Mechanical Engineers Part L – Journal of Materials – Design and Applications* **218** (L3) 183–192

Banks W.M., Pethrick R.A, Armstrong G.S., Crane R.L., and Hayward D, (2004) Dielectric and mechanical studies of the durability of adhesively bonded CFRP structures subjected to aging in various solvents, *Proceedings of the Institution of Mechanical Engineers Part L – Journal of Materials – Design and Applications* **218** (L4) 273–281

Bishopp J., (1997) The history of Redux (R) and the Redux bonding process 11, *International Journal of Adhesion and Adhesives*, **17** (4) 287–301

Bishopp J., (2005) Aerospace: A pioneer in structural adhesive bonding, *Handbook of Adhesives and Sealants*, Elsevier London, Volume 1, Pages 215–347

Boinard P., (2002) 'Dielectric Studies of Carbon Fibre Composite Bonded Systems', PhD Thesis University of Strathclyde

Boinard P., Banks W.M., and Pethrick R.A., (2002) Use of dielectric spectroscopy to assess adhesively bonded composite structures, Part I: water permeation in epoxy adhesive, *Journal of Adhesion* **78** (12) 1001–1026

Boinard P., Banks W.M., and Pethrick R.A., (2005) Changes in the dielectric relaxations of water in epoxy resin as a function of the extent of water ingress in carbon fibre composites, *Polymer* **46** (7) 2218–2229

Bucknall C.B., (2007) New criterion for craze initiation, *Polymer* **48** (4) 1030–1041

Bucknall C.B., and Partridge I.K., (1983) Phase separation in epoxy resins containing polyethersulphone, *Polymer* **24** 639–644

Bucknall C.B., Karpodinis A., and Zhangi X.C., (1994) A model for particle cavitation in rubber-toughened plastics, *Journal of Materials Science* **29** 3377–3383

Crane R.L., Hayward D., McConnell B., Pethrick R.A., Mulholland A.J., Mckee S., and MacKay C., (2006) Assessment of adhesively bonded aluminium joints by high frequency dielectric measurements, *Sampe Journal* **42** (5) 35–38

Garden L.H., Hayward D., and Pethrick R.A., (2007) Dielectric non-destructive testing approach to cure monitoring of adhesives and composites, *Proceedings of the Institution of Mechanical Engineers Part G – Journal of Aerospace Engineering*, **221** (G4) 521–533

Halliday S.T., Banks W.M., and Pethrick R.A., (1999) Dielectric studies of ageing in aluminium epoxy adhesively bonded structures: design implications, *Proceedings of the Institution of Mechanical Engineers Part L – Journal of Materials – Design and Applications* **213** (L2) 103–119

Hsieh T.H., Kinloch A.J., Taylor A.C., and Sprenger S., (2011) The effect of silica nanoparticles and carbon nanotubes on the toughness of a thermosetting epoxy polymer, *Journal of Applied Polymer Science* **119** 2135–2142

Kazilas M.C., and Partridge I.K. (2004) Temperature modulated dielectric analysis (TMDA), a new experimental technique for the separation of temperature and reaction effects in dielectric cure monitoring signals, *Measurement Science and Technology* **15** (5) L1–L4

Li Z.C., Joshi S., Hayward D., Gilmore R., and Pethrick R.A., (1997) High frequency electrical measurements of adhesive bonded structures – An investigation of model parallel plate waveguide structure, *NDT & E International* **30** (3) 151–161

MacKay C., Hayward D., Mulholland A.J., McKee S., and Pethrick R.A., (2005) Reconstruction of the spatial dependence of dielectric and geometrical properties of adhesively bonded structures, *Journal of Physics D – Applied Physics* **38** (12) 1943–1949

MacKay C., Hayward D., McKee S., Mulholland A.J., and Pethrick R.A., (2007) An inverse problem of reconstructing the electrical and geometrical parameters characterising airframe structures and connector interfaces, *Inverse Problems in Science and Engineering* **15** (3) 177–190

MacKinnon A.J., Jenkins S.D., McGrail P.T., and Pethrick R.A., (1992) A dielectric, mechanical, rheological, and electron microscopy study of cure and properties of a thermoplastic-modified epoxy resin, *Macromolecules* **25** 3492–3499

Mackinnon A.J., Jenkins S.D., McGrail P.T., and Pethrick R.A., (1993) Dielectric, mechanical and rheological studies of phase separation and cure of a thermoplastic modified epoxy resin: incorporation of reactively terminated polysulfones, *Polymer* **34** (15) 3252–3263

Mackinnon A.J., Jenkins S.D., McGrail P.T., and Pethrick R.A., (1994) Investigation of thermoplastic-modified thermosets: positron annihilation and related studies of an amine-cured epoxy resin, *Polymer* **35** (24) 5319–5326

Mackinnon A.J., Jenkins S.D., McGrail P.T., and Pethrick R.A., (1995) Cure and physical properties of thermoplastic modified epoxy resins based on polyethersulfone, *Journal of Applied Polymer Science* **58** (13) 2345–2355

Maistros G.M., and Partridge I.K., (1998) Monitoring autoclave cure in commercial carbon fibre/epoxy composites. *Composites B, Engineering* **29** (3), 245–250

Maxwell I.D., and Pethrick R.A., (1983) Low temperature rearrangement of amine cured epoxy resins, *Polymer Degradation & Stability* **5** (4) 275–301

Maxwell I.D., Pethrick R.A., and Datta P.K., (1981) Thermal modification of amine cured epoxy resins – thermal and mechanical studies, *British Polymer Journal* **13** (3) 103–106

Pethrick R.A., (1998) Overview of fundamental aspects of dielectric monitoring of cure of some resins used in composite manufacture, *Plastics Rubber and Composites Processing and Applications* **27** (6) 257–265

Pethrick R.A., (2010) *Polymer Science and Technology for Engineers and Scientists*, Whittles, Publishing, Dunbeath, UK

Pethrick R.A., Hollins E.A., McEwan I., MacKinnon A.J., Hayward D., Cannon L.A., Jenkins S.D., and McGrail P.T., (1996) Dielectric, mechanical and structural, and water absorption properties of a thermoplastic-modified epoxy resin: poly(ether sulfone)–amine cured epoxy resin, *Macromolecules* **29** (15) 5208–5214

Pethrick R.A., Amornsakchai T., and North A.M., (2010) *Introduction to Molecular Motion in Polymers*, CRC Press, Whittles Publishing Dunbeath, UK

Skordos A.A., and Partridge I.K., (2004) Determination of the degree of cure under dynamic and isothermal curing conditions with electrical impedence spectroscopy, *Journal of Polymer Science B, Polymer Physics* **42** (1) 146–154

9
Non-destructive evaluation (NDE) of aerospace composites: methods for testing adhesively bonded composites

B. EHRHART and B. VALESKE, Fraunhofer
Institute for Non-Destructive Test Methods (IZFP),
Germany and C. BOCKENHEIMER,
Airbus Operations GmbH, Germany

DOI: 10.1533/9780857093554.2.?

Abstract: This chapter aims to establish the state-of-the-art with regard to non-destructive testing (NDT) techniques of adhesively bonded composite materials, with a special focus on the aerospace industry, and composite substrates. Typical material defects, as well as quality management through NDT methods throughout the product life cycle, are introduced. A complete overview of existing NDT techniques is given, followed by a description of limitations in terms of bond performance assessment. Recent concepts and nascent developments ('Extended NDT') for the future are explored.

Key words: non-destructive testing (NDT), adhesive bond, defect, bond performance assessment.

9.1 Introduction

The aerospace industry benefits from experience with adhesive bonding on carbon fibre reinforced polymer (CFRP) technology. In addition, there is great potential for the manufacturing of high-loaded structures. Control over the quality of the adhesive bond is a prerequisite for such applications. Various non-destructive testing (NDT) techniques are sufficient for the characterisation of defects such as pores, delamination or disbond within adhesive bonds.

This chapter aims to establish the state-of-the-art for non-destructive characterisation of adhesively bonded composite materials, with a special focus on the aerospace industry. This therefore focuses on a system based on CFRP and adhesive bonded CFRP structures. After presenting the background and main influences on an adhesive bond, this chapter reviews the most frequently applied control methods in the aerospace industry. Future challenges for adhesive bonding and its performance quality assessment are introduced.

9.2 Adhesive bonding in the aerospace industry

Adhesive bonding is an assembly process which has developed rapidly in the last decades. The technique has numerous advantages over other fastening methods. The following sections introduce the history and advantages of adhesive bonding, while also specifying the prerequisites for a proper adhesive bonding process.

9.2.1 Evolution of bonding

Adhesive bonding is one of the oldest assembly processes. Its development accelerated acutely over the last century, especially in the aeronautic industry whose history is entwined with the two world wars. Fokker started using adhesive bonding in 1915, leading the way for other manufacturers. In 1940, adhesive bonds were used on wood spars for the DH 98 Mosquito aircraft (Adams and Cawley, 1988). In the 1960s, the aerospace industry also started using adhesive bonding technology, convinced it was an optimal solution for the manufacture of lightweight, resistant structures (Simon, 1994). Under constant development, adhesive bonding also reached civil aviation, where high-performance epoxy based resins are now widely used. As an example, Airbus began adhesive bonding in 1972. The Airbus aircraft family now has over 345 bonding features (Stöven, 2010). The future of adhesive bonding is promising, with the ever-present need for better performance keeping demand high for the technology, and generating new applications with their concurrent challenges.

9.2.2 Advantages of bonding over other assembling technologies

Adhesive bonding offers several advantages over other assembling processes. A structure assembled with adhesive bonding benefits from homogeneous stress distribution. This minimises any stress concentration that may appear with discontinuous assembly methods (riveting, screw fastening, point soldering, etc.) (Adams and Cawley, 1988; Light and Kwun, 1989; Simon, 1994; Cognard, 2002). Since additional fastening elements, such as bolts, are not required for adhesive bonding, this method enables design of light, strong and even complex structures. Uniform stress distribution confers high fatigue resistance on a structure, and thus also a longer service lifetime than those structures that had to be machined (e.g. bored, heated) prior to assembly. Bonding also allows the joining of two distinct materials. A continuous adhesive joint acts as a sealant, and has interesting properties in electrolytic corrosion protection, and vibration and sound damping. The process can be fully automated, highly relevant for its use on industrial assembly lines. These advantages suggest that adhesive bonding is an ideal manufacturing solution for the next generation of joined structures. This technology also retains development potential for broader implementation in other industries.

9.2.3 Prerequisite for adhesive bonding

Limitations to the wider application of adhesive bonding are numerous. Although bonding is a widespread assembly method in the industry, science has not yet explained how an adhesive actually functions (Cognard, 2002). There is a real lack of knowledge regarding the mechanisms of adhesion. For this salient reason, those working in the industry do not trust the method enough to use it for high-loaded, so-called 'primary', structures; instead, adhesive bonding is used for 'secondary' structures (spars, stringers, etc.) where a possible failure would not harm the integrity of the product directly.

Many parameters are decisive in the quality of an adhesive bond, starting with the bonding process. Briefly, proper bonding requires that a completely clean surface that presents a good affinity with the adhesive be prepared. Good wettability of the adherend surface is also necessary for optimal bonding. The bonding process, including surface preparation, materials handling, and the adhesive cure, require a high degree of control (Light and Kwun, 1989; Simon, 1994; Cognard, 2002). In 2003, the German Association for Soldering (DVS) introduced the German norm DVS 3310 'Quality requirements for adhesive bonding technology' with regard to the requirements of the EN ISO 9000:2000 norm, section 3.4.1, regarding adhesive bonding as soldering: a special process. Adhesive bonding is defined as a process where conformity of the resulting product cannot be easily or economically verified. Furthermore constant monitoring and/or observance of documented processing instructions is necessary to ensure that quality objectives are met. These process parameters are essential for the quality of adhesive bonds.

9.3 The role of non-destructive testing (NDT) in testing adhesive bonds

Even if the materials, processes and methods are designed to enable the highest possible quality, adhesive bonds can encounter defects at all phases of their life cycle, as indeed can any other part or material. Taking this risk into account, NDT is required not only for in-service or manufacturing, but throughout the life cycle of a part. Naturally, the bonding process itself plays a decisive role in quality assessment, and non-destructive evaluation is an asset for process monitoring.

The definition of the DVS, introduced in Section 9.2.3, specifies that constant control and/or observance of documented processing instructions ensures that quality objectives are met, preventing the occurrence of defects. For this reason, control methods for quality assessment are implemented for bonding processes in the aeronautic industry. These processes involve several critical steps requiring close inspection and control, governed by specifications proper to each manufacturer. For example, steps range from adherent surface preparation; time between pre-treatment and application; work life conditions of the adhesive and adherend material; and shop conditions during the bonding operation and preparation for curing.

In the next sections, an introduction to existing defects in adhesive bonds is given, followed by the objectives of NDT for adhesive bonding along the life cycle of an adhesive bonded part, from preparation to recycling.

9.3.1 General presentation of defects

Without detailing their causes, several typical defects exist. The simplest case of an adhesive bond between two composite laminates is considered first (see Fig. 9.1). Defects may affect:

- the adherend/substrate: cracks, dents, scratches, holes;
- the interface: disbonds, zero-volume disbonds or weak/kissing bonds;
- the adhesive layer: porosity, voids, cracks, poor cure (under-cure or over-cure), etc.

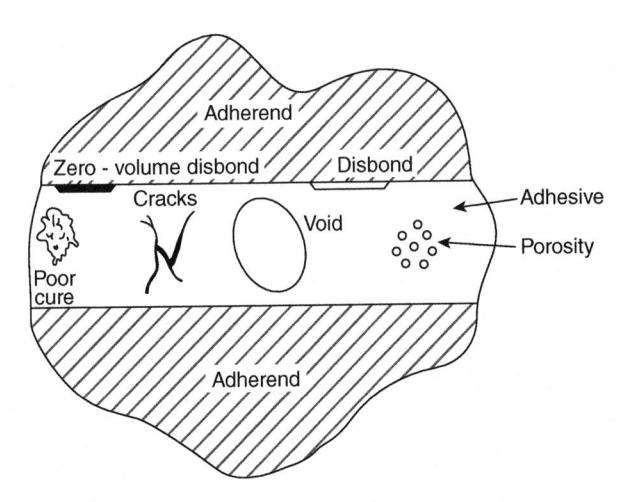

9.1 Typical defects in bonded joints (Adams and Cawley, 1988). Reprinted from *NDT International*, Volume 21/Issue 4, R. D. Adams, P. Cawley, A review of defect types and nondestructive testing techniques for composites and bonded joints, pp.208–222, Copyright August 1988, with permission from Elsevier.

In the case of a sandwich composite structure, particularly the honeycomb core sandwich, core cells are bonded to a CFRP skin, and the fillet of adhesive on the cell walls is crucial. Several other defects are characteristic (see Fig. 9.2):

- the honeycomb core may be damaged and shortened locally, preventing optimal contact with the skin (Fig. 9.2a).
- the CFRP skin may also be damaged (bumps, holes, cracks, etc.) preventing contact with the honeycomb core cells (Fig. 9.2b).
- missing adhesive at the interface between the honeycomb core and the CFRP skin resulting in the absence of adhesion and structural integrity (Fig. 9.2c).
- the adhesive fillet itself may have formed improperly, weakening the mechanical performances of the sandwich structure (Fig. 9.2d).

Other defects may appear during manufacturing, which are directly linked to manufacturing processes, and where process parameters are decisive:

- an ill-adapted or incorrect adhesive application (thickness of adhesive layer, geometry and quantity of adhesive, etc.);
- rheological deviations (temperature of storage, mixing ratio of adhesive components, etc.);
- curing deviations (temperature, pressure, etc.);
- defects resulting from outer sources such as foreign materials in adhesive layers.

The 'service life' of the part, where the adhesive bonded composite is exposed to the elements, also introduces characteristic defects, especially from hydrothermal ageing and corrosion, but also from impact (caused by any number of things such as ramps, tools, hail, stones, birds, etc.).

9.3.2 NDT for quality assessment prior to bonding processes

Surface preparation is a decisive step in the realisation of an adhesive bond, because it sets up the quality of the adherend. As mentioned above, control of the environment where bonding takes place is a decisive parameter for surface preparation treatment. Norms exist regarding the concentration of airborne particles, and even specify the temperature and humidity for optimal work conditions. The preparation itself should take place in a clean, dry and isolated area to prevent surface contamination. The surfaces must be clean and free of any inhomogeneity such as fatty substances, oils, foreign bodies or anything likely to affect the bonding operation. Cleanliness is usually assessed by means of close control and inspection before and

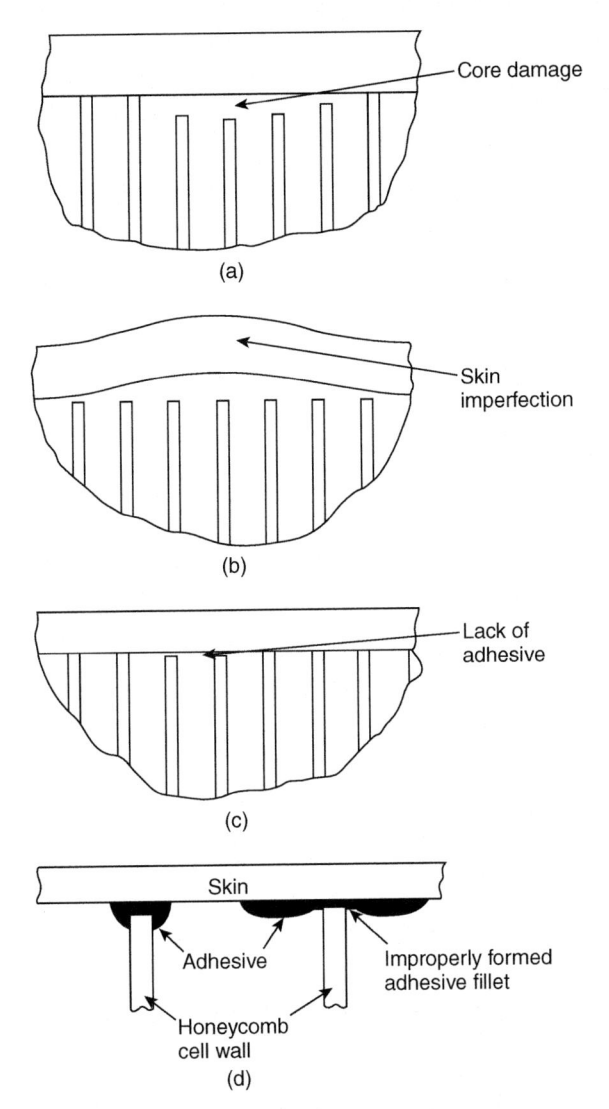

9.2 Defects in a bonded honeycomb/skin construction (Adams and Cawley, 1988). Reprinted from *NDT International*, Volume 21/Issue 4, R. D. Adams, P. Cawley, A review of defect types and nondestructive testing techniques for composites and bonded joints, pp. 208–222, Copyright August 1988, with permission from Elsevier.

after the numerous critical pre-treatment processes in serial production (including pre-treatment by abrasion or blasting with subsequent cleaning steps, water break testing and surface resistance checking).

The following methods are either analytic or mechanical, and allow the characterisation of the surface and bonded structures to obtain information related to the surface state (Stöven, 2006). These methods are integrated to ensure quality prior to bonding.

Surface characterisation prior to bonding is generally carried out by:

- *Visual inspection*: Visual inspection of prepared surfaces is performed for each part, to check whether pre-treatment was carried out homogenously on the entire surface. Visual inspection also establishes the homogeneity of treated areas (e.g. spots, shadows) and is performed in order to detect any damage to the component surfaces (e.g. abrasion, scratches), the dimensions of the parts, and any visible contamination or foreign bodies.
- *Surface resistance testing*: After pre-treatment by abrasion or blasting, a surface resistance test is performed if contaminated areas are still apparent.
- *Water break test*: After pre-treatment by abrasion or blasting, a water break test (WBT) is performed on the pre-treated adherend surfaces, to evaluate the presence of contaminants. This method, which relies on the visual inspection of a film of water flowing on the surface, is qualitative, and therefore subject to criticism for lack of sensitivity and objectivity (Parker and Waghorne, 1991).

9.3.3 NDT for quality assessment within bonding processes

Adhesive bonding is an assembly process. For instance, it is used for monolithic and sandwich composites, for joining stringers in the role of doublers and/or stiffeners, as well as for load transfer parts. The following three processes are established (Stöven, 2006):

- Secondary bonding: two solid cured parts are bonded together.
- Co-bonding: one uncured part is bonded to a cured solid part.
- Co-curing: two uncured parts are bonded together.

After the adherent surface preparation (described in Section 9.3.2), the adherend material must be processed in as little time as possible, again to avoid contamination. Alternatively, adherend parts must be stored and hermetically sealed. Once surface preparation is complete, the relevant and compatible adhesive is applied to the adherend material and the bonding operation may start. The bonding process is carefully controlled. From the autoclaves, curing ovens, presses and related equipment, parameters such as temperature, pressure, vacuum and time are recorded and controlled continuously to ensure that no procedural deviation occurs.

9.3.4 NDT for quality assessment of adhesive bonded parts

After the bonding process, mechanical destructive tests are recognised as the only valid way to evaluate the mechanical performances of the adhesive bond, and thus success of the adherent surface preparation and bonding process (ASTM International, 2008). Among the testing methods applied to bonded specimens, the most important is the Interlaminar Fracture Toughness Energy in mode I 'G1c' since its result is representative of the surface preparation prior to bonding. This cleavage test is performed with a precracked double cantilever beam specimen, and consists of opening by the application of stress to both sides of the precracked extremity. To be representative, the specimens are manufactured from the same material and under the same conditions, including surface treatment, as the original part. The absolute requirement is that failure in the crack propagation must never be due to the adhesive.

Other characteristic tests are the single lap shear strength and the interlaminar shear strength tests. In these, the specimens are not precracked and are tested in shear (mode II). The interlaminar shear strength test is a measurement of the resistance of the composite to delamination under shear forces parallel to the layers of the laminate, and so to the adhesive/adherent interface.

Alternatively, glass transition temperature with dynamic mechanical analysis (DMA) and degree of cure with dynamic scanning calorimetry (DSC) can be performed respectively on the cured laminate and adhesive to evaluate the cure cycle. These mechanical tests can provide values of stress related to the structure and history of the bonded part. Even if such tests are mandatory, their destructive character obviously means they cannot be applied to the final bonded product, reopening the question of how far their results can be transferred.

In conclusion, the methods introduced above are required for the control of the adhesive bonding process, in order to assess that adhesion is of high quality and that no contamination or other surface pollution has weakened the bond. For quality management purposes, NDT is of primary importance.

9.4 Non-destructive testing (NDT) methods

It is essential that any defects in the bonded composite product are detected throughout the product life cycle (Light and Kwun, 1989; Nottorf et al.,

2006). As well as metal products and metal bonded structures, NDT methods are used on adhesive joints for composite materials. The literature reports a large number of studies on NDT techniques for composites and bonded joints (Adams and Cawley, 1988). The following conventional NDT methods for adhesive bond quality control are presented in this chapter: visual inspection; the leak test; tap test and woodpecker; Fokker bond tester (FBT); acoustic emission (AE); ultrasonic (US) methods; X-ray/neutron radiography; shearography; and infrared (IR) thermography.

This section introduces the principle of each NDT method, its main features, and typical examples of applications in the inspection of adhesive bonded structures in aerospace. The content of the following paragraphs discusses general literature (see recommended websites) and the European adhesive engineer course documentation, from Bockenheimer and Valeske (2006).

9.4.1 Visual inspection

Visual testing is the most common non-destructive inspection method, relying on the naked eye. Several tools may also be used to augment the eye, such as lighting, magnifying glasses, microscopes, cameras and video recording. A visual inspection is easily performed, is economical, and usually requires little equipment. However, it is also evidently limited by the inspector's sight, and defects which do not reach the surface are invisible. Visual inspection is a standardised NDT technique in most aeronautic manufacturers. Typical applications range from pre-bonding inspection of the substrate surface, and the right adhesive application (geometry checks, fill-in, amount) to the post-bonding detection of defects in the side surfaces and surface of the bonded joints (cracks, scratches, blisters, voids, corrosion, degree of filling, foreign bodies, etc.).

9.4.2 Leak test

The leak test is performed by immersing a part, usually a sandwich composite structure, in a hot water tank. The temperature of the water induces the expansion of air in the structure, and if a crack or a delamination is present, gas bubbles escape the structure and are immediately detected by visual inspection. A leak test is easy to set up for small parts and only requires simple equipment. It is limited by the resolution of the human eye, and can only lead to the detection of an emerging surface crack or delamination. The leak test is a standardised NDT technique in most aeronautic manufacturers. Due to its limitations, basic applications include the inspection of delaminations and disbonds in sandwich composite structures, especially honeycomb core.

9.4.3 Tap tests and Woodpecker

The tap test is a resonance test. The surface is tapped with a tool producing a constant frequency and the sound of the intact adhesive bond is recorded as a reference. If a defective bond is tapped, the frequency shifts and hence the sound alters.

Several inexpensive tools have been designed for tap tests, from coins to hammers. For instance the Wichita tap hammer, electronic hammers with integrated diagnostics, and the Mitsui Woodpecker. The Woodpecker (see official website in 'Useful literature') is an automated hammer, offering selectable tapping intensity and frequencies ranging from 2 to 16 Hz for inspection. Defect detection is based on a qualitative evaluation of the emitted signal attenuation, comparing a safe bonded area, as a reference, and the part to be measured. Deviations in signal parameters are automatically recorded and displayed on the Woodpecker.

A classical tap test is easy to perform, and inexpensive. It relies on the experience of the inspector and is generally qualitative since it is limited by the capacities of the ear. It does not allow defects to be quantified either in position or size, and only works for defects close to the surface. The Woodpecker is also limited by composite skin thickness above the bonded area. The detectable delamination diameter is influenced by this thickness, which should not exceed 2 mm (see recommended websites for more information).

The tap test is a standardised NDT technique in most aeronautic manufacturers. It is used for the detection of voids, delaminations and disbonds, especially in the case of sandwich composite structure.

9.4.4 FBT

The FBT is also a resonance test. It was developed by Fokker in the 1970s. This technique focuses on the adhesive bond and is based on the damped spring-mass system. The substrate is considered as the mass and the adhesive layer, the weightless spring and damper. The technique relies on a transducer made of a piezoelectric oscillator, vibrating at a range of frequencies from 20 to 500 kHz. The excites the complete system to evaluate the resonance frequencies, so characterising the bonded structures. Two parameters are adjustable: the resonance frequency A and the amplitude B. The evaluation of both parameters gives important information about nature of the adhesive layer, substrate and adhesive layer thicknesses, presence of defects, and cohesion in the adhesive layer bondline. The high cost is a difficult calibration on several reference samples and a correlation to their related destructive tests. These are only relative and qualitative; the FBT sorts the measurement in quality classes depending on frequency A and amplitude B.

The FBT is a standardised NDT technique in most aeronautic manufacturers. It is still used for maintenance of older aircraft, for the inspection of metal to metal bonds, and of honeycomb core sandwich composite parts. The FBT enables the detection of delamination, voids, porosity, adhesive layer thickness and corrosion. It also allows the characterisation of cure degree, plasticisation, ageing (due to moisture, heat exposure, fatigue, etc.) of the bonded part, but fails in the detection of adhesion quality.

The FBT is still used for some applications. Major improvements including additional test modes have also been developed. One example: the *Bondmaster* includes a mechanical impedance analysis (MIA) mode and a pitch-catch mode (with Lamb waves) for applications such as water ingress inspection in honeycomb sandwich structures. The Bondmaster also offers improved handling flexibility.

9.4.5 AE

AE is based on the principle that a stressed defective bonded structure should generate an acoustic answer due to elastic energy, friction, dislocations or crack propagation. The acoustic events are mostly provoked by heating the bonded structure by hot air or IR lamps. Acoustic emissions are recorded by a sensor over 15 to 30 seconds and the signal is processed to monitor the acoustic answer.

Key parameters are: stress application parameters such as pulse duration, frequency and amplitude. AE is an inexpensive and swift technique sufficient for inspecting large structural bonded parts. The stress induced must, however, be applied cautiously to avoid any damage, and the outcome relies on the stability as well as the precision of the measuring device. Finally, AE may be applied for the inspection of defects, cracks, corrosion and also the ingress of moisture to core composite structures.

9.4.6 US methods

US methods are without a doubt the most versatile NDT method. A very large diversity of US methods are available, and in use, for the inspection of composite parts. The common principle of these techniques is the generation of a US pulse (1 to 25 MHz) which propagates into the material, to be reflected or attenuated by an obstacle (defects, inhomogeneities, interphases). The propagation behaviour of US waves into the material is linked directly to intermolecular forces and mechanical behaviour. US waves are particularly sensitive to the differences of impedance within the structure and so, for all US techniques involving contact of a transducer that functions as an actuator and/or sensor, the use of a coupling medium (such as water, oil or gel) is required to minimise signal loss.

Once the US wave has been reflected or transmitted, a sensor detects it, delivering information such as intensity (defect size, adhesive properties) and time of flight in the inspected part (depth of obstacle). The choice of this sensor is a compromise between test frequency and sensor size. The US field generated in the structure by the transducer must deal with a complex interference area in the near field of the transducer. Measurements must be done in the far field, starting at a usual depth of three times the near field size. The entire surface of bonded structures is generally scanned in order to provide information about the precise size and location of any defects.

US techniques offer many options for inspection. Two main modes exist:

- the pulse-echo mode where the actuator is also the sensor; and
- the transmission mode where a separate actuator and sensor are used.

Different coupling methods are available to suit all inspection needs:

- contact measurements with the sensor on the surface;
- immersion measurements (usually in water tanks);
- squirter measurements (flow of water onto the surface to avoid complete immersion);
- air-coupled US measurements (without a coupling medium); and
- laser stimulated US measurements.

Finally, a large variety of sensor sizes, orientation, and geometry, enable different US wave directions, and beam pressure to travel in particular directions in the bonded part. This variety of sensors enables the generation of longitudinal, transversal and Lamb waves for the detection of various defect orientations and geometries.

US techniques have benefited from numerical reconstruction processes to eliminate parasitic signal changes, especially due to coupling issues, thickness variations and composite plies arrangements. Spatial resolution and quantification has been improved in methods such as phased array or sampling phased array, where the combination of sensors can be electronically triggered for angular scanning and focusing variations.

US inspection is a standardised NDT technique in all aeronautic manufacturers. Its application ranges from the inspection of skin disbonds in honeycomb core sandwich structures of composites, porosity in bond lines, delaminations and disbonds in bonded parts, to the precise position, size and geometry of defects in the complete part.

However, despite the universal use of ultrasonics, this NDT technology is unable to detect changes in adhesive bond strength, a fact that is well known. The US waves can nevertheless yield information regarding the morphological and elastic features at the interface of adhesive bonds, by their behaviour in the inspected material.

9.4.7 X-ray radiography and computed tomography

X-ray radiography for the inspection of bonded parts functions as it does in the medical field. An X-ray source (25–75 kV or for computed tomography, microfocus X-ray tubes up to 450 kV) irradiates the bonded part and reveals the presence of various types of defect depending on the absorption of X-rays. Absorption is linked to the density of an element, and so high atomic number elements tend to absorb more X-rays than lighter ones such as organic compounds. The transmitted X-rays are measured behind the inspected part, either by a film plate or by modern detectors.

X-ray radiography successfully reveals the inner defects of bonded metallic and composite parts. Computed tomography enables an even finer detection of small defects (pores, voids, cracks, etc.) also displaying three-dimensional views of the inspected part. These two X-ray techniques are, however, very expensive in terms of equipment (entailing X-ray sources, secured area for the inspection, etc.) and time (from sample preparation to data processing and evaluation). The orientation of the bonded part to irradiation is a key parameter for an optimal result. X-ray radiography is a standardised NDT technique in most aeronautic manufacturers. It is usually selected for the detection of defects in metallic honeycomb core sandwich structures, such as water inclusions. It is also applied to the inspection of special X-ray opaque adhesives (for voids and porosity detection).

9.4.8 Shearography

Digital shearography uses a coherent light source (laser) to irradiate the inspected surface of the bonded part. The resulting diffuse reflection from the surface is captured by a shearing interferometer which delivers phase information about the interfering light waves: this is called the speckle pattern. It corresponds to a gradient of the surface profile in the direction of observation. A digital image acquisition system records a picture of the unstressed reference surface, also called a shearogram. The inspected surface is then deformed by applying a vacuum, traditionally, or a thermal or mechanic excitation, in order to increase the deformation around the defect. A second shearogram of the stressed surface is then captured. Finally, the difference of the stressed and unstressed shearograms is observed and reveals the deformation gradients, and by extension, allows the detection of defects.

Shearography is a very rapid inspection method, with the advantages that it can be contactless, automated and applied to large areas. The sensitivity of this technique is around the few nanometres mark. The appearance of little defects, however, requires a high stress solicitation, and only defects being related to the inspected surface are detectable. Shearography is a standardised NDT technique in most aeronautic manufacturers. It has been used for the inspection of large parts such as rotor blades, for the detection of bumps, delamination and disbonds.

9.4.9 Active IR thermography

The active thermography principle is excitation, either pulsed or periodic, of the bonded part in order to create a thermal flow. The detection of slight temperature increase on the surface part with an IR camera allows the monitoring of heat diffusion in the bonded part, which alters depending on the presence of defects in the part and bondline. In general, a defect-free part will appear homogeneous and cold, since no defects would generate a heat concentration by poor local heat transfer.

The inspection parameters are numerous, but each is essential. The sensitivity of the IR camera is usually of a few mK. With an image frequency higher than a few kHz, the pixel resolution is naturally of great importance. Several excitation sources exist and should be chosen according the defect orientation in the bonded part: contactless methods (flash lights, laser, hot air, IR lamps, induction) or mechanical stimulation (US, vibration). The nature and energy of excitation provide different response times, directly correlating to the depth reached in the bonded part, and the maximum size of defect detectable.

Thermography is an inspection technique suitable for large areas, and can be fully automated, but its use is conditioned upon a good knowledge of a part's thermal properties. In particular, the inspection of carbon fibre composite material is enhanced by the high thermal conductivity of the fibre, whereas in the case of glass fibre composite, its low thermal conductivity prevents any heat diffusion in the material.

Active thermography is a standardised NDT technique in most aeronautic manufacturers. It is used for the inspection of delaminations, disbonds, cracks and pores in the adhesive bonds. Defects are also revealed in the composite itself, or the skin of composite sandwich structures. Water can also be detected in sandwich composite with honeycomb core.

9.5 Challenges in non-destructive testing (NDT) of adhesive bonds

None of the 'conventional' NDT methods mentioned above is able to assess the mechanical performance of an adhesive bond. The literature unanimously concludes that although defects can be detected in adhesive composite joints, to date no statements can be made regarding the quality of adhesion, its strength, or its properties (Adams and Cawley, 1988; Baumann and Netzelmann, 2003; Valeske et al., 2008; Wilhelm et al., 2010).

9.5.1 Bond performance assessment

The term 'weak adhesive bond' must be explained in order to detail bond performance assessment. A weak bond is a defective adhesive bond presenting the following characteristics, as established in the literature (Marty et al., 2004; Decourcelle and Kellar, 2009):

- the strength measured with a lap shear test must be below 20% of the nominal bond strength;
- the mode of failure must be of adhesive type (i.e. purely at the interface between the adherend and the adhesive);
- the weak bond must be undetectable from normal bonds by using classical NDT techniques.

Another term referring to a weak bond, also widespread in the literature, is 'kissing bond' (Weise et al., 2002; Baumann and Netzelmann, 2003). The term 'kissing bond' is randomly attributed to all possible weak bonds, whereas in fact it relates to the particular case where the adherend surface and the adhesive bond are in intimate contact but without any adhesion at the interface. To prevent any confusion, the general term 'weak bond' will be used here.

The performance of an adhesive bond comprises its required *functionality* and *durability* and is governed by the following factors:

- *loadings* (chemical, mechanical and physical in-service loading);
- *defects* (disbonds, unbonds, pores, delaminations, cracks, etc.); and
- *properties* (adhesion, cohesion, network structure, interphases, molecular mobility, etc.).

In order to ensure the performance of adhesive bonds within a quality insurance concept, the adhesive bonds are designed in the most adequate way to manage loadings; inspection by means of conventional NDT is conducted to take care of defects; and finally, properties are controlled through a precise monitoring of materials and processes.

However, there is no method to assess the quality of the part which is to be bonded or the bond performance of the adhesive bonded part. Indeed, no evaluation of the minimum bond performance, and, by extension, no characterisation of weak adhesive is available to date through any NDT method (Ehrhart et al., 2010). This limitation of adhesive bonding technology is of utmost importance to its further development.

9.5.2 Extended NDT methods

If the quality of the adhesive bond can be defined as the absence of defects, durability in the service environment, and high mechanical performance (Light and Kwun, 1989), the quality is a parameter beyond current NDT technology. Unlike the quality of other assembly technologies, the quality of an adhesive bond requires the physical and chemical properties of the adhesion and interface to be characterised (Valeske *et al.*, 2008).

In this connection, a novel approach has recently been developed. This aims at 'extending' the use of conventional laboratory characterisation methods, and even NDT techniques, for the determination of physical and chemical properties. This approach is known as 'Extended NDT' for the characterisation of the adherent surface and the adhesive bond (Valeske *et al.*, 2008; Bockenheimer, 2011; Ehrhart *et al.*, 2011; Markus *et al.*, 2011).

Research on this topic is being conducted worldwide. For the characterisation of the adherend surface, a portable IR spectrometer and an X-ray fluorescence technique are the only commercial technologies to demonstrate success in detecting surface contamination, so far. Other methods are under development and may be implemented in manufacturing and in-service environments (Ehrhart *et al.*, 2011; Markus *et al.*, 2011). Regarding the characterisation of adhesive joint performances, a particularly promising approach known as a 'laser proof test' exists. This method involves the generation of a shock wave through a laser high energy and short pulse excitation. The shock wave travels back and forth in the bonded part, generating a tensile test with a precise load at the interface between the substrate and the adhesive layer (Bossi *et al.*, 2009; Ecault *et al.*, 2011). Such a technique should allow, once developed and calibrated, the NDT of a 'good' strong bond whereas a 'bad' weak bond would be disbonded by the shock.

All the above concepts and techniques for the quality assessment of the adhesive bond performances are currently only at the development stage, but their success should allow the broader application of adhesive bonding in the aeronautic industry, in the near future.

9.6 Conclusion

NDT methods present many advantages: they can be applied directly to the adhesive bonded part without harming product integrity. The maturity and experience of the industry with NDT techniques have led to the automation of inspections, both in their execution and evaluation. NDT methods are mandatory within the quality management of an adhesive bond. Their presence through the complete life cycle of the part, from its design phase to the bonding process and beyond, allows for constant quality monitoring and the early detection of process deviation or damage occurring during the in-service life of the final bonded part.

However, if the use of NDT methods is now well accepted, NDT inspection also involves strong requirements and norms for the test sequence itself. NDT techniques must be set up by trained and qualified inspectors who need experience in order to choose the most adequate NDT technique for a particular application. An NDT inspection is therefore demanding. To conclude, several conventional characterisation methods are in the process of being optimised and developed for a better detection capability towards assessment of bond performance, but to date no method can reliably and with good reproducibility detect any weak adhesive bond. Adhesive bonding on CFRP shows great development potential, especially in the field of NDT.

9.7 Sources of further information and advice

ASNT – The American Society for Nondestructive Testing
Website: *http://www.asnt.org/*

DVS – Deutscher Verband für Schweißen und verwandte Verfahren e.V.
Website: *http://www.die-verbindungs-spezialisten.de/*

FAA – Guidelines for Analysis, testing, and NDI of Impact Damaged Composite Sandwich Structures
Website: *http://www.tc.faa.gov/its/worldpac/techrpt/ar02-121.pdf*

Mitsui Woodpecker information
Website: *http://www.wp632.com/*

NDT.net – Open Access NDT Database
Website: *http://www.ndt.net/*

9.8 References

Adams, R.D., Cawley, P., 1988. A review of defect types and nondestructive testing techniques for composites and bonded joints. *NDT International* **21**, 208–222.

ASTM International, 2008. ASTM D_897-08 Standard test method for tensile properties of adhesive bonds.

Baumann, D.J., Netzelmann, D.U., 2003. Zerstörungsfreie Prüftechniken für Materialverbunde, Forschungszentrum Karlsruhe Gmbh, Germany (No. ISSN 0948-1427).

Bockenheimer, C., 2011. Novel Approaches to Non-destructive CFRP Bond Quality Assessment. Presentation at SAE 2011 Aerotech Congress (unpublished).

Bockenheimer, C., Valeske, B., 2006. Non-destructive tests for bonded joints – European Adhesive Engineer – EAE. Course material (unpublished).

Bossi, R., Housen, K., Walters, C., Sokol, D., 2009. Laser bond testing. *Materials Evaluation* **67**, 819–827.

Cognard, P., 2002. Collage des matériaux: Mécanismes & Classification des colles – Adhesive joining: mechanisms & adhesive classification. L'Entreprise industrielle, Collage des matériaux 4. From Internet database Techniques de l'ingénieur, Editions T.I., Paris.

Decourcelle, N., Kellar, E.J.C., 2009. *Development of a Methodology to Produce Samples and Ultrasonic Techniques for Kissing Bonds in Adhesive Joints*. Research Reports for Industrial Members of TWI 1–12.

Ecault, R., Boustie, M., Touchard, F., Berthe, L., Chocinski, L., Ehrhart, B., Bockenheimer, C., 2011. Damage of composite materials by use of laser driven shock waves. Presented at the Proceedings of the American Society for Composites 26th Annual Technical Conference/2nd Joint US–Canada Conference on Composites, Montreal, Quebec, Canada, p. 14.

Ehrhart, B., Muller, C.E., Valeske, B., Bockenheimer, C., 2010. Methods for the quality assessment of adhesive bonded CFRP structures – a resumé. Presented at the 2nd International Symposium on NDT in Aerospace 2010, DGZfP, Hamburg, Germany.

Ehrhart, B., Valeske, B., Ecault, R., Boustie, M., Berthe, L., Bockenheimer, C., 2011. Extended NDT for the quality assessment of adhesive bonded CFRP structures. Presented at the International Workshop Smart Materials, Structures & NDT in Aerospace Conference NDT in Canada 2011, Montreal, Quebec, Canada.

Light, G.M., Kwun, H., 1989. Nondestructive evaluation of adhesive bond quality: state of the art review (State of the art review No. SwRI Project 17-7958-838). Nondestructive Testing Information Analysis Center San Antonio, TX.

Markus, S., Tornow, C., Dieckhoff, S., Boustie, M., Ecault, R., Berthe, L., Bockenheimer, C., 2011. *Extended Non-Destructive Testing of Composite Bonds* (11 ATC-0492 No. 2011-01-2514). SAE International, Warrendale, PA.

Marty, P., Desaï, N., Andersson, J., AB, C., 2004. NDT of kissing bond in aeronautical structures. Presented at the 16th World Conference of NDT Proceedings, Citeseer.

Nottorf, E., Engelstad, S., Renieri, M., 2006. Certification Aspects of Large Integrated Bonded Structure. Presentation at ASIP Conference (unpublished).

Parker, B.M., Waghorne, R.M., 1991. Testing epoxy composite surfaces for bondability. *Surface and Interface Analysis* **17**, 471–476.

Simon, H., 1994. Assemblage par collage – Joining by adhesive bonding. Plastiques et composites, Traité Plastique et Composites A3758. From Internet database Techniques de l'ingénieur, Editions T.I., Paris.

Stöven, T., 2006. CFRP-bonding, Seminar Composites und Kleben.

Stöven, T., 2010. Rivetless Aircraft Assembly – a dream or feasible concept. Presentation at VDI EUCOMAS Conference (unpublished).

Valeske, B., Bockenheimer, C., Henrich, R., 2008. New NDT approach for adhesive composite bonds. *NDT Applications* **1**, 203–209.

Weise, V., Scruby, C., Birt, E., Jones, L., 2002. Report on 'Kissing Bond and Environmental Degradation Detection using Nonlinear Ultrasonics' (Report from METEOR Project No. QINETIQ/FST/CR023457), QinetiQ in Confidence. QinetiQ Ltd, Famborough.

Wilhelm, T., Hinnen, M., Schmidt, W., Montnacher, J., Verl, A., 2010. Produktionstaugliche Methode zur Quantifizierung der Adhäsionseigenschaften. *Adhäsion Kleben&Dichten JG* **54**, 14–21.

10
Non-destructive evaluation (NDE) of composites: assessing debonding in sandwich panels using guided waves

S. MUSTAPHA, University of Sydney, Australia and National ICT Australia (NICTA), Australia and L. YE, University of Sydney, Australia

DOI: 10.1533/9780857093554.2.238

Abstract: This chapter focuses on assessing debonding in sandwich carbon fibre/epoxy (CF/EP) composite structures with a honeycomb core using acoustic waves activated and captured by piezoelectric (PZT) elements. The fundamental anti-symmetric A_0 Lamb mode was excited at a low frequency. The debonding location was assessed using the time-of-flight (ToF) of damage-reflected waves, and the severity of debonding was evaluated using both the magnitude of the reflected wave signal and the delay in the ToF of Lamb wave signals. Also, an imaging algorithm based on time reversal was developed to detect debonding in composite sandwich structures using guided ultrasonic wave signals from an active sensor network. The correlation coefficients between the original and reconstructed time-reversed signals in the time or frequency domains were calculated to define the damage index (DI) for individual sensing paths, which were subsequently used to develop an imaging algorithm to identify the location of damage in the monitoring area enclosed by the active sensor network. The results demonstrated the effectiveness of guided waves activated and captured by surface-mounted PZT elements in detecting and assessing debonding/damage in composite sandwich structures.

Key words: debonding, sandwich carbon fibre/epoxy (CF/EP) composite structures, guided waves, time reversal, correlation coefficient.

10.1 Introduction

The high specific strength and the high bending stiffness of composite sandwich structures make them good candidates for application in various fields including aerospace, marine and civil structures. However, debonding between the skin and the core due to impact, cycling and fatigue loading or due to harsh environmental conditions such as extreme temperatures (Hsu, 2009) can substantially reduce the performance of these composite sandwich structures and may cause catastrophic failure during service life. The involvement of these types of structures in many applications generates the need for evaluation methods to access their integrity during service.

Damage detection in sandwich composite structures is particularly difficult because the skin–core configuration makes the damage invisible. Various methods based on non-destructive testing (NDT) such as the instrumented tap tester and air-coupled transmission ultrasonic C-scans (Hsu, 2009) are currently in use for inspection and evaluation of sandwich structures. However, over the past few decades an enormous amount of research has been directed toward developing structural health monitoring systems to be installed on various structures for on-going scrutiny of their integrity, with commensurate focus on the applicability of guided wave for detection and characterization of flaws and defects.

Guided waves are sensitive to both surface and embedded structural damage. They have proved their proficiency and have been widely used to develop various damage identification algorithms for assessing delamination, debonding, holes, cracks/notches and corrosion in both composite and metallic materials (Anton and Inman, 2009; Cuc et al., 2007; Mustapha et al., 2011; Seth et al., 2002; Su and Ye, 2004; Valle and Littles, 2002). Moreover, they have significant potential for decreasing the cost of inspection and for monitoring complex geometries.

Thwaites and Clark (1995) utilized phase velocity in the range of frequencies between 5 and 50 kHz to detect debonding in sandwich composite structures, accessing only one side of the structure. Castaings and Hosten (2001) studied the propagation of guided waves in composite sandwich-like structures and their use for traditional NDT using angle beam transducers. Osmont et al. (2001) monitored the health condition of sandwich plates using piezoelectric (PZT) sensors, based on analysis of the interaction of Lamb waves with damage, using finite element modeling (FEM). Diamanti et al. (2005a, 2005b) experimentally examined the potential for using low-frequency Lamb waves to detect damage in sandwich beams based on analysis of time-of-flight (ToF). Yang and Qiao (2005) looked at wave propagation and scattering, and also performed some experiments on aluminium beams, carbon fibre/epoxy (CF/EP) composite laminates, and sandwich composite beams to validate the effectiveness of guided waves in damage detection. Qi et al. (2008) studied the effect of debonding in a composite sandwich structure on ultrasonic guided waves. It was demonstrated that Lamb waves are effective in detection of debonding, giving a quantitative estimation of the debonding by calibrating the energy of wave signals received after interacting with the debonding. More work was carried out by Song et al. (2009, 2010) in an attempt to understand the guided wave propagation mechanism in honeycomb composite panels. They numerically solved the wave propagation by considering two finite element models; one included the geometry of individual cells of honeycomb whereas the other had an effective core of orthotropic properties. At a low frequency, good agreement was observed between the results of the two models. They also looked at single and multiple debonding in sandwich panels and calculated a damage index based on the intact and contaminated signals after a disbond was inserted. They were able to construct an image of the debonding based on probability analysis of the leaky guided wave.

This chapter focuses on the application of guided waves for detection of debonding in composite sandwich structures. Section 10.2 describes some of the signal processing tools used within our work and also describes the method used for data fusion for image generation. Section 10.3 briefly describes some aspects of numerical modelling of guided wave using finite element analysis (FEA) and the response when they encounter a debonding. Section 10.4 describes quantitative assessment in sandwich CF/EP composite beams with a honeycomb core using acoustic waves activated and captured by surface-mounted PZT elements. Section 10.5 introduces an imaging algorithm based on time-reversal, which was developed to detect debonding in composite sandwich structures using guided ultrasonic waves signals from an active sensor network. Section 10.6 presents some conclusions and suggestions for future studies.

10.2 Processing of wave signals

10.2.1 Selection of wave modes and locating of debonding

Lamb wave velocity depends on both the excitation frequency (f) and the thickness (h) of the plate combined in a frequency–thickness ($f \cdot h$) product. Once a central frequency of a transducer is selected, it imposes an upper limit on the thickness of the plate in which the fundamental Lamb wave mode can be excited without introducing high-order Lamb modes. Such a condition is critical in the use of Lamb waves for non-destructive evaluation (NDE) applications, since the excitation of a single Lamb mode can simplify the signal processing.

For the composite sandwich plate with plain woven composite face sheets and Nomex honeycomb core used in this study, preliminary experimental results showed that the A_0 mode was the most pronounced in magnitude in the low-frequency ranges at which the S_0 mode of the Lamb wave was barely visible, as shown in Fig. 10.1. Any required mode can be selected and it is preferable to select a frequency where only one mode exists, when the complexity in interpretation of the wave can be minimized.

In a sandwich composite beam with surface-bonded PZT actuator and sensors (Fig. 10.2), when the activated A_0 Lamb mode encounters a debonded area, it transmits through the debonding area with a reflection

Assessing debonding in sandwich panels 241

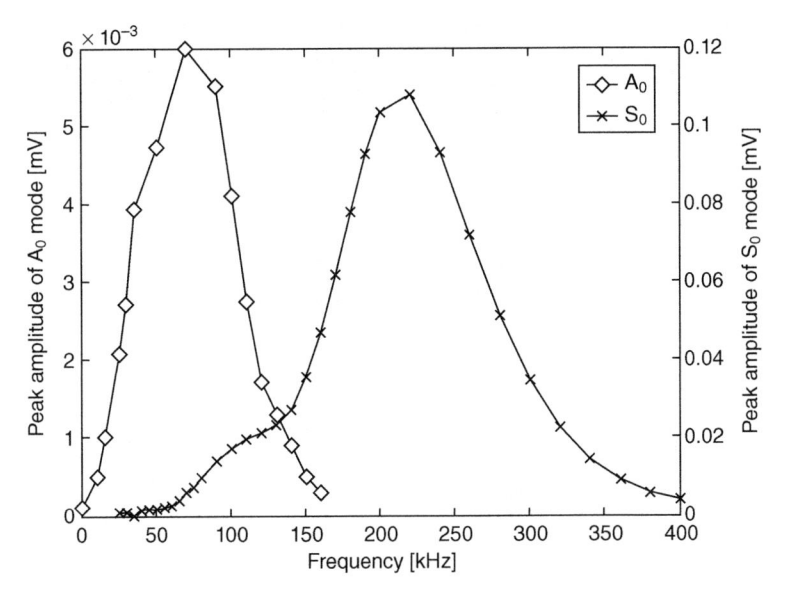

10.1 Magnitudes of A_0 and the S_0 modes activated in composite surface panel.

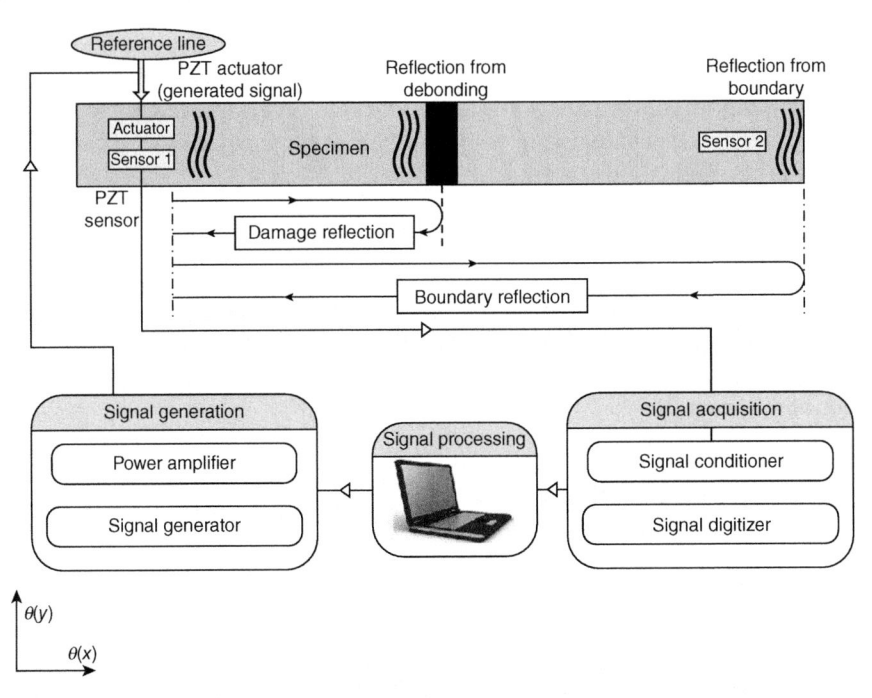

10.2 Schematic diagram of experimental set-up for debonding detection.

© Woodhead Publishing Limited, 2013

wave. With a 'pulse and echo' configuration of the wave actuator and Sensor 1, the debonding location can be identified by:

$$d = \frac{C_g \Delta T}{2} \qquad [10.1]$$

where d is distance from the centre of the sensor to the centre of the debonding; C_g and ΔT are the wave propagation velocity and the difference in ToF between the damage-reflected wave and the incident wave captured by Sensor 1, respectively.

10.2.2 Hilbert transform

To distinguish wave reflection from different sources, e.g. damage and structural boundaries, it is essential to determine a precise value for the ToF from the captured wave signals, and appropriate signal processing is of vital importance (Raghavan and Cesnik, 2007a, 2007b). Wave propagation in an elastic medium is the transportation of energy, and the interaction of waves with structural damage can significantly influence their propagation properties, accompanying energy reflection, transmission and mode conversion (Rose, 1999). Differences in the location and severity of damage can produce unique energy scattering phenomena, and the energy of the response signals can contain key information about the damage. The captured Lamb wave signals were thus evaluated in terms of their energy magnitude in the time domain through a Hilbert transform.

Accordingly, two damage-sensitive features, i.e. time delay, defined as the difference in ToF between an intact benchmark and a beam with debonding, and the reflection coefficient (i.e. the ratio of energy magnitude between the damage-reflected wave signal and the incident signal). Both damage signatures can be applied to different actuator-sensor configurations, either 'pulse and echo' (i.e. actuator and Sensor 1) or 'pitch and catch' (i.e. actuator and Sensor 2).

10.2.3 Damage image construction

Damage index (DI)

The damage index (DI) can be defined using the characteristics of wave signals in the time domain or the frequency domain. In the time domain, the changes between the original input wave signal V_A, $\{a_1, a_2, \ldots, a_n\}$ and the reconstructed signal after time reversal $V'_A \{b_1, b_2, \ldots, b_n\}$ of a sensing path can be measured quantitatively by the correlation coefficient, defined by Wang et al. (2010).

$$\rho_{a,b}(t) = \frac{n\sum a_i b_i - \sum a_i \sum b_i}{\sqrt{\left[n\sum a_i^2 - \left(\sum a_i\right)^2\right]\left[n\sum b_i^2 - \left(\sum b_i\right)^2\right]}} \qquad [10.2]$$

Before correlation, both V_A and V'_A are normalized to compensate for the dispersion in wave energy. The value of $\rho_{a,b}(t)$ indicates the degree of relationship between two signals V_A and V'_A. When each signal is perfectly approximated by the other, $\rho_{a,b}(t) = 1$. Furthermore, the correlation coefficient approaches unity if the sensing path is very distant from the damage (when the captured signal is not affected by the damage), but it decreases the closer the sensing path is to the damage, reflecting the fact that the time reversibility of the captured wave signal deteriorates after it interacts with damage.

The DI is defined as the similarity ($\rho_{a,b}(t)$) of the wave signals subtracted from unity (Wang et al., 2010):

$$DI = 1 - \rho_{a,b}(t) \qquad [10.3]$$

In this way, the greater the DI, the higher the possibility of the existence of damage or the closer the damage to the wave path, defined by the time reversibility of the input wave signal.

Distribution density function

For the existence of damage at any location near the wave path, the effect of the DI is confined in a bell shape with a normal distribution function, which means the effect of the DI is maximized on the wave path and decreases gradually as the distance from the wave path increases. Considering a normal distribution function with mean μ and standard deviation σ, the density function is given by

$$f(z) = \frac{1}{\sigma\sqrt{2\pi}} e^{-\frac{(z-\mu)^2}{2\sigma^2}} \quad \text{for} \quad -\infty < z < \infty \qquad [10.4]$$

where z is the distance to the path. The use of the function $f(z)$ highlights damage if it is close to the path but its effect decreases when it is further away, and this function determines the size of the affected zone for each individual sensing path. Determination of the affected zone is addressed later in this section. The variation of the σ value and its effect is shown in Fig. 10.3, as the σ increases the bell shape become wider and its influence covers a larger area.

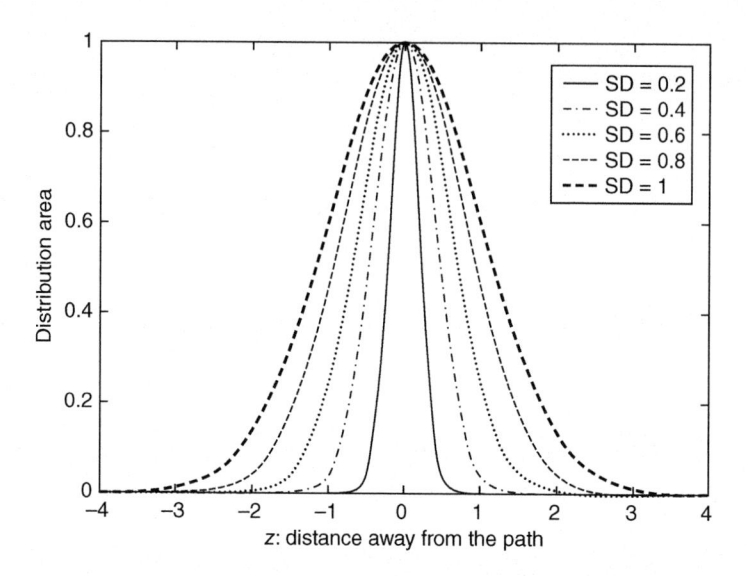

10.3 Representation of damage presence beside a sensing path.

The definition of σ was based on an experimental investigation in which a circular steel block 40 mm in diameter simulating damage was placed midway along an actuator-sensor path of 400 mm in length and then gradually shifted off the path in the direction perpendicular to it, as shown in Fig. 10.4(a).

An S_0 wave mode was excited at 150 kHz, and the captured wave signals before and after the Hilbert transform are displayed in Fig. 10.4, showing the influence of damage on the magnitude of S_0 mode. It can be seen that the wave signal was substantially affected by the damage as long as the distance from the centre of the damage to the path was less than 40 mm, confirming that the transmitted wave signal was sensitive to damage even when it did not lie precisely across the wave pathway. Therefore, in equation (10.4) μ and σ were set to 0 and 40 mm respectively.

Further, two different approaches were investigated to determine the value of z with a rectangular and an elliptical affected-zone area for each sensing path:

- $z = \dfrac{d}{D_k}$ where d is the normal distance from a node to the kth path and D_k is the distance between the actuator and the sensor of the kth path, and

- $z = \dfrac{D_a + D_s}{D_k} - 1$ where D_k is the distance between the actuator and sensor for the kth sensing path, D_a and D_s are the distances between a node to the actuator and the sensor respectively (Wang *et al.*, 2010).

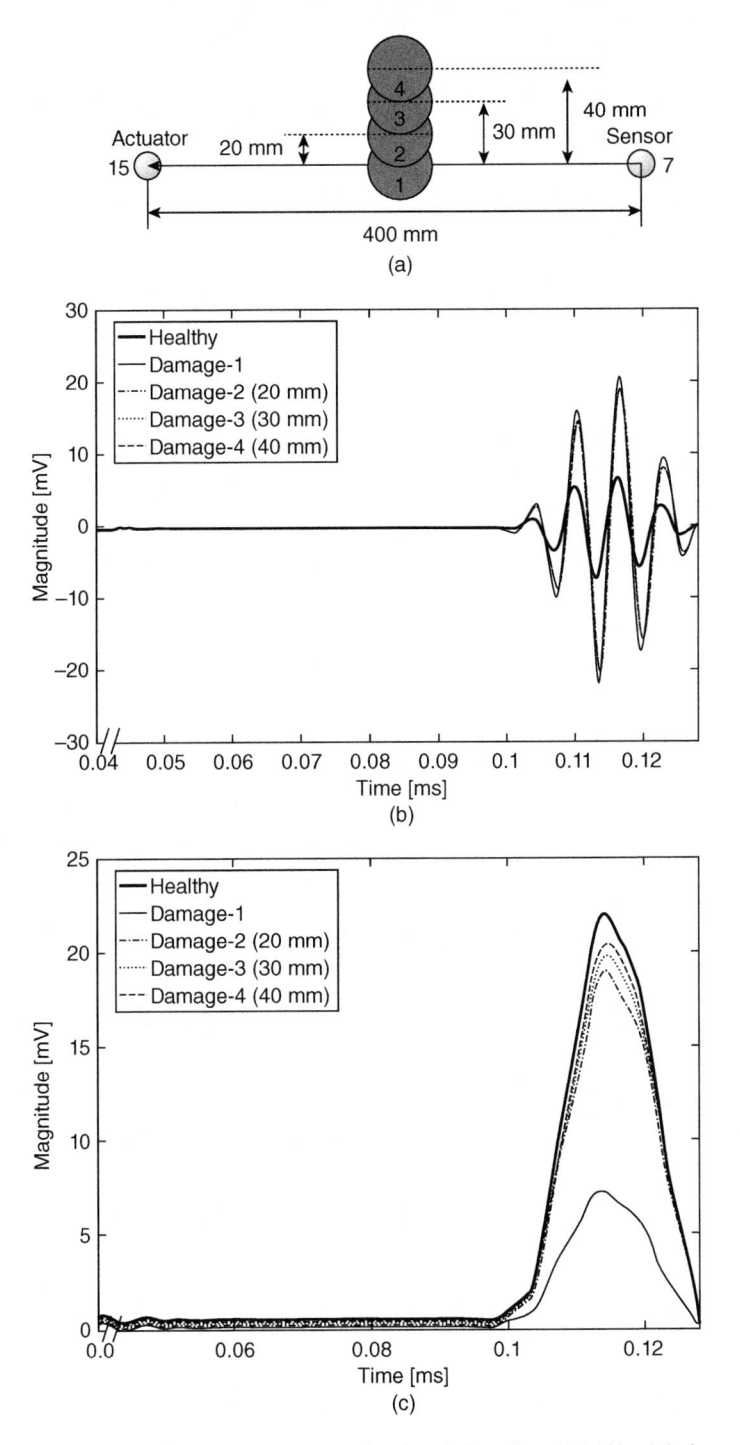

10.4 Effect of damage on magnitude of signal at 150 kHz: (a) damage position, (b) S_0 mode captured by sensor, and (c) S_0 mode after Hilbert transform.

Image construction

The monitoring area was meshed into uniform grids and the presence of damage at each grid was estimated by fusion of perceptions of the existence of damage from individual sensing paths. Assuming that there are N sensing paths for damage identification from the sensor network, estimation of the presence of damage at position (x, y) in the monitoring area can be written as (Wang *et al.*, 2009):

$$P(x, y) = \sum_{k=1}^{N} P_k(x, y) = \sum_{k=1}^{N} DI_k f_k(z) \qquad [10.5]$$

where DI_k and $f_k(z)$ are described in equations (10.3) and (10.4) respectively.

This process was applied to each individual path, and $P(x, y)$ at each grid was obtained for the area. In this way, perceptions of the damage were combined for all the paths in which the existence of damage was highlighted by an area with high values of $P(x, y)$ indicating the existence of damage.

10.3 Numerical simulation of wave propagation

Numerical modelling of the characteristics of elastic wave propagation is of a great interest to many researchers working in the area of structural health monitoring. Wave propagation in composite structures is relatively more complex due to the anisotropy in the material properties and multi-layered lay-up. Various numerical methods have been developed and used to calculate the propagation of elastic waves, including the finite element method (Ye *et al.*, 2006; Zienkiewicz and Taylor, 1989), the finite difference method (Strikwerda, 2004), finite strip elements (Liu and Achenbach, 1994; Liu and Xi, 2002), spectral element method (Doyle, 1989; Patera, 1984; Peng *et al.*, 2010), boundary element methods (Cho and Rose, 1996; Schafbuch *et al.*, 1993; Zhao and Rose, 2003), mass-spring lattice models (Hyunjune and Younghoon, 2000) and local interaction simulation approach (Lee and Staszewski, 2003a, 2003b). In this section, we aimed to numerically simulate elastic wave in the sandwich structure based on the FEM; the characteristics of the wave propagation are discussed and subsequently its interaction with debonding is investigated.

10.3.1 FEM for a sandwich structure

Modelling composite materials entails some difficulties and is not well implemented in commercial finite element software. The configuration of the plies limits the type of element that can be used. Shell elements are widely used for modelling thin wall structures and are ideal for laminated composites.

Assessing debonding in sandwich panels 247

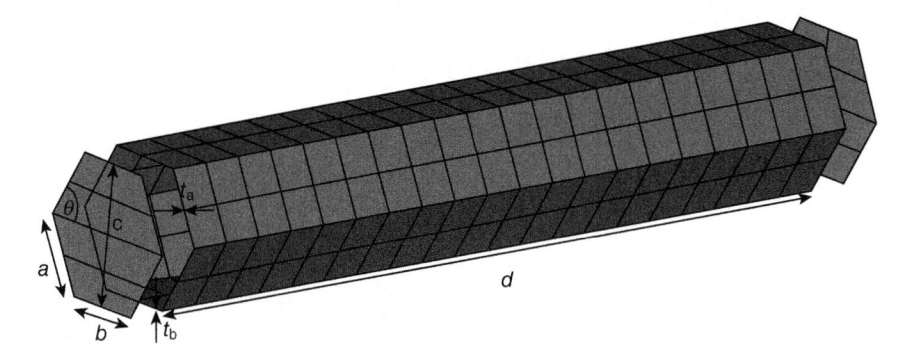

Dimensions (mm)	HRH-10-1/8-3
a	19.3
b	61.58
c	3.4
d	20.45
t_a	0.0496
t_b	0.0962
θ (degrees)	61.6

10.5 Mesh and geometry of the honeycomb unit cell.

Dynamic finite element simulation using ABAQUS was conducted to evaluate the wave propagation characteristics in the sandwich plate and to identify the sensitivity of the wave to debonding. The FEM of the sandwich consisted of shell elements for both the skin and the core. Note that the skin was modelled at the centre line, an offset of 0.44 mm was applied (S4R, a 4-node quadrilateral shell element with 6 DoF). The core geometry was included and details of the geometry are shown in Fig. 10.5.

The skin and the core were assembled together using node-to-node tie elements based on the master–slave formulation in ABAQUS/EXPLICIT®, where the skin is the master and the core is the slave. The master–slave formulation will allow penetration of the master surface into the slave surface, but the slave surface cannot penetrate into the master (Liyanapathirana and Kelly, 2010). To model the debonding between skin and core, the nodes in the debonded region were left untied (Burlayenko and Sadowski, 2009), but in order to prevent the debonded face sheet from overlapping with the core at the interfacial damage, 3D spring elements SPRINGA were introduced between nodes of the debonded region (Burlayenko and Sadowski, 2009). As the mesh for an FE model is dependent on the wavelength of elastic wave propagating in the model, it is recommended to refine the mesh such that six or more elements should exist per wave length for elastic wave propagation (Yang et al., 2006). In this model the adhesive was ignored and perfect bonding was assumed.

The mechanical properties of the laminate were obtained from Li et al. (2006), while the materials of the core were obtained from testing carried within our labs. Table 10.1 summarizes the properties that were used for both the panels and the core. Since the Nomex core is a non-woven sheet made of short aramid fibres (Nomex), its isotropy is assumed (Florens et al., 2006). The model was simply supported to simulate the same boundary conditions as in the experiments.

Lamb waves were excited by prescribing concentrated forces along nodes of the PZT element, and wave propagation was activated using a waveform of five sinusoidal cycle tone bursts endowed by the Hanning function, which effectively prevented wave dispersal. Dynamic simulation was accomplished using ABAQUS/EXPLICIT® code. The node displacements at the sensor location were acquired to determine the characteristics of the different Lamb modes. The simulated Lamb signals were acquired at a sampling frequency of 20.48 MHz by a sensor model (Lin and Yuan, 2001b), where the displacements of all the neighbouring nodes beneath the sensor were averaged and normalized respectively (Mustapha et al., 2011).

Table 10.1 Properties of CF/EP sandwich panel

(a) Properties of plain woven fabric CE/EP composite laminate

$E_{11} = E_{22}$	$G_{12} = G_{23} = G_{31}$	μ_{13}	μ_{12}	μ_{23}	Density
55.8 GPa	3.65 GPa	0.05	0.06	0.38	1600 kg/m^3

(b) Properties of Nomex core (HRH 10 1/8-3)

Cell diameter	Core thickness	Density
3.2 mm	20.4 mm	64 kg/m^3

(c) Properties of the Nomex material

E	v	Density
3.6 GPa	0.30	1380 kg/m^3

10.3.2 Results for wave propagation and interaction with damage

Sandwich beam

A refined mesh of the sandwich beam (560 mm × 35 mm) resulted in a total of 615 144 elements and 580 462 nodes, which consequently resulted in 3 382 992 variables. Two circular PZT elements (10 mm in diameter) functioning as actuator and sensor were placed 200 mm apart, symmetrical about the mid-span of the beam. A debonding 20 mm in length and running across the width of the beam was placed at mid-span between the two PZTs. Two frequencies of 20 and 150 kHz were selected to excite the guided wave in the beam during the numerical simulation.

Snapshots of the wave field for the excitation frequencies of 20 and 150 kHz are shown in Fig. 10.6, being representations in the time domain through transmission signals for both the intact and debonded state models. Figure 10.6(a) shows the excited wave at 20 kHz where the A_0 is dominant. The wave mode propagated through the whole thickness of the beam, treating the sandwich beam as single layer. However, the wave was partially

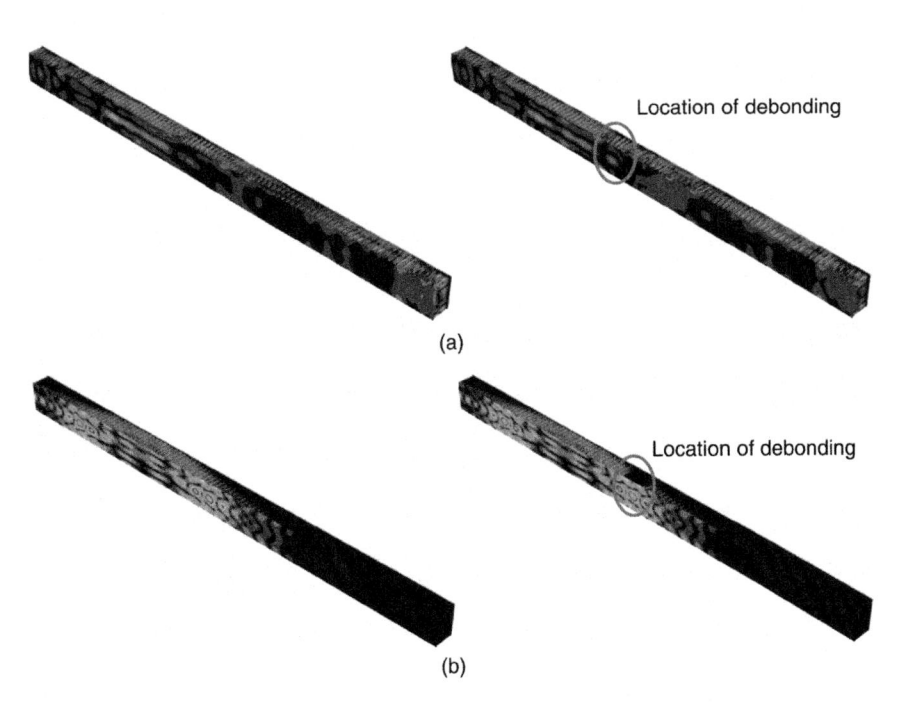

10.6 (a) A_0 at excitation frequency of 20 kHz and (b) S_0 at excitation frequency of 150 kHz.

restrained in the top half of the core when the S_0 mode was excited at 150 kHz, as in shown Fig. 10.6(b). The addition of debonding caused a disruption in the wave field as well as alteration in the magnitude of the signal for both the A_0 and the S_0 modes.

Sandwich plate

A total of 1 102 246 elements and 848 983 nodes formed the model of the sandwich plate (560 mm × 560 mm), resulting in 5 092 512 variables in the model. The actuator was placed in the centre of the plate and was excited at frequencies of 10 and 150 kHz. Figure 10.7(a) shows the propagation of S_0 in the in-plane direction on the sandwich plate. In Fig. 10.7(b) after we inserted debonding in the plate the interruption of the wave field is very apparent and is highlighted.

A sensor was placed in the FEM model in the position highlighted in Fig. 10.7(a), and signals were captured for the frequencies of 10 and 150 kHz, displayed in Fig. 10.8 and Fig. 10.9 respectively. Comparing them with the experimental results indicates very similar behaviour. The interaction of the S_0 excited at 150 kHz with the debonding resulted in reduction in amplitude whereas the A_0 mode excited at 10 kHz increased slightly in amplitude upon interaction with the debonding, these effects were due to the leaky and non-leaky behaviour of the guided waves.

Note that guided waves have leaky and non-leaky behaviour, depending on the selected mode and frequency. When a leaky wave propagates, it

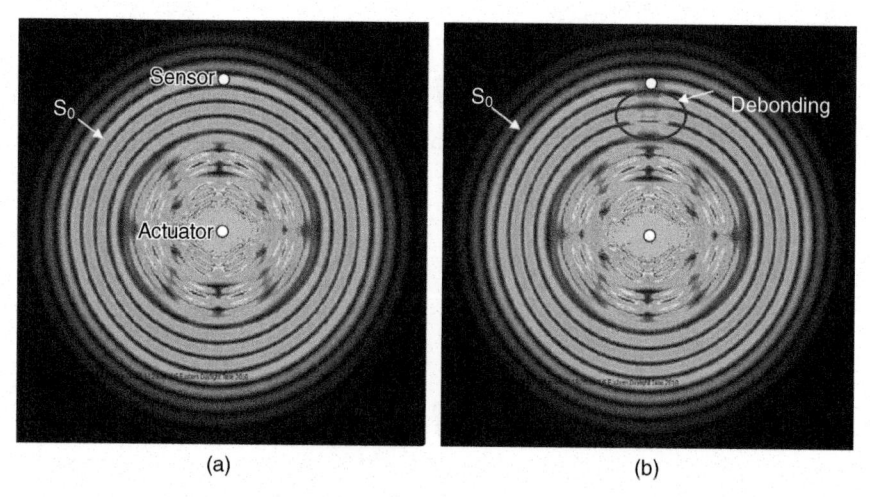

10.7 Excitation of Lamb wave at 150 kHz: (a) intact plate and (b) interaction with debonding.

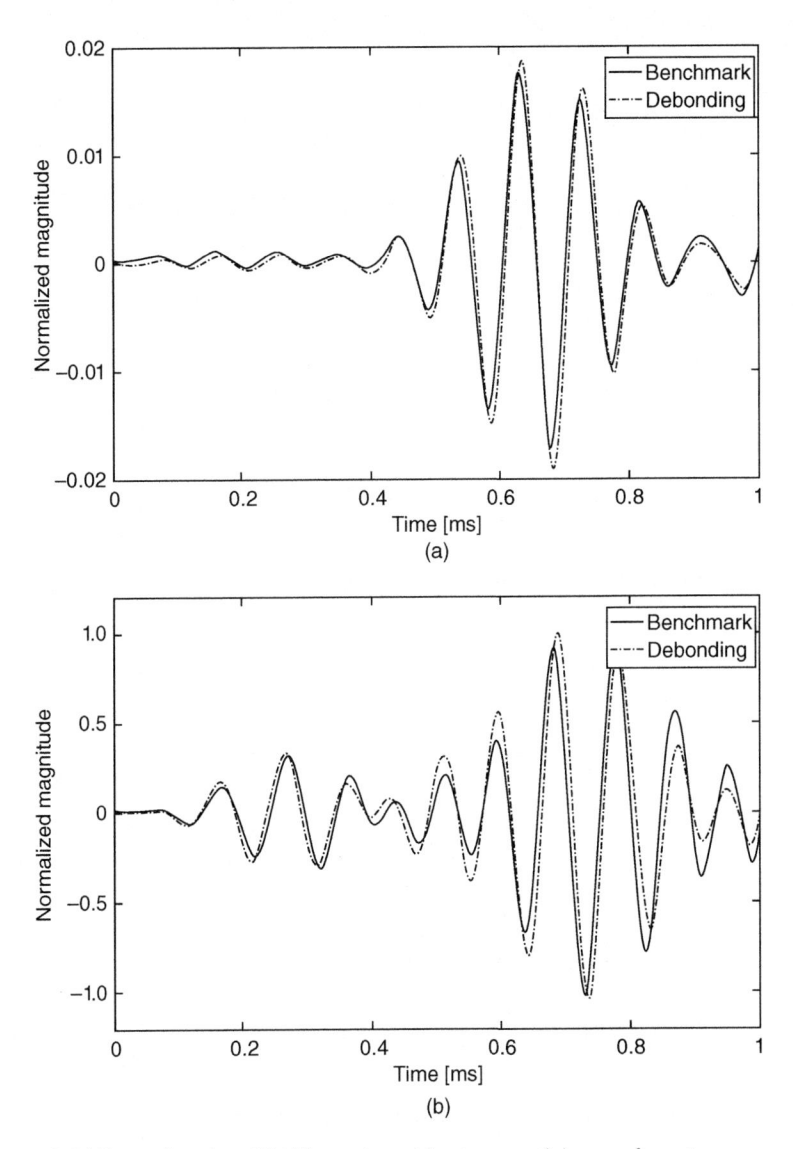

10.8 Wave signal at 10 kHz captured by sensor: (a) experiment, (b) simulation.

attenuates quickly because of losing energy to the half-space. A debonding or crack at the interface can block the leaky energy transmission into the half-space and hence reduce the attenuation, resulting in an increase in amplitude. In contrast, non-leaky wave energy is concentrated on the top surface, and on interacting with damage the wave will scatter, causing a drop in the energy level of the wave signal.

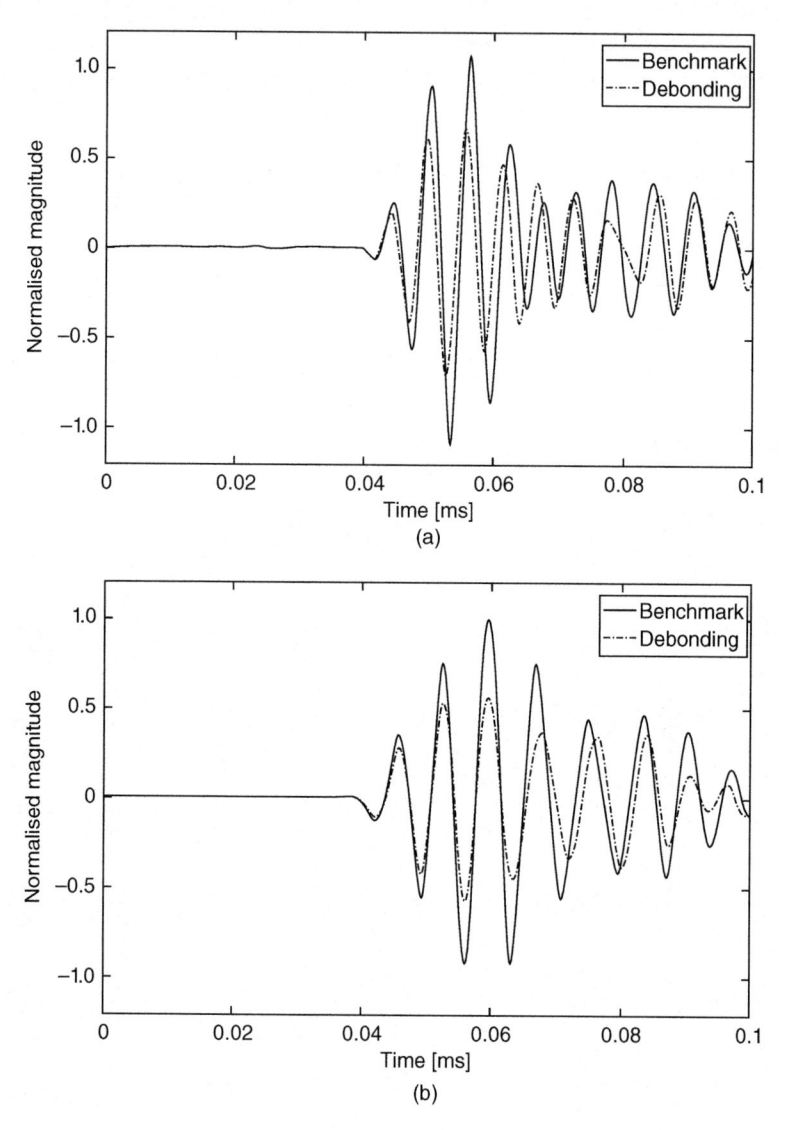

10.9 Wave signal at 150 kHz captured by sensor: (a) experiment, (b) simulation.

10.4 Debonding detection and assessment in sandwich beams

A quantitative assessment of debonding in sandwich CF/EP composite structures with a honeycomb core using acoustic waves activated and captured by surface-mounted PZT elements is presented. Debonding was introduced at different locations in sandwich CF/EP composite beams. The fundamental anti-symmetric A_0 Lamb mode was excited at a low frequency. The transmitted and reflected wave signals in both surface panels were captured by PZT elements after interacting with the debonding damage and specimen boundaries.

We quantitatively investigated the effects of the extent of debonding on the delay in ToF and the energy of the damage-reflected wave signal of the fundamental anti-symmetric A_0 mode at a low frequency in sandwich CF/EP composite beams with a honeycomb core, when the extent of debonding is increased gradually until the sandwich beam is fully debonded.

10.4.1 Sandwich beams

For the test specimens of sandwich composite beams, skin panels were manufactured using Cycom 970/T300 prepregs (Mikulik *et al.*, 2008) consisting of eight woven plies in a quasi-isotropic lay-up [±45, 0/90, ±45, 0/90]s. The core was HexWeb HRH10-1/8-4 (12.7 mm thick) (Hoo Fatt and Park, 2001). The skins and core were assembled via secondary bonding using FM 1515-3 film adhesive (Liljedahl *et al.*, 2008). The specimens were cut into 450 mm (length) × 30 mm (width) sections (shown in Fig. 10.10) with two skin panels each of 1.76 mm in thickness.

Four PZT (PI® PIC151, PQYY-0586) element electrodes, measuring 20 mm × 5 mm × 1 mm, were surface-mounted on each specimen, using a silver-epoxy adhesive (RiteLok SL65®). One PZT element served as the actuator to generate Lamb waves and the other three served as sensors to capture the incident signal (assuming it was infinitely close to the actuator), reflected wave signals (Sensor 1), and transmitted Lamb wave signals (Sensor 2) in the composite beams, as shown in Fig. 10.2. Sensor 3 was located beneath Sensor 1 on the opposite side of the beam, functioning

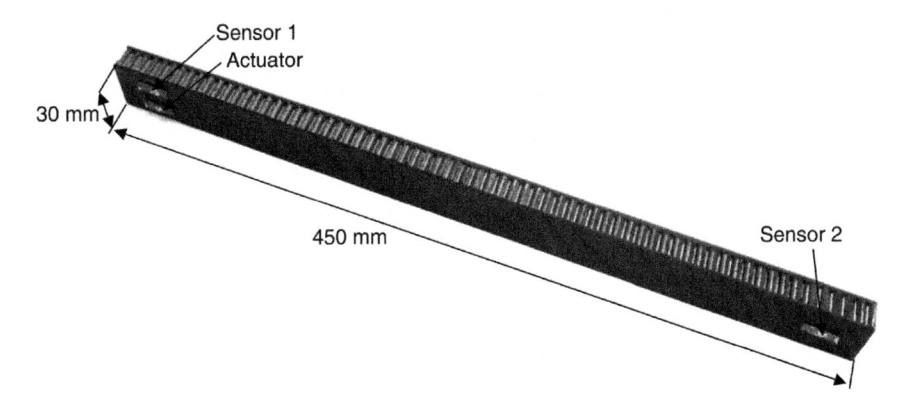

10.10 Geometry of the sandwich CF/EP beam.

Table 10.2 Summary of specimen configurations for sandwich beams

Sample	Debonding location	Debonding size
Benchmark	N/A	N/A
Debonded (1)	half-way	20 mm in span
Debonded (2)	3/4 way	20 mm in span

similarly to Sensor 1. One of the specimens was kept intact as a benchmark, and the others were damaged by introducing a cut at the interface between the skin and the core, using a thin sharp blade. Table 10.2 lists the details of the manufactured specimens. To understand the Lamb wave propagation in the composite panel and to study the effect of the core on the wave propagation behaviour, the composite skin panel was cut into 450 mm (length) × 30 mm (width) × 1.76 mm (thickness) beams, surface-mounted with PZT elements in similar configuration as for the sandwich beams.

10.4.2 Experimental set-up

With the actuator and sensor both located at one end of the beam (Fig. 10.2), activation and acquisition of Lamb wave signals were fulfilled using an active signal generation and data acquisition system developed on the VXI platform, consisting mainly of a signal generation unit (Agilent© E1441), signal amplifier (PiezoSys® EPA-104), signal conditioner (Agilent© E3242A) and signal digitizer (Agilent© E1437A). Five-cycle sinusoidal tone bursts enclosed in a Hanning window were generated and applied to the PZT actuator, and the Lamb wave signals were captured using the PZT sensor at a sampling rate of 20.48 MHz. The acquisition duration was set to ensure that at least the first reflected Lamb wave signal from the right-hand end of the beam was captured. The specimens (sandwich beam or skin panel) were supported at both ends without clamping. Any clamping of the specimens could cause a change in its stiffness, making it difficult to reproduce the experimental results unless the exact amount of applied torque could be controlled.

10.4.3 Results and discussion

Lamb wave propagation velocity

The magnitudes of the incident and boundary-reflected waves with a central frequency of 6.5 kHz in the surface panel and the sandwich beam are displayed in Fig. 10.11(a). The propagation velocity of the anti-symmetric A_0 mode was determined using the ToF and the total propagation distance of

Assessing debonding in sandwich panels 255

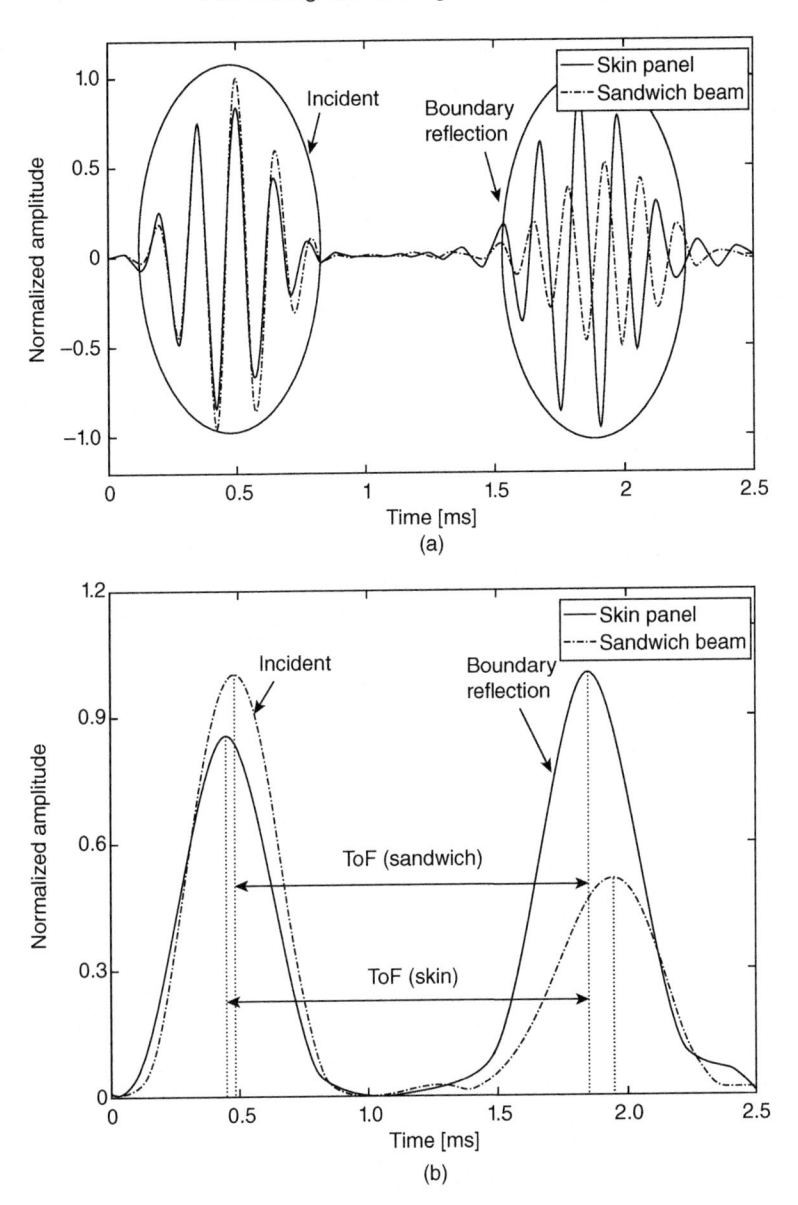

10.11 Simulated wave signals in composite skin panel and sandwich beams: (a) pulse-echo signal and (b) Hilbert transform.

Table 10.3 Comparison of propagation velocity in skin panels and sandwich beams

Case	Propagation velocity [m/s]	Error [%]
	Experiment	
Skin panel	609	~1
Sandwich beam	584	~1

the wave between the centre of Sensor 1 and the far right end of the specimen. To extract the precise ToF the Hilbert transform was employed on both the experimental and simulated Lamb wave signals, as explained in Section 10.2.2. The ToF was defined as the time interval between the peaks of the incident wave and the boundary reflection captured by Sensor 1, as shown in Fig. 10.11(b). Using equation (10.1), the propagation velocity of the A_0 mode at a central frequency of 6.5 kHz was 609 m/s in the skin panel and 584 m/s in the sandwich beam.

The difference in propagation velocity between the skin and the sandwich beam can be attributed to the energy scattering of the Lamb wave signal within the core. As shown in Fig. 10.11(b), there was a clear reduction in the magnitude of the wave reflection in the sandwich beam. The amplitude of the transmitted signals was reduced by about 40% after travelling for 800 mm. The energy loss may be attributable to the existence of the Nomex honeycomb core. The results from experiments and simulation show good agreement, though some minor differences exist, as shown in Table 10.3.

Debonding assessment

Figure 10.12(a) displays the Lamb wave signal captured by Sensor 1 from the experimental results, activated by the actuator at a central frequency of 6.5 kHz in the specimen with a section of debonding 20 mm in length in the middle of the span ($x = ½ L$). Using Equation (10.1) and the wave propagation velocity in the sandwich beam, the centre of the debonding was identified at 222 mm from the reference line (centre of the actuator and Sensor 1) shown in Fig. 10.2. Further analysis was undertaken for a debonding section 20 mm in length placed at three-quarters of the span ($x = ¾ L$). The results are shown in Fig. 10.12(b), where the centre of debonding was located at 328 m. This method shows the accuracy in detection of flaws in the structure, with an error of less than 5%. FEA simulation was also conducted, and the results showed good agreement with the experiments.

Interaction of Lamb waves with various boundaries can alter the characteristics of the wave signals, reducing the accuracy of the current approach.

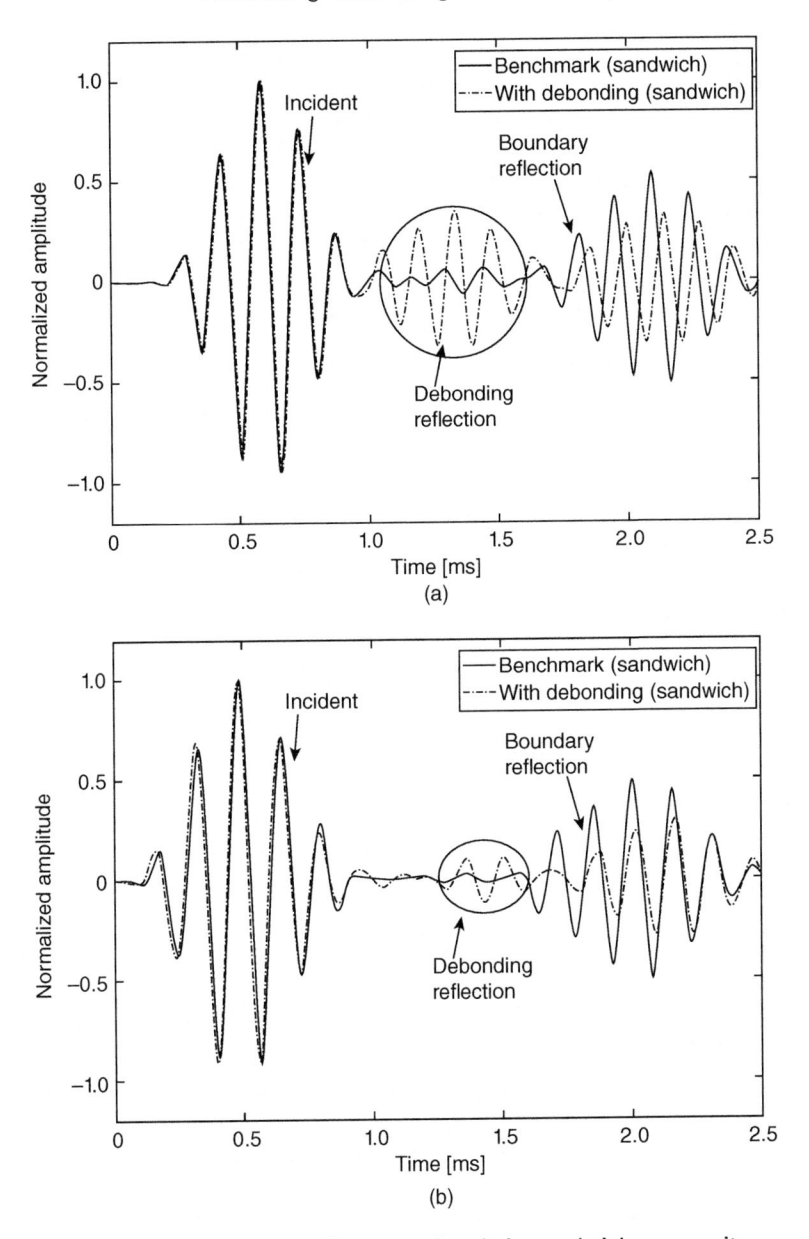

10.12 Experimental results for wave signals in sandwich composite beams with debonding (20 mm in length): (a) debonding at mid-span and (b) debonding at three-quarters along the span from the left-hand end.

Although the sensor may be placed at a distance from all boundaries (Kim-Ho Ip, 2004) to distinguish different reflected components of the wave signals, this becomes impractical in application. Considering the short span of the beam specimens in this study, the PZT actuator and Sensor 1 were placed near one end (the left-hand end of Fig. 10.2) to increase the ToF of the damage-induced component in the wave signals. However, with this configuration, the wave reflection from the left-hand end of the beam overlapped with the wave signals captured by Sensor 1. Moreover, it was observed that the amount of torque applied to fix the ends of the beam could also have a noticeable effect on the experimental results. When the torque was increased to a certain level, extra boundary reflections appeared. Therefore a free boundary arrangement may facilitate interpretation of the wave signals with consistent results.

Debonding severity

The delays in ToF for both the debonding-reflected signals captured by Sensor 1 in a 'pulse and echo' configuration and the transmitted wave signals captured by Sensor 2 in a 'pitch and catch' configuration were evaluated. In experiments the length of the debonding was increased gradually from 10 mm, using a thin sharp metal blade, until the entire skin was peeled off the honeycomb core, while the centre of the debonding remained constant (in the middle of the beam).

Debonding changes the local effective modulus of the structure and, as a consequence, the local speed of the Lamb wave varies. Each time the extent of debonding was increased the ToF was determined and the delay was calculated relative to the ToF for the beam without debonding. The time delay is plotted against the extent of debonding in Fig. 10.13.

The delay in the ToF for both reflected and transmitted wave signals showed the same trend. For Sensor 1, however, the captured wave had travelled double the distance and crossed through the debonding twice, compared with that captured by Sensor 2; therefore the time delay for Sensor 1 was halved before including it in Fig. 10.13. Furthermore, the time delay initially increased with the increasing extent of debonding. However, as the extent of debonding increased beyond a certain length (about 60 mm), the time delay began to decrease gradually. This phenomenon is attributed to the fact that the Lamb waves propagated in the skin panel when the extent of debonding became large, and the skin panel had a higher effective stiffness than the sandwich beam, allowing the Lamb waves to travel more quickly and thus reduce the time delay. Similarly, the variation of the reflection coefficient with debonding length was also evaluated. An important observation was that as the extent of debonding was increased, the magnitude of the reflection from the debonding captured by Sensor 1 increased

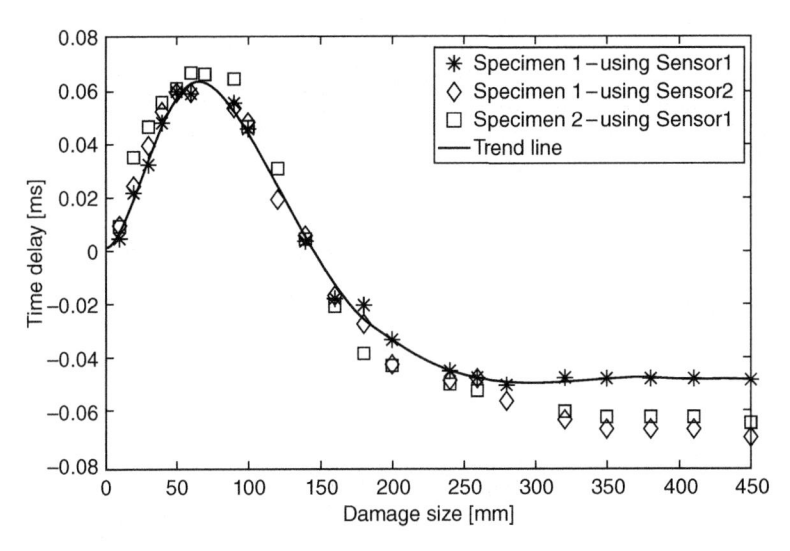

10.13 Correlation between debonding extent and delay in ToF in sandwich composite beams.

progressively when the extent of debonding was increased from 10 to 30 mm, accompanied by a reduction in the magnitude of the boundary reflection, shown in Fig. 10.14. However, as the debonding was increased beyond 35 mm, the magnitude of the reflection decreased, but it began to increase again when the extent of debonding reached 70 mm. This phenomenon is attributed to the competition of two influencing factors when the extent of debonding becomes large: the size of damage, which increases the magnitude of the reflection, and the decrease in travelling distance, which helps maintain the magnitude of the reflection. Figure 10.15 shows the results for the reflection coefficient versus the debonding size.

10.5 Debonding detection in sandwich panels using time reversal

Some attention has been paid recently to the application of time-reversal wave signals for damage detection (Lin and Yuan, 2001a). Ing and Fink (1998b, 1998a) applied the time-reversal process (TRP) to Lamb waves signals to refocus the energy in the time and spatial domains so as to compensate for the dispersive characteristics of Lamb waves. Kim and Sohn (2006) evaluated issues related to time reversal, and found that the bonding quality of surface mounted PZT elements affected the results; thus special care should be taken to control the quality of the bonding. Xu and Giurgiutiu (2007) investigated reversing A_0 and the S_0 modes either separately

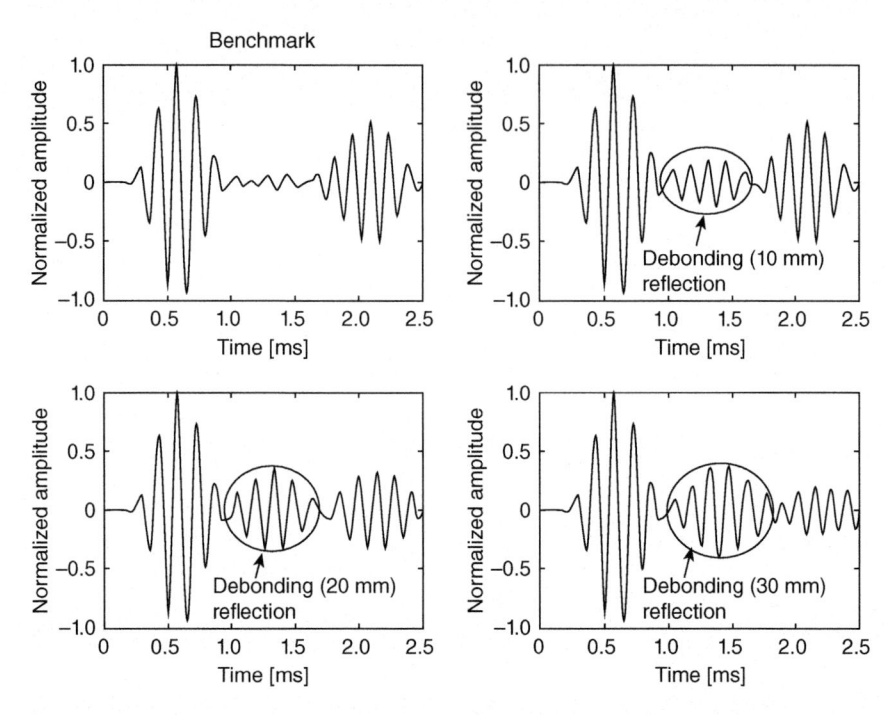

10.14 Increase in magnitude of debonding-reflected signals with extent of debonding.

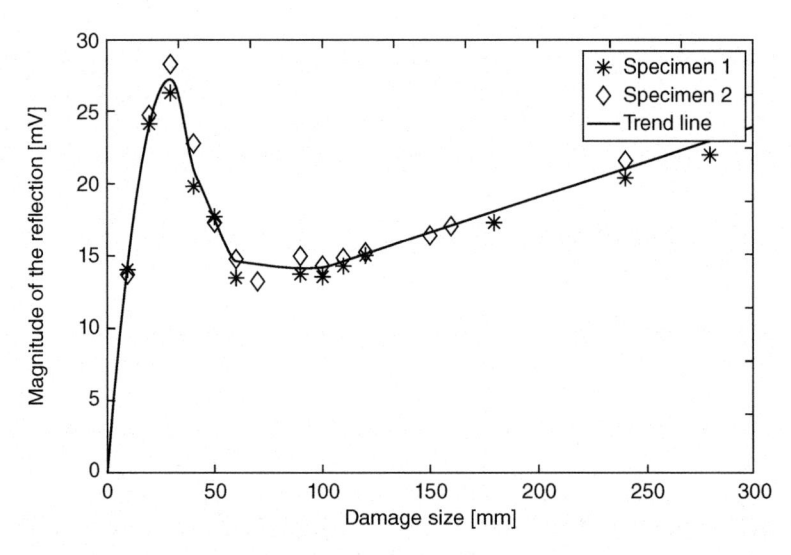

10.15 Correlation between debonding extent and reflection coefficient in sandwich composite beams.

or combined, in an aluminium structure, in order to reconstruct the original wave signal using narrow band tone burst excitation. Park *et al.* (2007) and Sohn *et al.* (2007a, 2007b) applied time-reversal to guided waves in order to detect defects in composite structures using methods of constitutive outlier analysis and wavelet-based signal processing. Gangadharan *et al.* (2009) investigated the use of time reversal in metallic structures, and experimentally tested A_0 and S_0 modes using narrowband and broadband excitation. It has been observed that for both S_0 and A_0 modes, as the number of cycles is increased in the tone-burst signal, the reconstructed time-reversed signal is closer to the input signal in shape. However, the method fails in some cases, as when Thien *et al.* (2006) used time reversal to detect cracks in a pipeline system, reconstruction of the original signal was unfortunately not very successful.

This part of the work investigated the use of time-reversal ultrasonic wave signals method for detecting debonding areas in composite sandwich structures. Experimental investigations using both A_0 and S_0 Lamb wave modes were conducted in detail using CF/EP sandwich composite panels (with and without damage). The effects of number of cycles and the central frequency selected on the reversibility of Lamb waves are discussed. An imaging algorithm based on the time-reversal ultrasonic wave signals was developed to define the locations of damage and debonding in a sandwich CF/EP composite panel with a honeycomb core.

10.5.1 Time reversal of wave signals

The principle of the time-reversal process in a two-dimensional plate is illustrated in Fig. 10.16, where a tone-burst of N cycles is applied to transducer A functioning as an actuator (step 1), activating a wave signal that is captured by transducer B acting as a sensor (step 2); the captured signal is time-reversed in the time domain (step 3) and reapplied to transducer B, which now has an actuating role, and then the wave signal at transducer A is collected and time-reversed as a reconstructed signal of the original one (step 4).

Based on Fig. 10.16, when a tone burst signal $V_A(\omega)$ at a central frequency ω is applied to transducer A, it activates a wave signal that propagates within the plate.

$$E_A(\omega) = k_A(\omega) V_A(\omega) \qquad [10.6]$$

with the response signal in the time domain at transducer A

$$V_A(t) = \frac{1}{2\pi} \int_{+\infty}^{-\infty} V_A(\omega) e^{i\omega t} d\omega \qquad [10.7]$$

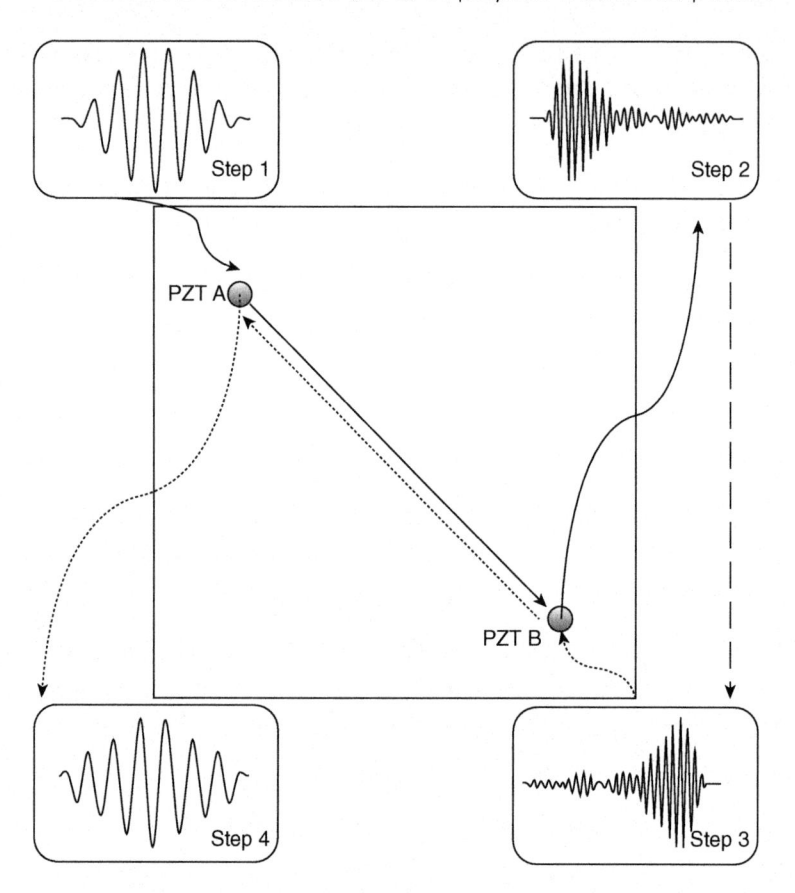

10.16 Illustration of the time-reversal process: generation and acquisition of signals.

where $E_A(\omega)$ is the wave energy and $k_A(\omega)$ is the electro-mechanical coefficient of transducer A. The wave energy that propagates to sensor transducer B is defined by

$$E_B(\omega) = G(\omega) k_A(\omega) V_A(\omega) \quad [10.8]$$

where $G(\omega)$ is the transfer function that relates the wave energy at transducer B to the input voltage at transducer A. On the other hand, the wave signal captured by transducer B can be defined by

$$V_B(\omega) = k_B(\omega)^{-1} G(\omega) k_A(\omega) V_A(\omega) \quad [10.9]$$

where $k_B(\omega)$ is the electro-mechanical coefficient of transducer B. If $k_B(\omega) = k_A(\omega)$ when the same transducers are used for actuator and sensor, this results in

$$V_B(\omega) = G(\omega) V_A(\omega) \quad [10.10]$$

When V_B is time-reversed and applied as a wave signal burst voltage to transducer B following the same procedure, a wave signal at transducer A is captured:

$$V'_A(\omega) = G(\omega)V_B(\omega) \quad [10.11]$$

Substituting the value of V_B from equation (10.10) into equation (10.11) results in

$$V'_A(\omega) = G^2(\omega)V_A(\omega) \quad [10.12]$$

Therefore the reconstructed time domain response signal at transducer A is defined by

$$V'_A(t) = \frac{1}{2\pi}\int_{+\infty}^{-\infty} V'_A(\omega)e^{i\omega t}d\omega$$
$$V'_A(t) = \frac{1}{2\pi}\int_{+\infty}^{-\infty} G^2(\omega)V_A(\omega)e^{i\omega t}d\omega \quad [10.13]$$

If there is no anomaly along the wave propagation path between transducers A and B, then there is no dispersion in wave central frequency, i.e. $G(\omega)$ is a scalar only, which represents the energy dissipation (or reduction in magnitude of the wave signal). Therefore, theoretically $V'_A(t)$ is similar to $V_A(t)$ in the time domain, and the shape of the reconstructed signal $V'_A(t)$ should be the same as the original input signal $V_A(t)$ after normalizing. However, a defect in (or near) a wave path results in a dispersion in the wave's central frequency, so that $G(\omega)$ is no longer just a scalar, and the shape of the reconstructed $V'_A(t)$ after wave interaction with damage will differ from that of the original $V_A(t)$ after normalizing, depending on the level of the wave form distortion. With the original wave signal $V_A(t)$ and the reconstructed wave signal $V'_A(t)$ after time reversal in the time domain, a DI can be defined by the difference between the two (Wang et al., 2009), and the perception of damage can be obtained without requiring the other baseline signals for comparison (Woo et al., 2009).

10.5.2 Experimental investigation

Two composite sandwich panels were manufactured using CF/EP woven fabric prepreg (Mikulik et al., 2008), consisting of surface CF/EP laminates in a quasi-isotropic configuration [±45, 0/90]s with a nominal thickness of 0.88 mm, and a core of Nomex honeycomb (HRH 10 1/8-3 with 20.4 mm in thickness) (Hoo Fatt and Park, 2001). The CF/EP laminate was initially cured in an autoclave following the required procedures. The surface CF/EP laminates and the core were bonded together with FM 1515-3 film adhesive (Liljedahl et al., 2008) using secondary bonding in the autoclave.

After bonding, the panels were trimmed to 560 mm × 560 mm in size (Fig. 10.17(a)). One of the panels was left intact (no damage), while the other had an embedded debonding of 40 mm × 40 mm introduced using a thermoplastic film (0.02 mm in thickness). The central position of the debonding is shown in Fig. 10.17(b). The panels were simply supported at the four edges during the experiment.

Experimental set-up

A sensor network was created consisting of 16 circular PZTs with a distance of 100 mm between elements, enclosing a square area. Elements (P1 to P16) measuring 10 mm in diameter and 1 mm in thickness, were surface-mounted on each of the sandwich panels. One additional PZT element (P17) was placed in the middle of the plate, as shown in Fig. 10.17(b). While one PZT element functioned as the actuator in order to activate wave signals, the others functioned as sensors to capture the wave signals, and the role of actuator alternated until all the PZT elements had functioned as the actuator. Sinusoidal tone bursts (90 V peak-to-peak) enclosed in a Hanning window were used as the input signal for the actuator. More details are provided in Section 10.4.2.

10.5.3 Results and discussion

Tuning of frequency and number of cycles

As multiple wave modes may appear concurrently when activating a wave signal, which complicates the time-reversal process (Ing and Fink, 1998a, 1998b; Woo *et al.*, 2009), it is crucial when using the time-reversal process to select the wave modes very carefully. It has been demonstrated that to properly reconstruct the wave signal using the time-reversal process, a single wave mode should be activated at a time (Xu and Giurgiutiu, 2007).

Based on the results shown in Fig. 10.1, frequencies of 25 and 140 kHz were selected, respectively, for the assessment. The use of narrowband excitation (Giurgiutiu, 2008) has been recommended to enhance the reconstructed time-reversed signal and to decrease the dependency of time reversal on frequency. Some initial experiments were carried out to determine the number of cycles in a tone-burst that suited the composite sandwich structure. An actuator-sensor path (i.e. P17–P11) 200 mm in length in the intact sandwich panel was used for this investigation, where the wave signals were reconstructed using the four symmetric tone-bursts of 4.5, 5.5, 6.5 and 7.5 cycles, respectively. It was observed that the similarity between the original and the reconstructed signals increased as the number of cycles in the tone-burst increased until the similarity of 99% (i.e. $\rho_{a,b} = 0.99$) was

Assessing debonding in sandwich panels   265

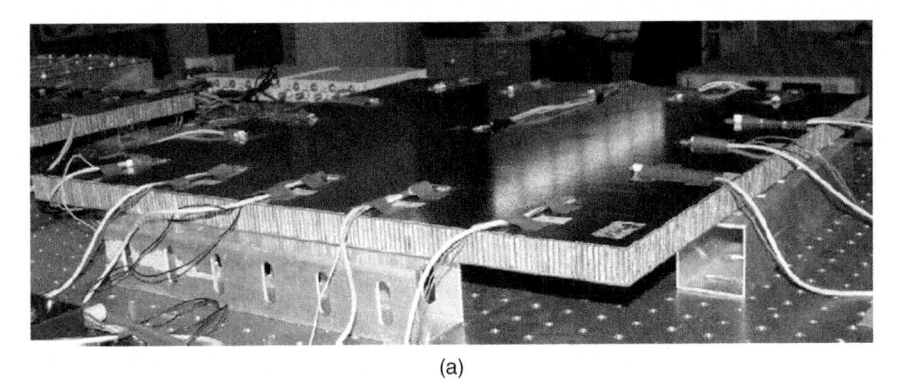

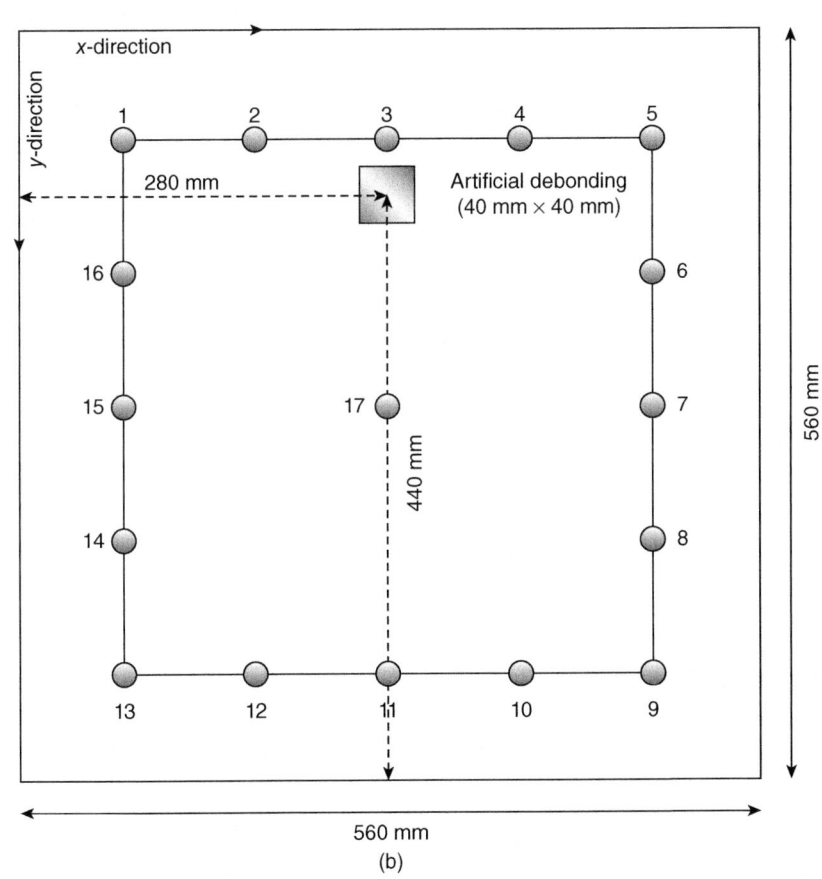

10.17 CF/EP sandwich composite panel: (a) experimental set-up and (b) sensor network on surface of panel.

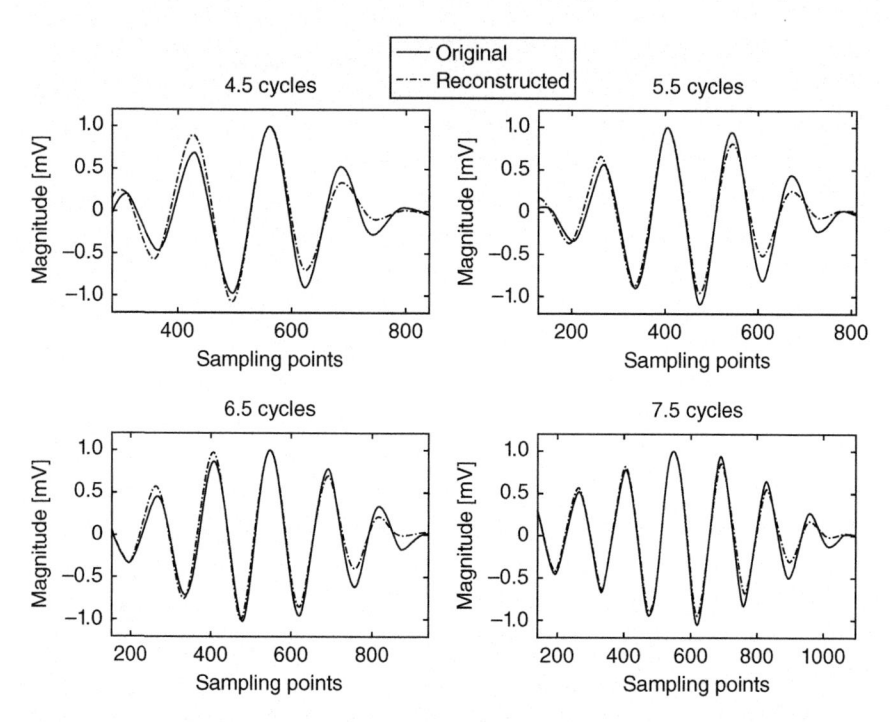

10.18 Similarity between reconstructed and original signals of S_0 mode for different tone-bursts excited at 140 kHz.

achieved for the tone-burst of 7.5 cycles, as shown in Fig. 10.18. On that basis the tone-burst of 7.5 cycles was selected for the study. Although a tone-burst with more cycles may help the reversibility of wave signals, it might not be suitable for short actuator-sensor paths because of the short time available for the whole wave packet to propagate from the actuator to the sensor.

Damage detection using A_0 and S_0 modes

Based on the wave propagation mechanism of A_0 and S_0 modes in a plate, the A_0 mode displays mostly out-of-plane displacement whereas the S_0 mode has predominantly radial displacement of particles (Su and Ye, 2009). Preliminary testing showed that the A_0 mode was not very sensitive to debonding at the interface between the skin panel and the core at the selected frequency of 25 kHz. Therefore, a steel block (40 mm in diameter and 147 grams in weight) was placed on top of the sandwich plate to simulate damage. A central frequency of 25 kHz was used for wave activation since the S_0 mode does not appear at this frequency (Fig. 10.1) and thus interpretation of the signals was less complex. Paths P6–P10 and P8–P1 in

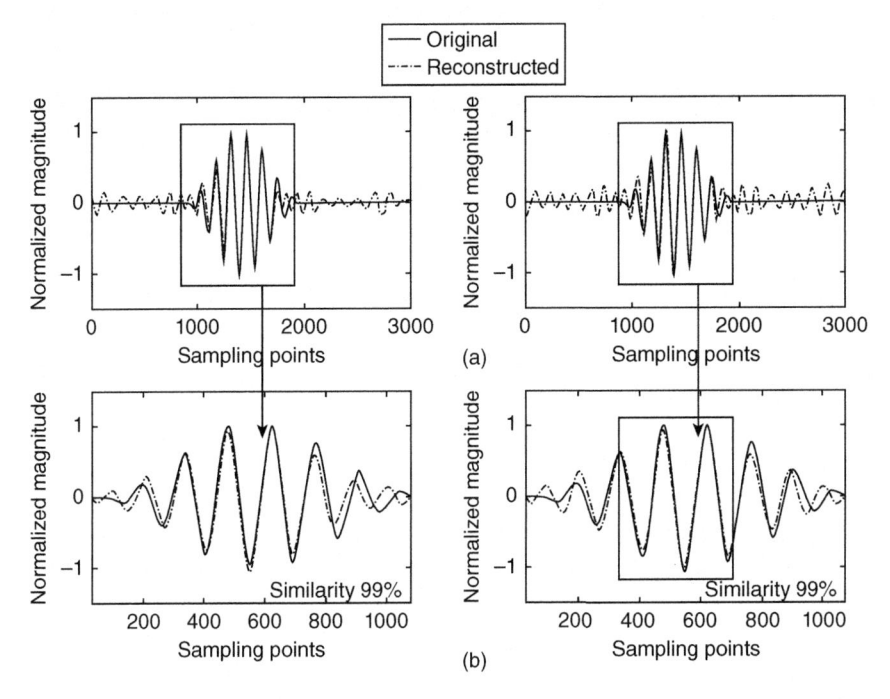

10.19 Time reversal of A_0 for paths P6–P10 and P8–P1: (a) full signal and (b) section for defining the damage index.

the intact sandwich panel were randomly selected for evaluation. The captured signals were successfully time-reversed, and the reconstructed signal achieved a similarity of 98% (i.e. $\rho_{a,b} = 0.98$), as shown in Fig. 10.19. In the time-reversal process of the A_0 mode the attenuation of the A_0 mode is most important since it propagates through the thickness of the sandwich plate (Song *et al.*, 2009). Also, when the path is long (e.g. more than 400 mm) the magnitude of the signal decreases significantly, resulting in complication of the time-reversal process and more high-frequency noise. In this case, a bandpass (Su and Ye, 2004) filter is used, which passes a signal in a certain bandwidth while attenuating those both above and below specified limits.

From the network of 16 PZT elements a total of 160 paths were available but with the dual function of the PZT elements they could be reduced to 80 paths (e.g. instead of using both paths P1–P2 and P2–P1, only P1–P2 was employed). As shown in Fig. 10.20(a), some sensing paths (with dashed lines) were very close to the edge of the plate, resulting in extensive boundary reflection and affecting the reconstruction process; therefore those wave paths were ignored. A total of 64 paths were left within the scanned area. Different groups of sensing paths were thus applied to construct the damage images. Groups I and II were established using 26 and 64 actual sensing paths provided by the 16 actual sensors (P1, P2, ..., P16), respectively, shown in Fig. 10.20(a) and (b).

The correlation coefficients between the original and the reconstructed signals were calculated and the DI was obtained using equation (10.3). Estimation of the presence of damage in every grid in the area covered by the sensing paths was obtained after fusion of the perceptions from individual paths (equation (10.5)). Figure 10.21(a) illustrates the images from Groups I and II sensing paths for the A_0 mode at the central frequency of 25 kHz, where the circle indicates the damage simulated by the steel block and the cross highlights the centre of estimation of damage. The values in each image were normalized. Overall, the results demonstrated that the method could identify the damage clearly.

The effectiveness of the S_0 mode in detecting damage was also evaluated in order to compare the results with those obtained using the A_0 mode, changing the central frequency to 140 kHz where the A_0 mode is less significant in magnitude (Fig. 10.1). The results were improved significantly for Group I, while minor improvement was achieved with Group II. The results obtained are shown in Fig. 10.21(b) using Groups I and II sensing paths. Both Groups I and II predicted the position of the damage (using A_0 and S_0 modes), and the accuracy of prediction of the position of damage using the A_0 mode improved dramatically as the number of paths was increased, as summarized in Table 10.4.

Table 10.4 Results of damage identification in CF/EP sandwich panels using guided wave signals for different defects

Description of defect	Actual position Coordinates [mm]	Predicted position Coordinates [mm]	Distance between actual and predicted positions [mm]
Group I			
Damage (using A_0 mode)	(150,250)	(63,309)	105
Damage (using S_0 mode)	(150,250)	(160,238)	15.6
Debonding (using S_0 mode)	(200,40)	(233,63)	40
Group II			
Damage (using A_0 mode)	(150,250)	(149,229)	21
Damage (using S_0 mode)	(150,250)	(145,238)	13
Debonding (using S_0 mode)	(200,40)	(212,55)	19

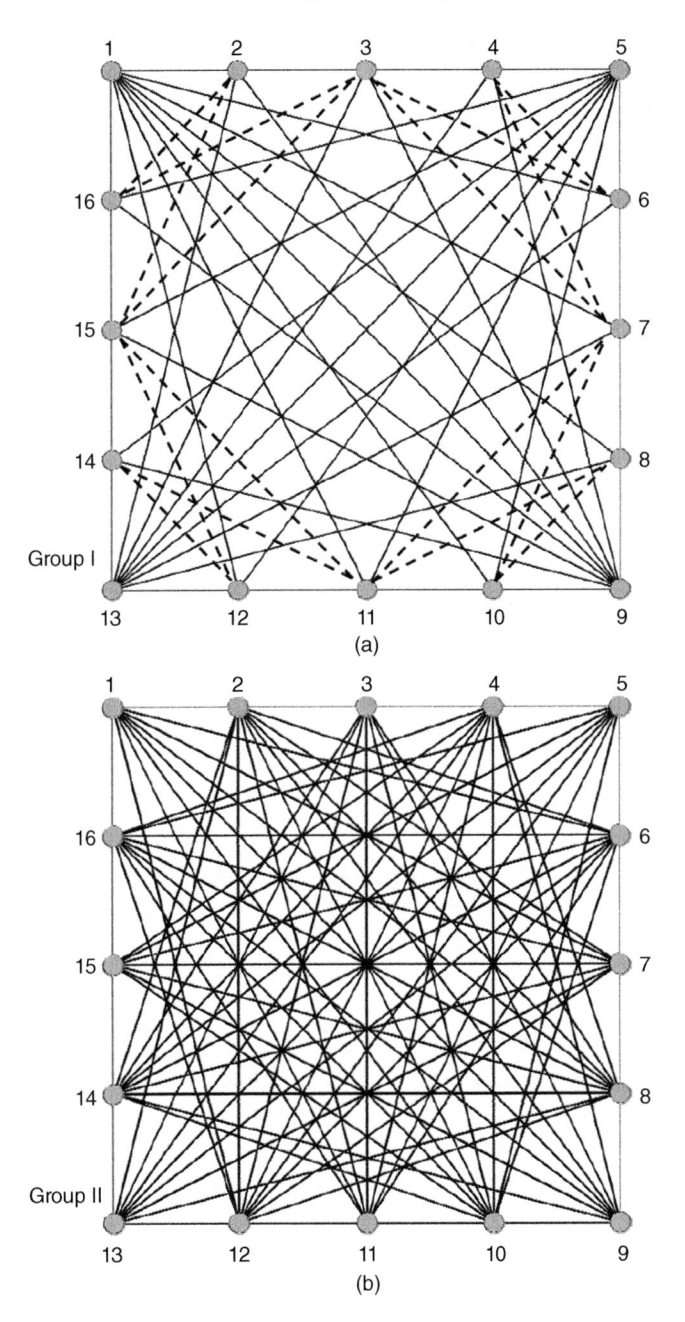

10.20 Sensor network paths: (a) Group I with 26 paths and (b) Group II with 64 paths.

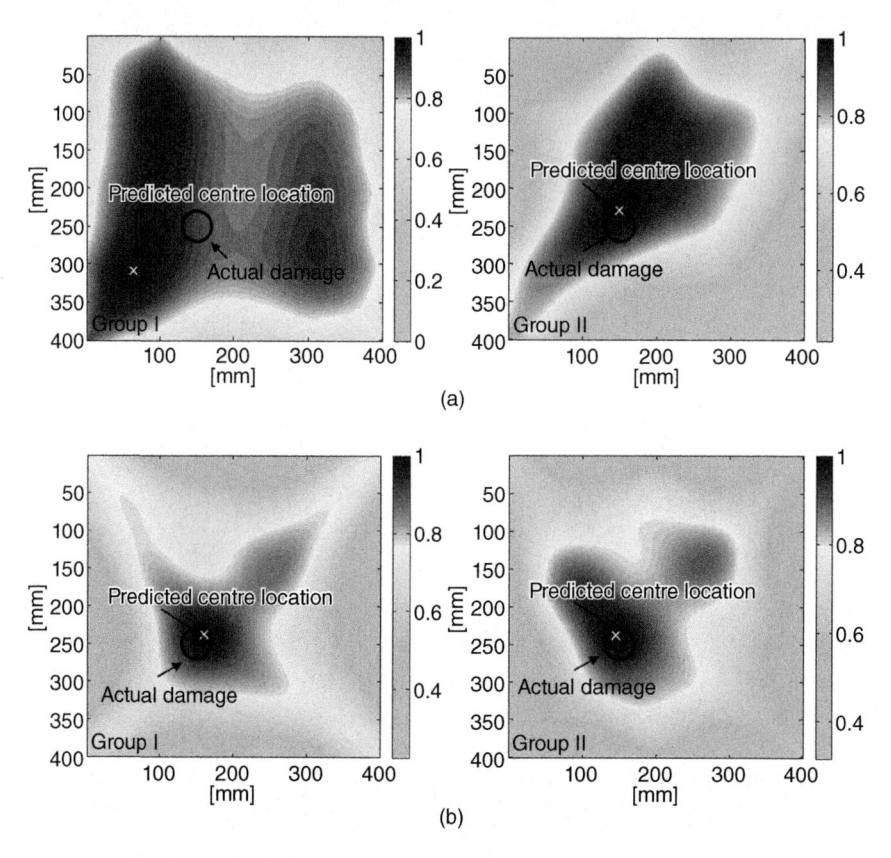

10.21 Image of the damage constructed with two groups of actuator-sensor paths using (a) A_0 mode and (b) S_0 mode.

Debonding detection using S_0 mode

Based on the tuning of the frequency and the selection of number of cycles in a tone-burst in the previous section, 140 kHz and 7.5 cycles were selected to activate the S_0 mode. Paths P6–P10 and P12–P16 in the intact CF/EP sandwich panel were randomly selected to evaluate the time-reversal process before proceeding to the image construction. The results of two reconstructed signals of the S_0 mode are shown in Fig. 10.22; the reconstruction was successful, achieving a similarity of more than 96% (i.e. $\rho_{a,b} = 0.96$). As the S_0 mode propagation was confined within the CF/EP laminate with the wave energy mostly propagating within the laminate, its attenuation was much less pronounced than that of the A_0 mode. Wave signals were captured and reconstructed from individual actuator-sensor paths of two different sensor networks (Group I and II), the original and the reconstructed signals were correlated for each path, and the damage index was defined.

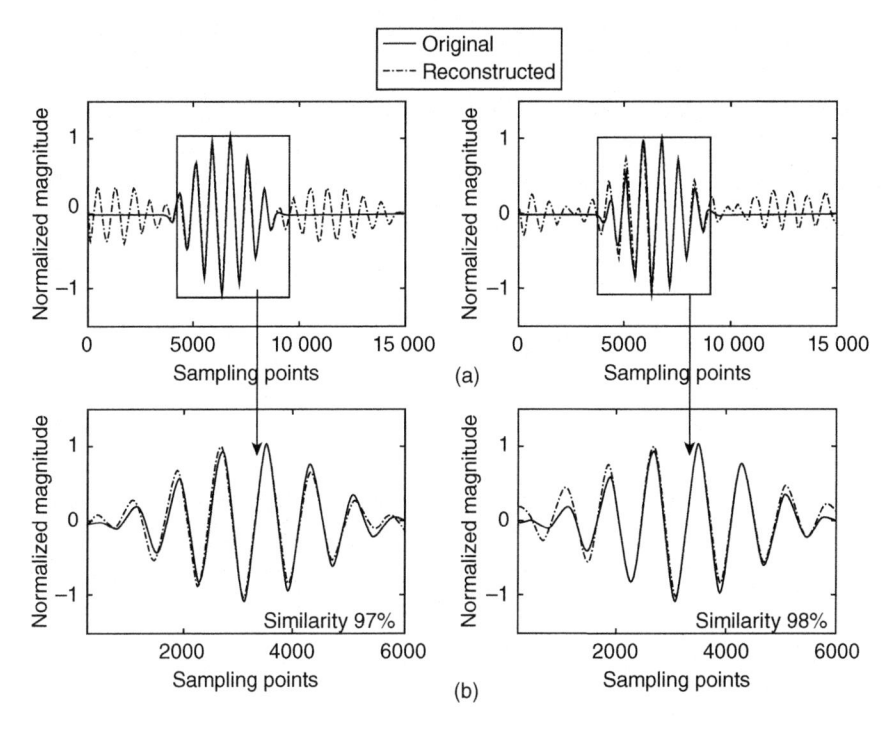

10.22 Time reversal of S_0 for paths P6–P10 and P12–P16: (a) full signal and (b) section for defining the damage index.

The distributions in two affected zones (rectangular and elliptical) were used in assessment and the rectangular affected zone showed better focus in the estimation, as shown in Fig. 10.23. The results for estimation of the presence of debonding are shown in Fig. 10.23, where good agreement between the estimated and the actual position of the debonding can be observed. The distance between the centre of debonding and the estimated location was 38.5 mm for Group I and 13 mm for Group II, as summarized in Table 10.4. The results for predicting the position of the debonding using the elliptical function are shown in Fig. 10.23(b); it appears that the use of an elliptical affected zone did not change the results dramatically.

10.6 Conclusion and future trends

The effects of debonding size on the delay in ToF and the energy of damage-reflected wave signals of the fundamental anti-symmetric A_0 mode of low frequency in sandwich CF/EP composite beams were investigated. The location of the debonding could be determined confidently using the ToF of the damage-reflected waves. An extensive assessment of the severity of the debonding was conducted using the time delay or reflection coefficient of

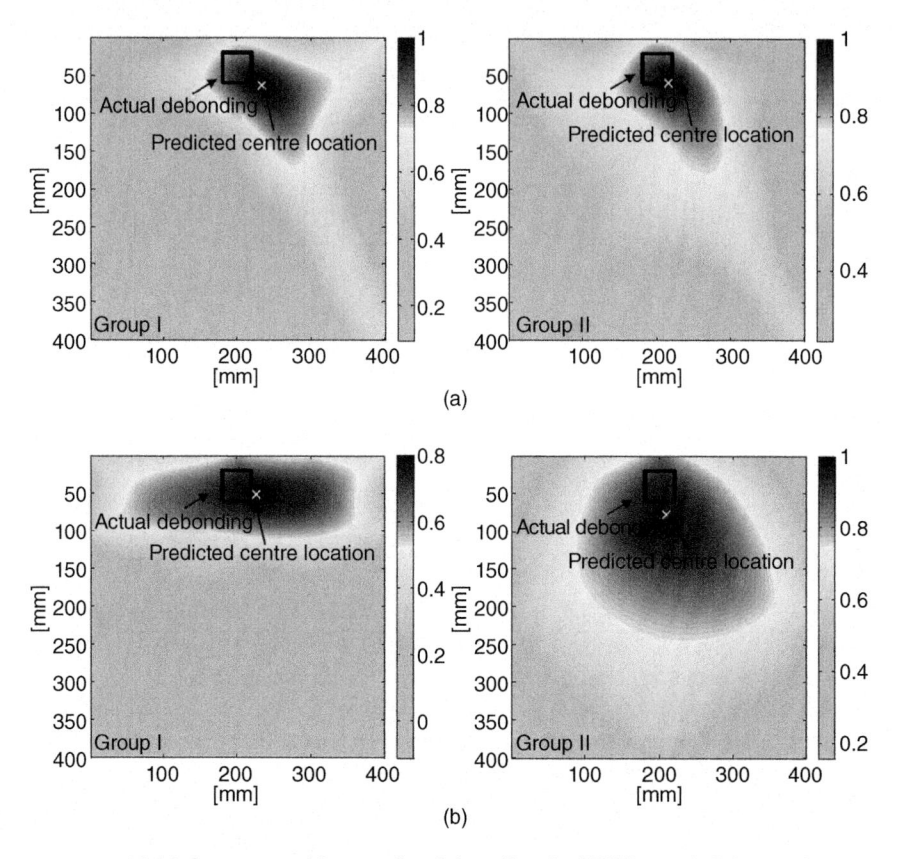

272   Non-destructive evaluation (NDE) of polymer matrix composites

10.23 Constructed image for debonding in CF/EP sandwich panel using S_0 mode with different affected zones; (a) rectangular shape and (b) elliptical shape.

debonding-reflected wave signals, and this approach proved capable of providing correlations between the extent of debonding and the time delay or reflection coefficient, though such correlations are not unique.

A time-reversal algorithm based on correlation analysis using ultrasonic wave signals from an active sensor network was also developed through experimental studies. For individual actuator-sensor paths, the changes in original and reconstructed signals were measured by correlation coefficients. The perceptions from individual paths were fused to obtain the final estimate of the presence of damage. A composite sandwich panel with imbedded debonding was tested. The predicted positions of the debonding agreed well with the actual positions. Both A_0 and S_0 modes were activated on two different occasions and their reversibility was demonstrated. The S_0 mode showed very good accuracy and ease in use whereas the A_0 mode resulted in a higher error and greater complexity in the wave signals. The identification results demonstrated that the time-reversal algorithm has good potential for the location of debonding in sandwich structures.

Our investigation proved the proficiency of guided waves for detection of flaws in sandwich composite structures but there are still many challenges to be further investigated in this field:

- The actuation and sensing of wave-generated signal using PZT via cables may set a limit on the implementation of this method for structural health monitoring. Cabling requires considerable planning for connection and could significantly increase the weight of the monitored structure. Therefore, there is a need for wireless actuation and sensing systems, to resolve the many issues associated with traditional cabling methods.
- The majority of current approaches employ the lowest-order symmetric (S_0) or anti-symmetric (A_0) modes, for convenience in signal processing. However, to characterize damage of small extent with acceptable accuracy, guided waves need to be excited at very high frequency in order to obtain a very short wavelength and thus higher sensitivity to miniature flaws. A possible choice could be the use of a high-order mode such as S_1 and A_1 which possess greater sensitivity to damage than the lowest-order modes.
- Most work done so far in the area of damage detection based on guided waves is still developed in the laboratory under controlled conditions. The effect of harsh environmental conditions on the activation and acquisition of signals is of great concern, and must be addressed.
- Most studies have focused on flat plates of uniform thickness or plates with little curvature, in spite of the fact that structural components in practice either vary in thickness or display a marked curvature of shape. The effect of anisotropy of medium (in both material and geometrical properties) on the propagation characteristics of Lamb waves, and the impact on algorithms of damage identification, has not been well addressed.
- The effectiveness of most existing approaches has been verified by identifying single instances of damage in a structure. In practice, however, engineering structures are likely to evidence multiple instances of damage simultaneously. Rather than being a simple expansion from the approaches developed for single damage, qualitative or quantitative identification of multiple damage is a highly challenging topic.
- Signal processing and data fusion play the most substantial role in delivering efficient and accurate identification of damage based on captured wave signals. The majority of the currently existing signal processing tools were developed for laboratory tests, and may not be very effective for application in the field. Thus signals captured in the field often present a very complex appearance, as a consequence of corruption by

background noise. To reduce dependence on the subjective judgement of individuals, fragmented signal analyses and manipulation should be avoided, and replaced by automated signal processing and data fusion with less involvement of the operator. Ideally, the processing and fusion exercise would be automatically triggered and accomplished when signals are available.
- States of damage and health are defined relatively; a damage event is not meaningful without a reference or comparison. Although it may not be explicitly stated, most existing approaches are primarily dependent on referencing or comparing the experimentally captured signals with baseline signals obtained theoretically or experimentally from a benchmark counterpart that is assumed to be free of damage or in a prior state. However, during this process, any changes in working conditions or measurement noise, if they are not included in the baseline signals, can lead to false identification. Furthermore, baseline signals are not always obtainable. Thus the excessive reliance of algorithms on a benchmark may narrow their applicability. It is of vital importance to minimize such reliance in further studies, through appropriate algorithms.
- The past two decades have witnessed an explosion of sensor technology. The majority of today's wave-based damage identification or SHM employs piezoelectric transducers, but their practical application may be limited due to particular features such as their brittleness and the significant degradation of their performance when the ambient temperature is high (over roughly half their Curie temperature) or low (in cryogenic environment). Recent advances in material science, manufacturing and bio technology have yielded encouraging prospects for overcoming these problems of piezoelectric materials, further benefiting the development of novel sensors and sensor networks for damage identification and structural health monitoring (SHM) in particular.

10.7 References

Anton, S. R. & Inman, D. J. 2009. Reference-free damage detection using instantaneous baseline measurements. *AIAA Journal*, **47**, 1952–1964. doi: 10.2514/1.43252

Burlayenko, V. N. & Sadowski, T. 2009. Free vibration of sandwich plates with impact-induced damage. *PAMM*, **9**, 179–180. doi: 10.1002/pamm.200910064

Castaings, M. & Hosten, B. 2001. The propagation of guided waves in composite, sandwich-like structures and their use for NDT. *Review of Progress in Quantitative Nondestructive Evaluation*, **557**, 999–1006.

Cho, Y. & Rose, J. L. 1996. A boundary element solution for a mode conversion study on the edge reflection of Lamb waves. *The Journal of the Acoustical Society of America*, **99**, 2097–2109. doi: 10.1121/1.415396

Cuc, A., Giurgiutiu, V., Joshi, S. & Tidwell, Z. 2007. Structural health monitoring with piezoelectric wafer active sensors for space applications. *AIAA Journal*, **45**, 2838–2850. doi: 10.2514/1.26141

Diamanti, K., Soutis, C. & Hodgkinson, J. M. 2005a. Lamb waves for the non-destructive inspection of monolithic and sandwich composite beams. *Composites Part A: Applied Science and Manufacturing*, **36**, 189–195. doi: 10.1016/j.compositesa.2004.06.013

Diamanti, K., Soutis, C. & Hodgkinson, J. M. 2005b. Non-destructive inspection of sandwich and repaired composite laminated structures. *Composites Science and Technology*, **65**, 2059–2067. doi: 10.1016/j.compscitech.2005.04.010

Doyle, J. F. 1989. *Wave propagation in structures*, New York, Springer-Verlag.

Florens, C., Balmés, E., Clero, F. & Corus, M. 2006. Accounting for glue and temperature effects in Nomex based honeycomb models. *International Conference on Noise and Vibration Engineering*, Leuven, Belgium.

Gangadharan, R., Murthy, C. R. L., Gopalakrishnan, S. & Bhat, M. R. 2009. Time reversal technique for health monitoring of metallic structure using Lamb waves. *Ultrasonics*, **49**, 696–705. doi: 10.1016/j.ultras.2009.05.002

Giurgiutiu, V. 2008. Piezoelectric wafer active sensors. *Structural Health Monitoring*, Burlington, MA, Academic Press. doi: 10.1016/B978-012088760-6.50008-8

Hoo Fatt, M. S. & Park, K. S. 2001. Dynamic models for low-velocity impact damage of composite sandwich panels – Part A: Deformation. *Composite Structures*, **52**, 335–351. doi: 10.1016/S0263-8223(01)00026-5

Hsu, D. K. 2009. Nondestructive evaluation of sandwich structures: a review of some inspection techniques. *Journal of Sandwich Structures and Materials*, **11**, 275–291. doi: 10.1177/1099636209105377

Hyunjune, Y. & Younghoon, S. 2000. Numerical simulation and visualization of elastic waves using mass-spring lattice model. *IEEE Transactions on Ultrasonics, Ferroelectrics and Frequency Control*, **47**, 549–558.

Ing, R. K. & Fink, M. 1998a. Self-focusing and time recompression of Lamb waves using a time reversal mirror. *Ultrasonics*, **36**, 179–186. doi: 10.1016/S0041-624X(97)00100-5

Ing, R. K. & Fink, M. 1998b. Time-reversed Lamb waves. *IEEE Transactions on Ultrasonics, Ferroelectrics, and Frequency Control*, **45**, 1032–1043.

Kim, S. B. & Sohn, H. 2006. Application of time-reversal guided waves to field bridge testing for baseline-free damage diagnosis. *Health Monitoring and Smart Nondestructive Evaluation of Structural and Biological Systems V*, San Diego, CA, USA Proc. SPIE.

Kim-Ho Ip, Y.-W. M. 2004. Delamination detection in smart composite beams using lamb waves. *Smart Materials and Structures*, **13**, 544–551. doi: 10.1088/0964-1726/13/3/013

Lee, B. C. & Staszewski, W. J. 2003a. Modelling of Lamb waves for damage detection in metallic structures: Part I. Wave propagation. *Smart Materials & Structures*, **12**, 804–814. doi: Pii S0964-1726(03)67508-7

Lee, B. C. & Staszewski, W. J. 2003b. Modelling of Lamb waves for damage detection in metallic structures: Part II. Wave interactions with damage. *Smart Materials & Structures*, 12, 815–824. doi: Pii S0964-1726(03)67509-9

Li, H. C. H., Beck, F., Dupouy, O., Herszberg, I., Stoddart, P. R., Davis, C. E. & Mouritz, A. P. 2006. Strain-based health assessment of bonded composite repairs. *Composite Structures*, **76**, 234–242. doi: 10.1016/j.compstruct.2006.06.032

Liljedahl, C. D. M., Fitzpatrick, M. E. & Edwards, L. 2008. Residual stresses in structures reinforced with adhesively bonded straps designed to retard fatigue crack growth. *Composite Structures*, **86**, 344–355. doi: 10.1016/j.compstruct.2007.10.033

Lin, X. & Yuan, F. G. 2001a. Detection of multiple damages by prestack reverse-time migration. *AIAA Journal*, **39**, 2206–2215. doi: 10.2514/2.1220

Lin, X. & Yuan, F. G. 2001b. Diagnostic Lamb waves in an integrated piezoelectric sensor/actuator plate: analytical and experimental studies. *Smart Materials & Structures*, **105**, 907–913. doi: 10.1088/0964-1726/10/5/307

Liu, G. R. & Achenbach, J. D. 1994. A strip element method for stress analysis of anisotropic linearly elastic solids. *Journal of Applied Mechanics*, **61**, 270–278. doi: 10.1115/1.2901440

Liu, G. R. & Xi, Z. C. 2002. *Elastic waves in anisotropic laminates*, New York, CRC Press.

Liyanapathirana, D. S. & Kelly, R. B. 2010. Interpretation of the lime column penetration test *Materials Science and Engineering*, **10**, 1–10. doi: 10.1088/1757-899X/10/1/012088

Mikulik, Z., Kelly, D. W., Prusty, B. G. & Thomson, R. S. 2008. Prediction of flange debonding in composite stiffened panels using an analytical crack tip element-based methodology. *Composite Structures*, **85**, 233–244. doi: 10.1016/j.compstruct.2007.10.027

Mustapha, S., Ye, L., Wang, D. & Lu, Y. 2011. Assessment of debonding in sandwich CF/EP composite beams using A_0 Lamb wave at low frequency. *Composite Structures*, **93**, 483–491. doi: 10.1016/j.compstruct.2010.08.032

Osmont, D. L., Devillers, D. & Taillade, F. 2001. Health monitoring of sandwich plates based on the analysis of the interaction of Lamb waves with damages *In:* Davis, L. P. (ed.) *Smart Structures and Materials 2001: Smart Structures and Integrated Systems*, Newport Beach, CA, SPIE.

Park, H. W., Sohn, H., Law, K. H. & Farrar, C. R. 2007. Time reversal active sensing for health monitoring of a composite plate. *Journal of Sound and Vibration*, **302**, 50–66. doi: 10.1016/j.jsv.2006.10.044

Patera, A. T. 1984. A spectral element method for fluid dynamics: laminar flow in a channel expansion. *Journal of Computational Physics*, **54**, 468–488. doi: 10.1016/0021-9991(84)90128-1

Peng, H., Ye, L., Meng, G., Mustapha, S. & Li, F. 2010. Concise analysis of wave propagation using the spectral element method and identification of delamination in CF/EP composite beams. *Smart Materials and Structures*, **19**, 085018. doi: 10.1088/0964-1726/19/8/085018

Qi, X., Rose, J. L. & Xu, C. 2008. Ultrasonic guided wave nondestructive testing for helicopter rotor blades. *17th World Conference on Nondestructive Testing*, Shanghai, China.

Raghavan, A. & Cesnik, C. E. S. 2007a. Guided-wave signal processing using chirplet matching pursuits and mode correlation for structural health monitoring. *Smart Materials and Structures*, **16**, 355–366. doi: 10.1088/0964-1726/16/2/014

Raghavan, A. & Cesnik, C. E. S. 2007b. Review of guided-wave structural health monitoring. *The Shock and Vibration Digest*, **39**, 91–114. doi: 10.1177/0583102406075428

Rose, J. L. 1999. *Ultrasonic waves in solid media*, New York, Cambridge University Press.

Schafbuch, P. J., Rizzo, F. J. & Thompson, R. B. 1993. Boundary element method solutions for elastic wave scattering in 3D. *International Journal for Numerical Methods in Engineering*, **36**, 437–455. doi: 10.1002/nme.1620360306

Seth, S. K., Spearing, S. M. & Constantinos, S. 2002. Damage detection in composite materials using Lamb wave methods. *Smart Materials and Structures*, **11**, 269–278. doi: 10.1088/0964-1726/11/2/310

Sohn, H., Park, H. W., Law, K. H. & Farrar, C. R. 2007a. Combination of a time reversal process and a consecutiv outlier analysis for baseline-free damage diagnosis. *Journal of Intelligent Material Systems and Structures*, **18**, 335–346. doi: 10.1177/1045389x0606629

Sohn, H., Park, H. W., Law, K. H. & Farrar, C. R. 2007b. Damage detection in composite plates by using an enhanced time reversal method. *Journal of Aerospace Engineering*, **20**, 141–151. doi: 10.1061/(ASCE)0893-1321(2007)20:3(141) (11 pages)

Song, F., Huang, G. L. & Hudson, K. 2009. Guided wave propagation in honeycomb sandwich structures using a piezoelectric actuator/sensor system. *Smart Materials and Structures*, **18**, 125007 (8pp). doi: 10.1088/0964-1726/18/12/125007

Song, F., Huang, G. L., Kim, J. H. & Haran, S. 2010. Multiple debondings' detection in honeycomb sandwich structures using multi-frequency elastic guided waves. Proceedings of the SPIE, **7560**, 765018-765018-12. doi: 10.1117/12.848144

Strikwerda, J. 2004. *Finite Difference Schemes and Partial Differential Equations*, Philadelphia, PA, Society for Industrial and Applied Mathematics.

Su, Z. & Ye, L. 2004. Fundamental Lamb mode-based delamination detection for CF/EP composite laminates using distributed piezoelectrics. *Structural Health Monitoring*, **3**, 43–68. doi: 10.1177/1475921704041874

Su, Z. & Ye, L. 2009. *Identification of damage using lamb waves: from fundamentals to applications*, Berlin, Heidelberg, Springer-Verlag.

Thien, A. B., Puckett, A. D., Gyuhae, P. & Farrar, C. R. 2006. The use of time reversal methods with Lamb waves to identify structural damage in a pipeline system. *In:* Kundu, T. (ed.) *Health monitoring and smart nondestructive evaluation of structural and biological systems V*, San Diego, CA, USA Proc. SPIE.

Thwaites, S. & Clark, N. H. 1995. Non-destructive testing of honeycomb sandwich structures using elastic waves. *Journal of Sound and Vibration*, **187**, 253–269. doi: 10.1006/jsvi.1995.0519

Valle, C. & Littles, J. W. 2002. Flaw localization using the reassigned spectrogram on laser-generated and detected Lamb modes. *Ultrasonics*, **39**, 535–542. doi: 10.1016/S0041-624X(02)00249-4

Wang, D., Ye, L. & Lu, Y. 2009. A probabilistic diagnostic algorithm for identification of multiple notches using digital damage fingerprints (DDFs). *Journal of Intelligent Material Systems and Structures*, **20**, 1439–1450. doi: 10.1177/1045389x09338323

Wang, D., Ye, L., Su, Z., Lu, Y., Li, F. & Meng, G. 2010. Probabilistic damage identification based on correlation analysis using guided wave signals in aluminum plates. *Structural Health Monitoring*, **9**, 133–144. doi: 10.1177/1475921709352145

Woo, P. H., Bum, K. S. & Sohn, H. 2009. Understanding a time reversal process in Lamb wave propagation. *Wave Motion*, **46**, 451–467. doi: 10.1016/j.wavemoti.2009.04.004

Xu, B. & Giurgiutiu, V. 2007. Single mode tuning effects on lamb wave time reversal with piezoelectric wafer active sensors for structural health monitoring. *Journal of Nondestructive Evaluation*, **26**, 123–134. doi: 10.1007/s10921-007-0027-8

Yang, C., Ye, L., Su, Z. & Bannister, M. 2006. Some aspects of numerical simulation for Lamb wave propagation in composite laminates. *Composite Structures*, **75**, 267–275. doi: 10.1016/j.compstruct.2006.04.034

Yang, M. & Qiao, P. 2005. Modeling and experimental detection of damage in various materials using the pulse-echo method and piezoelectric sensors/actuators. *Smart Materials and Structures*, **14**, 1083–1100. doi: 10.1088/0964-1726/14/6/001

Ye, L., Yang, C., Su, Z. & Bannister, M. 2006. Some aspects of numerical simulation for Lamb wave propagation in composite laminates. *Composite Structures*, **75**, 267–275. doi: 10.1016/j.compstruct.2006.04.034

Zhao, X. & Rose, J. 2003. Boundary element modeling for defect characterization potential in a wave guide. *International Journal of Solids and Structures*, **40**, 2645–2658. doi: 10.1016/S0020-7683(03)00097-0

Zienkiewicz, O. C. & Taylor, R. L. 1989. *The finite element method*, London, McGraw-Hill.

11
Non-destructive evaluation (NDE) of composites: detecting delamination defects using mechanical impedance, ultrasonic and infrared thermographic techniques

B. S. WONG, Nanyang Technological University, Singapore

DOI: 10.1533/9780857093554.2.279

Abstract: The non-destructive tests in this chapter are focused on delamination defect detection in solid plate composites and disbonds in honeycomb structures. The following topics are presented: mechanical impedance inspection of disbonds, ultrasonic inspection of delaminations and thermographic inspection of delaminations. These three techniques are described in detail individually, and generally major advantages of each are: mechanical impedance inspection does not require a liquid couplant between the probe and specimen, ultrasonics is usually considered to be the most sensitive technique and thermographic can test large areas quickly. A comparison will be made between the techniques and the following factors will be discussed: sensitivity in terms of defect size detectable, ability to size a defect and determine its shape, ability to locate the position and depth of a defect and ability to test a large area quickly.

Key words: mechanical impedance inspection of disbonds, ultrasonic inspection of delaminations, thermographic inspection, defect size and location detectability and large area inspection.

11.1 Introduction

Fibre reinforced laminated composite (FRC) materials are widely used in aircraft, wind turbines and other structures requiring light weight. Defects in these materials can be produced during new manufacture and during use of the materials. The main mechanism by which defects are produced during use is impact damage. For example, aircraft and wind turbines can be subjected to bird strikes. Typical defects produced by impact damage are matrix cracking, fibre fractures, delaminations between the composite prepreg sheets in solid composite plates and debonding defects in honeycomb structures between the face sheets and honeycomb structures. Fibre fractures and matrix cracking do affect the tensile strengths, but delamination defects are of most concern because they reduce the compression strength significantly [1]. Disbond defects in honeycomb structures will also have a similar effect.

Delaminations are produced because of the low transverse and interlaminar shear strength of FRC materials. The delaminations will tend to grow outwards from the point of contact of the impacting object. The depth which the delaminations are produced varies and it is important for non-destructive tests to determine this depth if possible. Therefore the non-destructive tests in this chapter are focused on delamination defect detection in solid plate composites and disbonds in honeycomb structures. The following topics will be presented:

(a) mechanical impedance inspection of disbonds;
(b) ultrasonic inspection of delaminations;
(c) thermographic inspection of delaminations.

The three techniques will be described in detail individually later, but generally a major advantage of each is:

(a) mechanical impedance inspection does not require a liquid couplant between the probe and specimen;
(b) ultrasonics is usually considered to be the most sensitive technique;
(c) thermographic can test large areas quickly.

The word defect will usually be used for disbonds in case (a) and delaminations in cases (b) and (c). A comparison will be made between the techniques and the following factors will be discussed:

- sensitivity in terms of detectable defect size;
- ability to assess the size of defect and determine its shape;
- ability to locate the position and depth of a defect;
- ability to test a large area quickly.

A focus of the ultrasonic and thermographic work is to investigate defect sizes from 1 to 11 mm in diameter at several depths below the testing surface in order to determine the limits of sensitivity of these techniques in testing FRCs.

Comparisons have been made by previous workers between ultrasonics and thermography. Amenabar *et al.* [2] used specimens with large defects >30 mm in size up to a depth of 0.25 mm. Pulse and lock-in thermography techniques both performed similarly and very well on these defects. Ultrasound using a 5 MHz probe found all defects and provided a good compromise between material penetration and defect resolution, as well as providing good depth measurement. Garnier *et al.* [3] also compared optical methods with these techniques. Some important conclusions were that ultrasound determines defect depth accurately and thermography is most suited to large specimens. The advantages and disadvantages were listed. The work reported in this chapter on ultrasonic and thermographic tests utilized a larger number of defects which are smaller and deeper, hence assessing a wider range of defect size and depth assessment.

11.2 Using mechanical impedance: disbonding in aluminium honeycomb structures

As mentioned above, the mechanical impedance is attractive because the low frequencies utilized mean the technique can be used with dry contact between the interrogating probe and the specimen, unlike pulse echo ultrasound which requires a liquid couplant. The technique is also quick and easy to use and portable, enabling it to be used for on-site applications. This section describes tests carried out using an MIA 3000 mechanical impedance instrument produced by Inspection Instruments Ltd. A probe is utilized to apply a localized forced oscillation to the area under investigation. A defective area can be recognized from the different response it produces when compared with a non-defective area. A difference would be a reduced resonant frequency due to the reduced stiffness of the defective area. These responses can be deduced from the frequency spectrum displays produced by the instrument. The probe can be used manually or attached to an x–y table to produce 'C' scan displays.

Cawley [4] conducted investigations with the technique on disbonds in stiff honeycomb structures. A skin–core disbond 10 mm in diameter in a honeycomb with a 0.5 mm thick carbon fibre-reinforced (CFR) plastic skin was detectable, as was a disbond 20 mm in diameter in a 1 mm thick skin. The defects investigated in this chapter were also disbonds, for example as indicated in Fig. 11.1, which also shows the mechanical impedance probe. These disbonds will be referred to as defects in this chapter.

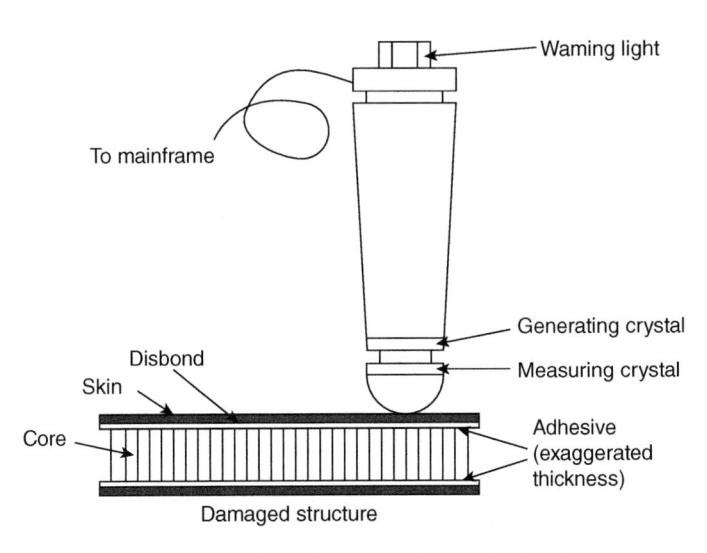

11.1 The experimental arrangement for evaluating honeycomb specimens using the MIA 3000 probe.

The skins of the honeycombs were aluminium and the cores were made of Nomex. The fabrication process for the specimens, which were manufactured, is identical to that used on military aircraft. This structure is used in the F16, for its leading edge flaps, its fixed trailing edges and flaperons and its ventral fin. The honeycomb structure is also a common component of other aircraft types. Disbonds between the skin and core caused by poor fabrication or damage during use for example by an impacts due to bird strikes will lead to severe weakening of the structure and possibly failure of the aircraft. Testing of the honeycomb structures with techniques such as mechanical impedance is therefore of paramount importance and the technique is used by many aircraft users throughout the world.

11.2.1 Test specimens

The specimens were all cured in an autoclave. For the first specimens to be described which had round defects in them, the materials used were:

- Aluminium sheets (7075 T6) of thickness 0.80 mm, 1.25 mm, 1.57 mm and 2.25 mm.
- Nomex honeycomb core of thickness 12.50 mm and core size 4.68 mm.
- Adhesive, FM 73M.
- Defects simulated using ten layers of high-temperature plastics.

A typical specimen fabricated is shown in Fig. 11.2. Four of these specimens were fabricated with the skin thicknesses indicated above.

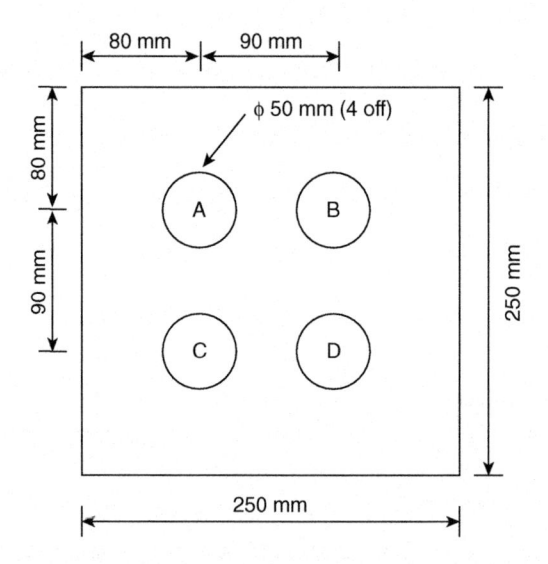

11.2 Dimensions of a typical specimen.

The aluminium used was AL0.80/N12.5/S0:

- AL refers to the skin material, i.e. aluminium;
- 0.80 refers to the skin thickness in mm;
- N refers to the core material, i.e. Nomex;
- 12.5 is the honeycomb thickness in mm;
- S0 refers to the specimen code no.

For the irregular shaped defect specimens (which are discussed on pages 286–291), the skin materials were either three layers of glass fibre (code GF) or three layers of boron fibre (code BF). The defects were made of three layers of peel ply impregnated between the skin and the core. The core material was Nomex (code N3) or aluminium (code AL) of thickness 12.5 mm. The ellipse defects reported were of size 80 mm by 60 mm and the elongated defect of size 120 mm by 40 mm.

11.2.2 Test procedure

The MIA 3000 is an electronic transmitter and receiver, which acquires data from the material under test, analyses the data and presents it in graphic form on a cathode ray tube. To perform these functions the instrument uses a probe containing two piezo-electric crystals, one of which converts electrical signals into vibrations and, by the reverse procedure, the modified vibrations on the other crystal are converted back into electrical signals. See Fig. 11.1. Typical displays for a non-defective (good) and a defective area (bad) are shown in Fig. 11.3.

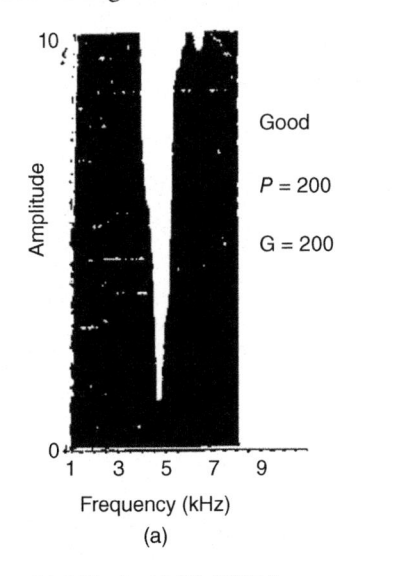

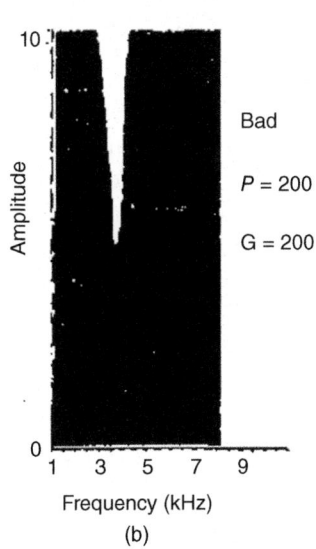

11.3 Typical MIA 3000 frequency spectra.

The resonating frequency can be determined from the position of the peak of the amplitude trace. In a structure, the layer of material above a disbond or delamination may be regarded as a plate that is being restrained around its edges. If the plate is being excited, it can resonate with the first mode being the membrane resonance. From the vibration of a plate [5], the first mode of the membrane resonant frequency for a plate fixed around its boundary is given by:

$$Fr = \frac{0.47h}{r^2} \sqrt{\frac{E}{\rho(1-v^2)}} \qquad [11.1]$$

where h = depth of the defect (in this case the skin thickness)
  r = defect radius
  E = flexural modulus of the layer above the defect
  ρ = density of the material
  v = Poisson's ratio of the material above the defect

The above equation represents the resonant frequency at the centre of the thin (membrane) circular plate. As a general guide, the defect diameter must be at least 20 times the skin thickness for this equation to be valid. The constant, 0.47, is the result of using a fixed boundary condition. Hence it is obvious that for a given structure the membrane resonant frequency, Fr, at the centre of the defect is a function of the depth and size of the defect and its boundary conditions.

This experiment investigated the relationship between the depth of the defect and its resonant frequency at the centre of the defect as measured by the MIA 3000. The theoretical resonant frequencies were calculated using equation (11.1) with the following properties for aluminium:

- flexural modulus, E = 71 Gpa;
- density of aluminium, ρ = 2950 kg/m^3;
- Poisson's ratio, v = 0.33.

11.2.3 Test results

Resonant frequency versus depth

Experiments were conducted to measure the resonant frequencies over the round defect specimens described in Section 11.2.2 ([6]). Good agreement was found between the experimental and theoretical curves as shown in Fig. 11.4.

Resonant frequency versus size

The data can be replotted as shown in Fig. 11.5. Resonant frequencies are plotted against defect sizes. As expected the theoretical resonant frequency

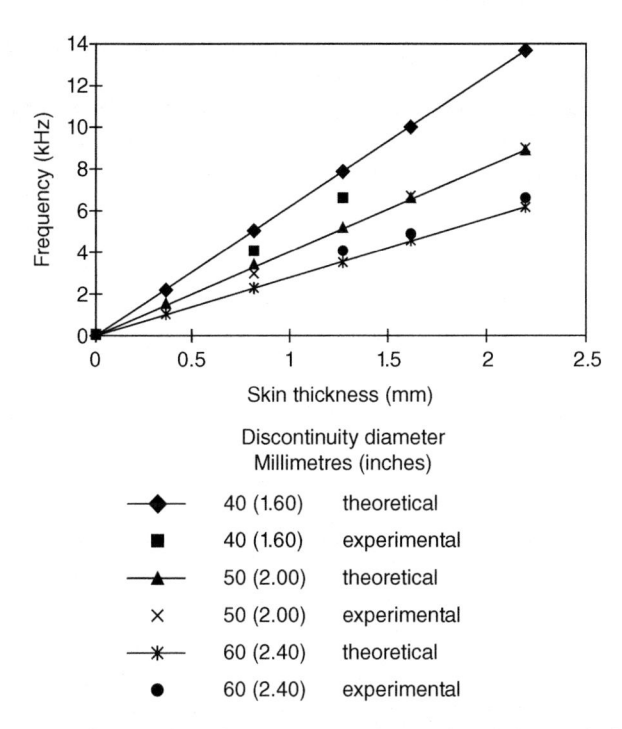

11.4 Comparison between experimental and theoretical resonant frequency plotted against skin thickness.

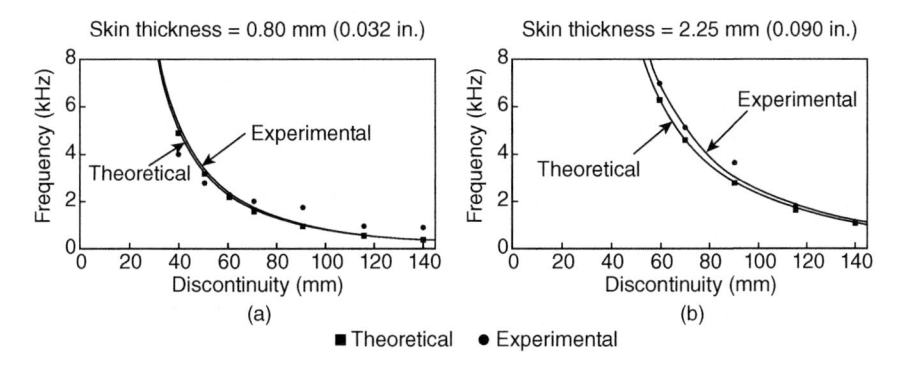

11.5 Resonant frequency versus defect size.

derived from equation (11.1) agree well with the experimental values. The good agreement implies that for the autoclave cured aluminium skin specimens used in this chapter (which is homogeneous and relatively stiff) it is possible to determine the diameter of the defect from the resonant frequency recorded and the use of equation (11.1). This assumes that the defect would be circular or near-circular in shape. Experiments carried out with cold cured glass fibre reinforced skinned specimens have revealed more variation of the resonant frequency for a constant defect size because of the inhomogeneity of the skin and the variations on the specimen caused by the cold cure process. Errors would result from the sizing of defects from the resonant frequency using specimens with inhomogeneous skin structures.

Evaluating irregularly shaped defects

It was concluded that resonant frequency at a delamination area were lower than the resonant frequency of a good area. Therefore, it is possible to detect defects within the specimens by monitoring the drop in the resonant frequency as the probe is swept across the entire test area. Although some scanning points were positioned along the defect's edges, it was noted that the dip in resonant frequency was only evident at positions within the defects. This means that the defect's size would be smaller than the actual simulated delaminations. This part of the experiment also investigated the response of the defects as the probe was swept through different directions. Figures 11.6, 11.7 and 11.8 show the different scanning directions and resonant frequency responses for different defects.

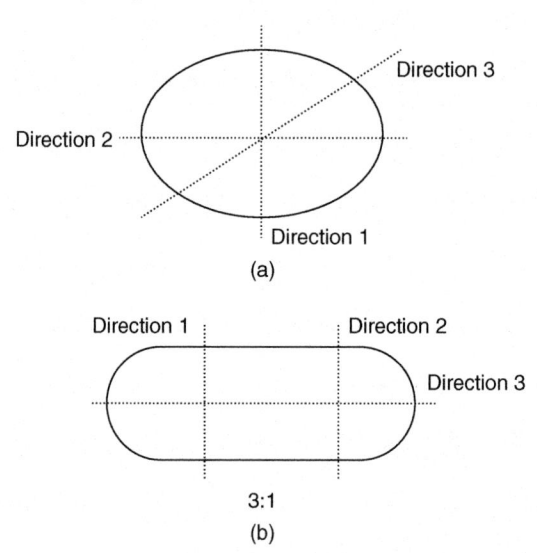

11.6 Scanning directions for different defects: (a) ellipse, (b) 3:1.

Detecting delamination defects 287

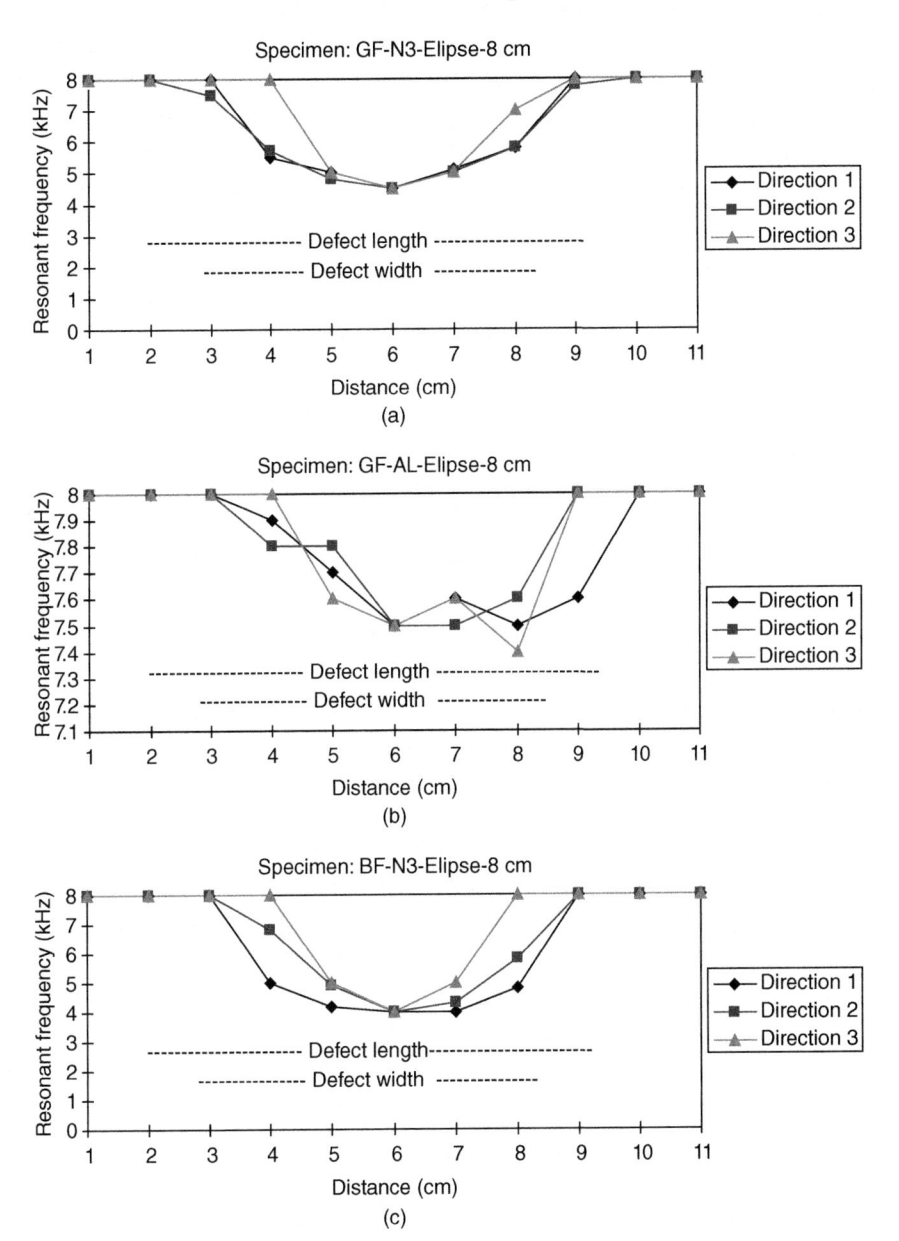

11.7 Resonant frequency against distance for ellipse shaped defects.

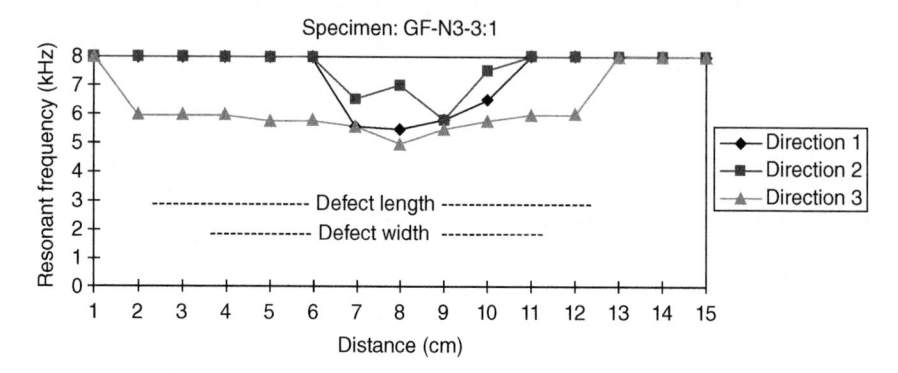

11.8 Resonant frequency against distance for 3:1 shaped defect.

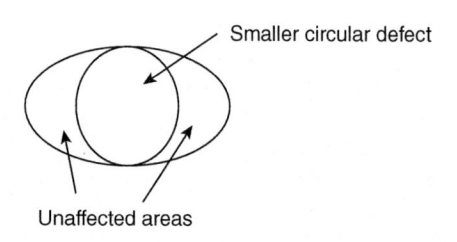

11.9 Illustration of an elliptical defect behaving as a smaller circular defect.

From the resonant frequency against distance plot for the various defects, the graphs show that the resonant frequency along the defects decreases as the probe is moved towards the centre. For all the defects, the decrease in resonant frequency is only evident within the defect's area. This means that the defect's size will tend to be smaller than the actual simulated. It was noted that the different scanning directions for the same defects, like the circle, produced the same results.

For the larger and odd-shaped defects, like the ellipse, the resonant frequency dropped across scanning, directions 2 and 3 producing the same results as direction 1. This indicates that although the defect had a 'longer length' along that direction, it behaved like a smaller defect. Hence the frequency response was similar to a defect having the smaller diameter. Figure 11.9 shows the 'unaffected' area for the elliptical defects. The frequency response of the ellipse behaved like a smaller circular defect.

For the other defects simulated, similar results were observed. The results indicated that the resonant frequency response for an odd-shaped defect behaved like a plate with the shortest dimension 'clamped'. This was indicated by the unaffected area that was not showing any response when the direction was scanned.

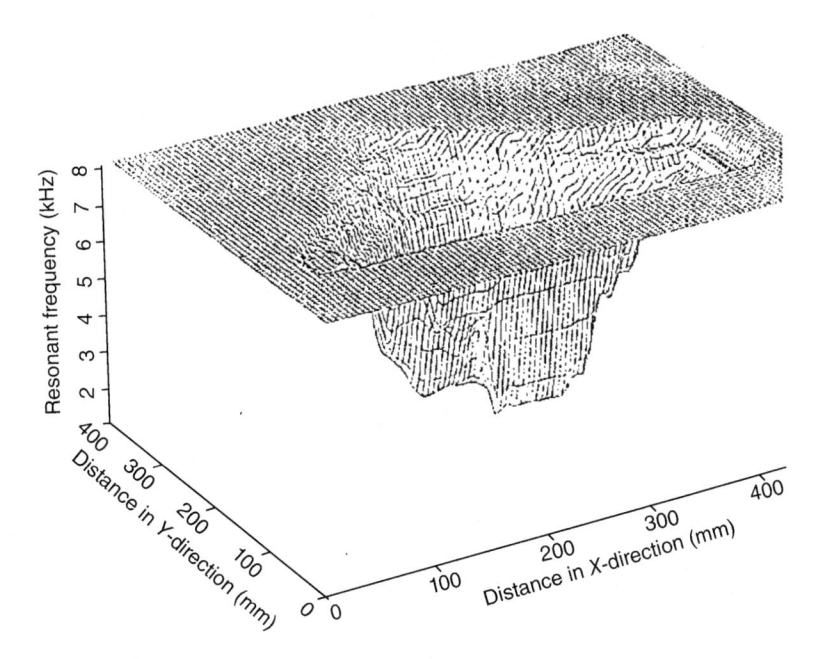

11.10 Resonant frequency plot. Specimen AL090/AL62.5/S1 (instrument power = 70 and gain = 50). All figures are resonant frequency values in kHz.

Resonant frequency change across parts of an irregularly shaped defect

A specimen AL090/AL62.5/S1 was used for the main part of this section. The core was made of aluminium and was 62.5 mm thick in order to simulate an actual specimen from an aircraft. Tests were confined to resonant frequency examinations. Figure 11.10 shows the resonant frequency plot of the specimen. It can be seen that the resonant frequency measured changes across the main defect. However, this change in resonant frequency was not uniform throughout the defect. The rate of change varied at different parts of the defect.

Figure 11.11 shows a schematic outline of the defect. Plots of the resonant frequency versus distance in one direction were made for several distinct regions to evaluate the effect of the defect's extremity curvature and the effect of the main part of the defect on the resonant frequency values. Plots were made of the four directions shown in Figure 11.11. For example from direction 1, the edge of the defect has a smaller radius as compared with direction 5. The plots all start 20 mm from outside the edge of the defect and ended 80 mm into the defect.

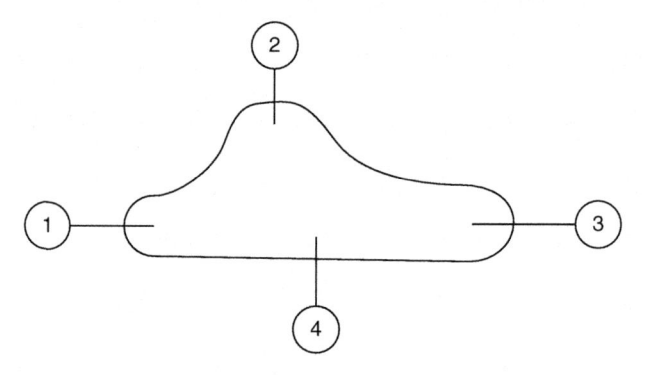

11.11 Outline of defect and different directions considered.

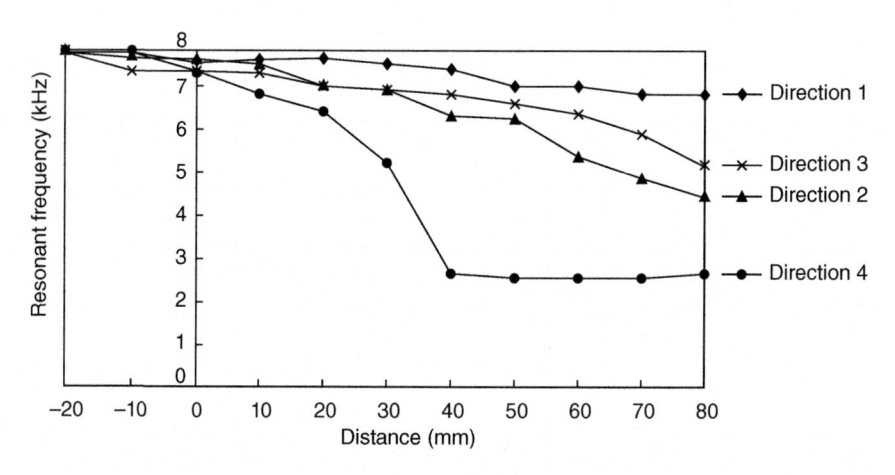

11.12 Frequency variations in the various directions.

The following observations can be made about each direction:

- **Direction 1:** The edge had the smallest radius here and therefore the structure was most stiff here. As expected, the rate of reduction in frequency here was the smallest in all the four directions.
- **Direction 2:** The radius of the edge here was greater than at direction 1. Therefore, as expected the reduced stiffness increased the rate of reduction in frequency compared with direction 1.
- **Direction 3:** This direction showed a frequency reduction rate, which was intermediate between directions 1 and 2 even though the radius was greater than for direction 2 and hence a greater frequency drop rate would be expected because of reduced stiffness. However from examination of direction 2 from Fig. 11.12 it could be seen that the central mass of the defect was reached sooner than was the case for direction 3. Therefore the main mass where the lowest stiffness and lowest

resonant frequency exist would have greater effect on direction 2 and would reduce its frequency reduction rate to more than for direction 3.
- **Direction 4:** The reduction in frequency at this flat edge of the defect was the most drastic. This was due to the maximum possible radius of curvature, essentially infinity, and also closer proximity to the central defect mass than directions 1, 2 and 3.

These results show that the rate of frequency reduction over the extremity of a defect is affected by two variables:

- The radius of curvature of the defect extremity. Also this radius would need to exist over about a semicircle as in the above cases before this variable becomes effective. The greater the radius, the greater the frequency reduction to be expected.
- The proximity of the edge to the central area of the defect. The closer the central area of the defect, the greater the frequency reduction to be expected.

These varying rates of frequency change over the defect's extremity cause difficulties in deriving a procedure for evaluating the exact extremity of the defect edge. It is necessary that an empirical approach be developed to determine the defect's extremities. It has been found so far that for skin thicknesses of 3.6 mm, as in the above specimen, the extremities of a large irregular defect with radii between about 100 mm to infinity can be delineated reasonably accurately from the fact resonant frequency begins to drop at or close to the defect's extremity. However, for a thicker specimen of 6.4 mm as is the case for a practical structure on a plane the resonant frequency would not drop until some way into the defect, making defect delineation more difficult.

11.2.4 Conclusions

The following conclusions are all deduced from tests utilizing the MIA 3000 mechanical impedance instrument, although they should also be applicable to any mechanical impedance instrument which produces the appropriate amplitude–frequency displays shown in Fig. 11.3 and has a similar single frequency capability. The main conclusions are as follows:

- Resonant frequency decreases as the centre of a defect is approached.
- For the aluminium-skinned specimens the defects were found to be reasonably accurately represented by a vibrating plate clamped at its extremities. This enabled the theoretical equation for the latter to predict the resonating frequency of the former. Therefore for delaminations in aluminium skinned honeycomb specimens it would be possible

determine the defect size from the resonant frequency if the defect is circular in shape.
- The resonant frequency is generally affected by the nearest boundary. An important result of this is that an ellipsoidal defect, which is a common shape, would be assessed as a circular defect of radius equal to the minor diameter of the ellipse.
- The rate of resonant frequency drop across a defect extremity is affected by the radius of the extremity at that point and also by the proximity of the point to the main central area of the defect.

11.3 Using ultrasonic 'C' scanning: carbon fibre-reinforced (CFR) composites

Conventional ultrasonic immersion 'C' testing of CFR structures is widely used. The procedure usually uses a focused immersion compression wave probe which directs ultrasonic pulses normally into the specimen. The probe is attached to an x–y table and is moved in a raster fashion over the specimen. 'A' scan data is collected as the probe moves and software is able to generate plan view images of any defects detected. Many references [1–3] report work on larger defects and this section focuses on evaluation in terms of sensitivity, resolution and sizing of smaller defects (1–11 mm in diameter) at varying depths in solid laminar composites and how to determine how much detail can be revealed on a subsurface delamination caused by impact damage.

11.3.1 Test specimen

The first specimen utilized is shown in Fig. 11.13. It contains delamination defects of various sizes ranging from 1 to 11 mm in diameter. These defects are reproduced at various depths below the testing surface ranging from 0.28 mm (2 layers of prepreg) to 2.8 mm (20 layers of prepreg). Each layer of prepreg was 0.14 mm. The defects were made of peel ply. Each defect consisted of three layers of peel ply placed on top of each other in order try to ensure that an air gap existed at least between each of the three layers. All defects were round expect the 1 mm defect which was square in view of the difficulty of producing a round defect of 1 mm diameter.

11.3.2 Test results

Figure 11.14 shows 'C' scan images of plastic insert defects in a CFR specimen shown in Fig. 11.13. The defects were made by inserting two layers of peel ply between the second and third prepreg layers. The air gaps between each ply layer served as defects reflecting the ultrasound back to the probe.

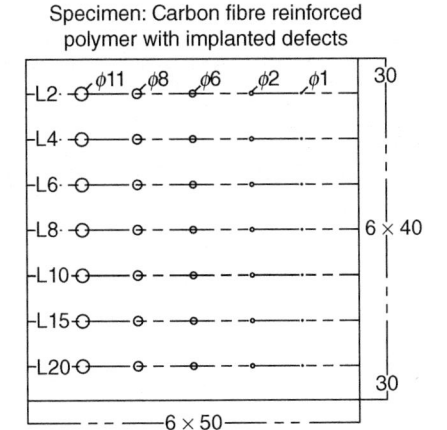

11.13 A composite test piece (made of peel ply) with implanted delamination defects of known sizes and depths. 'L' refers to how many prepreg layers below the top of the test piece the defect is located. The total thickness of the test piece was 4.2 mm or 30 layers of prepreg. The specimen represents part of the tail fin of an F16 aircraft.

Since each prepreg layer was 0.15 mm, the depth of each defect was 0.3 mm from the top surface of the specimen. The specimens and probe were immersed in water. The energy from the probe then travelled through the water before entering the specimen, hence providing very efficient transfer of energy between the probe and specimen. A highly damped 10 MHz frequency probe was used and the energy from the probe was focused with a lens on the front of the probe to a small area on the specimen.

Figure 11.14 (a) clearly shows the ability of a 'C' scan to resolve the square shape of the 1 mm wide defect. The defects in Fig. 11.14 (b) to (e) are clearly shown to be circular and the focused probe enabled quite accurate sizing of the diameters of the defects and determination of the defect shapes. It is important to note that all the defects shown in Fig. 11.13 were detectable using this 'C' scan system, and the depths, sizes and shapes of the defects could be determined.

Table 11.1 shows comparisons between the ultrasonic image sizes shown in Fig. 11.14 compared with the actual sizes of the manufactured plastic insert defects. The image sizes were measured simply by taking the widths of images as viewed in the figure (an infinity drop technique) and using the distance scale shown in the images. As expected, oversizing occurs due to beam spread effects, but at least an approximate indication of the defect size can be obtained.

Results from this same ultrasonic system on a carbon fibre skinned, nylon core honeycomb specimen are next described. Figure 11.15 shows the

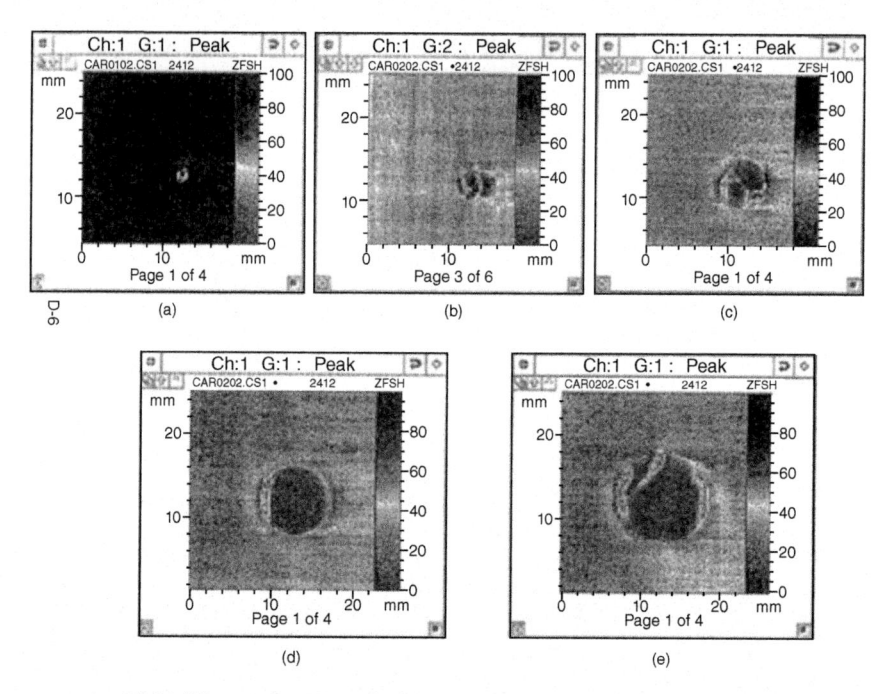

294   Non-destructive evaluation (NDE) of polymer matrix composites

11.14 'C' scan images of plastic insert defects in carbon fibre-reinforced specimens. All defects were 0.3 mm below the top surface of the specimen. All defects were circular except (a): (a) 1 mm width, square shaped; (b) 4 mm diameter; (c) 6 mm diameter; (d) 8 mm diameter; (e) 11 mm diameter.

Table 11.1 Comparisons between actual defect sizes and ultrasonic 'C' scan size measurements for defects at 0.28 mm depth in specimen shown in Fig. 11.13. All figures are in mm

Actual defect size	1	2	6	8	11
Ultrasonic 'C' scan measured size	1.5	4	8	9	12

specimen and some 'C' scan images. Figure 11.15(a) shows a cross-section of a real delamination defect in the top skin of the honeycomb specimen created by an impact on the top of the specimen surface. It was possible to determine that this defect occurred between the third and fourth layers because the image '(3)' produced in Fig. 11.15(b) was with the gate set to detect defects between layers 3 and 4. The image '(10)' shows the honeycomb cell structure since the gate was set to detect defects between the top skin and honeycomb core. Some concluding remarks on this section are included in Section 11.5.

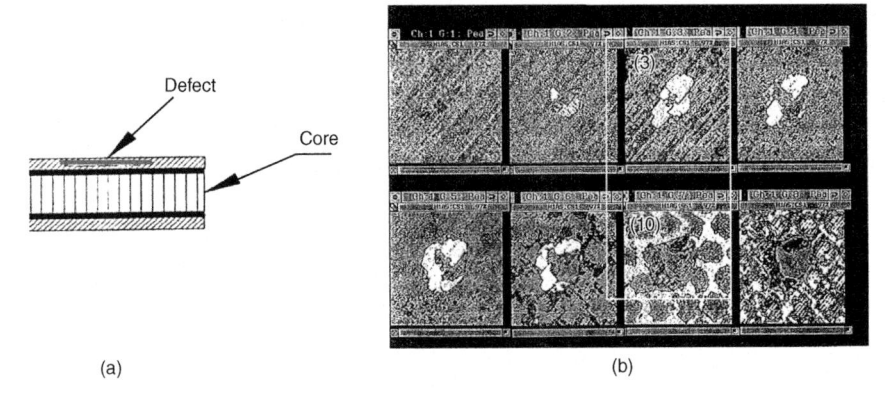

11.15 A honeycomb composite consisting of two solid composite skins each 1.5 mm thick and each made of 10 layers of 0.15 mm thick prepreg attached to a honeycomb core. (a) A 'real' (not plastic insert) delamination defect created by impact damage. This defect is positioned between the third and fourth layers. (b) The '(3)' shows a 'C' scan image of this defect which was achieved by positioning the 'gate' to detect signals between the third and fourth layers. The '(10)' shows a 'C' scan image with the gate positioned to detect defects between the skin and top of the core. The honeycomb cell structure is clearly shown. The other images in this figure show images from other gate positions.

11.4 Using infrared thermography

Thermographic non-destructive testing (NDT) utilizes electromagnetic waves which have a wavelength longer than light. The most common wavelength used is between 8 and 12 microns because light at this wavelength has low absorption and good transmission through the atmosphere. This is important because most non-destructive tests are conducted by active thermography, in which the heat from a source, such as a large lamp, is directed to the specimen to be tested and hence the heat needs to travel through the atmosphere. Thermography is mainly suitable for defects just a few millimetres below the surface of a specimen. Its main advantage is that it can test a large area quickly, unlike other NDT techniques like ultrasound which can only test small areas with a probe in one position at once, and the time-consuming process of probe movement is required to test large specimens.

11.4.1 Pulse thermography

Figure 11.16 illustrates the principle involved. A short pulse of heat (termed a light wave in Fig. 11.16) from a source is directed to the specimen's surface.

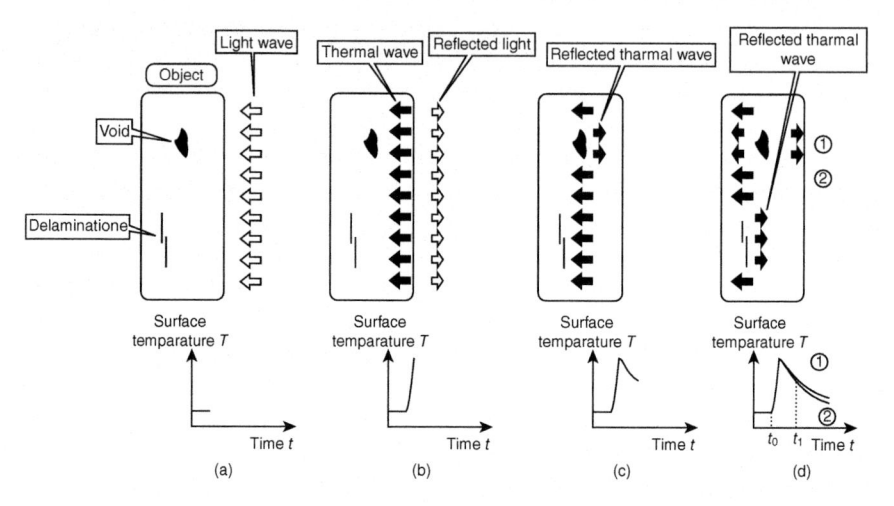

11.16 Various stages of pulsed thermography.

The heat travels to the surface by radiation. The surface is heated and the heat then travels into the specimen by conduction and is termed a thermal wave. When the thermal wave strikes a defect such the void in Fig. 11.16(a), most of it is reflected from the defect because the defect is usually air since air has a poor heat conductivity. The reflected wave then returns to the specimen surface at position 1 (Fig. 11.16(d)), where it heats the surface to a higher temperature than the surrounding specimen surface. Heat then leaves the surface by radiation. A thermal camera is able to receive the heat emitted by the surface and converts this into a temperature profile of the surface. Position 1 is at a higher temperature than position 2 and hence position 1 is viewed as a hot spot (Plate I between pages 296 and 297). Faint images of other defects can also be seen in this Plate. Figure 11.15 also shows the time temperature profiles. The highest temperature shown in Fig. 11.16(d) is where the heat source has been switched off. The figure also shows that at any time after switching off the source, the temperature at position 1 is always higher than that at position 2. The heat profile for the delamination defect is not shown.

Non-metals such as fibre-reinforced composites can be tested more effectively than metals. This is because metals have high conductivity. The profile shown in Plate I between pages 296 and 297 would only exist for a very short time in metals because the heat from the hot area would be conducted away very rapidly. A disadvantage of the pulse thermography system is that noise signals due to heat radiation being directly reflected off the specimen surface can lead to spurious signals being detected. This heat radiation is not due to internal defects. A technique to overcome this problem is lock-in thermography, which is described next.

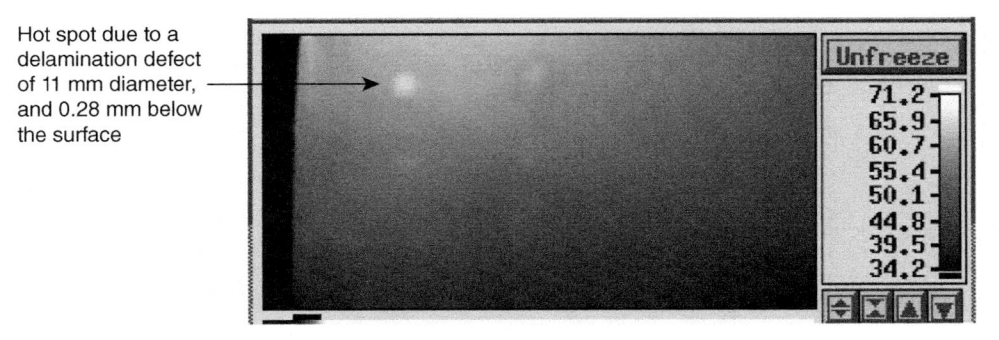

Plate I (Chapter 11) A temperature profile image from a thermal camera of the surface of the specimen in Fig. 11.13. The hot spot due to a delamination defect of 11 mm diameter, and 0.28 mm below the surface is shown. The temperature scale on the right-hand side is in °C.

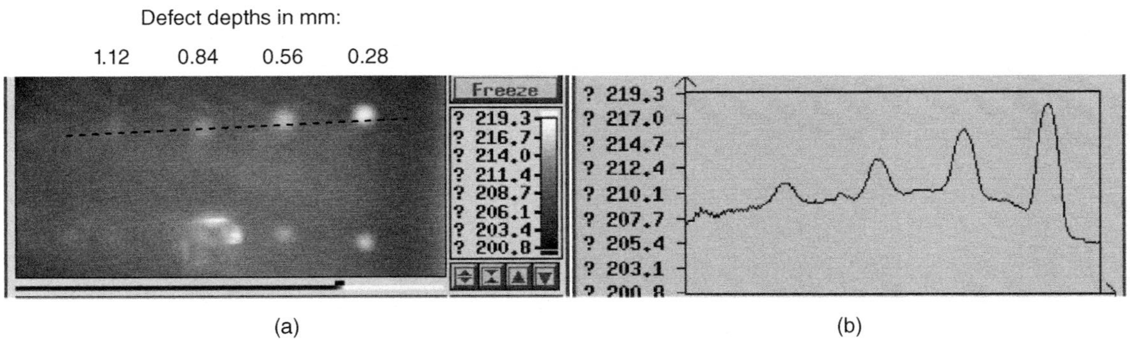

Plate II (Chapter 11) (a) Lock-in phase difference images. The top row of images are from delamination defects of 11 mm diameter. Defect depths in mm are shown. Defects created by plastic inserts. (b) Plot of phase difference versus distance across the specimen along the dotted line shown in (a).

(a) Frequency = 0.0037 Hz

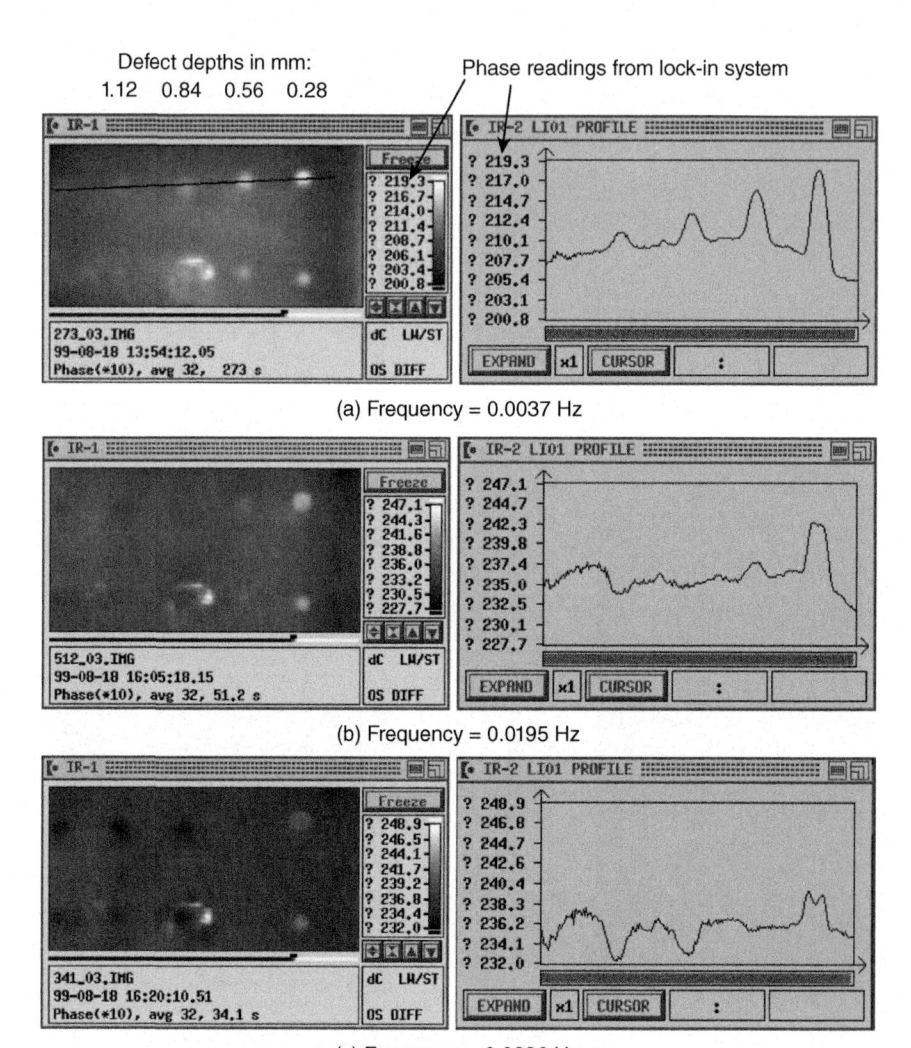

(b) Frequency = 0.0195 Hz

(c) Frequency = 0.0296 Hz

Plate III (Chapter 11) Thermograms and phase profile plots of 11 mm defects of specimen.

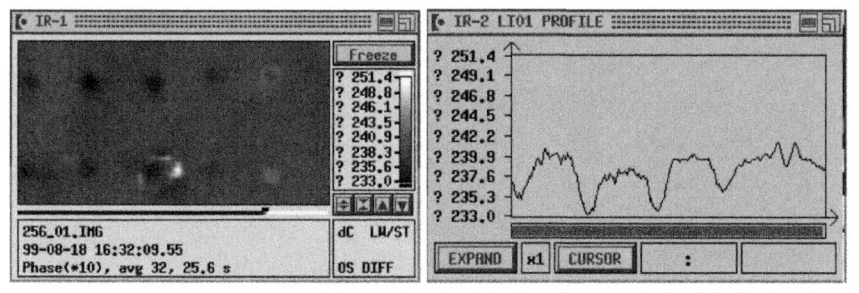

(d) Frequency = 0.0391 Hz

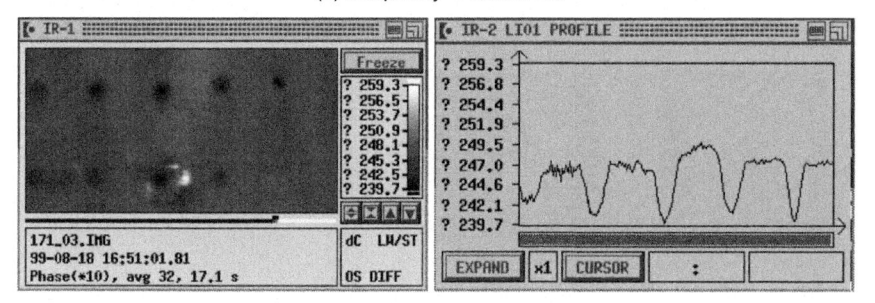

(e) Frequency = 0.0585 Hz

Plate III Continued

Plate IV (Chapter 11) Lock-in thermographic image of specimen shown in Fig. 11.15(a).

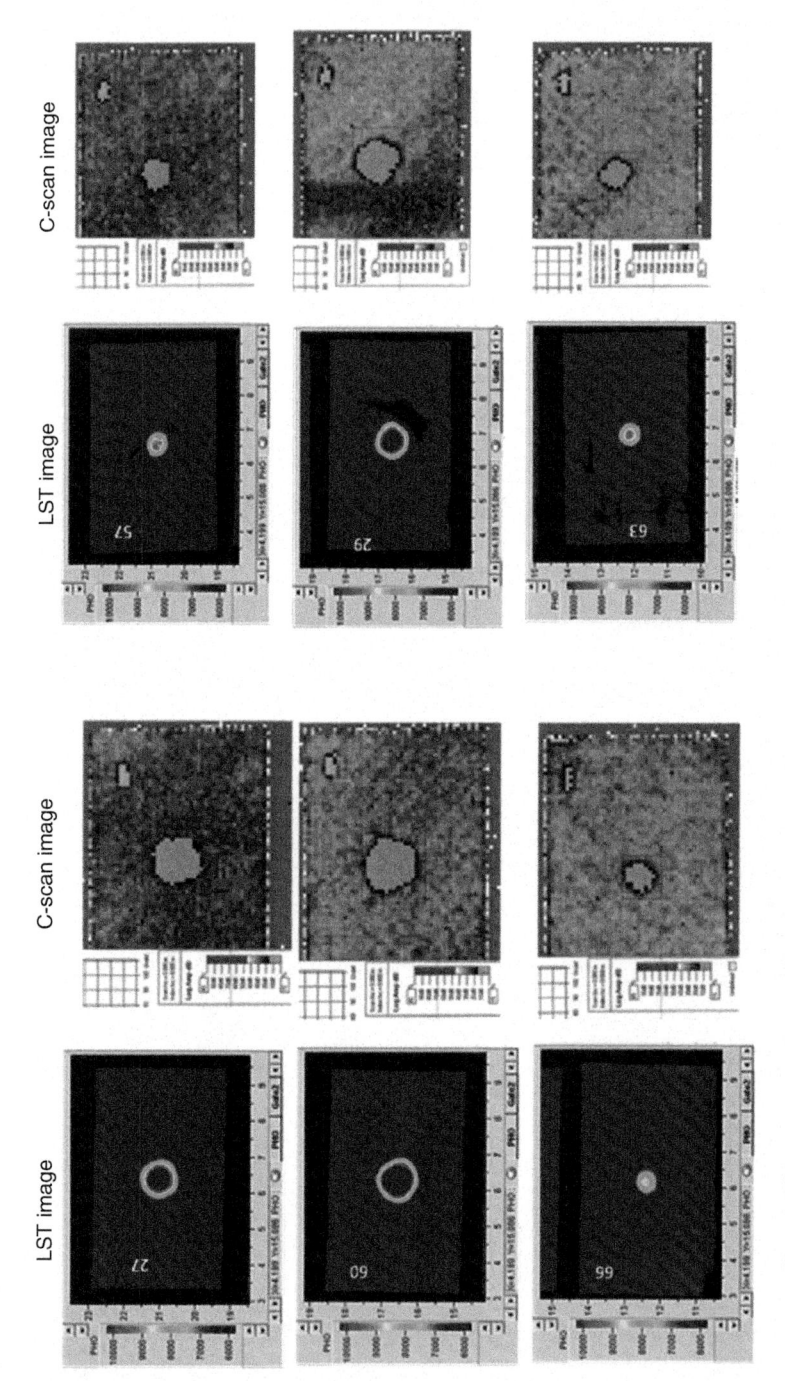

Plate V (Chapter 12) Comparison of UT and LST scans of honeycomb sandwich panels with impact damage from various loading levels. It was consistently observed that areas detected by LST were slightly higher than those found using immersion UT.

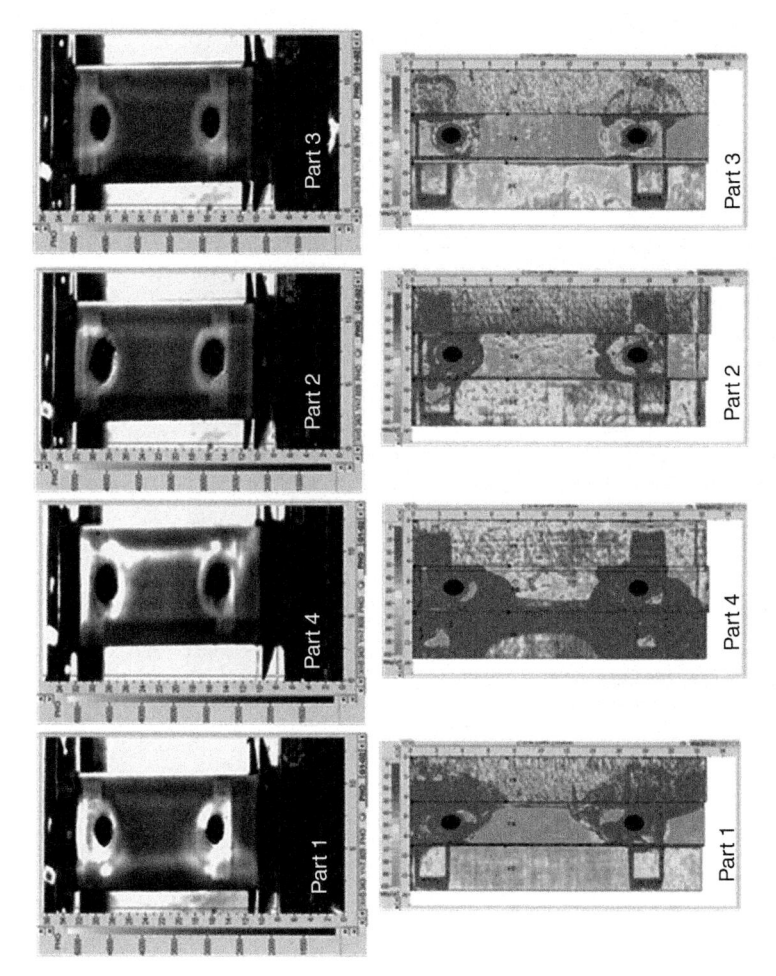

Plate VI (Chapter 12) Top: LST-scans of the four hardbacks studied. Parts 1 and 4 show larger and more intense hot spots, which are expected to be associated to damage produced by impact. All the images presented correspond to the image difference between two observation times ($t = 25$ and 57 s). Bottom: C-scan images of the four hardbacks. Comparison of LST and UT images show corresponding areas. Note that the LST images show the mirror image of the part or area scanned.

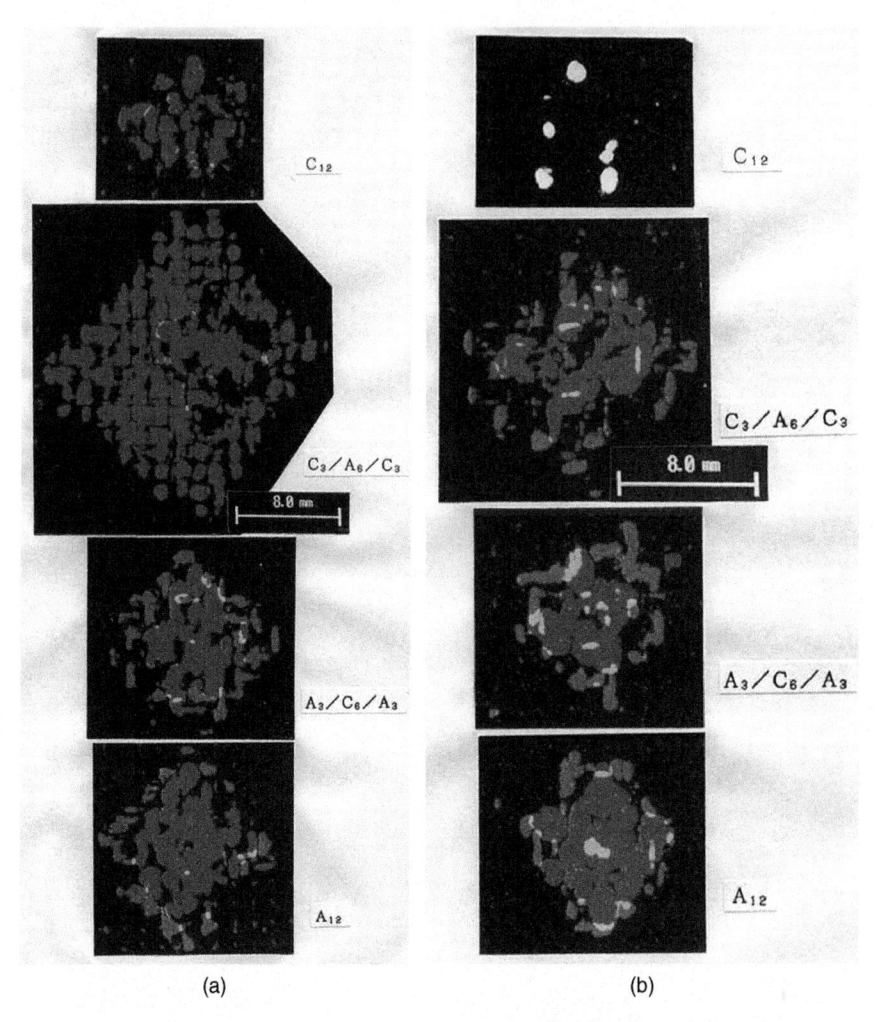

Plate VII (Chapter 16) Acoustic images showing delaminations at interfaces. (a) Impactor: diameter $\Phi = 5$ mm, velocity $V = 113$ m/s. (b) Impactor: diameter $\Phi = 10$ mm, velocity $V = 40$ m/s. Frequency = 30 MHz.

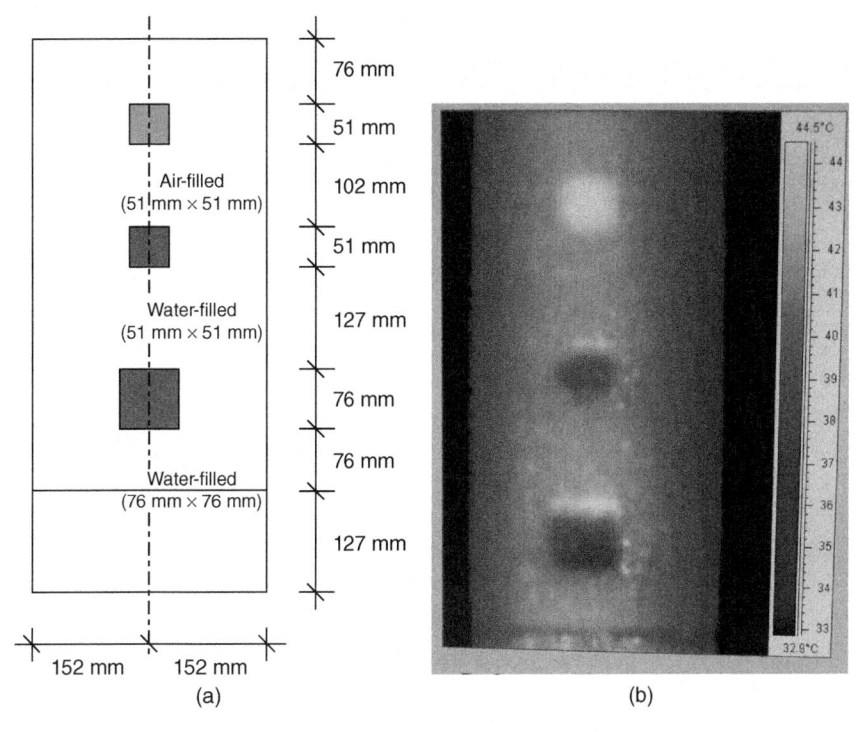

Plate VIII (Chapter 18) (a) Schematic view, and (b) infrared image of bridge deck specimen BD2 showing the air-filled and water-filled debonds.

Plate IX (Chapter 18) (a) Photograph of a pile, and (b) infrared image showing debonds.

© Woodhead Publishing Limited, 2013

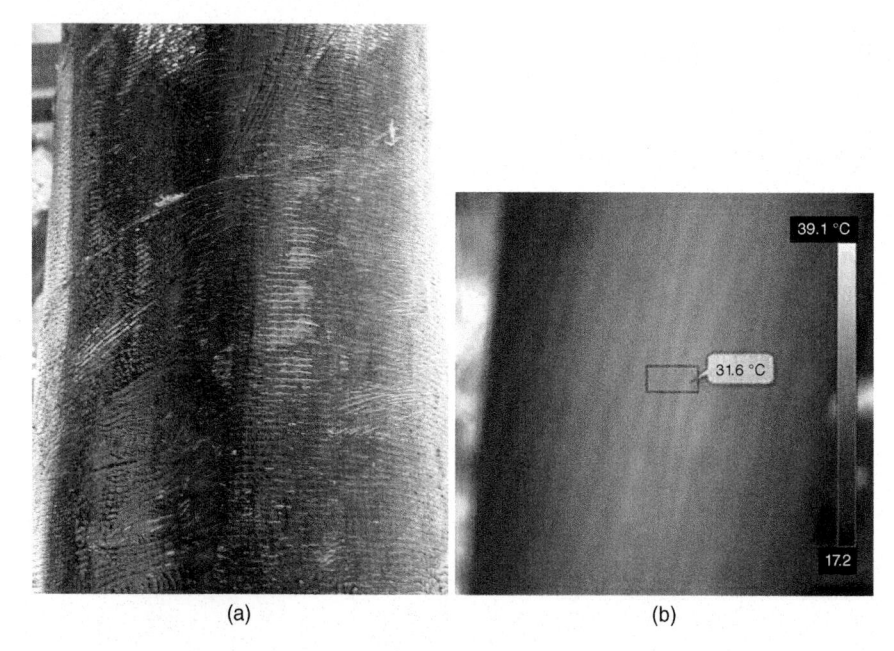

Plate X (Chapter 18) (a) Photograph of a pile, and (b) infrared image showing good bond.

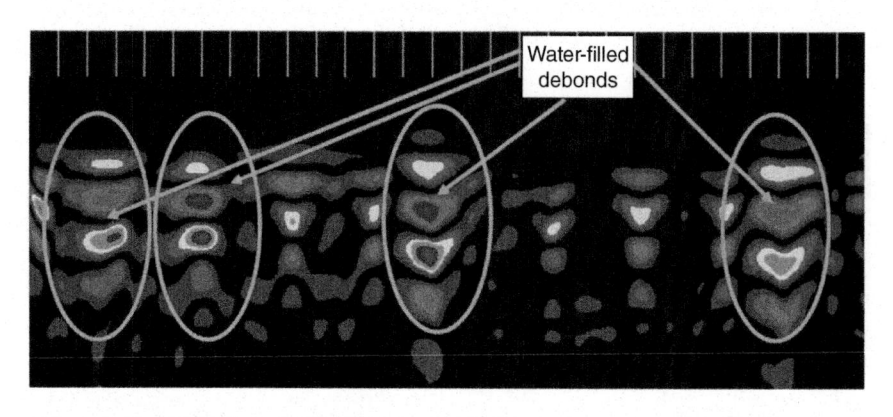

Plate XI (Chapter 18) Results of GPR test showing water-filled debonds.

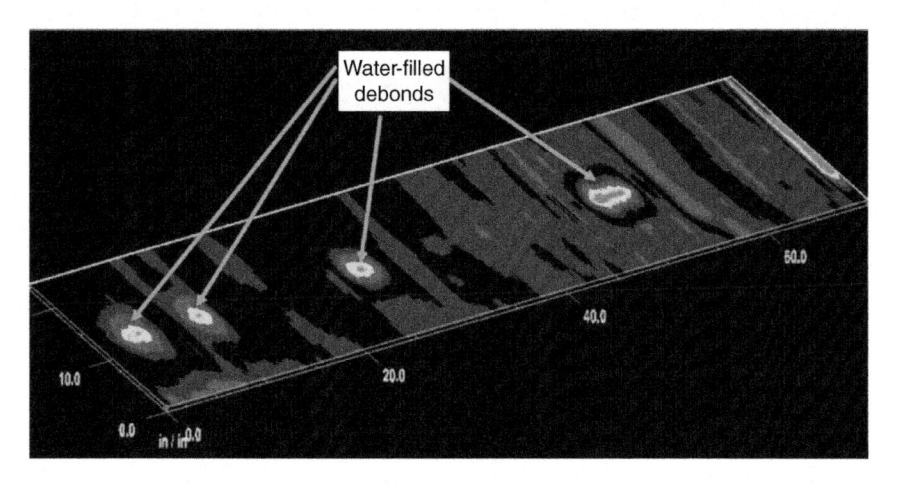

Plate XII (Chapter 18) Three-dimensional GPR image showing location of water-filled debonds.

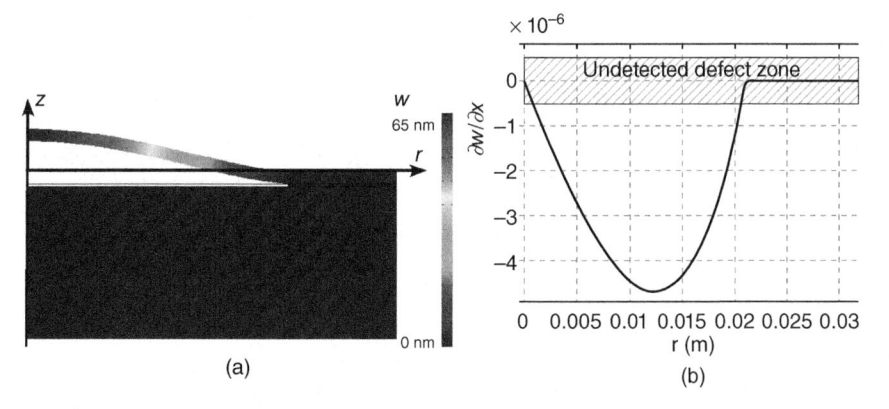

Plate XIII (Chapter 20) Results of the finite element simulations for a defect diameter equal to 40 mm and an applied depressure $\Delta P = 100$ Pa.

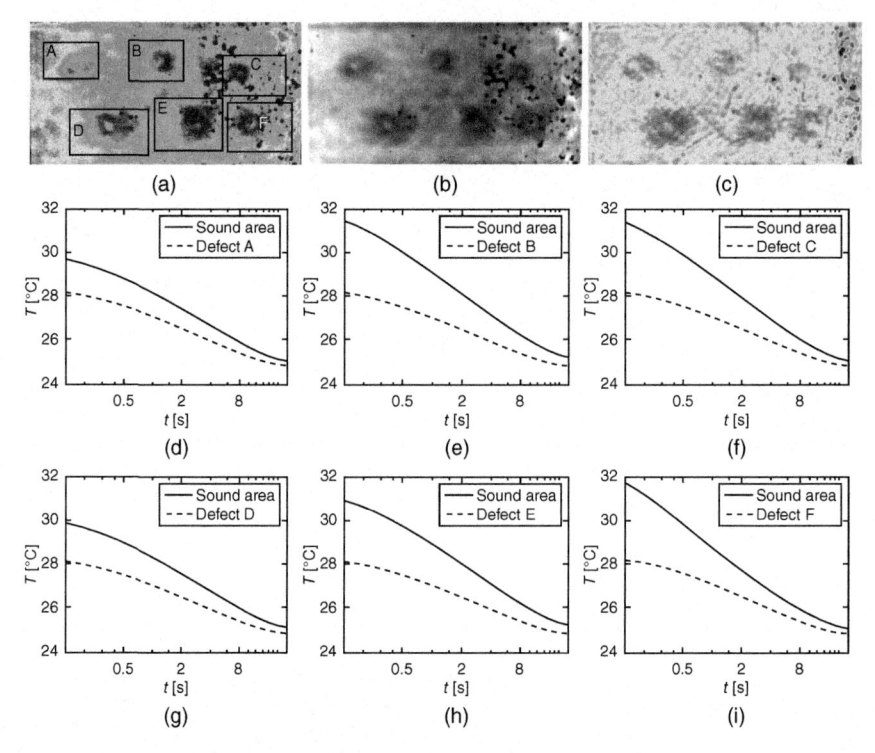

Plate XIV (Chapter 24) First time derivatives from a fifth degree polynomial at $t = 19.1$ seconds: (a) colour map 'jet', (b) gray levels, and (c) phase at $f = 0.043$ Hz and thermal profiles for defects.

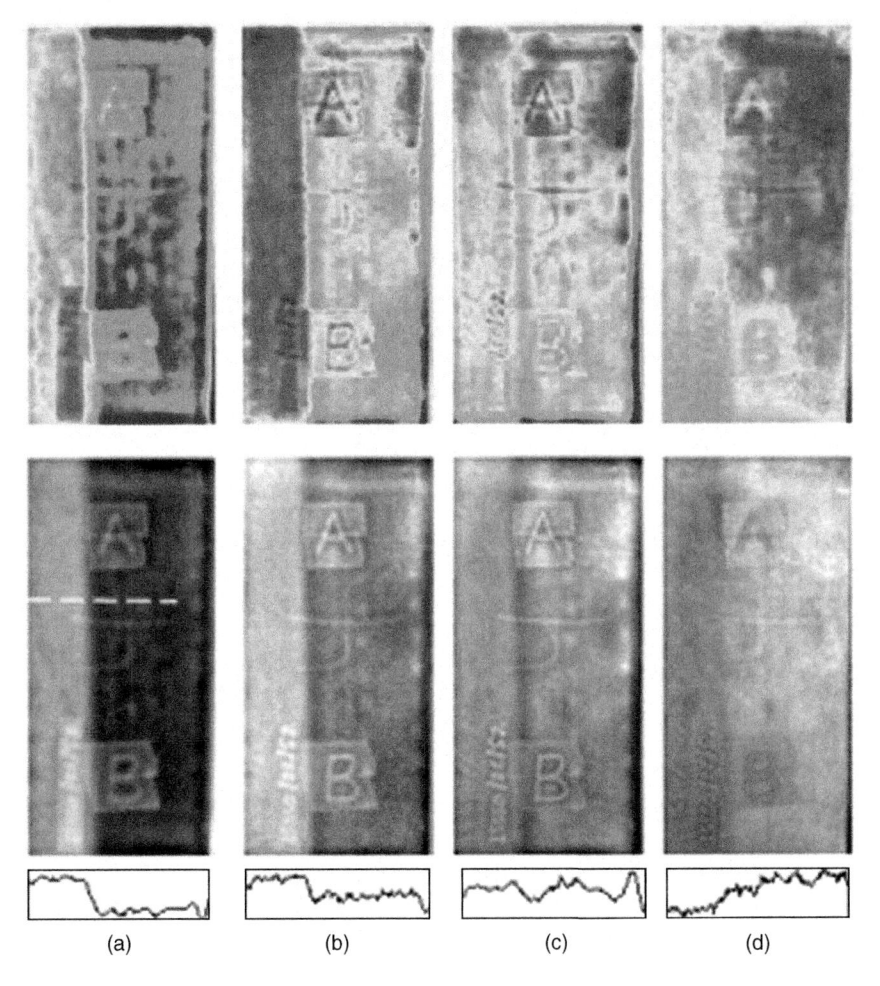

Plate XV (Chapter 24) 40 s heating with five lamps (1000 W), 2385 frames at 1.1 Hz: jet (top) and grey (bottom) phasegram at $f =$ (a) 0.0059, (b) 0.0118 (c) 0.0378 and (d) 0.1796 Hz. Spatial profiles passing through the specimen centre are also shown.

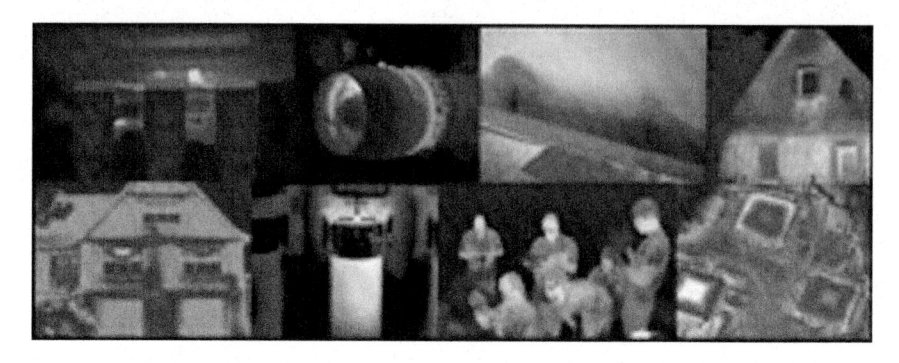

Plate XVI (Chapter 25) Different thermography applications.

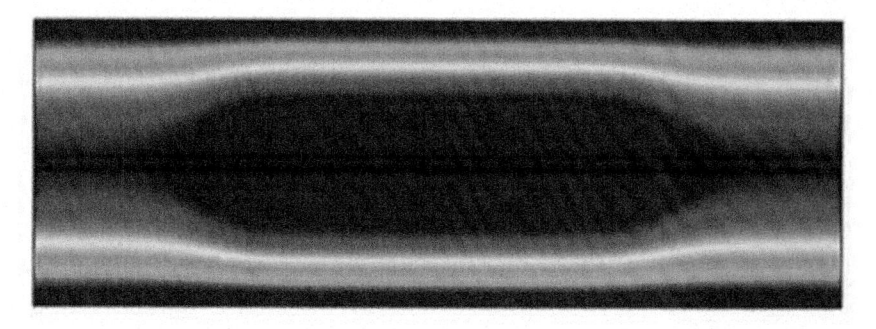

Plate XVII (Chapter 25) Thermal response of delamination defect.

11.4.2 Lock-in thermography

Figure 11.17 illustrates a lock-in thermography system. Rather than directing a single pulse of heat to the specimen as described in Section 11.4.1, a continuous sinusoidal heat wave is used. This is produced by varying the intensity of the lamp source sinusoidally. The system controller and module in Fig. 11.17 send the required signal to the lamp. The sinusoidal heat wave enters into the specimen as with the pulse thermography technique. The wave is reflected by any defects and returns to the specimen surface where it interferes with incoming thermal waves. This interference pattern is monitored by the thermal camera which can produce phase and amplitude data. This phase and amplitude data is compared with the sine heat wave which leaves the lamp source, producing phase and amplitudes differences. The name lock-in is used because the system lamp and camera are locked together and the software in the system can compare the data from the camera with the lamp source sine wave. An example of phase difference thermal images is shown in Plate II (between pages 296 and 297). The specimen used was that shown in Fig. 11.13, the same as that used for the pulse thermography tests described in Section 11.4.1.

Plate II(a) shows phase difference images from the specimen shown in Fig. 11.13. The defect images are approximately 11 mm in diameter showing the ability to size the defects. The images are less visible as the defect depth increases owing to reduced contrast because of increasing attenuation with depth, and the fact that many of the thermal waves will not move perpendicularly into the specimen as required to produce the phase difference images. The reduction in contrast with defect is quantified in Plate II(b), which is a plot of phase difference against distance along the dotted line shown in (a). The numbers on the vertical scale indicate absolute phase difference values. This figure indicates the possibility of determining the defect depth from the phase difference values if the defect size is determined from an image. The defect images in the bottom part of Plate II(a) are from 6 mm diameter defects, again varying in depth as with the top row of defects. The bright image shown in the bottom part of the image is probably due an actual defect produced unintentionally, i.e. not due to a plastic insert.

Table 11.2 shows a comparison between the actual defect size and lockin thermographic size from the top row of defects indicated in Plate II(a) and the phase plot versus distance in Plate II(b). The horizontal alignment of the fibres in the composite specimen would mean that the heat flow travelling through the specimen at right angles to the surface would tend to spread out slightly due to the greater heat conductivity along the fibres rather than at right angles to the fibres. This would result in oversizing of the defects. Hence the defects were sized by taking the widths of the

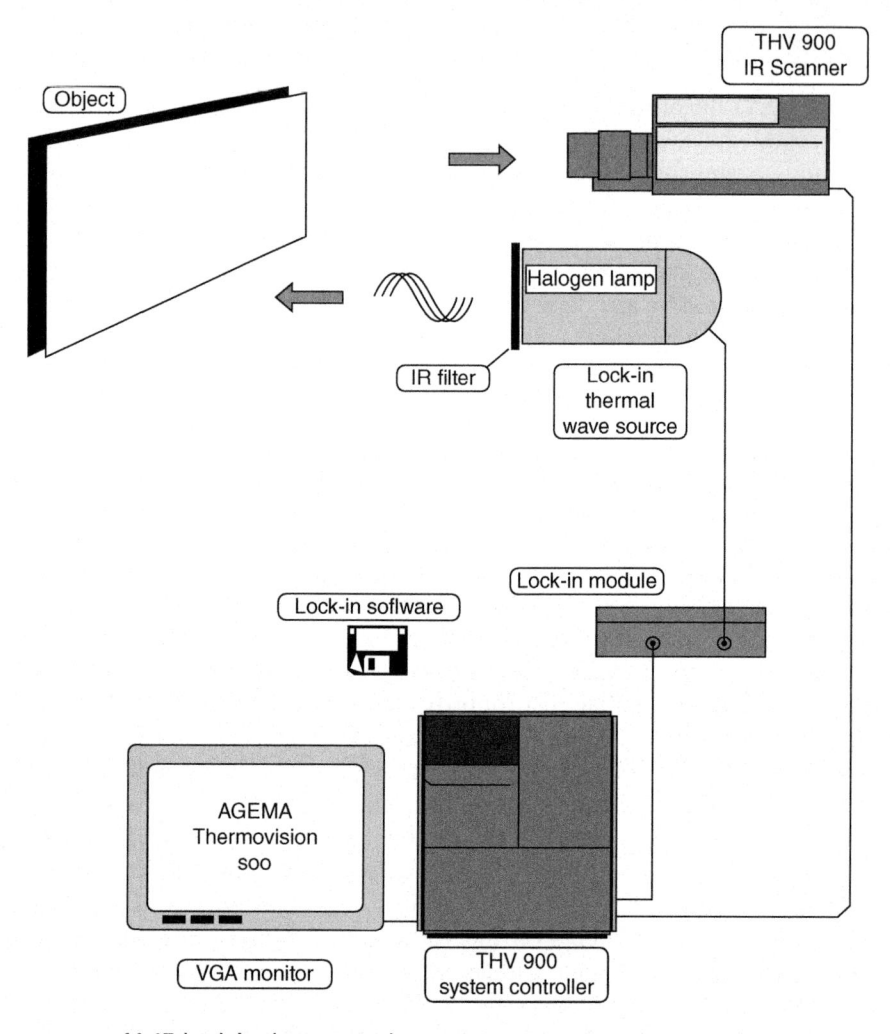

11.17 Lock-in thermography system.

Table 11.2 Comparisons between actual defect sizes and lock-in thermographic size measurements for defects at various depths in specimen shown in Plate II. All figures are in mm

Defect depth	1.12	0.84	0.56	0.28
Actual defect size	11	11	11	11
Lock-in thermographic measured size	12.1	13.5	10.5	13.5

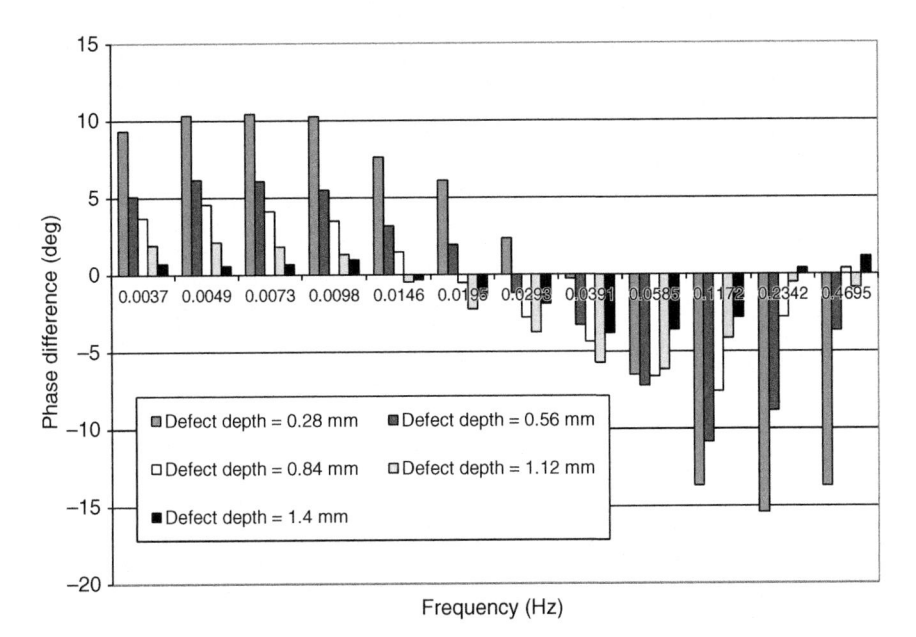

11.18 Phase differences caused by 11 mm defects at different depths versus testing frequency.

respective phase signals in Plate II(b) at the half-amplitude heights. Reasonably accurate measurements resulted with no definite trends at different depths.

The selection of the testing frequency is important in lock-in thermographic testing. Plate III (between pages 296 and 297) show how the phase contrast values of the 11 mm defects at different depths change as the test frequency changes. Figure 11.18 summarizes these values. The general features observable from both these figures are as follows. The phase differences of the defects can be positive (light coloured) or negative (dark coloured) with respect to the backgrounds. Either positive or negative values can be used for defect detection. To select the best frequency to detect the defect at a particular depth, the frequency giving the maximum phase contrast should be selected and the frequency with the lowest phase contrast (the 'blind' frequency) must be avoided. For example to test the defect at 0.28 mm depth and frequency of 0.0037 Hz could be used to produce the maximum positive signal or a frequency of 0.2342 Hz could be used to produce the maximum negative signal (not shown in Plate III). The blind frequency to be avoided is about 0.0293 Hz.

A theoretical model of phase difference against testing frequency for defects of various depths was developed by Bai and Wong [7]. The advantage of this model is that it enabled the prediction of the phase differences

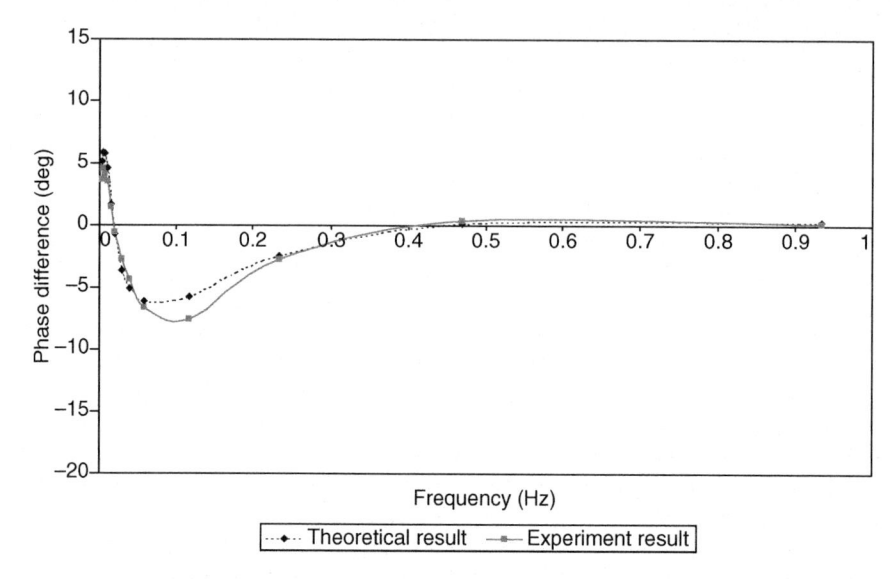

11.19 Theoretical and experimental results of phase difference versus testing frequency. The defect depth was 0.84 mm.

for various defect sizes without having to produce a test specimen to determine the optimum frequencies to be used. For example Fig. 11.19 shows that the optimum testing frequency to be used would be about 0.1 Hz and blind frequencies to be avoided would be 0.15 and 0.4 Hz. The model was suitable only for large defects since it considered one-dimensional heat flow only. However it would be accurate for defects of size 11 mm.

Plate IV (between pages 296 and 297) shows a lock-in thermographic image of the specimen shown in Fig. 11.15(a) which was a delamination defect caused by impact damage. When compared with the ultrasonic 'C' scan image in Fig. 11.15(b) it can be seen that the defect is detected with thermography showing a similar shape and size. However, the resolution detail appears to be superior using ultrasonics.

11.4.3 Conclusions

Figure 11.20 shows defect detectability in the carbon fibre specimen illustrated in Fig. 11.13. Defects in the upper left hand part of the chart are detectable because they have large diameters and shallow depths (i.e. near the specimen surface). Defects in the bottom right hand part of the chart are not detectable because they are of small diameters and are deep in the specimen. The conventional pulse method (solid line and above) can be seen to be the least sensitive technique, with low signal-to-noise ratios. The

Detecting delamination defects 301

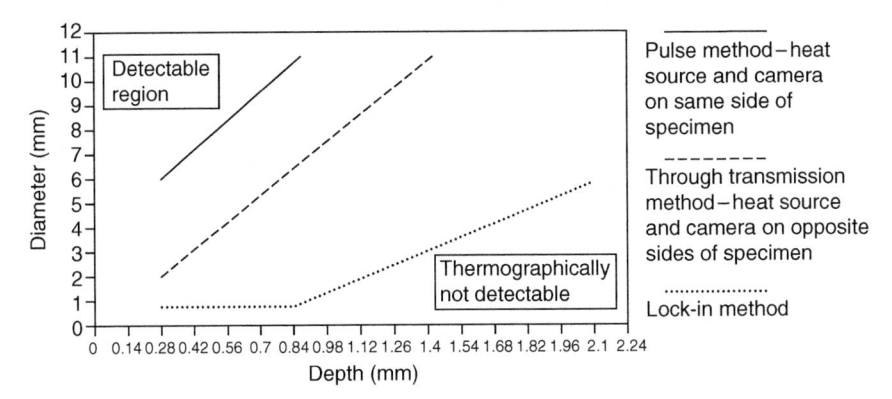

11.20 Defect detectability of thermographic techniques in a carbon fibre reinforced composite specimen.

heat source and thermographic camera are on the same side of the specimen in this case. Results for a through transmission technique are indicated as the dashed line and above. The heat source is used to direct the heat through the specimen and the camera is placed on the opposite side of the specimen. This procedure is clearly more sensitive than when source and camera are on the same side. This is because in through transmission the heat only needs to traverse one way through the specimen; it does not return after reflecting from the defect. A higher percentage of the heat energy is more likely to traverse, perpendicularly to the defects and specimen surfaces.

The lock-in phase analysis method (dotted line and above) is clearly shown to be the best method with the highest signal-to-noise ratio of all three techniques. The production of phase data from interfering signals which removes unwanted noise from surface reflection data is clearly the preferred procedure. These results show generally that thermography is more sensitive to defects with high diameter (size) to depth ratios. Large and shallow defects will be more readily detected. It is important to note that ultrasonic 'C' scan testing can detect all the defects represented in Fig. 11.20 above 1 mm in size, whatever the depth.

11.5 Conclusion: comparing different techniques

Table 11.3 summarizes the techniques covered in this chapter. For disbond defects in honeycomb structures, mechanical impedance is the fastest for measuring a point location on the specimen and it requires no couplant. Ultrasonic 'C' scan requires couplant and is reasonably fast for scanning very small areas. Large areas, e.g. several square metres, can take an hour or more, depending on the scan pitch resolution. 'C' scan has the advantage

Table 11.3 Comparisons between mechanical impedance, ultrasonic 'C' scan and thermography for testing FRC materials from data in this chapter

	Mechanical impedance	Ultrasonic 'C' scan	Pulse (reflection) thermography	Lock-in thermography
Practical advantage/ disadvantage	No couplant required. Portable. A small probe placed on specimen surface.	Couplant required. Portable systems exist.	No contact required with specimen. Portable heat source and camera can be one metre away from specimen. Slightly curved surfaces can be tested.	No contact required with specimen. Portable heat source and camera can be one metre away from specimen. Slightly curved surfaces can be tested.
Testing time	Fast for a point location.	Reasonably fast for very small location but requires considerable time for a large area.	Fast technique for large area.	Fast technique for large area but longer than pulse thermography.
Ability to test a large area quickly	Time consuming due to 'C' scan raster movement.	Time consuming due to 'C' scan raster movement.	Most effective NDT technique.	Effective NDT technique. Slightly longer inspection time than pulse thermography.

Type defect most suited to detect	Disbond defect in honeycomb structures between skin and core.	Delamination defects in laminar composites. Disbonds in honeycomb structures.	Near surface delamination defects in laminar composites.	Near surface delamination defects (but deeper than pulse thermography) in laminar composites.
Defect sensitivity	Suitable for large defects. 40 mm dia or larger in current work.	Most sensitive technique. All 1 mm defects detected up to depth of 2.8 mm.	Only large near surface defects detectable, e.g. 6 mm dia. defect not detectable deeper than 0.28 mm.	Greater sensitivity to near surface defects than pulse reflection, e.g. 6 mm defect detectable to 2.1 mm depth.
Defect sizing ability	Reliable for large defects.	Slightly enlarged sizing for defects 1–11 mm dia.	Reliable for detectable defects.	Reliable for detectable defects.
Defect shape assessment	Reliable for large defects.	Reliable for defects 1–11 mm dia.	Reliable for detectable defects.	Reliable for detectable defects.
Defect x–y position location	Can be determined accurately.	Can be determined accurately.	Can be determined accurately.	Can be determined accurately.
Defect depth location	For disbond measurement, it is possible to measure the skin thickness from resonant frequency values, if the defect size is measured first.	Can be determined accurately.	Approximate depth could be determined.	Could be determined from a database.

of being sensitive to delamination defects in solid laminar composites and disbonds in honeycomb structures. Figure 11.15(b) '(10)' shows the ability to visualize the honeycomb cell structure and this would not be visible if disbond occurs.

Pulse and lock-in thermography would not be able to detect disbond defects because the honeycomb wall structure is thin, which creates a very low ratio between the wall thickness and depth below the testing surface. However, pulse and lock-in thermography are the most desirable techniques to test for delamination defects very close to the specimen surface because they can both test a large area (e.g. one metre square) quickly (perhaps several minutes). Lock-in thermography can even match the sensitivity of ultrasound for very near defects. Also no contact between the specimen and heat source and camera is required for the thermographic techniques. In addition both the heat source and camera can be positioned at oblique angles to the specimen, meaning the set-up procedure is easier and hence quicker. Also slightly curved surfaces such as aircraft wings and fuselages do not seriously reduce the thermographic sensitivity.

In terms of sensitivity, mechanical impedance is only effective for large disband defects. Ultrasonic is the most sensitive technique being able to detect all defects from 1 to 11 mm down to 2.8 mm depth. Generally the thermographic techniques can match 'C' scan for very near surface defects but not deep defects. All the techniques are reliable for the defect sizing and shape determination for the defects described in this chapter because of the generally large size to depth ratio. Also, for a similar reason, all the planar x–y positions of all defects in this chapter were accurately determined.

As for defect depth measurement, for mechanical impedance the skin thickness can be measured by measuring the defect size first by a scanning technique and then using the resonant frequency measured at the centre of the defect as a measure of skin thickness (Fig. 11.4). Ultrasonic 'C' scan is the most accurate technique for defect depth measurement and the delamination defect position in terms of in which two ply layers it exists can be determined. Pulse thermography depth measurement has not been discussed in this chapter but approximate defect depth measurements can be made by determining the temperature versus time profiles of defect indications. Deeper defect indications take longer to appear after the pulse heat has been applied and also take longer to disappear. Lock-in phase measurements could be used to determine depth measurements. If the defect size is determined by image size measurements the phase amplitude values could be used for depth measurement (Plate II(a) and (b)).

The results described in this chapter are appropriate for specimens made of the same composite materials as the test specimens described. However the same qualitative trends should be applicable to specimens made of different materials, e.g. lock-in thermography should be sensitive to more deeper defects than pulse thermography. In terms of which technique to select for an application then the sensitivity to the defects which need to be detected is of prime consideration. For large disbond defects, mechanical impedance should be used, but for smaller defects ultrasonic 'C' scan would be required. To test very near surface delamination defects quickly over a large area of laminar composite, then one of the thermographic techniques would suffice. For slightly deeper defects, lock-in is preferable, and for even deeper defects ultrasonic, 'C' scan is required.

11.6 References

1. Perez MA, Gil L, Oller S. Non-destructive evaluation of low velocity impact damage in carbon fiber-reinforced laminated composites. *Ultragarsas (Ultrasound)* 2011; **66**(2): 21–26.
2. Amenabar I, Mendikute A, Lopez-Arraiza A, Lizaranzu M, Aurrekoetxea J. Comparison and analysis of non-destructive testing techniques suitable for delamination inspection in wind turbine blades. *Composites: Part B* 2011; **42**: 1298–1305.
3. Garnier C, Pastor ML, Eyma F, Lorrain B. The detection of aeronautical defects *in situ* on composite structures using non destructive testing. *Composite Structures* 2011; **93**: 1328–1336.
4. Cawley P. The sensitivity of the mechanical impedance method of nondestructive testing. *NDT International* 1987; **20**(4): 209–215.
5. Cawley P. The impedance method of non-destructive inspection. *NDT International* 1984; **17**(2): 59–65.
6. Wong BS, Tui CG. Mechanical impedance inspection of composite structures. ASNT Spring conference 1993.
7. Bai W, Wong BS. Photothermal models for lock-in thermographic evaluation of plates with finite thickness under convection conditions. *Journal of Applied Physics, USA* 2001; **89**(6): 3275–3282.

Part III
Non-destructive evaluation (NDE) techniques in aerospace applications

Part III

12
Non-destructive evaluation (NDE) of aerospace composites: application of infrared (IR) thermography

O. LEY and V. GODINEZ, Mistras Group Inc., USA

DOI: 10.1533/9780857093554.3.309

Abstract: In aerospace applications fiber-reinforced composites offer lightweight, high strength and corrosion resistance, among other advantages. Despite these advantages, composite structures are sensitive to impact damage, and other factors that compromise the integrity of the composite structure. Non-destructive testing techniques capable of providing an accurate damage assessment of the composite structure are critical to prevent premature failure and extend service life of the part. This chapter deals with the application of dynamic thermography as an efficient methodology to scan wide areas rapidly and detect impact damage, delaminations and other defects and problems.

Key words: active thermography, fiber composite, delamination, thermal contrast, heat propagation.

12.1 Introduction: thermography as a non-destructive evaluation (NDE) technique

Non-destructive evaluation or NDE refers to an inspection method to examine a part, structure, or system without impairing its future usefulness. The requirements for NDE are driven by the need for low-cost methods and instruments with great reliability, sensitivity, user friendliness and high operational speed as well for applicability to increasingly complex materials and structures. The techniques that fall within the non-destructive category are radiography, ultrasound, eddy current, acoustic emission, penetrant, vibration analysis and thermography, among others. The main reason driving the continuous development of NDE techniques is to improve safety, eliminate failures, reduce the need of unscheduled maintenance and extend the service life of parts and structures, such as those manufactured using composite materials.

In the aerospace industry, composites are being used to fabricate primary and secondary structures, and radomes, as well as to fabricate rockets, shafts, motorcases and hardbacks for military applications. The necessity for more efficient and cost-effective aircraft has led to the development of innovative testing and evaluation techniques. Smart and cost-effective methods for evaluating the integrity of aircraft structures are necessary to reduce both manufacturing costs and out of service time of aircraft due to maintenance. Nowadays, in aerospace applications, thermal NDE techniques are frequently used in the assessment of both composite materials and metal components. In particular, thermography has been successfully used in the last few years to inspect for:

- water ingress or moisture which can degrade the mechanical properties of some resins, or lead to freezing inside the part, causing ice expansion forces during the flight;
- disbond or delamination in composite structures, which will result in the part having low strength;
- metallic or non-metallic inclusions which can reduce strength by kinking the composite fibers around the inserted material;
- finding notches under multi-ply composite patching or identifying regions of delamination between two plies on a multi-ply composite repair;
- finding drilling-induced defects on multi-ply laminates of carbon fiber composites;
- detecting barely visible impact damage on carbon fiber-reinforced plastic (CFRP) panels and honeycomb sandwich structures produced by low-velocity impact;
- through-skin sensing assessment of CFRP, to assess thickness changes, delaminations, moisture ingress, the presence of voids or inclusions.

NDE techniques are characterized as active or passive, and as surface, near surface or volumetric. Active NDE is where energy is introduced into the system under inspection and an observable change in the energy is detected due to the presence of an anomaly. Passive techniques monitor the system in steady state when subjected to normal conditions; and a change in the normal or expected behavior is associated to the presence of a defect or malfunction.

The selection of an NDE technique for the detection of a defect or failure mechanism depends on the flaw type and characteristic, material properties, accessibility, sensitivity required, as well as time available to perform the inspection; in many cases, given the complexity in shape and/or material composition of the part or structure and a single NDE technique might not be sufficient to inspect a system. In any case, several techniques are tested using specifically designed defect panels. This chapter focuses on dynamic thermography, and its goal is to provide general information of when application of thermal imaging is feasible as NDE and the limitations of this technique. Thermal NDE methods involve the measurement or mapping of surface temperatures as heat flows through the object of interest.

Thermography is an NDE technique that presents the following key advantages in comparison with other NDE available technologies, namely:

- It is non-contact and non-invasive.
- It can inspect relatively large areas in a single snapshot.
- The data is in nature pictorial, which might help the inspector reach rapid decisions.
- The current price of infrared (IR) cameras is now competitive with other non-destructive testing (NDT) equipment.
- The security of personnel is guaranteed when compared with radiography.

12.1.1 Fundamentals of thermography: passive and active thermography

Thermal measurement methods have a wide range of uses. They are used by the police and military for night vision, surveillance and navigation aid; by firemen and emergency rescue personnel for fire assessment, and for search and rescue; by the medical profession as a diagnostic tool for tumor detection, cutaneous blood flow measurement, as well as fever and local inflammation detection. In the construction industry thermography is used for energy audits; and in general manufacturing or power industry thermography is used as a tool for preventative maintenance, processes control and non-destructive testing (NDT).

The basic premise of thermographic NDT is that a cold or hot spot in the surface of interest is indicative of a problem. Depending on the thermal property changes between the defect and the sound area, the defect can be observed either as a hot or bright spot; or as a cold or dark spot. The local surface temperature change is because the flow of heat from the surface of a solid is affected by internal flaws such as disbonds, voids or inclusions; or because a certain area is experiencing increased heat generation due to failure.

There are two basic types of thermography: passive and active; in both cases a thermal imager or IR detector is used to observe the temperature distribution of the surface under investigation. Thermal imaging systems are instruments capable of producing images utilizing the IR radiation being emitted by any body or part at temperature higher than absolute zero. The use of thermal imaging systems allow surface temperature information to be rapidly collected over a wide area and in a non-contact fashion, which makes its use desirable in NDE applications (Holst, 2000). Selection of the IR camera or imager for a given application depends on the magnitude of the temperature change expected due to the defect and the speed at which such change takes place during a dynamic type approach. There are two main types of IR cameras: those with cooled IR image detectors and those with uncooled detectors. Cooled detectors operate in the 60 to 100K requiring external cooling; these detectors have greater sensitivity and use IR radiation in the 3 to 5 micrometers range. Uncooled detectors are mostly based on pyroelectric and ferroelectric materials or microbolometer technology detecting radiation between 8 and 12 micrometers. Further details and classification of IR imaging systems available in the market can be obtained elsewhere (Kruse, 2002; Kaplan, 2007).

In passive thermography, the camera is simply pointed at the test piece and from the thermal image a temperature map is constructed. The thermal image is compared with that obtained from a fault-free system. Applications using passive thermography are:

- Plant condition monitoring and predictive maintenance, based on the fact that an erratic or deviant thermal behavior or operating equipment is the precursor of costly and ineffective operation and then to failure due to mechanical friction, electrical overheating, valve or pipe blockage, or inefficient insulation.
- Process monitoring and control where the condition (quality, consistency, etc.) of the product is correlated to the temperature variations recorded during the process. It is used in industrial processes, where the temperature needs to be measured and controlled, such as those in the food, chemical and petroleum industry, among others.
- Medical imaging to actively or passively look at metabolic activity of near to the skin tissues, such as tumors, or to assess perfusion or blood flow changes that can indicate early onset of disease, stress conditions on an individual, inflammation or to monitor core temperature.
- Aerial thermography to look for vapor or water leaks or to assess roof or building insulation and roof moisture detection.
- Night vision and surveillance, which uses IR radiation in the long wavelength 8 to 12 μm to detect and identify targets through atmosphere in the dark and in bad weather.

In passive thermography, the camera is simply pointed at the test piece and from the thermal image a temperature map is constructed. Active thermography involves heating the surface of the object rapidly using an external heat source and observing how the temperature decays with time. Flaws in the material show up by variations in the temperature decay rate.

Active thermography, on the other hand, involves heating the surface of the object using an external heat source and observing how the surface temperature decays with time as the heat deposited penetrates or diffuses into the material. Internal flaws in the material show up by variations in both the surface temperature distribution and decay rate. Depending on the protocol followed to deposit energy of the surface the active thermography technique is further classified. The imaging modality combined with the possibility of detecting and characterizing flaws as well as determining material properties makes active thermography a fast and robust testing method even in industrial or production environments. Nevertheless, depending on the kind of defect (thermal properties, size, depth) and sample material (CFRP, metal, glass fiber) or sample structure (honeycomb, composite layers, foam), active thermography can sometimes produce equivocal results or might not produce information about the defect or the sensitivity required by the application in hand.

12.1.2 Types of dynamic or active thermography techniques

In active or dynamic thermography heat is deposited in the surface or the interior of the material in a controlled manner and for a certain period of time. This is done via optical heating using various types of lamps; using jets of hot or cold air or water; or using infrared heaters and/or heating blankets. Also microwaves, ultrasonic waves or eddy currents are commonly used to generate heat inside a material. The duration of the heat deposition, as well as frequency at which pulses of energy are deposited, and the way the surface temperature changes are analyzed, have been used to distinguish between dynamic thermography techniques. Among these techniques one can find: lock-in thermography (LT), pulse thermography (PT), step heating (SH), vibrothermography and line scanning thermography (LST), as shown in Fig. 12.1. Every material responds differently to thermal excitation depending on the stimulation protocol. Thermography based on optical techniques, in general, provides very good defect resolution (depending on

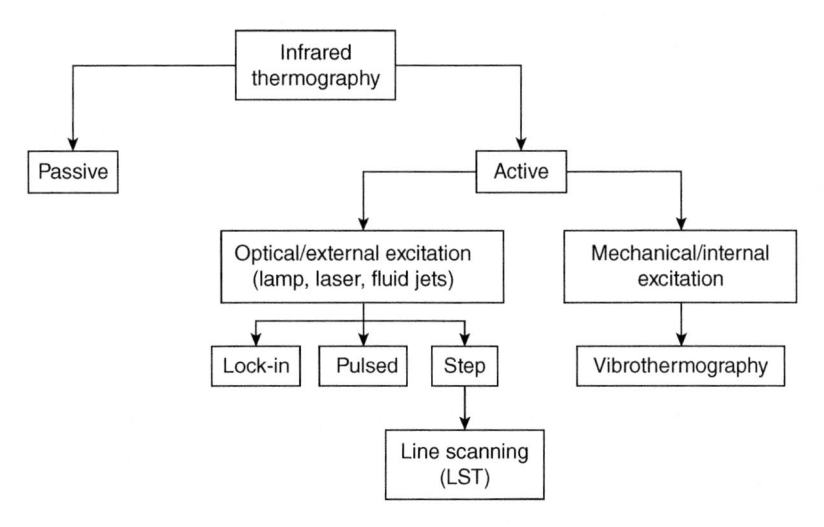

12.1 General classification of thermography techniques. This diagram shows the distinction between passive and active techniques and the classification or types of active techniques commonly used in NDE.

the IR sensor size, the optics, and the distance between the imager and the surface of interest). However, results can be strongly affected by surface features.

Most dynamic thermography techniques utilize a stationary thermal imager and are limited by the analysis of a section of the sample at a time (Ley *et al.*, 2009a). These stationary techniques can only inspect a portion of the region of interest at a time, increasing the inspection time of large areas due to repositioning of the heat source and IR camera. To inspect large regions, the thermal imager is commonly placed further away from the surface. This solution limits the thermal image resolution or the number of pixels in the image that are used to represent a defect. In this chapter we briefly describe the types of dynamic or active thermography, but focus on the use of LST.

LT is based on thermal waves generated inside the specimen under study by submitting it to a periodic (sinusoidal) thermal stimulation. In the case of a sinusoidal temperature stimulation of a specimen, highly attenuated and dispersive waves, known as 'thermal waves', are found inside the material (in a near surface region) (Ibarra-Castanedo *et al.*, 2007).

PT is one of the most popular thermal stimulation methods in IR thermography. The thermal stimulation pulse can have a duration from a few milliseconds for high thermal conductivity material inspection to a few seconds for low thermal conductivity specimens.

SH involves surface temperature monitoring during and after the heating process; the heating pulse is longer in duration than in PT. This technique (SH) is sometimes referred to as time-resolved IR radiometry or TRIR.

Vibrothermography is an active IR thermography technique where, hot spots in the surface of study are generated due to the effect of mechanical vibrations induced externally to the structure at a few fixed frequencies. In this technique heat is released by friction at the crack tip or surfaces (Han, 2007).

The LST technique (Cramer *et al.*, 1999) deposits heat along a thin line which is swept from edge to edge of the surface under inspection. In LST, the IR camera moves in tandem with the heat source at a set speed and is able to capture the thermal profile of the sample after the heat deposition took place. A diagram of the basic set-up is shown in Fig. 12.2, where one can observe that the field of view of the camera is restricted to an area of the sample surrounding the heat application region. During the scan, the temperature of the region swept by the heat source increases, whereas the surface temperature of the region in front of the heat application remains constant. It is important to note that LST requires the scanning speed and heat intensity to be optimized to match the heat diffusion in the material being studied. A thin material with good thermal conductivity will require fast scanning speed and significant heat deposition, on the other hand, a

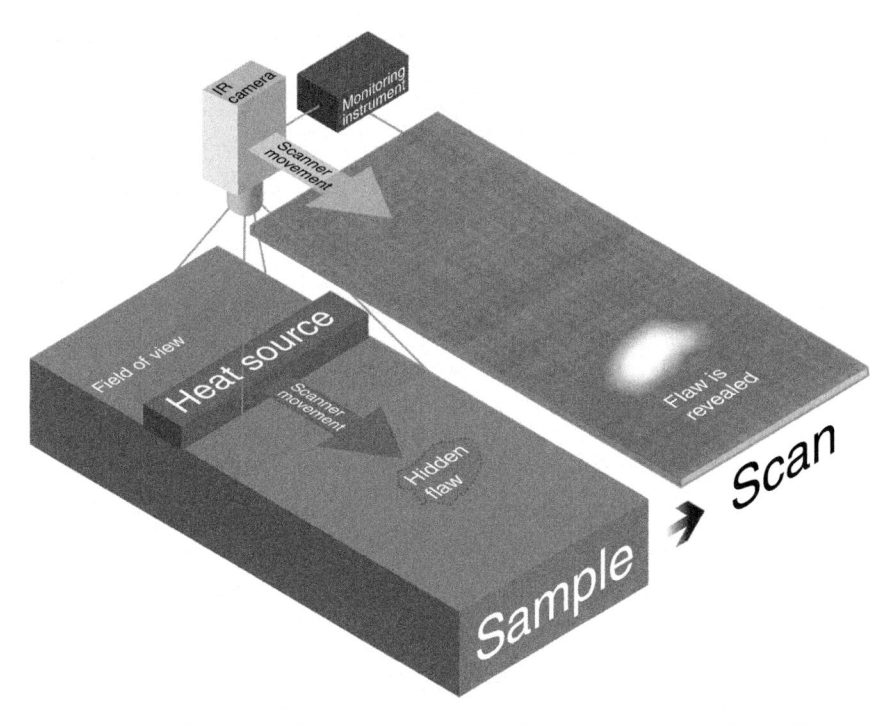

12.2 Set-up of LST where camera and heat source move in tandem through the surface to be inspected. The image on the left shows a side view of the heat source, IR camera and the surface being studied. The image on the right shows the LST thermal image generated by stacking a selected pixel line captured in every frame.

thick material or a material with lower thermal conductivity will require a slower scan with reduced heat deposition intensity (Ley *et al.*, 2009a; Ley and Godinez, 2011).

12.2 Heat propagation in dynamic thermography

The dynamic surface temperature variations recorded during an active thermography test are mainly the result of the heat diffusion process taking place inside the object or system studied and at its surface. Radiation dominates the transfer of thermal energy from the object's surface to the IR detector, as well as the absorption of the heat within the material. Conduction dominates the transfer of heat deposited on the material and the transient temperature variations observed in the surface. Propagation of the heat deposited in the surface is controlled by a diffusion process, which is described by the transient heat transfer equation discussed in detail elsewhere.

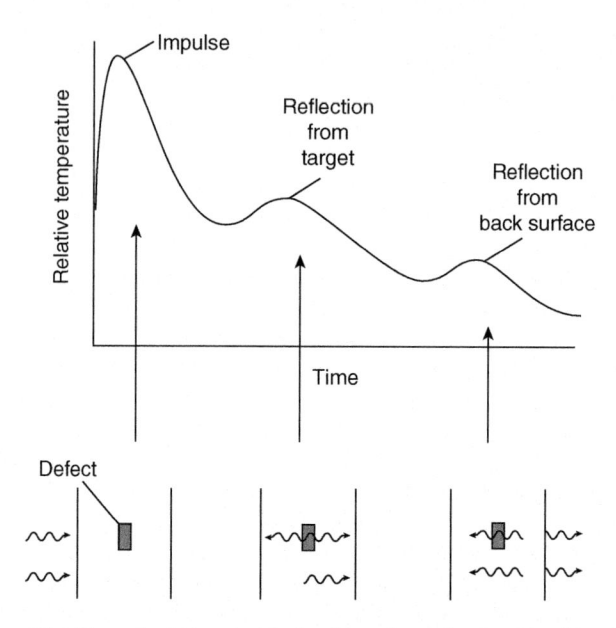

12.3 Transient thermal behavior of a point in the object's surface after heat deposition. When the energy reaches an interface where the material properties and the rate of heat propagation or diffusion change.

Dynamic thermography is based upon the following premises: (1) the rate at which a material transfers heat depends on its thermal properties, namely heat capacity, thermal conductivity, density, thermal diffusivity and thermal effusivity; (2) following surface heat deposition, the surface temperature of the object drops in time following Newton's cooling law; and (3) regional (volumetric) variations in the thermal properties of the body or material in question are observed by spatial and temporal surface temperature changes. Figure 12.3 shows the general transient thermal behavior of a point in the object's surface, where it is observed that the temperature rises after heat deposition, then it drops following Newton's cooling law until the energy reaches an interface where the material properties and the rate of heat propagation or diffusion change.

12.2.1 Thermal response of material interfaces, delaminations, disbonds and substructures

A material interface formed by either an inclusion or delamination or by the presence of a substructure can be observed as a region where the local temperature changes. When an inclusion is present within the base material (for the purposes of this chapter a composite), the appearance of either a cold or hot spot depends on the material property changes between the base material and the inclusion. If the thermal diffusivity of the inclusion is higher than that of the base material, the heat will diffuse faster on the inclusion, making the surface temperature in the area of the inclusion drop with respect to the surrounding area. This is the common response of water or moisture ingress. On the other case, if the thermal diffusivity of the inclusion material is lower than that of the base material, the surface will reveal a hot spot. A delamination commonly behaves as a hot spot, because the thermal diffusivity of air is generally lower than that of the composite which increases the resistance to heat propagation, and slows down the rate at which the energy flows from the surface to the material. The same hot spot behavior is observed when the thickness of a given region is reduced, e.g. plies of a composite separate or an air pillow is inserted between layers of the composite. Finally, a substructure is observed as a darker area. This is due to the fact that the substructure provides extra material through which the heat deposited should diffuse, which will be seen as a reduction in the surface temperature.

It is expected that the damage generated as a result of low-velocity projectile impact in a laminate composite will produce delaminations that will affect heat propagation and generate discontinuities or a region of increased heat resistance. Consequently, when a dynamic thermography technique is used, the region affected will be observed as a hot spot. The temperature rise associated with a delamination is the result of the material properties, the severity of the discontinuities, the amount of heat applied and the time at which the temperature is recorded after heat application.

Projectile impact can generate large areas of delamination, fiber fracture, fiber buckling and matrix cracking (Almond *et al.*, 1996; Cantwell *et al.*, 1983, 1984). In most cases, fiber composites are able to absorb the impact energy and transfer it to the matrix. A low-velocity impact is known to produce delaminations between the layers with no visible surface manifestation. Such delaminations may grow in service, causing severe stiffness reduction in the structure. Testing in composite coupons subjected to low-energy projectile impacts has shown reductions in residual tensile and compression strength of up to 50% (Cantwell *et al.*, 1983, 1984), conditions resulting in a complex fracture process. As a result, impact damage is one of the most significant damage types, since this can initiate delaminations which greatly reduce the compressive and fatigue strengths of a composite component. The development of non-destructive inspection techniques has been pushed forward by the aerospace industry where composites are used in many critical applications.

12.2.2 Factors limiting the success of thermography inspections

The ability to detect a void, delamination, inclusion or in general the presence of an interface within an object using dynamic thermography depends on the following factors:

- The variation in the thermal properties among the materials constituting the interface. This is defined by a parameter called thermal mismatch γ, defined as $\gamma = \dfrac{e_{material} - e_{flaw}}{e_{material} + e_{flaw}}$, where e_i (with i = material or i = flaw) denotes the thermal effusivity of each materials forming the interface or the material and the flaw. e_i is a term associated to the thermal inertia of the material and it is given by $e_i = \sqrt{k\rho C_p}$, where k is the thermal conductivity, ρ is the material density and C_p is the heat capacity of the material under consideration. The thermal mismatch (γ) is equivalent to the acoustic impedance (widely used in ultrasonics) and if γ is larger than 0.5, then the interface can be easily detected by thermal methods. In composites, the thermal properties are commonly unknown, and depend strongly on the fiber concentration, as well as the material constituting the matrix; also fiber orientation produces anisotropies in the thermal properties that will affect heat diffusion and defect detection.
- The relative size of the defect within the material compared to its depth. The known limit to determine the minimum size defect that it can be detected using dynamic thermography is given by $2d \geqslant z$, where z is the depth of the defect, and d is the defect diameter. This limitation is independent of the minimum number of pixels needed to identify a defect, as that depends on the working distance (distance between the imager and the surface under study) and the type of lens used in the imager.
- The magnitude of the surface temperature change – also known as thermal contrast – produced by the transient heating pulse compared with the minimum temperature difference that the IR images can detect. As part of a dynamic thermography technique the thermal contrast should be optimized, that is, the time at which the temperature difference between the sound region and the defect area is chosen to be the maximum possible value. That depends on the material properties and defect size or interface depth; and also on the camera sensitivity or its capability of detecting small temperature changes. When the temperature change between the defect and the sound area is very small, one could change the heat deposition by either increasing the magnitude or duration or number of heating pulses deposited on the surface of the object under study. In this case one should ensure that the material is not damaged by the heat deposition, which depends on its thermal diffusivity or the ratio between the ability of the material to store heat and to trans-

fer such heat. IR imaging systems are characterized by the noise equivalent temperature difference (NETD), which is defined as the amount of incident signal temperature that would be needed to match the internal noise of the detector such that the signal-to-noise ratio is equal to one. The characteristic NETD values of cooled cameras is between 20 and 30 mK, and for microbolometers is between 50 and 80 mK.

Active thermography presents some limitations fundamentally due to an exponential rate of attenuation of the defect signature with its depth – a consequence of the dependence of thermography on the heat conduction processes to convey information about internal structural anomalies. Defect enhancement techniques must be applied in order to produce a successful thermal inspection (Shepard, 2007). Conventional image processing techniques such as the time-derivative approach by thermographic signal reconstruction (TSR) (Shepard et al., 2002), the automatic algorithm based on differential absolute contrast (DAC), and the application of the Fourier transform, known as pulsed phase thermography (PPT) (Maldague 2001) have been applied to pulsed thermographic inspection showing both advantages and disadvantages.

12.2.3 Parameter optimization in LST inspections

In LST, heat is deposited in the form of a thin line along the length of the scan. The heat source is always on during the scan and the camera moves in tandem with the heat source, always observing the same area with respect to the heat deposition location. In LST the user has to set scan velocity, index resolution, scan length and heat deposition intensity. The scan length determines the total length of the thermal image that will be generated using the LST protocol as shown in Fig. 12.2. The vertical resolution of the LST image is given by the camera frame rate, the index resolution and the speed of the scan (i.e. the velocity at which the heat source and IR camera move along the surface scanned).

The LST technique is able to produce a series of images of the whole area scanned. Each image in the series shows the surface temperature distribution at a given time after heat deposition. The images are generated by defining an observation window or a given pixel line from all frames acquired from the camera during the scan. The final image or image of the whole area scanned is formed by stacking the selected pixel line from all the frames captured. When using images with a sensor resolution of 240 × 320 pixels, a maximum of 240 images of the whole area can be constructed. The time elapsed between consecutive pixel lines depends on the scanning speed and the camera frame rate. Figure 12.4 shows an example of the images that can be generated using the LST technique following heat

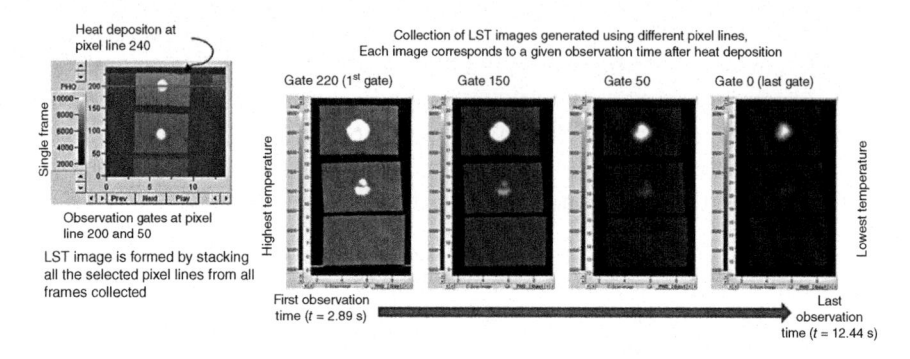

12.4 Panel showing the observation gate selection with respect to the heat deposition location. The LST thermal image is generated by stacking the selected observation lines from all frames recorded during the scan. The panel on the right shows a collection of LST thermal images generated using different observation gates. The images show the same scale and represent how the surface temperature drops after heat deposition.

deposition. The images show the same scale and were generated using different observation windows. Each image in the series shows the surface temperature distribution of the whole area scanned at a given time after heat deposition; the time is defined by the distance between the heat application and the observation gate, and the speed at which the scan is set.

Observation of a defect using LST requires proper optimization of the scanning parameters (scan velocity, index resolution, scan length and heat deposition intensity), as these determine the section of the cooling curve that will be observed, and it should match the speed at which the energy travels within the material; and the amount of heat deposited over the surface should be sufficient to produce a thermal gradient between the defect and the sound area. Particularly, when scanning thin materials displaying good thermal conductivities, the scanning speed should be set higher than the speed used on materials that have lower thermal diffusivities. Scanning at high speeds provides observation of earlier times after heat deposition, and scanning at lower speeds provides images corresponding to latter observation times. In addition one should consider that if the heat deposition is kept the same (Q) in two scans performed at speed V_o and $2V_o$, then the scan performed at speed $2V_o$ is depositing half the amount of heat that the scan performed at speed V_o. With this in mind, in LST when scanning materials that have low thermal diffusivities one should use lower speeds and low heat deposition intensities, and for scanning thin materials or materials with high thermal diffusivity one should use high heat deposition intensities and scanning speeds.

12.3 Thermography in aerospace composites

To reduce weight and improve strength in the aerospace industry, composite structures have gained popularity as a replacement for conventional materials and structures. With the large increase in the use of composite materials and honeycomb structures, the need for high speed, large area inspection for fracture critical, sub-surface defects in aircraft, missiles and marine composites led to broad acceptance of IR-based NDT methods.

The use of advanced composite materials in the aerospace industry has drastically increased over the last two decades. An increasing number of safety critical parts are now being manufactured from these advanced composites and this trend is expected to continue. Damage to composites is not always visible to the naked eye and the extent of damage is best determined for structural components by suitable NDE techniques. One of the main problems associated with composite materials is low-velocity impact damage caused by various incidents (i.e. a knock from a tool during manufacturing or repair work, etc). Low velocity impacts can produce barely visible impact damage (BVID) (Almond *et al.*, 1996). The poor resistance of composites to impact damage is due to low strength in non-fiber reinforced directions and low plastic deformation in the area local to the impact. Cyclic compressive and tensile stresses can cause rapid growth of BVID to a point where damage can reach a critical level and the component may fail catastrophically. BVID can remain invisible until component failure.

Smart methods for assessing the integrity of a composite structure are essential to both reduce manufacturing costs and out of service time of the structure due to maintenance (Garnier *et al.*, 2011). Nowadays, thermal NDT is commonly used for assessing composites. Next a test using LST on various composite structures and materials commonly used in the aerospace industry will be presented. Only thermal images employing LST will be shown as it is the technique that the authors work with and research into. There are several published studies using other types of dynamic thermography techniques in these same types of composites published elsewhere (Ibarra-Castanedo *et al.*, 2009, 2008, 2007).

12.3.1 Case studies in aerospace applications

As examples showing the use of thermography, and particularly LST, for the NDE of composite parts and structures commonly employed in the aerospace industry, we will show: (1) the thermal behavior of honeycomb sandwich panels with inserts, and experiencing impact damage; (2) the thermal response of substructures, inserts, thickness changes and delaminations in solid composite panels with varying number of plies; and (3) some composite blades used in light aircraft. Most of the tests or cases presented correspond to test coupons fabricated to assess the feasibility and sensitivity of thermography as an inspection technique.

The LST scans presented here were obtained using our laboratory set-up which performs vertical scans of up to 1.5 m (5 feet) in length and 40 cm (16 inches) of width. The system employs a cooled infrared camera working in the mid-wave IR range (3–5 micrometers). The lamp employed in the system corresponds with a quartz lamp that is 40 cm long and able to deposit 3200 W of energy across the length when set to its maximum operating power. During the LST scans, energy is deposited over the surface of interest in the form of a line 6 mm (0.25 inches) thick. The heat deposition is controlled as a percentile value of the maximum deliverable power (3200 W). The surface to be inspected was positioned facing the heat source as required by the LST technique (Fig. 12.2). The samples were positioned vertically in the direction of the scan.

Detection of thickness variations, embedded defects and presence of substructures in solid laminate composites

We studied four flat composite panels with bonded substructures in the back wall or showing tapering or thickness variations at the edges due to reduction in the number of plies. The test coupons had various types of inserts, such as flat bottom holes, pull tabs, Teflon pillow inserts and Grafoil inserts. The thermal image of a part containing a substructure of the same material as the composite panel will show in general a darker region in the area occupied by the substructure. Figure 12.5 shows an example of a thermal image of a stitched composite. The figures shows the thermal response at two different times after heat deposition. In this set of IR images one observed the darker areas as the frame where the composite skin is stitched; as time passes following heat deposition, the heat penetrates and diffuses inside the frame or substructure and is not able to diffuse as quickly in the areas where there is only composite skin, making them appear at higher temperature.

Figure 12.6 shows an example of a test part containing flat bottom holes and a substructure. In this case the substructure varies in thickness and it has flat bottom holes at different depths. This figure shows the map of defects, as well as LST thermal images generated at different observation times or using different scanning parameters (scanning speed and heat deposition intensity), as marked in each figure. At $v = 0.75$ in/s (19 mm/s) all defects are observed except for the deeper rectangular defects at the substructure. More heat is needed to be able to develop some of the rectangular defects. At $v = 1$ in/s and 1.5 in/s (25 and 38 mm/s) all defects embedded in the sample, as well as the deep rectangular defects are observed.

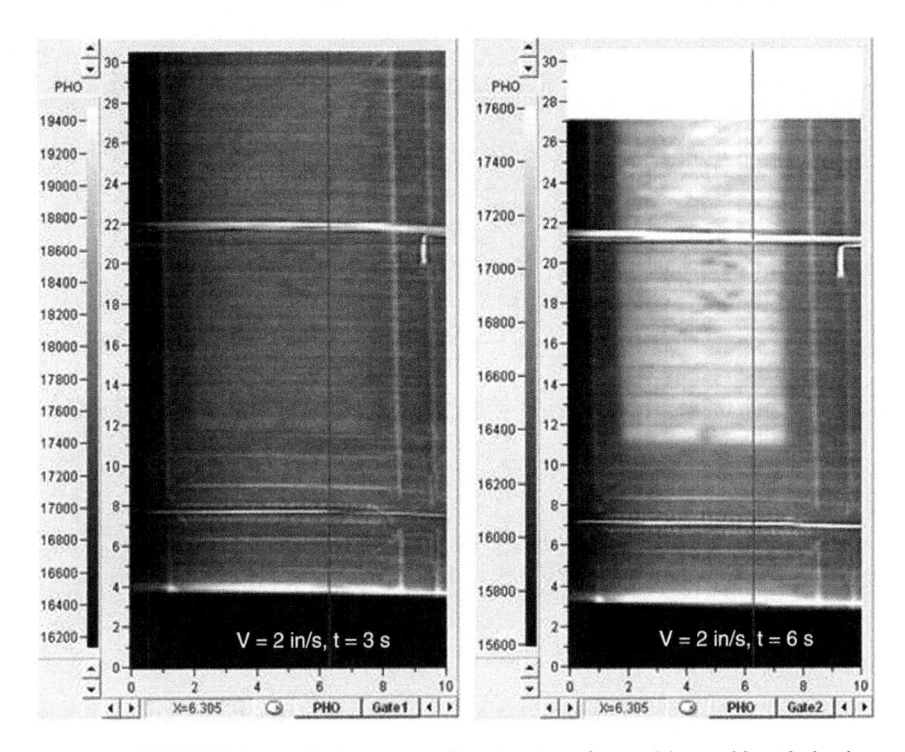

12.5 LST thermal of a composite structure formed by a skin stitched to a substructure. Two different times after heat application are shown. Details about the surface are observed initially and later the presence of the substructure is observed as a darker area.

Figure 12.7 shows another test coupon with flat bottom holes (FBH), pillow inserts (PI) and Grafoil (Gf) inserts embedded in the material. In this case the FBH, the PI and the substructure are observed; a hot spot is produced by the presence of FBH and PI The Gf inserts are observed as a darker area, but with limited contrast. In this test piece, an increase in the signal or surface temperature is observed at one edge due to the tapering or gradual thickness reduction in the sample. The darkening in the region where the substructure is present due to the increase in material thickness. The top row of thermal images shown in Fig. 12.7 corresponds to a fast scan ($v = 1.5$ in/s) and earlier times after heat deposition, for that reason the substructure is not apparent in these images; as a slower scan was performed (bottom line of images in Fig. 12.7) and latter observation times are generated by the scan, the substructure and signal increase due to the thickness reduction starts to be observed.

Finally, Fig. 12.8 shows a section of a fabricated panel with (from left to right) pulled tabs (PT), flat bottom holes (FBH), and Grafoil (Gf) and

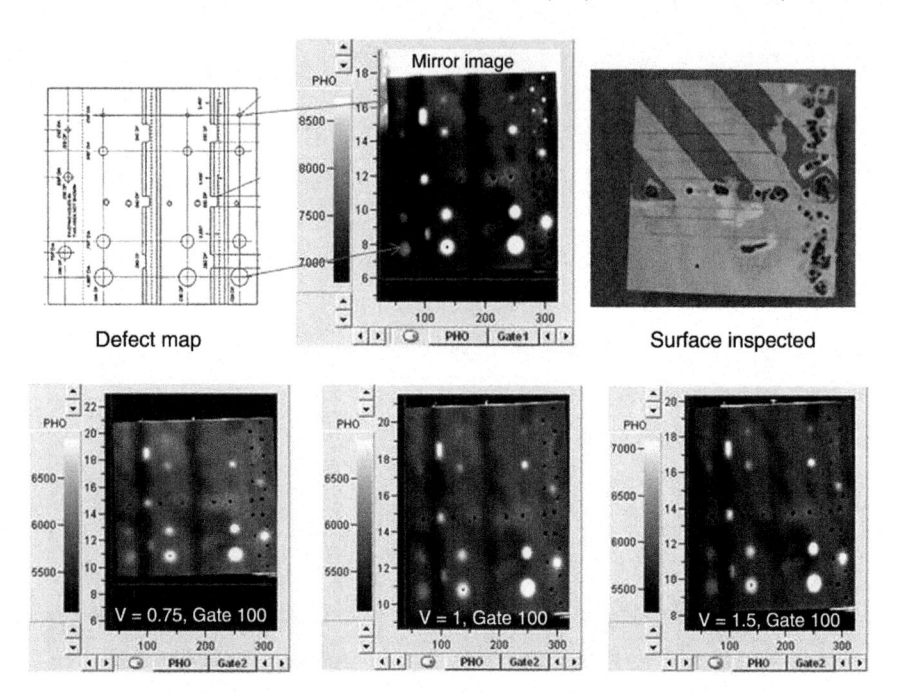

324 Non-destructive evaluation (NDE) of polymer matrix composites

12.6 Carbon composite test coupon containing flat bottom holes (FBH) of different sizes and depths and substructural elements. The top part of the figure shows the defect map, a characteristic thermal image generated using LST, and a picture of the surface inspected. The thermal image corresponds to the mirror image of the defect map. The bottom images show scans at different velocities (0.75 in/s, 1 in/s and 1.5 in/s; 19, 25 and 38 mm/s). Note that faster scanning speeds produce images of the surface at earlier times of the cooling curve. The effect of diffusion is also observed as the speed is lowered.

variable thicknesses along the vertical direction (see item (d) in the figure). In this case one can observe the different thermal responses depending on the defect/inclusion types, and the response of the thicker part (top) and the thinner part (bottom) of the sample. Item (d) shows how the signal changes along the tapered area.

Laminate carbon/epoxy sandwich composite with Nomex honeycomb core

With the development of new materials technology, honeycomb sandwich structure composite materials are increasingly used in the aerospace, civil engineering and other fields because of its good performance in lightweight, high strength and stiffness, resistance of fatigue. In the process of manufacture and application, there is likely to produce disbond, flattening sandwich

Application of infrared (IR) thermography   325

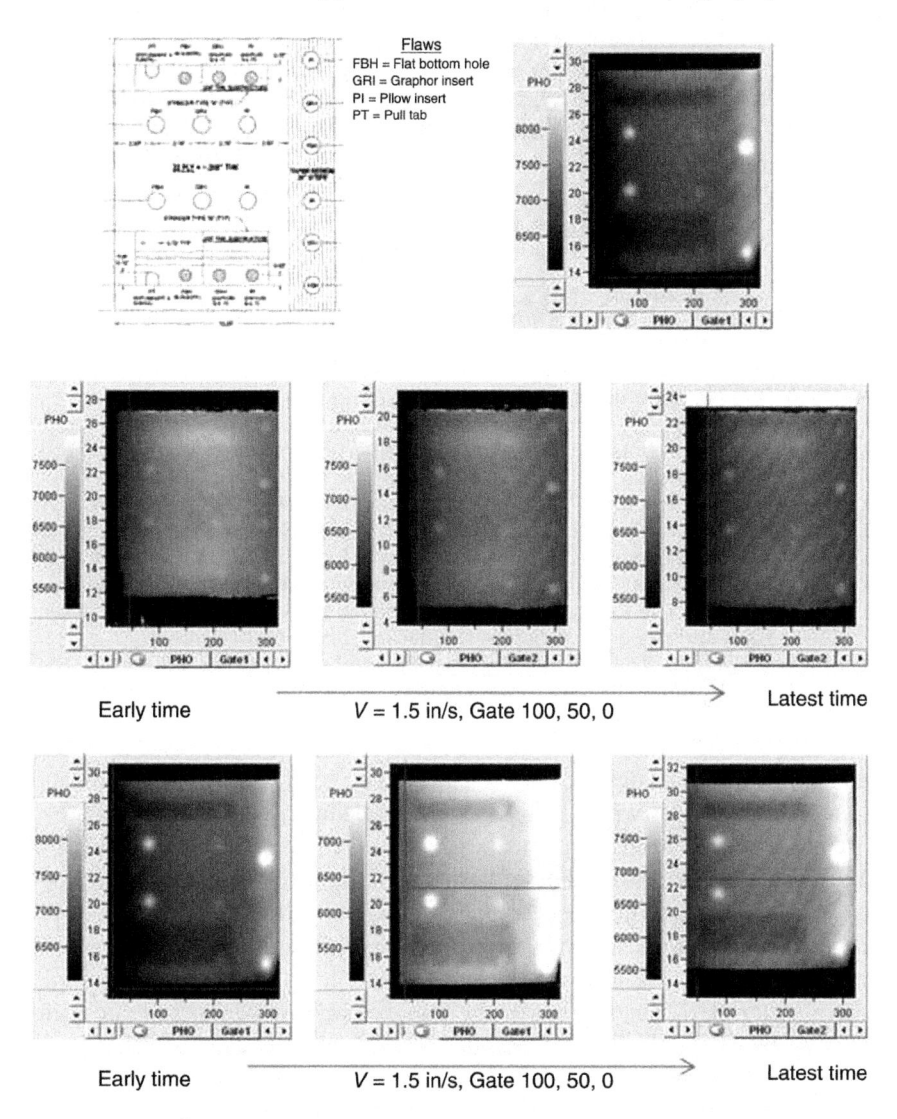

Early time V = 1.5 in/s, Gate 100, 50, 0   Latest time

Early time V = 1.5 in/s, Gate 100, 50, 0   Latest time

12.7 Carbon composite test coupon containing bonded substructure, flat bottom holes (FBH), pillow inserts (PI) and Grafoil (Gf) inserts embedded in the laminate. The top row shows the defect map and a characteristic thermal image. The second row shows images from earlier observation times where the FBH and PI embedded in the laminate are detected. The third row shows late observation times where the substructure, observed as a darker region, starts to be visible.

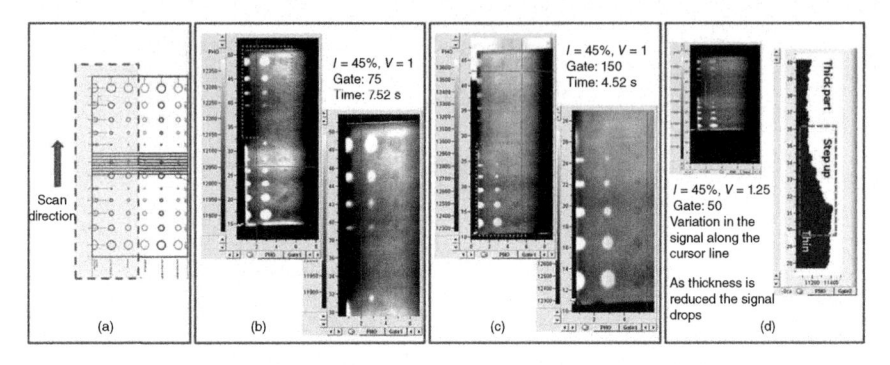

12.8 Flat laminate composite with varying number of plies in the vertical direction and containing, from left to right, pulled tabs (PT), flat bottom holes (FBH), and Grafoil (Gf); as well as variable thicknesses along the vertical direction (see item (d) in the figure). From left to right, Panel (a) shows the defect map; Panel (b) shows the response of the thicker part (top), and Panel (c) shows the response of the thinner part (bottom) of the sample. Panel (d) shows how the signal changes along the tapered area, where the number of plies changes. The top part is thicker and shows lower temperature (darker area) and the bottom part is thinner as shows higher signal magnitude. The PT and FBH are observed as hot spots and the Gf inserts are observed as cold spots.

and other internal injuries between the honeycomb sandwich panels and sandwich structures, which will result in the decline of the mechanical properties in material strength (or stiffness) and affect the safety of the structure. Therefore, how to assess the defects of the composite materials with honeycomb sandwich structure become the most important concern in production, application and research departments. At present, the X-ray detection method used in the honeycomb structure can detect the defects of the pores, extra material, disbond between the honeycomb sandwich and edge component. However, the X-ray method is not only strict to the installation and safety, but also difficult to perform in-service.

Next we present scans of different test coupons with embedded defects, core spliced, different number of plies and with impact damage. Figure 12.9 shows the interaction of the heat front with the embedded defects for a composite containing three plies. In this set of thermal images the potted core was observed as a dark spot and it was visible during the early stages after heating; the pillow insert indication was observed as a hot or bright spot in the early stages, and for later times, the indication turned into a dark spot. Finally, the machine-core defect was observed as a bright spot that showed the least contrast with respect to the sound region. As the thickness of the carbon composite increased, as shown in Fig. 12.10, the time necessary for heat penetration increased. This required to reduce scanning speed

Application of infrared (IR) thermography 327

○ Machined core
◉ Pillow insert
● Potted core

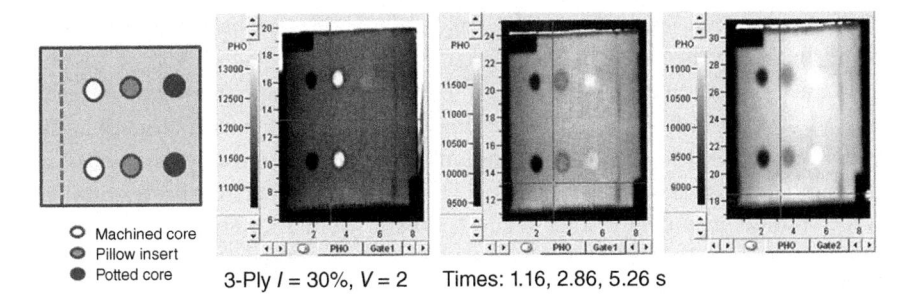

3-Ply *l* = 30%, *V* = 2   Times: 1.16, 2.86, 5.26 s

12.9 Thermal images of composite sandwich panels with Nomex core. The composite panels have three plies, and the sample contains four types of fabricated defects: machined core, pillow inserts, potted core and spliced core. The defect map is shown on the left, and the right-hand side images show the thermal response at different times, indicating both the difference thermal response of the fabricated defects and the time require to obtain a contrast change in the region of the defect.

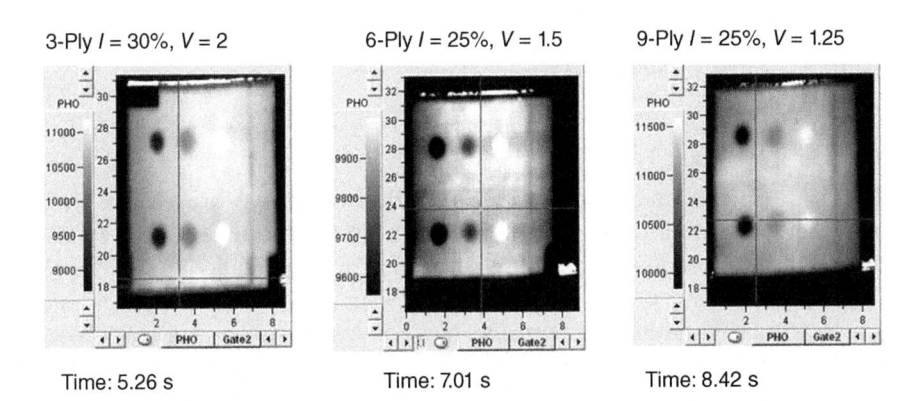

3-Ply *l* = 30%, *V* = 2   6-Ply *l* = 25%, *V* = 1.5   9-Ply *l* = 25%, *V* = 1.25

Time: 5.26 s   Time: 7.01 s   Time: 8.42 s

12.10 Thermal images of composite sandwich panels with Nomex core with fabricated defects (as shown in Fig. 12.9). The images show the thermal response of test coupons where the number of plies in the laminate composite varied from three to nine. The progression of figures shows the need to increase the observation time as the composite thickness increases.

© Woodhead Publishing Limited, 2013

and increase heat deposition. The spliced core line is observed as a dark spot, similar to the potted core.

Plate V (between pages 296 and 297) shows the region affected by impact on 152 mm^2 (6 in-by-6 in) laminate sandwich composites with Nomex honeycomb core. A total of nine specimens were studied on both sides using LST and immersion ultrasonic test (UT). For the LST scans, three samples were studied in a single scan set to a 70 cm (28 in) length. The scanning speed used for the LST scans ranged between 50 and 6 mm/s (2 in/s and 0.25 in/s). The fast scans allowed observation times of up to 6 s after heat deposition for a scanning speed of 50 mm/s (2 in/s) to 50 s when the 6 mm/s (0.25 in/s) scanning speed was selected. The slower scans were performed to try to observe information from the aluminum honeycomb; however, it was observed that most of the damage generated was very superficial and a speed of 25 mm/s (1 in/s) was selected for the inspection of all test coupons. Plate V shows both the LST images and the UT C-scans obtained for six of the samples scanned. The areas showing delamination in the LST and UT scans were calculated using a cluster analysis algorithm. It was consistently observed that areas detected by LST were slightly larger than those found using immersion UT. In this case, the region affected in the C-scans was defined when the UT amplitude fell below 15%. Similar studes comparing various NDE techniques on the assessment of sandwich panels have been performed (Ibarra-Castanedo et al., 2009).

Composite propellers for light aircraft

There are several manufacturers around the world that are redesigning propellers for light aircraft to contain blades incorporating advanced structural composites or aluminum alloy, depending on weight and life-cycle cost requirements. Some of the composites used for composite blade fabrication are Kevlar or graphite; a composite blade is composed of a metal blade shank retention section onto which is molded a low-density foam core that supports built-up layers of composite laminate. Advantages of composite blades include longer service life and substantial weight reduction over conventional aluminum blades. The composite blade service life is longer because most damage to composites can be repaired and returned to service, while damage to an aluminum blade requires removal of material and it ultimately falls under dimensional limits for strength and must be retired from service.

Figure 12.11 shows a thermal image of a Kevlar composite blade. The images show indications in the same areas identified as problem areas and marked with an oval. The regions with different gray level or brightness show areas of either interfaces between materials, the presence of wrinkles and the presence of delaminations. The base of the blade is darker because

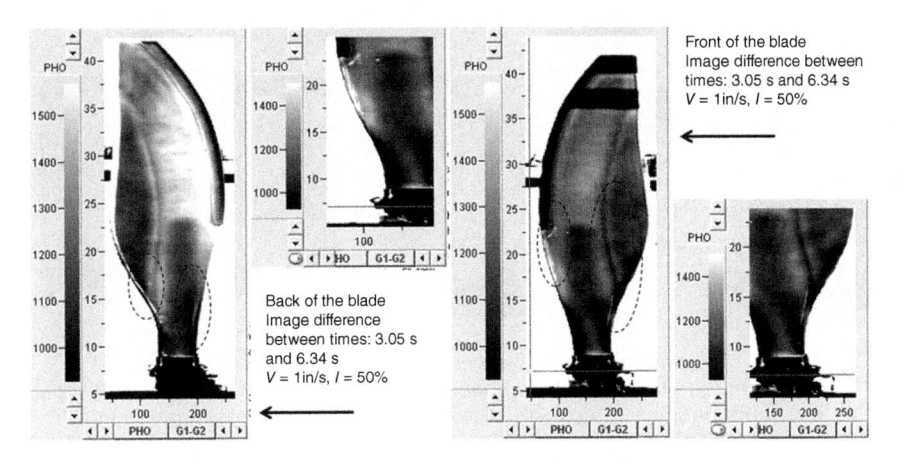

12.11 Thermal response of a composite blade used in light aircraft. The images correspond to a difference of two thermal images obtained at two different times after heat deposition. The front and back of the blade are shown and the sections showing an unexpected response are marked with an oval.

of the increased thickness of the composite layer. To scan these blades two scans were performed perside, and the inspection required about 3 minutes per side. Comparing the scan with that of a blade without defects allows the regions with problems to be found.

12.3.2 Composites outside aerospace applications

As noted before, thermography is very effective at detecting disbonds or delaminations. Figure 12.12 shows two examples of disbond detection in applications outside the aerospace industry. Composite patching has proven its effectiveness and cost benefits in the aerospace industry for several years now and composite material patching is a very promising method for repairing and/or reinforcing both concrete and metallic structures, which has found applications in both marine and civil engineering Infrastructure. Figure 12.12a shows a hot spot due to a delamination in a composite patch repair; and Fig. 12.12b shows the adhesive distribution between the shell and the spar of a wind turbine blade.

In addition, Figure 12.13 shows the thermal image of a composite made of materials commonly used in the fabrication of wind turbine blades; this sample corresponds to the leading edge of a blade, where the two parts of the shell come together. The sample has a laminate layer, a layer of PVC, and adhesive. Some defects on the form of flat bottom holes are milled within the adhesive layer. In this case a multiple scan protocol was followed

330 Non-destructive evaluation (NDE) of polymer matrix composites

(a) Composite repair over concrete

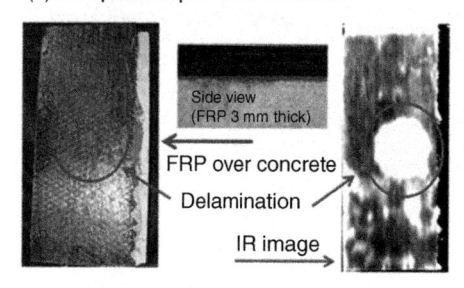

(b) Bond between shell and spar

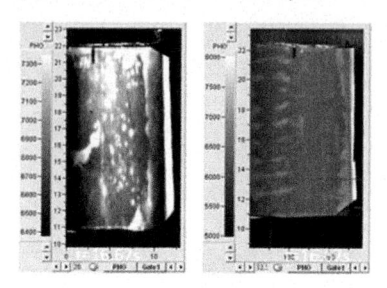

12.12 Example thermal images showing the feasibility of the LST technique to detect delamination on structures containing composite skins bonded to a substructure. Part (a) shows a delamination in a fiber-reinforced plastic (FRP) composite bonded to concrete; and part (b) shows a composite skin bonded to a spar in a wind turbine blade. The hot spots in the thermal image show the presence of voids in the adhesive between the shell and the spar indicating the presence of delaminations in the sample.

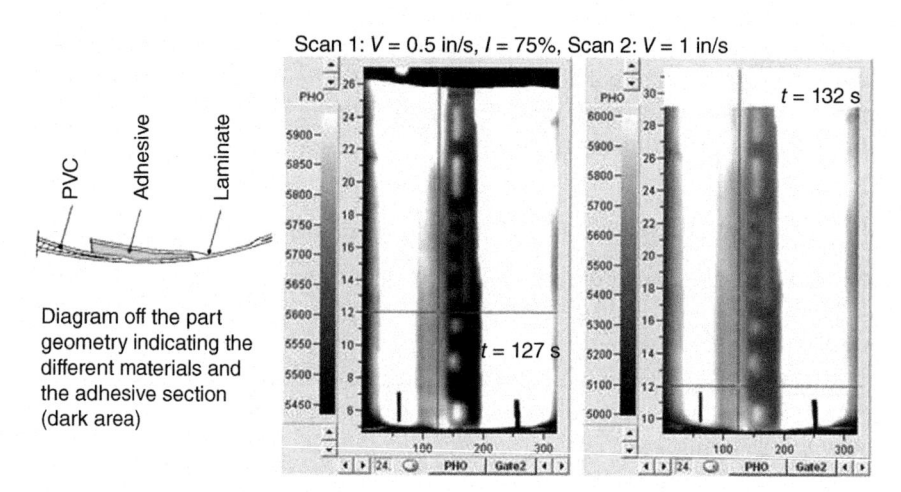

Scan 1: $V = 0.5$ in/s, $I = 75\%$, Scan 2: $V = 1$ in/s

Diagram off the part geometry indicating the different materials and the adhesive section (dark area)

$t = 127$ s

$t = 132$ s

12.13 LST thermal images of a section of a leading edge of a wind turbine blade. The sample has fabricated defects in the area of the adhesive. The LST thermal images show the observation of millings in the back of the sample.

to be able to observed the defects. The first scan is only used to deposit the heat (v = 0.5 in/s (12 mm/s)) and the second scan v = 1 in/s (25 mm/s) is done after enough time for heat diffusion was allowed. More details about the multiple scan protocol are described in (Ley and Godinez, 2011).

12.3.3 Comparison of thermography inspections with other NDE techniques

As noted before, the factors that favor the use of thermography over other NDE techniques depend on the magnitude of the signal produced by the defect, the sensitivity or minimum defect size detection compared to the engineering requirements of the part or structure under consideration, the ease of the inspection deployment, the speed at which the inspection can be performed and the density or resolution of data captured during the inspection. In thermography the defect signal is dictated by the ratio of thermal properties of the materials forming the interface in question; in many instances the thermal property variation will be higher than the acoustic property variation (impedance), in such cases the use of thermography is preferred over ultrasonic techniques.

Regarding speed of inspection and deployment it is convenient to note that sound propagation in a solid material is faster than heat diffusion; however, thermography is inherently associated with a wide field of view compared with ultrasonic techniques, where to obtain a C-scan map the sensors should raster in two directions. Finally, the ease and speed of deployment in manufacturing, in service or in the field are important factors that affect the selection of thermography over other NDE techniques. In addition, regarding the resolution the inspection technique, thermography can in general offer high-resolution information of the area scanned; whereas ultrasonic inspections are typically carried out via point-by-point measurement techniques which can be time consuming, and more importantly, are susceptible to missing possibly important information if the sampling gridwork is too coarse.

As an example of testing the feasibility of different NDE techniques, Plate VI (between pages 296 and 297) shows LST thermal images and ultrasonic C-scans of a composite hard back that constitutes part of a missile launcher system used in attack helicopters. The launcher system includes rails and a hardback, which can be subject to impact damage from accidental tool drops, routine operation and/or ballistic threats. The composite hardback and the launch rails both have complex geometries that can challenge the inspection process. The goal was to establish an inspection method that could quickly and accurately assess damage extent in order to minimize service time and return the missile system back into the field. Four hardbacks, that constitute a missile launcher in an attack helicopter, were inspected visually and by different NDT techniques, namely:

- LST;
- immersion ultrasound;
- Acousto ultrasonics.

The NDT techniques employed were capable of detecting the region affected by impact. From the techniques used, immersion UT seem to be the most sensitive but it requires contact with the surface, and immersing the part to be studied underwater. Using the portable acousto ultrasonic system seemed to be the easiest to deploy in the field, but given the size of the scanner and the size and geometry of the hardback, as well as the presence of the ridges in the surface due to the reinforcements near the holes, performing the scan required significant effort. Finally, LST was one of the easiest methods to implement and seems capable of detecting, with a single scan lasting 200 seconds, the defective regions in the hardback surface. Similar multi-technique comparisons are commonly performed when trying to determine the best NDE inspection protocol as can be observed in Amenabar *et al.* (2011).

12.4 Conclusion

In general, thermography and the LST technique provide ways to inspect large areas in a short time. The success of LST depends on proper optimization of the scanning parameters, namely heat deposition and scanning speed, which should be set to match the heat diffusion in the material. Many different NDT techniques can be used to inspect a system; the final selection depends of sensitivity, ease of deployment and allotted time for the inspection. Testing on test coupons with and without fabricated defects is necessary to compare NDE techniques and to select the most appropriate for the inspection; however, in some circumstances multiple inspection techniques might be necessary.

12.5 References and further reading

Almond DP, Delpech P, Beheshtey MH, and Peng W (1996), 'Quantitative determination of impact damage and other defects in carbon fibre composites by transient thermography', *SPIE Proc.*, **2944**, 256–264.

Amenabar I, Mendikute A, Lopez-Arraiza A, Lizaranzu M, and Aurrekoetxea J (2011), 'Comparison and analysis of non-destructive testing techniques suitable for delamination inspection in wind turbine blades', *Composites: Part B*, **42**, 1298–1305.

Cantwell W, Curtis P, and Morton J (1983), 'Post impact fatigue performance of carbon fibre laminates with non-woven and mixed-woven layers', *Composites*, **14**(3), 301–305.

Cantwell WJ, Curtis PT, and Morton J (1984), 'Impact and subsequent fatigue damage growth in carbon fibre laminates', *Int. J. Fatigue*, **6**(2), 113–118.

Cramer KE, Winfree WP, Reid D, and Johnson J (1999), 'Thermographic detection and quantitative characterization of corrosion by application of thermal line source', SPIE Conference on Nondestructive Evaluation of Utilities and Pipelines III, 277–768.

Cramer KE, and Winfree WP (1999), 'Method and apparatus for the portable identification of material thickness and defects using spatially controlled heat application', US Patent 6000844.

Garnier C, Pastor ML, Eyma F, and Lorrain B (2011), 'The detection of aeronautical defects *in situ* on composite structures using non destructive testing', *Composite Structures*, **93**, 1328–1336.

Han X (2007), 'Sonic infrared imaging: A novel NDT technology for detection of cracks delamination/disbonds in materials and structures', *Ultrasonic and advanced methods for nondestructive testing and material characterization*, edited by Chen CH, University of Massachosetts Dartmouth, USA, 369–384.

Holst GC (2000), *Common sense approach to thermal imaging*, SPIE, 265–309.

Ibarra-Castanedo C, Genest M, Piau JM, Guilbert S, Bendada A, and Maldague XPV (2007), 'Active infrared thermography techniques for the nondestructive testing of materials', *Ultrasonic and advanced methods for nondestructive testing and material characterization*, edited by Chen CH, University of Massachosetts Dartmouth, USA, 325–348.

Ibarra-Castanedo C, Grinzato E, Marinetti S, Bison PG, Avdelidis NP, Grenier M, Piau JM, Bendada A, and Maldague XVP (2008), 'Quantitative assessment of aerospace materials by active thermography techniques', in the 9th Quantitative Infrared Thermography Conference (QIRT), Krakow, Poland.

Ibarra-Castanedo C, Piau JM, Guilbert S, Avdelidis NP, Genest M, Bendada A, and Maldague XPV (2009), 'Comparative study of active thermography techniques for the nondestructive evaluation of honeycomb structures', *J. Rev. Nondestructive Evaluation*, **20**, 1–31.

Kaplan H (2007), *Practical applications of infrared thermal sensing and imaging equipment*, SPIE.

Kruse PW (2002), *Uncooled thermal imaging arrays, systems and applications*, SPIE.

Ley O, and Godinez V (2011), 'Feasibility of using line scanning thermography in NDE of wind turbine blades', SPIE Smart Structures/NDE, San Diego, CA (7983-65, Session 7B).

Ley O, Chung S, Godinez V, Schuttle J, and Bandos B (2009a), 'Use of line scanning thermography for wide area non-destructive/non-contact inspection of fiber reinforced composite structures', Aging Aircraft, Kansas City, Missouri.

Ley O, Chung S, Schuttle J, Dunne K, Ciazzo A, Bandos B, Valatka T, and Godinez V (2009b), 'Analysis of impact damage using line scanning thermography', presented at the QNDE conference in Kingston, RI.

Maldague XPV (2001), *Theory and practice of infrared technology for non-destructive testing*, John Wiley & Sons.

Maldague XPV, and Moore PO (2001), 'Infrared and thermal testing', *Nondestructive Handbook on Infrared Technology*, Volume 3, ASNT Handbook Series, 3rd edition, ASNT Press, 30–54.

Shepard SM (2007), 'Flash thermography: the final frontier', *NDT Magazine*, April.

Shepard SM, Lhota JR, Rubadeux BA, Ahmed T, and Wang D (2002), 'Enhancement and reconstruction of thermographic NDT data', *Thermosense XXIV*, Proc. SPIE, Eds: Maldague XPV, Rozlosnik AE, Vol. 4710, pp. 531–535.

Shepard SM, Lhota JR, and Ahmed T (2007), 'Flash thermography contract model based on IR camera noise characteristics', *Nondestructive Testing Eval.*, **22**, (2–3), 113–126.

Wu C, Wang W, Yuan Q, LI Y, Zhang W, and Zhang Y (2011), Infrared thermogrphy non-destructive testing of composite materials, *Adv. Mater. Res.*, **291–294**, 1307–1310.

13
Non-destructive evaluation (NDE) of aerospace composites: flaw characterisation

N. RAJIC, Defence Science and Technology Organisation, Australia

DOI: 10.1533/9780857093554.3.335

Abstract: Infrared thermography has the potential to efficiently address some of the unique maintenance challenges posed by thin-skinned polymer–composite structural assemblies which are increasingly used in the construction of civil and military airframes. This chapter provides an overview of the quantitative capabilities of active infrared thermography, with a particular focus on flash thermography and its application to the characterisation of delamination, impact damage and porosity.

Key words: infrared thermography, flash thermography, delamination, impact damage, porosity.

13.1 Introduction

Infrared thermography is a rapid broad-field non-destructive inspection (NDI) technique with a growing role in aerospace composite manufacturing and in-service maintenance. The imperative to continually improve aircraft performance and fuel efficiency has led to various innovations in structural design and an increasing use of advanced materials. This, in turn has driven a trend in aircraft manufacturing towards thin-skinned composite construction, exemplified by aircraft such as the NH-90. Such aircraft pose a new set of maintenance challenges that are not necessarily well served by many of the traditional methods of NDI. Infrared thermography offers considerable promise, particularly for the inspection of thin-skinned composite aircraft components. The reason is two-fold. Firstly, a thin composite laminate is technically well suited to inspection by infrared thermography, by virtue of certain fundamental thermal properties. Secondly, the inspection method is inherently rapid so a diagnostic result is typically achieved at a far lower labour cost relative to that of more conventional NDI methods, such as tap testing [1] and manual ultrasound [2]. This second point is especially significant in relation to through-life maintenance support for an airframe. Because of the vulnerability of composite materials to impact damage, inspections for usage-related or in-service damage are likely to be more frequent for a composite-skinned aircraft than for a metallic equivalent. An added maintenance burden obviously entails a financial cost; however, in the context of military aircraft any increase in maintenance down-time also constitutes a loss in operational availability, which has a less tangible but arguably more significant cost in terms of reduced capability. In this context, speed of inspection has strategic importance.

NDI has two distinct but related objectives; the first is to detect structural defects, the second is to furnish a quantitative description of those defects. Both objectives are important, but not all NDI techniques are able to achieve both. Active infrared thermography, while capable of achieving both in theory, is rarely applied in practice to the second aim. The reasons for that will be discussed shortly, but it is worth noting that without quantitative information relating to the size, depth, orientation and composition of a defect, its structural significance cannot be properly assessed. In that sense the mere detection of a flaw raises more questions than it answers.

Although thermography can be applied to quantitative NDI, its capabilities in that regard are limited, and with a few exceptions, are generally accepted as inferior to those of ultrasonic techniques. The reason is fundamental. A thermal inspection derives structural information from heat diffusion, a process that disperses energy across both time and space, making the retrieval of information germane to a quantitative assessment relatively difficult. Nonetheless, the task is not impossible. The purpose of this chapter is to provide an overview of the quantitative capabilities of active infrared thermography, focusing in particular on structural defects such as delamination, impact damage and porosity. As a short overview the treatment given in this chapter is required to be brief. In order to ensure the coverage is as instructive as possible the bulk of the discussion addresses fundamental principles of quantitative analysis with the view to furnishing the practitioner with enough knowledge to apply those fundamentals to particular problems.

The number of thermographic inspection methods available commercially has grown considerably over the past few decades; however, all employ the same basic principle of detection. A hidden defect is revealed by inducing a perturbation in heat flow through the component that is large enough to produce a temperature contrast measurable by infrared radiometry. Differences between methods are largely confined to the manner in which the interrogating heat flow is induced. The most basic distinction between methods is drawn on the basis of whether the heat flow is either induced naturally, for example through free convection or phase transformation, or forced. When the flow occurs naturally the inspection is said to be passive. In general, passive thermography is not well suited to quantitative inspection and therefore will not be considered here.

Instead, this chapter will focus on active methods, where the flow of heat is forced by the controlled application of a thermal stimulus. The thermal stimulus in an active method can come from a variety of energy sources;

however, radiative heating is the most widely used and the most amenable to quantitative inspection. In flash or pulsed thermography, the source is typically a flash discharge tube driven by a high-energy capacitor. That arrangement is particularly attractive for quantitative inspection since the discharge event, and therefore the applied thermal stimulus, is relatively short – in the order of 10 ms. When the stimulus occurs over a time/scale that is short relative to the time constants for diffusion through the structure (for composite materials time constants in the order of seconds are common), the inspection furnishes, to a good approximation, the thermal impulse response of that structure. Such a response provides in one inspection the richest possible source of thermal information for quantitative NDI. For that reason, flash thermography is arguably the most important quantitative thermographic technique and is therefore given the greatest coverage in this chapter. Other radiative techniques are also briefly discussed, including the optically excited lock in thermography.

The strengths and limitations of thermographic inspection are largely determined by a few key properties of heat diffusion, so an appropriate place to start the coverage is with a basic review of the physics of heat diffusion.

13.2 Fundamentals of heat diffusion

In practice, every thermal inspection involving a radiative excitation will at some stage involve heat flow in three spatial dimensions. However, providing the material behaves homogeneously, the period immediately following an excitation event is dominated by heat flow normal to the illuminated surface, which admits a simplified one-dimensional analysis of the heat diffusion problem. Accordingly, over that period, the temperature response at the surface is given by:

$$\rho C \frac{\partial \theta}{\partial t} - K \frac{\partial \theta^2}{\partial x^2} = 0 \qquad [13.1]$$

where θ is the temperature change at time t, x is the through-thickness coordinate and K, C and ρ are respectively the thermal conductivity, heat capacity and mass density of the material.

13.2.1 Half-space: harmonic excitation

Solving Eqn 13.1 for the case of a periodic excitation defined by $Q_0[(1 + \cos(\omega t))]$, where Q_0 is the peak energy density, acting on the surface of a semi-infinite solid or half space yields [3]:

$$\theta(x,t) = Q_o \frac{\mu}{2K} e^{-x/\mu} \cos\left(\omega t - \frac{x}{\mu} - \frac{\pi}{4}\right) \qquad [13.2]$$

$$= \frac{Q_0}{\varepsilon\sqrt{\omega}} e^{-x/\mu} \cos\left(\omega t - \frac{x}{\mu} - \frac{\pi}{4}\right) \qquad [13.3]$$

Here, ω is the circular frequency, $\varepsilon = \sqrt{K\rho C}$ is the thermal effusivity, and μ is the thermal diffusion length, which is defined as:

$$\mu = \sqrt{\frac{2\alpha}{\omega}} \qquad [13.4]$$

where $\alpha = K/\rho C$ is the thermal diffusivity. Note that the temperature response $\theta(x, t)$ depends on two material properties, the thermal effusivity and thermal diffusivity. These involve the same three material constants but their roles in the heat diffusion process are very different and are worth clarifying. Effusivity represents the resistance of a material to a change in temperature for a given thermal stimulus*. Equation 13.3 shows that a decrease in effusivity has the same effect as an increase in the stimulus, that is, it produces a stronger response. This obviously assists thermal inspection which explains in part why infrared thermography works especially well on polymer composite materials. Thermal diffusivity on the other hand governs the rate of heat flow through a body, which influences both the scale and phase of the temperature fluctuations as illustrated by Eqn 13.3. Although its role in the heat diffusion physics is more complex than thermal effusivity, notice that for the particular case of a half-space, diffusivity has no influence on either the amplitude or phase of the surface response (i.e. at $x = 0$). Compared with metals, composite materials have a low thermal diffusivity resulting in greater depth-wise attenuation as well as an increased phase lag.

The exponential decay in Eqn 13.3 is key to explaining an important practical limitation of thermal inspection. The term gives expression to the fact that the temperature oscillations** are intrinsically highly damped. For example, at a depth corresponding to one diffusion length the amplitude of temperature oscillations is a factor e^{-1} or $\approx 37\%$ lower than on the surface. At five diffusion lengths that figure drops to 1%. In summary, thermal waves have a characteristically shallow penetration depth for modulation frequencies that are practical for inspection, which explains why the performance of thermal inspection tends to be relatively poor on thick structures (i.e. ≥ 5 mm in polymer composites).

The intrinsic damping is not entirely a disadvantage for inspection. Although clearly a hindrance to penetration depth, damping can aid the

---

*Effusivity is sometimes referred to as thermal inertia.
**Or thermal waves, a term also employed in the present chapter; however, these oscillations lack many of the properties commonly associated with wave propagation.

inspection of thin composite components and highly textured laminates, leading to potentially better performance relative to ultrasonic methods for these types of applications. Recall that at the scale of the resin and reinforcing fibres, polymer composite materials are inhomogeneous. For high-frequency ultrasonic inspection, the large acoustic impedance mismatch between those material components can produce a complex back-scattered field which has the potential to obscure structural defects from detection. The equivalent thermal impedances are similarly mismatched, but the damping associated with heat diffusion is significant at the short length scale of the matrix/fibre structure so the impact of the material heterogeneity on the detectibility of defects is comparatively smaller. An experimental example is shown later that illustrates this aspect.

13.2.2 Half space: impulse excitation

The theory covered to this point addresses a sinusoidal excitation, yet flash thermography, the primary focus of this chapter, is applied with a short pulse. The two cases are in fact closely related and an extension of the harmonic theory to a pulsed excitation is straightforward.

It was argued previously that a flash illumination excites what amounts to a thermal impulse response of the structure. For composite materials that assumption is particularly valid since the flash duration is often two orders or more shorter than the thermal transients in a typical composite laminate. Recall that the amplitude spectrum of an impulse is flat over the infinite set of frequencies. An impulse response can therefore be expressed as an integral function of the harmonic response, which reduces [3] to the following:

$$\theta(x,t) = \frac{Q_0}{\varepsilon\sqrt{\pi t}} e^{-\frac{x^2}{4\alpha t}} \qquad [13.5]$$

At the surface of the body,

$$\theta(x=0,t) = \frac{Q_0}{\varepsilon\sqrt{\pi t}} \qquad [13.6]$$

Equation 13.6 defines a log linear time decay with a slope of 0.5. This observation has important implications for quantitative analysis, as will be explained shortly. Another important aspect of the expression is that it shows that the temperature decays exponentially as a function of depth squared, which is simply the impulse equivalent of the damping term noted previously for the harmonic case. It serves to emphasise that the penetration depth of thermal inspection is relatively shallow no matter what form the excitation takes. Indeed, the diffusion length defined by Eqn 13.4 for a harmonic source has a pulsed analogue, called the effective diffusion length

(μ_e), which is defined by Lau et al. [4] as the depth at which the temperature is a fraction e^{-1} of the surface response, viz.

$$\mu_e = 2\sqrt{\alpha t} \qquad [13.7]$$

The similarity in form of Equations 13.4 and 13.7 underscores that both relate to the same physical process. Both expressions describe a similar dependence between penetration depth and time – that is the diffusion length grows with elapsed time from an excitation pulse, or equivalently, with a drop in modulation frequency.

13.2.3 Finite slab: impulse excitation

Although clearly an abstraction, the notion of a semi-infinite body or half space serves an important practical purpose in developing a framework for quantitative inspection since the finite dimensions of a body manifest a departure in behaviour from that ideal. The key parameter is the timing of the departure relative to the excitation event.

Consider an insulated slab of thickness L. The temperature response can be derived from the harmonic solution for a semi-infinite body through a modification that accounts for the reverberation of thermal waves between the boundaries of the slab. The problem was dealt with by Green in 1930 [5] who expressed the solution as the integral:

$$\theta(x,t) = \frac{Q_0 \alpha}{\pi K} \int_0^\infty e^{-\alpha \lambda^2 t} \frac{\cos(\lambda x) + A e^{-i\lambda L} \cos(L - \lambda x)}{1 - A^2 e^{-2i\lambda L}} d\lambda \qquad [13.8]$$

where,

$$i\lambda = \sqrt{\frac{i\omega}{\alpha}} \qquad [13.9]$$

Equation 13.8 represents a summation of forward and backward travelling wave trains where each reflection contributes to the amplitude and phase of the temperature response at the surface. The integral is solved using the residue theorem, which for a slab with adiabatic boundaries ($A = 1$), i.e. an ideal reflector, yields

$$\theta(x,t) = \frac{Q_0}{\rho C L}\left[1 + 2\sum_{n=1}^\infty e^{-\alpha \beta^2 t} \cos(\beta x)\right] \qquad [13.10]$$

where

$$\beta = \frac{n\pi}{L} \qquad [13.11]$$

It is helpful for later purposes to introduce a characteristic time:

$$t_c = \frac{L^2}{\alpha \pi^2} \qquad [13.12]$$

The response at the slab surface can then be written as:

$$Q(x=0,t) = \frac{Q_0}{\rho C L}\left[1+2\sum_{n=1}^{\infty} e^{-n^2 \frac{t}{t_c}}\right] \qquad [13.13]$$

When $t \ll t_c$ the response defined by Eqn 13.13 approaches that of a half-space, i.e. Eqn 13.6. However, unlike the half-space the slab eventually attains a thermal steady state (defined by the first term in Eqn 13.13), so the rate of decay in its surface temperature must decrease as the body approaches equilibrium. The behaviour is illustrated in Fig. 13.1, which traces $\ln(\theta)$ for arbitrary values of L. Each trace begins with a log linear slope of −0.5. This is consistent with Eqn 13.6, and indicates that reflections from the back-wall are yet to contribute significantly to the response at the front wall. When it occurs, the departure from the half-space ideal is loosely analogous to the arrival of a reflected pulse in ultrasonic inspection. Indeed, like the time-of-flight in ultrasound, the timing of the departure depends on the distance between the source and the reflector. Although very helpful, the analogy with ultrasonics should be used with caution as the two phenomena differ in fundamentally important ways. For example, where an elastic wave arrival corresponds to a well-defined pulse or wave packet, the analogous breakaway in a thermal response does not, appearing instead as an asymptotic feature, as illustrated in Fig. 13.1.

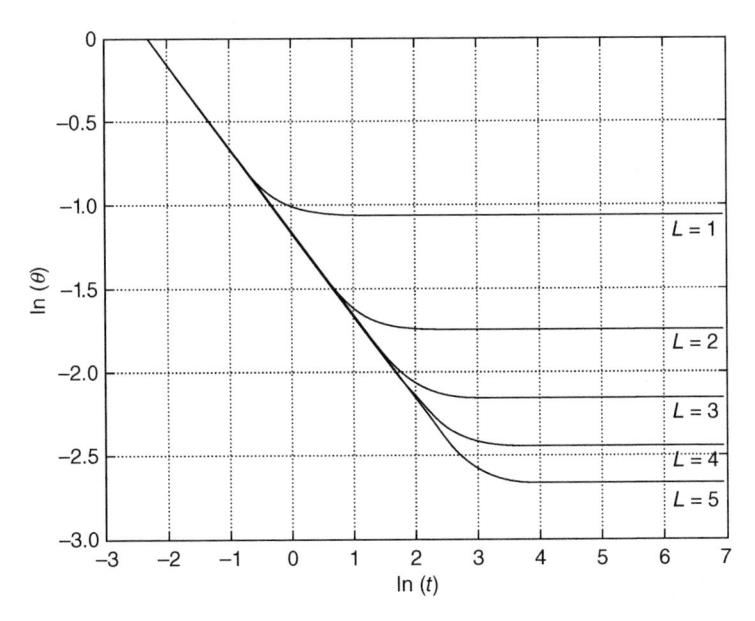

13.1 Temperature response for an arbitrary slab of varying thickness exposed to impulse heating at $t = 0$.

13.2.4 An estimate for slab thickness

Despite the indistinct transition, the existence of two well-defined asymptotic states allows for the identification of a characteristic time that can be correlated to the back-wall distance or slab thickness. That time is defined by the intersection of the asymptotes corresponding to the steady-state and log-linear responses [6], and falls within the transition between those two states as shown in Fig. 13.2. That intersection can be expressed as:

$$\frac{Q_0}{\varepsilon\sqrt{\pi t^*}} = \frac{Q_0}{\rho C L} \qquad [13.14]$$

which leads to a simple expression for the breakpoint time,

$$t^* = \frac{L^2}{\alpha \pi} \qquad [13.15]$$

The relation is significant. It connects a measurable point in the recorded temperature response to the slab thickness and thermal diffusivity, and provides therefore an experimental basis for the estimation of either value. Note also that although L is by strict definition the slab thickness, it can be viewed more generally as a characteristic length defining the distance between the heated surface and a reflector, and can therefore be assigned to represent the distance to a defect. The assignment is plausible if one

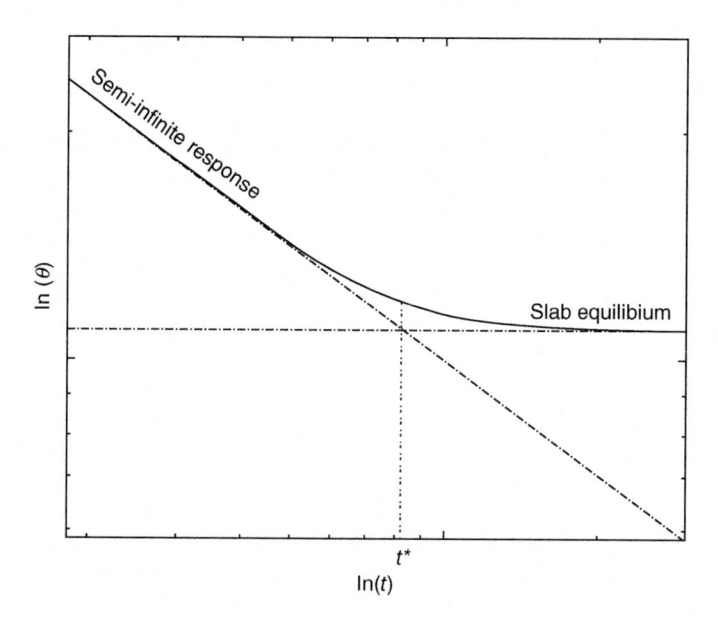

13.2 Response of a slab to impulse heating showing two asymptotic states and the breakpoint time t^*.

considers that a planar defect with a large thermal impedance will cause a termination of the log-linear decay process in much the same way that the back-wall of an ideal slab does. The chief difference between the two cases is that a structural defect has finite lateral dimensions, which prompts lateral heat flow. That of course violates the conditions under which Eqn 13.13 was derived; however, the effect is mitigated somewhat by the fact t^* occurs relatively early in the transition from one-dimensional heat flow.

In practice, a constant temperature is seldom achieved at steady state, even for a uniform slab, much less an arbitrary structure containing a real defect. Consequently, t^* cannot normally be measured by way of intersecting asymptotes, as suggested in Fig. 13.2. Fortunately, an alternative definition for t^* is available that provides a more useful basis for its measurement. The definition emerges from an analysis of the second derivative of the logarithmic time history for a slab. Figure 13.3 plots that derivative as a function of time which is non-dimensionalised with respect to the characteristic time t_c, which was defined previously. The peak in the derivative is shown to occur at a time

$$t = \pi t_c = \frac{L^2}{\alpha \pi} \qquad [13.16]$$

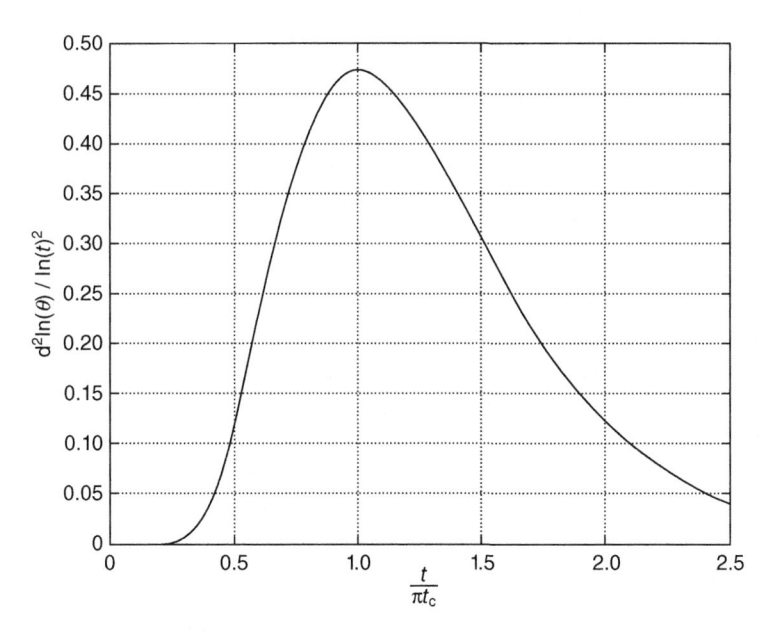

13.3 Second derivative of the response to impulse heating as a function of the characteristic time t_c.

which one will note is equivalent to the breakpoint time t^* defined previously. The challenge in using that approach in practice revolves around the computation of a robust second derivative, which is not trivial given that infrared detectors produce data with an intrinsically low signal-to-noise ratio (relative to visible detectors). Simple finite difference approximations are seldom feasible. A robust approach to this problem was developed by Shepard et al. [7] as part of a broader processing framework called the thermographic signal reconstruction (TSR) method. The central idea is to replace the measured response by a polynomial defined in the logarithmic time domain [7], viz.

$$\ln[\theta(t)] = \sum_{n=0}^{N} a_n [\ln(t)]^n \qquad [13.17]$$

In practice, the assignment of N needs to be done with care. The aim is to assign the lowest possible order that results in an accurate description of at least the second derivative at the breakpoint time t^*. Additional terms risk coupling experimental noise into the function. Once the coefficients a_n of the polynomial have been determined, the derivatives of the response can be expressed analytically and are then easily and quickly evaluated.

Figure 13.4 plots the second derivative of the response curves shown in Fig. 13.1. The derivatives are Gaussian in shape and centred at the breakpoint time which corresponds to the slab thickness, as defined by Eqn 13.15. It is also readily confirmed from Fig. 13.4, and required by Eqn 13.15, that

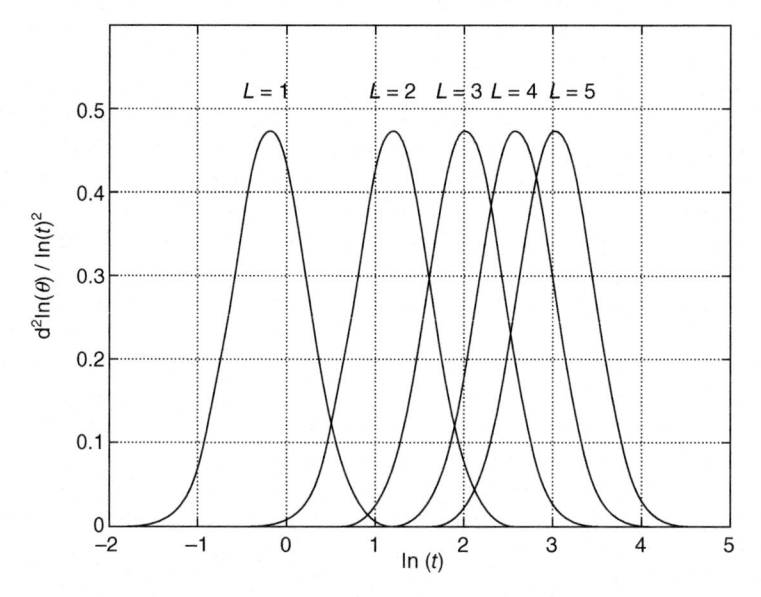

13.4 Second derivative of the response of a slab to impulse heating at $t = 0$.

the peak derivative shift in time as a function of L^2. This points to an important phenomenological difference with ultrasonics where the time of flight of an elastic wave is proportional to L.

The preceding discussion describes an effective approach to the measurement of slab thickness (or thermal diffusivity). It is not, however, the only available approach, and neither is it necessarily the most effective. A relationship similar to Eqn 13.15 was developed earlier by Ringermacher et al. [8] using a slightly different perspective. In that approach the starting point was a through-transmission problem and an expression describing the temperature response at the back-wall of a slab after applying a heat pulse to the front wall. The inflection (peak slope) in the back-wall response was related to the inflection of the front-wall contrast response, using as a reference the response at a location where the structure is known to be much thicker. Where that thickness is infinite, i.e. the half-space ideal, the inflection in the contrast response occurs at [8]:

$$t_i = \frac{3.64 L^2}{\pi^2 \alpha} \qquad [13.18]$$

Equation 13.18 has an identical functional form to Eqn 13.15 but the inflection point is seen to be delayed slightly with respect to the breakpoint time. Although the difference is not negligible, neither is it particularly significant. Both Eqn 13.15 and 13.18 yield excellent estimates for slab thickness (and thermal diffusivity) when the experimental conditions are reasonably consistent with the key assumptions, i.e. one-dimensional heat flow, a short excitation pulse width and in the case of Eqn 13.18 an appropriate reference. However, some difference in performance is likely where lateral heat flow is a factor, for instance where a flaw has relatively small lateral dimensions. Neither approach furnishes a precise estimate for flaw depth in that case; however, since Eqn 13.15 targets an earlier stage in the temperature response the assumption of one-dimensional heat flow should be slightly better met, and as a consequence that expression might yield better accuracy. Arguably, the most significant drawback of the contrast-based approach is the need for a valid reference, which for a variety of reasons [9] is not always available. That requirement was later removed [9] by substituting an experimentally tuned half-plane response for the reference.

The approach of targeting a single event in the transient response as the basis for quantitative evaluation, as embodied by Eqns 13.15 and 13.18, is effective, but it fails to exploit the significant amount of quantitative information available in the transient response in the lead-up to the breakpoint. In that period the assumption of one-dimensional heat flow is better met than at the breakpoint time, meaning that Eqn 13.13 provides a particularly good description of experimentally observed behaviour. That fact can be used to frame an optimisation problem where the values L and α

in Eqn 13.13 are tuned so as to produce in a least squares sense the best possible fit to the measured data. This approach has two particularly attractive characteristics: (i) an excellent robustness to experimental noise, and (ii) by targeting the breakaway at its earliest stages it adheres well to the assumption of one-dimensional heat flow. The optimisation problem is easily solved using non-linear least squares, as demonstrated in [10] and [11].

Another interesting approach to depth estimation, reported by Winfree and Zalameda [12], involves a reduction of the thermal inspection data to a set of eigenvectors or empirical orthogonal functions. These are known to provide a compact description of the variances caused by thermal constrasts relating to sub-surface defects (see [13]). Winfree and Zalameda used a one-dimensional double layered heat transfer model to generate theoretical temperature responses spanning a relatively large domain of possible defect thicknesses. Coefficients of the eigenvector representation of the theoretical data set were then compared with the equivalent coefficients determined for the experimental data and a best fit approach used to determine the likely defect thickness. The method was successfully validated on an experimental test case involving a 10-ply quasi-isotropic composite panel with synthetic defects. Thickness estimates were obtained at 256×256 discrete measurement points in under 10 seconds, confirming good computational efficiency.

A common thread running through the quantitative methodologies described thus far is the assumption of one-dimensional heat flow. Experience has shown that such approaches are valid for many practical purposes; however, alternative methodologies are available which eliminate that constraint. Examples include finite element based optimisation [14], artificial neural networks [15] and mathematical inversion [16]. Such approaches are generally more difficult to apply in practice than the methods already outlined and are seldom used outside of the research domain.

The emphasis shifts now to a consideration of specific composite defects and the quantitative methods available to address them.

13.3 Non-destructive evaluation (NDE) of delaminations and planar inclusions

Delamination involves a separation of adjacent plies in a laminate. Its detection is considered one of the key objectives for composite NDI largely because delamination is not uncommon and can significantly degrade the mechanical properties of a composite structure, particularly its buckling performance. Delaminations most often occur as a consequence of a structural load exceeding the interlaminar strength of the composite; for instance, where a load exceeds the design allowable, i.e. a static overload or a dynamic load caused by an impact event, or where the interlaminar strength has been degraded as a result of poor manufacturing practice during lay-up and cure of the laminate, i.e. due to contamination. Manufacturing processes such as cutting and drilling can also cause delaminations if tools exert a sufficiently high peel stress on the laminate.

13.3.1 Characterisation of depth

Precisely how a delamination responds to thermal inspection is governed largely by the morphology of the delamination surface, and specifically on whether the surfaces are physically separated or in contact. From the viewpoint of inspection an ideal delamination is one where the surfaces are well separated, since in that case the near-surface approaches an adiabatic condition and the resulting thermal signature is at a relative maximum. That situation is also consistent with the theory underpinning Eqn 13.15, which is strictly valid only for an infinitely vast adiabatic planar inclusion, i.e. a perfect reflector, so is an appropriate limiting case for a delamination. For that case an estimate for the delamination depth is relatively straightforward. If the lateral dimensions of the delamination are large compared to its depth below the surface, the assumption of one-dimensional heat flow holds reasonably well and Eqn 13.15 is expected to yield a good estimate for the depth. This will be demonstrated by means of an example shortly. The greater practical interest, however, is in what occurs when the delamination strays from the idealised case, that is when the interface permits some heat transfer, and the lateral dimensions of the delamination are not vast in comparison to the depth.

Numerical simulation provides an effective way of investigating the influence of these factors. Consider the following example. A 5 mm thick transversely isotropic carbon-epoxy laminate, with the thermal properties listed in Table 13.1, is exposed to a heat pulse 5 ms in duration. Eight structural configurations of the laminate are examined: a defect-free baseline, a blind hole defect, and finally a number of planar inclusions with a range of dimensions and material properties that represent cases of practical interest. Relevant details are provided in Fig. 13.5 along with Table 13.2. Simulations were carried out using the finite element method (FEM). Without loss of generality the defects are assumed to have a circular disc geometry; the symmetry allowing a 2D analysis of the problem. A structured mesh was used in the near surface layer to ensure an adequate level of refinement to properly describe higher-order transients. Implicit time-integration was employed to advance the solution.

Figure 13.6 plots the 2nd derivative of the surface response at the point of axial symmetry for each of the eight configurations listed in Table 13.2.

Table 13.1 Material thermal properties

Material	K_x (W/mK)	K_y (W/mK)	ρ (kg/m^3)	C (J/kg K)	ε (J/m^2 K s$^{0.5}$)	R
Carbon-epoxy	0.56	3.18	1530	950	902.20	–
Air	0.02	0.02	1.2	1000	4.89	0.989
Teflon	0.25	0.25	2200	1050	759.93	0.086

Table 13.2 Case descriptions and summarised results

Case	Description	a (mm)	h (mm)	t^* (s)	$\hat{d}$ (mm)	Error %
a	Ideal slab	30	2.5	5.19	2.51	0.34
b	Blind hole	5	2.5	3.92	2.18	−12.78
c	Adiabatic insert	5	0.005	3.75	2.13	−14.65
d	Air insert	5	0.005	3.42	2.04	−18.56
e	Teflon insert	5	0.005	3.41	2.03	−18.65
c$^+$	Adiabatic insert	15	0.005	5.12	2.49	−0.31
d$^+$	Air insert	15	0.005	4.21	2.26	−9.63
e$^+$	Teflon insert	15	0.005	3.83	2.16	−13.75

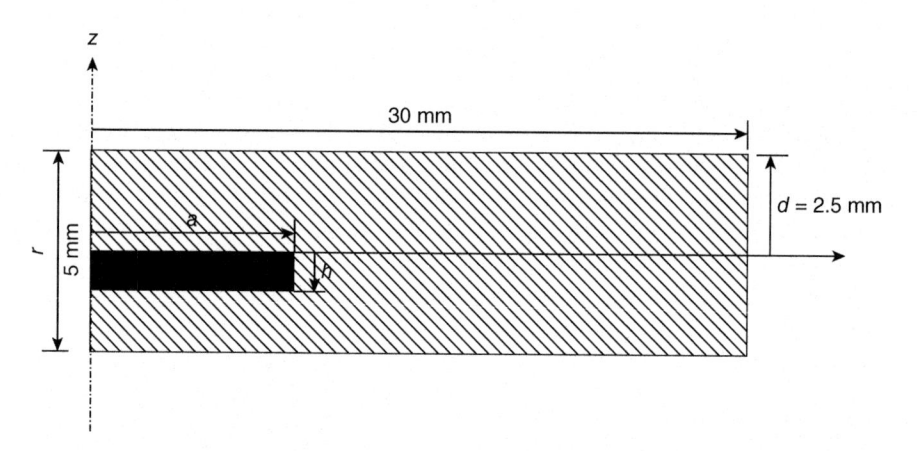

13.5 Modelled axisymmetric structure where z defines the axis of symmetry. Hatched region represents a carbon-epoxy laminate with orthotropic properties (see Table 13.1). The size and material properties of the solid region are varied to model different problems, as described in Table 13.2.

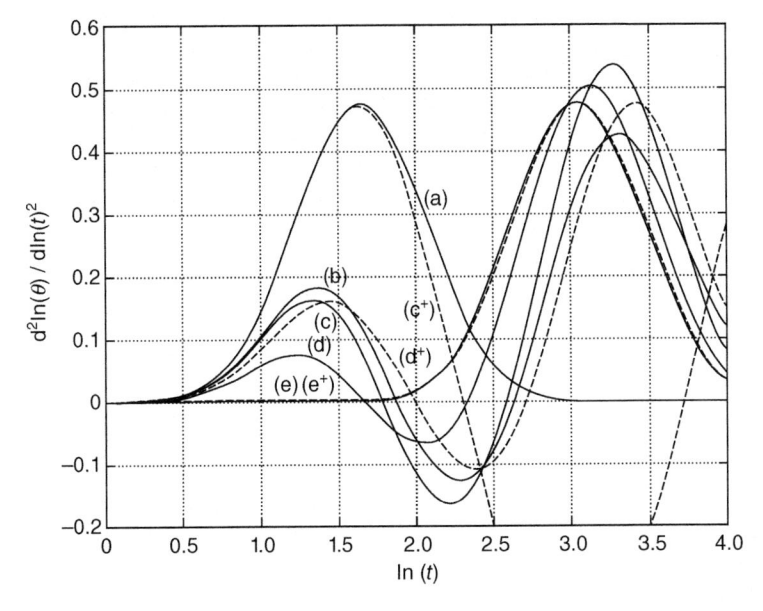

13.6 Second derivatives for the cases described in Table 13.2.

The first aspect to consider is the extent of variation in the position of the first peak, which is noteworthy because the depth L is constant (2.5 mm). Note that only case (a) is entirely consistent with the assumptions of Eqn 13.13, which is readily confirmed by comparison with Fig. 13.1, as well as by the low error in the corresponding estimate for L in Table 13.2. Accordingly, that curve serves as our baseline for comparison.

The larger adiabatic insert (case c^+) most closely approaches the baseline result. But even that response diverges in time, with the appearance of a trough first, followed then by a second peak associated with the back-wall of the slab. As the properties of the insert are varied and its lateral size reduced the variation from the baseline is seen to grow significantly, and that in turn translates to a growing bias in the estimate for L. Interestingly, the estimate is always conservative, i.e. Eqn 13.15 consistently underpredicts L.

Consider now the effect of lateral size in isolation. As the size decreases two things occur: an advance in the timing of the peak and a drop in its magnitude. The first effect has a profound influence on the accuracy of the depth estimate which was previously remarked on and relates directly to the growing influence of lateral heat flow. The second effect also has an impact on accuracy but in a less direct way. A decline in the strength of the peak implies a weaker thermal response, which in view of the relatively low SNR (signal-to-noise ratio) levels associated with infrared detection makes a robust calculation of the second derivative in practice all the more challenging.

Consider next the influence of the thermal properties of the insert. Before we discuss that aspect it is appropriate that a justification be given for the use of a material insert as a substitute for a true delamination. First it is worth noting that such an approach mirrors a practice occasionally used to manufacture NDI standards for delamination defects. As remarked previously the morphology of the delamination surface is an important factor in determining the response of a delamination to inspection. However, the morphology depends on a large range of factors including the material composition of the laminate, the circumstances that produced the delamination, the residual stress state in the laminate, and so on. Consequently, the response to inspection in practice is highly variable, potentially falling anywhere between the extremes of a near-adiabatic reflector when the surfaces of the delamination are widely separated, and a near-perfect thermal transmitter when the surfaces are in intimate contact. The two insert materials listed in Table 13.2 produce behaviours that represent close to these extremes and are therefore useful in defining limits to the response that might be expected for a real delamination.

The effect of the insert material on the thermal response is determined by the difference in thermal effusivity of the insert and host materials, which can be expressed in terms of the thermal wave reflection coefficient; a property of the interface defined by:

$$R = \frac{\varepsilon_h - \varepsilon_i}{\varepsilon_h + \varepsilon_i} \qquad [13.19]$$

where the subscripts i and h refer to the insert and host materials respectively. A coefficient of 1 describes a perfect reflector, while a value of 0 describes a perfectly matched insert (i.e. thermally transparent). Table 13.1 lists values of the coefficient for the insert materials relevant to this exercise. Returning to Fig. 13.6, the reflection coefficient is seen to affect both the breakpoint time and the peak magnitude, advancing the former and reducing the latter as the coefficient declines. The practical implications are identical to those outlined for a decreasing lateral size, so need not be discussed again. However, the behaviour of Teflon (cases e and e$^+$) merits a little more attention since it highlights a limitation in the use of a Teflon film insert as a surrogate for a delamination in NDI test standards. Unless practical measures are taken to ensure such inserts are detached from the adjacent ply in the cured laminate, the insert is likely to produce a relatively low thermal contrast that will represent only one extreme of the spectrum of possible behaviour for a true delamination. Note also that the second derivatives for cases e and e$^+$ are barely discernible at the scale used in Fig. 13.6, which raises the point that in practice, where experimental noise

13.3.2 Lateral sizing

As thermography is a visual technique the sizing of defects should be relatively straightforward. This, however, is true only to a limited extent. An approximate size can be deduced very quickly by simply measuring the size of the defect signature, with appropriate corrections for the various (generally known) geometrical transformations introduced by the optical imaging system. However, a challenge arises when a precise determination of size is required. This is because the image of the defect produced by thermographic inspection is a thermal artefact rendered by a heat diffusion process, and accordingly the size of that artefact is a biased estimator for the true size. That fact is illustrated in Fig. 13.7 which shows simulated surface temperature profiles for cases d and d⁺ (Table 13.2) at various stages during an inspection. Notice that at no stage is there a sharply defined feature corresponding to the edge of the defect and indeed the gradient of the

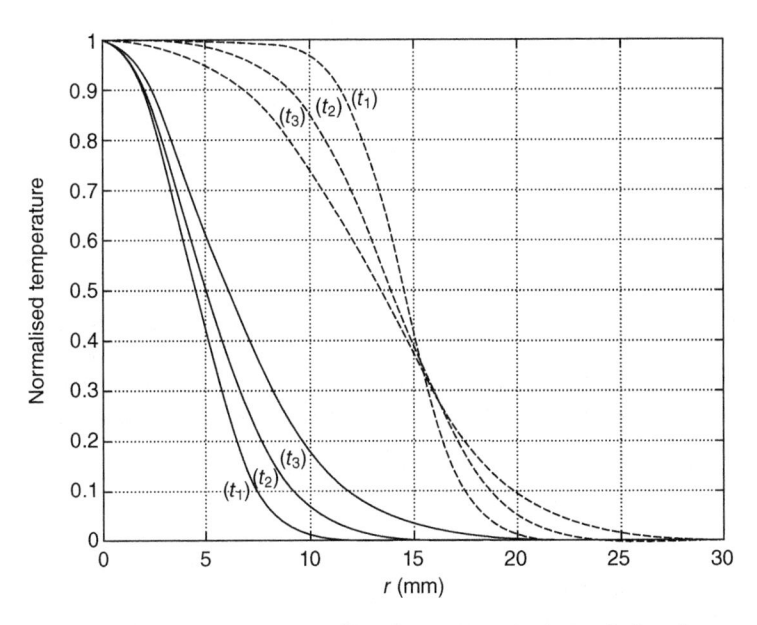

13.7 Surface temperature profiles from numerical simulation for cases d (solid line) and d⁺ (dashed line), as described in Table 13.2, at the following delays after the heat pulse: 2.80 s (t_1), 6.84 s (t_2) and 14.30 s (t_3).

temperature profile decreases as time progresses, as required by heat diffusion. The practical effect is broadly equivalent to that of a poorly focused lens in conventional imaging where the focus degrades the longer the object is viewed.

Since the effect of heat diffusion is deterministic, it is in theory possible to retrieve the size of a defect by deconvolving the effect of diffusion from the thermal signature. However, in practice, that approach leads to an ill-conditioned inverse mathematical problem, which whilst not intractable (see Favro *et al.* [17]) is not trivial to solve when the data are contaminated with noise, i.e. sourced from an experimental measurement.

In practice, defects are normally sized using an approach that is a little less rigorous but on the whole quite effective. The approach entails the identification of a distinct feature in the temperature profile that correlates with defect size, much like the approach taken to depth estimation described previously. One such feature is the position of the steepest gradient at a time corresponding to the peak temperature contrast. Prior work by Krapez and Cielo (see [6]) has shown that this location occurs near the lateral boundary of the defect. This was verified by the present author for the data given in Fig. 13.7, where using the steepest gradient approach the insert boundary location was estimated to within 10% of the known value.

The computation of a gradient is straightforward for synthetic data but far more challenging for real experimental data. An alternative metric that avoids the need for an evaluation of a gradient is the full width at half maximum (FWHM) of the contrast signature. That parameter was first applied to sizing by Wetsel and McDonald [18] in relation to photothermal imaging, a thermal inspection technique employing a harmonically modulated laser beam. Almond and Lau [19] later developed a theoretical basis for the application of the FWHM in flash thermography. They showed that for the limiting case of a perfect reflector with relatively large lateral dimensions the FWHM differs from the defect size by a known time-dependent error. That error was defined as $-0.54\mu_e$, indicating a conservative estimate. It is worth briefly examining how that proposition relates to the simulated contrast profiles shown in Fig. 13.7. For the dashed curves the full width at the half max point (a normalised amplitude of 0.5) reduces as time advances, which is consistent with theory. However, the converse is true for the solid curves corresponding to the smaller defect. The reason is two-fold. Firstly, the theory is limited to the case where the diffracted thermal wave fields from opposing edges of a defect do not interfere, which is only met when the defect is relatively large and the diffusion times relatively short. Secondly, the theory is developed for an isotropic material whereas the case considered here involves a transversely isotropic material with a relatively high in-plane thermal diffusivity.

Notwithstanding the limitations of the underpinning theory the FWHM still provides, for practical purposes, a useful estimate for defect size. Applied to case d in Fig. 13.7, i.e. the solid traces, the FWHM yields a maximum error of 15%, while for case d$^+$, i.e. the dashed traces, the error is roughly 11%. Saintey and Almond [20] note that the highest predictive accuracy is achieved by measuring the FWHM at the earliest possible time following the heating pulse, which is largely consistent with the results shown in Fig. 13.7.

The FWHM approach can also be applied to processed inspection data. For instance, Hung *et al.* [21] in a study on layered panels with artificial bonding defects found that FWHM measurements taken from a second derivative profile furnished by the TSR method correlated well with the known size. In applying the FWHM approach to processed data one needs to be mindful that a transform to a derivative or a phase map [22] can introduce oscillations in the contrast profile, leading to possible ambiguity in the FWHM definition.

13.3.3 Lateral size from depth profiling

Any one of the depth analysis methods described in the previous section can be used to furnish a depth map, from which lateral sizing of defects is a relatively straightforward exercise, involving little more than a length measurement. However, it is important to note that such size measurements are still only approximate since the depth map is sourced from an analysis reliant on assumptions that need to be justified. Consider for example the assumption of one-dimensional heat flow, which while reasonably well met at the centre of a defect invariably deteriorates as an edge is approached. Yet, sizing relies on information derived from an edge. Again, a numerical simulation provides a convenient means of illustrating the point. Figure 13.8 plots the depth profile determined for case d using the second derivative of the log response to establish t^* in Eqn 13.15. Recall from Table 13.2 that in this case the estimated depth at the point of axial symmetry $r = 0$ was 2.04 mm or approximately 18% lower than the known value. That error was ascribed in large part to the effect of the violation of the assumption of one-dimensional through-thickness heat flow, which occurs because of the relatively small size of the defect in relation to its depth. Note the variation in error as the edge of the defect is approached and then crossed. Although the profile in Fig. 13.8 raises several interesting questions, the main point is merely to demonstrate that a depth profile determined this way is likely to produce a biased estimate for defect size.

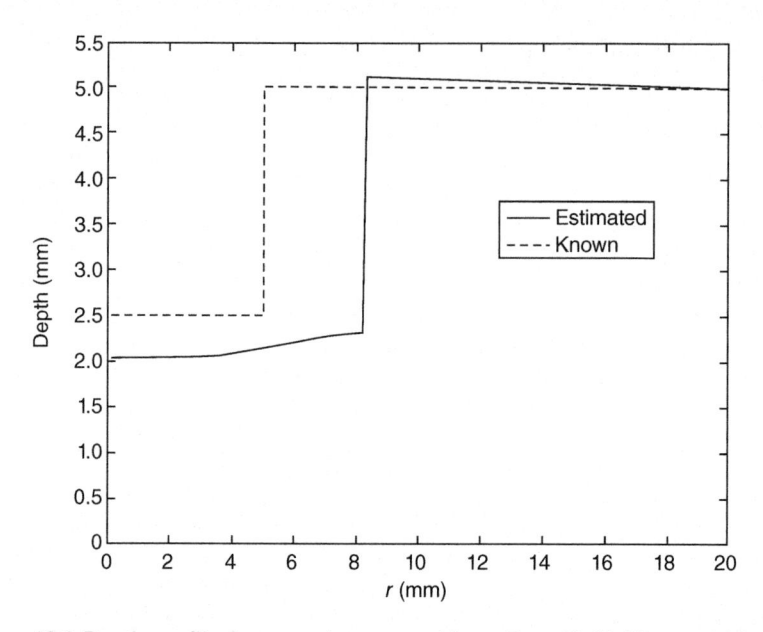

13.8 Depth profile for case d computed from Eqn 13.15. The transition at a radial position of approximately 8.3 mm denotes the point where the peak in the second derivative corresponding to the air insert could no longer be detected. In practice that location is likely to occur nearer the physical edge of the defect.

13.4 Non-destructive evaluation (NDE) of impact damage

A hard impact on a laminate can cause delamination, matrix cracking, fibre failure and/or fibre–matrix separation [23]. Perforation is also possible if the impact velocity is sufficiently high (e.g. ballistic); however, the primary need for NDI in relation to impact damage occurs in the lower velocity regime. Lower impact velocities have the potential to create barely visible impact damage (BVID), which describes a situation where potentially significant internal structural damage is produced with little if any visible evidence of damage at the surface. In that event a serious decline in the mechanical properties of the structure may go unnoticed.

The delaminations and interlaminar matrix cracks caused by low-velocity impact tend to be distributed through the laminate in a pattern that resembles the branches of a pine tree [24]. Adjacent to the impact site the damage is confined to a relatively small region, but this region extends in size with increasing depth and typically reaches a maximum size near the back-wall of the laminate. The distribution of damage through the laminate also depends strongly on the lay-up. For example, delaminations tend only to occur between plies of different fibre direction (because of a mismatch in bending stiffness), and develop preferentially along the direction of the fibre in the lower laminate, leading to the oblong shaped indications often seen in NDI scans of impact damage.

Figure 13.9 shows a sequence of raw thermographs obtained from a flash inspection of the front (impact) and back faces of a 50-ply quasi-isotropic carbon-epoxy laminate damaged by a controlled low-velocity impact. The case is an exemplar for BVID as the internal damage is extensive yet the only evidence of damage at the impact site is a barely visible indentation. Several of the aforementioned characteristics are apparent in the thermographs. The first row pertains to the near surface structure of the damage and confirms both a large size variation between the front and rear side delaminations as well as a dominant 45° orientation which is consistent with the known lay-up sequence. Shortly after, the front side inspection reveals evidence of a longer delamination with a 0° orientation, which develops a strong thermal contrast 0.5 seconds into the sequence (third row). At that stage, the rear surface response also shows evidence of a deeper delamination with a −45° orientation. The results in the last row reveal yet deeper structure, particularly on the impact side where the lateral size of the indication has grown significantly, now almost matching the size revealed in the back face inspection. Incidentally, it is worth noting the absence of fine detail in this last set of results, which illustrates the time-dependent nature of the damping caused by heat diffusion.

Providing the lay-up and ply thickness are known, observations of this type can be readily used to determine depth. A more formal analysis can also be applied to such inspection results, keeping in mind that the scope for that will be curtailed somewhat by the multilayered nature of impact damage. The key difficulty is that the measured contrast is a product of contributions from damage at various layers which are not easily deconvolved by analysis. This means that only the shallowest delamination, which produces the earliest breakpoint, is readily handled by the depth analysis approaches described previously. These approaches can be extended to deeper layers providing the delamination footprint extends beyond the overlap zone, e.g. the −45° indication in the third row of Fig. 13.9, and the analysis is performed sufficiently far from any overlap to avoid detecting a breakpoint from a shallower delamination.

Lateral sizing can be done using the FWHM approach described previously, applied to a thermal profile bisecting the damage; however, as with a depth analysis, account needs to be taken of the multilayered structure of the damage, and in particular the ambiguity that arises in defining an appropriate 'maximum' for the FWHM measurement where the indication is a product of overlapping delaminations at different depths. In this situation

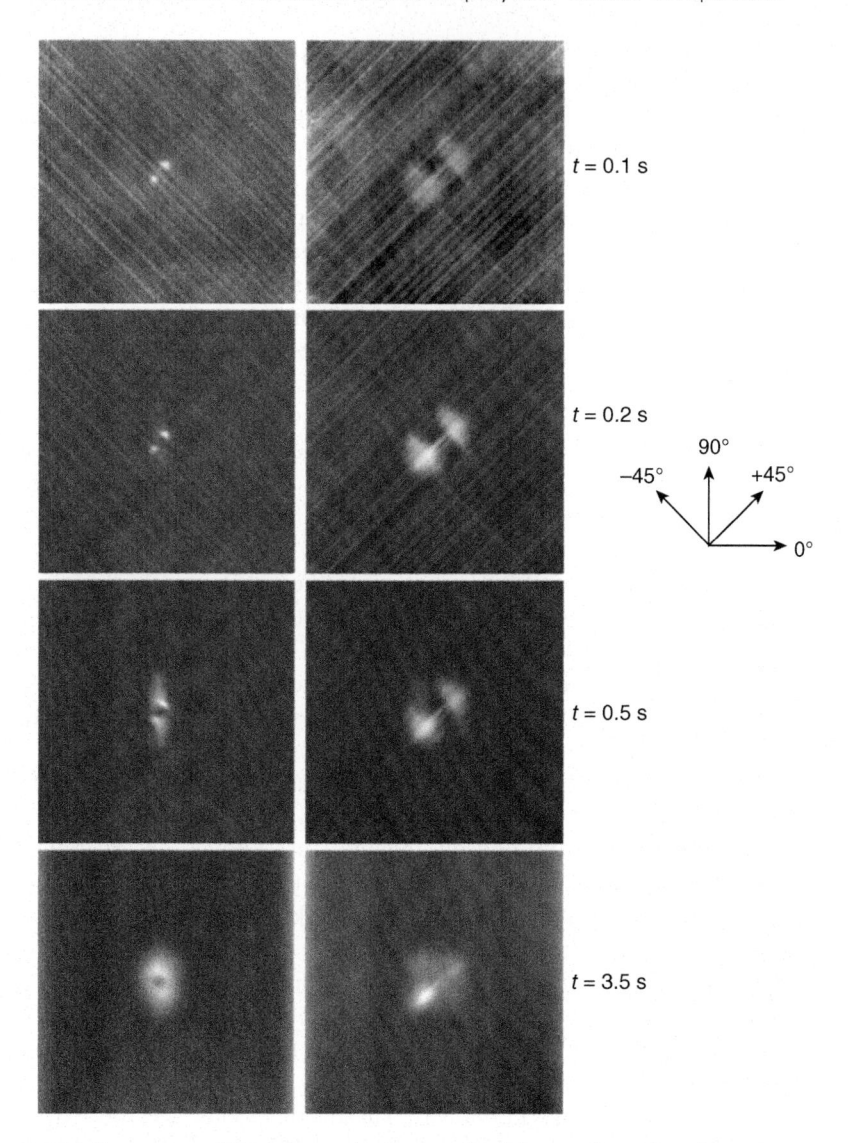

13.9 Sequence of thermographs from a flash inspection of an impact-damaged composite laminate.

the peak spatial derivative might be worth considering as an alternative measure.

13.5 Non-destructive evaluation (NDE) of porosity

Birt and Smith [25] provide a concise definition for porosity, describing it as '... a large number of microvoids, each of which is too small to be of structural significance or to be detected individually by a realistic inspection technique, but which collectively may reduce the mechanical properties of a component to an unacceptable degree.' The detection of an individual microvoid is not feasible for conventional thermal inspection since the thermal diffusion lengths associated with normal inspection practice are typically much too large*. The best prospect for detection lies in the measurement of a material bulk property like thermal diffusivity which is known to be a strong function of porosity. Work by Zalameda and Winfree [26], Connolly [27] and Hendorfer et al. [28] has shown that for carbon–epoxy composites the diffusivity varies as a largely linear function of volume fraction in the range of ≈1 – 7%. The dependence on porosity stems from separate effects on the thermal mass ρC and thermal conductivity K of the matrix. The dependence of the former is expressed by the law of mixtures (LOM), viz.

$$\rho_H C_H = v_m \rho_m C_m + v_v \rho_v C_v \qquad [13.20]$$

where the subscript H denotes a heterogeneous material made up of a matrix denoted by the subscript m, with a distribution of voids denoted by the subscript v, and v is the volume fraction. The thermal conductivity of a heterogeneous structure has been expressed analytically as a function of the properties of the matrix and void material by several authors. For instance, Kerrisk [29] found that where the scale of the heterogeneity of the material is small compared with the length scale of the thermal disturbance, a condition he calls the steady-state limit, the conductivity can be expressed as follows:

$$K_H^K = K_m \frac{K_v(1+2v_v)+2K_m v_m}{K_m(2+v_v)+K_v v_m} \qquad [13.21]$$

Ringermacher et al. [30] provides a different expression which he developed using a dielectric analogy borrowed from electromagnetics. The result is:

$$K_H^R = K_m + v_m(K_v - K_m)\left[\frac{K_m}{K_m + \eta_L(K_v - K_m)}\right] \qquad [13.22]$$

Here, ηL is called the 'dethermalization factor', a parameter that depends on the spheriodal aspect ratio of the microvoids. Although the two expressions are clearly different, they are consistent in suggesting that thermal conductivity should decrease with increasing volume fraction of porosity**. This raises an interesting point. Thermal mass also declines with increasing porosity, which results in a potential ambiguity with respect to the net effect

---

*Compared with the length scale of an individual void.
**Assuming the density, specific heat and thermal conductivity of a microvoid are negligibly small compared with the values for the matrix.

of porosity on thermal diffusivity. However, it is generally accepted (see Grinzato et al. [31]) that for carbon fibre composite materials the conductivity is reduced proportionately more than the thermal mass, so the net effect is a decline in diffusivity, consistent with the experimental findings of Connolly [27] and others, and, it can be shown, with Eqns 13.20 and 13.21.

Note that these expressions describe only the conductivity of the matrix, and not the laminate. Springer and Tsai [32] adopted a cylindrical filament model to develop an expression for the laminate conductivity, which they gave as:

$$K_L = K_H \left(1 - 2\sqrt{\frac{v_f}{\pi}}\right) + \frac{K_H}{B}\left[\pi - \frac{4}{\sqrt{1-B^2\frac{v_f}{\pi}}} \tan^{-1}\left(\frac{\sqrt{1-B^2\frac{v_f}{\pi}}}{1+\sqrt{B^2\frac{v_f}{\pi}}}\right)\right] \qquad [13.23]$$

where the subscript f denotes the fibre and

$$B \equiv 2\left(\frac{K_H}{K_f} - 1\right) \qquad [13.24]$$

Zalameda [33] reports reasonable agreement between experimental measurements and predictions furnished by the cylindrical filament model, particularly for porosity levels below 2%.

The estimates of diffusivity presented in the work of Zalameda [33] and Connolly [27] were derived by fitting a theoretical one-dimensional heat-flow model to experimental data acquired from the back-wall of a sample in a through-transmission inspection, reminiscent of the method pioneered by Parker et al. [34]. In Parker's method, diffusivity is calculated from the time required for the back-wall temperature to reach half of the maximum rise ($t_{\frac{1}{2}}$), which is inversely proportional to the thermal diffusivity:

$$\alpha = \frac{1.38L^2}{\pi^2 t_{\frac{1}{2}}} \qquad [13.25]$$

This simple approach is well proven, and applied routinely in laboratories around the world. Unfortunately, a through-transmission arrangement is impractical for most in-service applications where often only one side of a component is accessible. That situation is easily handled using a single-sided variant of Parker's approach, which follows trivially from Eqn 13.15. The approach performs well in practice, as reported by Shepard et al. [35] who found good correlation with Parker's method. It is worth noting that the model-fit approach used by Zalameda [33] and Connolly [27] could similarly be adapted to a single-sided inspection.

For the assessment of porosity in large structures the high speed of inspection provided by thermography offers a potential advantage over more commonly applied ultrasonic methods. Indeed, the practice of thermal diffusivity imaging has been successfully applied for some time [30], [36] and should continue to benefit from advances in computational hardware. Hendorfer *et al.* [28] observed in their study that thermography was both more sensitive and provided better spatial resolution than an ultrasonic method.

13.6 Experimental demonstration

Theoretical models and numerical simulations are useful in illustrating fundamental concepts and properties, but offer only limited insight into the practical feasibility of a particular approach to analysis. Models always involve some element of simplification or approximation, which if poorly considered can seriously erode fidelity and the instructive value of a simulation. For that reason an experimental demonstration is arguably more instructive and convincing.

Consider the experimental case study shown in Fig. 13.10. The sample is a carbon-epoxy laminate with three rows of artificial defects varying in

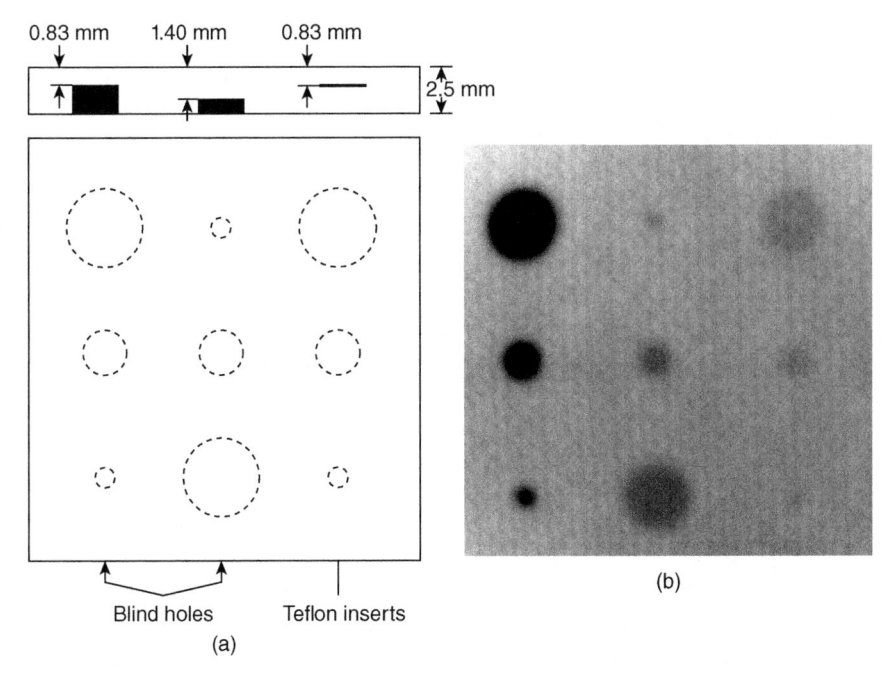

13.10 An experimental case study involving a carbon-epoxy laminate sample with three rows of artificial defects of varying depth and size.

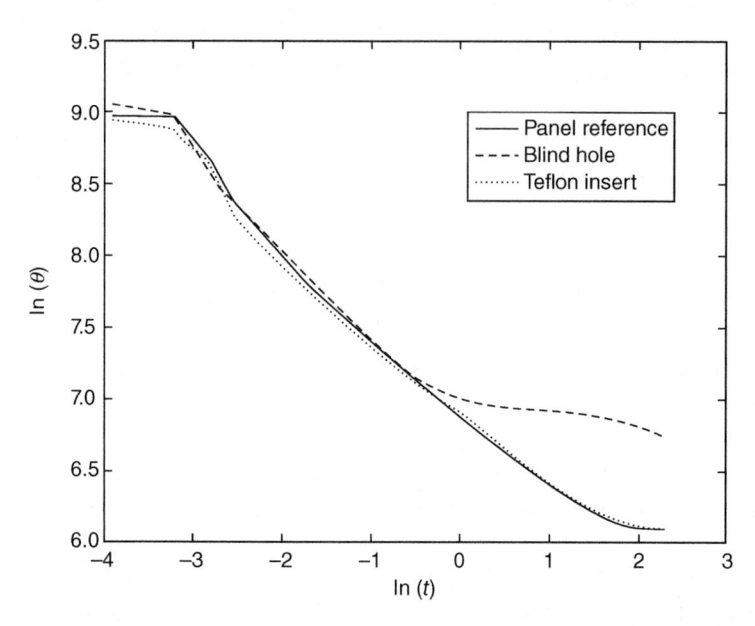

13.11 Log–log decay of the temperature response at three locations on the panel.

depth and size. The laminate was inspected using flash thermography, with the infrared imagery acquired using a cryogenically cooled commercial 640 × 512 indium antimonide detector, at a frame rate of 50 Hz. The flash duration was approximately 20 ms. Shown alongside the schematic is a raw thermograph taken 2.2 seconds after the flash trigger. All of the defects produce noticeable thermal contrasts. Human perception of image contrast is, however, particularly acute, so the fact an indication can be resolved does not necessarily imply a good basis for quantitative analysis. Figure 13.11 illustrates that point. Three time traces are shown here. The solid line serves as our reference, corresponding to a point on the panel remote from any defect. It has a breakpoint at $\ln(t) \approx 1.5$, which corresponds to the back-wall signature (i.e. the delay corresponds to the laminate thickness of 2.5 mm). The dashed line is the decay measured at the centre of the deepest blind hole (upper left defect in the schematic). Its breakpoint occurs much earlier, as expected. Note, however, the dotted trace, which corresponds to the largest of the Teflon inserts. The decay is almost indistinguishable from the reference trace, with no noticeable evidence of a breakpoint, which, in theory, should occur at approximately the same time as for the blind hole.

The implications for quantitative analysis will be discussed shortly, but first it is worth pointing out several other features that distinguish experimental inspection results from simulated ones. Note that the early-time

Flaw characterisation 361

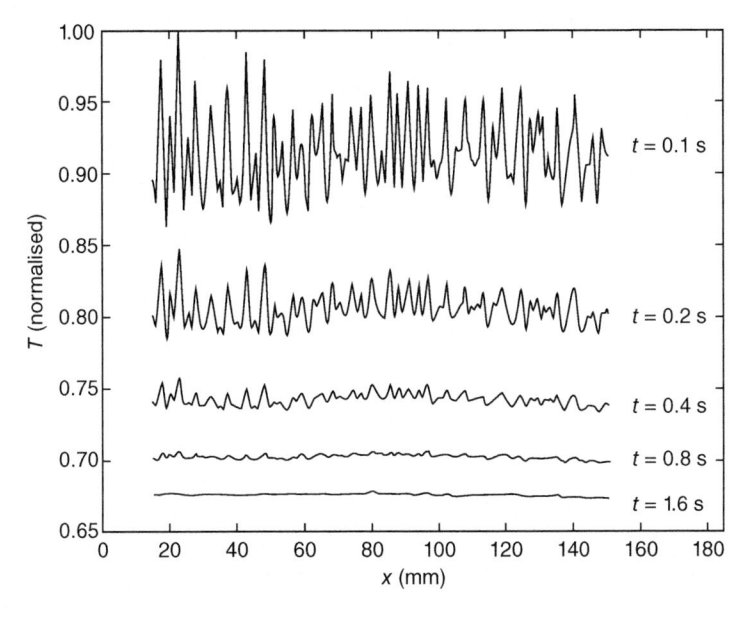

13.12 Time variation in the temperature profile along a straight line positioned mid-way between the upper two rows of defects.

behaviour is profoundly at odds to the log-linear decay predicted by theory (cf. Fig. 13.1). The flat response in the period $\ln(\theta) < -3$ is caused by saturation of the detector. Since the duration of the flash discharge is of the order of the frame period, even if the detector were not saturated the response is technically forced and therefore violates a key assumption of the theory. The radiant emittance eventually falls within the dynamic range of the detector allowing measurement of the decay; however, that decay is evidently not log-linear, particularly early on in the response. This is partly caused by a persistence in the excitation from the flash discharge which is neither quenched nor shuttered so the response remains partly forced. Another reason is that the effective diffusion length in that early time phase is sufficiently short to resolve the heterogeneity in the surface ply, which has a woven bidirectional structure resulting in complex heat flow between the resin and fibre components. Figure 13.12 shows large periodic variations in temperature across the laminate which are consistent with this explanation.

Consider now the determination of defect depth, which we tackle using the approach embodied by Eqn 13.15. In order to mitigate some of the impact of the heterogeneity of the laminate, the analysis is applied to measurements that are spatially averaged over a 5×5 cluster of adjacent pixels. Depth estimates are sought at five locations, corresponding to four defects;

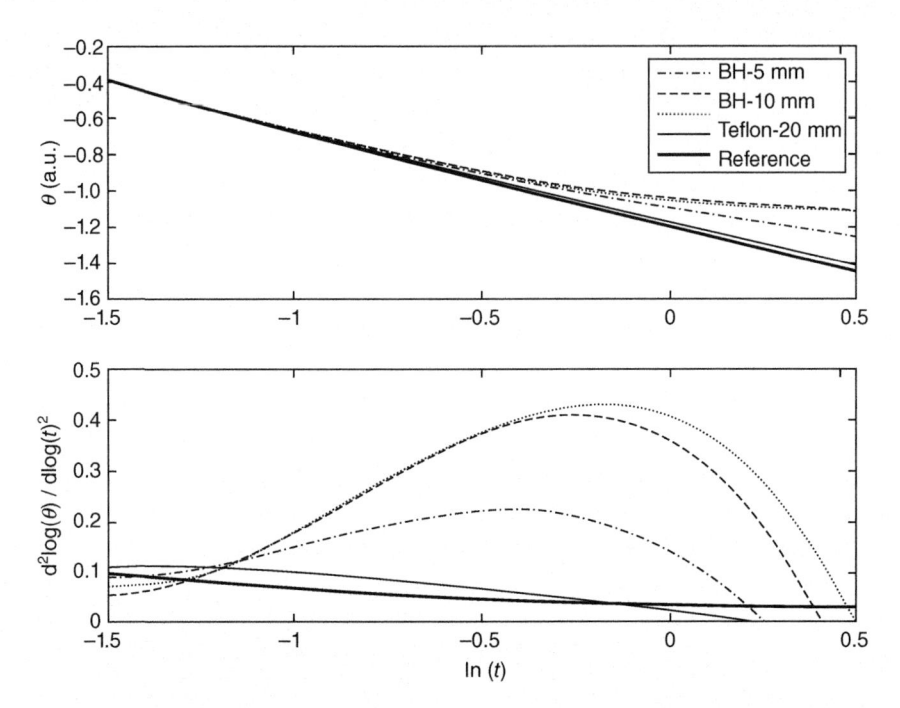

13.13 Traces of the log decay (upper) and second derivative (lower) at five locations on the laminate.

three blind holes and one Teflon insert, all at the same nominal depth of 0.83 mm, as well as an arbitrary reference away from the defects. The subplots in Fig. 13.13 show the log decay and its second derivative for the five cases, all obtained from a polynomial fit to the averaged data set, less the saturated portion of the sequence which was excluded to minimise the model order. The absence of an asymptotic zero second derivative affirms the previous remarks about experimental deviations from the log-linear ideal. For the blind hole defects these deviations are largely insignificant since the breakpoint time t^* is still distinct. This is not true for the Teflon insert case, where background variations in the second derivative rule out any prospect of a meaningful depth estimate.

A diffusivity value of 2.7×10^{-7} m^2 s^{-1} was used in the calculation. This value was determined by substituting the breakpoint time measured for the 20 mm diameter blind hole into Eqn 13.15 where L was set to 0.83 mm, the known depth. The breakpoint times measured for the 10 and 5 mm diameter holes consequently produce depth estimates of 0.81 and 0.75 mm respectively. Two aspects of this result are consistent with the findings of the previous case study involving simulated data: (i) the depth is underpredicted, and (ii) the error grows inversely with the lateral size of the defect.

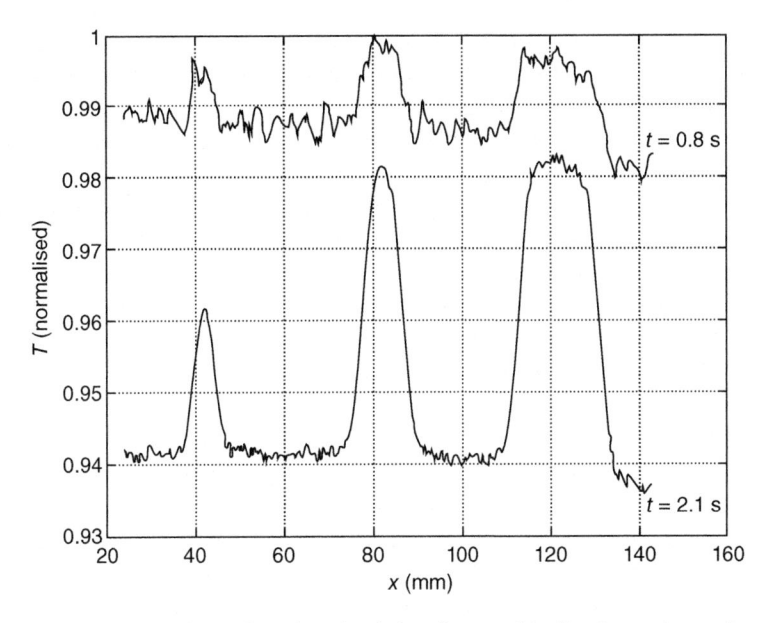

13.14 Line profiles bisecting the left column of indications shown in Fig. 13.10.

Consider now, as a final task, the calculation of lateral size. Figure 13.14 plots the temperature profile along a line bisecting the centres of the left column of blind holes shown in Fig. 13.10. Note the profile at $t = 0.8$ s and in particular the oscillations with a short length scale, which are produced by inhomogeneities in the laminate, as well as the variation in the background level, which is caused by uneven illumination from the flash tubes. Both features have the potential to bias the FWHM estimate. The second profile is more amenable to analysis, but recall from previous remarks that as diffusion progresses the potential for bias in the estimate grows. Estimates deduced from the latter profile for the three indications are 18.6, 9.3 and 5.4 mm, which are within 10% of the known values. Such accuracy is acceptable for many practical purposes.

13.7 Future trends

The use of infrared thermography should grow steadily within the aerospace sector particularly for in-service maintenance support of composite structures. Its remarkable speed advantage over other methods is impetus enough for an increased uptake irrespective of any future advancement in quantitative capability.

Unfortunately, the main impediment to a significant improvement in capability is, fundamentally, the immutable constraint imposed by the physics of heat diffusion. Although advancements in infrared detector technology and computational hardware will certainly help improve capability to some extent, one should not expect thermography to ever rival an advanced ultrasonic testing (UT) method, such as phased array, for detailed quantitative NDE.

Notwithstanding those fundamental constraints, with continued growth in computing performance there are prospects for a significant shift in the approach to quantitative characterisation, which has the potential to improve both the scope and quality of information obtained from thermal inspection. The enabler is evolutionary computing, in the form of methods such as genetic algorithms and particle swarm optimisation, which provide a means of solving problems that were previously considered intractably complex. Evolutionary algorithms combined with numerical modelling tools such as FEM should eliminate the need for many of the fundamental assumptions inherent in current approaches, and would thereby provide a potentially more powerful basis for quantitative analysis. A further attraction of such an approach is that by furnishing a detailed FEM model of a defect it should help close the gap that currently exists between diagnostic and prognostic assessments of a defect.

Another possible future trend is diagnostic baselining for high-value structures such as aircraft. The aim is to produce a reference data set against which future inspection results can be compared. The potential advantages are significant such as an enhanced probability of detection and increased scope for degradation monitoring and long-term durability assessment. That the practice is currently uncommon is partly a result of the prohibitive cost. The cost, however, is largely a function of the speed of inspection, and therefore might be justified for a rapid technique such as infrared thermography.

13.8 References

1. P. Cawley and R. D. Adams. The mechanics of the coin-tap method of non-destructive testing. *Journal of Sound and Vibration*, **122**(2):299–316, 1988.
2. R. D. Adams and P. Cawley. A review of defect types and nondestructive testing techniques for composites and bonded joints, *NDT International*, **21**(4):208–222, 1988.
3. H. S. Carslaw and J. C. Jaeger. *Conduction of Heat in Solids*. Oxford University Press, Oxford, 2nd edition, 1959.
4. S. K. Lau, D. P. Almond, and P. M. Patel. Transient thermal wave techniques for the evaluation of surface coatings. *Journal of Physics D: Applied Physics*, **24**(3):428, 1991.
5. G. Green. Some problems in the conduction of heat. *Philosophical Magazine Series 7*, **9**(56):241–260, 1930.
6. J. C. Krapez and P. Cielo. Thermographic nondestructive evaluation: data inversion procedures Part 1: 1D analysis, *Research in Nondestruction Evaluation*, **3**(2): 81–100, 1991.

7. S. M. Shepard, J. R. Lhota, B. A. Rubadeaux, D. Wang, and T. Ahmed. Reconstruction and enhancement of active thermographic image sequences. *Optical Engineering*, **42**(5):1337–1342, 2003.
8. H. I. Ringermacher, D. J. Mayton, D. R. Howard, and B. Cassenti. Towards a flat-bottom hole standard for thermal imaging. *Review of Progress in Quantitative Nondestructive Evaluation*, 17, 1998.
9. H. I. Ringermacher and D. R. Howard. Synthetic thermal time-of-flight(sttof) depth imaging. *AIP Conference Proceedings*, **557**(1):487–491, 2001.
10. N. Rajic. A quantitative approach to active thermographic inspection for material loss evaluation in metallic structures. *Research in Nondestructive Evaluation*, **12**:119–131, 2000.
11. J. G. Sun. Analysis of pulsed thermography methods for defect depth prediction. *Journal of Heat Transfer*, **128**(4):329–338, 2006.
12. W. P. Winfree and J. N. Zalameda. Thermographic depth profiling of delaminations in composites. *AIP Conference Proceedings*, **657**(1):981–988, 2003.
13. N. Rajic. Principal component thermography for flaw contrast enhancement and flaw depth characterisation in composite structures. *Composite Structures*, **58**(4):521–528, 2002.
14. N. Rajic. Genetic algorithm for flaw characterization based on thermographic inspection data. In X. P. Maldague and A. E. Rozlosnik, editors, *Proceedings of SPIE 4710 -Thermosense XXIV*, pages 522–530, Orlando, Florida, USA, 2002.
15. M. B. Saintey and D. P. Almond. An artificial neural network interpreter for transient thermography image data. *NDT and E International*, **30**(5):291–295, 1997.
16. D. J. Crowther, L. D. Favro, P. K. Kuo, and R. L. Thomas. Inverse scattering algorithm applied to infrared thermal wave images. *Journal of Applied Physics*, **74**(9):5828–5834, 1993.
17. L. D. Favro, D. J. Crowther, P-K. Kuo, and R. L. Thomas. Inversion of pulsed thermal-wave images for defect sizing and shape recovery. In J. K. Eklund, editor, *Proceedings of SPIE 1682 – Thermosense XIV*, pages 178–181, Orlando, Florida, USA, 1992. SPIE.
18. G. C. Wetsel and F. A. McDonald. Subsurface-structure determination using photothermal laser-beam deflection. *Applied Physics Letters*, **41**(10):926–928, 1982.
19. D. P. Almond and S. K. Lau. Defect sizing by transient thermography. i. An analytical treatment. *Journal of Physics D: Applied Physics*, **27**(5):1063, 1994.
20. M. B. Saintey and D. P. Almond. Defect sizing by transient thermography. ii. A numerical treatment. *Journal of Physics D: Applied Physics*, **28**(12):2539, 1995.
21. M. Y. Y. Hung, Y. S. Chen, S. P. Ng, S. M. Shepard, Y. Hou, and J. R. Lhota. Review and comparison of shearography and pulsed thermography for adhesive bond evaluation. *Optical Engineering*, **46**(5):051007–1, 2007.
22. X. Maldague and S. Marinetti. Pulse phase infrared thermography. *Journal of Applied Physics*, **79**:2694–2698, 1996.
23. F. Chen and J. M. Hodgkinson. Impact damage. *Wiley Encyclopedia of Composites*. Wiley, 2011.
24. S. Abrate. *Impact on Composite Structures*. Cambridge University Press, 1st edition, 1998.
25. E. A. Birt and R. A. Smith. A review of nde methods for porosity measurement in fibre-reinforced polymer composites. *Insight*, **46**(11):681–686, 2004.

26. J. N. Zalameda and W. P. Winfree. Thermal diffusivity measurement on composite porosity samples. *Review of Progress in Quantitative Nondestructive Evaluation*, **9**:1541–1548, 1990.
27. M. P. Connolly. The measurement of porosity in composite materials using infrared thermography. *Journal of Reinforced Plastics and Composites*, **11**:1367–1375, 1992.
28. G. Hendorfer, G. Mayr, G. Zauner, M. Haslhofer, and R. Pree. Quantitative determination of porosity by active thermography. *AIP Conference Proceedings*, **894**(1):702–708, 2007.
29. J. F. Kerrisk. Thermal diffusivity of heterogeneous materials. ii. limits of the steady-state approximation. *Journal of Applied Physics*, **43**(1):112–117, 1971.
30. H. I. Ringermacher, D. R. Howard, and R. S. Gilmore. Discriminating porosity in composites using thermal depth imaging. *AIP Conference Proceedings*, **615**(1):528–535, 2002.
31. E. G. Grinzato, S. Marinetti, and P. G. Bison. NDE of porosity in CFRP by multiple thermographic techniques. In Xavier P. Maldague and Andres E. Rozlosnik, editors, *Proceedings of SPIE 4710 – Thermosense XXIV*, pages 588–598, 2002.
32. G. S. Springer and S. W. Tsai. Thermal conductivities of unidirectional materials. *Journal of Composite Materials*, 1:166–173, 1967.
33. J. N. Zalameda. Measured through-the-thickness thermal diffusivity of carbon fiber reinforced composite materials. *Journal of Composite Technology & Research*, **21**:98–102, 1999.
34. W. J. Parker, R. J. Jenkins, C. P. Butler, and G. L. Abbott. Flash method of determining thermal diffusivity, heat capacity, and thermal conductivity. *Journal of Applied Physics*, **32**(9):1679–1684, 1961.
35. S. M. Shepard, Y. Hou, T. Ahmed, and J. R. Lhota. Reference-free interpretation of flash thermography data. *Insight*, **48**(5): 298–307, 2006.
36. W. P. Winfree and D. M. Heath. Thermal diffusivity imaging of aerospace materials and structures. *Proceedings of SPIE, Thermosense, XXIII*, **3361**:282–290, 1998.

14
Non-destructive evaluation (NDE) of aerospace composites: detecting impact damage

C. MEOLA and G. M. CARLOMAGNO,
University of Naples Federico II, Italy

DOI: 10.1533/9780857093554.3.367

Abstract: This chapter gives an overview of the use of infrared thermography to investigate the impact damage of composites mainly used for aerospace applications. The chapter concentrates on low-energy/low-velocity impact, difficult to detect since it may cause significant material degradation without perforation and so without visible damage. Infrared thermography is used with a twofold function. The first consists of monitoring the material thermal response to an impact event useful for design purposes. The second refers to the use of infrared thermography as a non-destructive evaluation (NDE) technique for detecting buried impact damage.

Key words: composites for aerospace applications, low-velocity impact, infrared thermography, on-line monitoring of the material thermal behaviour under an impact event.

14.1 Introduction

Composite materials offer several advantages over metals. These include high specific stiffness and high specific strength combined with a significant reduction in weight (Jones, 1975; Hull and Clyne, 1996; Soutis, 2005). Composites are attractive for many industrial applications. One of the most important fields of application is aerospace, in which lightness is of key importance. Indeed, the aircraft industry has been and continues to be involved in the development of high performance light materials.

Composites entered military aviation in 1960 and civil aviation in 1970. During the 1980s, composites were used for a variety of secondary components (Shütze, 1996) such as rudder and wing trailing edge panels, which require directional reinforcement (Piening *et al.*, 2002). In the 2000s composites were also introduced into the fabrication of primary load-carrying structures, such as in the A380, the world's largest passenger aircraft, of which they cover 25% by weight. The central wing box of the aircraft is entirely made of carbon fibre-reinforced polymer (CFRP). An ever greater percentage of these materials is used in the Boeing Dreamliner (Boeing

787), which has been conceived with almost 50% composite materials, bringing both a fuel saving and reduction in CO_2 emission of almost 20%.

Because of their low intralaminar strength, fibre composites are susceptible to delamination during their lifetime. In fact, composites are able to absorb impact energy thanks to a polymeric matrix that distributes energy through the material. In this way, a low-energy/low-velocity impact does not produce perforation, but rather delamination between the layers, with no visible surface manifestation. Such delamination may grow in service, causing severe reduction of stiffness in the structure, and likely leading to catastrophic failure. Since the high probability of impact-damaging in composite materials during manufacture, service and maintenance, knowledge of impact processes and response of the material is of paramount importance. However, impact damage in composites is a very complex mechanism, involving matrix cracking, surface buckling, delamination, fibre shear-out and fibre rupture (Richardson and Wisheart, 1996; Park and Jang, 2001; Shyr and Pan, 2003; Elder *et al.*, 2004; Baucom and Zikry, 2005; Caneva *et al.*, 2008).

The scientific community is substantially divided into two main groups when it comes to impact damage, with one group mainly interested in the dynamics of impact load (DIL), or in the comprehension of physical phenomena associated with an impact event, for design and preventive purposes, and the second driven instead towards the detection of damage produced, or to non-destructive inspection (NDI) for assessment of defect-free parts and/or maintenance purposes. It is worth considering that:

- on one hand, it is very important to understand material behaviour under impact relating to the matrix, the fibres and the stacking sequence for design purposes;
- on the other hand, it is important to be able to assess the integrity of a part, which also means having techniques capable of discovering the damage in its incipient stage.

Computer simulation is widely adopted by industry for modelling an impact event because it is faster and cheaper than physical testing, However, test validation is generally compulsory. The primary objective of numerical simulation in composites is to predict the delamination threshold load (DTL) (Kachanov, 1987; Lemaitre and Chaboche, 1990; Elder *et al.*, 2004; Iannucci and Willows, 2006). An impact event is accompanied by complex phenomena which depend on several factors, mainly on target geometry (Schoeppner and Abrate, 2000; Cantwell, 2007; Minak and Ghelli, 2008) and impactor shape (Kim and Goo, 1997; Mitrevski *et al.*, 2005). In particular, Kim and Goo (1997) investigated the impact response of composite laminates through the dynamic contact problem for an elliptical impactor. They found that an impactor with a more 'blunt' shape produced the highest force-indentation. Mitrevski *et al.* (2005) later confirmed this with non-destructive techniques.

The main objective of NDI is to individuate techniques which comply with the standards of a specific field (such as Aeronautical Standards). An ideal NDI technique should be effective in detecting buried slight delaminations, and should be simple, fast and cheap, as well as safe (Gros and Takahashi, 1999; Pitkänen *et al.*, 2000). Ultrasound is the most popular NDI technique (Krautkramer and Krautkramer, 1990; Blitz and Simpson, 1996; Gros and Takahashi, 1999; Pitkänen *et al.*, 2000; Mitrevski *et al.*, 2005; Zhang *et al.*, 2009). Ultrasound supplies information about material characteristics (thickness, density, stiffness, porosity), but requires a lot of time to scan large surfaces. Infrared thermography (IRT) is a faster technique (Meola and Carlomagno, 2004; Meola and Carlomagno, 2006). IRT has proved useful, and in some cases more effective than the most widely used methods (Meola *et al.*, 2008).

This chapter gives an overview on the use of IRT for investigating impact damage in composites. Key examples are reported which replicate the most important results obtained at the Department of Aerospace Engineering of the University of Naples Federico II, to which the authors belong.

14.2 Effectiveness of infrared thermography

Within the impact damage of composites an infrared imaging system is used with a twofold function:

- Non-destructive testing performed before the impact to check for manufacturing defects, and after impact to detect any impact damage.
- On-line monitoring performed during an impact event to record thermal behaviour of composites under impact.

14.2.1 Materials

Several specimens of different composite types were fabricated, including both different types of matrix and fibre reinforcement, and different stacking sequences. Some specimen details are listed below:

- CFRP|$_u$: includes unidirectional carbon fibres embedded in an epoxy matrix and overlaid following the stacking sequence [0/45/90/–45]$_s$ to obtain laminates of thicknesses 2.4 and 4.7 mm. The thinner laminate was cut into several specimens of width 100 mm and length 130 mm, named CFRP-U$_{th}$, which were impacted with a modified Charpy pendulum at energy E_i between 0.5 and 2.8 J. From the thicker laminate a specimen CFRP-U$_{tk}$, of side about 40 cm, was obtained and was impacted with a drop test machine at two points with E_i respectively equal to 15 and 30 J.

- CFRP$|_f$: was obtained by overlaying epoxy adhesive pre-impregnated carbon twill woven fabrics to get a thickness of about 2 mm; specimens are called CFRP-F. Two specimens of width 220 mm and length 620 mm were prepared and impacted with a modified Charpy pendulum at several different points with E_i varying from 0.5 up to 4 J.
- CFRP$|_s$: is a sandwich composed of two carbon/epoxy skins 1 mm thick and of Nomex® honeycomb core of cell size about 3 mm and thickness 15 mm. This was impacted with a drop test machine at 5 J.
- GFRP$|_u$: is made of E-glass fibres embedded in an epoxy matrix. The stacking sequence is $[0_2,90_2]_s$ with an overall laminate thickness of 2.9 mm. Each specimen is 100 mm wide and 130 mm long. Specimens were impacted with a modified Charpy pendulum at energy E_i between 4 and 25 J.
- FML: is a fibre metal laminate specimen that includes four aluminium layers with two glass/epoxy layers between them. It is 130 × 100 mm^2 and 1.5 mm thick, and was impacted with a modified Charpy pendulum at energy E_i in the range 2.7 J–4.5 J.

These specimens were either used for impact tests with simultaneous monitoring of their thermal response, or impacted and then inspected using NDI. Some of the results obtained will be shown later in this chapter. For more information the reader is addressed elsewhere (Meola and Carlomagno, 2009, 2010a, 2010b; Meola et al., 2009a, 2009b, 2011; Carlomagno et al., 2011).

14.2.2 Instrumentation and test set-up

Two infrared cameras were used (the SC3000 and the SC6000 (FLIR Systems)), both equipped with a Stirling cooled Focal Plane Array (FPA) Quantum Well (QWIP) detector working in the long wave infrared band 8–9 μm. For the SC3000, the standard acquisition rate is 60 Hz full frame, which is 320 × 240 pixels, but can reach up to 900 Hz with a reduced field of view (16 horizontal lines). The SC6000 has a higher spatial resolution, 640 × 512 pixels full frame, and is equipped with a windowing option linked to frequency frame rate and temperature range.

For non-destructive testing, the infrared system is coupled with the lock-in option and IRLockIn© software. In this work, optical lock-in thermography (OLT) was used in which (Fig. 14.1) thermal stimulation is performed with halogen lamps. One or two lamps were utilized depending on surface area. Tests were performed by varying the heating modulation frequency in order to evaluate the conditions of the material at different depths through its thickness, according to the basic lock-in equation:

$$\mu = \sqrt{\frac{\alpha}{\pi f}} \qquad [14.1]$$

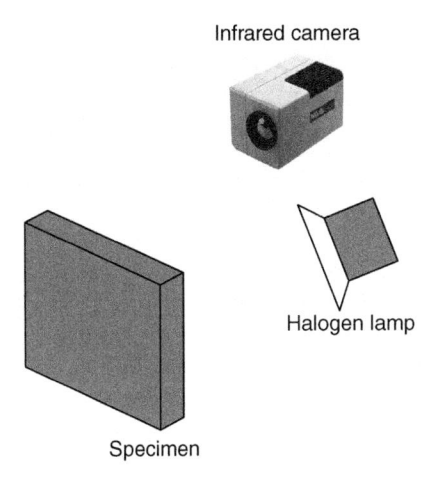

14.1 Lock-in thermography test set-up.

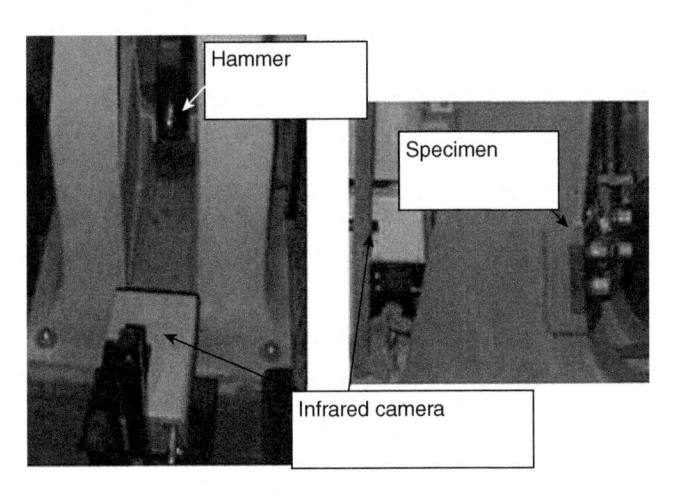

14.2 Set-up for impact tests.

where μ is the thermal diffusion length and α is the material thermal diffusivity coefficient. Impact tests were carried out with a modified Charpy pendulum, which includes enough room for positioning of the infrared camera (Fig. 14.2). The impact was performed by using two different hammers, one with an ogival nose 18 mm in diameter and the other hemispherical 24 mm in diameter. Specimens were held from the shorter sides, while free to move along the two longer sides. The impact energy value was set by varying the starting height of the Charpy arm. As shown in Fig. 14.2, the infrared camera was positioned to view the specimen surface opposite to the impacted one. During an impact event, thermal images were acquired in time sequence at 900 or 300 Hz with the SC3000 and at 96 Hz with the SC6000. The frame rate was varied to account for both temporal and spatial thermal phenomena. Some tests were performed with a drop test machine, but mainly for non-destructive evaluation (NDE) purposes.

14.3 On-line monitoring

Under low-energy impact a material may experience only an elastic phase without permanent deformation; the impact energy is released when the material recovers its undeformed status. For higher energy the material also undergoes a plastic phase in which some energy is dissipated as heat with consequent permanent material modifications. The impact pushing force causes local bending of the surface. Initially, by the side opposite to impact (the side in view of the infrared camera, see Fig. 14.2) the surface undergoes a convex curvature, meaning local material expansion. From the thermal point of view, this generally entails a cooling down effect (thermoelastic effect; Biot, 1956; Thomson, 1978). Depending on impact energy E_i, the material may behave differently (Meola and Carlomagno, 2009, 2010a, 2010b; Meola *et al.*, 2009a, 2009b, 2011; Carlomagno *et al.*, 2011). For low E_i once the force is removed, the surface attempts to recover its undisturbed position, and its initial ambient temperature. As E_i increases above the elastic limit, damage occurs to the material, associated with energy dissipation (Meola *et al.*, 2009a, 2009b; Meola and Carlomagno, 2010a). By analysing local surface warming, it is possible to acquire information about the damage.

14.3.1 Qualitative analysis

As examples, some thermal images are reported in Figs 14.3–14.6. The images shown in Fig. 14.3 are extracted from a sequence taken at 900 Hz with the SC3000 infrared camera during impact at 1.2 J of the specimen CFRP-F1 (i.e., belonging to the second group), while those in Fig. 14.4 refer to an impact at 2.7 J of the specimen FML-1 (i.e., belonging to the fifth group).

In Figs 14.5 and 14.6 thermal images for two specimens belonging to the GFRP|$_u$ type respectively GFRP-A and GFRP-B both impacted at 12.3 J, are shown. These thermal images were taken with the SC3000 at 900 Hz (Fig. 14.5) and at 300 Hz (Fig. 14.6). Notice that, for this camera, the field of view reduces to 16 horizontal lines and to 48 horizontal lines for frame rate at respectively 900 Hz and 300 Hz.

For all sets of images, the surface viewed displays an almost uniform colour (uniform temperature distribution) before impact (Figs 14.3a, 14.4a, 14.5a, 14.6a). Suddenly, after the impact, it is possible to see (especially for

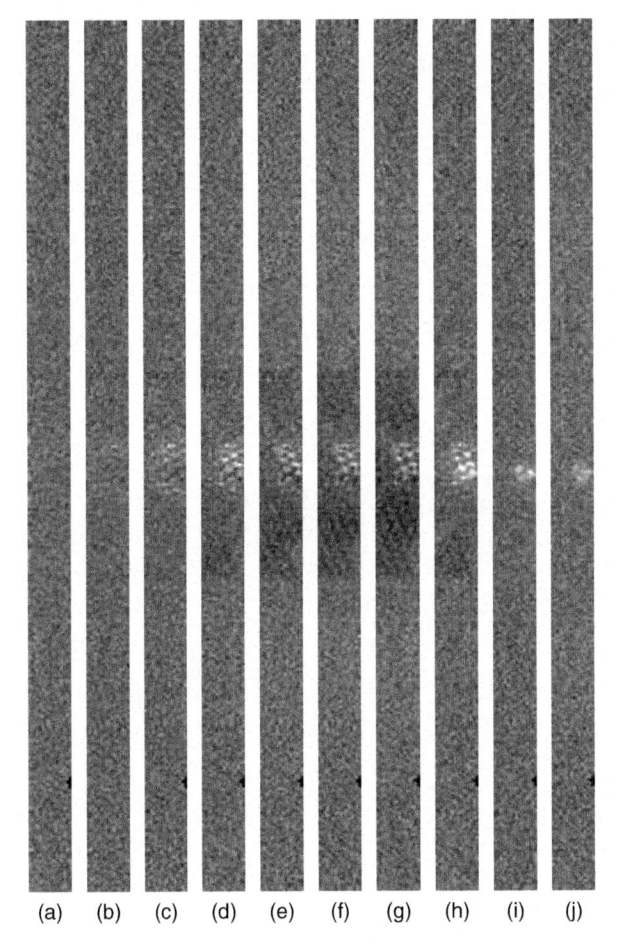

14.3 Images in time sequence (900 Hz) while the specimen CFRP-F1 is being impacted at 1.2 J: (a) before impact; (b) 0.0011 s after impact; (c) 0.0022 s after impact; (d) 0.0044 s after impact; (e) 0.0077 s after impact; (f) 0.0088 s after impact; (g) 0.011 s after impact; (h) 0.021 s after impact; (i) 0.029 s after impact; (j) 0.382 s after impact.

images taken at 900 Hz) zones of darker colour, which means cooling down due to the thermo-elastic effect. Temperature variations linked to thermo-elastic phenomena evolve very quickly. In fact, in the elastic phase, the surface cools down, or heats up, as it assumes a convex (under impact), or a concave (as the impact force is removed and the surface goes back) shape. Such thermo-elastic effects can only be visualized when taking images at a high frame rate. Conversely, the temperature increase due to dissipation of energy under impact is slower and it can be visualized by taking images at

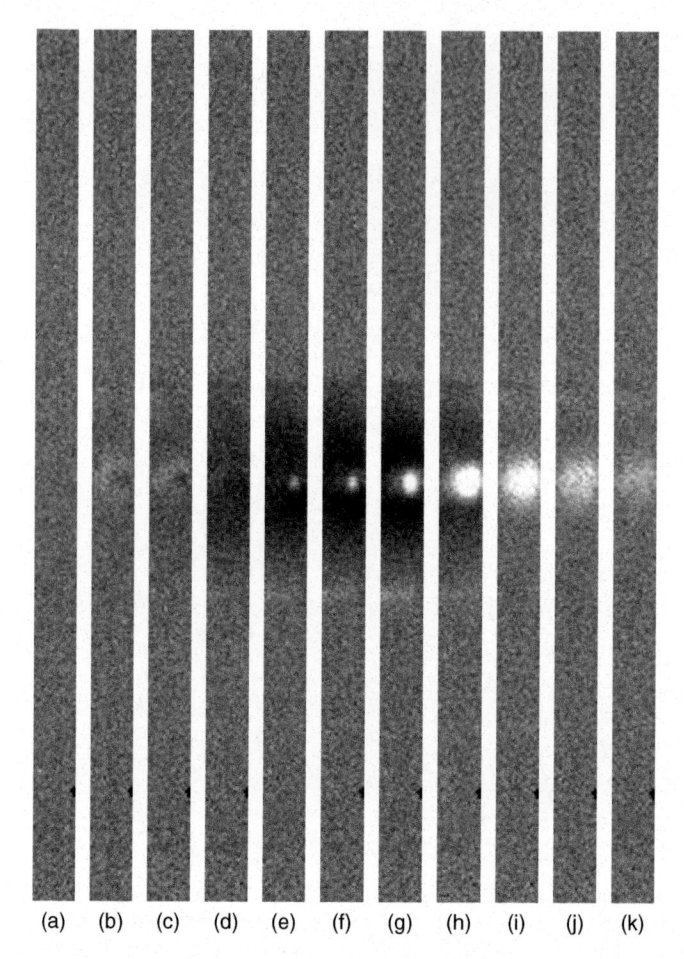

14.4 Images in time sequence (900 Hz) while the specimen FM-1 is being impacted at 2.7 J: (a) before impact; (b) 0.0011 s after impact; (c) 0.0022 s after impact; (d) 0.0044 s after impact; (e) 0.0077 s after impact; (f) 0.0088 s after impact; (g) 0.016 s after impact; (h) 0.021 s after impact; (i) 0.171 s after impact; (j) 0.377 s after impact; (k) 1.694 s after impact.

a lower frame rate. In fact, it is possible to see, as the hot stain remains visible for some time after removal of the impact load.

14.3.2 Quantitative analysis

In order to include all thermal phenomena from an impact event for each test, acquisition began a few seconds before the impact, and lasted for some time after. To better account for the material thermal behaviour, the first

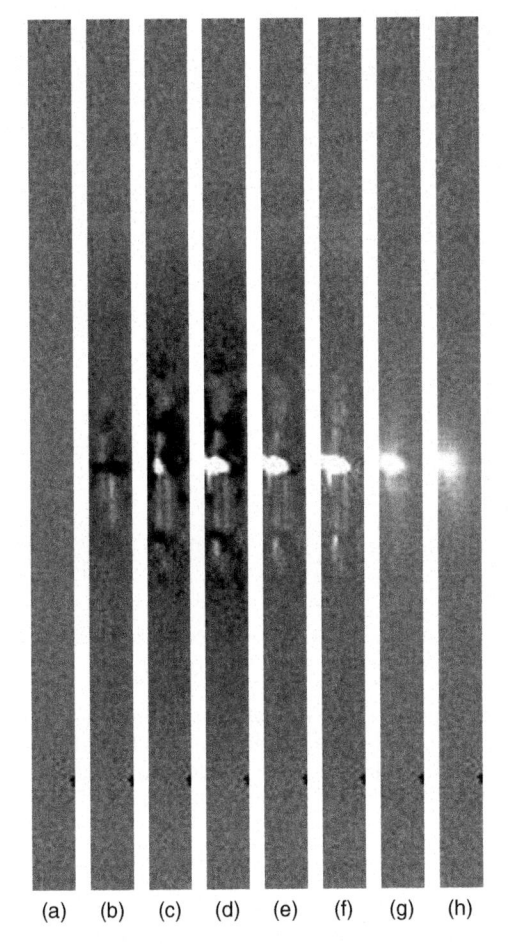

14.5 Images in time sequence (900 Hz) while the specimen GFRP-A is being impacted at 12.3 J: (a) before impact; (b) 0.0011 s after impact; (c) 0.0022 s after impact; (d) 0.0044 s after impact; (e) 0.0077 s after impact; (f) 0.022 s after impact; (g) 0.55 s after impact; (h) 1.63 s after impact.

image ($t = 0$) of the sequence (i.e., the specimen surface (ambient) temperature before impact) is subtracted from each subsequent image so as to generate a sequence of temperature difference ΔT maps:

$$\Delta T = T(i, j, t) - T(i, j, 0) \qquad [14.2]$$

with i and j representing lines and columns of the surface temperature map. As an example, plots of ΔT against time for three spots over the specimen surface (as indicated by arrows) are reported in Fig. 14.7. More specifically, the number of images (taken in time sequence at 900 Hz) is indicated in

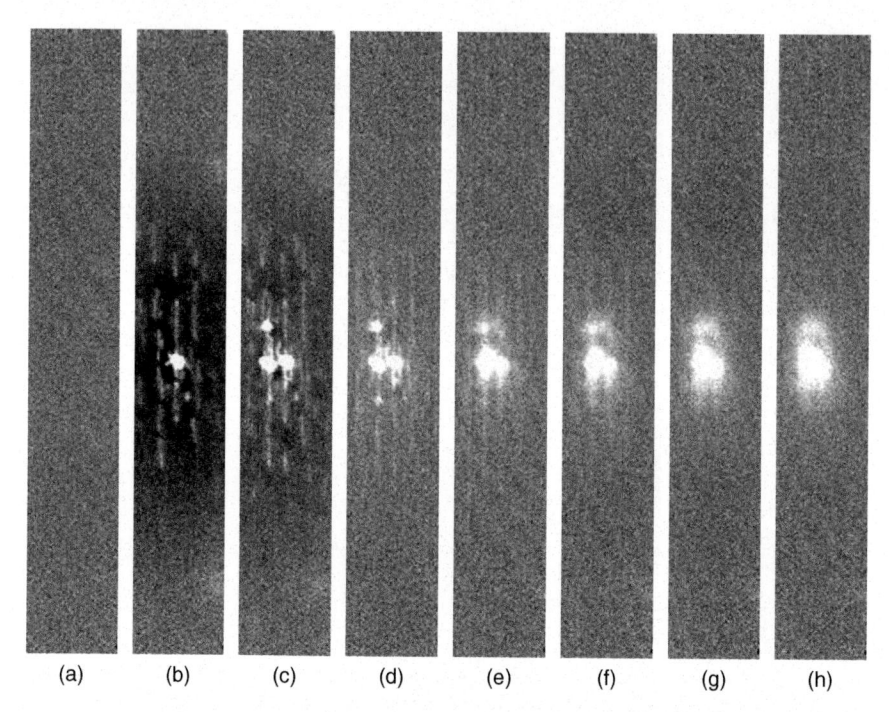

14.6 Images in time sequence (300 Hz) while the specimen GFRP-B is being impacted at 12.3 J: (a) before impact; (b) 0.0033 s after impact; (c) 0.0066 s after impact; (d) 0.033 s after impact; (e) 0.33 s after impact; (f) 0.5 s after impact; (g) 0.92 s after impact; (h) 1.35 s after impact.

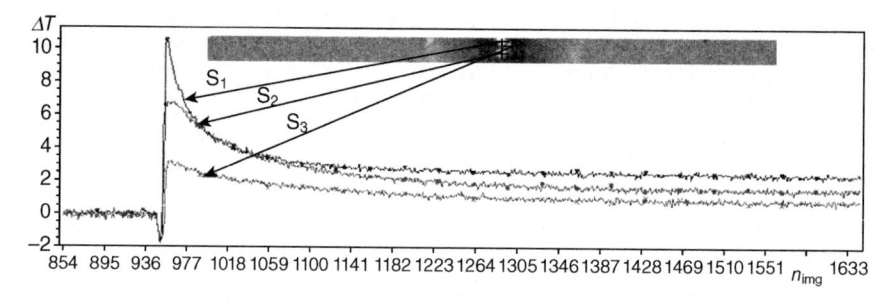

14.7 ΔT variation with time (number of images in time sequence) in three points over the surface of the specimen GFRP-C impacted at 9.7 J.

the abscissa. The specimen is the GFRP-C impacted at 9.7 J. From this figure it is possible to appraise the ΔT, in all the three spots as zero before impact, because the specimen surface at an almost uniform ambient temperature, suddenly goes below zero upon impact, and soon after rises to over 10 degrees for the S_1 spot, over 6 degrees for the S_2 and about 3 degrees for the S_3. Then, there is at first a rapid temperature decrease for each hot point, whose slope depends on the temperature peak and material thermal diffusivity, followed by a more gradual decrease towards room temperature.

To gain more detail about the behaviour of GFRP under impact, minima (negative values with respect to the initial ambient temperature) and maxima ΔT values over the impacted specimen surface are extracted from each image sequence for the different test conditions, and analysed separately. As an example, minima and maxima ΔT, evaluated over the surface of specimens impacted respectively at 7.5, 9.9 and 12.3 J, are plotted against time in Fig. 14.8(a, b). The lower minimum, which corresponds to the highest material expansion, is attained at the impact instant. After that, the temperature raises towards $\Delta T = 0$ (Fig. 14.8a) while a secondary minimum is displayed about 0.02 s later. The latter is most probably due to a shaking effect of the specimen. More specifically, the specimen does not recover its flat aspect immediately after the impact load, but its surface alternates from convex (i.e., the shape assumed under the pushing force) to concave, and again to convex until completely at rest. However, this vibration is very swift since it is completely extinguished in about 0.04 s. The maximum ΔT, as shown in Fig. 14.8b, is reached at about 0.005 s, lasting until about 0.02 s and then decreasing at first quickly and then slowly towards ambient temperature. Both lowest minima (in absolute value) and highest maxima are plotted against the impact energy E_i in Fig. 14.9. As can be seen, minima ΔT_{min} show an almost linear increase with E_i; conversely, maxima ΔT_{max} group into two distinct regions being E_i lower/equal, or greater than about 10 J. Such a distinction was attributed by Meola and Carlomagno (2010b) to the damaging mechanisms. In fact, micro-cracks form first in the material, at a low impact energy, and are then followed by fibre breakage as the energy reaches and/or passes a critical value.

The hot spots (e.g., Fig. 14.6c) coincide with heat generation loci which form in correspondence of local energy dissipation, or better where delamination and/or breakage occurs during the impact event. Evidently, temperature increase is proportional to energy released locally as a consequence of the damage. The ΔT difference between the three hot spots indicates the type and importance of local damage. Furthermore, the number of hot spots may vary depending on the type of material, the impact energy, and the damaging modes, amongst other factors. In fact, as shown in Fig. 14.10, a CFRP may display no hot spots for $E_i = 1.9$ J (Fig. 14.10a), or either a

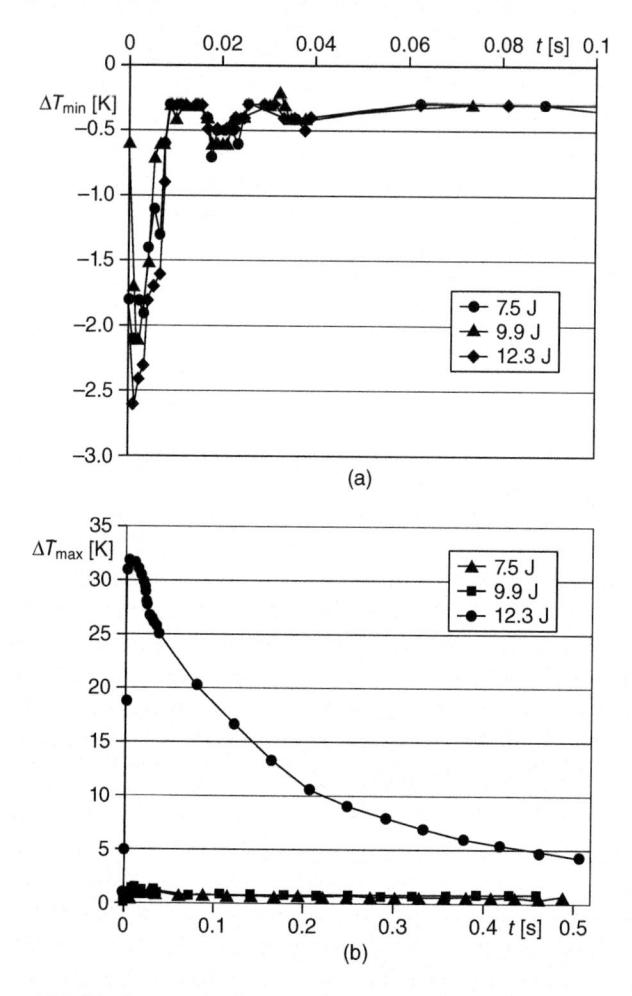

14.8 Maxima and minima ΔT values over the overall surface of GFRP specimens impacted with three different impact energies: (a) minima ΔT values; (b) maxima ΔT values.

vertical (Fig. 14.10b), or a horizontal (Fig. 14.10c) hot stripe for $E_i = 2.8$ J, depending on the fibre direction being vertical for Fig. 14.10b and horizontal for Fig. 14.10c. Conversely, a GFRP may display practically no hot spots for $E_i = 4.5$ J (Fig. 14.10d), or three (Fig. 14.10e), or four (Fig. 14.10f) hot spots for the impact energy increasing up to 9.7 J for Fig. 14.10e and to 15.9 J for Fig. 14.10f respectively.

Indeed, being able to visualize onset and evolution of hot spots represents a very important possibility, giving information on onset and propagation of impact damage, of prime importance in the aerospace industry. This may be useful for material design purposes, better choice of the type of

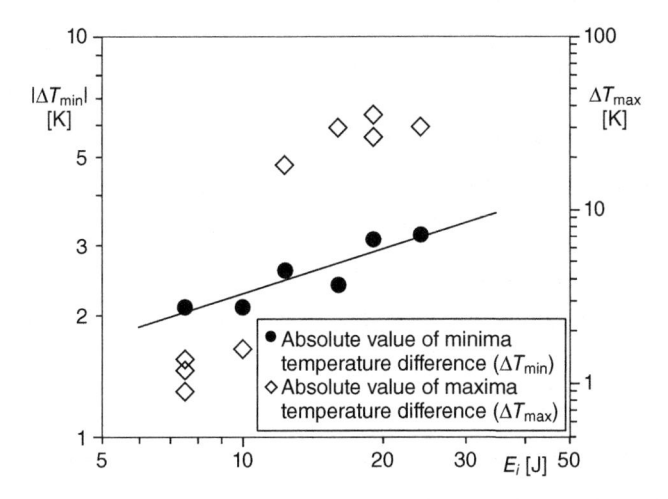

14.9 Variation of maxima and minima Δ*T* values with the impact energy.

material for a given part, and for setting a maintenance timetable. It is worth noting that the 16-line image, obtained with frame rate set at 900 Hz, allows for the cooling down (thermo-elastic) effect to be visualized, but does not allow for exhaustive measurement of the whole damaged area. This is a key task when specimen fibre direction is vertical (e.g., Figs 14.10b, d, e) and this is not known in advance. Such a drawback can be overcome by a low frequency frame rate and/or by using infrared imaging devices equipped with the windowing option. Some thermal images, which were taken at 96 Hz with the infrared camera SC6000 and which refer to GFRP specimens impacted both at 12 J, are reported in Figs 14.11 and 14.12. The only difference between the two specimens is that fibres on the external surface, viewed by the infrared camera, are directed vertically for the GFRP-F and horizontally for the GFRP-G as indicated by the alignment of the hot spots. As can be seen, independently of the fibre direction, it is possible to account for the whole damaged area extension. In addition, it is also worth noting that, thanks to the higher spatial and thermal resolutions, a high-contrast picture is obtained which allows for a clear discrimination of contiguous fibres and, in turn, for identification of the breakage points. This is evident by comparing, for example, Figs 14.11b and 14.10e.

14.3.3 Damage evolution and heat transfer mechanisms

As recently demonstrated by Meola and Carlomagno (2010b), the rise in temperature observed over the material surface during an impact event, can be exploited to understand onset and propagation mechanisms of

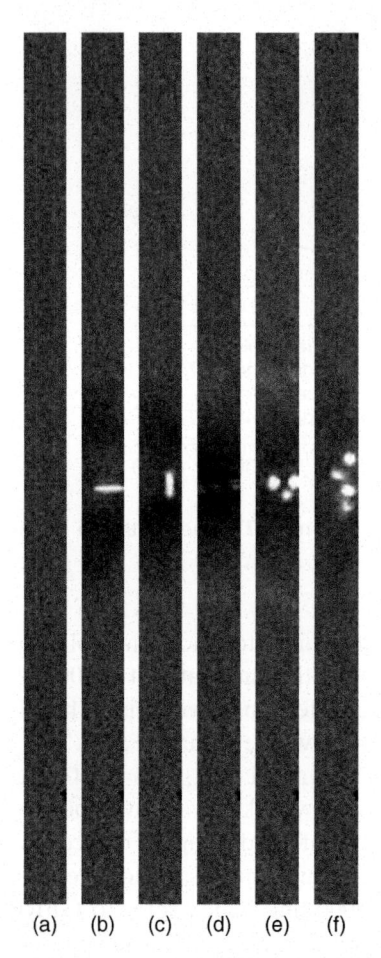

14.10 Comparison between thermal images taken at the same time of 0.017 s after impact for varying the reinforcing fibre and the impact energy: (a) CFRP- $U_{th}1$, E_i = 1.9 J; (b) CFRP- $U_{th}2$, E_i = 2.8 J; (c) CFRP- $U_{th}3$, E_i = 2.8 J; (d) GFRP-D, E_i = 4.5 J; (e) GFRP-C, E_i = 9.7 J; (f) GFRP-E, E_i = 15.9 J.

impact damage. In fact, an increase in temperature is entailed by dissipated impact energy. The mechanisms of energy absorption by a composite during impact depend on several factors, including: impact velocity, geometrical parameters, and material inherent characteristics (i.e., brittle or ductile) (Abrate, 1991; Kang and Kim, 2000; Naik *et al.*, 2006). For low-energy/low-velocity impacts, the energy absorbed by the specimen E_a is generally regarded as the sum of: membrane energy E_m, bending energy E_b and damage energy E_d.

$$E_a = E_m + E_b + E_d \qquad [14.3]$$

Detecting impact damage 381

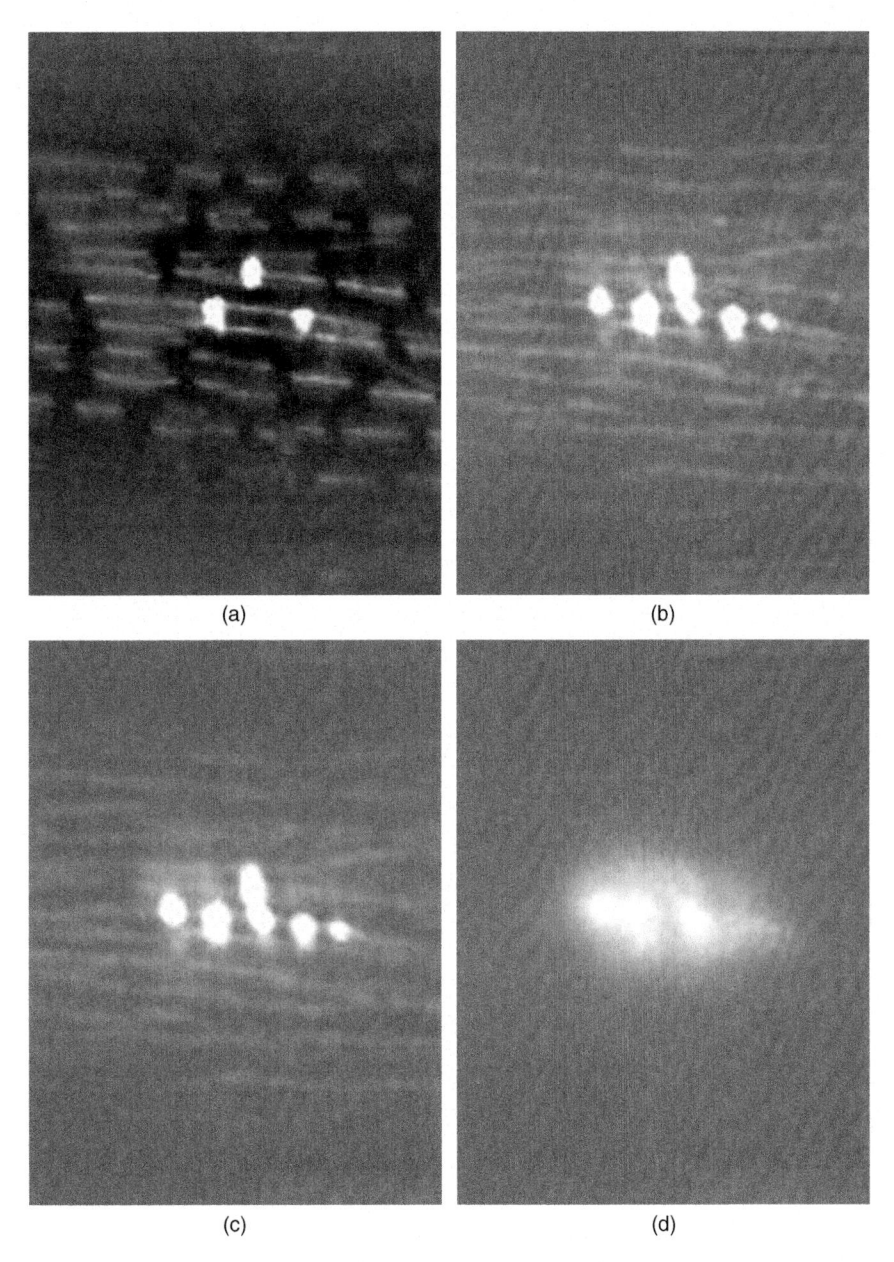

14.11 Some thermal images taken at 96 Hz of the specimen GFRP-F impacted at 12 J: (a) $t = 0.0104$ s; (b) $t = 0.1248$ s; (c) $t = 0.2392$ s; (d) $t = 2.4752$ s.

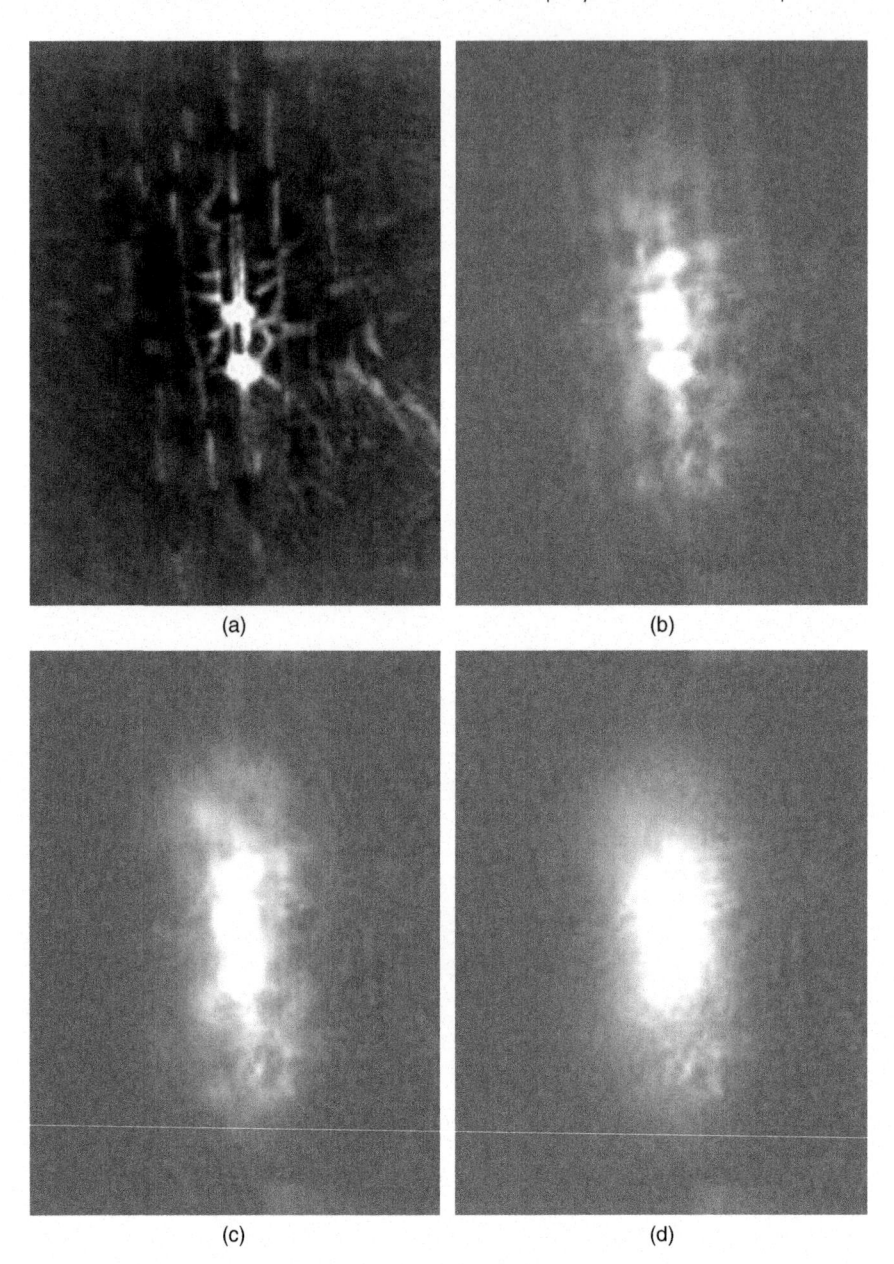

14.12 Some thermal images taken at 96 Hz of the specimen GFRP-G impacted at 12 J: (a) $t = 0.0104$ s; (b) $t = 0.364$ s; (c) $t = 0.9672$ s; (d) $t = 4.1392$ s.

The importance of each contribution depends on material properties. For a brittle material E_d includes two terms, one accounting for fibre breakage and the other for delamination. For a ductile material, for which no fibre breakage occurs and the energy is predominantly spent in deformation (E_m and E_b) and delamination perpendicular to the impactor axis, E_d coincides with delamination energy (Kang and Kim, 2000). There is a general belief that the damage process is initiated by matrix cracks, which in turn lead to delamination at ply interfaces. A numerical simulation of the damage initiation process in E-glass/epoxy laminates was carried out by Chang and co-workers using the 3DIMPACT code (Wu and Chang, 1989; Choi *et al.*, 1991; Choi and Chang, 1992). In particular, for low-velocity impact, they proposed the matrix failure criterion to predict the occurrence of critical matrix cracks (Choi *et al.*, 1991) and the impact-induced delamination growth criterion (Choi and Chang, 1992). Later, Aslan *et al.* (2003), using the 3DIMPACT transient dynamic finite element code and experimental tests, found, for an E-glass/epoxy laminate 4.8 mm thick, the largest delamination arising at the back (i.e., opposite the impact side) and depending on the specimen in-plane dimension. They assumed delamination to be induced by matrix cracks formed under surface bending.

Meola and Carlomagno (2010b) provided new insights into the impact damage mechanisms of GFRP, considering that most of the absorbed energy is dissipated as heat and, taking into account the material thermal properties and the involved heat transfer mechanisms. In particular, they assert that the impact energy, dissipated as heat, is spent in the formation of matrix microcracks, delamination and the rupture of fibres. For $E_i < 7$ J energy dissipates to produce micro-cracks in the matrix, and delamination. This causes a local temperature rise detected by the infrared device during impact monitoring. However, at this stage the associated delamination is too weak to be detected by non-destructive techniques. As the impact energy overcomes 7 J, the associated damage becomes more significant, giving rise to the abrupt temperature increase visualized during the monitoring phase (see figures 5, 6, 9, 10 in Meola and Carlomagno, 2010b) and through the anomaly detected by lock-in thermography (see figure 16 in Meola and Carlomagno, 2010b).

In addition, Meola and Carlomagno (2010b) found that the hot spots, initially appearing in correspondence of fibres breakage, enlarge with time and coalescence to form a unique warm zone named A_w. A_w can be measured from thermal images, by choosing a limiting ΔT value between sound and delaminated material and by knowing the lens spatial resolution. The distribution of A_w with time for impact energies of 19 and 15.9 J is shown in Fig. 14.13. As can be seen, A_w increases with impact energy. For each energy level A_w increases quickly until about 0.3–0.4 s, and then remains virtually constant with only small fluctuations around the average value.

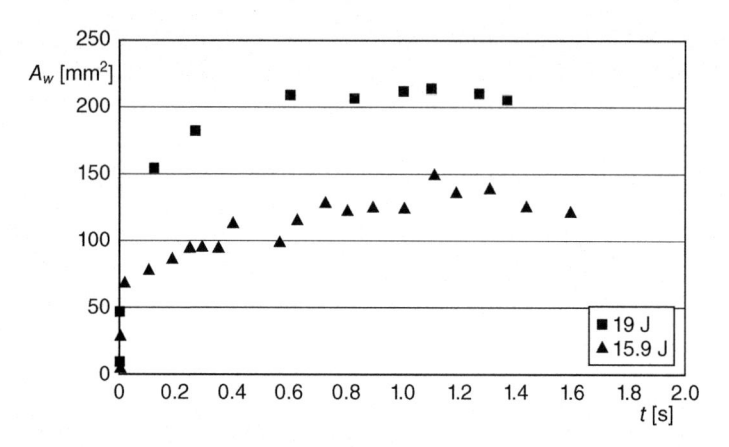

14.13 Variation of the warm area with time.

Indeed, the final A_w value coincides with the area damaged during impact. The latter was detected both through ND and visual inspection after impact, which was possible thanks to the translucent appearance of the panels. An important result from Meola and Carlomagno (2010b) is that the damage extension can be seen as a function of the impact energy and area of contact A_{HS} between the specimen surface and the hammer head:

$$A_w = f(A_{HS}, E_i) \quad [14.4]$$

However, as recently observed by Meola *et al.* (2011), the hot spots developing in both carbon and glass reinforced specimens lie within the hammer nose diameter. Thus, it seems that damage enucleates within the surface struck by the impactor, with no damage arising outside the directly impacted surface. Further, the final warm area accounting for damage extension, is caused mainly by delamination rather than fibre breakage. It should also be noted that hot spots form at a critical energy level dependent on many factors involving type and thickness of the material, as well its thermal and mechanical properties. For a CFRP 2.5 mm thick Carlomagno *et al.* (2011) observed a strong temperature increase at an impact energy of 2.8 J, much lower than the 10 J previously mentioned for the GFRP. Therefore, from the analysis of infrared images it is possible to gain information for every material, about onset, propagation and the final extension of the impact damage.

14.4 Non-destructive evaluation (NDE) of different composite materials

NDI is performed with OLT to detect either manufacturing defects, and/or impact damage. In this chapter only examples of impact damage detection

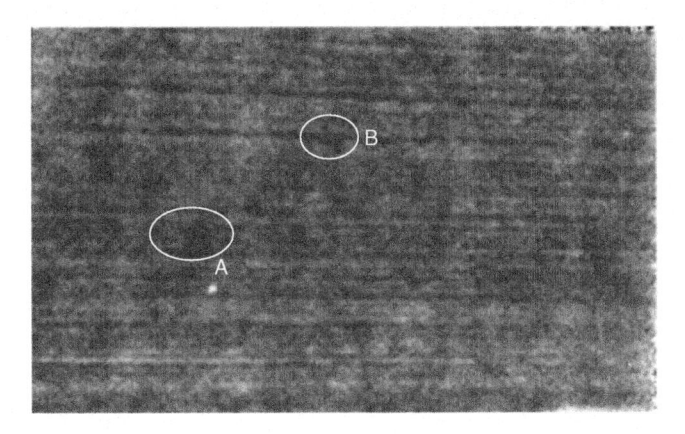

14.14 Phase image taken at $f = 0.5$ Hz of the specimen type CFRP-U_{th}2 after impact at 2.8 J.

induced through laboratory impact tests with either a modified Charpy pendulum, or a drop test machine, are reported. For all tests, unless otherwise specified, the rear side (i.e., opposite to the percussion) was viewed during non-destructive tests.

14.4.1 CFRP materials

As a first example, Fig. 14.14 shows a phase image taken at $f = 0.5$ Hz of the specimen CFRP-U_{th} after impact at 2.8 J. The specimen was impacted at two different points at the same energy, but by varying the direction of the holding force with respect to fibre length. In particular, the impact was directed to zone A, with the holding sides perpendicular to fibre direction (i.e., specimen horizontal, as in figure, with fibres grasped at their extremes). Zone B was instead impacted with the holding sides parallel to fibre direction (i.e., the specimen rotated by 90°). In the phase image, a darker zone is clearly evident in correspondence with the first impact A, accounting for some damage which occurred there. Conversely, it is quite difficult to recognize the mark produced by the second impact B since in this second case delamination mainly occurs parallel to fibre direction. In fact, the impact energy was mostly spent in a combination of delamination and fibre rupture at the first time (A zone) and in delamination between contiguous fibres at the second time (B zone).

Figure 14.15a displays the results of the inspected specimen CFRP-F1. This specimen was impacted at two points with two different energies of 2.5 and 4 J. To acquire information about the extension of the damage, tests were carried out with a close-up view, at a higher spatial resolution. An example of increased damage visibility is given in Fig. 14.15b which shows

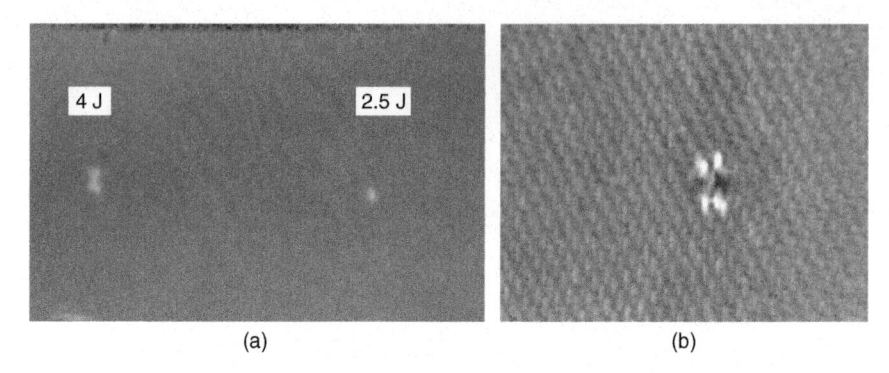

14.15 Phase images of the specimen CFRP-F1: (a) full view of the specimen impacted in two points with two different energies 2.5 and 4 J; (b) close up view to the zone impacted at 2.5 J.

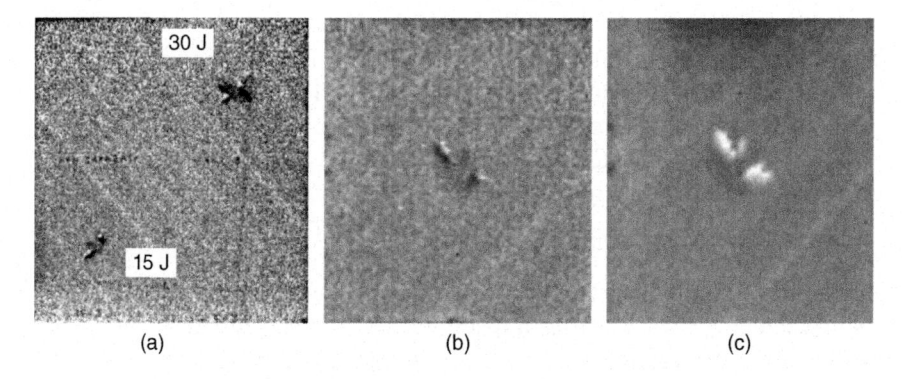

14.16 Phase images of the specimen CFRP-U$_{tk}$: (a) full view for $f = 0.75$ Hz; (b) close view to the specimen one-quarter impacted at 15 J for $f = 0.5$ Hz; (c) close view to the one-quarter specimen surface impacted at 15 J for $f = 0.05$ Hz.

the close-up view image of the zone impacted at 2.5 J. It is clearly distinguishable the lobed appearance of the impact damage in the CFRP material.

Figure 14.16a shows a phase image taken of the specimen CFRP-U$_{tk}$, at $f = 0.75$ Hz, presenting two impacts at 15 and 30 J. Both impacts are well identified as almost two-lobed appearance zones. To account for the depth of damage, for the zone (about ¼ of the specimen surface) including the impact at 15 J, Figs 14.16b and c show two phase images taken at 0.5 and 0.05 Hz respectively. As shown, the lobed structure becomes less pronounced as the heating frequency is reduced from 0.75 to 0.5 Hz, and undergoes inversion from dark to white at $f = 0.05$ Hz. This needs to be explained.

Generally a defect in a phase image is visualized through the phase contrast $\Delta\phi$ which may be expressed as:

$$\Delta\phi = \phi_d - \phi_s \qquad [14.5]$$

with subscripts d and s respectively indicating defect and sound material. It must be born in mind that $\Delta\phi$ may change sign by varying wave frequency ($\Delta\phi$ may become positive, zero, or negative) (Meola and Carlomagno, 2004, 2006). More specifically, when the thermal wave (of suitable frequency value) touches the top surface of the defect, the defect appears, but is poorly contrasted (only a mark) and $\Delta\phi$ could be either positive or negative. Generally, by decreasing the frequency, the contrast enhances and becomes highest as the thermal wave approaches the defect midplane. The contrast then worsens once more, vanishing ($\Delta\phi = 0$) when the thermal wave reaches the bottom of the defect. The corresponding frequency is called *blind frequency* f_b. Afterwards, the defect appears again but with changed colour (change of $\Delta\phi$ sign). A typical variation of $\Delta\phi$, with an initial negative value, by varying thermal wave frequency is represented in Fig. 14.17.

Going back to interpreting the data from Eq. (14.2), considering that defect depth is equal to 1.8 μm and assuming a thermal diffusivity of 0.007 cm²/s for sound CFRP material, it is possible to evaluate depth values of about 1, 1.2 and 4 mm respectively, corresponding to the three heating frequencies of 0.75, 0.5 and 0.05 Hz. Thus, for the phase images in Figs 14.16a and b, which refer to a specimen of overall thickness 4.7 mm, we are visualizing the impact damage with $\Delta\phi$ assuming negative values for $f < f_b$ (see the plot in Fig. 14.17). This also means that some damage is present at a depth of 1 mm (appears darker in Fig. 14.16a) and extends deeper than 1.2 mm. In fact, as f approaches the blind frequency f_b, the phase angle contrast $\Delta\phi$ worsens towards zero and so defects appear lighter as evidenced by Fig. 14.16b compared with Fig. 14.16a. In the phase image

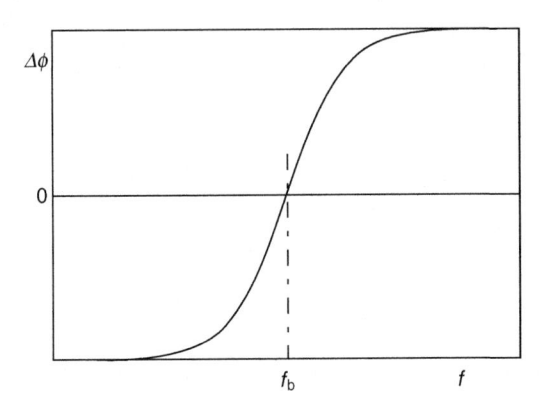

14.17 $\Delta\phi$ distribution with f.

shown in Fig. 14.16c where we see darker and lighter zones, instead we see the condition of a material layer at a depth of 4 mm. Therefore, looking at Fig. 14.17, the damage is again visible, already detected as darker lobes in Fig. 14.16a for $f = 0.75$ Hz, but now lighter since it is visualized at a frequency $f > f_b$ for which there is an inversion of $\Delta\phi$. Conversely, the darker zone in the centre between the lighter lobes indicates local delamination at a depth of 4 mm.

14.4.2 GFRP materials

Some phase images for GFRP specimens are reported in Figs 14.18–14.20. In particular, Fig. 14.18 shows phase images taken of the GFRP-F specimen impacted at $E_i = 12$ J at heating frequencies of 0.67 (a), 0.14 (b) and 0.045 Hz (c) (see Fig. 14.11). Tests were performed by viewing the specimen's front side (i.e., the side struck by the impactor). As can be seen, some damage has already appeared at the higher heating frequency of 0.67 Hz which allows for a depth of approximately 0.4 mm assuming a thermal diffusivity equal to 0.00116 cm^2 s^{-1} (Meola et al., 2002). In Fig. 14.18c the damage appears more clearly outlined, and it is visible because the damage shape practically coincides with the warm area present in Fig. 14.11d (note that the figures are not to scale).

Figure 14.19 collects phase images taken at $f = 0.14$ Hz from the rear side (the side viewed by the infrared camera during impact monitoring) of different specimens at varying conditions of impact. More specifically, Figs 14.19a and b refer to specimens GFRP-F and GFRP-G respectively, both impacted at $E_i = 12$ J. The only difference between the two specimens during impact tests, as already evidenced, lies in fibre direction in the external layer (that viewed by the camera during impact), which is vertical for the GFRP-F and horizontal for the GFRP-G. However, this means that damage caused by the impact might have been affected by the distribution of the grasping force. In fact, the damage stain takes a better clear two-lobed appearance on the GFRP-G specimen (Fig. 14.19b) while restricted to a more slender oval on the GFRP-F specimen. In the latter case, because of the external layer, the grasping force is parallel to fibre direction and allows slipping effects of contiguous fibres; the impact energy is mostly spent in delamination between contiguous fibres. Instead, for the GFRP-G specimen, for which the holding force is perpendicular to fibre direction, the impact energy is spent in a combination with delamination, and ruptures with a consequently wider extension of the damaged area. Figure 14.19c refers to the specimen GFRP-H impacted at $E_i = 9.7$ J and Fig. 14.19d refers to the GFRP-I specimen impacted at $E_i = 7.5$ J. Both are characterized by horizontal direction of fibres on the external layer. As can be seen, there is a reduction in extension of the damage area, by reducing impact energy.

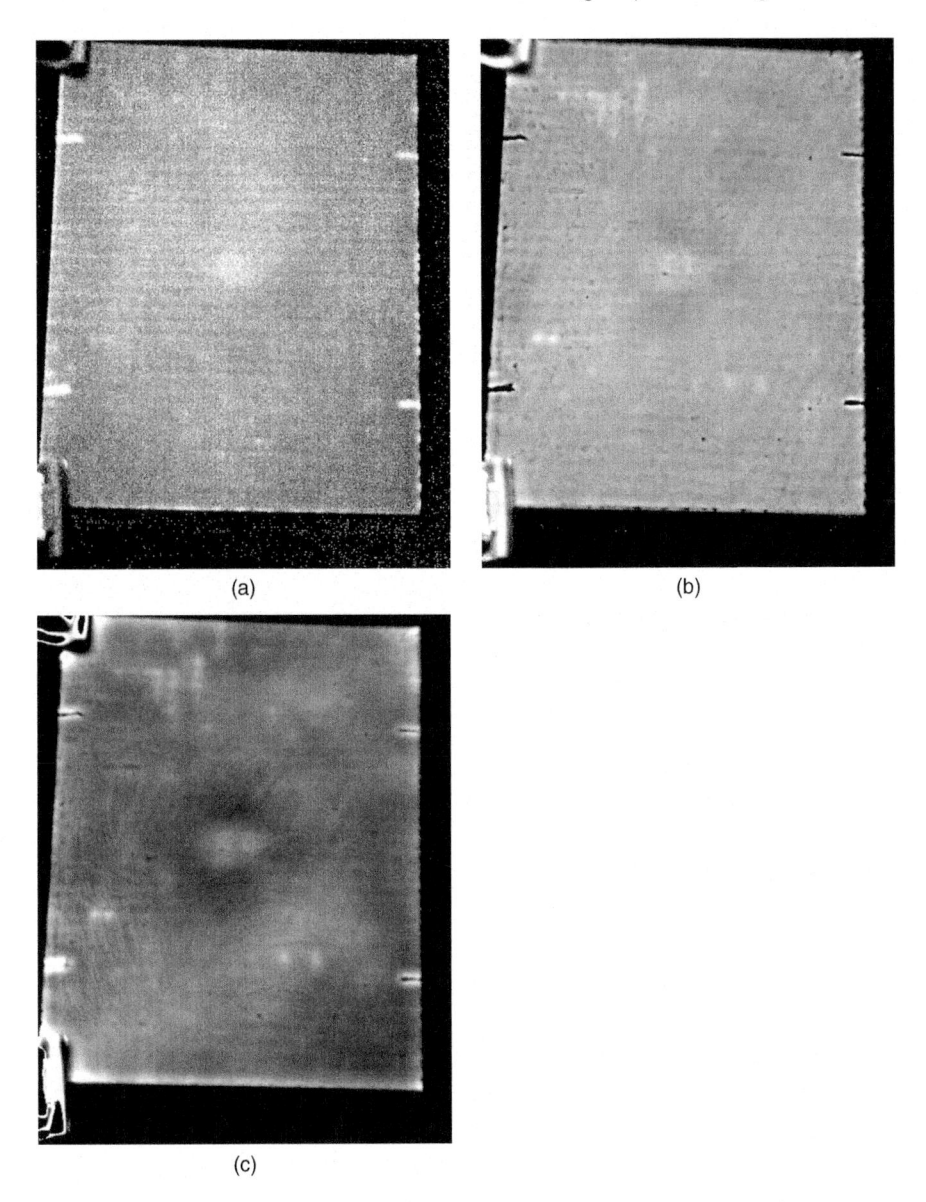

14.18 Phase images of the GFRP-F specimen taken from the front side: (a) $f = 0.67$ Hz; (b) $f = 0.14$ Hz; (c) $f = 0.045$ Hz.

(a) (b)

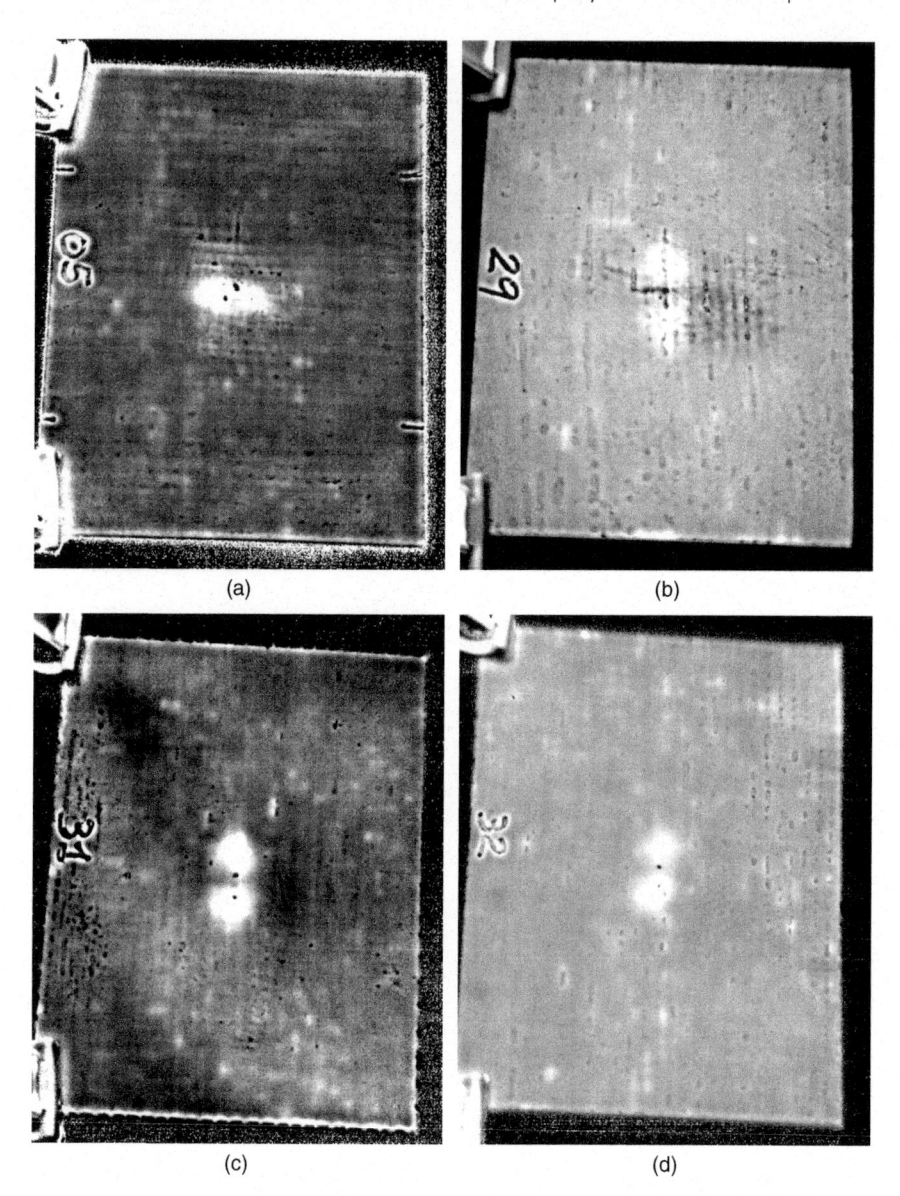

(c) (d)

14.19 Comparison of phase images taken at $f = 0.14$ Hz from the rear side of: (a) GFRP-F, $E_i = 12$ J; (b) GFRP-G, $E_i = 12$ J; (c) GFRP-H, $E_i = 9.7$ J; (d) GFRP-I, $E_i = 7.5$ J.

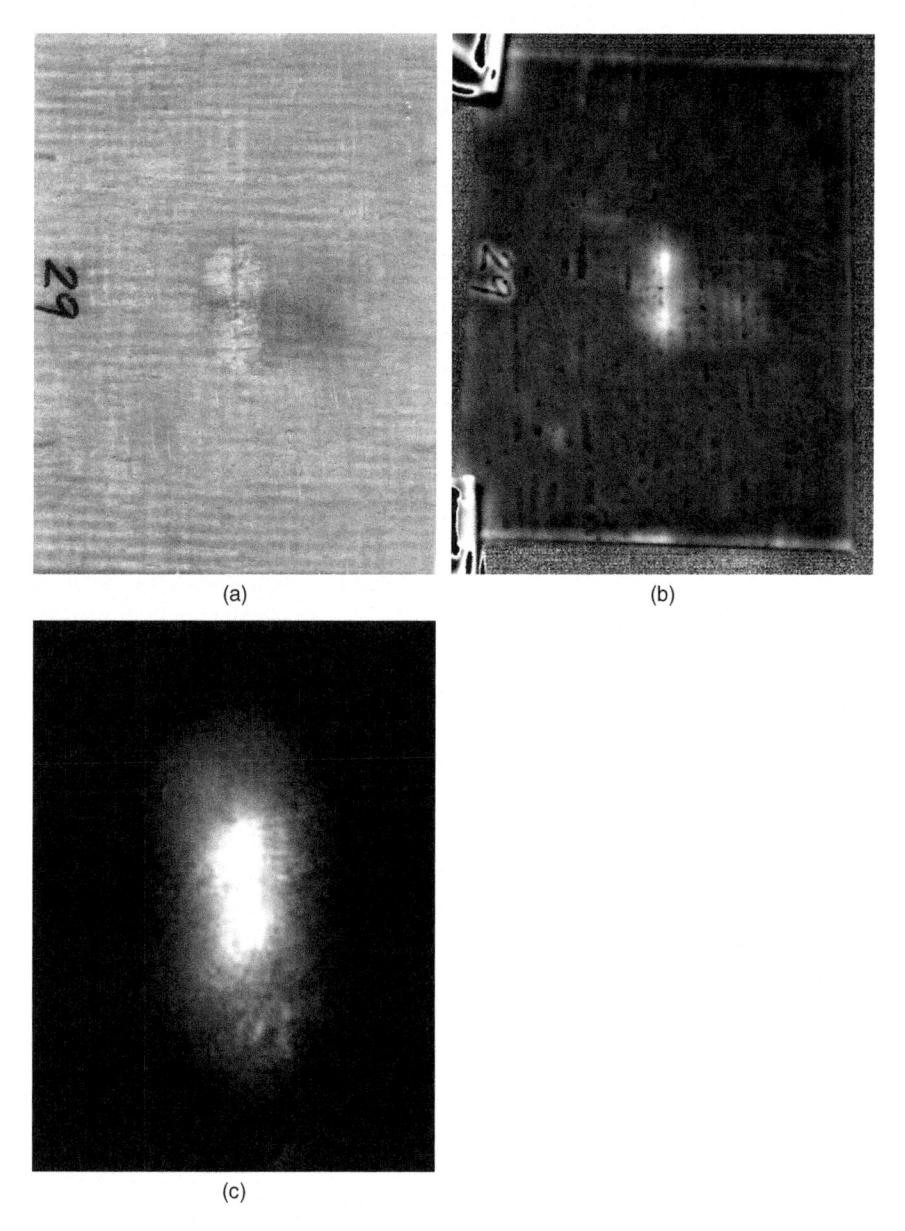

14.20 Comparison between images of the specimen GFRP-G: (a) visible picture taken after impact; (b) phase image taken at $f = 0.045$ Hz; (c) thermal image taken 4.15 s after impact.

For greater detail a photo of impacted specimen GFRP-G is compared with a phase image and a thermal image in Fig. 14.20. Both Fig. 14.20a and b include the entire specimen surface, while in Fig. 14.20c only the portion which extends horizontally from the marks, visible on the specimen top and bottom edges in Fig. 14.20a is reported. At first sight, good agreement is found between the damage stain present on the three images. Of course, the damage stain is visible to the naked eye because of the translucent characteristic of the tested GFRP material. On a thorough analysis, it is possible to note that in Fig. 14.20c (the thermal image) the damage stain is surrounded by a lighter halo. This may be due to either slighter delamination, or lateral heat conduction through sound material, or both, This effect needs further consideration by ad hoc tests.

14.4.3 Sandwich structures

A phase image of a CFRP/Nomex® sandwich, taken at $f = 0.5$ Hz, is shown in Fig. 14.21. In this case, the shape of the impact damage appears as a bowl, closer to the shape in metals rather than that occurring in composites. However, the CFRP skin is thin and the impact damage starts in the Nomex® core beneath. It is the core, then, that drives the damage shape.

14.5 Conclusion and future trends

It has been shown that a remote infrared imaging device may be helpful in the investigation of impact damage in composites. A significant advantage of using such a device for on-line monitoring of the material surface during impact, is being able to follow the material thermal behaviour, and gain information towards improved understanding of the damaging mechanisms. In fact, one can visualize and record the reaction of the material surface to impact. From a correct interpretation of the temperature signatures it is then possible to identify the damage origin, and to follow its successive growth and propagation. Importantly, it is possible to have, by a remote imaging device and so without altering the surface under inspection, a clear picture of what it is happening, during an impact event, both over the surface and within the material. This is of utmost importance in optimizing the material design as well as for preventative measures and maintenance of parts in service for the safe life of the aircraft.

As a general comment, for all the tested materials, the light temperature signatures bear witness to the formation of micro-cracks at interplay regions. At this stage the associated delamination is too weak to be detected by non-destructive techniques. As the impact energy overcomes 1.2 J for CFRP and 7 J for GFRP, the associated damage becomes more important. This is supported by the abrupt temperature rise visualized during online monitoring;

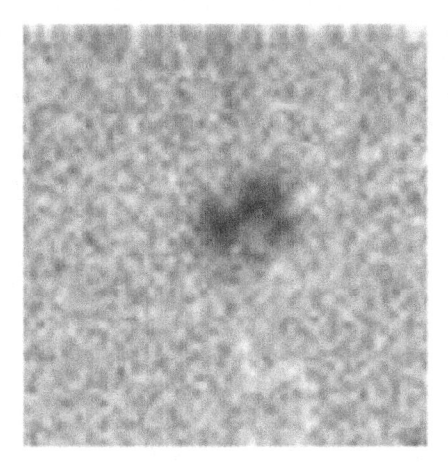

14.21 Phase image of a sandwich panel taken at 0.5 Hz.

some damage also becomes visible through non-destructive testing. However, the results shown above, and in the work by Meola *et al.* (2010a, 2010b), are mainly for a translucent GFRP material, which was chosen since it is suitable for validation of results through visible observation.

To the authors' knowledge, notwithstanding the huge amount of literature dealing with the impact topic, apart from the work by Meola *et al.*, there is no document specifically concerned with heat generation at damage onset. In the future, we will continue the investigation on the impact load with a more consistent variation of material types and test conditions for a better characterization of different types of composites.

14.6 Acknowledgements

Impact tests were performed with the support of Prof. G. Caprino Dr V. Lopresto, Prof. F. Ricci, Dr S. Boccardi, Dr F. De Falco and Dr V. Grasso.

14.7 References

Abrate S (1991), 'Impact on laminated composite materials', *Applied Mechanics Review*, **44**, 155–190.

Aslan Z, Karakuzu R, Okutan B (2003), 'The response of laminated composite plates under low-velocity impact loading', *Composite Structures*, **59**, 119–127.

Baucom JN, Zikry MA (2005), 'Low-velocity impact damage progression in woven E-glass composite systems', *Composites Part A*, **36**, 658–664.

Biot MA (1956), 'Thermoelasticity and irreversible thermodynamics', *Applied Physics*, **27**, 240–253.

Blitz J, Simpson G (1996), *Ultrasonic Methods of Nondestructive Testing*, Chapman & Hall, London.

Boeing 787 Dreamliner. Program Fact Sheet. http://www.boeing.com/commercial/787family/program facts.html (accessed on: March 22, 2011).

Caneva C, De Rosa IM, Sarasini F (2008), 'Monitoring of impacted aramid-reinforced composites by embedded PVDF acoustic emission sensors', *Strain*, **44**, 308–316.

Cantwell WJ (2007), 'Geometrical effects in the low velocity impact response of GFRP', *Composites Science and Technology*, **67**, 1900–1908.

Carlomagno GM, Meola C, Ricci F (2011), 'Infrared thermography and piezoelectric patches for impact damage detection in composite structures', in Proc. of the IWSHM 2011 – International Workshop on Structural Health Monitoring 2011, Stanford University, Stanford, CA, September 13–15, 2011.

Choi HY, Chang FK (1992), 'A model for predicting damage in graphite/epoxy laminated composites resulting from low-velocity point impact', *Journal of Composite Materials*, **26**, 2134–2169.

Choi HY, Wu HYT, Chang FK (1991), 'A new approach toward understanding damage mechanisms and mechanics of laminated composites due to low-velocity impact: part II – analysis', *Journal of Composite Materials*, **25**, 1012–1038.

Elder DJ, Thomson RS, Nguyen MQ, Scott ML (2004), 'Review of delamination predictive methods for low speed impact of composite laminates', *Composite Structures*, **66**, 677–683.

Gros XE, Takahashi K (1999), 'On the efficiency of current NDT methods for impact damage detection and quantification in thermoplastic toughened CFRP materials' *NDT.net*, **4** (3).

Hull D, Clyne TW (1996), *An introduction to Composite Materials*, Cambridge University Press, Cambridge.

Iannucci I, Willows ML (2006), An energy based damage mechanics approach to modeling impact onto woven composite materials – part I: numerical Models, *Composite Part A*, **37**, 2041–2056.

Jones RM (1975), *Mechanics of Composite Materials*, Hemisphere Publishing Corporation, New York.

Kachanov LM (1987), *Introduction to Continuum Damage Mechanics*, Martinus Nijhoff Publishers, Boston, MA.

Kang TJ, Kim C (2000), 'Impact energy absorption mechanism of largely deformable composites with different reinforcing structures', *Fibers and Polymers*, **1**, 45–54.

Kim SJ, Goo NS (1997), Dynamic contact responses of laminated composite plates according to the impactor's shapes; *Computer & Structures*, **65**, 83–90.

Krautkramer J, Krautkramer H (1990), *Ultrasonic Testing of Materials*, 4th Edition, Springer Verlag, Berlin.

Lemaitre J, Chaboche JL (1990), *Mechanics of Solids Materials*, Cambridge University Press, Cambridge.

Meola C, Carlomagno GM (2004), 'Recent advances in the use of infrared thermography', *Measurement Science and Technology*, **15**, 27–58.

Meola C, Carlomagno GM (2006), 'Application of infrared thermography to adhesion science' *Journal of Adhesion Science and Technology*, **20**, 589–632.

Meola C, Carlomagno GM (2009), 'Infrared thermography to impact-driven thermal effects', *Applied Physics A*, **96**, 759–762.

Meola C, Carlomagno GM (2010a), 'Infrared thermography in health monitoring of CFRP for aeronautical applications', *Aerotecnica The Journal of Aerospace Science Technology and Systems*, **89**, 179–186.

Meola C, Carlomagno GM (2010b), 'Impact damage in GFRP: new insights with Infrared Thermography', *Composites Part A*, **41**, 1839–1847.

Meola C, Carlomagno GM, Squillace A, Giorleo G (2002), 'Non-destructive control of industrial materials by means of lock-in thermography', *Measurement Science and Technology*, **13**, 1583–1590.

Meola C, Carlomagno GM, Di Foggia M, Natale O (2008), 'Infrared thermography to detect residual ceramic in gas turbine blades', *Applied Physics A*, **91**, 685–691.

Meola C, Carlomagno GM, Lopresto V, Caprino G (2009a), 'Impact damage evaluation in composites with infrared thermography', 3rd European Conference for Aerospace Science EUCASS, Versailles (France) 6–9 July, 2009.

Meola C, Carlomagno GM, Ricci F, Lopresto V, Caprino G (2009b), Analisi del comportamento all'impatto di compositi con termografia all'infrarosso, Atti del 13° Congresso AIPnD Conferenza Nazionale sulle Prove non Distruttive Monitoraggio Diagnostica, Rome, 15–17 October 2009, CD-ROM, www.ndt.net/search/docs.php3?MainSource=-1&id=8199

Meola C, Carlomagno GM, Ricci F, Lopresto V, Caprino G (2011), 'Investigation of Impact Damage in Composites with Infrared Thermography', in Proceedings of the 6th NDT in Progress, Prague (Czech Republic) 10–12 October 2011, P. Mazal, Editor, pp 175–182.

Minak G, Ghelli D (2008), 'Influence of diameter and boundary conditions on low velocity impact response of CFRP circular laminated plates', *Composites Part B*, **39**, 962–972.

Mitrevski T, Marshall IH, Thomson R, Jones R, Whittingham B (2005), 'The effect of impactor shape on the impact response of composite laminates', *Composite Structures*, **67**, 139–148.

Naik NK, Shrirao P, Reddy BCK (2006), 'Ballistic impact behavior of woven fabric composites: Formulation', *International Journal of Impact Engineering*, **32**, 1521–1552.

Park R, Jang J (2001), 'Impact behaviour of aramid/glass fiber hybrid composite: evaluation of four-layer hybrid composites', *Journal of Materials Science*, **36**, 2359–2367.

Piening M, Pabsch A, Sigle C (2002), Structural element of high unidirectional rigidity, US 6355337.

Pitkänen J, Hakkarainen T, Jeskanen H, Kuusinen P, Lahdenperä K, Särkiniemi P, Kemppainen M, Pihkakoski M (2000), 15th WCNDT 2000 *NDT.net*.

Richardson MOW, Wisheart MJ (1996), 'Review of low-velocity impact properties of composite materials'. *Composites Part A*, **27**, 1123–1131.

Schoeppner GA, Abrate S (2000), 'Delamination threshold loads for low velocity impact on composite laminates', *Composites Part A*, **31**, 903–915.

Shütze R (1996), Aircraft wings, US5496002 (1996).

Shyr TW, Pan YH (2003), 'Impact resistance and damage characteristics of composite laminates', *Composite Structure*, **62**, 193–203.

Soutis C (2005), 'Fiber reinforced composite in aircraft construction', *Progress in Aerospace Science*, **41**, 143–151.

Thomson W (1978), 'On the thermoelastic, thermomagnetic and pyroelectric properties of matters', *Philosophical Magazine*, **5**, 4–27.

Wu HYT, Chang FK (1989), 'Transient dynamic analysis of laminated composite plate subjected to transverse impact'. *Computers and Structure*, **31**, 453–466.

Zhang J, Fox BL, Gao D, Stevenson AW (2009), 'Inspection of drop-weight impact damage in woven CFRP laminates fabricated by different processes', *Journal of Composite Materials*, **43**, 1939–1946.

15
Non-destructive evaluation (NDE) of aerospace composites: ultrasonic techniques

D. K. HSU, Iowa State University, USA

DOI: 10.1533/9780857093554.3.397

Abstract: This chapter discusses the defects in aerospace composite structures and the ultrasonic techniques used for their non-destructive detection and characterization. The defects discussed include manufacturing defects such as porosity, errors in ply layup or stacking sequence, ply waviness, and service-induced delamination, disbond, crushed core, and microcracks. Ultrasonic inspection of both solid laminates and honeycomb sandwiches is discussed. The techniques addressed include novel processing of C-scan images in water-coupled ultrasonic testing, application of air-coupled ultrasound, and electromagnetically generated shear waves to exploit their strong interaction with fiber direction. In terms of instrumentation, this chapter discusses laboratory systems, stationary systems used in manufacturing environment, and portable inspection systems for field applications.

Key words: ultrasonic non-destructive evaluation (NDE), aerospace structures, defects in composites, solid laminate, honeycomb sandwich.

15.1 Introduction

Composites have been used on aircraft structures in civil aviation for several decades; their use on military aircraft has an even longer history. These composite structures have delivered their proclaimed advantages of light weight and non-corrosion. They have been largely safe and reliable structures and served the aircraft industry well. As far as non-destructive evaluation (NDE) and non-destructive inspection (NDI) are concerned, composite structures are typically inspected intensively before they are assembled onto the aircraft. Once in service, they are usually only inspected by eye. For honeycomb sandwich structures, the tap test is the most commonly used technique. In terms of the three main non-destructive testing (NDT) modalities, namely, ultrasound, X-ray, and electromagnetic methods, ultrasound has been by far the main staple in the inspection of composites. X-ray is also used but most electromagnetic methods do not apply to composites due to their non-conducting nature.

In the NDE of composite aircraft components in a manufacturing environment for quality assurance purposes, the use of ultrasound has matured to a stage where large composite structures are routinely inspected using large ultrasonic scanning equipment and facility. The anisotropic and inhomogeneous nature of composites makes the ultrasonic NDE of composites an interesting and challenging research topic. Over the years there has been ample literature on this topic; examples include the work of Summerscales (1987, 1990), Armstrong and Barrett (1998), and Smith (2009). The fundamentals of ultrasonic interaction with composites are treated in a recent book by Rokhlin *et al.* (2011).

15.2 Inspection of aerospace composites

In order for NDE to serve as an essential component in the life-cycle management of aerospace composite structures, it must meet the needs for manufacturing process control and quality assurance, as well as in-service damage inspection and characterization. Ultrasonic methods have been developed for cure monitoring of polymer resin in the manufacturing of composite laminates and for monitoring the resin flow in resin transfer molding of dry fiber preforms of composite structures. The quality issues of aerospace composites include the quantification and characterization of porosity in solid laminates, the verification of ply orientation and stacking sequence, and the detection and characterization of delaminations, foreign object inclusions, and fiber waviness or 'marcel'. The primary need for NDT for in-service inspection is the ability to detect, quantify, and characterize impact damage, including barely visible impact damage, of solid laminates and honeycomb sandwich structures. NDT techniques are also needed to detect disbond caused by exposure to localized heating, ingression of hydraulic fluid, and encounter with anomalous overload. For space structures, NDT methods are also needed for detecting and characterizing matrix microcracking caused by repeated excursions to cryogenic temperatures. Examples of some of these defects and damage in aerospace composites will be described in this section.

Porosity in solid laminates has been a persistent quality issue in aerospace composites. The term porosity here refers to distributed small voids in the resin matrix of polymer-based composites. They are typically the result of non-ideal curing process, such as inadequate evacuation of the volatile gas released during the cure cycle that led to the entrapment of microscopic or macroscopic gas bubbles in the resin matrix. Figure 15.1 shows a typical micrograph of a cut and polished cross-section of a porous carbon fiber composite. The porosity within a ply is typically a distribution of small voids, of the order of 10–30 µm, scattered in the matrix resin and between the reinforcing fibers. In the resin-rich layer between the plies, or in the resin pockets at the intersections of fiber tows in a woven composite laminate, the porosity tends to take the form of larger and longer voids, of the order of 100 µm or greater. There is a tendency for the void to elongate along the

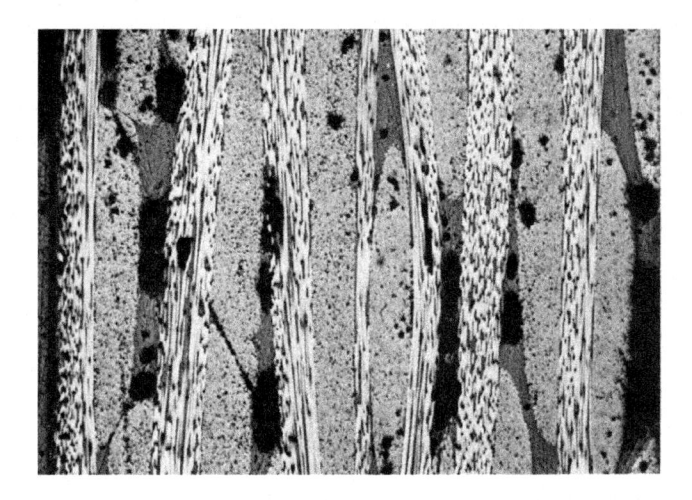

15.1 The cross-sectional view of a woven carbon composite containing porosity. The photo shows an area of approximately 1.5 × 2.1 mm², with wavy plies containing fiber tows going in different directions. Small porosities in the plies are small dark dots and large porosities between the plies and at the weave intersections are dark blobs.

fiber directions at the ply interfaces. Ultrasound is the primary NDT tool for the detection and characterization of porosity in aerospace composites. Specific methods are described in Section 15.4.1.

The verification of the ply layup or stacking sequence in a solid laminate is very important. Errors in layup or stacking sequence can lead to the rejection of a part. Destructive validation of the laminate layup can be done by cutting and polishing a cross-sectional surface and examining the fiber orientations with optical microscopy. Non-destructively, the reflected ultrasound signal from a certain depth in the interior of the laminate is sensitive to the orientation of the fibers in the plies. The detection of layup errors and verification of stacking sequence can therefore be done non-destructively using ultrasound. Several innovative ultrasonic methods for mapping out the stacking sequence and for detecting layup errors of a solid laminate are described in Section 15.4.2.

To reduce the large amount of labor required in hand layup and to avoid human errors, composite laminates are increasingly manufactured by tape-laying machine and fiber-placement machines. However, defects and imperfections such as gaps and overlaps of fiber strips may still exist in composites made by machines. Minute, isolated imperfections are not easily detected by NDT techniques but they do not pose as a threat unless there is a distribution of such defects. The size, distribution, and severity of defects that must be detected in a composite manufactured with tape-laying or fiber-placement machines depend on the design and application.

In a fiber-reinforced composite laminate, the waviness or 'marcel' of the fiber tows, either in the vertical direction (out-of-plane) or in the horizontal direction (in-plane) can degrade the stiffness and strength of the designed structure. The wrinkles of the plies can be the result of thermal expansion due to temperature profile anomalies during curing. Being an elastic wave, the propagation of the ultrasound can be strongly affected by the geometric distortion associated with the waviness of the plies. Several ultrasonic methods, including one-sided, pitch-catch contact mode ultrasound and air-coupled ultrasound, have been used to detect the waviness in fiber plies.

The three types of defect described above are examples related to the manufacturing of the composites. Once the composite structures are deployed in the field, they may suffer damage through load, handling, and environment. The primary concern with in-service NDT of composite structures is low-velocity impact damage of solid laminates and honeycomb sandwiches due to hail damage, tool drop during maintenance, bird strike, and impacts of runway debris. Impact damage of a solid laminate typically occurs in the form of a series of delaminations that increases in size as the depth increases from the impact surface. At each ply interface, the delaminations often consist of two lobes aligned with the fiber direction in one of the plies. The shape of the internal delaminations has been photographed and documented using a resin burn-off and de-ply procedure after the impacted region of the laminate was first infused with a gold chloride/ether solution (Blodgett et al., 1986). Unlike metals, an impact damage of a solid laminate of composite can exhibit little or no visual indication. Such 'barely visual impact damage' (BVID) could pose a risk for further damage growth and potential failure as the component remains in service. The detection and characterization of impact damage are therefore of great interest in the NDT of composite structures. The ultrasonic reconstruction of impact-induced delamination damage zone has received considerable attention from NDE researchers over the years and has led to much success.

Adhesively bonded composite structures such as honeycomb sandwiches are also susceptible to impact damage. Low-velocity impacts of honeycomb sandwiches with composite laminate facesheets can lead to crushed or buckled core, disbond between the facesheet and the core, and delaminations in the facesheet. The morphology of the damaged core usually assumes the shape of an upwardly curving arc, as shown schematically in Fig. 15.2. A photograph of the cell wall deformation, taken after the damage zone was sectioned, is also included in Fig. 15.2 for comparison. It is not unusual for the honeycomb core fracture or buckle line to occur at a depth that is a substantial fraction of the core thickness. It is also not unusual to have the damaged core remain bonded to the facesheet after the impact. Brittle

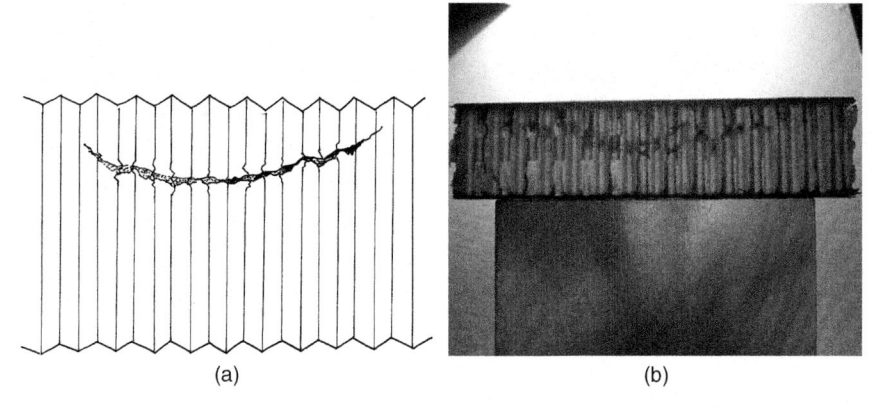

15.2 Damage morphology of impacted honeycomb sandwich: (a) the upwardly curving line of fracture or buckle of the honeycomb cell walls; (b) an example of buckled cell wall in a CFRP/Nomex sandwich.

honeycomb core materials, such as fiberglass cells or Nomex cells, tend to fail by fracture and crushing, whereas ductile honeycomb cells such as aluminum cells fail by an accordion-style buckle collapse. In a low-velocity impact on an aluminum honeycomb sandwich with a carbon or glass composite facesheet, the permanent deformation of the aluminum honeycomb cells can easily result in a separation or disbond between the deformed core and the facesheet. In the NDI of impact damage on honeycomb sandwiches, contact mode ultrasonic testing is generally ineffective, but a water-coupled or air-coupled transmission mode ultrasonic scan can readily map out the damage zone.

The adhesive bond between the facesheet and the honeycomb core may also fail if the surface of the sandwich structure is exposed to sudden heating of a localized area, such as that of hot exhaust gas or a fire. The sudden heating of a localized area can produce a large stress in the facesheet. Since the facesheet is rigidly supported by the cell walls underneath, the stress is released by buckling upward and creating a disbond. Figure 15.3 shows the morphology of a facesheet to honeycomb core disbond of an aluminum slat wedge on the leading edge of an aircraft wing. The disbond was induced by applying heat to a local area of the facesheet. The micrographs of the inside of the facesheet and the top view of the honeycomb cell wall shown in Fig. 15.3 indicated a mixture of adhesive failure and cohesive failure. Regions with no adhesive on the disbonded honeycomb cell wall corresponded to adhesive failure, whereas regions with adhesive on both the facesheet and the honeycomb corresponded to cohesive failure. In addition to transmission ultrasonic scans, disbond in a honeycomb

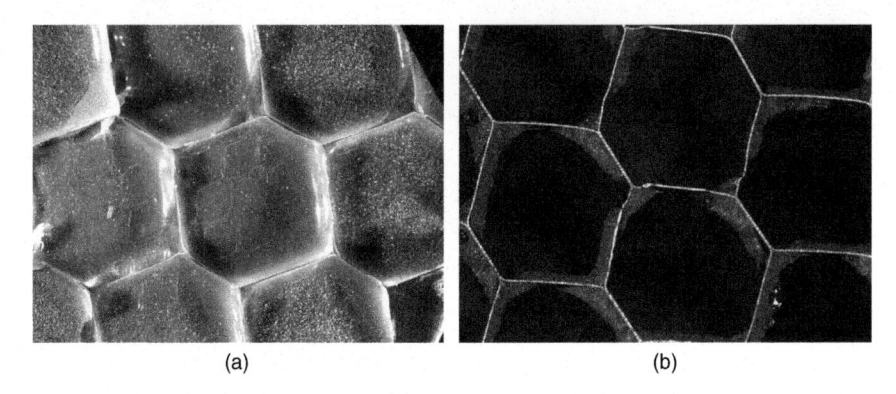

15.3 Disbond failure between facesheet and honeycomb core showing a mixture of cohesive and adhesive failure: (a) the facesheet; (b) the top of the honeycomb core. The disbond was induced by applying heat to a local area on the facesheet.

sandwich can also be imaged with a number of other NDT methods, including thermography, shearography, and tap test imaging.

15.3 Ultrasonic inspection methods for aerospace composites

15.3.1 Normal incident longitudinal waves

Normal incident longitudinal waves are the staple of ultrasonic inspection of composites. In this configuration the wave propagation is independent of the fiber directions in the plies of a composite laminate. The interaction of the longitudinal waves with the characteristic ply thickness of a laminate (typically 120 µm) is generally weak since the wavelength at the commonly employed testing frequency in the single megahertz range is much greater than the ply thickness. However, in special cases where higher frequencies are used, 30 MHz and above, distinct echoes from the ply interfaces can appear (Ourak *et al.*, 1998). Furthermore, ultrasonic testing is typically carried out with broadband pulses, so that any resonance phenomena such as mechanical resonance of the ply thickness or stop band and pass band are smeared out due to the broad frequency bandwidth.

With pulsed longitudinal waves, the ultrasonic testing of composites falls into two categories: material property evaluation and flaw detection. In material property evaluation, ultrasonic velocity measurements are used to determine the elastic moduli of the anisotropic composite material (Zimmer and Cost, 1970; Papadakis *et al.*, 1991; Rokhlin *et al.*, 2011). Both ultrasonic attenuation and velocity measurements are useful techniques for the NDE of porosity in composites. The presence of porosity increases the attenuation and decreases the velocity of ultrasound. The changes in attenuation and velocity depend on frequency, the volume fraction of porosity, and the detailed morphology of the voids, such as their size, shape, and orientation in the composite laminate.

Ultrasonic detection of defects in composites such as foreign object inclusions, delaminations, and disbond of adhesively bonded interfaces is perhaps the primary NDE method for composite inspection and continues to expand in its scope of applications. The presence of a defect is indicated by the appearance of an echo in a pulse-echo test and by the disappearance or reduction of the transmitted pulse in a through-stransmission ultrasonic (TTU) test.

15.3.2 Interaction of shear waves and fiber directions

Owing to the high degree of anisotropy of fiber-reinforced composites, the fiber direction in the composite plies interacts strongly with the in-plane vibration of a normal incident shear wave. When the shear vibration of the incident wave is at an angle with the fiber axis, the shear wave is decomposed into two components parallel and perpendicular to the fiber direction. The parallel component propagates in the thickness direction at a greater velocity than the perpendicular component. The time of flight of a shear wave pulse through an anisotropic composite laminate depends on the ply layup of the laminate. When using shear waves in the transmission mode, the 'cross-polarized' configuration, where the polarization of the transmitting transducer is maintained perpendicular to that of the receiving transducer, is highly sensitive to any layup or stacking sequence errors (see also Section 15.4.2).

Longitudinal waves are easily coupled from the transducer to the specimen under testing by the application of a liquid or gel couplant. The coupling of shear waves is more difficult and requires a highly viscous layer of couplant. In practice, honey serves as a convenient and washable shear wave couplant. In metals shear waves can be readily generated using electromagnetic acoustic transducers (EMAT) (Thompson, 1990). In non-conducting composites such as polymer matrix composites reinforced by glass or Kevlar fibers and in weakly conducting carbon fiber-reinforced plastic (CFRP) composites, ultrasonic waves cannot be generated by the direct application of EMAT. However, ultrasonic waves can be first generated using EMAT in an aluminum adhesive tape attached to a composite, the waves generated in the aluminum layer can then propagate into the composite. The aluminum tape is then peeled off after testing. For stacks of composite prepregs laid up for curing, the entire stack may be sandwiched between metal blocks and shear waves may be generated in the metal blocks with EMAT. The waves can then propagate through the prepreg stack by exploiting the tack of the un-cured prepregs. This method has been employed in the detection of errors in ply layup and stacking sequence (Fei and Hsu, 2002).

15.3.3 Effects of anisotropy and acousto-ultrasonics

Even though the effects of in-plane anisotropy of a fiber-reinforced composite laminate do not strongly influence the normal incident longitudinal ultrasonic signals in pulse-echo or through-transmission tests, the elastic anisotropy can certainly affect the beam profile and spreading, which can in turn affect the size and shape of the flaw images in ultrasonic scanning of defects (Minachi *et al.*, 1993). Due to the spreading of the beam, the waves generated by a longitudinal transducer placed at one location on the composite laminate can be detected by another transducer placed at some distance away. For thin plates and large transducer separation, the wave phenomenon is dominated by guided waves or Lamb waves, but in thick laminates and relatively short transducer separation distances, the received signals can be understood based on the various wave models allowed by the elastic symmetry of the laminate and their associated slowness surfaces. As an example of the oblique wave propagation in a thick (25 mm) unidirectional carbon composite laminate, two longitudinal wave contact mode transducers were placed on the surface, one as a transmitter and the other as a receiver, along the fiber direction. The separation distances were varied from 10 to 110 mm and the received pulses were analyzed to identify their mode based on the time of flight. The modes observed were quasi-longitudinal wave (QL), quasi-shear vertical wave (QSV), and a pure shear horizontal wave (SH). The echoes reflected from the bottom surface were QL → QL, QL → QSV, QSV → QSV, and SH → SH. As the separation distance between the transducers was increased, the time of flight of the echoes each increased in its own way. Using the slowness surface of the QL, QSV, and SH modes, the changes of the time of flight as a function of the transducer separation were calculated. The measured time of flight of the various echoes agreed well with the calculated values (Hsu and Margetan, 1992).

The configuration of coupling two transducers, separated by a distance that is large compared with the transducer size, to the same surface of a structure to investigate the condition of the material between the transducers is traditionally referred to as the acousto-ultrasonic (AU) technique (Duke, 1988; Vary, 1990). The AU technique was widely applied to the NDE of composite components and the results were correlated to the material conditions, presence of defects, and damage. The AU signals are typically too complex to be analyzed in terms of individual modes involved, like that described above, and are in fact processed using approaches similar to those in acoustic emission analysis. The AU technique, as a phenomenological approach, is useful for evaluating the material conditions between the transmitting and the receiving transducers.

15.3.4 Water-coupled ultrasonic testing

In manual ultrasonic inspection of structures using portable flaw detectors in a spot check fashion, small amounts of oil or water serve to couple the ultrasonic waves from the transducer into the structure. In scanning mode, water is the primary medium to provide low attenuation, consistent coupling from the transducer to the component. The two main scanning modes of ultrasonic inspection are the immersion mode and the squirter mode. Immersion tanks equipped with scanning frame or gantry are widely used in shops and laboratories. Squirter systems are used for scanning large aerospace composite structures. Squirters are typically used in the through-transmission mode of inspection. Specially designed squirters provide a jet of water column that carries the ultrasonic beam generated by a transducer behind the column. The two water jets are then aimed at each other and precisely aligned; the structure to be inspected is then placed in the path of the beam. A scan gantry moves the pair of aligned squirters in unison over the surface of the part.

Water-coupled ultrasonic inspection is also carried out in a number of variations. In a bubbler system, water is continuously fed to the space between a transducer and the part surface, using a water recirculation system. Specially designed transducer holders with foam or brush are used to contain the coupling water during the inspection. A combination of vacuum and pump have also been used to make the water coupling system a 'dripless bubbler' (Patton and Hsu, 1994). For localized inspection, leakers or weepers using permeable membrane can also be attached to the transducer to provide coupling water at a slow feed rate.

15.3.5 Air-coupled ultrasonic testing

The development of air-coupled ultrasonic NDE began in earnest in the early 1980s. Over the last 30 years the technology has matured considerably and is now applied by aircraft manufacturers in production inspection. Air-coupled ultrasound eliminates the need for a liquid couplant and the associated water handling system. It therefore enjoys a great advantage in terms of portability and convenience, and can be applied to such structures as honeycomb sandwich with perforated facesheets. However, the greatest challenge in air-coupled ultrasonics is the enormous loss of ultrasonic energy at the air–solid interface where most of the energy of the wave is reflected and only a tiny fraction of the energy is transmitted. Therefore, the development of air-coupled ultrasound has been basically a development of the transducer technology. For the piezoceramic type of air-coupled transducer, special material layers are used to bridge the high acoustic impedance of the transducer element and the low impedance of air (Buckley, 2000; Bhardwaj, 2001). Commercially available piezoceramic air-coupled transducers and ultrasonic systems today have sufficient penetration power and sensitivity for NDE of aerospace composites. An alternate approach to the fabrication of air-coupled ultrasonic transducers is the micromachined capacitance transducer, which can have a high dynamic range and broad bandwidth (Schindel and Hutchins, 1995).

Owing to the high attenuation of ultrasound in air (approximately 1.6 dB/cm at 1 MHz), air-coupled ultrasonic tests for aerospace composites are typically conducted in the frequency range of 100–400 MHz. With piezoceramic transducers driven by a long tone-burst, the preferred mode of operation is through-transmission. Despite the great impedance mismatch between air and composites, air-coupled ultrasound is able to penetrate a wide variety of composite laminates and bonded structures. For solid laminates, air-coupled ultrasound at 100 kHz can penetrate as much as 50 mm of CFRP using a portable ultrasonic flaw detector. Owing the low frequency of air-coupled ultrasound, the wavelength of sound and the laminate thickness can be of the same order of magnitude so that resonance effects are often prominent even using tone-burst signals. The transmission of air-coupled ultrasound as a function of the sample thickness may contain a number of resonance peaks. Since the piezoceramic transducers are relatively narrowband, the frequency used should be chosen judiciously for efficient transmission (Hsu, 2006). In thin laminates of composites (a few millimeters thick), air-coupled ultrasound at 400 kHz can readily detect and image imbedded flaws of 5 mm diameter. Because of resonance, the transmission of air-coupled ultrasound through a thin plate is actually most efficient when the plate is tilted at an angle with respect to the sound beam (Bar *et al.*, 2010). The guided wave phenomenon of air-coupled ultrasound in composite laminates has been a topic of many extensive studies (Kazys *et al.*, 2006).

Air-coupled ultrasound has been proven highly effective for NDE of honeycomb sandwich structures. In the TTU configuration, the honeycomb cell walls serve as conduits for guided wave transmission. An air-coupled ultrasonic scan has one of the highest detection sensitivity for flaws in honeycomb sandwiches compared with other NDE methods.

15.4 Ultrasonic inspection of solid laminates

15.4.1 Ultrasonic NDE of porosity

The most commonly used NDE method for detecting and evaluating porosity in solid composite laminates is based on the measurement of the extra attenuation attributable to the presence of porosity (Stone and Clarke, 1975). A common practice is to set an accept/rejection criterion based on the extra attenuation of a porous region over and above the baseline attenuation of a porosity-free region. It is important to recognize, however, that the value of the 'rejectable attenuation' depends on many factors, most importantly the effects of frequency. Ideally one would like to have a simple linear relationship between the porosity volume fraction and the extra attenuation it introduces to the ultrasonic waves produced by a transducer of a given nominal frequency. In reality, however, ultrasonic attenuation is a frequency-dependent quantity that is affected by the size, shape, orientation, and distribution of the voids in the laminate. Since the frequency spectral profiles of transducers with the same nominal frequency can vary considerably depending on the pulser/receiver circuitry, the use of a simple fixed 'dB drop' value for rejecting components for excessive porosity is too much of an over-simplification. Investigations of porosity and its effects on ultrasonic attenuation in composites have generally showed that, for a given porosity volume percent, the ultrasonic attenuation versus frequency curve is usually approximately linear and that the frequency slope of the attenuation, i.e., $d\alpha/df$, showed a more robust correlation with porosity volume fraction. Figure 15.4 shows an example of frequency-dependent attenuation data of CFRP laminates with porosity volume fractions from essentially zero to about 8%.

The advantages of using the frequency slope of the attenuation is that the amplitude changes of the ultrasonic signal due to variations of

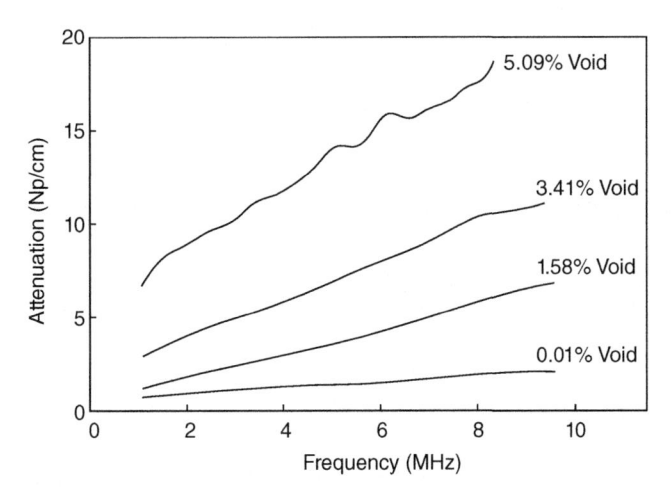

15.4 Ultrasonic attenuation as a function of frequency for CFRP laminates containing increasing amount of porosity.

transducer bandwidth and pulser/receiver circuitry are included in the analysis. This practice, however, requires the extra step of performing an FFT (fast Fourier transform) spectral analysis of the ultrasonic data. Detailed investigations of ultrasonic attenuation and porosity in several composite systems, including laminates of different layup and woven laminates, showed that the correlation between attenuation slope $d\alpha/df$ and porosity volume fraction was reasonably robust (Jeong and Hsu, 1995). Although the constant of proportionality between $d\alpha/df$ and porosity volume fraction may require determination or verification for material systems of substantially different construction (for example, woven versus unidirectional layup) and porosity of characteristically different morphology (spheroidal voids versus elongated ribbon-like voids), the method of using attenuation slope still provides a more stable correlation with the porosity volume fraction.

In a porous solid, both the attenuation and the velocity are expected to be frequency dependent and depend on the void content and morphology. Although used less frequently than attenuation, ultrasonic velocity can sometimes be a simpler and useful method in the evaluation of porosity in composites. Depending on the material characteristics and the morphology of the ply interfaces and porosity, the velocity has been found to be less sensitive to frequency than the attenuation. When ultrasonic velocity is used in the correlation with porosity content, it is important that the precise thickness of the laminate is known since the change in velocity for the porosity content of interest may be only a few percent. It is also advantageous to use the relative change of velocity, $\Delta v/v$, with respect to a porosity-free specimen in the evaluation of the porosity content. In cases where the frequency dependence of ultrasonic velocity is not strong, the FFT spectral analysis step may not be needed.

15.4.2 Detection and mapping of layup and stacking errors

The ply layup and stacking sequence of a composite laminate may be verified destructively on sacrificial pieces using a resin burn off and un-stacking procedure. The stacking sequence of a laminate may also be verified using optical microscopy on cut and polished edge surfaces. Non-destructive ultrasound may be used to detect ply layup and stacking errors and to map the layup and stacking sequence of laminates.

The ply interface in a composite laminate typically does not give rise to distinct reflection echoes to a normal incident ultrasonic pulse unless the frequency of the ultrasound is very high, such as 30 MHz (Ourak *et al.*, 1998). However, C-scan images of composite laminates using time ultrasonic signals time gated to corresponding ply interfaces often reveal the fiber directions in the adjacent plies. Time gated ultrasonic C-scan images were first used to map out the delaminations of an impact damage at different ply interfaces (Moran and Buynak, 1989). More recently the full waveform ultrasonic data were processed to extract the ply layup and stacking sequence of composite laminates (Smith and Clarke, 1994). As stated earlier the propagation of normal incident longitudinal waves are not sensitive to the fiber directions of the plies, the fact that fiber orientation of the plies can be extracted from the scan image based on time domain signal of a longitudinal pulse was made possible mainly by the imperfections present at the ply interfaces. These imperfections included voids, resin-rich, or resin-starved regions, and variations of the interfacial thickness. The imperfections tend to align themselves along the fiber direction; as a result, the C-scan image at a depth corresponding to a ply interface invariably contained features (streaks) that reflected the fiber direction. It was observed experimentally that the fiber direction of the upper ply of the interface often showed up more predominantly in the interface C-scan image. To facilitate the identification of the fiber orientation in the C-scan image, a spatial 2D FFT was performed on the scan image (Hsu et al., 2002). An angular distribution plot was then generated by summing over all the frequency components, i.e., summing over the distance from the origin of the 2D FFT image, at each angle. This amplitude versus angle plot will then reveal the fiber directions involved; the largest peak usually occurred at the angle of the fiber direction of the upper ply at the interface. Figure 15.5(a) shows a C-scan image of a 20° ply intentionally placed at the midplane of a $[(90_3/0_3)_2, 90_3, 20_3, 90_3, (0_3/90_3)_2]$ CFRP laminate as a test of layup error detection. The scan image showed streaks tilted at a 20° angle and some faint streaks at 0° and 90°. The angular distribution plot generated from the 2D spatial FFT of the scan image showed clearly peaks at 0°, 20°, and 90°, with the dominating peak at 20° (Fig. 15.5(b)). The full waveform block of ultrasonic data, containing A-scan traces at all (x, y) position of the scan, can be processed in this manner by moving a narrow time gate with a width of approximately one or two plies, through the block of the data. Figure 15.6 shows the determination of the ply orientation in a $[-45/45/0_4/-45/45]_s$ CFRP laminate.

Shear wave ultrasonics has not been fully exploited for NDE of composites, but it holds distinct advantages for detecting ply orientation and stacking sequence errors of composite laminates. Owing to the highly anisotropic nature of the fiber plies in a composite laminate, the ply orientation and stacking sequence interact strongly with the in-plane vibration of a normal incident shear ultrasonic wave propagating through the thickness. The transmitted signal in a 'cross-polarized' configuration, with the transmitting and receiving transducers oriented perpendicular to each other, was found to be particularly sensitive to any ply orientation anomaly and stacking sequence errors. To evaluate the fiber orientation in a solid laminate, the transmitting transducer and the receiving transducer are rotated

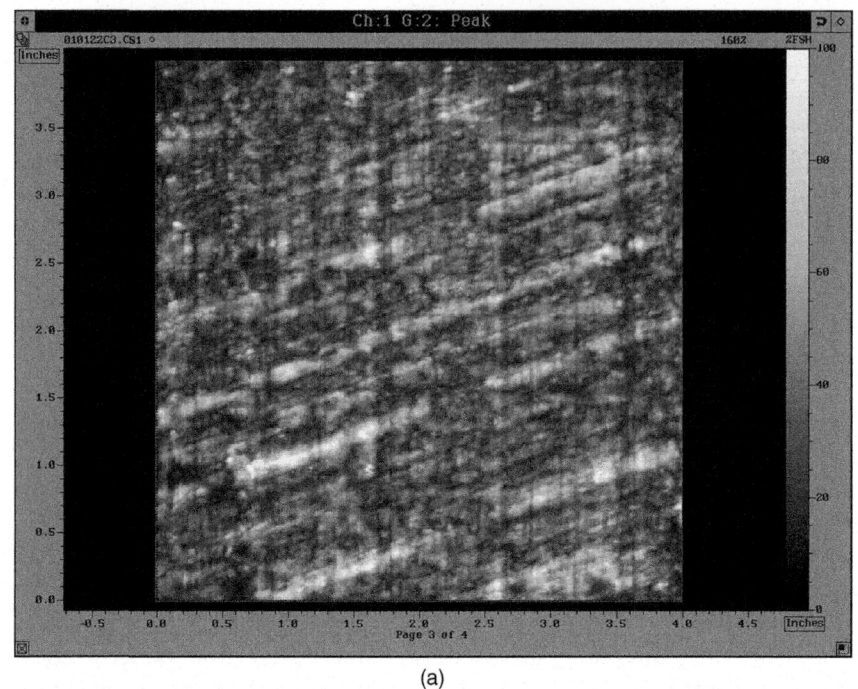

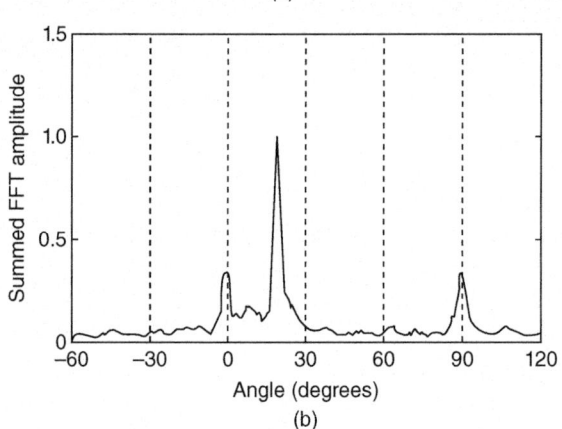

410   Non-destructive evaluation (NDE) of polymer matrix composites

15.5 (a) Ultrasonic C-scan image of the 18th interface between 20° and 90° plies in [(90$_3$/0$_3$)$_2$, 90$_3$, 20$_3$, 90$_3$, (0$_3$/90$_3$)$_2$] CFRP laminate. (b) The angular distribution plot deduced from the 2D spatial FFT of the scan image identified the presence of the 20° ply layup error.

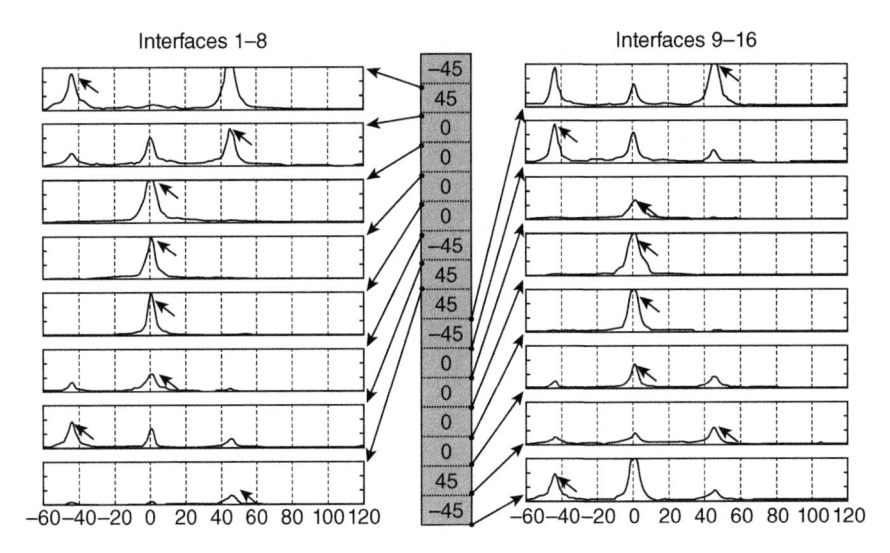

15.6 Angular distribution plot of the ply interfaces in a [−45/45/0₄/−45/45]ₛ CFRP laminate where all internal ply orientations were correctly identified.

simultaneously in an 'azimuthal scan' while maintaining the orthogonality of the two transducers. The scan data are presented in a polar plot of the received peak to peak amplitude of the transmitted RF signal as a function of the transducer angle. Anomalies in the solid laminate, such as misoriented ply or an error in stacking sequence, would exhibit themselves as pronounced changes from the azimuthal scan of an error free laminate. For example, a single misoriented ply in a 24-ply laminate can be readily detected. The solid curve in Fig. 15.7 shows the cross-polarized azimuthal scan of shear wave through the laminate [(0/90/45/−45)₃]ₛ. The dashed curve shows the azimuthal plot of the scan signal when the 12th ply was intentionally placed at +45° instead of at −45°. The change produced by the layup error on the azimuthal plot was quite striking.

The behavior of the shear wave propagation through a solid laminate containing plies oriented at different angles has been modeled based on vector decomposition of the shear vibration into components parallel to the fibers and perpendicular to the fibers. The model was used to simulate a number of ply layup and stacking sequence anomalies and the technique was found to be very effective (Hsu *et al.*, 1997). In addition, full waveform of the RF signal (A-scan trace) was used to display the azimuthal scan data as an image of transducer angle versus propagation time. This image, similar to an ultrasonic B-scan, can serve as a 'fingerprint' of a specific layup. For example, model simulation of the azimuthal scan showed very different fingerprints for the symmetric layup [(0/45/90/−45)₃]ₛ and the asymmetric

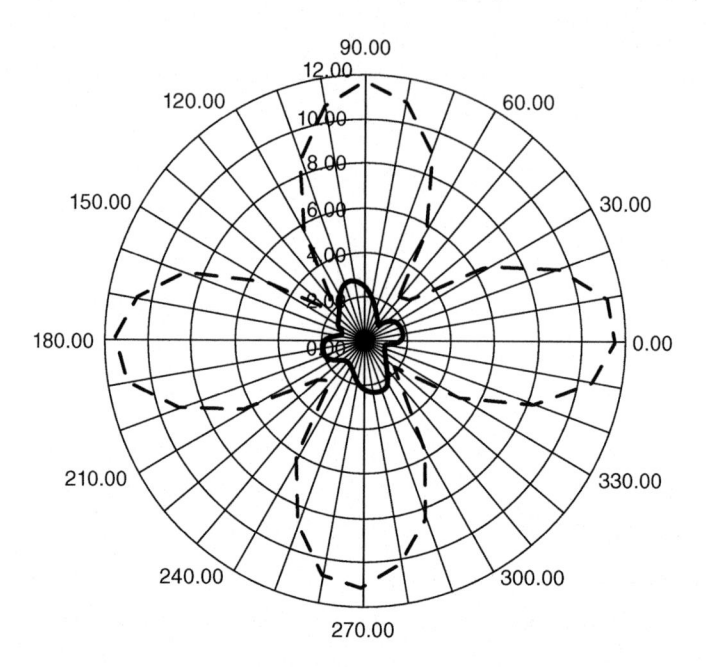

15.7 Azimuthal scan of transmitted shear wave amplitude for a 24-ply laminate with a ply orientation error. The solid curve was obtained on the error-free [(0/90/45/−45)$_3$]$_s$ laminate and the dashed curve was the result when the 12th ply was mistakenly placed at +45°.

layup [(0/45/90/−45)$_6$]$_T$. These results showed considerable promise for future use of shear ultrasonic waves as an NDT tool for fiber composite laminates.

One of the difficulties in applying shear ultrasonic waves is the coupling mechanism. It is difficult to maintain the viscous shear couplant constant, especially when the transducers are rotated on the surface of a sample as in an azimuthal scan. One way to avoid the changing coupling condition is to use EMAT as the transmitter and the receiver for shear waves on composites. Since the polymer matrix composites (even with carbon fibers) do not have adequate electrical conductivity for EMAT to function directly, a metallic layer, such as that in an aluminum tape, is applied to the surface and the shear waves are generated in the metal layer, from which they propagate into the composite laminate. With the aid of the metallic layer, the EMAT probes can be rotated at ease and consistent data can be obtained. For an un-cured stack of composite prepregs, the entire stack may be sandwiched and squeezed between two metal blocks, while EMAT transducers are used to generate and receive shear waves for interrogating the ply layup of the prepreg stack. The use of EMAT for the shear wave NDE on cured and un-cured composite laminates has been investigated extensively, both in modeling and in experiment (Fei and Hsu, 2002). This approach can potentially be applied in a manufacturing setting for in process monitoring of composite laminate structures.

15.4.3 Ultrasonic imaging of impact damage

The morphology of impact damage in solid composite laminates has been the subject of many studies. An impact on the surface typically produces a series of interlaminar delaminations that increase in size with depth before fading out. Buynak and Moran (1987) and Moran and Buynak (1989) first mapped out impact delaminations with ultrasonic scan by gating the time-domain waveform in a pulse-echo test. Since the delaminations at a given ply interface tend to occur in pairs, and aligned with one of the fiber directions, the delamination pattern would appear to rotate in step of 90° in a quasi-isotropic layup. This 'flower petal' pattern of impact delamination is well known in the ultrasonic NDE of composites (Gosse and Hause, 1988). Since the delaminations are basically physical discontinuities that can block the propagation of ultrasonic waves, the deeper delaminations are partially obscured by the presence of shallower delaminations in a pulse-echo scan. Due to the strong reflection of the delaminations, impact damage is not difficult to detect with conventional longitudinal wave ultrasound, the challenge in practical implementation arises when there are large areas of surfaces to inspect for BVID. The need for inspecting known vulnerable areas for impact damage has led to the development of handheld ultrasonic devices (see Section 15.6).

For a given impactor configuration and boundary condition of the composite structure, delaminations occur when a certain threshold impact energy is exceeded. However, the morphology of low energy (pre-delamination) impact damage is also of interest and concern, particularly with regard to the potential growth of damage under load and fatigue. Laboratory studies of low-energy impact in carbon and glass composite laminates showed that, below the delamination threshold energy, the damage zone consisted mainly of a distribution of microcracks in the resin matrix. The orientation of the microcracks is tilted toward the impact site on the surface. Photomicrographs of a polished cross-sectional surface cut through the impact site showed that microcracks to the right of the impact site were all tilted to the left, and similarly microcracks to the left of the impact point were tilted to the right. When there were stacks of plies with the same fiber orientation in the laminate, the tilted microcracks were found to extend through the thickness of the stack and terminate at the ply interfaces on its ends.

The behavior of the microcracks under fatigue loading has been investigated (Hsu *et al.*, 2011). Specimens with impact-induced microcracks were loaded in the laboratory in three-point fatigue tests and the changes of the

15.8 Morphology of microcracks and delaminations of a 24-ply CFRP laminate after fatigue failure. The tilted microcracks were caused by the impact and the delaminations were caused by the subsequent fatigue test.

microcrack morphology were followed and observed with optical microscope during the experiment. Some of the tilted microcracks were found to increase in width and began to initiate horizontal delaminations on their ends at the ply interfaces as the number of fatigue cycles increased. The delaminations were initially localized but grew in length and eventually joined with other delaminations. As a result, a pyramid of delaminations was formed with the tilted microcracks serving as the connecting 'jogs'.

Figure 15.8 shows the morphology of a failed impact zone in a 24-ply $[(0/90)_6]_S$ CFRP laminate. The impact point is at the top center. The thick continuous horizontal lines are the delaminations and the tilted jogs are the original impact-induced microcracks before the fatigue test. The specimen failed at the end from the propagation of the delaminations. In laminates made of unidirectional prepregs, the propagation of the delaminations at the ply interfaces was generally unimpeded. In contrast, the propagation of impact induced delaminations in laminates of woven prepregs can be easily interrupted by intervening fiber tows in the path of their propagation and growth.

In terms of ultrasonic testing of the incipient failure, the cloud of microcracks and the initial short localized delaminations were generally not easily detected with conventional ultrasonic testing; however, the delaminations became easily detectable once their length grew to several millimeters.

15.4.4 Detection of ply waviness in laminates

In the layup and curing of composite laminates, nonuniformity in ply stiffness or thermal expansion anomalies can cause one or more of the plies to buckle and result in an interior wrinkle or 'waviness'. Such ply waviness may occur without any indication on the outer surface. These localized ply waviness can be detrimental to the stiffness and strength of the laminate and therefore need to be detected. A subtle ply waviness with a large aspect ratio (length to height ratio of the 'wave') does not scatter the ultrasound strongly and can be difficult to detect. It may be discernible on a B-scan image, mainly by the presence of a region behind the waviness that contained disturbed waves. Because of the characteristics shape of the ply waviness, they are somewhat easier to detect using oblique incident waves. A pitch-catch configuration of two Rayleigh wave transducers (for steel) held together in a head-to-head fashion can be used to generate oblique incident waves in CFRP composite laminates. This approach was found to be more effective than normal incident longitudinal waves in detecting the ply waviness (Hsu et al., 2007).

Ply waviness is also a defect of concern in the manufacturing of wind energy composite turbine blades, especially those occurring in the thick glass composite skin (up to 40 mm) near the trailing edge of the blade. Due to the large thickness and high attenuation of the glass composite structure, low-frequency air-coupled ultrasound in the 100–200 kHz range was used in the detection of waviness in wind turbine trailing edge. The oblique incidence pitch-catch configuration was again found to be more effective in detecting the ply waviness in thick glass composites. Two air-coupled transducers tilted approximately 13° from the normal incidence direction were mounted symmetrically in a pitch-catch configuration with the wave propagation direction oriented perpendicular to the length direction of the ply waviness. The two transducers were then moved in unison to perform a manual or motorized scan of the sample surface. In air-coupled ultrasonic testing, the specular reflection from the top surface of the sample can easily overwhelm the signal associated with the interior ply waviness. The specular reflection signal must therefore be thoroughly blocked in order to detect the flaw signal. A simple approach of eliminating the specular reflection is to mount the two transducers in a block of foam, with two channels drilled at the desired angle and separated appropriately in distance. Figure 15.9 shows the air coupled pitch-catch set-up and a scan image of a ply waviness in a 25 mm thick glass composite obtained at a frequency of 120 kHz.

15.5 Ultrasonic inspection of sandwich structures

15.5.1 TTU scan

Honeycomb sandwich structures are widely used on aircraft flight control surfaces such as rudder, aileron, spoiler, and flap. The facesheet of honeycomb sandwiches are typically thin composite laminates and the interior of

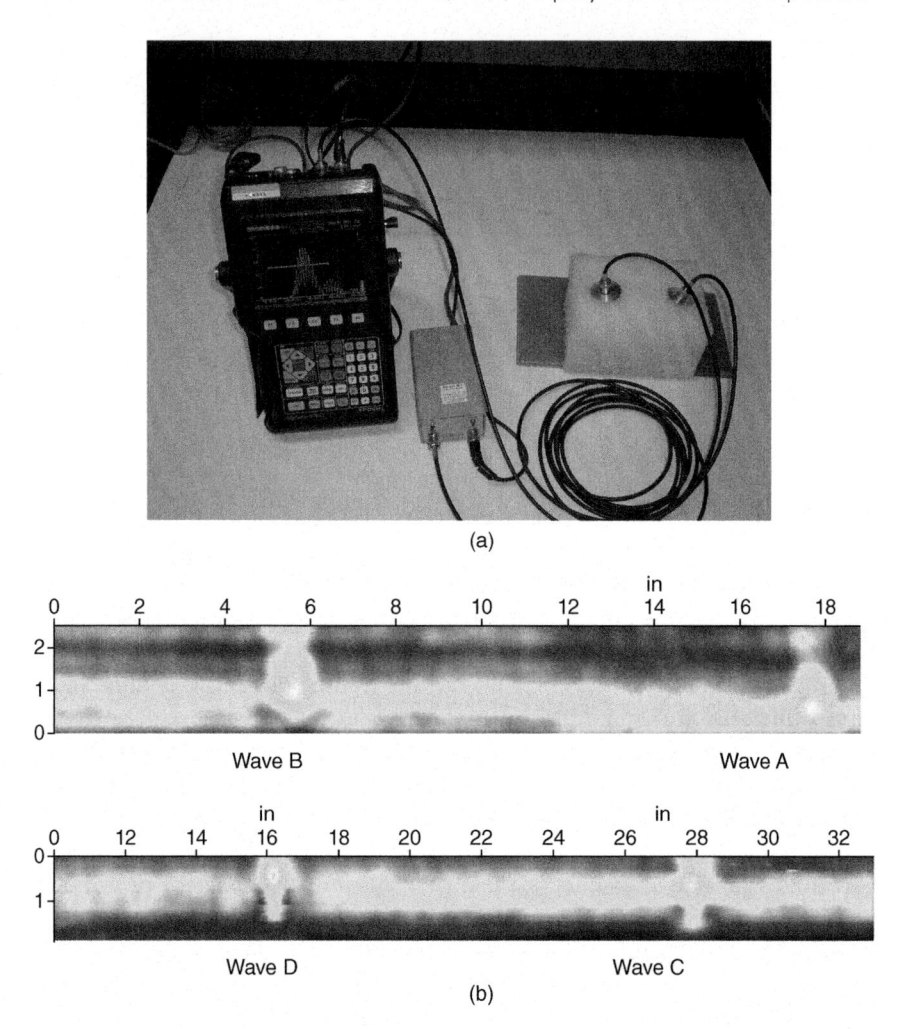

416   Non-destructive evaluation (NDE) of polymer matrix composites

15.9 (a) A manual air coupled pitch-catch set-up for detecting ply waviness in thick composite laminates, and (b) an ultrasonic pitch-catch scan image of a ply waviness in a wind turbine blade training edge obtained with a motorized scanner.

the sandwich is mostly honeycomb cell walls made of Nomex, fiberglass, or aluminum. The hollow structure of honeycomb sandwiches does not lend itself to effective ultrasonic pulse-echo inspection; instead the TTU scan has been the most widely used and effective mode for inspecting flaws and damage in honeycomb or foam sandwich structures. The light weight and large size of sandwich flight control structures are not conducive to immersion ultrasonic scan, so the most widely used mode for implementing the TTU inspection for aerospace sandwich structures has been the water squirter system (see also Section 15.6).

15.5.2 Air-coupled ultrasonic inspection of sandwiches

Air-coupled TTU scan has proven highly effective in the NDE of aerospace honeycomb sandwiches. The typical frequency range for air-coupled TTU inspection is 100–400 kHz using planar or focused transducers separated by a typical distance of 100 mm. The technique has been applied at a transducer separation of as much as 250 mm on thick and curved structures such as sections of engine nacelle. The performance of air-coupled TTU detection of flaws and damage in honeycomb sandwiches has been investigated extensively in a comparative with other NDT techniques. Air-coupled TTU scans were made on a large set of 42 honeycomb panels containing various engineered flaws to establish the probability of detection (POD) data.

15.5.3 Inspection of composite sandwich with perforated facesheet

Some honeycomb composite sandwiches on aircraft have perforated facesheet for acoustic noise control purposes; a notable example is the engine nacelle. The inspection of honeycomb sandwich with perforated facesheet cannot be done with water squirters. Current industrial practice is to waterproof the perforated facesheet by applying acrylic tapes (about 50 mm wide) over the entire surface. This manual operation of taping and un-taping is tedious, labor intensive, and therefore costly. Perforated honeycomb structures may be inspected with air-coupled ultrasound without any surface preparation. To demonstrate the effectiveness of air-coupled ultrasonic detection of flaws in perforated honeycomb sandwich, experiments were performed on a test panel of $3/8$ in (10 mm) sized honeycomb cells and a facesheet with perforation spacing of approximately 3 mm. The test panel contained engineered disbonds of 12 and 25 mm diameters. The panel was scanned with a pair of 120 kHz air coupled transducers in the through-transmission mode. The scan image clearly showed the disbonds, as well as the honeycomb pattern inside the sandwich. The presence of the perforation on the facesheet did not interfere with the ultrasonic detection and imaging of the disbond flaws and the honeycomb cells. To illustrate the results, the scanned image was printed on a transparency and then laid over the perforated facesheet, as shown in Fig. 15.10.

It has been demonstrated in the laboratory that sections of engine nacelle with perforated facesheet can be inspected with air-coupled TTU scan in the 100–200 kHz frequency range at large standoff with the transducers separated by up to 250 mm to clear the curvature of the part. An air-coupled ultrasonic scan can effectively detect flaws in perforated honeycomb sandwiches, but the low frequency and the long tone-burst length can limit the pulse repetition rate and therefore the throughput of the scan. Further

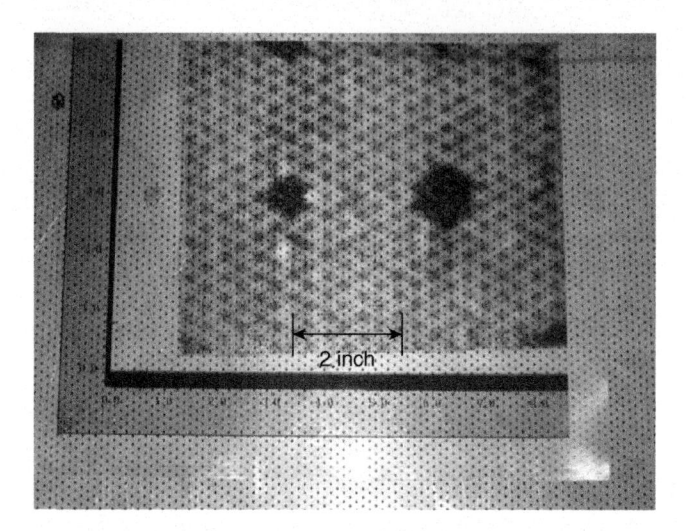

15.10 Air-coupled ultrasonic scan image printed on transparency and laid over a perforated facesheet to show the two engineered disbond flaws with 12 and 25 mm diameter and the interior honeycomb pattern. The perforation had a spacing of approximate 3 mm.

development is needed in applying air-coupled ultrasound to high-speed industrial production inspection of aerospace parts.

15.5.4 Field inspection of honeycomb sandwich on aircraft

The primary challenge of field inspection of structures on aircraft is the large area involved. Boeing and Airbus require only visual inspection to detect impact damage in honeycomb composites. The NDI of honeycomb sandwich flight control surfaces has traditionally been carried out with a tap test. Even today a hearing-based manual tap test is still widely used in the field. A tap test can be subjective and sometimes unreliable, but instrumented tap tests have been developed. Examples are the Mitsui 'Woodpecker' and the WichiTech RD3. In addition, a semi-automated tap test with imaging capability has also been developed (Peters *et al.*, 2000). Generic ultrasonic flaw detectors are of limited use on aircraft honeycomb structures in the field, but air-coupled ultrasound has been developed and tested on aircraft in recent years (Peters *et al.*, 2004).

In order to apply air-coupled ultrasound on aircraft honeycomb structures, the transducers are mounted on a yoke so that TTU inspection can be performed on flight control surfaces with two-sided access. A portable air-coupled ultrasonic set-up with transducers on a yoke is adequate for cursory manual survey for damage on the structure. However, more quantitative inspection and the detection of more subtle defects would require

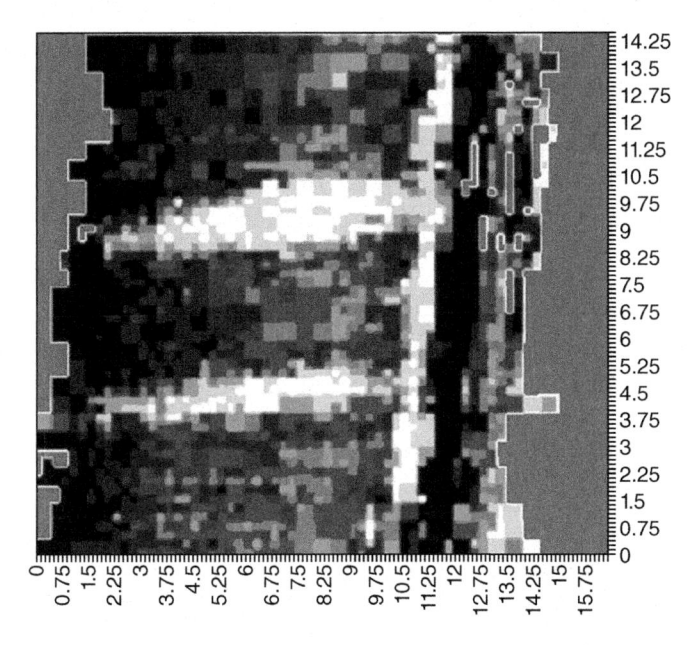

15.11 Transverse tree-strike damage on a Black Hawk rotor blade were imaged with a manual air-coupled ultrasonic scan system using a magnetic position encoder.

a system with scanning and imaging capability. Computer controlled and motorized scanners have their advantages but such systems are often too complex and cumbersome for field use. For this reason, efforts were made to develop a simple manual scanner with position encoding to apply air coupled TTU inspection of honeycomb sandwich structures (Barnard *et al.*, 2005). Using the portable hand-scan system, damage and repairs on rotor blades, flaps, and trailing edges were imaged. Figure 15.11 shows a manual scan image of damages on a helicopter rotor blade.

15.6 Ultrasonic non-destructive testing (NDT) instruments for aerospace composites

The ultrasonic instruments used for the NDI of aerospace composites range from small portable manual ultrasonic flaw detectors to large fully automated factory scanning systems. They can perhaps be divided into three categories: handheld flaw detectors for point measurement, the intermediate system portable enough for on aircraft application and capable of scanning a local area to produce images, and the stationary large scan gantries for high throughput factory inspection of whole composite components. Various models of portable ultrasonic flaw detectors are available from most NDT instrument venders; one example is the Epoch 4 of Olympus. Large stationary ultrasonic scanning systems using water jet (squirters) such as the AUSS-5 have been developed by major aircraft manufacturers. To meet the need for intermediate systems that are portable enough for on-aircraft application and can provide quantitative scan data, there have been a number of development efforts. One example is the mobile automated scanner (MAUS) system developed by the former McDonnell Aircraft Company (now Boeing). There have also been efforts to develop portable instruments for acquiring ultrasonic scan images manually without a motorized scanner by combining the position data from simple position encoders and the ultrasonic data from flaw detectors (Barnard *et al.*, 2005). With the recent development and advances in ultrasonic phased arrays, the needs for portability and inspection speed in field applications are being increasingly met.

15.7 Conclusion

Ultrasonic NDE methods have been the primary inspection techniques for aerospace composite structures. The intimate relationship between ultrasonic parameters and the elastic and mechanical properties of the composite material makes ultrasound the most effective probing field for defects and damage of composite structures. Conventional ultrasonic methods will continue to be important and new ultrasonic testing modalities are also coming to the fore. These include increasingly wide use of ultrasonic phased arrays and laser ultrasound. The ability of an ultrasonic passed array to electronically steer and focus the beam has distinct advantages in the inspection of anisotropic composites, in addition to the benefits of greater scan speed and imaging capability. The technologies of laser generated and detected ultrasound have matured to large-scale industrial application. The progress of ultrasonic NDE for aerospace composite materials and structures is expected to move forward, both in the fundamental development of the field–flaw interaction but also in the technological development of implementation and application.

15.8 References

Armstrong K B and Barrett R T (1998), *Care and Repair of Advanced Composites*, Warrendale, PA, Society of Automotive Engineers, Inc.

Bar H N, Dayal V, Barnard D J and Hsu D K (2010), 'Plate wave resonance with air coupled ultrasonics,' *Review of Progress in Quantitative Nondestructive Evaluation*, **29**, 1069–1076.

Barnard D J, Peters J J and Hsu D K (2005), 'Towards a generic manual scanner for nondestructive inspection,' *Review of Progress in Quantitative Nondestructive Evaluation*, **24**, 1669–1676.

Bhardwaj M (2001), 'Non-contact ultrasound: The last frontier in nondestructive testing and evaluation,' in *Encyclopedia of Smart Materials*, Biderman A ed., New York, John Wiley & Sons.

Blodgett E D, Miller J G and Freeman S M (1986), 'Correlation of ultrasonic polar scatter with de-ply technique for assessment of impact damage in composite laminate,' *Review of Progress in Quantitative Nondestructive Evaluation*, **5**, 1227–1238.

Buckley J (2000), 'Air coupled ultrasound – A millennial review,' *15th World Conference on Nondestructive Testing*, Rome, Italy.

Buynak C F and Moran T J (1987), 'Characterization of impact damage in composites,' in *Review of Progress in Quantitative NDE*, Thompson D O and Chimenti D E (eds), Vol. 6, pp. 1203–1211, Plenum Press, New York, NY.

Duke J C Jr. (1988), *Acousto-ultrasonics: Theory and applications*, New York, Plenum Press.

Fei D and Hsu D K (2002), 'A model and experimental study of fiber orientation effects on shear wave propagation through composite laminates,' *Journal of the Acoustical Society of America* **111**(2), 840–855.

Gosse J H and Hause L R (1988), 'A quantitative nondestructive evaluation technique for assessing the compression-after-impact strength of composites plates,' *Review of Progress in Quantitative Nondestructive Evaluation*, **8**, 1011–1020.

Hsu D K (2006), 'Nondestructive testing using airborne ultrasound,' *Ultrasonics*, **44**, Suppl. 1, 1019–1024.

Hsu D K and Margetan F J (1992), 'Analysis of acousto-ultrasonic signals in unidirectional thick composites using the slowness surfaces,' *Journal of Composite Materials*, **26**(4), 1050–1061.

Hsu D K, Fischer B A and Koscamp M (1997), 'Shear wave ultrasonic technique as an NDE tool for composite laminates before and after curing,' *Review of Progress in Quantitative Nondestructive Evaluation*, **16**, 1975–1982.

Hsu D K, Fei D and Liu Z J (2002), 'Ultrasonically mapping the ply layup of composite laminates,' *Materials Evaluation*, **60**(9), 1099–1106.

Hsu D K, Dayal V, Barnard D J and Im K H (2007), 'Ultrasonic inspection of solid composite laminates using a one-sided pitch-catch contact mode,' *Review of Progress in Quantitative Nondestructive Evaluation*, **26**, 1021–1028.

Hsu D K, Dayal V, Subramanian A, Im K H, and Barnard D J (2011), 'A study of microcracks and delaminations in composite laminates,' *Int. Sym. Nondestrutive-Testing of Materials and Structures*, May 15–18, 2011, Istanbul, Turkey.

Jeong H and Hsu D K (1995), 'Experimental analysis of porosity-induced ultrasonic attenuation and velocity change in carbon composites,' *Ultrasonics*, **33**(3), 195–203.

Kazys R, Demcenko A, Zukauskas E and Mazeika L (2006), 'Air-coupled ultrasonic investigation of multi-layered composite materials', *Ultrasonics*, **44**, 819–822.

Minachi A, Margetan F J and Hsu D K (1993), 'Delamination sizing in composite materials using a Gauss-Hermite beam model,' *Ultrasonics*, **31**(4), 237–243.

Moran T J and Buynak C F (1989), 'Correlation of ultrasonic imaging and destructive analysis of low energy impact events,' *Review of Progress in Quantitative Nondestructive Evaluation*, **8**, 1627–1634.

Ourak M, Ouaftouh M, Duquennoy M and Xu W J (1998), 'Characterization of ply boundaries in carbon epoxy composite materials using longitudinal ultrasonic waves,' *Review of Progress in Quantitative Nondestructive Evaluation*, **17**, 1147–1154.

Papadakis E P, Patton T, Tsai Y M, Thompson D O and Thompson R B (1991), 'The elastic moduli of a thick composite as measured by ultrasonic bulk wave pulse velocity,' *Journal of the Acoustical Society of America*, **89**(6), 2753–2757.

Patton T C and Hsu D K (1994), 'Doing focused immersion ultrasonics without the water mess,' *Review of Progress in Quantitative Nondestructive Evaluation*, **13**, 701–708.

Peters J J, Barnard D J, Hudelson N A, Simpson T S and Hsu D K (2000), 'A prototype tap test imaging system: initial field test results,' *Review of Progress in Quantitative Nondestructive Evaluation*, **19**, 2053–2060.

Peters J J, Barnard D J and Hsu D K (2004), 'Development of a fieldable air coupled ultrasonic inspection system,' *Review of Progress in Quantitative Nondestructive Evaluation*, **23**, 1368–1375.

Rokhlin S I, Chimenti D E and Nagy P B (2011), *Physical Ultrasonics of Composites*, New York, Oxford University Press.

Schindel D W and Hutchins D A (1995), 'Application of micromachined capacitance transducers in air-coupled ultrasonics and nondestructive evaluation,' *IEEE Transactions on Ultrasonic Ferroelectric Frequency Control*, **42**, 51–58.

Smith R A (2009), 'Composite defects and their detection,' Materials Science and Engineering, Vol. III, *UNESCO Encyclopedia of Life Support System (EOLSS)*, Ramsey, Isle of Man, EOLSS Publishers Co. Available from http://www.eolss.net/Eolss-sample AllChapter.aspx [accessed 8 November 2011].

Smith R A and Clarke B (1994), 'Ultrasonic C-scan determination of ply stacking sequence in carbon fiber composites,' *Insight – Journal of the British Institute of Non-destructive Testing*, **36**(10), 741–747.

Stone D E W and Clarke B (1975), 'Ultrasonic attenuation as a measure of void content in carbon fiber reinforced plastics,' *Non-destructive Testing*, **8**(3), 137–145.

Summerscales J editor (1987, 1990), *Non-destructive Testing of Fiber-reinforced Plastics Composites*, Vol. I, Vol. II, New York, Elsevier Applied Science Publications.

Thompson R B (1990), 'Physical principles of measurements with EMAT transducers,' in Thurston R N and Pierce A D, *Ultrasonic Measurement Methods, Physical Acoustics XIX*, San Diego, Academic Press, 157–199.

Vary A (1990), ' Acousto-ultrasonics,' in *Non-destructive Testing of Fiber Reinforced Plastics Composites*, Vol. 2 (J. Summerscales, ed.), New York, Elsevier Applied Science, 1–54.

Zimmer J E and Cost J R (1970), 'Determination of the elastic constants of a unidirectional fiber composite using ultrasonic velocity measurements,' *Journal of the Acoustical Society of America*, **47**, 795–803.

16
Non-destructive evaluation (NDE) of aerospace composites: acoustic microscopy

B. R. TITTMANN, C. MIYASAKA and M. GUERS,
The Pennsylvania State University, USA, H. KASANO,
Takushoku University, Japan and H. MORITA,
Ishikawajima – Harima Heavy Industries Co. Ltd, Japan

DOI: 10.1533/9780857093554.3.423

Abstract: This chapter presents two case studies on the use of acoustic microscopy for the characterization of carbon fiber-reinforced epoxy composites (CFRP) and aramid fiber composites (AFRP). The emphasis is on impact damage ranging from hard impacts (bullets, micrometeorites, hail, etc.) to soft impacts (bird strike) typically encountered in aerospace applications. The chapter ends with challenges and future trends, such as the embedding of sensors in composite panels.

Key words: non-destructive evaluation (NDE), carbon fiber/aramid fiber (CF/AF) interply, hybrid laminates, impact-induced damage, compression after impact, low-velocity impact, healing delaminations, scanning acoustic microscopy (SAM), C-scan, AS4/PEEK (polyetheretherketone) composite, finite element method (FEM), CAPA, embedded sensors, structural health monitoring.

16.1 Introduction

This chapter is a review of some recent results of current work in addition to some previous work [1–5]. The impact-induced damage is the most serious type of damage for carbon fiber-reinforced plastic (CFRP) composites and aramid fiber plastic (AFRP) composites [6]. In particular, delamination caused by impact loading is known to produce significant reductions in the residual compression strength (i.e., compression-after-impact (CAI) strength) of CFRP laminated composites, which is a major problem associated with the structural integrity of composite compression components. Although the aramid fiber has similar density and tensile strength to the carbon fiber, the tensile modulus and elongation of the aramid fiber are approximately half and double of the carbon fiber, respectively.

Much effort has therefore been devoted to the study of the effects of impact-induced damage on the load-bearing capacities of these components in the past 30 years [7–11]. In the course of these studies, attempts have also been made to investigate the damage mechanisms [12–20]. However, there are few reports of the impact-induced damage analysis for composite laminates.

Acoustic microscopy [1] is uniquely suited to imaging and characterizing damage below the surface, nondestructively. The heart of an acoustic microscope is a high numerical aperture (NA) transducer which sends and receives ultra-high-frequency (UHF) acoustic waves into a focal spot. The focal spot is adjusted to an arbitrary depth below the surface and then scanned parallel to the surface to provide a high-resolution image of the focal plane.

16.2 Case study: damage analysis using scanned image microscopy

16.2.1 Introduction

In this study, first, we completed impact tests with the carbon fiber/aramid fiber (CF/AF) hybrid fabric laminates; second, non-destructively visualized internal damage (e.g., delaminations) with scanning acoustic microscopy; and third, sectioned the laminates, and observed the cross-sectional surfaces with scanning electron microscopy (SEM) to characterize impact damage and hybrid effects. The experimental investigation was conducted to evaluate the impact loading damage within hybrid fabric laminates comprising carbon and aramid fibers. The experiments were undertaken on a series of interplay hybrid specimens with different stacking sequences of prepregs. The spherical steel balls, having diameters of 5.0 and 10.0 mm and velocities of 40 and 113 m/s respectively, were propelled normal to the center of the surface of the specimens via an apparatus having an impact mechanism similar to that of an air-gun, leading to impact damage to the specimens. The surface damage (e.g., micro-cracks, debondings) of the specimens was visualized by SEM, and internal damage (e.g., delamination) of these was visualized as a horizontal cross-sectional image by a scanning acoustic microscope (SAM) to evaluate impact-induced damage.

16.2.2 Sample preparation

Table 16.1 shows the properties of fibers and matrices of the composites used in this study, and Table 16.2 shows the specifications of the composites. Although the aramid fiber has similar density and tensile strength to the carbon fiber, the tensile modulus and elongation of the aramid fiber are approximately half and double of the carbon fiber, respectively. Table 16.3 shows the stacking sequences of the specimens. 'C' denotes the carbon fiber laminates and 'A' the aramid fiber laminates. The hybrid specimens comprising the carbon and aramid laminates are denoted 'CAC' or 'ACA',

Table 16.1 Properties of fiber and matrix

Constituent	Carbon fiber	Aramid fiber	Epoxy resin
Type	T300-3000	Kevlar 49	Toray # 2500
Specific weight	1.74	1.45	1.24
Tensile strength (MPa)	3040	2765	78
Tensile modulus (GPa)	225.55	131.12	3.96
Elongation (%)	1.3	2.4	2.0

Table 16.2 Specification of fabric reinforcement

Type	Torayca cloth	Kevlar cloth
No.	#6343	K281
Texture	Fabric with plain weave	Fabric with plain weave
Density	5/cm	6.8/cm
Weight	200 g/cm^2	178 g/cm^2
Thickness	0.27 mm/ply	0.285 mm/ply

Table 16.3 Stacking sequence of specimens

Specimen	Stacking sequence
C	C_{12}
CAC	$C_3/A_6/C_3$
ACA	$A_3/C_6/A_3$
A	A_{12}

respectively, depending on the order of the stacking sequence. A subscript number refers to the number of prepregs. The shape of the specimen was rectangular and its X-Y-Z dimensions were 180 mm × 100 mm × 3 mm.

16.2.3 Impact testing

Impact tests were performed using an air-gun type of apparatus (see Fig. 16.1). Air was bled from a supply line into a cylindrical reservoir. Two steel balls, 5.0 and 10.0 mm in diameter, were used as the impactors. The steel balls were accelerated by compressed air discharged from the reservoir. The impact machine had a velocity-measuring device at the end of the gun barrel. As the steel ball traveled through this device, laser beams emitted by photodiodes were interrupted, triggering an electronic counter. The

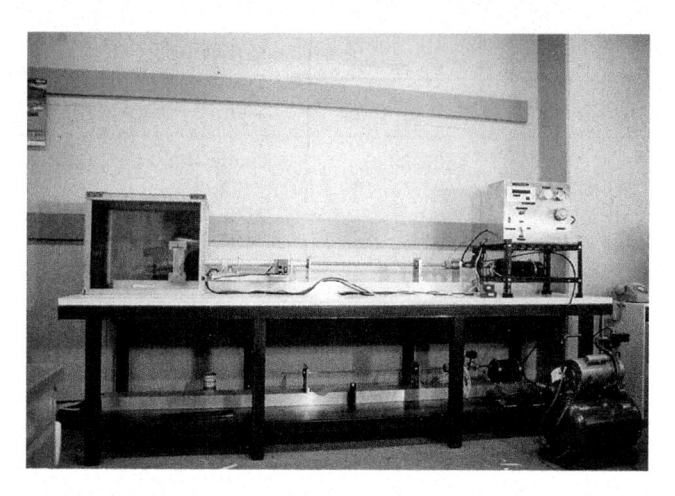

16.1 An air-gun type of impact apparatus.

impact velocity was calculated from the distance between the photodiodes and the time intervals between beam interruptions. The impact velocity was controlled by adjusting the air pressure in the reservoir. The test specimens were mounted in the fixture during testing and the edges were clamped by the frames. In the impact test, the specimens were damaged by the steel impactor ball propelled normal to the center of the surface at the desired impact velocity or energy level.

The specimens were impacted under two different test conditions. The steel balls were propelled at velocities of 40 and 113 m/s. The diameter and weight of the balls were 5.0 mm, 0.5 g and 10.0 mm, 4.1 g, respectively, in order produce the same kinetic energy, i.e., the same impact energy of 2 J.

16.2.4 Acoustic microscopy

The impact-damaged specimens were non-destructively inspected by the pulse-wave mode of the mechanical scanning acoustic reflection microscope (hereinafter called simply 'SAM'). Figure 16.2 is the schematic diagram of the SAM (Olympus Corporation; model: UH Pulse 100). An electrical signal is generated by a transmitter/receiver. The electrical signal is transmitted to a piezoelectric transducer (i.e., $LiNb_2O_3$) located on the top of a buffer rod (i.e., fused quartz) through the single-pole-double-throw (SPDT) switch. The electrical signal is converted to an acoustic signal (i.e., ultrasonic plane wave) at the transducer. The ultrasonic plane wave travels through the buffer rod to a spherical recess (hereinafter called simply the 'lens') located at the bottom of the buffer rod. The lens converts the ultrasonic

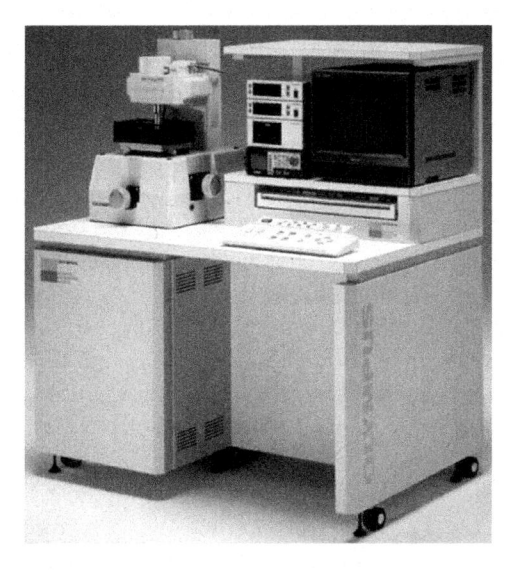

16.2 A mechanical scanning acoustic reflection microscope (pulse-wave mode).

plane wave to an ultrasonic spherical wave (i.e., ultrasonic beam). The ultrasonic beam is focused within the specimen, and reflected from the specimen. The reflected ultrasonic beam, which carries acoustic information of the specimen, is again converted to an ultrasonic plane wave by the lens.

Suppose a pulse-wave emitted from an acoustic lens through a coupling media (e.g., water) is focused onto the back surface of the specimen (i.e., CF/AF hybrid fabric-reinforced plastic laminates). Suppose the pulse-wave is strong enough to travel through the specimen and reflect back to the acoustic lens. Since the specimen is a layered material, the timing of each reflection is different because the traveling distance is different. An oscilloscope may monitor both the amplitude and the delay of the pulse-wave reflected from each plane of the specimen. Note that an output of an electrical pulse signal generated from a transmitter/receiver is approximately 50–300 V. The pulse wave can be focused through the specimen of interest for obtaining an optimized acoustic image of good quality. For example, when the first interface needs to be clearly visualized to detect a defect such as a delamination, the pulse-wave is focused onto the first interface for maximizing the amplitude of the pulse wave reflected from the first interface. Then, the reflected pulse wave is electrically gated out for visualizing the first interface as a horizontal cross-sectional image by horizontally scanning the acoustic lens.

16.2.5 Imaging analysis: delamination

Plate VII (between pages 296 and 297) displays acoustic images at 30 MHz showing delaminations at interfaces of specimens. From the images it is clear that the delaminations at interfaces B of the hybrid specimens are larger than that of C. Tables 16.4 and 16.5 provide delamination areas calculated from the acoustic images. Table 16.4 shows that the delamination area of ACA is almost equal to that of A. The delamination area at interface B of CAC is extremely large. The crack created within the back portion of C of CAC probably propagated toward A located in the center of CAC.

The impacted specimens were sectioned. Figures 16.3 and 16.4 are the surfaces of the sectioned specimens observed by SEM. Cracks are seen near the impact area and the portion close to the back surface in C. The cracks caused by the separations between fibers and matrix can be clearly noticed at the back surface in A. The lateral and matrix cracks along the orientation of fibers can be seen in CAC. The behavior of cracks in ACA and A seem similar. The cracks of A and ACA are related to the characteristics of the aramid fibers (elongation is large), and the cracks of C and CAC are related to the carbon fibers (elongation is small).

Table 16.4 Damage area calculated from horizontal cross-sectional images obtained by SAM (impactor: diameter $\phi = 5$ mm, velocity $V = 113$ m/s)

Specimen	Interface A	Interface B
C	25.68	40.48
CAC	30.66	199.04
ACA	41.36	73.00
A	46.46	71.59

Table 16.5 Damage area calculated from horizontal cross-sectional images obtained by SAM (impactor: diameter $\phi = 10$ mm, velocity $V = 40$ m/s)

Specimen	Interface A	Interface B
C	*	4.82
CAC	32.62	56.54
ACA	18.54	40.42
A	36.69	51.75

*Damage area too small to be measured.

Acoustic microscopy 429

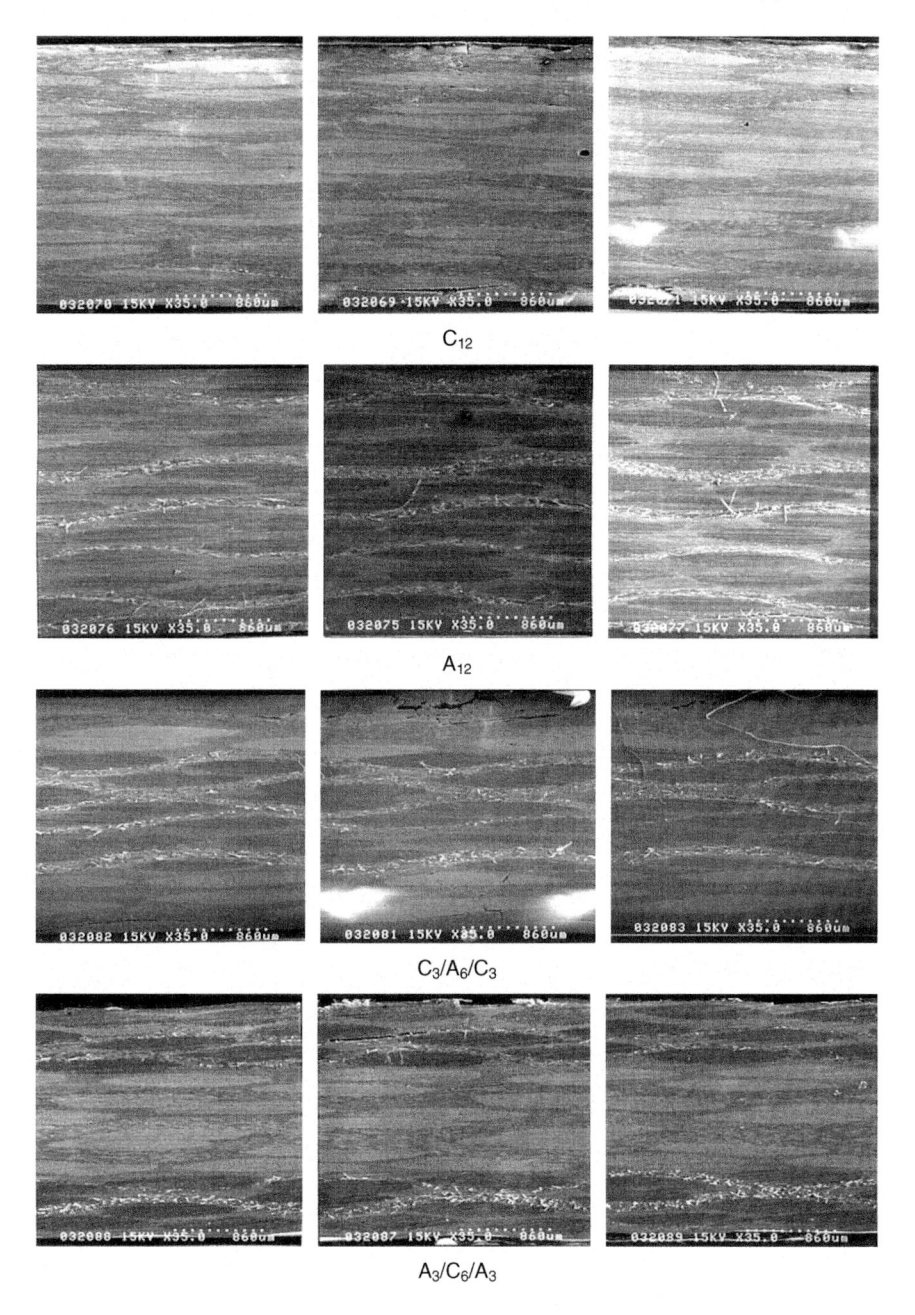

C_{12}

A_{12}

$C_3/A_6/C_3$

$A_3/C_6/A_3$

16.3 SEM images (impactor: diameter $\phi = 5$ mm, velocity $V = 113$ m/s).

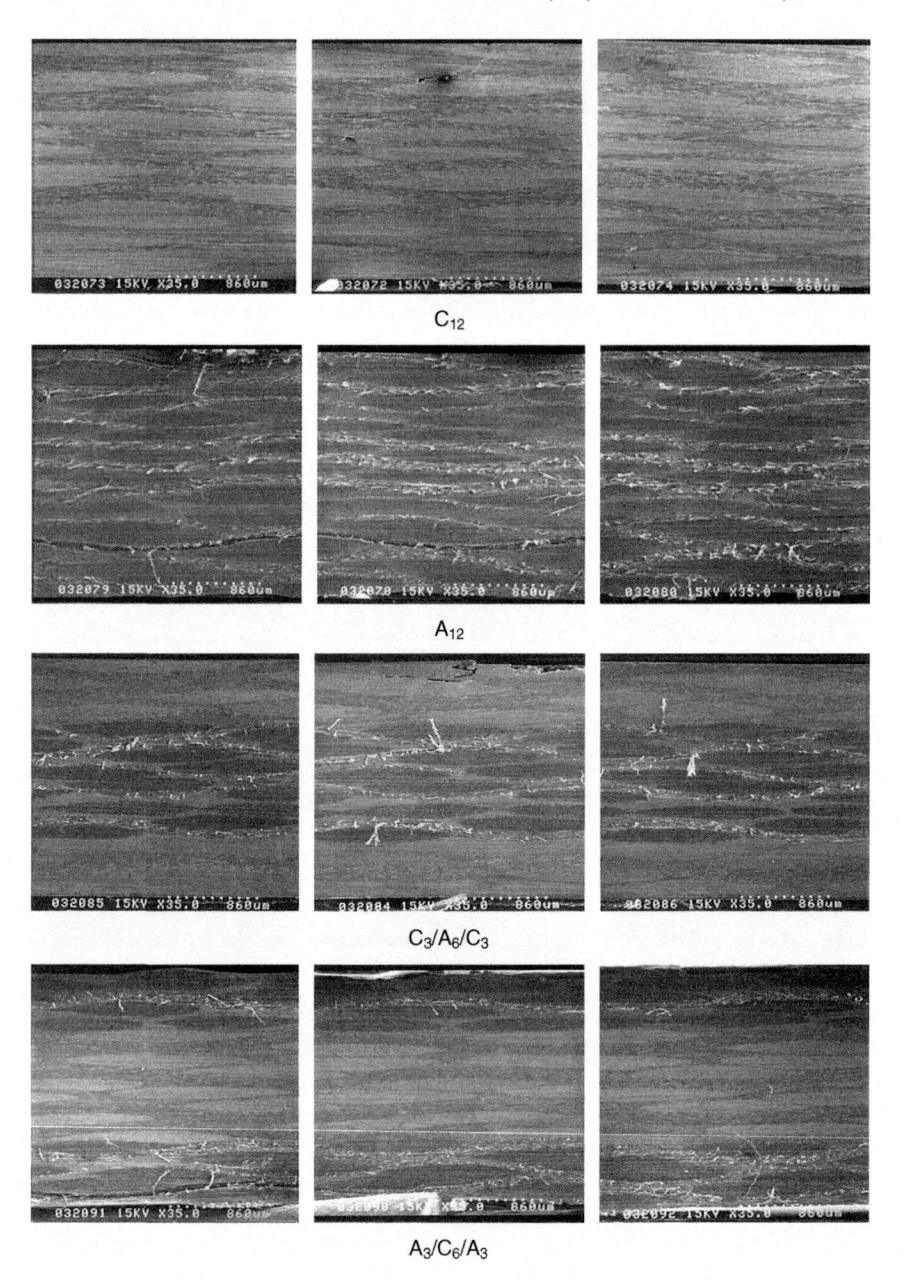

C_{12}

A_{12}

$C_3/A_6/C_3$

$A_3/C_6/A_3$

16.4 SEM images (impactor: diameter $\phi = 10$ mm, velocity $V = 40$ m/s).

16.2.6 Discussion

The shape of a delamination is typically in the shape of a diamond. The diagonal lines of the diamond give the orientation of fibers. The areas of the delaminations of the hybrid specimens (i.e., CAC and ACA) and the AFRP are larger than that of the CFRP. When the portion of the AFRP is internally located within the hybrid specimen (i.e., CAC), the impact tends to generate large delaminations at interface B. When the portion of the AFRP is located at the surfaces of the hybrid specimen (i.e., ACA), the impact tends to generate the delamination at the interface B in the same way as that of A. Cracks observed within the AFRP are caused by the separation of the aramid fibers and the matrix, while cracks observed within the CFRP are caused by fracture of the matrix, the separation of the carbon fibers and the matrix, and the disconnections of the carbon fibers.

16.3 Case study: damage analysis using acoustic microscopy

16.3.1 Introduction

In order to use composite materials in aeronautical turbo engines, the resistance to impact damage caused by soft bodies (birds) must be understood. In this work the subperforation flat-wise soft body impact resistance of two kinds of carbon fiber/PEEK (polyetheretherketone) systems were evaluated using a gelatin projectile. Tested systems were AS4/PEEK (APC-2/AS4, ICI-Fiberite) and AS4/PEEK+IL, which consists of APC-2 prepreg and PEEK film inserted between layers as an interleave (IL). To investigate the effects of the stacking sequence on resistance, three lay-ups- (0/+30/0/–30)s, (0/+60/0/–60)s, and (0/+45/90/–45)s – were tested. A gas-gun system was used for the impact tests, with a velocity range between 90 and 190 m/s. The mass of the projectile was about 3 g. The projected damage areas were measured with an ultrasonic C-scan system. The characteristics of the soft body impact damage are evaluated by comparisons of the damage caused by the aluminum and gelatin projectiles. The relative impact resistance was discussed earlier with the data measured in the previous hard body and gelatin impact tests. The impact damage caused by the gelatin projectile was almost the same as the damage shape caused by the aluminum projectile, but the gelatin impact resulted in a smaller projected damage area. The relationship between damage area (DA) and impact energy (IE) is linear for both the gelatin impact and the aluminum bullet impact. The DA/IE of gelatin impact was less than that for the aluminum bullet impact, and the gelatin impact indicated higher impact energy thresholds.

The basic deformation mode of the aluminum and the composite specimens considered were pure bending at a lowest natural frequency without node diameters or circles, but in the frequency domain, the behavior of composites was more complicated than that of aluminum plates. The effects of stacking sequence system (T200/#3900) indicated a different relationship from that for the CF/PEEK systems. The impact resistance can be assessed with the proposed non-dimensional stacking parameter $\beta T2$ and the value of $(DA/IE)/\beta T2$ which is independent of both fiber angle and lamina thickness. The CF/PEEK systems indicated a good proportional correlation of $(DA/IE)/\beta T2$ between the high-velocity impact to the low velocity and the gelatin impact.

16.3.2 Background

Bird impact damage in aeronautical turbo engine components made out of polymer matrix composites is a major design concern. To consider the invisible impact damage on composite air frames, the impact tolerance of the residual compression strength is the major issue. However, in the case of the bird impact on engines, since the projectile has the same order of size as the target, and since the damage can significantly alter the shape of the target or reduce its stiffness, impact damage resistance is the primary design issue. Impact resistance is dependent on the choice of material system. Many high-resistance systems, such as toughened epoxy, thermoplastics, interleaved structures, stitched structures, and 3D composites have been developed. To use these new systems, the damage resistance must be measured as an aid in material screening. The stress field caused by an impact loading is complex and damage results from many failure modes, e.g., delaminations, fiber breakage, and matrix cracking. An impact damage propagation process was reported [1] whereby the intraply matrix cracking caused by shear or bending was the initial damage mode, and delaminations initiated from the matrix cracking propagated into the nearby interface. A failure prediction model which consists of the three-dimensional finite element method (FEM) analysis and an interlaminar failure criterion was also proposed [19]. However, such a method is too complex for assessing the damage resistance of material systems in the actual design process, and a general but simple parameter to determine absolute impact resistance is not available at this time. Therefore, comparing the overall damage caused by a coupon impact test is a reasonable way to assess the resistance of the candidate material systems.

In a previous study [21], the impact resistance of CF/PEEK and CF/toughened epoxy systems were investigated with hard body impact tests using a drop weight impactor (low-velocity impact) and an air-gun impactor (high-velocity impact). Both tests used the same specimens and supporting conditions. The projectile for the low-velocity test was a steel hemispherical tip, and the projectile for the high-velocity test was an aluminum hemispherical bullet. However, a bird is a soft body projectile. The hard projectile behavior during impact can be considered as a rigid body, but the soft body impactor shows a liquid-like behavior. These differences in impactor characteristics change the contact force history and the contact region. There are relatively few papers that report findings regarding soft body impact on composites. Therefore, it is necessary to evaluate the effects of such differences of projectile behavior on impact damage. A gelatin impactor can be used as a model of an actual bird. Shooting a bare gelatin bullet with a small bore gun used in coupon tests is difficult because the gelatin bullet can easily become stuck in the barrel. A very light holder to hold the gelatin bullet may be useful to avoid this situation.

In this work the subperforation flat-wise soft body impact resistance of carbon fiber/PEEK systems was evaluated in a high-velocity impact test with gelatin projectiles held in lightweight holders. The specimens and supporting conditions were the same as the previous hard body impact tests. The characteristics of soft body impact damage were evaluated by comparing the damage caused by the gelatin and the aluminum projectiles. The relative impact resistance was discussed earlier with the data measured in the hard body and the gelatin impact tests.

16.3.3 Experimental procedure

Specimens

The specimens were tested in previous high-velocity impact tests with aluminum bullets. The systems tested were AS4/PEEK (APC-2/AS4, ICI-Fiberite) and AS4/PEEK+IL, which consists of APC-2 prepreg and PEEK film (TALPA2000C, Mitsui Toatsu Chemical) inserted between layers as an interleave. APC-2/AS4 was in the form of pre-impregnated unidirectional sheets. Its modulus properties were also found. The APC-2 prepreg is a well-known thermoplastic composite system using a PEEK matrix. Some previous studies discussed the high interlaminar toughness and impact resistance of this system. The PEEK film interleaving significantly reduces delamination caused by impact loading. To investigate the effects of a stacking sequence on resistance, three lay – ups- (0/+30/0/–30)s, (0/+60/0/–60)s and (0/45/90/–45)s – were tested. All laminates were consolidated in an autoclave according to manufacturers' specifications. Flat specimens were cut to 100×100 mm^2 and they were approximately 3 mm thick. Ultrasonic C-scans were performed to ensure that samples were free of gross defects prior to the impact tests.

Gas-gun system and gelatin projectile

The schematics of the gas-gun system are as follows. The specimen was clamped between the two frame plates which have a central circular cutout with a 76.2 mm diameter. The specimen was aligned so as to be impacted at the center of the plate. The supporting conditions were the same as the previous hard body impact tests. The total mass of the projectile was approximately 3 g. The impact velocity ranged from 90 to 190 m/s, and the corresponding impact energy was 12 to 55 J.

The gelatin bullets were composed of a foam material gelatin holder and the gelatin itself. The holders were machined from foam material blocks to the final cup-like shape. The gelatin was poured into the holder and consolidated in a refrigerator. The holder and the gelatin struck the specimens together. The foam material used in the holder was very brittle and the mass fraction of the holder was less than 15%. Thus, the effect of the holder on impact loading was considered to be relatively low. Under the high pressure of the gas-gun, the bond between the gelatin and the holder were broken in the barrel. To obtain sufficient impact energy, some improvements of the gelatin bullet structure were performed with a series of shooting tests. As for the results of the pilot tests, inserting an intermediate epoxy bonding layer in the bottom section of the holder was effective for a higher pressure limit. A pre-bonded foam block, consisting of two pieces of foam blocks bonded together with an epoxy adhesive, was machined to the cup-like holder shape. The upper limit of the gas pressure on the final design of the bullet (Design F) was 0.3 MPa (45 psi), which resulted in an impact velocity of 190–210 m/s and an impact energy of 52–68 J. The deviation of velocity was about 10%, which was almost the same value in former impact test results with the high velocity (HV) gas gun.

Acoustic microscopy

After impact the specimens were evaluated for a damage profile and a projected area using a digital ultrasonic C-scan system (Ultran NDC700). A 10 MHz transducer with a focal length of 12 mm and diameter of 9.6 mm was used. The Sonix Software of Flex Scan 4.0 was used with a Sonix 100 MHz digital board. The bridge precision was 12.7 μm. The scan and step axes both had a length of 81.3 mm (3.2 inches) and an increment of 0.5 mm (0.02 inches). The scan acceleration was 104.6 mm/s^2 (4.12 in./sec^2) and the scan velocity was 43.8 mm/s (1.7247 in./sec). The projected damage area discounts delaminated areas that lie on top of one another.

16.3.4 Gelatin impact damage

Damage profile on acoustic images

The impact damage shapes caused by the gelatin projectile and shown by the C-scan images were almost the same as the damage shapes caused by the aluminum projectile. However, the gelatin impact resulted in smaller projected damage areas. Some of the C-scan images of the impacted specimens, the incident energy of which ranged from 30 to 40 J, are shown in Fig. 16.5. The gelatin impact damage on AS4/PEEK+IL specimens was compared with the damage caused by the aluminum bullets. The damage profile consists of wedge-shaped delaminations bounded between the fiber orientation of the plies above and below the laminae interface.

However, some specimens, especially those with a stacking sequence of (0/30/0/−30)s, indicated a characteristic split damage profile as shown in Fig. 16.6. In the hard body impact, the damage was concentrated around the impact point. The gelatin impact loading, which has a wider contact area and relatively flat contact stress distribution, may be the cause of the split damage. The effects on the unconcentrated loading were also shown in the external appearances of impacted specimens. The hard body impacted specimens have an indentation mark at the impact point and a distinguished matrix cracking at the back surface. The gelatin impacted specimens had no indentations on the impact side and some hair-line cracking on the back side.

Relation between damage area and impact energy

The projected damage areas measured from the C-scan for both gelatin impact and aluminum bullet impacts are shown in Fig. 16.7 as a function of the incident impact energy. The overall level of damage increased with increasing energy. The relationship between DA and IE is linear for both impact conditions in each of the material systems. The ratio of the DA/IE indicates the impact resistance of each specimen, both relatively and quantitatively. The lines in Fig. 16.7 are least-square linear fits. The DA/IE for gelatin impact was less than that of the aluminum bullet impact, and the gelatin impact indicated higher impact energy thresholds, which were 5 to 20 J, shown as the intercepts on energy axis. The value of DA/IE was also changed with stacking sequence, the lower the difference of the fiber angle between adjacent layers, the lower the value of DA/IE.

16.3.5 Discussion

Effects of stacking sequence, stacking parameter $\beta T2$

To consider the effect of layup on the value of DA/IE, a stacking parameter β, based on the mismatching coefficient *mmc*, was proposed. Coefficient

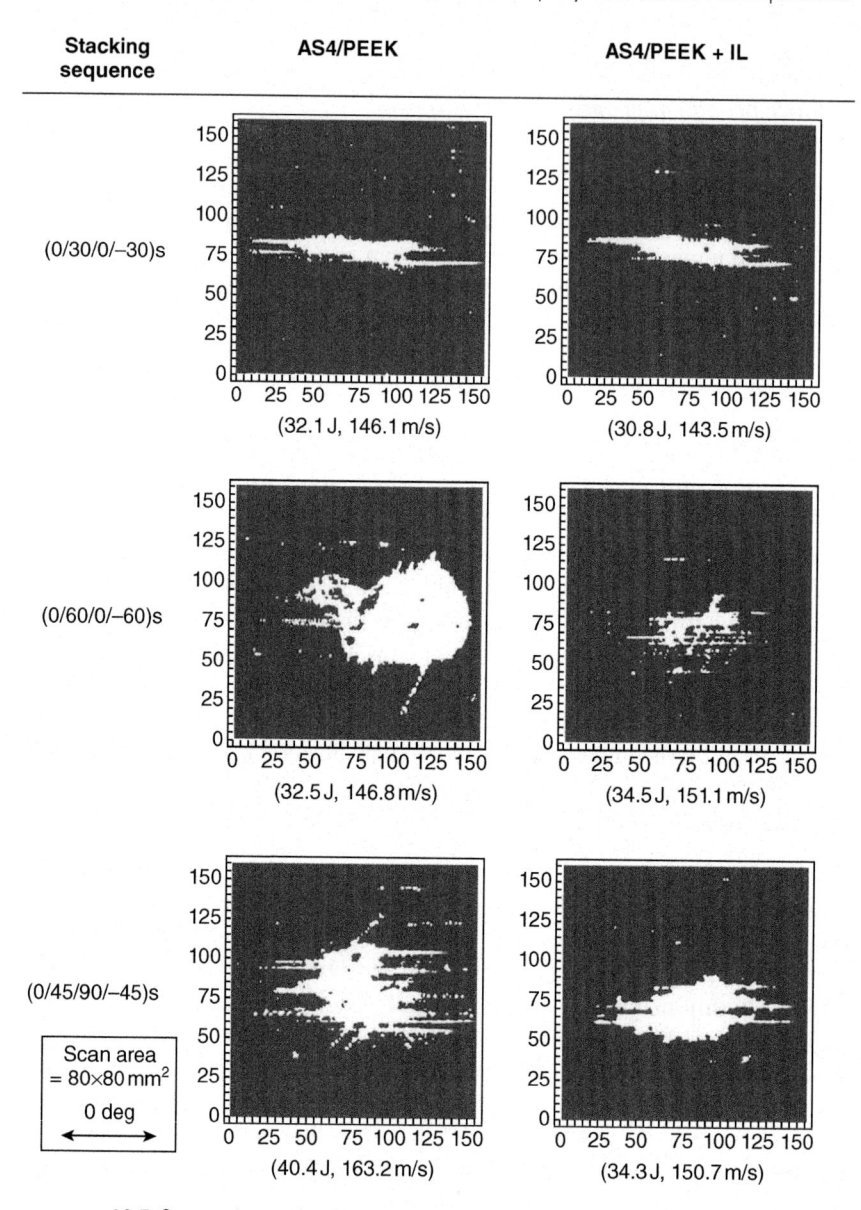

436 Non-destructive evaluation (NDE) of polymer matrix composites

Stacking sequence	AS4/PEEK	AS4/PEEK + IL
(0/30/0/−30)s	(32.1 J, 146.1 m/s)	(30.8 J, 143.5 m/s)
(0/60/0/−60)s	(32.5 J, 146.8 m/s)	(34.5 J, 151.1 m/s)
(0/45/90/−45)s	(40.4 J, 163.2 m/s)	(34.3 J, 150.7 m/s)

Scan area = 80×80 mm² 0 deg

16.5 C-scan images of gelatin damage.

Acoustic microscopy 437

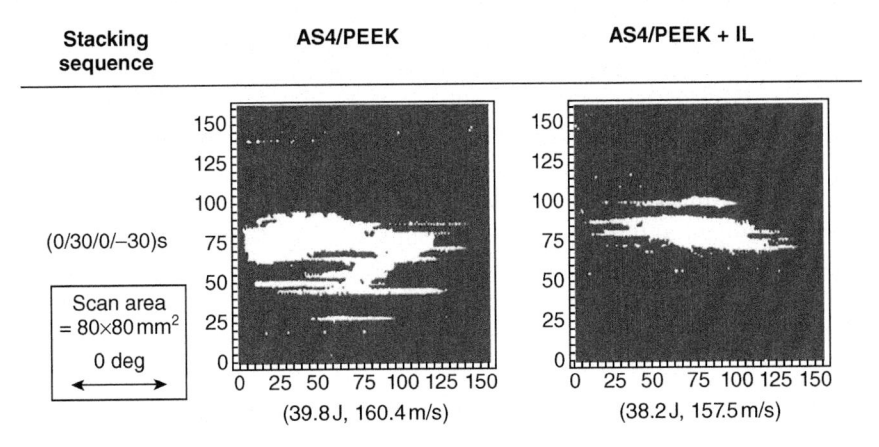

Stacking sequence	AS4/PEEK	AS4/PEEK + IL
(0/30/0/−30)s Scan area = 80×80 mm² 0 deg	(39.8 J, 160.4 m/s)	(38.2 J, 157.5 m/s)

16.6 Split damage caused by gelatin impact.

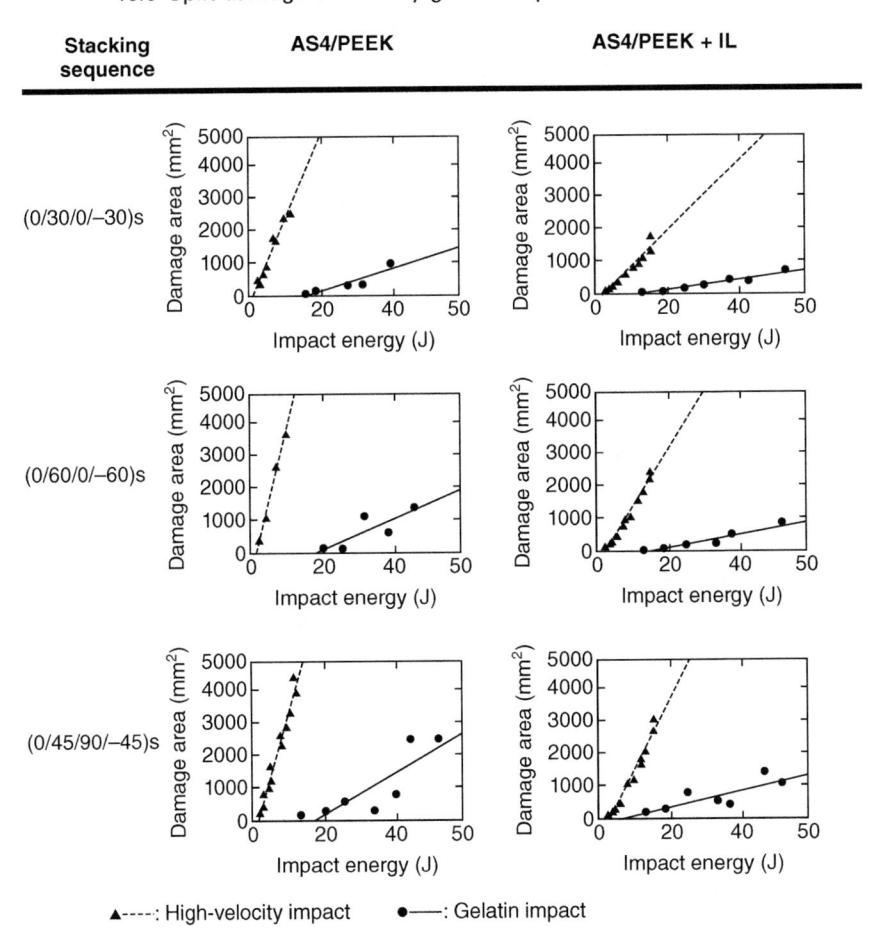

16.7 Projected damage area versus impact energy.

▲----: High-velocity impact   ●——: Gelatin impact

mmc indicates the mismatching of bending stiffness between two adjacent layers, and the delamination area is proportional to the value of *mmc*. However, despite the fact that a unidirectional laminate has no interlaminar interface where adjacent fiber angles are mismatched, the value of *mmc* for an unidirectional laminate is not 0 because it is calculated from the difference of bending stiffness. To resolve this contradiction, the parameter β accounts for the difference of the in-plane stiffness between the adjacent layers as follows. It is assumed that a laminated beam, which has its longitudinal axis aligned with the direction q, is cut from the impacted specimen. This beam is loaded by bending moment M. The maximum stagger offset of bending stress $(\Delta\sigma(\theta))_{max}$ at the interlaminar interface can be presented as a value divided by the bending moment M, namely:

$$\frac{[\Delta\sigma(\theta)]_{max}}{M} = \frac{[\Delta Q_{11}(\theta) \cdot T(\theta)]_{max}}{D_{11}(\theta)} \quad [16.1]$$

where $\Delta Q_{11}(\theta)$ is the difference in in-plane stiffness Q_{11} between two adjacent layers in direction θ, $T(\theta)$ is the distance from the laminate neutral surface to the interface, and $D_{11}(\theta)$ is the bending stiffness of the entire laminate in the direction θ. The parameter β is defined as an integrated value, with respect to θ, of Equation (16.1). That is:

$$\beta = \left(\frac{1}{2}\pi\right)\int_{0}^{2\pi}\{[\Delta Q_{11}(\theta) \cdot T(\theta)]_{max}/D_{11}(\theta)\} \quad [16.2]$$

The stacking parameter β indicates the amount of the bending stress discontinuity at interlaminar interface and it is considered empirically that the larger the parameter β, the larger the impact damage area. A linear relation between DA/IE and β was obtained in each material system for high and low-velocity impact tests with hard body projectiles, and the value of (DA/IE)/β indicated the impact resistance, which is independent of fiber angle. However, specimens which had different lamina thicknesses indicated different (DA/IE)/β, and were found to be dependent on the dimension of β which is (length)$^{-2}$.

To resolve the disadvantage of β, a nondimensional stacking parameter $\beta T2$, modified from β is proposed as follows:

$$\beta T2 = \beta \cdot (t^2) \quad [16.3]$$

where t is the thickness of the lamina. The value of DA/IE, measured in both the previous low- and high-velocity hard body impact tests and the present gelatin impact test, and stacking parameter β and $\beta T2$ of tested specimens were found. In the case of AS4/PEEK+IL specimens, the interleaved layers were ignored in order to calculate the parameters. The thickness of each specimen was assumed to be 3.1 mm, and the thickness of the lamina was interpolated from the total thickness of the specimen. In the low-velocity impact test, the ratio of (DA/IE)/βT2 in the four-layer specimens, which were (0/30)s, (0/60)s and (0/90)s, were larger than eight-layer specimens, which were (0/30/0/–30)s, (0/60/0/–60)s and (0/45/90/–45)s. However, the relation between DA/IE and βT2 were not only proportional, but the data of the four- and eight-layer specimens were on the same fitting line. Thus the effects of a stacking sequence on impact resistance can be assessed with stacking parameter βT2 and the value of (DA/IE)/βT2, indicating the impact resistance, which is independent of both fiber angle and lamina thickness. The (DA/IE)/βT2 values of the tested material were also found.

Relative impact resistance on various impact conditions

The relative impact resistance, measured with three kinds of impact coupon tests (low- and high-velocity hard body impact and gelatin impact), will be compared in this section. These three tests used the same specimens and supporting conditions, but used different projectiles and impact velocities. The impact resistance of tested specimens can be indicated with the values of DA/IE and (DA/IE)/βT2 in each test.

The DA/IE, measured in the low-velocity impact and the gelatin impact, can be shown as a function of the corresponding DA/IE on the high-velocity impact. In the overall correlation, the DA/IE on the high-velocity impact increased with the increasing of both the DA/IE on the low-velocity and the gelatin impact. Therefore, in general, a high-resistance material system, measured with high-velocity test, will also indicate higher resistance in low-velocity and gelatin impact tests. Each material system has three data points with different stacking sequences. The scatter of these three points can be reduced to one point with the (DA/IE)/βT2 values. The CF/PEEK systems (with and without interleave) indicated a good relation of (DA/IE)/βT2 between the high-velocity impact to the low-velocity and the gelatin impact, and it means that the same relative impact resistance on these two systems was obtained through the three different impact conditions. The (DA/IE)/βT2 value of the AS4/PEEK+IL system was about 50–60% less than the AS4/PEEK system in the three kinds of test.

However, the toughened epoxy system had relatively low (DA/IE)/βT2 on the high-velocity impact, but high (DA/IE)/βT2 on the low-velocity impact. Thus, as a result, the relative impact resistance of the different materials was changed with impact conditions, and this phenomenon should be considered carefully in a material screening process.

16.3.6 Conclusions

In this work, the subperforation flat-wise soft body impact resistance of carbon fiber/PEEK systems was evaluated with comparisons of the damage caused by gelatin projectiles and aluminum projectiles. The extent of damage was measured with a digital ultrasonic C-scan system. The relative impact resistance was discussed with the data measured in the previous hard body impact tests and the present gelatin impact test.

The relation between DA and IE is linear for both the gelatin impacts and the aluminum bullet impacts. The ratio of the DA/IE indicates the impact resistance of each specimen, both relatively and quantitatively. The DA/IE of the gelatin impacts was less than that of the aluminum bullet impacts, and the gelatin impacts indicated higher impact energy thresholds.

16.4 Future trends: using embedded ultrasonic sensors for structural health monitoring of aerospace materials

The structural health monitoring of structures during active use has long been of interest to the NDE community. One technique uses embedded ultrasonic sensors and actuators. However, the interpretation of the ultrasonic signals can be difficult, for example in the case of carbon fiber-reinforced composite when the operator tries to distinguish between the growth of harmless micro-cracks and the development of delaminations. This study focuses on two types of structures, aluminum plates such as used in wing structures in aircraft and graphite plates such as encountered in aircraft disc brakes where carbon–carbon composite is used [22–24].

The objective in this work is a source influence study. The technical approach is to use FEM to simulate the ultrasonic emissions from sources represented by piezoelectric wafers embedded in the composites. A special FEM code entitled CAPA was used to model both transmitter and receiver piezoelectrics as well as the guided wave propagation in flat panels of graphite and aluminum alloy. The ultrasonic waveforms were modeled from both transverse and longitudinal embedded piezoelectric sources. The results show distinct differences in the amplitudes, durations and frequency content, creating a potential avenue for embedding ultrasonic sensors for structural health monitoring.

16.4.1 Background

The rapid progress in the performance of personal computers has been a major factor supporting the application of computationally intensive numerical techniques. FEM has been established as one of the standard methods of computer-aided engineering for a large variety of physical systems. Its advantages include a great versatility concerning the model geometry and material composition, the availability of efficient numerical solvers, and the capability of handling coupled-field problems within a single model [25].

FEM solves the governing partial differential equations with the appropriate boundary conditions numerically. Therefore, it enables an accurate computation of complex dynamic behavior, including wave propagation in isotropic and anisotropic solids, and fluids [26, 27]. The complexity of wave propagation in a plate is such that many dispersive modes may exist. If piezoelectric structures such as ultrasonic transducers are considered for simulation, the electric field and its interaction with the mechanical field must also be taken into account. The first application of FEM to modeling of the piezoelectric effect dates back to 1970s [28]. Nowadays, advanced 2D and 3D finite element techniques for simulating piezoelectric transducers are available [29]. Consequently, FEM can be used for modeling wave propagation in plates including the transducers applied in a real-world setup.

16.4.2 Technical approach

The theoretical background of the FEM has been treated extensively in the literature [25–29]. A number of FEM software systems are available. In this section, we perform transient FEM simulations of the complete set-up using the coupled-field FEM system CAPA [25]. The experimental set-up makes use of finite cylindrical transducers and, therefore, requires solution in all three spatial dimensions. Nevertheless, as discussed above, the x-dimension is insignificant for characterization of the physical nature of the waveguide. This allows an efficient 2D FEM analysis, which can be performed on a standard PC. A portion of the 2D models is shown in Fig. 16.8(a)(b).

The commercial broadband transducers applied in the experimental set-up are self-contained components with a complex inner structure involving various materials. The associated material and geometrical parameters are difficult to obtain and represent major uncertainty in the modeling process. For this reason, the transducers are modeled in a simplified manner as homogeneous piezoelectric elements made of the well-known ceramic PZT-5A. Their thicknesses are selected to achieve the appropriate center frequencies of the receiver (~100 KHz). Their bandwidth is controlled by means of manipulating the material damping parameters of PZT-5A in order to match the bandwidth of the real-world transducers.

For generation of a structured 2D finite element mesh (see Fig. 16.8(a)), a discretization rate of at least 12 linear elements per wavelength of the wave is selected, representing a compromise between model size and accuracy [28]. At a frequency of 50 KHz, this discretization rate results in a model consisting of 280 000 finite elements (for $h_1 = h_3 = 1.25$ mm, $h_2 = 1.6$ mm). A transient analysis with 9000 time steps requires about 160

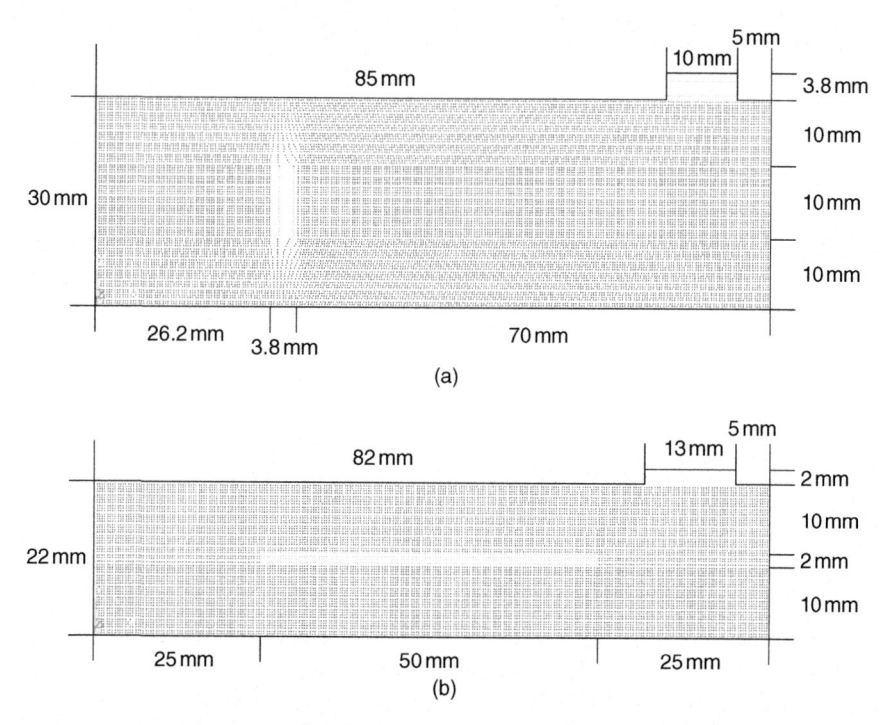

16.8 (a) FEM mesh for source placed perpendicular to the plate. Note that the receiving transducer is in the upper right-hand corner. (b) FEM mesh for source parallel to plate. Note that the detecting transducer is in the upper right corner.

minutes of computation time on a state-of-the-art PC. The effect of the discretization rate on accuracy and size of the model is further discussed in the next section. Finite element simulations were performed as a start on a very simplified basis. The objective was to compare two waveforms: one from a source with displacement in the plane of the sample and the second from a source with displacement perpendicular to the plane of the sample. For this purpose a piezoelectric element was postulated to be placed within an extended plate in the two orthogonal orientations. Each of the elements were considered subjected to voltage spike excitations. Figures 16.8(a)(b) show the two configurations, including dimensions and the meshings used in the simulations.

16.4.3 Results

The results are shown in Figs 16.9–16.12. The simulations are presented in several different ways: as displacement versus time, as fast fourier transform

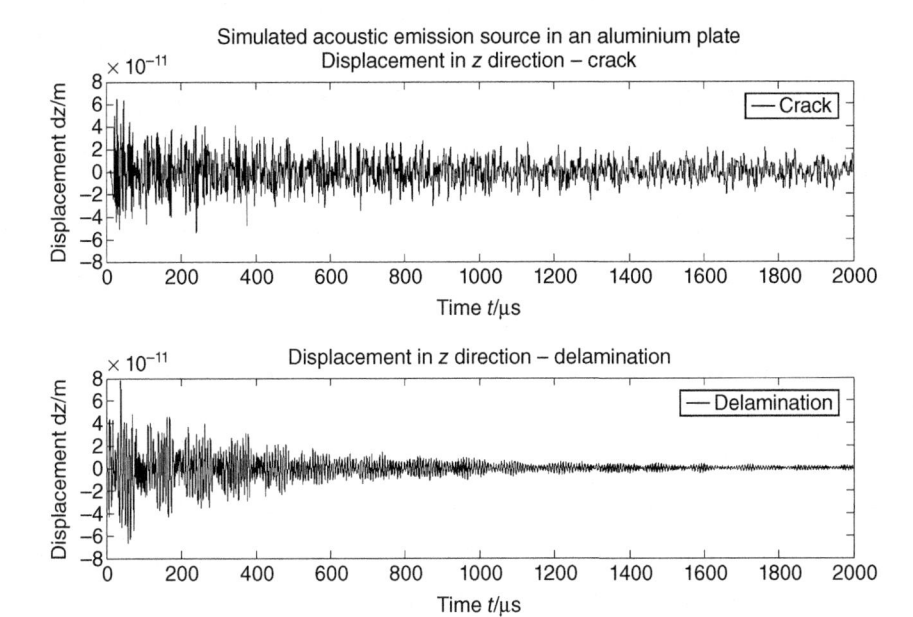

16.9 Simulations of waveforms for two orthogonal orientations of the piezoelectric generating source in an aluminum plate: (a) waveforms for generation perpendicular to the panel, labeled crack; (b) waveforms for generation parallel to the panel, labeled delamination. The time scales are the same.

(FFT) versus frequency, as ultrasonic waveforms versus time and as FFT versus frequency. Figure 16.9 is a plot of the displacement just below the center of the piezoelectric element on the top of the structure at the far right, a likely place for the positioning of a receiving transducer. The simulation shows a strong burst of acoustic energy at the beginning of the time trace followed by a long train of signal pulses.

Another result is shown in Fig. 16.10 displaying the FFT of the waveforms of Fig. 16.9. The frequency spectra for the two are different: (a) waveforms for generation perpendicular to the panel showing a broad spectrum and (b) waveforms for generation parallel to the panel showing a few somewhat narrow bands across the frequency spectrum. The frequency scales are the same.

Similar conclusions could be drawn for an analogous plate of graphite. Figures 16.11 and 16.12 show the displacements in the z-direction under the transducer and electrical potential as an oscilloscope would record it for a graphite panel.

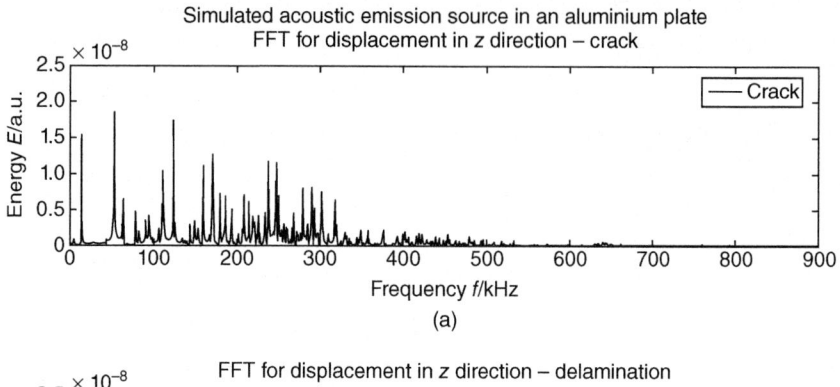

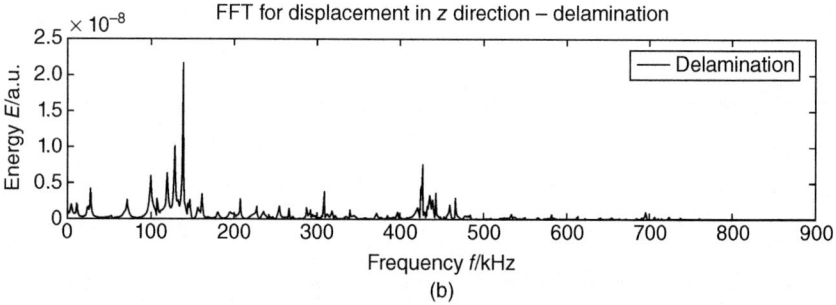

16.10 Simulations of waveform Fourier transforms for two orthogonal orientations of the piezoelectric generating source in an aluminum plate: (a) generation perpendicular to the panel, labeled crack; (b) generation parallel to the panel, labeled delamination. The frequency scales are the same.

16.5 Conclusion

This chapter reports on the use of finite element techniques to predict the source influence signals expected from plates of aluminum and graphite [30]. The simulation for a delamination shows a strong burst of acoustic energy at the beginning of the time trace; whereas the microcrack simulation is seen to produce a long train of signal pulses. This points to a significant difference in the type of acoustic signals and could serve as a guide in recognizing delaminations using sensors embedded in structures.

Two case studies on the use of acoustic microscopy for the characterization of carbon fiber reinforced epoxy composites (CFRP) and aramid fiber composites (AFRP) have been presented. The emphasis is on impact damage ranging from hard impact (bullets, micrometereorites, hail, etc.) to soft impact (bird strike) typically encountered in aerospace applications. Although much effort has been devoted to the study of the effects of

Acoustic microscopy 445

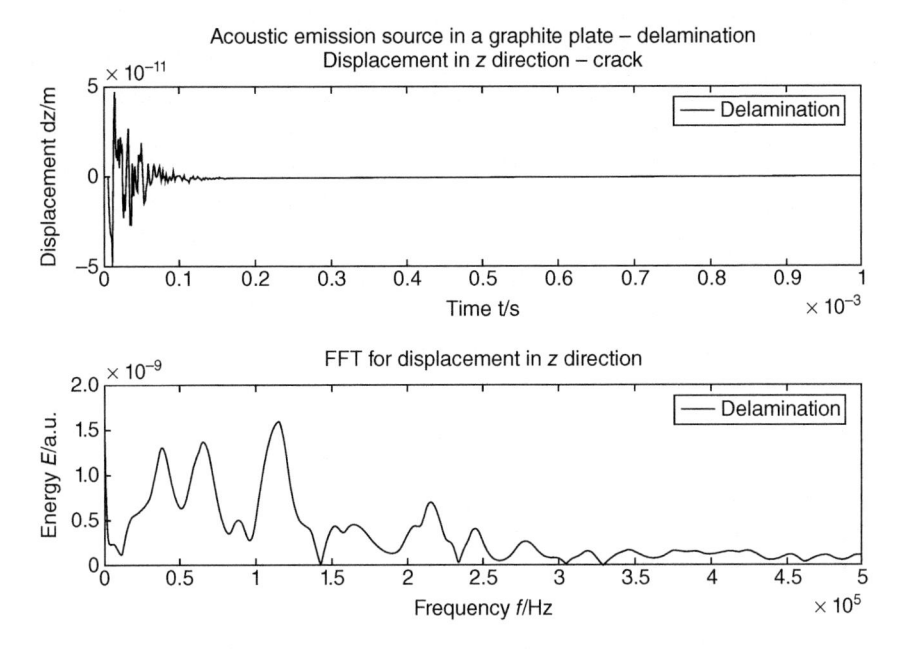

16.11 Simulations of waveform and its FFT for generation parallel to the panel, labeled delamination.

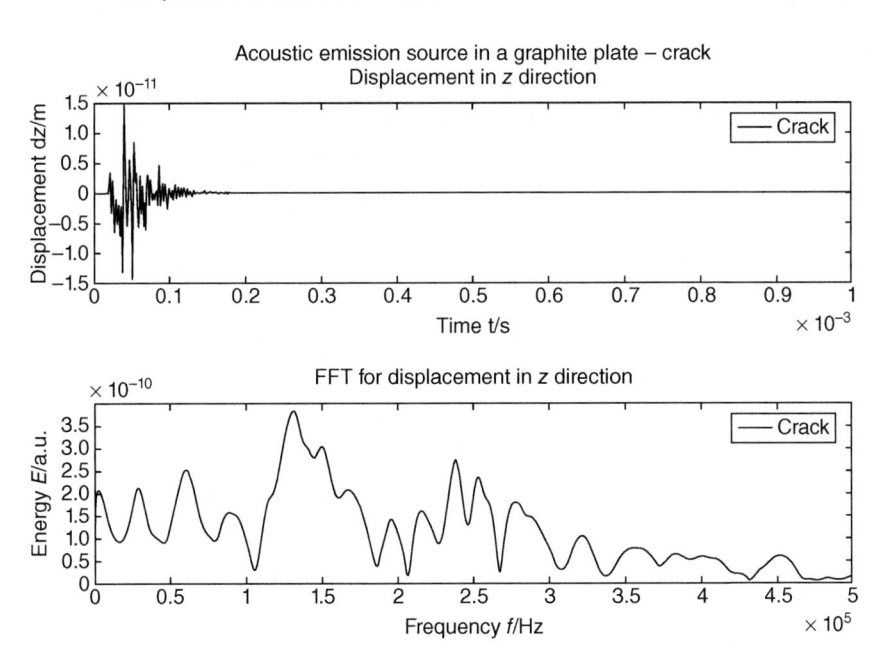

16.12 Simulations of waveform and its FFT for the piezoelectric generating source in an graphite plate perpendicular to the panel, labeled crack.

impact-induced damage on the load-bearing capacities of components in the past 30 years [31] there are a few reports of the impact-induced damage analysis for composite laminates.

Impact-induced damage is the most serious type of damage for carbon fiber reinforced composites (CFRP) and aramid fiber composites (AFRP). In particular, delamination caused by impact loading is known to produce significant reductions in the residual of carbon fiber reinforced plastic laminated composites, which is a major problem associated with the structural integrity of composite compression components. Although aramid fiber has similar density and tensile strength to carbon fiber, tensile modulus and elongation of the aramid fiber are approximately half and double of the carbon fiber, respectively.

Acoustic microscopy is shown here to be uniquely suited to image and characterize damage below the surface, nondestructively. Finite element modeling is an interesting approach to the propagation of ultrasonic waves in composite structures.

16.6 References

1. Tittmann, B.R. and Miyasaka, C., Scanning acoustic microscopy, in *Encyclopedia of Imaging Science and Technology*, J.P. Hornack, Ed., pp. 1228–1248. New York: Wiley. 2002.
2. Miyasaka, C., Kasano, H. and Tittmann, B.R., Experimental analyses on impact-induced damage in CFRP laminated composites with scanned image microscopy. Presented at the 17th Annual International Conference Composites/Nano Engineering (ICCE-17), July 26–31, 2009, Hawaii.
3. Morita, H. and Tittmann, B.R., 'Ultrasonic characterization of soft body impact damage on CF/PEEK laminates with gelatin projectiles,' in *Review of Progress in Quantitative Nondestructive Evaluation*, Vol. 16, D.O. Thompson and D.E. Chimenti, Eds, Plenum Press, NY, 1997.
4. Tittmann, B.R. and Crane, R., 'Ultrasonic inspection of composites,' in *Comprehensive Composite Materials*, A. Kelly and C. Zweben, Eds, Vol. 5, Chapter 12, pp. 259–320, Elsevier Science, New York, 2000.
5. Miyasaka, C. and Tittmann, B.R., 'Principles and applications of practical shear wave lens at low frequencies for scanning acoustic microscopy,' in *Proceedings of 15th World Congress on Nondestructive Testing in NDT for Materials Properties – Methods and Instrumentation*. Rome, Italy, 2001.
6. Challenger, K.D., The damage tolerance of carbon fiber reinforced composites – a workshop summary, *Composite Structures*, **6**, 295–318, 1986.
7. Starnes, J.H., Jr., Rhodes, M.D. and Williams, J.G., Effect of impact damage and holes on the compressive strength of a graphite/epoxy laminate, in *Nondestructive Evaluation and Flaw Criticality for Composite Materials*, ASTM STP696, R.B. Pipes, Ed., 1979, pp. 145–171.
8. Sharma, A.V., Low-velocity impact tests on fibrous composite sandwich structures, in *Test Methods and Design Allowables for Fibrous Composites*, ASTM STP734, C.C. Chamis, Ed., 1981, pp. 54–70.

9. Labor, J.D., Impact damage effects on the strength of advanced composites, in *Nondestuctive Evaluation and Flaw Criticality for Composite Materials*, ASTM STP696, R.B. Pipes, Ed., 1979, pp. 172–184.
10. Williams, J.D. and Rhodes, M.D., Effect of resin on impact damage tolerance of graphite/epoxy laminates, in *Composite Materials: Testing and Design (Sixth Conference)*, ASTM STP787, I.M. Danial, Ed., 1982, pp. 450–480.
11. Sharma Avva, V., Effect of specimen size on the buckling behavior of laminated composites subjected to low-velocity impact, in *Compression Testing of Homogeneous Materials and Composites*, ASTM STP808, R. Chair and R. Papirno, Eds, 1983, pp. 140–154.
12. Rhodes, M.D., Williams, J.G. and Starnes, J.H., Jr., Low velocity impact damage in graphite-fiber reinforced epoxy laminates, *Polymer Composites*, 2(1), 36–44, 1981.
13. Takeda, N., Sierakowski, R.L. and Marvern, L.E., Microscopic observations of cross-sections of impacted composites laminates, *Composite Technology Review*, 41(1), 40–44, 1982.
14. Malvern, L.E., Sierakowski, R.L. and Ross, C.A., 'Impact failure mechanisms in fiber reinforced composite plates,' *Proc. IUTAM*, 1977, pp. 24–27.
15. Chai, H., Knauss, W.G. and Babcock, C.D., Observation of damage growth in compressively loaded laminates, *Experimental Mechanics*, 23, 329–337, 1983.
16. Joshi, S.P. and Sun, C.T., Impact-induced fracture in a laminated composite, *Journal of Composite Materials*, 9, 51–56, 1986.
17. Cantwell, W.J., Curtis, P.T. and Morton, J., An assessment of the impact performance of CFRP reinforced with high strain carbon fibers, *Composites Science and Technology*, 9(2), 40–46, 1987.
18. Joshi, S.P. and Sun, C.T., Impact-induced fracture in a quasi-isotropic laminate, *Journal of Composites Technology & Research*, 9(2), 40–46, 1987.
19. Clark, G., Modeling of impact damage in composite laminates, *Composites*, 20(3), 209–214, 1989.
20. Choi, H.Y. and Chang, F.K., A model for predicting damage in graphite/epoxy laminated composites resolting from low-velocity point impact, *Journal of Composite Materials*, 26, 2134, 1992.
21. Morita, H., Adachi, T., Tateishi, Y. and Matsumoto, H., Charaterization impact damage resistance of CF/PEEK and CF/toughened epoxy laminates under low velocity impact tests, in *Proceedings of the American Society for Composites, Tenth Technical Conference*, 376–385, 1995.
22. Wang, C.J., The effect of resin thermal degradation on thermostructural response of carbon-phenolic composites and the manufacturing process of carbon-carbon composites, *Journal of Reinforced Plastics and Composites*, 15(11), 1111–1129, 1996.
23. Chapman, L.R. and Holcombe Jr., C.E., 'Carbon material and process of manufacturing,' U.S. patent 68309812, 2004.
24. McManus, H.L.N. and Springer, G.S., High temperature thermomechanical behavior of carbon-phenolic and carbon-carbon composites, II. Results, *Journal of Composite Materials*, 26(2), 230–235, 1992.
25. Kaltenbacher, M., *Numerical Simulation of Mechatronic Sensors and Actuators*. Germany: Springer Verlag, 2004.
26. Hughes, T.J.R., *The Finite Element Method*. Dover Publications, Mincola, NY, 2000.

27. Zienkiewicz, O.C. and Taylor, R.L., *The Finite Element Method*. McGraw-Hill, London, 1998.
28. Allik, H. and Hughes, T.J.R., Finite element methods for piezoelectric vibration, *International Journal of Numerical Methods in Engineering*, **2**, 151–157, 1970.
29. Lerch, R., Simulation of piezoelectric devices by two- and three-dimensional finite, elements, *IEEE Transactions on Ultrasonic, Ferroelectrics and Frequency Control*, **37**(3), 233–247, 1990.
30. Wojcik, G.L., Vaughan, D.K., Abboud, N. and Mould, Jr., J., Electromechanical modeling using explicit time-domain finite elements, in *Proc. IEEE Ultrasonics Symposium*, 1993, pp. 1107–1112.
31. Masters, E. and Reifsnider, K.L., *Damage in Composite Materials*, ASTM Publication 775, ASTM, Philadelphia, PA, pp. 40–62, 1982.

17
Non-destructive evaluation (NDE) of aerospace composites: structural health monitoring of aerospace structures using guided wave ultrasonics

M. VEIDT and C. K. LIEW, The University of Queensland, Australia

DOI: 10.1533/9780857093554.3.449

Abstract: Monitoring safety-critical structures over their lifetime in order to improve reliability and availability and reduce life-cycle costs is of vital importance in many technical fields. Aerospace, with zero catastrophic failure tolerance, is the leader in the development of structural health monitoring (SHM) systems and considering the constantly increasing use of composite materials in aerospace applications, the development of SHM systems for aerospace composite materials and structures is of particular interest. This chapter outlines the present progress of health monitoring of aerospace composite structures by compiling the technologies used and summarising significant contributions made. First the three major transducer systems used are briefly reviewed. This is followed by a detailed discussion of the current status in four main areas of guided waves ultrasonic SHM systems development, namely transducer integration, simulation techniques, transducer network optimisation and signal processing. The results show that SHM will be one of the key technologies to ensure the structural integrity of ageing and future composite aircraft structures. At the same time it is recognised that a step-by-step implementation strategy is required to ensure direct technological and economical benefits, in which SHM systems with increasing complexities are taking over more and more responsibility.

Keywords: aerospace structures, composite materials, health monitoring, guided wave ultrasonics.

17.1 Introduction

Monitoring safety-critical structures over their lifetime in order to improve reliability and availability, as well as reduce life-cycle costs, is of great interest in many technical fields, including aerospace, civil and mechanical engineering. Hence, the developments and number of published works concerning structural health monitoring (SHM) has increased rapidly in recent years (e.g. [1–4]).

The aerospace industry, with its particular requirement of zero tolerance for catastrophic failure, has always been the leader in the development of

SHM systems, and it is recognised that SHM will be one of the key technologies to ensure the structural integrity of future aircraft structures. The use of SHM will contribute to (i) reduced structural weight by changing design principles; (ii) reduced maintenance costs; and (iii) increased availability, but it is recognised at the same time that SHM technologies will be introduced only if they have direct economic benefits [5–9]. A step-by-step SHM implementation approach is thus applied, in which different SHM systems with increasing complexities are building up on each other and advancing the technology to meet these industry requirements. Consequently, following the increasing use of composites in aerospace applications, the development of SHM systems for these materials and structures is of particular interest (e.g. [10–12]).

A SHM system is defined as an integrated part of the structure to be monitored and should ideally be able to detect, locate and evaluate damage in the structure, estimate its severity and monitor its evolution with time. Many non-destructive testing (NDT) methods exist for damage detection and evaluation, but not all of them are well suited for SHM, owing to integration and cost issues. The main approaches for SHM in aerospace composite structures are systems using optical fibre sensors to measure strains at several positions along the fibre or piezoelectric transducers which use structural vibrations or propagating guided waves to determine the integrity of the structure [11–15]. A third, relatively new SHM transducer technology that is starting to become popular in aerospace applications is comparative vacuum monitoring, in which pressure changes in micro cavities are monitored to indicate damage [16,17]. All these methods allow large areas to be monitored without the need of scanning processes and the devices can be permanently attached or integrated into the structure. In addition, the piezoelectric transducers can be used not only as sensors but also as transmitters, which allows the development of active monitoring systems.

The most widely used SHM methodology for composite aerospace structures is guided wave (GW) ultrasonics, in which the analysis of stress waves propagating along the component is used to identify damage (e.g. [4,14,15,18–20]). GW ultrasonics is applied as an active as well as passive monitoring techniques [21], where the latter is comparable to an autonomous acoustic emission measurement system. But the majority of GW SHM systems in development use active transducers to excite specific stress wave modes to interrogate the component. The wave modes are selected based on their propagation behaviour within the structure and their sensitivity to particular damage types. In this configuration wave scattering and wave mode conversion are analysed to detect and characterise damage.

Major design aspects that govern any SHM system are (i) changes of the monitored physical properties due to damage; (ii) transducer–structure interaction to reliably measure the expected changes with the required resolution; and (iii) signal processing and analysis tools to extract the required information on structural integrity. Although continuous advancements have been made in all these areas over the years as illustrated, for example, in the references listed above, the transition of SHM to become common practice in design of new and maintenance of existing aerospace platforms is relatively slow, the main reasons being high development costs and uncertainties regarding performance and reliability of the SHM transducers and measurement systems.

The aim of this review is to report on the current status of the application of SHM to aerospace composite structures by collating recent significant contributions in the areas of SHM design outlined above. The next section provides a brief overview of the major SHM transducer systems used for aerospace composite structures and their key performance characteristics. In Section 17.3, the current status in four areas of GW SHM system development is discussed and the current major challenges identified. The essential findings of the review are summarised in the final section.

17.2 Structural health monitoring (SHM) transducer systems

17.2.1 Piezoelectric transducer systems

Smart materials technology incorporating piezoelectric transducers was pioneered in the aerospace industry when there was a need to control vibrations of large structures like satellites and solar array panels as well as for isolation and noise control of high-precision equipment for radio frequency applications and in optical and microgravity experiments [22–26].

In the last two decades, piezoelectric materials have gained significant interest in SHM. The SHM systems developed with attached or integrated piezoelectric transducers are aimed at both detecting and characterising in-service damage and manufacturing flaws before they grow and propagate into critical failure modes. The defects targeted with piezoelectric SHM systems are diverse and exist in a wide range of materials, e.g. fatigue cracks in concrete, corrosion in metals, delamination in monolithic laminates, and skin–core disbond in honeycomb sandwich structures. The prospect of the application of piezoelectric transducer systems in SHM thus spans into a number of major industries including aerospace, transport, civil, mining and energy production [27–34]. There are also research activities that investigate the application of the technology to monitor the integrity of composite strengtheners and repair patches [35–38].

Piezoceramics, especially those made from lead zirconium titanate (PZT), are currently the most widely used and reliable transducers in GW SHM

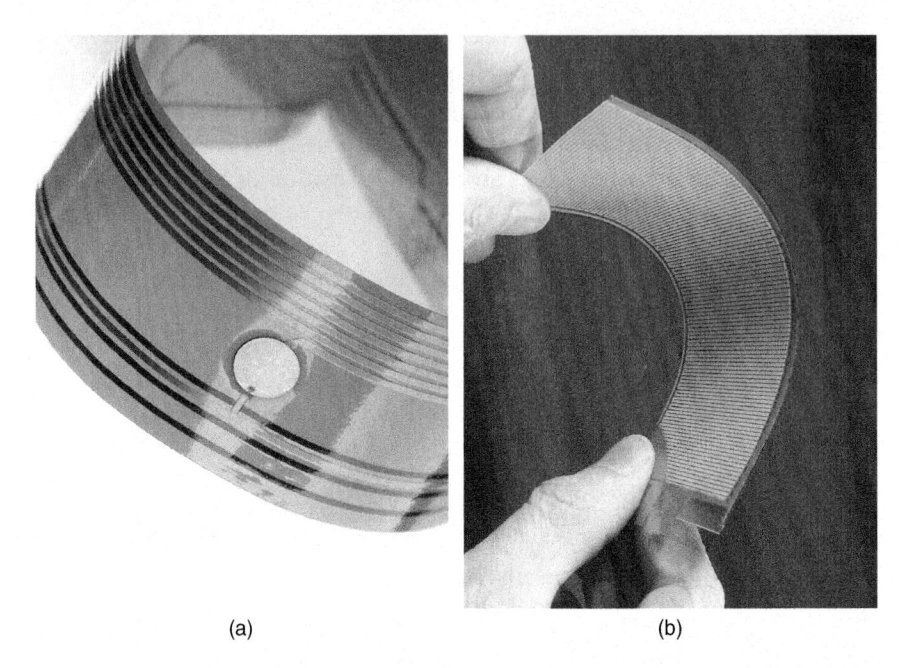

452 Non-destructive evaluation (NDE) of polymer matrix composites

17.1 Piezoceramic SHM transducer solutions: (a) SMART Layer® from Acellent Techologies, reproduced with permission from Acellent Techologies (www.acellent.com); (b) Active Fibre Composites from Materials Systems Inc. Reproduced with permission from MSI (www.matsysinc.com).

applications. Industry SHM solutions like SMART Layer and Active Fibre Composite (AFC) shown in Fig. 17.1 are derived from the application of piezoceramics but with different transducer constructions.

The SMART Layer is made of piezoceramic transducers distributed on polyimide film with a printed circuit [39] and has been reported in use for a number of composite applications [40,41]. AFCs consist of unidirectional piezoceramic fibres sandwiched between interdigitated electrodes printed on polyimide film [42,43]. AFC is mainly used for shape and vibration control of structures but has recently been recognised for SHM applications [44,45]. Other types of piezoelectric transducers are, for example, DuraAct patches made of a layer of piezoelectric foil sandwiched with conductive electrodes and encapsulated in a polyimide membrane [46] or low-profile flexible piezopolymers particularly polyvindylidene fluoride (PVDF) thin films [47–49], which are especially considered as GW sensors.

17.2.2 Fibre optic Bragg grating systems

Apart from piezoelectric elements, optical fibres are also reliable and robust yet lightweight types of sensor that are commonly used in SHM transducer

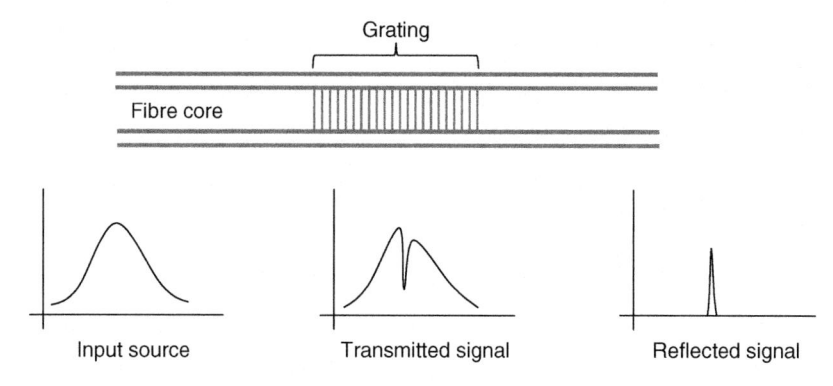

17.2 Fundamental operating principle of fibre Bragg grating.

systems. Among the different styles of fibre optic sensors, fibre Bragg grating (FBG) receives the most attention for SHM purposes. This is due to its low cost, ease of implementation and resistance to harsh environments. In addition, it is a reference-free measurement technique contrary to interferometric-type fibre optic sensors. The grating is a series of close parallel lines printed into the core of the fibre, which when loaded exhibit different reflection and transmission properties due to a change in the grating spacing, as schematically illustrated in Fig. 17.2.

Strain in a structure is identified as a shift in the reflected signal from the FBG. The optical sensor is thus effectively a strain measuring sensor. The advantages in SHM applications compared with conventional resistance strain gauges are that FBG are more durable, possess a lower profile and no electrical wiring is required. In addition, they have multiplexing capabilities, i.e. multiple measurement locations along the fibre can be created by varying the grating spacing.

For damage detection, the FBG identifies this change as a deformation in shape and reduction in amplitude of the reflected spectrum due to the presence of non-uniform strains and loss in the structure's load-carrying ability. In addition, FBG can also replace thermocouples to measure thermal load profiles, as well as monitor the quality of cure and adhesive bondline of repair patches [50,51]. Most studies observe the strain directly to monitor for damage [52–54] or directly measure the light intensity and power in the reflection spectra from FBG [55–57].

Because of the advantages of FBG over strain gauges, these optical sensors have received significant interest for operational load monitoring in aerospace structures [58,59]. The composite laminate construction also facilitates the embedment of these low-profile sensors where no strength degradation of the primary structure has been observed [60]. To prolong the lifespan of these sensors, stringent practices are applied in manufacturing the fibre gratings while a protective coating is necessary for long-term reliability [61]. Integration of FBG for SHM also considers important issues like temperature compensation, interrogation and fibre encapsulation techniques [62].

17.2.3 Comparative vacuum monitoring systems

One of the relatively new SHM technologies that has gained accreditation by major commercial aircraft industries is comparative vacuum monitoring (CVM) developed by Structural Monitoring Systems Ltd (SMS) [63]. The construction of the CVM sensor is simple but robust and durable. The sensor is typically made of a flexible elastomer that can conform to a curved surface if required and attachment is via a pressure-sensitive adhesive applied under atmospheric conditions or integrated directly into the construction of the host structure. Inside the sensor, there are fine cavities known as 'galleries' which are alternately filled with low vacuum and atmospheric pressure (Fig. 17.3). When a flaw develops under the sensor, this causes air to flow to the vacuum galleries to stabilise the pressure difference. This change from a reference can be measured by a fluid flow meter connected to the sensor to infer the presence of the flaw. SMS offers instruments to facilitate inspections at workshop and *in situ* environments.

The CVM system has been successfully used for crack monitoring at hot spots particularly near rivet joints in aluminium structures since its early stage of development [16,17]. The sensor displays a very high sensitivity to the crack size and its growth but this sensitivity is compromised when more than one flaw is present under the one sensor patch [16]. Sensors can be manufactured to small footprints to mitigate this problem while large area inspection is preserved by linking the sensors. Recently, CVM has also

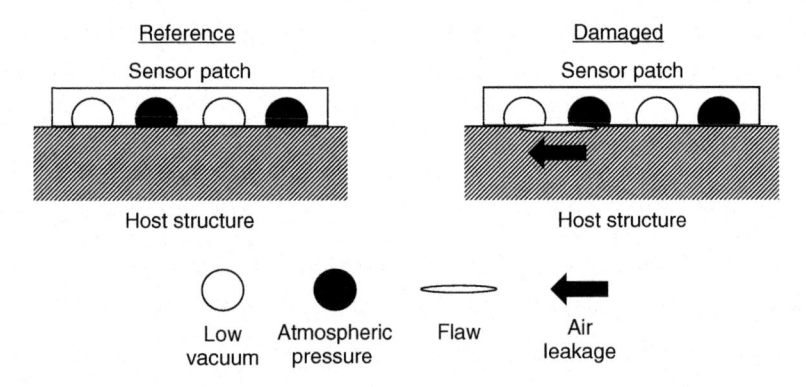

17.3 Simple schematics showing the operating principle of the CVM sensor.

demonstrated potential in composite inspection by monitoring for disbond, low-velocity impact damage and crack growth in repair patches [64,65].

However, there are integration issues when the sensor is embedded within a laminate. Experimental investigations show that embedded galleries can affect the strength and stiffness of the host structure but these effects can be minimised by applying galleries below critical sizes, as well as oriented along the loading direction and normal to the damage propagation path [66,67].

17.3 Guided wave (GW) structural health monitoring (SHM) systems for composite structures

As mentioned in Section 17.1, the most widely investigated SHM approach in aerospace composite structures is the application of ultrasonic GWs. GW is a term used to represent a wave propagating along a confined space, and whose propagation characteristics are dependent on the structural boundaries. Examples of guided waves are Rayleigh, Love, Stoneley and Lamb waves (e.g. [68,69]). Lamb waves are especially popular in SHM for composite inspection due to the ability to propagate large distances along a thin structure like composite laminates. Unlike bulk waves used in conventional ultrasonic testing, these GWs have wavelengths much larger than the thickness of the structure and hence energy is constrained within its boundaries. This becomes the motivation of extensive developments in this field as large area inspection offered by GWs can save inspection time, allows monitoring of structural regions with limited access and overcomes limitations of through-thickness conventional ultrasonic testing for thin structural components (e.g. [14,15,18–20]).

GWs contain rich information regarding the mechanical properties along their path of propagation and possess high sensitivity to changes in material and geometric inhomogeneities due to damage. But the dispersive behaviour of GWs and the existence of an infinite number of wave modes creates major challenges in the interpretation of GW response signals for damage detection and characterisation. Dispersion effects cause phase shifts due to velocity variations as a function of frequency while mode conversion may occur as a result of interaction with defects, discontinuities and boundaries. Although signal processing algorithms exist to help reducing dispersive effects and compensate for environmental changes, which may result in alterations of the interrogation wave pulse [19,70,71], extracting quantitative information from GW signals requires thorough understanding of the physical processes related to propagation and scattering of elastic waves in solids. If the interaction of GWs with damages is fully understood, analysing time or frequency domain signal components can be used to determine physical damage characteristics. Detecting and locating damage is the primary focus for the diagnostics stage of SHM. Major techniques for this purpose are time-of-flight based methods such as beamforming [72–74] or time reversal methods [75–77]. More quantitative information on damage size and severity is required for the prognostics aspect of SHM, which has the aim of predicting residual strength or life time of a component. Diffraction tomography [78–82] is one of the most promising techniques to determine quantitative damage characteristics. In all these techniques apart from time reversal reconstruction the signals used to infer the presence of damage are calculated by methods like baseline subtraction where signals are compared to a reference. The main problem with this method is the requirement of a good reference signal and accurate comparison. This can be a challenging task as there are many variables which can affect the overall transient wave signals, such as environmental effects like noise and temperature [83,84] or long-term transducer durability and signal transfer consistency [85].

The next sections discuss four major aspects of GW SHM system design in more detail and identify the major challenges with a special focus on aerospace composite structures. Section 17.3.1 considers transducer and system reliability and durability. In Section 17.3.2 the simulation tools currently used to predict the system performance are discussed and Section 17.3.3 investigates the particular problem of optimal transducer network design. Finally the major methods of processing SHM sensor signals are reviewed in Section 17.3.4.

17.3.1 Transducer performance and integration

Primary and secondary aerospace structures are required to carry loads which may exert up to 3000 microstrains, while design standards typically require composite parts to support much higher strains [86]. Apart from loading, these structures can also encounter a wide range of environmental effects such as extreme temperatures and moisture. These conditions impose significant degradation effects on sensors integrated into the structure for SHM.

Some of the early works gauge the durability of embedded piezoceramic sensors by simply measuring their output voltage during mechanical testing of the substrate specimen to detect for any loss in performance [87,88]. Electrical and temperature cyclic loadings have also been considered together with static and fatigue loadings in the tests [89–91]. In general, it has been found that these sensors can presently sustain actuation and sensing performance up to 2000 microstrains. Above this strain level, failures attributed to sensor fracture, depolarisation and interface debonding can occur. Sensor voltage can drop by more than 10% if thermal cyclic loading is introduced to the specimen. Similar results have been found for the durability of piezoceramic fibres in integrated AFCs where fibre fragmentation has been observed above 2000 microstrains [92,93].

Electrical impedance or admittance is another effective way of measuring piezoceramic transducer durability as the method considers both its electrical and mechanical properties. The impedance spectra have shown high sensitivity to a broad range of ultrasonic frequencies in order to capture local or near-field dynamics around the transducer [94,95]. According to the admittance curve, interface debonding of the transducer is related to a relative increase in the gradient and moves towards the gradient of an unbonded element while a drop in this gradient is an indication of transducer damage (Fig. 17.4). At present, the method has been successfully used to determine the durability of surface-bonded sensors where similar transducer performance outcomes as voltage measurements have been obtained [97,98].

Although piezoelectric devices have yet to achieve the stringent requirements of aerospace composite applications, understanding the operating conditions and service loads of the host structure is the first step of achieving a robust ultrasonic transducer design. Selection of suitable embedment techniques, matching interfacial materials and profile conformation are some of the range of design aspects that need to be considered carefully. Protection of the electrodes and the adherences of interfaces are also essential to ensure continuous electrical connection to the transducers for SHM operation. Embedding piezoceramic elements at the neutral axis of the

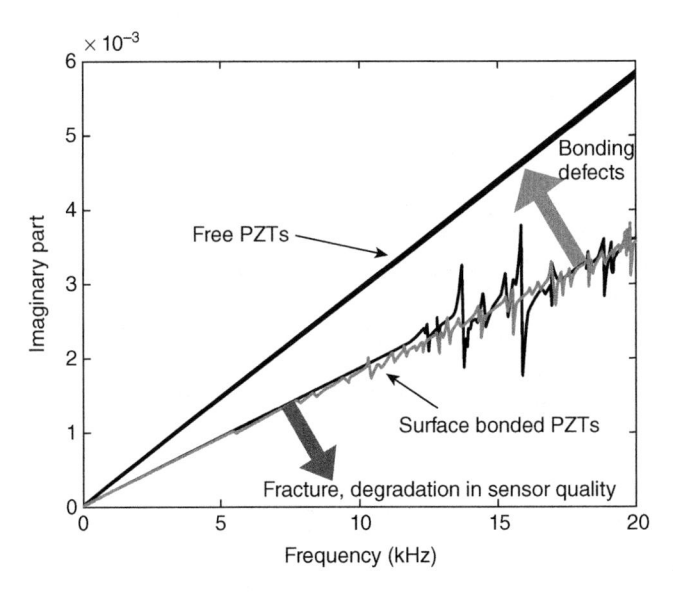

17.4 Electrical admittance versus frequency plots to measure the quality of bonded PZT sensors (from [96]).

structure where bending stress is at minimum can be applied to reduce transverse strain transfer to the transducer. However, increasing the number of plies above the transducer can reduce acoustic transduction efficiency [98]. A multilayer transducer construction can also be used to provide some buffer in strain transfer [99] but again considerations must be given to ensure adequate actuation performance. For greater flexibility, AFCs and piezopolymers can serve as alternative sensors to provide more resistance to bending loads compared to monolithic piezoceramic wafers. A stand-off element has also recently been introduced as an interface between surface-mounted piezoceramic wafers and the host structure to reduce transducer stress by local out-of-plane bending [100]. Integration issues are discussed in more detail in the following sub-section.

For FBG, there are two main concerns with respect to sensor performance and durability. The first issue is the overall optical fibre strength. This is usually not a problem for glass silica used as the base material for the fibres, as this can withstand over 5000 microstrains. However, the FBG fabrication process including fibre drawing and grating writing can introduce internal stresses and surface damages that significantly degrade the mechanical strength of the optical fibre [101,102]. The second issue is that the stability of the grating is susceptible to high temperatures. Continuous exposure to temperatures above room conditions can cause the grating to degrade with time and thus affect the reflectivity response of the sensor [103]. Careful manufacturing processes and the application of coating technologies are important to improve the lifespan of FBG sensors especially for those used in harsh environments. At the end-user, selection of suitable optical fibres depending on the application is vital for the long-term reliability of the FBG system as an SHM tool.

Transducer integration

In comparison to surface mounted sensors, embedded sensors possess the benefit of being shielded from external damage and direct environmental effects while preserving the aerodynamic profile of the structure. However, this comes with additional design considerations in the manufacturing process of the host structure to embed the sensor. A sensor element can be integrated into a laminate host structure either by direct embedment or by a cut-out method. In both ways, the sensor is included in the ply layup prior to curing with the only difference that the cut-out method requires a volume of plies to be removed to accommodate the sensor, Fig. 17.5.

Direct embedment results in stiffness increase due to additional volume of material present and creates resin-rich regions around the edges of the sensor. The cut-out method introduces discontinuity in addition to surrounding resin-rich pockets. These can be a source of stress concentration,

17.5 (a) Direct and (b) cut-out methods for embedding a monolithic PZT sensor (■) in a composite laminate.

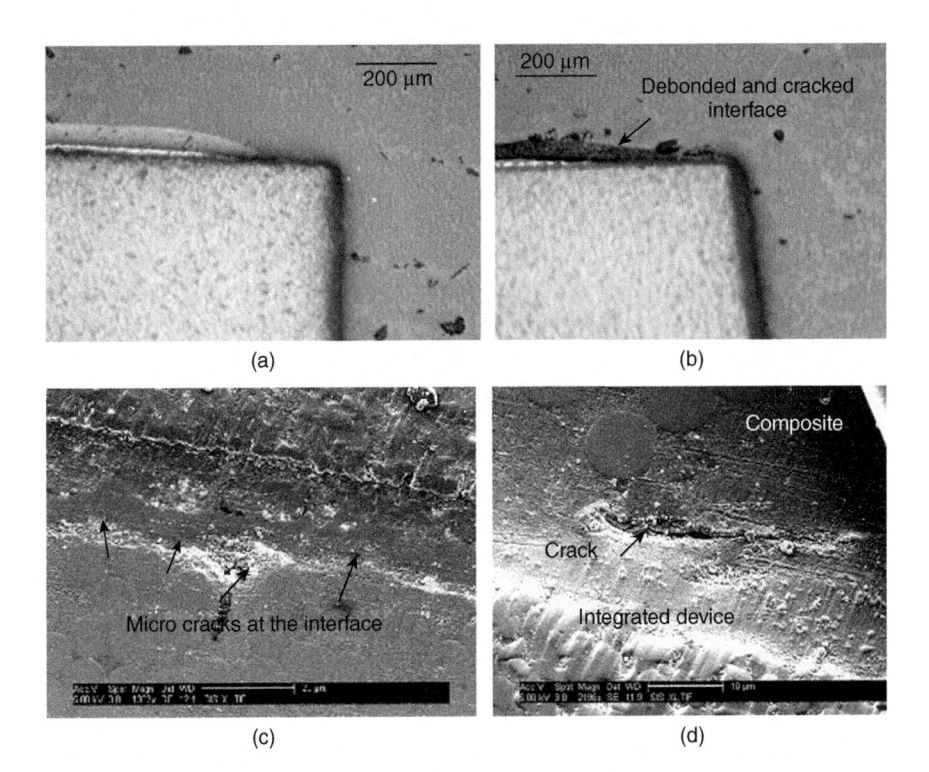

17.6 Micrographs around a corner of an integrated sensor in glass fibre epoxy: (a) an unstressed sample, (b) a sample tensile tested to 400 MPa in the horizontal direction along the plane of the images, (c) and (d) micro-cracking at the interface between integrated sensor and composite (from [109]).

crack initiation, interface debonding and overall reduction in the strength of the host structure [104,105]. These detrimental effects are dependent on its shape, size and position, as well as the laminate ply stacking sequence. However, there can be little to negligible impact if the geometry of the host structure is unaffected and size of resin-rich pockets are insignificant in the case of thin sensors, thick laminate or sensor characteristic length aligned to the fibre direction [106–108] (Fig. 17.6).

In static loading, typically less than 5% strength degradation is observed for piezoceramic sensors embedded directly at the mid-plane of the host laminate structure [109]. This figure can increase up to 15% for fatigue loading with degradation observed especially in the shear mode [110,111]. Increasing tensile strength degradation can also be observed for sensor elements located further from the mid-plane of the host structure due to increasing non-uniform stress distribution through the laminate thickness. Critical failure modes for thin monolithic sensor embedments are debonding and micro-cracks at the sensor-laminate interface which is different for fibre optic sensors where failure occurs at the tip of the resin-rich pocket [105,112].

Instead of a volume cut-out, plies can also be interlaced around the embedded sensor for load redistribution and thus delaying the onset of damage and improving the strength of the structure [113–115]. A less sophisticated approach is to either insert or replace a large ply area with a thin layer consisting sensor elements, as demonstrated in the SMART Layer technology [39–41]. When inserting a sensor layer, it needs to be ensured that any slight increase in laminate thickness does not affect the function and load-bearing capability of the original structure while a sensor layer which replaces a large ply area must be able to bear the load supported by the replaced ply. In both cases, the transfer of load via the sensor layer is important and this depends on the interface attachment between the host structure and the sensor layer, as well as their construction and materials.

Integrated piezoceramic transducer requires encapsulation to shield it from electrical interference. This is of particular importance to composites with conductive fibres, especially carbon fibres which may short circuit the sensors. Non-conductive glass or polyester fibre layers can be applied as insulating interface ply layers. A more practical approach is the encapsulation of sensors with polyimide film or Kapton tape used commonly in commercial sensors. The issue with the use of polyimide film is its low adherence properties which can directly affect the load-bearing capability within the region of the encapsulated sensor. To improve the adhesion of the film to the composite matrix, polyimide can be treated either using dry or wet processes for better wettability, chemical bonding and interlayer diffusion [116]. Dry methods such as plasma activation have been shown to significantly increase peel resistance of polyimide to epoxy [117] while wet methods using a curing agent like potassium hydroxide can introduce functional groups on the polyimide surface for better adhesion properties [118].

Wiring between transducer and instrumentation such as function generator, signal converter and amplifier is an important design parameter in sensor embedment. Wiring requirements are different for different sensor constructions. For monolithic piezoceramics, a wire lead needs to be attached either by bonding or soldering directly to the sensor. Interdigitated electrodes are used in AFC [43] while the SMART Layer has a polyimide printed circuit [39]. Locating sensors at the edge or thickness side of the composite structure is a recommended design to ensure minimal embedded lead wires running through the composite which are sources of stress concentration, a known problem faced by FBG sensors. Embedding of wires can be via either direct or cut-out methods, similar to the embedment of the sensor.

There have been a lot of recent advances in wireless technology for SHM to minimise the amount of wiring from sensors to data acquisition and processing devices [119–122]. Most of these developments are targeted at civil structures but similar concepts and hardware designs can be applied for aerospace composites. The fundamental design of this technology involves the use of a wireless sensing unit connected in close proximity to the sensor which functions to drive the sensor as well as to transmit signals to a central data collection infrastructure. The sensing unit can support multiple embedded sensors and can be installed with software algorithms to process the data on-the-fly and deliver damage detection results, thus automating the monitoring process and saving time [123,124]. The wireless sensing unit can be connected to the sensor via the wiring techniques while powering can be by a battery pack or by energy harvesting methods [125,126].

17.3.2 Simulating system performance

The capability to simulate the performance of SHM systems is essential to design optimised applications for particular platforms due to the complex characteristics of individual system components and their interaction. The simulation of two design aspects are considered: (i) propagation of GW and their interaction with structural features and damage; and (ii) transducer–structure interaction for excitation and detection of GW. The topic of the design of transducer networks for optimal performance regarding probability of detection and robustness are discussed in Section 17.3.3.

Guided wave propagation and scattering

Selecting a wave mode with optimal propagation characteristics in regard of dispersive behaviour and sensitivity to the particular damage type is an essential component in the SHM system design process. Many research groups have developed their own numerical tools to calculate dispersion relations for different materials and structures. The major commercial analytical tool is DISPERSE, developed by researchers at Imperial College [127]. The program can handle different geometries and types of materials including single and multilayer, elastic and viscoelastic anisotropic

462 Non-destructive evaluation (NDE) of polymer matrix composites

laminates. However, complicated material systems require significant amount of guidance from the user to trace all wave modes in the entire frequency space. In addition, the Global Matrix Method [128] implemented in DISPERSE requires longer computation time as the number of layers increases. Recently it has become more common to use semi-analytical finite element (SAFE) or other numerical simulation techniques to generate dispersion curves and determine through-thickness displacement fields and energy propagation directions [129–131].

Analytic solutions for GW propagation problems only exist for a very limited number of simple cases. But for real engineering structures with complex geometries, joints and varying thicknesses and for advanced engineering materials such as composite laminates with anisotropic characteristics, numerical techniques are required to simulate the propagation behaviour of guided waves. Explicit finite element (eFE) analysis has served as primary computational tool for understanding stress wave propagation and scatter at inhomogeneities. Use of eFE modelling has been encouraged by competitive advantages for modelling geometrical complexity combined with material anisotropy, when compared with other numerical simulation methods such as finite difference or boundary element methods. eFE analysis has successfully been used to simulate GWs propagation and scattering in composite laminates (e.g. [132–134]). 2D and 3D simulations have been performed and simulation results for scattering at delaminations have been validated experimentally (Fig. 17.7).

However, eFE computational costs for 3D problems are very high, and it is necessary to carefully test and optimise simulation procedures for every structural component and damage type to guarantee accurate predictions, although accepted best practice recommendations exist regarding mesh

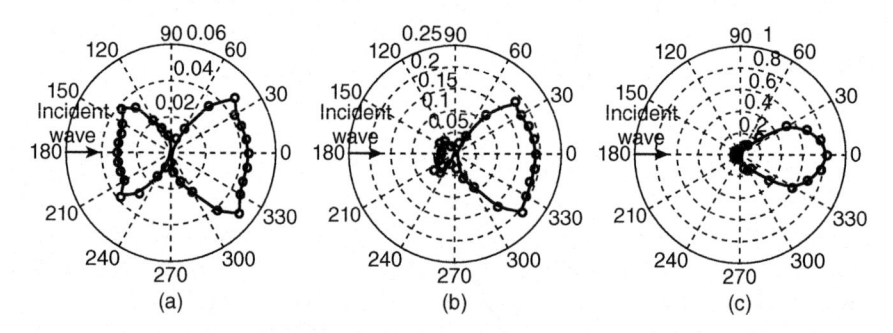

17.7 Scatter directivity patterns from explicit finite element simulation for 140 kHz A_0 Lamb waves from (a) 3 mm, (b) 5 mm, and (c) 11 mm diameter delaminations located between the 4th and 5th plies (centre) of a [45/–45/0/90]$_s$ composite laminate (from [135]).

density, optimisation for stability and robustness, and damage modelling. Recent developments using global–local simulation techniques and absorbing boundary layers address the problem of high computational costs and try to increase the simulation area for complex composite structures [136–139]. Another development in the last 10 years is the application of time domain spectral element (TDSE) analysis as a computationally more efficient numerical method for solving GW propagation and scattering problems in complex structures, e.g. [140,141]. But all these new methods and concepts require the development of special purpose (sub-) programs and their use as efficient tools for GW SHM system design for complex composite structures still remains to be demonstrated.

Transducer–structure interaction

Understanding the transducer–structure interaction for the excitation and detection of GWs is essential to designing optimised SHM systems. In almost all simulation studies discussed above, GWs are produced by applying simple piston-type displacement or pressure boundary conditions under the transducer area. However, piezoelectric actuators generally create complex waves particularly in anisotropic substrates and their dynamic characteristics vary greatly with the type of attachment to the host structure. Hence, it is desirable to know the complete 3D stress and displacement fields generated by the transducer and to calculate the GW field of a transmitting transducer with the aim to understand for which cases these 1D models are good approximations.

Rajic [142] presents a comprehensive analysis for axisymmetric configurations, i.e. for piezoelectric circular disc wafer transducers adhesively bonded on plates. The simulation is using a finite difference (FD) formulation and includes viscoelastic behaviour as well as transverse isotropy of the piezoceramic transducer, adhesive layer and substrate. A comparison of experimental and calculated far-field waveforms shows that uniform pressure piston excitation is not a good model for a 500 kHz narrow band toneburst excitation of a 20 mm diameter, 1 mm thick Pz27 transducer on a 0.65 mm thick aluminium plate. Including the viscoelastic layer is essential and the predictive accuracy of the model greatly depends on the quality of the material property data (Fig. 17.8).

FE simulation methods have been by used by Dumas *et al.*, [143], Velichko and Wilcox [144] and Gao and Rose [145]. A hybrid implicit/explicit FE approach is presented by Bottai *et al.* [146], where an implicit dynamic model handles the coupled electro-mechanical constitutive equations and explicit time integration handles the wave propagation. In Raghavan and Cesnik [147] explicit FE simulation is used to validate an analytical method to calculate the far-field solution to the 3D Green's function for isotropic

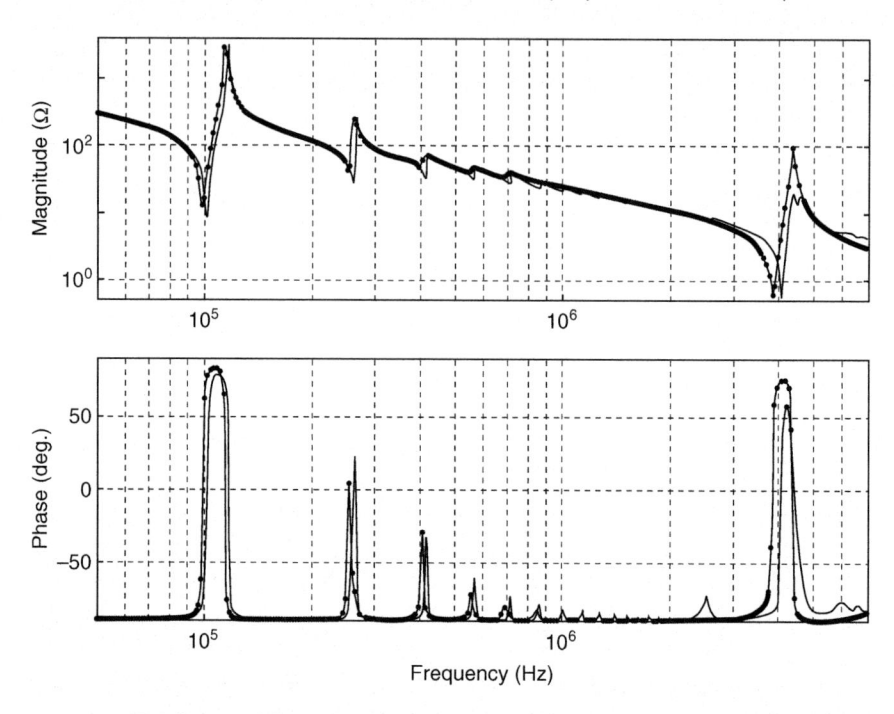

17.8 Comparision of measured (solid line) and calculated (line with symbols) impedance spectra for a 20 mm diameter 0.5 mm thick Pz27 transducer (from [142]).

and anisotropic plates. A very similar approach is presented by Ditri and Rose [148] where a semi-analytical normal mode expansion technique is applied to simulate GWs excited from surface mounted piezoelectric transducers and the predicted wave fields are compared with results from standard FE calculations.

TDSE analysis has recently been extended to include piezoelectric excitation and detection [149,150] using a hybrid TDSE formulation, which takes advantage of the high-order spectral element formulation in the in-plane direction and applies a linear FE formulation in the thickness direction. This hybrid approach maintains accuracy with increased numerical efficiency, especially for low-frequency wave modes.

17.3.3 Optimal transducer networks

The development of reliable damage detection and evaluation methods is not the only key technology required to successfully monitor the integrity and safety of structures. Without an optimised and robust transducer network, the measured data may not contain enough information for damage diagnosis, or detection performance may deteriorate over time.

Deciding how many transducers to use and where to place them is one of the most important decisions associated with the design of smart materials in general and GW SHM systems in particular. However, this is a very challenging problem because of the large number of possible transducer locations and complexity of structures and damage. A transducer network needs to provide effective coverage of the inspected structure with minimum interference from boundary effects. At the same time, sensors must not be integrated in a way that can compromise the original strength and integrity of the structure.

A promising technique to address this issue is optimal transducer placement (OTP), which is being actively researched for aerospace structures as well as other contexts (e.g. [151,152]). OTP is a general automated technique that determines the optimal or near-optimal configuration of transducer networks by solving an optimisation problem with concurrent satisfaction of specified design and functional objectives. In addition, it is essential to include as prior knowledge engineering expertise regarding critical areas, expected damage characteristics as well as functional requirements of the structure [151]. Optimisation techniques can then be used to help decide on the ideal positions and optimal number of transducers, including the important concept of fail-safe networks [153], recognising that transducer failure and degradation is an important problem in multi-transducer architectures. Constraint optimisation problems have a great degree of complexity and various tools based on biological and physical analogies have been developed in recent years to solve multi-variable, constraint optimisation systems such as genetic algorithms, simulated annealing, and tabu search, which are well documented (e.g. [154,155]).

The latest techniques of optimal transducer network design for GW SHM systems apply probabilistic optimisation techniques with the aim of including uncertainties of different SHM system processes [156,157]. The process in Guratzsch and Mahadevan [156] includes combining numerical simulation of wave propagation and scattering with probabilistic analysis to estimate the variability of structural response and sensor output. Based on this, a fitness function is calculated and optimised in order to maximise the reliability of damage detection. In Flynn and Todd [157] the theoretical framework of Bayes risk is used to calculate an objective function that is either maximised for probability-of-detection (given allowable false alarms) or minimised for false alarm rate (given a target probability-of-detection) (Fig. 17.9).

17.3.4 Signal processing

Typical raw signals collected in GW SHM are in the time domain. It has been shown that damage in composite materials can be identified directly

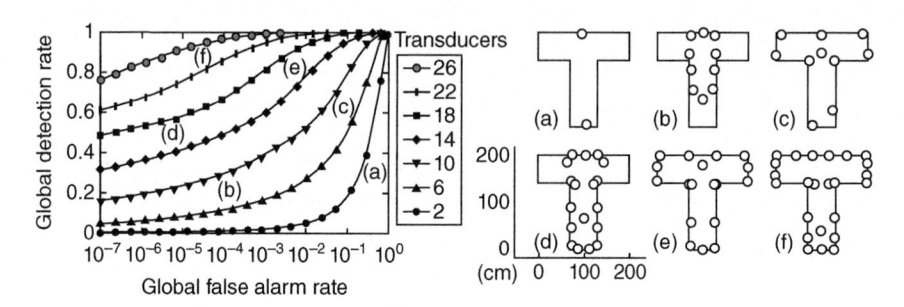

17.9 Transducer network optimisation results for a T-shaped plate by balancing between global detection rate and global false alarm rate (from [157]).

from time series analysis (e.g. [1,4]). However, the difficulty remains in making quantitative estimates of physical damage properties and isolating defect-generated GW responses from other influences such as environmental changes, dispersive changes of different propagating wave modes and interferences of reflections from boundaries or structural features.

Frequency analysis is an established approach to examine guided wave signals (e.g. [158]). Transforming the time signals to the Fourier space, the dispersion curves can be obtained while the wave modes can be separated. In this approach time information is lost and thus frequency analysis does not readily permit the determination of the defect location. Recently, time-frequency analysis, in particular wavelet transform, has been used to overcome this problem [159–162]. The wavelet transform convolutes the time signal against different scales and translation of a wavelet function to generate a map of coefficients that preserves both the information of time and frequency. This rich information then allows an elaborate comparison between reference and damaged structures to quantify the flaw present (Fig. 17.10).

Wavelet analysis has also been used as a tool for feature extraction of sensitive parameters to be used as inputs of intelligent systems, e.g. fuzzy logic and neural networks, which in turn produce outputs which are able to identify the damage [163,164] (Fig. 17.11).

17.4 Conclusion

This review of the current status and trends of health monitoring of aerospace composite structures shows that extended efforts are undertaken in all fields of SHM research and constant progress is made to make SHM one of the key technologies to ensure the structural integrity of ageing and future composite aircraft structures. The results also show that a step-by-

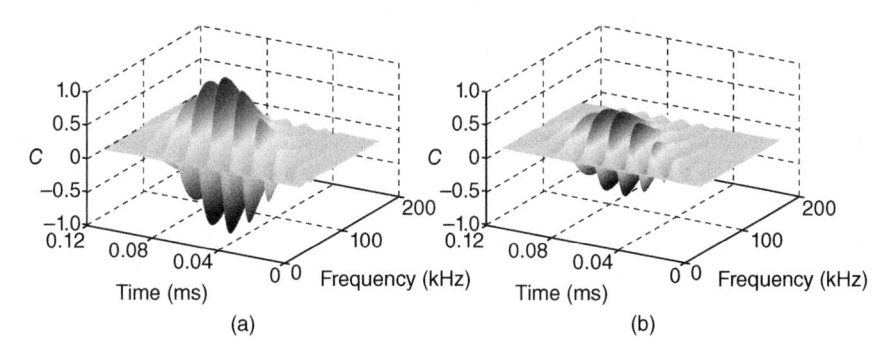

17.10 Wavelet coefficients C of 80 kHz A_0 Lamb waves measured on an (a) undamaged reference and (b) a delaminated laminate plate. Sensor measurement is 200 mm from a Pz27 actuator with a 20 mm delamination located centrally in this straight line wave path on the damaged plate (from [162]). Reproduced with permission from Liew, CK & Veidt, M 2011, 'Damage Evaluation in Smart Materials using Time-Frequency Analysis', *Advanced Materials Research*, Vol. 313, pp. 2299–2306, Trans Tech Publications.

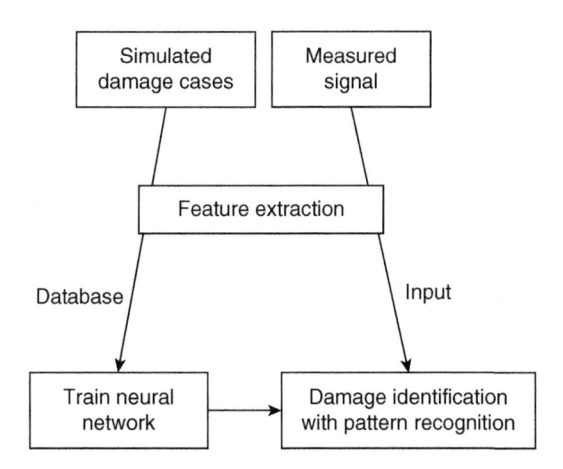

17.11 A typical pattern recognition scheme using neural networks. A large representative damage case database obtained from simulation is used to train a neural network. The trained neural network is then used to identify the damage from a measured signal. Both simulation and measurement data undergo feature extraction using tools like wavelet analysis to obtain sensitive parameters for damage identification.

step implementation approach is most promising, in which SHM systems with increasing complexities build up on each other to advance the technology and to meet the fundamental requirement of direct economic benefits that the industry has for the implementation of any SHM system.

Piezoelectric transducers, fibre optic Bragg grating and comparative vacuum monitoring systems are currently the major types of integrated transducers for SHM applications, and GW ultrasonics is the main inspection methodology for aerospace composite structures. Four design aspects have been identified where additional development is required to further advance SHM of composite structures and enable the transition to practical applications, namely (i) transducer performance and integration; (ii) system simulation; (iii) network optimisation; and (iv) signal processing.

The current literature shows that it is still a challenge to achieve the stringent requirements of aerospace composite applications regarding SHM system durability and reproducibility. Environmental influences and operating conditions may result in failure of embedded or surface mounted transducers and their mechanical and electrical connections, or create inconsistent system responses. The latter is a special challenge for comparative measurement techniques which are using baseline response signals to identify and characterise damage. The development of new manufacturing processes and optimal designs of transducer systems are vital to solve the problem of long-term reliability of high-fidelity transducers.

The capability to simulate the performance of SHM systems is recognised to be essential to enable the design of optimised applications. The two fundamental system characteristics of particular interest are (i) the propagation of guided waves in composite structures and their interaction with structural features and damage; and (ii) transducer–structure interaction for excitation and detection. The capability to efficiently and accurately simulate the system response are basic requirements to select optimal excitation signals, determine number and locations of transducers for maximum probability-of-detection with minimum false alarms and develop signal processing methodologies. Although constant progress is made using explicit finite element and time domain spectral element analysis, the efficiency of simulating large area complex components needs to be further improved.

In summary, this review has shown the potential and importance of SHM for aerospace composite structures and has identified fundamental issues regarding system sensitivity, robustness and design efficiency that need to be resolved until it can play its role as a key technology to ensure the structural integrity of safety critical, high-value aerospace infrastructure assets.

17.5 References

1. Boller, C, Chang, FK & Fujino, Y 2009, *Encyclopedia of Structural Health Monitoring*, John Wiley & Sons Inc., Hoboken.
2. Balageas, D, Fritzen, CP & Guemes, A 2006, *Structural Health Monitoring*, ISTE Ltd, London.
3. Farrar, CR & Worden, K 2007, 'An Introduction to Structural Health Monitoring', *Philosophical Transactions of the Royal Society A*, Vol. 365, pp. 303–315.

4. Staszewski, WJ, Boller, C & Tomlinson, GR 2004, *Health Monitoring of Aerospace Structures: Smart Sensor Technologies and Signal Processing*, John Wiley & Sons Ltd, Chichester.
5. Kapoor, H, Boller, C, Giljohann, S & Braun, C 2010, 'Strategies for Structural Health Monitoring Implementation Potential Assessment in Aircraft Operational Life Extensions Considerations', *2nd International Symposium on NDT in Aerospace*, Hamburg.
6. Giurgiutiu, V 2008, 'Structural health monitoring and NDE – An Air Force Office of Scientific Research Structural Mechanics Program Perspective', *Proceedings of the 4th European Workshop on Structural Health Monitoring*, Cracow.
7. Speckmann, H & Daniel, JP 2004, 'Structural Health Monitoring for Airliner, from Research to User Requirements, a European View', *Collection of Technical Papers – CANEUS 2004 – Conference on Micro-Nano-Technologies for Aerospace Applications*, California.
8. Arnaiz, A, Buderath, M & Ferreiro, S 2009, 'Condition Based Operational Risk Assessment for Improved Aircraft Operability', *Safety, Reliability and Risk Analysis: Theory, Methods and Applications – Proceedings of the Joint ESREL and SRA-Europe Conference*, Prague.
9. Balageas, DL 2002, 'Structural Health Monitoring R&D', *European Research Establishments in Aeronautics (EREA), Aerospace Science and Technology*, Vol. 6, pp. 159–170.
10. Herszberg, I, Bannister, MK, Verijenko, V, Li, HCH & Buderath, M 2009, 'Integration of Structural Health Monitoring for Composite Structures into the Aircraft Health Management and Maintenance Systems', *17th ICCM International Conference on Composite Materials*, Edinburgh.
11. Takeda, N 2008, 'Recent Developments of Structural Health Monitoring Technologies for Aircraft Composite Structures', *26th International Congress of the Aeronautical Sciences*, Anchorage, USA.
12. Diamanti, K & Soutis, C 2010, 'Structural Health Monitoring Techniques for Aircraft Composite Structures', *Progress in Aerospace Sciences*, Vol. 46, pp. 342–352.
13. Montalvao, D, Maia, NMM & Ribeiro, AMR 2006, 'A Review of Vibration-based Structural Health Monitoring with Special Emphasis on Composite', *The Shock and Vibration Digest*, Vol. 38, pp. 295–324.
14. Raghavan, A & Cesnik, CES 2007, 'Review of Guided-wave Structural Health Monitoring', *The Shock and Vibration Digest*, Vol. 39, pp. 91–114.
15. Su, Z & Ye, L 2009, *Identification of Damage using Lamb Waves: From Fundamentals to Applications*, Springer Verlag, Berlin.
16. Wishaw, M & Barton, DP 2001, 'Comparative Vacuum Monitoring: A New Method of *in situ* Real-time Crack Detection and Monitoring', *10th Asia-Pacific Conference on Non-Destructive Testing (APCNDT)*, Brisbane, Australia.
17. Roach, D 2009, 'Real Time Crack Detection using Mountable Comparative Vacuum Monitoring Sensors', *Smart Structures and Systems*, Vol. 5, pp. 317–328.
18. Clarke, T & Cawley, P 2011, 'Enhancing the defect localization capability of a guided wave SHM system applied to a complex structure', *Structural Health Monitoring*, Vol. 10, pp. 247–259.

19. Croxford, AJ, Wilcox, PD, Drinkwater, BW & Konstantinidis, G 2007, 'Strategies for Guided-wave Structural Health Monitoring', *Proceedings of the Royal Society A: Mathematical, Physical and Engineering Sciences*, Vol. 463, pp. 2961–2981.
20. Ng, CT & Veidt, M 2009, 'Lamb Wave Based Imaging of Damage in Composite Laminates', *Smart Materials and Structures*, Vol. 18, 074006.
21. Staszewski, WJ, Mahzan, S & Traynor, R 2009, 'Health Monitoring of Aerospace Composite Structures – Active and Passive Approach', *Composites Science and Technology*, Vol. 69, pp. 1678–1685.
22. Bailey, T & Hubbard, EJ 1985, 'Distributed Piezoelectric-Polymer Active Vibration Control of a Cantilever Beam', *Journal of Guidance, Control and Dynamics*, Vol. 8, No. 5, pp. 605–611.
23. Scheulen, D & Baier, H 1994, 'Smart Materials for Active Vibration Damping and Shape Control', *International Symposium on Advanced Materials for Lightweight Structures*, Netherlands, pp. 485–490.
24. Resch, M, Berg, H & Elspass, WJ 1995, 'System Identification and Vibration Control for Composite Structures with Embedded Actuators', *Smart Structures and Materials 1995: Smart Structures and Integrated Systems*, San Diego, pp. 511–519.
25. Quenon, D, Boyd, J, Buchele, P, Self, R, Davis, T, Hintz, T & Jacobs, J 2001, 'Miniature Vibration Isolation System for Space Applications', *Smart Structures and Materials 2001: Industrial and Commercial Applications of Smart Structures Technologies*, Newport Beach, CA, pp. 159–170.
26. Damaren, CJ & Oguamanam, DCD 2004, 'Vibration Control of Spacecraft Box Structures using a Collocated Piezo-Actuator/Sensor', *Journal of Intelligent Material Systems and Structures*, Vol. 15, No. 5, pp. 369–374.
27. Munn, TE, Kent, RM, Bartolini, A, Gause, CB, Borinski, JW, Dietz, J, Elster, JL, Boyd, C, Vicari, L, Cooper, K, Ray, A, Keller, E, Venkata, V & Sastry, SC 2002, *Health Monitoring for Airframe Structural Characterization*, Technical Report CR-2002-211428, NASA Langley Research Center.
28. Staszewski, W, Boller, C & Tomlinson, G 2004, *Health Monitoring of Aerospace Structures: Smart Sensor Technologies and Signal Processing*; John Wiley & Sons Ltd, Chichester.
29. Xiaoling, Z, Huidong, G, Guangfan, Z, Ayhan, B, Fei, Y, Chiman, K and Rose, JL 2007, 'Active Health Monitoring of an Aircraft Wing with Embedded Piezoelectric Sensor/Actuator Network: I. Defect Detection, Localization and Growth Monitoring', *Smart Materials and Structures*, Vol. 16, No. 4, pp. 1208–1217.
30. Brownjohn, JWW 2007, 'Structural Health Monitoring of Civil Infrastructure', *Philosophical Transactions of the Royal Society A*, Vol. 365, pp. 589–622.
31. Song, G, Mo, YL, Otero, K & Gu, H 2006, 'Health Monitoring and Rehabilitation of a Concrete Structure using Intelligent Materials', *Smart Materials and Structures*, Vol. 15, No. 2, pp. 309–314.
32. Song, G, Gu, H, Mo, YL, Hsu, TTC & Dhonde, H 2007, 'Concrete Structural Health Monitoring using Embedded Piezoceramic Transducers', *Smart Materials and Structures*, Vol. 16, No. 4, pp. 959–968.
33. Mijarez, R, Gaydecki, P & Burdekin, M 2007, 'Flood Member Detection for Real-Time Structural Health Monitoring of Sub-Sea Structures of Offshore Steel Oilrigs', *Smart Materials and Structures*, Vol. 16, No. 5, pp. 1857–1869.

34. Kim, MH 2006, 'A Smart Health Monitoring System with Application to Welded Structures using Piezoceramic and Fiber Optic Transducers', *Journal of Intelligent Material Systems and Structures*, Vol. 17, No. 1, pp. 35–44.
35. Chiu, WK, Galea, SC, Koss, LL & Rajic, N 2000, 'Damage Detection in Bonded Repairs using Piezoceramics', *Smart Materials and Structures*, Vol. 9, No. 4, pp. 633–641.
36. Saafi, M & Sayyah, T 2001, 'Health Monitoring of Concrete Structures Strengthened with Advanced Composite Materials using Piezoelectric Transducers', *Composites Part B*, Vol. 32, No. 4, pp. 333–342.
37. Wu, Z, Ghosh, K, Qing, X, Karbhari, V & Chang, FK 2006, 'Structural Health Monitoring of Composite Repair Patches in Bridge Rehabilitation', *Smart Structures and Materials and Materials 2006: Sensors and Smart Structures Technologies for Civil, Mechanical, and Aerospace Systems*, San Diego, pp. 1–9.
38. Baker, A 2008, *Structural Health Monitoring of a Bonded Composite Patch Repair on a Fatigue-Cracked F-111C Wing*, DSTO-RR-0335.
39. Lin, M & Chang, FK 2002, 'The Manufacture of Composite Structures with a Built-in Network of Piezoceramics', *Composites Science and Technology*, Vol. 62, pp. 919–939.
40. Qing, X, Beard, SJ, Kumar, A, Ooi, TK & Chang, FK 2007, 'Built-in Sensor Network for Structural Health Monitoring of Composite Structure', *Journal of Intelligent Material Systems and Structures*, Vol. 18, pp. 39–49.
41. Qing, X, Beard, SJ, Kumar, A, Chan, HL & Ikegami, R 2006, 'Advances in the Development of Built-in Diagnostic System for Filament Wound Composite Structures', *Composites Science and Technology*, Vol. 66, pp. 1694–1702.
42. Bent, AA & Hagood, NW 1997, 'Piezoelectric Fiber Composites with Interdigitated Electrodes', *Journal of Intelligent Material Systems and Structures*, Vol. 8, pp. 903–919.
43. Nelson, LJ 2002, 'Smart Piezoelectric Fibre Composites', *Materials Science and Technology*, Vol. 18, pp. 1245–1256.
44. Brunner, AJ, Barbezat, M, Huber, C & Flueler, PH 2005, 'The Potential of Active Fiber Composites made from Piezoelectric Fibers for Actuating and Sensing Applications in Structural Health Monitoring', *Materials and Structures*, Vol. 38, pp. 561–567.
45. Brunner, AJ, Birchmeier, M, Melnykowycz, MM & Barbezat, M 2009, 'Piezoelectric Fiber Composites as Sensor Elements for Structural Health Monitoring and Adaptive Material Systems', *Journal of Intelligent Material Systems and Structures*, Vol. 20, pp. 1045–1055.
46. PI Ceramic 2007, *DuraAct – Piezoelectric Patch Transducers for Industry and Research*, Physik Instrumente GmbH & Co.
47. Rathod, VT, Mahapatra, DR, Jain, A & Gayathri, A 2010, 'Characterization of a Large-Area PVDF Thin Film for Electro-mechanical and Ultrasonic Sensing Applications', *Sensors and Actuators A: Physical*, Vol. 163, pp. 164–171.
48. Caneva, C, De Rosa, IM & Sarasini, F 2008, 'Monitoring of Impacted Aramid-reinforced Composites by Embedded PVDF Acoustic Emission Sensors', *Strain*, Vol. 44, pp. 308–316.
49. De Rosa, IM & Sarasini, F 2010, 'Use of PVDF as Acoustic Emission Sensor for *in situ* Monitoring of Mechanical Behaviour of Glass/Epoxy Laminates', *Polymer Testing*, Vol. 29, pp. 749–758.

50. Botsev, Y, Gorbatov, N & Tur, M 2004, 'Fiber Bragg Grating Sensing in Smart Composite Patch Repairs for Aging Aircraft', *Proceedings of SPIE*, Vol. 5502, pp. 100–103.
51. Ben-Simon, U, Kressel, I, Botsev, Y, Green, AK, Ghilai, G, Gorbatov, N, Tur, M & Gali, S 2007, 'Residual Strain Measurement in Bonded Composite Repairs for Aging Aircraft by Embedded Fiber Bragg Grating Sensors', *Proceedings of SPIE*, Vol. 6619.
52. McKenzie, I, Jones, R, Marshall, IH & Galea, S 2000, 'Optical Fibre Sensors for Health Monitoring of Bonded Repair Systems', *Composite Structures*, Vol. 50, pp. 405–416.
53. Jones, R & Galea, S 2002, 'Health Monitoring of Composite Repairs and Joints using Optical Fibres', *Composite Structures*, Vol. 58, pp. 397–403.
54. Tasmasphyros, GJ, Kanderakis, GN, Furnarakis, NK & Marioli-Riga, ZP 2003, 'Detection of Patch Debonding in Composite Repaired Cracked Metallic Specimens using Optical Fibers and Sensors', *Proceedings of SPIE*, Vol. 5145, pp. 128–136.
55. Sekine, H, Fujimoto, SE, Okabe, T, Takeda, N & Yokoburi Jr, T 2006, 'Structural Health Monitoring of Cracked Aircraft Panels Repaired with Bonded Patches using Fiber Bragg Grating Sensors', *Applied Composite Materials*, Vol. 13, pp. 87–98.
56. Takeda, S, Yamamoto, T, Okabe, Y & Takeda, N 2007, 'Debonding Monitoring of Composite Repair Patches using Embedded Small-diameter FBG Sensors', *Smart Materials and Structures*, Vol. 16, pp. 763–770.
57. Li, HCH, Beck, F, Dupouy, O, Herszberg, I, Stoddart, PR, Davis, CE & Mouritz, AP 2006, 'Strain-based Health Assessment of Bonded Composite Repairs', *Composite Structures*, Vol. 76, pp. 234–242.
58. Ecke, W, Latka, I, Willsch, R, Reutlinger, A & Graue, R 2001, 'Fibre Optic Sensor Network for Spacecraft Health Monitoring', *Measurement Science and Technology*, Vol. 12, pp. 974–980.
59. Murayama, H, Kageyama, K, Naruse, H, Shimada, A & Uzawa, K 2003, 'Application of Fiber-optic Distributed Sensors to Health Monitoring for Full-scale Composite Structures', *Journal of Intelligent Material Systems and Structures*, Vol. 14, pp. 3–13.
60. Satori, K, Fukuchi, K, Kurosawa, Y, Hongo, A & Takeda, N 2001, 'Polyimide-coated Small-diameter Optical Fiber Sensors for Embedding in Composite Laminate Structures', *Proceedings of SPIE*, Vol. 4328, pp. 285–294.
61. Foote, P, Breidne, M, Levin, K, Papadopolous, P, Read, I, Signorazzi, M, Nilsson, LK, Stubbe, R & Claesson, A 2004, 'Operational Load Monitoring Using Optical Fibre Sensors', in Staszewski, W, Boller, C & Tomlinson, G, *Health Monitoring of Aerospace Structures*, John Wiley & Sons Ltd, Chichester, pp. 75–123.
62. Majumder, M, Gangopadhyay, TK, Chakraborty, AK, Dasgupta, K & Bhattacharya, DK 2008, 'Fibre Bragg Gratings in Structural Health Monitoring – Present Status and Applications', *Sensors and Actuators A: Physical*, Vol. 147, pp. 150–164.
63. Davey, K, Kristensen, PE, Sharp, PK & Clark, G 1997, 'Evaluation of an Innovative NDI Method for Detection and Monitoring of Cracking', *International Aerospace Congress – IAC97*, Melbourne, Australia.

64. Walker, L 2004, 'Real Time Structural Health Monitoring – Is It Really This Simple?', *SAMPE 2004*, Long Beach, CA, USA.
65. Black, S 2008, 'Structural Health Monitoring: Composites Get Smart', *High-Performance Composites*, CompositesWorld.
66. Kousourakis, A, Mouritz, AP & Bannister, MK 2006, 'Interlaminar Properties of Polymer Laminates Containing Internal Sensor Cavities', *Composite Structures*, Vol. 75, pp. 610–618.
67. Kousourakis, A, Bannister, MK & Mouritz, AP 2008, 'Tensile and Compressive Properties of Polymer Laminates Containing Internal Sensor Cavities', *Composites: Part A*, Vol. 39, pp. 1394–1403.
68. Graff, KF 1991, *Wave Motion in Elastic Solids*, Dover Publications, New York.
69. Rose, JL 1999, *Ultrasonic Waves in Solid Media*, Cambridge University Press, Cambridge.
70. Wilcox, PD 2003, 'A Rapid Signal Processing Technique to Remove the Effect of Dispersion from Guided Wave Signals', *IEEE Transactions on Ultrasonics, Ferroelectrics, and Frequency Control*, Vol. 50, pp. 419–427.
71. Croxford, AJ, Wilcox, PD, Lu, Y, Michaels, J & Drinkwater BW 2008, 'Quantification of Environmental Compensation Strategies for Guided Wave Structural Health Monitoring', *Proceedings of SPIE*, Vol. 6935, doi 10.1117/12.776362.
72. Veidt, M, Ng, CT, Hames, S & Wattinger, T 2008, 'Imaging laminar damages in plates using Lamb wave beam forming', International Conference on Multifunctional Materials and Structures, 28-31 July 2008, Hong Kong, China, in: *Advanced Materials Research*, Vols. 47–50, pp. 666–669.
73. Ihn, J-B & Chang, F-K 2008, 'Pitch Catch Active Sensing Methods in Structural Health Monitoring for Aircraft Structures', *Structural Health Monitoring*, Vol. 7, No. 1, pp. 5–15.
74. Zhao, X, Gao, H, Zhang, G, Ayhan, B, Yan, F, Kwan, C & Rose, JL 2007, 'Active Health Monitoring of an Aircraft Wing with Embedded Piezoelectric Sensor/Actuator network: I. Defect Detection, Localization and Growth Monitoring', *Smart Materials and Structures*, Vol 16, pp. 1208–1217.
75. Wang, CH, Rose, JT & Chang, F-K 2004 'A Synthetic Time-reversal Imaging Method for Structural Health Monitoring', *Smart Materials and Structures*, Vol. 13, pp. 415–423.
76. Sohn, H, Park, HW, Law, KH & Farrar, CR 2007, 'Damage Detection in Composite Plates by Using an Enhanced Time Reversal Method', *Journal of Aerospace Engineering*, Vol. 20, pp. 141–151.
77. Gangadharan, R, Murthy, CRL, Gopalakrishnan, S & Bhat, MR 2011, 'Time Reversal Health Monitoring of Composite Plates using Lamb Waves', *International Journal of Aerospace Innovations*, Vol. 3, pp. 131–142.
78. Rohde, AH, Veidt, M, Rose, LRF & Homer, J 2008, 'A Computer Simulation Study of Imaging Flexural Inhomogeneities Using Plate-wave Diffraction Tomography', *Ultrasonics*, Vol. 48, pp. 6–15.
79. Rohde, AH, Rose, LRF, Veidt, M & Wang, C 2009, 'Two Inversion Strategies for Plate Wave Diffraction Tomography', 2nd Asia-Pacific Workshop on Structural Health Monitoring, 2–4 December 2008, Melbourne, Australia, *Materials Forum*, Vol. 33, pp. 489–495.
80. Ng, CT & Veidt, M 2009, 'Lamb Wave Based Imaging of Damage in Composite Laminates', *Smart Materials and Structures*, Vol. 18, 074006

81. Simonetti, F & Huang, L 2008, 'From Beamforming to Diffraction Tomography', *Journal of Applied Physics*, Vol. 103, 103110.
82. Belanger, P, Cawley, P & Simonetti, F 2010, 'Guided Wave Diffraction Tomography within the Born Approximation', *IEEE Transactions on Ultrasonics, Ferroelectrics and Frequency Control*, Vol. 57, No. 6, pp. 1405–1418.
83. Konstantinidis, G, Drinkwater, BW & Wilcox, P 2006, 'The Temperature Stability of Guided Wave Structural Health Monitoring Systems', *Smart Materials and Structures*, Vol. 15, pp. 967–976.
84. Park, CY & Jun, SM 2010, 'Guided Wave Damage Detection in Composite Plates Under Temperature Variations, *Materials and Manufacturing Processes*, Vol. 25, pp. 227–231.
85. Tsoi, KA & Rajic, N 2009, Durability and Acoustic Performance of Integrated Piezoceramic Transducer Elements under Cyclic Loading, *Materials Forum*, Vol. 33, pp. 214–219.
86. Baker, A, Dutton, S & Kelly, D 2004, *Composite Materials for Aircraft Structures*, 2nd ed., AIAA Institute, Reston, VA.
87. Mall, S & Coleman, JP 1998, 'Monotonic and Fatigue Loading Behavior of Quasi-isotropic Graphite/Epoxy Laminate Embedded with Piezoelectric Sensor', *Smart Materials and Structures*, Vol. 7, No. 6, pp. 822–832.
88. Mall, S 2002, 'Integrity of Graphite/Epoxy Laminate Embedded with Piezoelectric Sensor/Actuator under Monotonic and Fatigue Loads', *Smart Materials and Structures*, Vol. 11, No. 4, pp. 527–533.
89. Mall, S & Hsu, TL 2000, 'Electromechanical Fatigue Behavior of Graphite/Epoxy Laminate Embedded with Piezoelectric Actuator', *Smart Materials and Structures*, Vol. 9, pp. 78–84.
90. Yocum, M, Abramovich, H, Grunwald, A & Mall, S 2003, 'Fully Reverse Electromechanical Fatigue Behavior of Composite Laminate with Embedded Piezoelectric Actuator/Sensor', *Smart Materials and Structures*, Vol. 12, pp. 556–564.
91. Bronowicki, AJ, McIntyre, LJ, Betros, RS & Dvorsky, GR 1996, 'Mechanical Validation of Smart Structures', *Smart Materials and Structures*, Vol. 5, pp. 129–139.
92. Wickramasinghe, VK & Hagood, NW 2004, 'Material Characterization of Active Fiber Composites for Integral Twist-actuated Rotor Blade Application', *Smart Materials and Structures*, Vol. 13, pp. 1155–1165.
93. Melnykowycz, M, Kornmann, X, Huber, C, Barbezat, M & Brunner, AJ 2006, 'Performance of Integrated Active Fiber Composites in Fiber Reinforced Epoxy Laminates', *Smart Materials and Structures*, Vol. 15, pp. 204–212.
94. Bhalla, S & Soh, CK 2003, 'Structural Impedance Based Damage Diagnosis by Piezo-Transducers', *Earthquake Engineering and Structural Dynamics*, Vol. 32, pp. 1897–1916.
95. Giurgiutiu, V, Zagrai, A & Bao, J 2004, 'Damage Identification in Aging Aircraft Sensors with Piezoelectric Wafer Active Sensors', *Journal of Intelligent Material Systems and Structures*, Vol. 15, pp. 673–687.
96. Park, G, Farrar, CR, Scalea, FLD & Coccia, S 2006, 'Performance Assessment and Validation of Piezoelectric Active-Sensors in Structural Health Monitoring', *Smart Materials and Structures*, Vol. 15, pp. 1673–1683.

97. Durr, J, Herold-Schmidt, U, Zaglauer, HW & Arendts, FJ 1998, 'Integration of Piezoceramic Actuators in Fibre-reinforced Structures for Aerospace Applications', *Proceedings of SPIE*, Vol. 3326, pp. 81–92.
98. Tsoi, KA & Rajic, N 2009, 'Durability and Acoustic Performance of Integrated Piezoceramic Transducer Elements under Cyclic Loading', *Materials Forum*, Vol. 33, pp. 214–219.
99. Favier, WG, Veidt, M & Hou, M 2010, 'Finite Element Analysis of Strain Transfer from a Mechanically Loaded Substrate to a Surface Mounted Piezoceramic Structural Health Monitoring Transducer', *6th Australasian Congress on Applied Mechanics (ACAM 6)*, Perth, Australia.
100. Rajic, N 2010, 'A Strategy for Achieving Improved Piezoceramic Transducer Durability under High Structural Loading', *Smart Materials and Structures*, Vol. 19, art. No. 105005.
101. Lemaire, PJ, Atkins, RM, Mizrahi, V & Reed, WA 1993, 'High Pressure H_2 Loading as a Technique for Achieving Ultrahigh UV Photosensitivity and Thermal Sensitivity in GeO_2 Doped Optical Fibres', *Electronics Letters*, Vol. 29, pp. 1191–1193.
102. Limberger, HG, Varelas, D & Salathe, RP 1996, 'Mechanical Degradation of Optical Fibers Induced by UV Light', *Proceedings of SPIE*, Vol. 2841, pp. 84–93.
103. Baker, SR, Rourke, HN, Baker, V & Goodchild, D 1997, 'Thermal Decay of Fiber Bragg Gratings Written in Boron and Germanium Codoped Silica Fiber', *Journal of Lightwave Technology*, Vol. 15, pp. 1470–1477.
104. Ghasemi-Nejhad, MN, Russ, R & Pourjalali, S 2005, 'Manufacturing and Testing of Active Composite Panels with Embedded Piezoelectric Sensors and Actuators', *Journal of Intelligent Material Systems and Structures*, Vol. 16, pp. 319–333.
105. Surgeon, M & Wevers, M 2001, 'The Influence of Embedded Optical Fibres on the Fatigue Damage Progress in Quasi-isotropic CFRP Laminates', *Journal of Composite Materials*, Vol. 35, pp. 931–940.
106. Dasgupta, A, Wan, Y & Sirkis, JS 1992, 'Prediction of Resin Pocket Geometry for Stress Analysis of Optical Fibers Embedded in Laminated Composites', *Smart Materials and Structures*, Vol. 1, pp. 101–107.
107. Huang, Y & Nemat-Nasser, S 2007, 'Structural Integrity of Composite Laminates with Embedded Micro-Sensors', *Proceedings of SPIE*, Vol. 6530, doi: 10.1117/12.715221.
108. Ghezzo, F, Huang, Y & Nemat-Nasser, S 2009, 'Onset of Resin Micro-cracks in Unidirectional Glass Fiber Laminates with Integrated SHM Sensors: Experimental Results', *Structural Health Monitoring*, Vol. 8, pp. 477–491.
109. Paget, CA & Levin, K 1999, 'Structural Integrity of Composites with Embedded Piezoelectric Ceramic Transducers', *Proceedings of SPIE*, Vol. 3668, pp. 306–313.
110. Mall, S 2002, 'Integrity of Graphite/Epoxy Laminate Embedded with Piezoelectric Sensor/Actuator under Monotonic and Fatigue Loads', *Smart Materials and Structures*, Vol. 11, pp. 527–533.
111. Melnykowycz, MM, Kornmann, X, Huber, C, Brunner, AJ & Barbezat, M 2005, 'Integration of Active Fiber Composite (AFC) Sensors/Actuators into Glass/Epoxy Laminates', *Proceedings of SPIE*, Vol. 5761, pp. 221–232.

112. Shivakumar, K & Bhargava, A 2005, 'Failure Mechanics of a Composite Laminate Embedded with a Fiber Optic Sensor', *Journal of Composite Materials*, Vol. 39, pp. 777–798.
113. Singh, DA & Vizzini, AJ 1994, 'Structural Integrity of Composite Laminates with Interlaced Actuators', *Smart Materials and Structures*, Vol. 3, pp. 71–79.
114. Shukla, DR & Vizzini, AJ 1996, 'Interlacing for Improved Performance of Laminates with Embedded Devices', *Smart Materials and Structures*, Vol. 5, pp. 225–229.
115. Hansen, JP & Vizzini, AJ 2000, 'Fatigue Response of a Host Structure with Interlaced Embedded Devices', *Journal of Intelligent Material Systems and Structures*, Vol. 11, pp. 902–909.
116. Pocius, AV 2002, *Adhesion and Adhesives Technology*, 2nd ed., Munich, Hanser.
117. Yun, HK, Cho, K, Kim, JK, Park, CE, Sim, SM, Oh, SY & Park, JM 1997, 'Effects of Plasma Treatment of Polyimide on the Adhesion Strength of Epoxy Resin/Polyimide Joints', *Journal of Adhesion Science Technology*, Vol. 11, pp. 95–104.
118. Kim, SH, Lee, DW, Chung, KH, Park, JK, Jaung JY & Jeong, SH 2002, 'Improvement in the Adhesion of Polyimide/Epoxy Joints using Various Curing Agents', *Journal of Applied Polymer Science*, Vol. 86, pp. 812–820.
119. Tanner, NA, Wait, JR, Farrar, CR & Sohn, H 2003, 'Structural Health Monitoring using Modular Wireless Sensors', *Journal of Intelligent Material Systems and Structures*, Vol. 14, pp. 43–56.
120. Lynch, JP, Sundararajan, A, Law, KH, Soh, H & Farrar, CR 2004, 'Design of a Wireless Active Sensing Unit for Structural Health Monitoring', *Proceedings of SPIE*, Vol. 5394, pp. 157–168.
121. Lynch, JP 2007, 'An Overview of Wireless Structural Health Monitoring for Civil Structures', *Philosophical Transactions of the Royal Society A: Mathematical, Physical and Engineering Sciences*, Vol. 365, pp. 345–372.
122. Zhao, X, Qian, T, Mei, G, Kwan, C, Zane, R, Walsh, C, Paing, T & Popovic, Z 2007, 'Active Health Monitoring of an Aircraft Wing with an Embedded Piezoelectric Sensor/Actuator Network: II. Wireless Approaches', *Smart Materials and Structures*, Vol. 16, pp. 1218–1225.
123. Wang, Y, Lynch, JP & Law, KH 2007, 'A Wireless Structural Health Monitoring System with Multithreaded Sensing Devices: Design and Validation', *Structure and Infrastructure Engineering*, Vol. 3, pp. 103–120.
124. Lynch, JP, Sundararajan, A, Law, KH, Kiremidjian, AS, Kenny, T & Carryer, E 2003, 'Embedment of Structural Monitoring Algorithms in a Wireless Sensing Unit', *Structural Engineering and Mechanics*, Vol. 15, pp. 285–297.
125. Guyomar, D, Jayet, Y, Petit, L, Lefeuvre, E, Monnier, T, Richard, C & Lallart, M 2007, 'Synchronized Switch Harvesting Applied to Selfpowered Smart Systems: Piezoactive Microgenerators for Autonomous Wireless Transmitters', *Sensors and Actuators A*, Vol. 138, pp. 151–160.
126. Park, G, Rosing, T, Todd, MD, Farrar, CR & Hodgkiss, W 2008, 'Energy Harvesting for Structural Health Monitoring Sensor Networks', *Journal of Infrastructure Systems*, Vol. 14, pp. 64–79.
127. Pavlakovic, B & Lowe MJS 2003, *Disperse Software Manual Version 2.0.16B*, Imperial College London, UK.

128. Lowe, M 1995, 'Matrix Techniques for Modeling Ultrasonic Waves in Multilayered Media', *IEEE Transactions on Ultrasonics Ferroelectrics and Frequency Control*, Vol. 42, pp. 535–542.
129. Castaings, M & Lowe, M 2008, 'Finite Element Model for Waves Guided along Solid Systems of Arbitrary Section Coupled to Infinite Solid Media', *The Journal of the Acoustical Society of America*, Vol. 123, pp. 696–708.
130. Bartoli, I 2006, 'Modeling Wave Propagation in Damped Waveguides of Arbitrary Cross-section', *Journal of Sound and Vibration*, Vol. 295, pp. 685–707.
131. Barbieri, E, Cammerano, A, De Rosa, S & Franco, F 2009, 'Waveguides of a Composite Plate by Using the Spectral Finite Element Approach', *Journal of Vibration and Control*, Vol. 15, No. 3, pp. 347–367.
132. Yan, F, Xue, Q, Rose, JL & Hasso, W 2010, 'Delamination Defect Detection Using Ultrasonic Guided Waves in Advanced Hybrid Structural Elements', *36th Annual Review of Progress in Quantitative Nondestructive Evaluation (QNDE)*, Vol. 1211, pp. 2044–2051.
133. Basri, R & Chiu, WK 2004, 'Numerical Analysis on the Interaction of Guided Lamb Waves with a Local Elastic Stiffness Reduction in Quasi-isotropic Composite Plate Structures', *Composite Structures*, Vol. 66, pp. 87–99.
134. Velichko, A & Wilcox, P 2010, 'A Generalized Approach for Efficient Finite Element Modeling of Elastodynamic Scattering in Two and Three Dimensions', *Journal of the Acoustical Society of America*, Vol. 128, pp. 1004–1014.
135. Ng, CT & Veidt, M 2011, 'Scattering of the Fundamental Anti-symmetric Lamb Wave at Delaminations in Composite Laminates', *Journal of the Acoustical Society of America*, Vol. 129, pp. 1288–1296.
136. Srivastave, A & Scalea, FL 2010, 'Quantitative Structural Health Monitoring by Ultrasonic Guided Waves', *Journal of Engineering Mechanics*, Vol. 136, No. 8, pp. 937–944.
137. Givoli, D 2004, 'High-order Local Non-reflecting Boundary Conditions: A Review', *Wave Motion*, Vol. 39, No. 4, pp. 319–326.
138. Sprenger, H, Raman, SR & Gaul, L 2011, 'Absorbing Boundary Conditions for Solid Waveguides', *Mechanics Research Communications*, Vol. 38, No. 3, pp. 158–163.
139. Drozdz, M, Lowe, M, Cawley, P, Moreau, L & Castaings, M 2006, 'Efficient Numerical Modeling of Absorbing Regions for Boundaries of Guided Waves Problems', *Review of Progress in Quantitative Non Destructive Evaluation*, Vol. 820, pp. 126–133.
140. Kudela, P, Zak, A, Karawczuk, M & Ostachowicz, W 2007, 'Modelling of Wave Propagation in Composite Plates Using the Time Domain Spectral Element Method', *Journal of Sound and Vibration*, Vol. 302, pp. 728–745.
141. Kim, Y, Ha, S & Chang FK 2008, 'Time-domain Spectral Element Method for Built-in Piezoelectric Actuator-induced Lamb Wave Propagation Analysis', *AIAA Journal*, Vol. 46, No. 3, pp. 591–600.
142. Rajic, N 2006, 'A Numerical Model for the Piezoelectric Transduction of Stress Waves', *Smart Materials and Structures*, Vol. 15, No. 5, pp. 1151–1164.
143. Dumas, D, Lani, F, Monnier, T & Lallart, M 2008, 'Simulation of Lamb Wave Propagation to Predict Damage Detection in Thin Walled Composite Structures' *4th European Workshop on Structural Health Monitoring*, Cracow, pp. 474–481.

144. Velichko, A & Wilcox, PD 2007, 'Modeling the Excitation of Guided Waves in Generally Anisotropic Multilayered Media', *Journal of the Acoustical Society of America*, Vol. 121, No. 1, pp. 60–69.
145. Gao, H & Rose, JL 2007, 'Impact of Ultrasonic Guided Wave Transducer Design on Health Monitoring of Composite Structures', *Proceedings of SPIE*, Vol. 6532, Article 653208.
146. Bottai, GS, Chrysochoidis, NA, Giurgiutiu, V & Saravanos, DA 2007, 'Analytical and Experimental Evaluation of Piezoelectric Wafer Active Sensors Performances for Lamb Waves Based Structural Health Monitoring in Composite Laminates', *Proceedings of SPIE*, Vol. 6532, Article 65320N.
147. Raghavan, A & Cesnik, CES 2005, 'Finite-dimensional Piezoelectric Transducer Modeling for Guided Waves Based Structural Health Monitoring', *Smart Materials and Structures*, Vol. 14, pp. 1448–1461.
148. Ditri, JJ & Rose, JL 1994, 'Excitation of Guided Waves in Generally Anisotropic Layers Using Finite Sources', *Journal of Applied Mechanics*, Vol. 61, pp. 330–338.
149. Kim, Y, Ha, S & Chang, FK 2008, 'Time-domain Spectral Element Method for Built-in Piezoelectric-actuator-induced Lamb Wave Propagation Analysis', *AIAA Journal*, Vol. 46, No. 3, pp. 591–600.
150. Ha, S & Chang, FK 2010, 'Optimizing a Spectral Element for Modeling PZT-induced Lamb Wave Propagation in Thin Plates', *Smart Materials and Structures*, Vol. 19, No. 1, Article 015015.
151. Staszewski, WJ & Worden, K 2001, 'An Overview of Optimal Sensor Location Methods for Damage Detection', *Proceedings of SPIE*, Vol. 4326.
152. Giurgiutiu, V & Cuc, A 2005, 'Embedded Non-destructive Evaluation for Structural Health Monitoring, Damage Detection, and Failure Prevention, *The Shock and Vibration Digest*, Vol. 37, pp. 83–105.
153. Staszewski, WJ, Worden, K, Wardle, R & Tomlinson, GR 2000, 'Fail-safe Sensor Distributions for Impact Detection in Composite Materials', *Smart Materials and Structures*, Vol. 9, No. 3, pp. 298–303.
154. Rao, SS 2009, *Engineering Optimisation: Theory and Practice*, John Wiley & Sons, Hoboken, NJ.
155. Aldick, R 2006, *Applied Optimization: Formulation and Algorithms for Engineering Systems*, Cambridge University Press, Cambridge.
156. Guratzsch, RF & Mahadevan, S 2010, 'Structural Health Monitoring Sensor Placement Optimization under Uncertainty', *AIAA Journal*, Vol. 48, No. 7, pp. 1281–1289.
157. Flynn, EB & Todd, MD 2010, 'A Bayesian Approach to Optimal Sensor Placement for Structural Health Monitoring with Application to Active Sensing', *Mechanical Systems and Signal Processing*, Vol. 24, No. 4, pp. 891–903.
158. Alleyne, DN & Cawley, P 1992, 'The Interaction of Lamb Waves with Defects', *IEEE Transactions on Ultrasonics, Ferroelectrics, and Frequency Control*, Vol. 39, pp. 381–397.
159. Lemistre, M & Balageas, D 2001, 'Structural Health Monitoring System Based on Diffracted Lamb Wave Analysis by Multiresolution Processing', *Smart Materials and Structures*, Vol. 10, pp. 504–511.
160. Su, Z & Ye, L 2004, 'Fundamental Lamb Mode-based Delamination Detection for CF/EP Composite Laminates using Distributed Piezoelectrics', *Structural Health Monitoring*, Vol. 3, pp. 43–68.

161. Liew, CK & Veidt, M 2011, 'A Wavelet-based Damage Detection Approach for Acousto-Ultrasonic *in-situ* Monitoring Systems', *Key Engineering Materials*, Vol. 472, pp. 809–814.
162. Liew, CK & Veidt, M 2011, 'Damage Evaluation in Smart Materials using Time-frequency Analysis', *Advanced Materials Research*, Vol. 313, pp. 2299–2306.
163. Su, Z & Ye, L 2005, 'Digital Damage Fingerprints (DDF) and its Application in Quantitative Damage Identification', *Composite Structures*, Vol. 67, pp. 197–204.
164. Liew, CK & Veidt, M 2009, 'Pattern Recognition of Guided Waves for Damage Evaluation in Bars', *Pattern Recognition Letters*, Vol. 30, pp. 321–330.

Part IV
Non-destructive evaluation (NDE) techniques in civil and marine applications

18
Non-destructive evaluation (NDE) of composites: techniques for civil structures

U. B. HALABE, West Virginia University, USA

DOI: 10.1533/9780857093554.4.483

Abstract: This chapter presents the basic theory and applications of modern NDE techniques for testing of composite structural components such as beams, columns, and bridge decks. The use of infrared thermography and ground penetrating radar is discussed in detail using results from several laboratory and field tests. In addition, localized testing using digital tap hammer is discussed. The chapter also presents discussions on issues and challenges on the use of NDE for civil structures, in addition to future trends and sources for further information.

Key Words: non-destructive evaluation (NDE), infrared thermography, ground penetrating radar (GPR), tap testing, digital tap hammer, civil infrastructure, structures, structural, fiber-reinforced polymer (FRP) composites, composites.

18.1 Introduction

Non-destructive evaluation (NDE) plays an important role in the annual inspection and maintenance of civil structures, which is crucial for ensuring their continued structural integrity and serviceability. While the customary practice for field inspection of civil structures is to rely primarily on periodic visual inspections, there is a growing need for modern NDE techniques which can detect subsurface defects at their early stages of formation before they grow to critical sizes and show up on the surface. Early detection of subsurface defects in structural components can enable timely repair and maintenance, thus prolonging the service life of the various structural components. This aspect has become very important in the current era of aging infrastructure, growing cost of new construction, and very tight budgetary constraints.

The rehabilitation of existing structures often utilizes innovative materials such as fiber-reinforced polymer (FRP) composites. For example, old concrete bridge decks can be replaced by lighter and stronger FRP composite decks, which also offer a significantly longer lifespan than their concrete counterpart (Liang and GangaRao, 2004). Deteriorating concrete and timber beams and columns in bridges, buildings, and parking garages can be rehabilitated by wrapping them with glass or carbon-reinforced FRP composite fabrics. These fabrics not only enhance the strength of the structural components, but also enhance their ductility (ability to deform under large wind/hurricane, earthquake, or blast loading), leading to increased safety under harsh loading conditions. Non-destructive monitoring of these rehabilitated structural components is very important for quality control during the initial construction, and periodic monitoring is needed to detect subsurface debonds that could be formed between the composite fabric and underlying material. Timely repair of the debonded locations can help to ensure the enhanced strength and ductility is continuously realized over the lifespan of the structure.

Modern NDE methods for application to civil infrastructure include techniques such as infrared thermography and ground penetrating radar (GPR). These techniques utilize commercially available portable equipment and offer the capability of high-speed data acquisition in order to scan large areas in field structures. Moreover, significant reduction in equipment prices over the last decade coupled with the development of user-friendly data analysis procedures and software have resulted in increasing use of these techniques by highway agencies and field inspectors. This chapter presents the basic methodology and laboratory/field testing results for both infrared thermography and ground penetrating radar techniques. In addition, a localized but effective testing technique such as digital tap testing is also discussed. The following three NDE techniques that are widely used for condition assessment of civil infrastructure are described:

- infrared (IR) thermography;
- GPR;
- digital tap testing.

18.2 Infrared thermography

18.2.1 Basics of infrared thermography testing

Infrared thermography is a non-contact technique which is used to generate a surface temperature profile of the test area in the form of gray scale or color infrared images. Brighter areas in the infrared image typically signify high-temperature zones while darker areas signify lower temperatures. The first step in infrared thermography testing is to create a thermal gradient in the test object by applying either a heating or cooling source to the specimen's surface. The rate of heat conduction through the specimen in regions consisting of subsurface defects is different from that of the surrounding defect-free area (Maldague, 1993, 2000, 2001; Maldague and Moore 2001). This produces variations in the surface temperature of the specimen. The second step is to remove the heating or cooling source and immediately acquire a sequence of infrared images from the specimen's surface in order to detect the subsurface defects. This acquisition is done using infrared cameras that are sensitive to the infrared energy emitted by the test object. Modern infrared cameras can store digital infrared images using a memory card or a laptop computer which can house the image acquisition and processing software.

The underlying principle behind infrared thermography can be found in several textbooks and other publications (Kaplan, 1999; Maldague, 1993, 2000, 2001; Maldague and Moore 2001). Infrared thermography measures the surface temperature of an object by detecting the amount of energy emitted by the object in the infrared band. The visible spectrum refers to the portion of the electromagnetic spectrum that is visible to the human eye and falls within wavelengths of 0.4–0.75 µm (1 µm or micrometer is one-millionth of a meter and about one-fiftieth of the diameter of a human hair). The infrared portion of the electromagnetic spectrum ranges from 0.75 to 100 µm wavelength. This infrared spectrum is divided into various sub-regions, which are: near infrared (0.75–3 µm), middle infrared (3–6 µm), far infrared (6–15 µm), and extreme infrared (15–100 µm) (Clark *et al.*, 2003). It should be noted that most practical measurements can be made to only about 20 µm (Kaplan, 1999).

As the temperature of an object reduces, the frequencies for the energy emitted by the object are lower, which means the wavelengths are higher. For example, the sun with a surface temperature of close to 6000 K, radiates most of its energy in the visible spectrum, which is what the human eye can detect. The tungsten filament glowing inside a light bulb has a temperature of around 2800–3300 K, with a significant portion of the energy in the visible spectrum. At temperatures typically found on the Earth's surface (i.e., in the vicinity of 300 K ≅ 30 °C), most of the energy is radiated in the infrared spectrum with peak energy close to 10 µm. Energy in this range is invisible to the human eye or a photographic camera, but can be detected using an infrared camera.

The lens of an infrared camera is made of special material (germanium) which is transparent in the infrared spectrum. Use of such lens material and special infrared detectors make the price of infrared cameras much higher than regular photographic cameras. The infrared detectors sense the infrared energy at various pixel locations, and the camera electronics convert the image into a visual color or gray scale image, where bright (white) colors indicate higher temperatures and dark (black) colors depict lower temperatures. These images (called thermograms or infrared images) show the surface temperature distribution of the viewed area. While old analog models allowed storage of the infrared images in a video tape, the stored image only showed the relative temperature profile and the actual temperature value was lost. The modern digital infrared cameras allow storage of full radiometric images in digital form in a computer's hard drive, CD/DVD disks, or other media (such as memory card in the camera itself), and the images can be viewed later using associated software, which also enables display of temperature data at every pixel in the image. Such radiometric data can be utilized for digital image processing and for developing automated image analysis procedures and algorithms (Shepard et al., 1999).

Whenever an object is thermally stimulated on its surface by heating (or cooling) the surface, conduction takes place through the thickness. The rate of thermal conduction through a defect is usually different from the conduction rate through defect-free material. If the defect is air-filled, it acts as an insulator, thus causing build-up of heat (storage of thermal energy) in the region above the defect (assuming the surface of the object is heated, not cooled). This phenomenon leads to differences in surface temperatures over defective areas compared with surrounding defect-free regions. The different parts of the surface radiate different amounts of thermal energy based on the surface temperatures in accordance with the Stefan–Boltzmann law (Kaplan, 1999). The radiated thermal energy is detected using an infrared camera, which produces thermal images showing defective areas with different surface temperatures compared with the surrounding defect-free areas. If the surface of the object is heated using a heater (an active source), and then infrared images are acquired after removal of the heat source; the images will show air-filled defects as hot spots. This phenomenon is altered if a cooling source is used instead of a heat source in order to provide thermal stimulation, or if the defect is water-filled instead of air-filled. It is important to note that consistent infrared measurements can only be made if the surface heating or cooling is uniform. In the case of bridge decks, it is often convenient to conduct infrared thermography testing under solar heating.

Commercial infrared cameras are typically available in two spectral ranges (~1–3 μm and 7–14 μm). The near infrared cameras (1 to 3 μm range) are typically used for night vision. It should be noted that the air medium between the test object and the infrared camera acts as a filter of thermal energy, often distorting the measured energy spectrum. The atmospheric transmission happens to be quite uniform in the 7 μm to 14 μm range (Kaplan, 1999); hence this range is preferred for most scientific measurements.

Figure 18.1 shows a typical infrared testing set-up in the laboratory using an infrared camera that is capable of acquiring infrared images at a capture speed of up to 60 frames/s. The data acquisitions of such cameras are typically controlled by a laptop computer-based software. These cameras are expensive and the sequence of images acquired by these cameras can be used in conjunction with advanced image processing algorithms (Shepard et al., 1999). The thermal sensitivity of such cameras typically ranges from

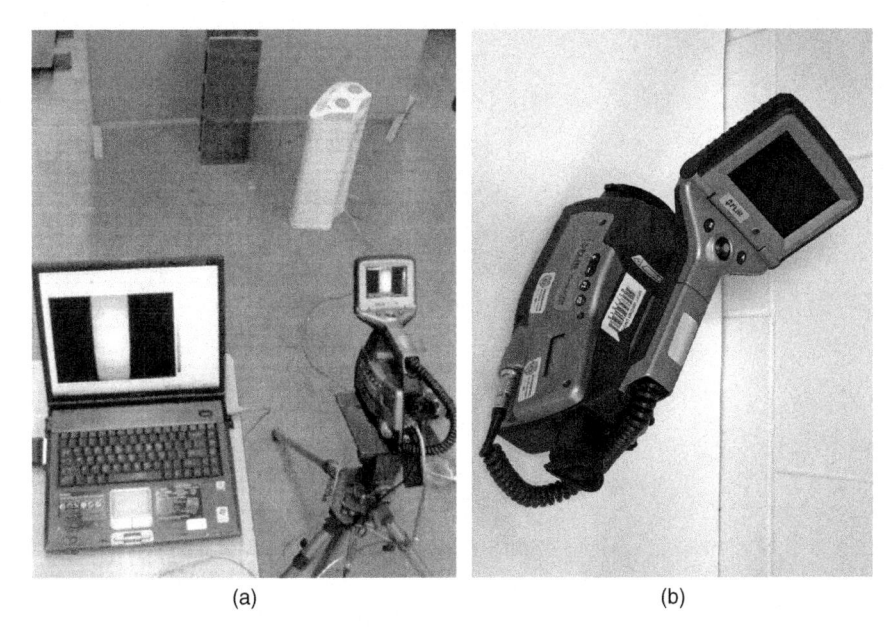

18.1 (a) Experimental set-up using digital infrared camera, and (b) close-up view of the camera.

0.01 to 0.06 °C at 30 °C. Figure 18.2 shows an infrared camera that produces radiometric JPEG still images, which can be stored in a SD memory card in the camera. Such cameras are typically low-cost (with thermal sensitivity of about 0.1 °C at 30 °C) and can be conveniently used for field inspections of structural components. The radiometric JPEG images from these cameras can be used to measure the temperature of any spot or average temperature over an area using computer software. Figure 18.3 shows various heating and cooling sources that can be used for thermal stimulation of a test object that is not typically heated by the sun (e.g., beams and columns under the bridges). For testing of bridge decks, it is convenient to use the sun as the heat source. For housing inspections, the thermal difference between the heated or cooled house and the ambient air is often utilized to determine areas of air and water leakage or low insulation.

18.2.2 Infrared thermography testing for glass fiber-reinforced polymer (GFRP) bridge decks

This section illustrates the infrared thermography testing of glass fiber-reinforced polymer (GFRP) bridge decks in the laboratory as well as field settings. The specimens used in the laboratory study consisted of GFRP

18.2 Infrared camera with SD card that stores radiometric JPEG images.

bridge deck modules of different sizes (Halabe *et al.*, 2007). The GFRP decks were composed of E-glass fibers and either polyester or vinyl-ester resin combination. The glass fibers were continuous strand roving and tri-axial fabrics. The fiber volume fractions for the polyester and vinyl-ester decks were about 45% and 50% respectively. The top surface of some of the decks was coated with 3/8″ (9.5 mm) thick wearing surface, which consisted of an epoxy-based overlay system made of a specially selected blend of resins, hardener, powder, and aggregates.

Debonds of different sizes were made by joining two polypropylene sheets with an enclosed air pocket in between them. The water-filled debonds were made by enclosing water within plastic pouches. The simulated debonds were placed on the top surface of the GFRP deck, which was followed by the application of 9.5 mm (3/8″) thick wearing surface on the entire top surface of the deck. The specimen BD1 was a bridge deck module of plan size 610 mm × 305 mm (24″ × 12″), flange thickness of 12.7 mm (0.5″), and web thickness of 8.9 mm (0.35″). This deck specimen had an overall depth of 203 mm (8″) and is shown in Fig. 18.4. The geometric cross-section, wearing surface thickness, and material composition of this specimen are the same as that used in field construction of many GFRP bridges. One side of the specimen, BD1a, consisted of two air-filled debonds of sizes 51 mm × 51 mm (2″ × 2″) and 76 mm × 76 mm (3″ × 3″) and the other side BD1b, consisted of two air-filled debonds of sizes 25 mm × 25 mm (1″ × 1″) and 13 mm × 13 mm (½″ × ½″) between the wearing surface layer and the deck. All these air-filled debonds had a thickness of 1.6 mm (1/16″). Since

Techniques for civil structures 489

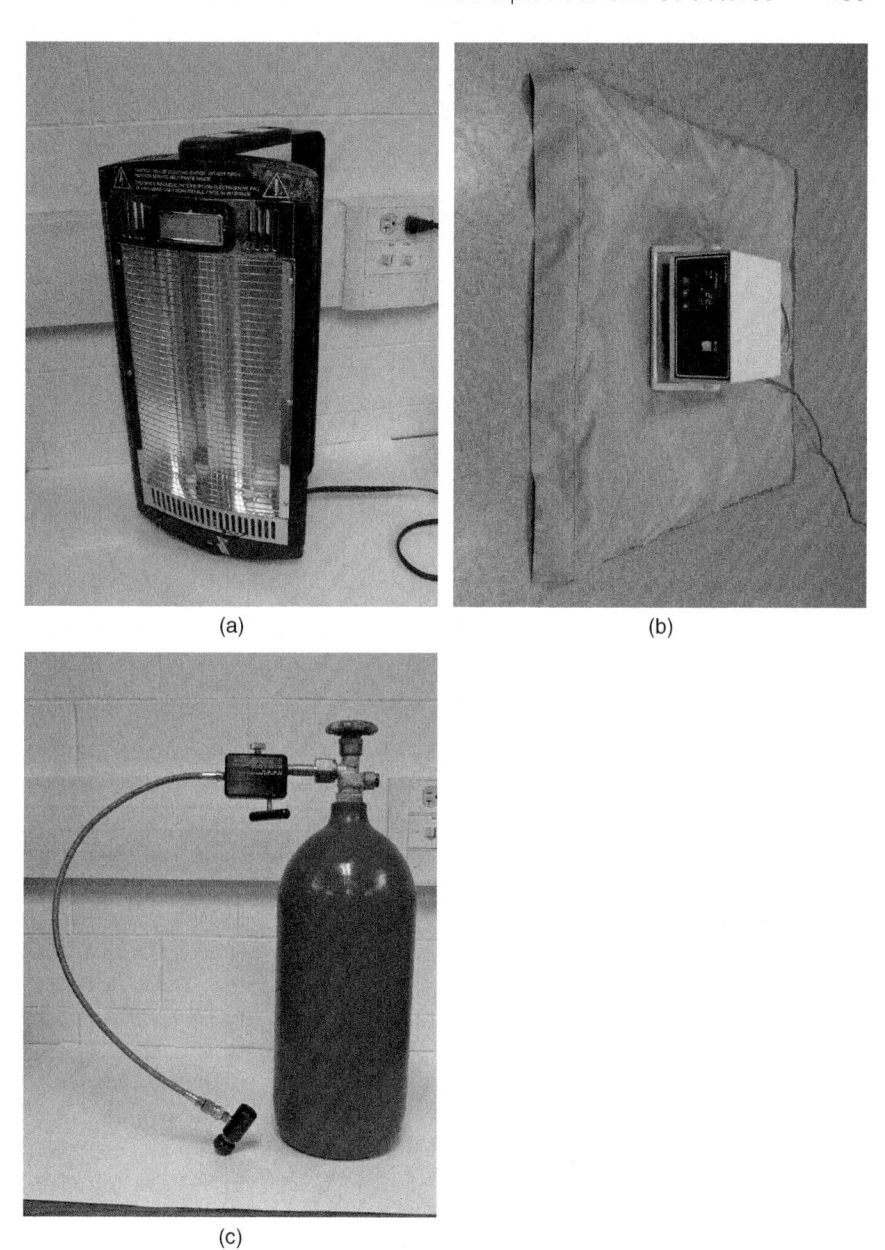

18.3 Various active heating and cooling sources: (a) 1500 W quartz heater, (b) 1500 W heating blanket, and (c) liquid carbon dioxide tank.

18.4 Front and cross-sectional views of the GFRP bridge deck specimen BD1 with wearing surface.

this was a laboratory specimen, both sides were used for this infrared testing study, but it should be noted that both sides (with wearing surface) represented the top flange of field decks. For the infrared thermography tests conducted in this research, the top and bottom flanges essentially acted as two separate specimens.

It is important to note that the polymer concrete wearing surface overlay placed on field GFRP bridge decks is an inhomogeneous mixture of coarse gravel and resin, which results in natural surface temperature variation on bridge decks. While no standard exists for thermographic testing of GFRP bridge decks, ASTM D 4788 – 03 (2007) standard for thermographic testing of steel reinforced concrete bridge decks states that the surface temperature difference must be at least 0.5 °C before an anomalous region is considered to be a debond or delamination. ASTM D 4788 – 03 (2007) also requires the use of 'imaging infrared scanner having a minimum thermal resolution of 0.2 °C.' The radiometric infrared camera used in this research had a sensitivity of 0.06 °C, which is more than adequate considering that only anomalies over 0.5 °C temperature difference (compared with the surrounding defect-free area) are of interest. Also, this camera was an uncooled model which is quite convenient for testing of GFRP bridge decks in the field.

A 1500 W quartz heater was used as the main heating source in laboratory testing. Specimen BD1 was uniformly heated by placing the quartz

Techniques for civil structures 491

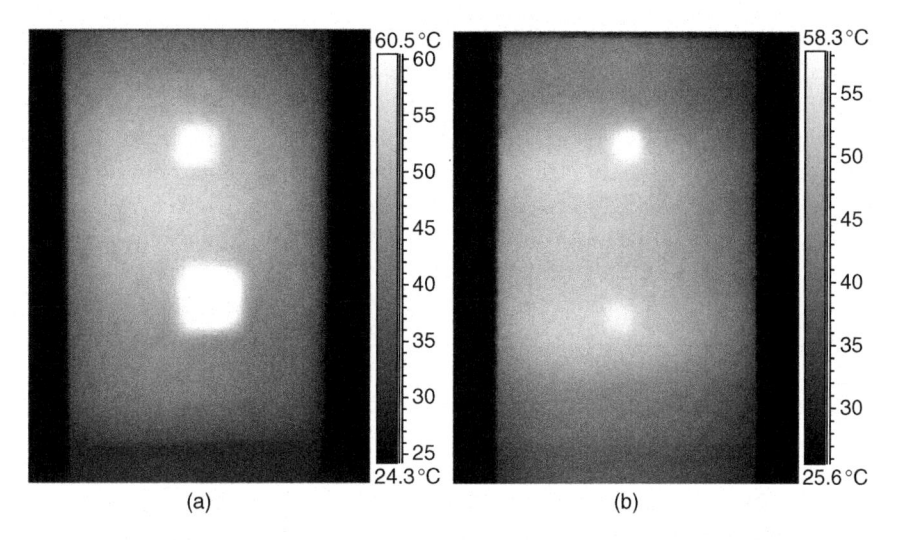

18.5 Infrared image of specimen BD1 (a) with air-filled debonds of sizes 76 mm × 76 mm × 1.6 mm (3″ × 3″ × 1/16″) and 51 mm × 51 mm × 1.6 mm (2″ × 2″ × 1/16″), and (b) with air-filled debonds of sizes 25 mm × 25 mm × 1.6 mm (1″ × 1″ × 1/16″) and 13 mm × 13 mm × 1.6 mm (½″ × ½″ × 1/16″).

heater at a distance of about 0.2 m from the specimen surface for a duration of 150 s. Figure 18.5 shows the infrared image of the 76 mm × 76 mm (3″ × 3″), 51 mm × 51 mm (2″ × 2″), 25 mm × 25 mm (1″ × 1″), and 13 mm × 13 mm (½″ × ½″) debonds, all with a debond thickness of 1.6 mm (1/16″). The infrared image of the specimen revealed the location of the debonds as white regions (areas with higher temperature when compared to the surrounding defect-free regions). The gray scale showing the temperatures of the deck is also available with the infrared image. The difference between the surface temperatures over the 76 mm × 76 mm (3″ × 3″) and 51 mm × 51 mm (2″ × 2″) debonded areas with respect to the surrounding defect-free area were about 11 °C and 8 °C respectively. The boundaries of the 76 mm × 76 mm (3″ × 3″), 51 mm × 51 mm (2″ × 2″), and 25 mm × 25 mm (1″ × 1″) debonds are well defined in the infrared image shown in Fig. 18.5. Though a well-defined boundary for the 13 mm × 13 mm (½″ × ½″) debond was not visible, the area shows sufficient contrast which confirmed the presence of the debond. The difference between the surface temperatures of the 25 mm × 25 mm (1″ × 1″) and 13 mm × 13 mm (½″ × ½″) debonded areas with respect to the surrounding defect-free area were about 8 °C and 4 °C, respectively.

To establish the surface temperature–time curves, the bridge deck specimen BD1 was heated using the 1500 W quartz tower heater and the infrared images were recorded over a period of time. The infrared images from the digital infrared camera were captured at 10-second intervals. The first infrared image was captured immediately after removing the heating source from the specimen's surface. The temperatures of the defective and defect-free areas at the 10-second time intervals were tabulated and curves were drawn for the temperatures of both the defective and defect-free areas along the *Y*-axis and time over the *X*-axis. An infrared image showing the most prominent boundary of the delamination would be obtained when the temperature difference between the defective and the defect-free area is at its maximum. The curves for all the four debonds were similar in nature, but with different values. It was observed that the maximum temperature difference between the debonded and defect-free areas for all the four debond sizes was obtained almost immediately (within 30 to 40 seconds) after the heating source was taken away. Figure 18.6 depicts the surface temperature-time curves for the 25 mm × 25 mm (1″ × 1″) and 13 mm × 13 mm (½″ × ½″) debonds. The graph also shows that the curves for the debonds and defect-free area converges. The decrease in the temperature difference is very slow for the first few minutes, after which it becomes more rapid. Images with reasonably good contrast between the defective and defect-free areas were available for about 300 s duration for the 76 mm × 76 mm (3″ × 3″) and 51 mm × 51 mm (2″ × 2″) debonds and about 200 s duration for the 25 mm × 25 mm (1″ × 1″) and 13 mm × 13 mm (½″ × ½″) debonds, after the heating source had been removed. The time available for good contrast in the infrared image reduces with the size of the defect. After this period of time the deck approaches thermal equilibrium which

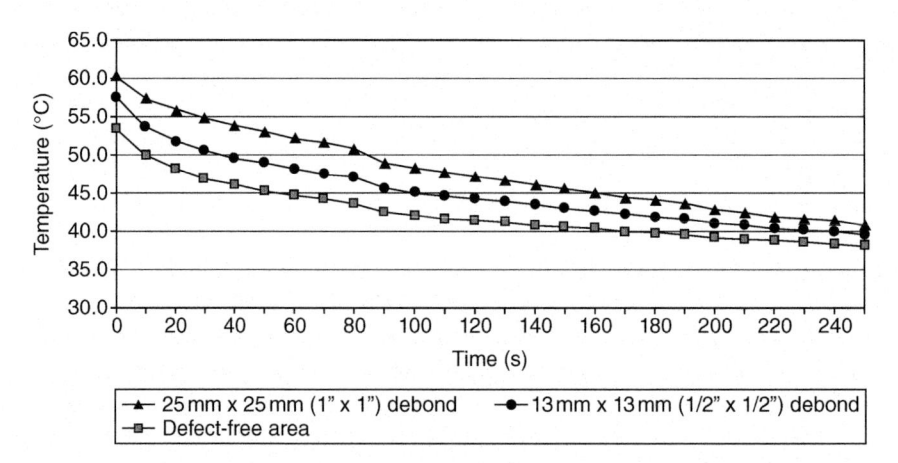

18.6 Surface temperature-time curves for air-filled debonds of sizes 25 mm × 25 mm × 1.6 mm (1″ × 1″ × 1/16″) and 13 mm × 13 mm × 1.6 mm (½″ × ½″ × 1/16″).

is facilitated by heat conduction from the higher-temperature regions (debonded areas) to lower temperature regions (defect-free areas).

Another bridge deck specimen was used to demonstrate the difference between the behavior of air-filled and water-filled debonds. This specimen, BD2 (shown in Fig. 18.7), had a plan size 610 mm × 305 mm (24″ × 12″), flange thickness of 11.4 mm (0.45″) and web thickness of 10.2 mm (0.4″). The overall depth of this deck was 102 mm (4″), which is typical of 'low-profile' GFRP bridge decks used in some of the recent field construction. Two water-filled debonds of sizes 51 mm × 51 mm × 1.6 mm (2″ × 2″ × 1/16″) and 76 mm × 76 mm × 3.2 mm (3″ × 3″ × 1/8″), and an air-filled debond of size 51 mm × 51 mm × 1.6 mm (2″ × 2″ × 1/16″) were placed on the surface of the GFRP deck specimen before it was overlaid with the 9.5 mm (3/8″) wearing surface layer to create simulated air-filled and water-filled debonds.

This specimen BD2 was also heated using the quartz heater. Plate VIII (between pages 296 and 297) shows the schematic view of the debond locations and the corresponding infrared image of the bridge deck. The surface temperature above the air-filled debond (white area in infrared image) is higher than the temperature above defect-free area (temperature difference: 5.7 °C). On the other hand, the temperatures above the two water-filled debonds are lower compared to the surrounding defect-free area, with a temperature difference of −3.9 °C for the 76 mm × 76 mm (3″ × 3″) debond and −2.5 °C for the 51 mm × 51 mm (2″ × 2″) debond. Thus, the surface

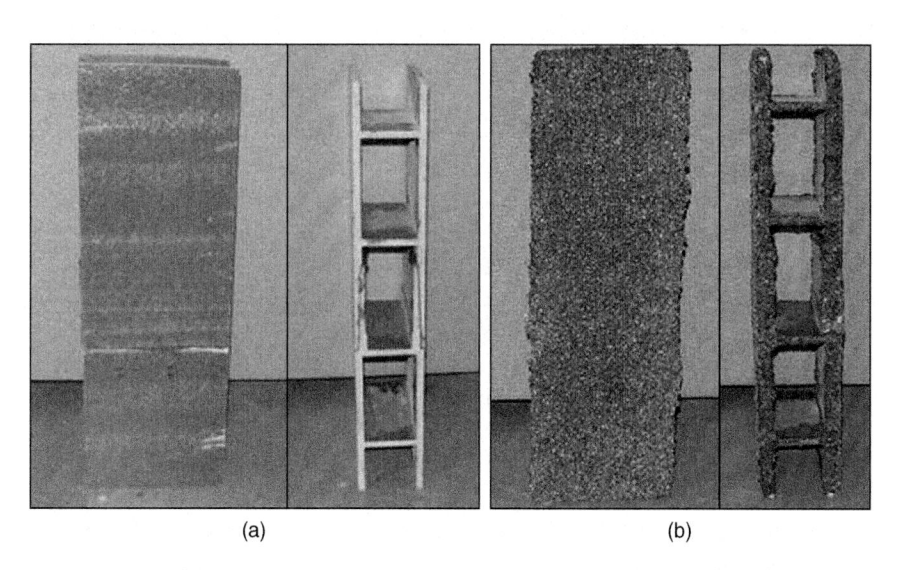

18.7 Front and cross-sectional views of the GFRP bridge deck specimens (a) without wearing surface overlay, and (b) with wearing surface overlay.

temperature difference for water-filled debond is lower in magnitude compared with the air-filled debond. This is because the thermal diffusivity of water (0.14×10^{-6} m^2 s^{-1}) is close to that of the GFRP bridge deck whose thermal diffusivity is about 0.13×10^{-6} m^2 s^{-1} along the flange depth (i.e., perpendicular to the fibers). On the other hand, air-filled debond has a significantly higher thermal diffusivity (33×10^{-6} m^2 s^{-1}) (Maldague, 1993) compared with the GFRP deck, which results in higher surface temperature difference. It can also be observed from Plate VIII that the top portion (6.3 mm × 76 mm or ~0.25″ × 3″ in plan) of the two water-filled debonds showed a higher temperature as represented by small white area above the dark area. This was due to the presence of small air bubbles inside the plastic pouch that was used to make the water-filled debond. The infrared images revealed that the boundaries of the air-filled debond were more clearly defined compared with the water-filled debonds. From visual observation of the infrared images, it is more difficult to detect the presence and mark the boundaries of water-filled debonds compared to air-filled debonds.

The behavior seen in the above laboratory tests have also been observed during field testing on the GFRP deck of La Chein bridge, which is a two-lane bridge located in Monroe County, West Virginia, USA. The bridge deck had 9.5 mm (3/8″) thick polymer concrete overlay, which showed signs of debonding as evident from surface cracks and unevenness. The objective of infrared testing was to detect debonds present at the interface between the wearing surface and the underlying GFRP deck and to quantify the debonded area (Halabe *et al.*, 2003b). The wearing surface overlay covered an area of 7.9 m × 12.5 m (26′ × 41′) in plan. For the purpose of infrared testing, the bridge was divided into numerous small grids and infrared images as well as digital photographs were acquired. The distance of the infrared camera from the bridge deck was about 2 m. For most part of the day, solar radiation was used as the heating source. The temperature of the deck was as high as 50 °C and the ambient air temperature was around 28–30 °C for most of the day. The test was started around noon when the sun had already heated the deck to a sufficiently high temperature to conduct the infrared testing. Typically, solar radiation is at its peak at noon while the air and deck temperatures are at their peak at about 2 p.m. The ideal time window to conduct infrared testing is from 11 a.m. to about 3 p.m. During late afternoon (around 4 p.m.), when the sun could not heat the deck efficiently, two 1500 W heating blankets of plan size 0.91 m × 0.91 m (3′ × 3′) each and shown in Fig. 18.3(b), were used as the heating source for the deck.

Some representative digital photographs and infrared images identifying subsurface debonds are shown in Fig. 18.8. During field testing, it is important to use digital cameras and take photographs of the areas for which infrared images are acquired. Since the bridge was hot and dry, the debonds

Techniques for civil structures   495

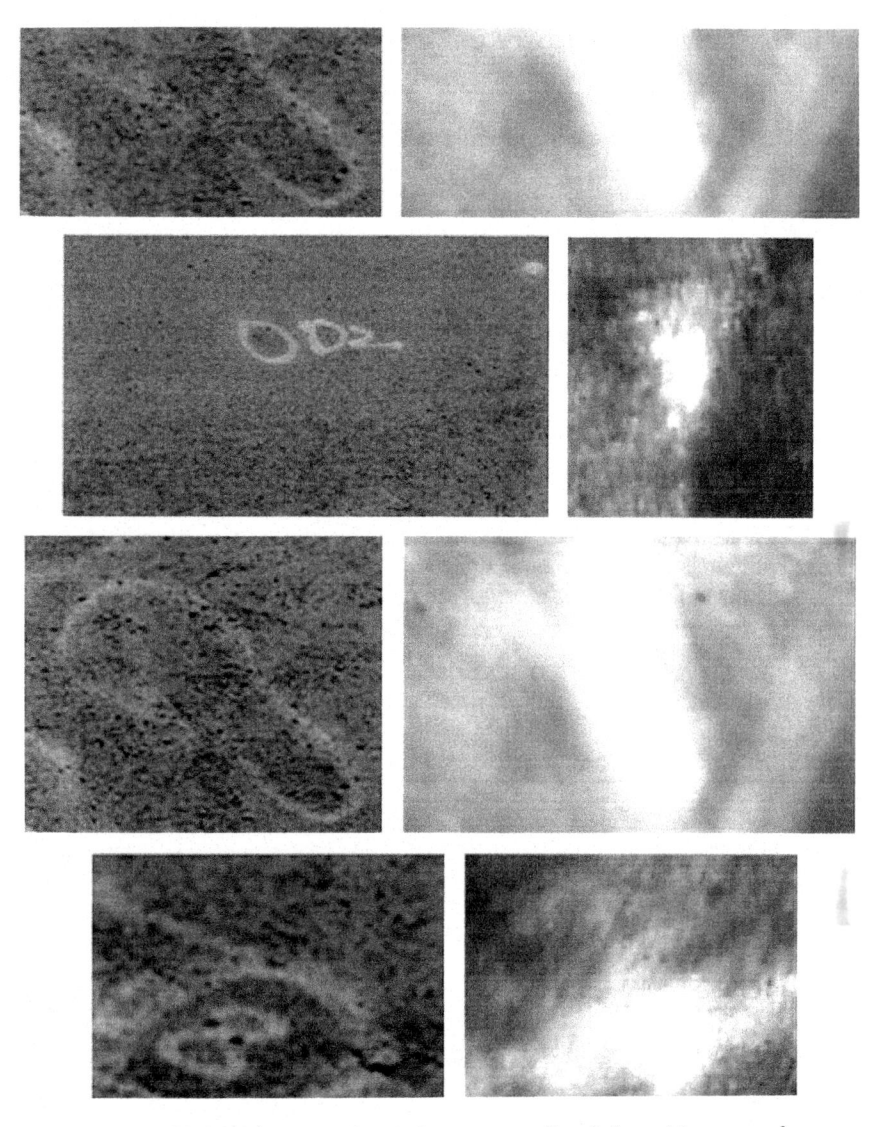

18.8 Digital photographs and corresponding infrared images of various debonded areas in the La Chein GFRP bridge deck.

were air-filled and appeared as bright white spots (high-temperature areas) in the infrared image. The corresponding debonded areas were marked on the bridge deck using white spray paint as seen in the digital photographs in Fig. 18.8. The infrared testing helped to map the shape and plan sizes of the various debonds, and based on these results it was concluded that approximately two-thirds of the deck wearing surface had defective areas (Halabe *et al.*, 2003b). Therefore, the bridge was recommended for wearing surface replacement. The debonding in this bridge was caused due to placement of the wearing surface during very hot weather (outside the temperature range specified by the resin manufacturer) which adversely affected the curing of the resin in the wearing surface layer.

Some portions of the bridge deck were under shadows of the guard rail. The infrared images of these portions (Fig. 18.9) revealed areas with lower temperature (with darker shades) that were caused due to the shadow. Such areas should not be misinterpreted as debonded areas. Some other areas of the deck had dirt particles and loose gravel on the surface that had separated from the polymer concrete overlay. An infrared image of a surface with loose gravel is shown in Fig. 18.10. Proper care should be taken when analyzing the infrared images to avoid erroneous classification of the areas under the shadows of guard rail, trees, or adjacent buildings, and areas with loose gravel/dirt as debonded areas.

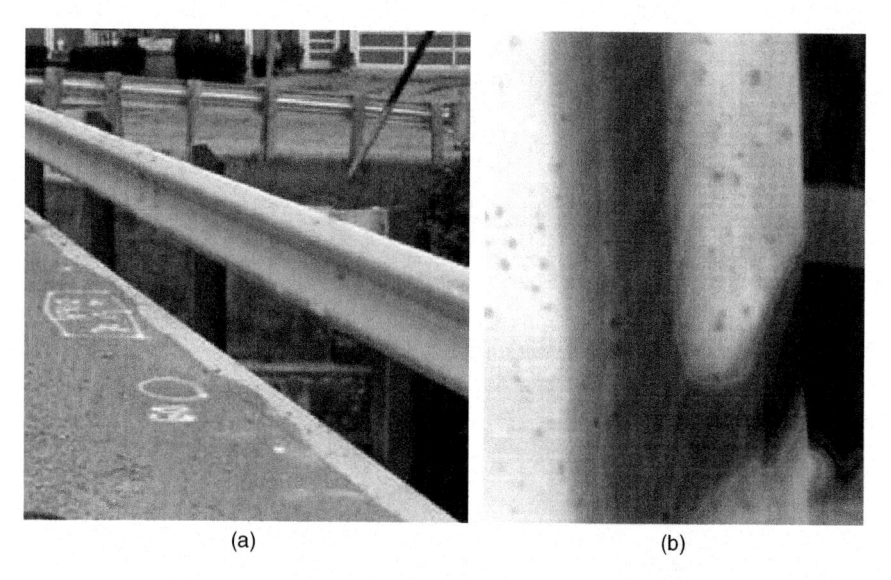

18.9 (a) Photograph of the guard rail area and (b) infrared image showing the area subjected to shadow of the guard rail.

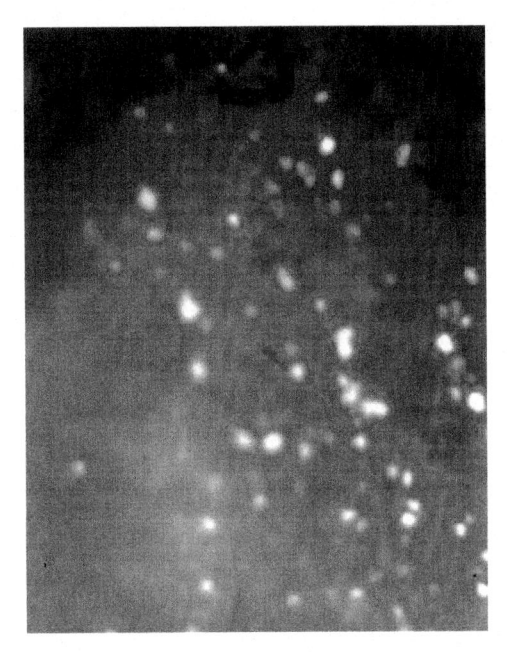

18.10 Infrared image showing loose gravel on the surface of the GFRP bridge deck.

The above discussions have focused extensively on the detection of debonds between the wearing surface and the underlying composite bridge deck. The use of infrared thermography has also been investigated for detecting air-filled and water-filled delaminations within flanges of GFRP composite deck specimens in the laboratory as well as under solar heating, with more promising results in case of air-filled delaminations (Halabe *et al.*, 2006, 2007; Hing and Halabe, 2010; Vasudevan, 2004). The term 'delamination' is used here to indicate a defect or separation within a composite layer, while 'debond' refers to a separation between dissimilar layers. However, some researchers have used the terms debonds and delaminations interchangeably in published literature without making a clear distinction between the two types of subsurface defects.

18.2.3 Infrared thermography testing for FRP wrapped columns

This section discusses the use of infrared thermography testing of concrete or timber columns wrapped with FRP fabric. The fabrics typically consist of glass or carbon fibers and are termed GFRP or CFRP fabrics.

Although both fabrics have been used to rehabilitate civil infrastructure, GFRP fabrics offer lower cost. However, CFRP fabrics offer a greater increase in strength and ductility properties of the structural component than GFRP fabrics (Dutta, 2010). The subsurface debonds that could form between the fabric and the underlying concrete or timber member at the time of initial wrapping or during service could lead to reduction in localized confinement, thus adversely affecting the strength and ductility of the rehabilitated member. Therefore, it is important to use techniques such as infrared thermography for condition assessment during initial rehabilitation work and for periodic monitoring of the structure in order to detect subsurface debonds and repair them by injecting resin into the debonds. If the debonds are large, then one may need to replace the fabric wrap in accordance with established standards (e.g., ACI 440.2R-08, 2008).

A study was conducted by Halabe *et al.* (2010) to evaluate the effectiveness of infrared thermography in detecting subsurface debonds in 152.4 mm × 304.8 mm (6″ × 12″) concrete cylinders wrapped with CFRP and GFRP fabrics. The specimens were prepared with pre-inserted debonds between the fabric and the underlying concrete specimens, and were heated using 1500 W quartz heater for about one minute. The infrared images were acquired immediately upon removal of the heating source. The distance of the infrared camera to the test specimen was 0.9 m (3′). The laboratory test results (Figs 18.11–18.14) showed that infrared thermography was successful in detecting subsurface debonds in both GFRP and CFRP wrapped cylinders. It should be noted from the infrared images in Figs 18.13 and 18.14 that the water-filled debonds also showed up as hot spots similar to the air-filled debond cases in Figs 18.11 and 18.12. According to the principle of conductive heat transfer, a water-filled debond should show up as

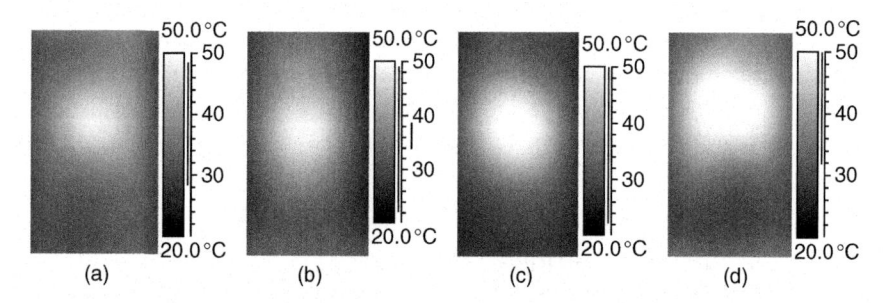

18.11 Infrared images of GFRP wrapped cylinders with air-filled debonds of sizes (a) 25.4 mm × 25.4 mm, (b) 35.6 mm × 35.6 mm, (c) 50.8 mm × 50.8 mm and (d) 76.2 mm × 76.2 mm.

Techniques for civil structures 499

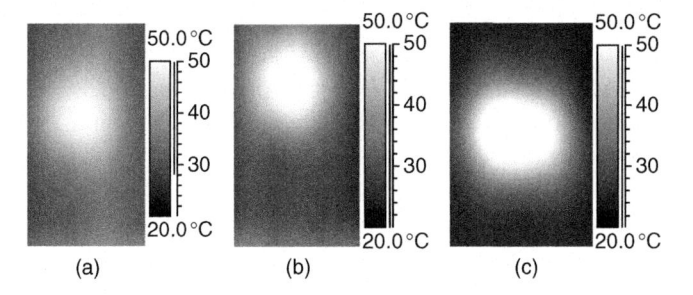

18.12 Infrared images of CFRP wrapped cylinders with air-filled debonds of sizes (a) 35.6 mm × 35.6 mm, (b) 50.8 mm × 50.8 mm and (c) 76.2 mm × 76.2 mm.

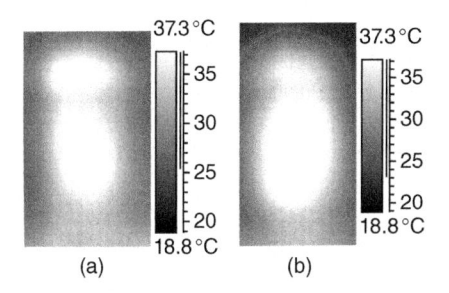

18.13 Infrared images of GFRP wrapped cylinders with water-filled debonds of sizes (a) 50.8 mm × 50.8 mm and (b) 76.2 mm × 76.2 mm

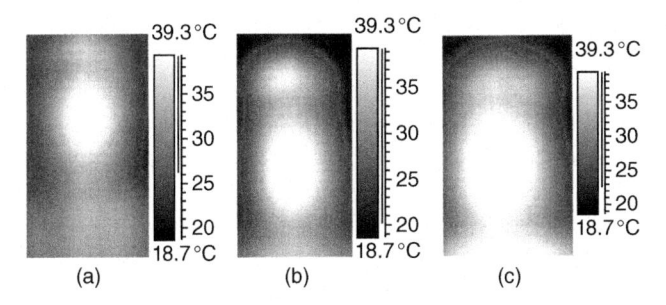

18.14 Infrared images of CFRP wrapped cylinders with water-filled debonds of sizes (a) 35.6 mm × 35.6 mm, (b) 50.8 mm × 50.8 mm and (c) 76.2 mm × 76.2 mm.

© Woodhead Publishing Limited, 2013

a cold spot or in other words, surface temperature of the region above the water-filled debond should be lower than the surface temperature of the region above the defect-free areas. This phenomenon has been observed in thicker planar composites (Plate VIII between pages 296 and 297) and is attributed to the fact that water has a higher thermal conductivity value (0.6 Wm^{-1} K^{-1}) than GFRP and CFRP composites (~0.17 Wm^{-1} K^{-1}). Thus, water-filled debonds conduct heat at a faster rate through the thickness compared to the defect-free area. However, in this study, the water-filled debonds showed up as hot spots similar to the case of air-filled debonds. This is because of the small thickness of the wrapping fabric coupled with high specific heat value of water (~4180 J/(kg K)). Thus, water stored a significant amount of heat which was then released during the cooling cycle when the infrared images were taken. This phenomenon becomes very prominent because of the small thickness of the FRP wraps. Also, it is important to note that air-filled debonds show a more clear boundary with higher surface temperature compared to the water-filled debond cases.

When conducting infrared tests with a heater, it is important to note that the heater should be at least 200 mm (8″) away from the specimen surface and moved from side to side while applying the heat. This ensures uniform heating of the test surface, otherwise one may see a pattern corresponding to the two heating rods in the infrared image instead of the subsurface defects. Also, in the case of vertical members, solar heating results in higher temperature of the top area and lower temperature of the bottom area due to naturally occurring convection (hotter air moving up and colder air moving down), thus masking the effect due to the subsurface debonds (Halabe et al., 2003a). This implies that solar radiation cannot be relied upon as a heat source for infrared testing of vertical members, although it is commonly used as an effective heat source for infrared testing of horizontal members like bridge decks. The testing of vertical members can be carried out using active heating sources such as heaters, halogen lamps, or heating blankets.

Plates IX and X (between pages 296 and 297) show the digital photographs and infrared images of two timber piles wrapped with GFRP fabric. The infrared images were obtained immediately after heating the test area with a 1500 W heater for about one minute. The infrared image in Plate IX(b) shows a number of debonds that were formed during wrapping because the surface of the underlying timber pile was not smooth. In contrast, the infrared image in Plate X(b) shows a very uniform surface temperature profile, which indicates a good bond between GFRP wrap and the underlying timber pile (Halabe and Periselty, 2011). Additional field testing results for GFRP wrapped timber piles and pile caps (beams) can be found in Halabe et al. (2005).

18.3 Ground penetrating radar (GPR)

18.3.1 Basics of GPR testing

GPR equipment typically consists of a mainframe system whose settings are controlled by a data acquisition laptop computer. The mainframe is connected to devices known as antennas to propagate short pulse (1 to 2 nanoseconds) electromagnetic waves into a medium and receive echo signals back from subsurface layer interfaces formed by debonds, delaminations, and inclusions such as steel reinforcing bars in case of reinforced concrete structural components. The wave velocity, signal amplitude, and wave attenuation characteristics of electromagnetic waves depend on the complex dielectric permittivity of the propagation medium (Halabe et al., 1993). For a medium with low electrical conductivity, the wave velocity (v) in the medium is given by the following equation:

$$v = \frac{c}{\sqrt{\varepsilon_r}} \qquad [18.1]$$

Where c is the velocity of electromagnetic waves in vacuum ($= 3 \times 10^8$ m/s) and ε_r is the relative dielectric constant of the propagation medium. The depth of a subsurface interface or inclusion (d) can be obtained using the following equation:

$$d = \frac{vt}{2} \qquad [18.2]$$

where t is the total round-trip time taken by the electromagnetic pulse to travel from the antenna to the subsurface feature and back to the antenna. Equation (18.2) assumes that the radar antenna is placed on the surface of the test specimen and the same antenna transmits and receives the echo signal (pulse-echo mode). Since the waves attenuate as they propagate through a medium, it is customary to use linear or exponential gain (increasing with travel distance) during radar data acquisition. The wave attenuation coefficient increases with antenna frequency and moisture content of the propagation medium (Halabe et al., 1993). While low-frequency antennas (15–900 MHz) are typically used for subsurface investigation in pavements and soils, higher-frequency antennas (1–2.5 GHz) are typically used in cases of concrete and composite structural components where the required penetration depth is much lower (few cm to 1 m). The higher frequencies offer shorter wavelengths and better resolution, so smaller defects can be detected. Some researchers have successfully used very high-frequency systems known as microwave or millimeter wave systems for composite applications (e.g., Akuthota et al., 2004; Fallahpour et al., 2011; Kharkovsky et al., 2008).

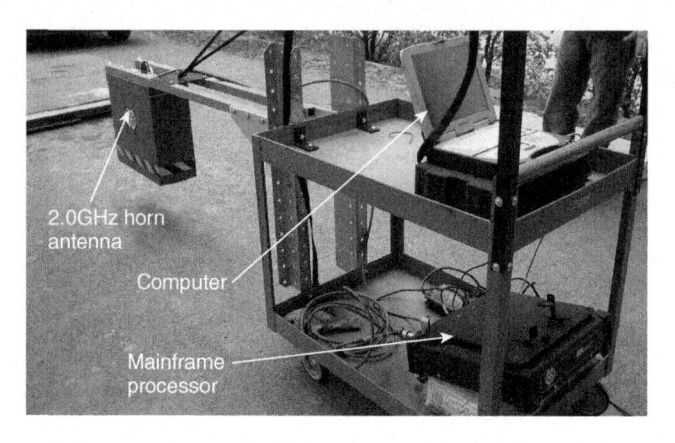

18.15 Components of a GPR system along with an air-launched radar antenna.

GPR antennas are typically of two types, air-launched and ground-coupled. Air-launched antennas (1–2 GHz) can be attached to the back of a vehicle and kept 0.15–0.3 m above the road or bridge surface. This allows for faster scanning of a pavement or bridge, with vehicle speeds ranging from ~30–65 kph (20 to 40 mph). An air-launched antenna mounted on a push cart housing a GPR system is shown in Fig. 18.15. The ground-coupled antennas (15 MHz to 2.5 GHz) are dragged over the test surface (using attachment handles or belts) and transmit higher amount of energy into the ground, which is required for deeper penetration (>1 m). Also, high-frequency (1–2.5 GHz) ground-coupled antennas (Fig. 18.16) typically have a smaller footprint (as small as 0.1 × 0.2 m), which results in a higher-resolution scan where small subsurface debonds and rebar layout are clearly visible.

In GPR applications, it should be noted that defects which have a higher dielectric contrast than the surrounding medium are easier to detect. This means water-filled defects and metallic inclusions (e.g., steel rebars) are much easier to detect than air-filled defects (Halabe *et al.*, 1995). Also, the electromagnetic pulses transmitted by commercial GPR antennas are linearly polarized. This means a conductor such as a rebar produces much higher signal amplitude if it is oriented along the polarization direction of the antenna. While the electromagnetic waves propagate easily through GFRP composites, they cannot penetrate conducting media such as CFRP composites owing to the high electrical conductivity of the carbon fibers. Some researchers have shown that electromagnetic waves can be propagated into CFRP composites if the polarization direction of the radar antenna is oriented perpendicular to the carbon fibers (Akuthota *et al.*,

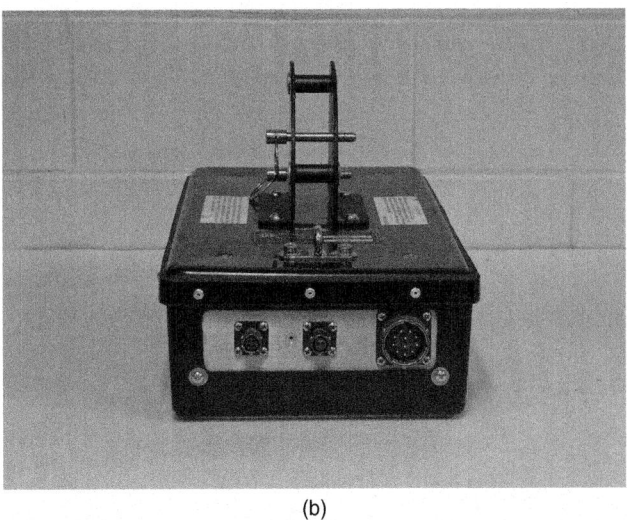

18.16 High-frequency ground-coupled radar antennas: (a) 1.6 GHz ground-coupled antenna mounted on a hand-held cart and (b) 0.9 GHz ground-coupled antenna.

504 Non-destructive evaluation (NDE) of polymer matrix composites

2004; Kharkovsky *et al.*, 2008). This approach can be effectively used for testing unidirectional CFRP composites. However, the CFRP fabrics used in civil structures often consist of multidirectional fabrics with fiber orientation in 0° and 90°, ±45°, or a combination of all four directions. Such multidirectional CFRP fabrics offer better ductility and earthquake resistance, but they can be tested only using other NDE techniques such as infrared thermography, but not GPR (Halabe *et al.*, 2010).

18.3.2 GPR testing for GFRP bridge decks

Figure 18.17 shows a low-profile (102 mm thick) GFRP composite deck, similar to the one shown in Figure 18.7. The deck size was 1830 mm × 915 mm in plan. Several simulated air-filled and water-filled debonds were inserted between a 9.5 mm polymer concrete wearing surface and the underlying deck (Hing and Halabe, 2010). Figure 18.18 shows the schematic layout of the subsurface debonds. A radar scan was acquired by moving a 1.5 GHz ground-coupled antenna along the line of the subsurface debonds. This longitudinal GPR scan (vertical cross-section), also known as B-scan, is shown in Plate XI (between pages 296 and 297). The water-filled debonds

18.17 GFRP bridge deck with simulated debonds inserted between the wearing surface and the underlying GFRP composite deck.

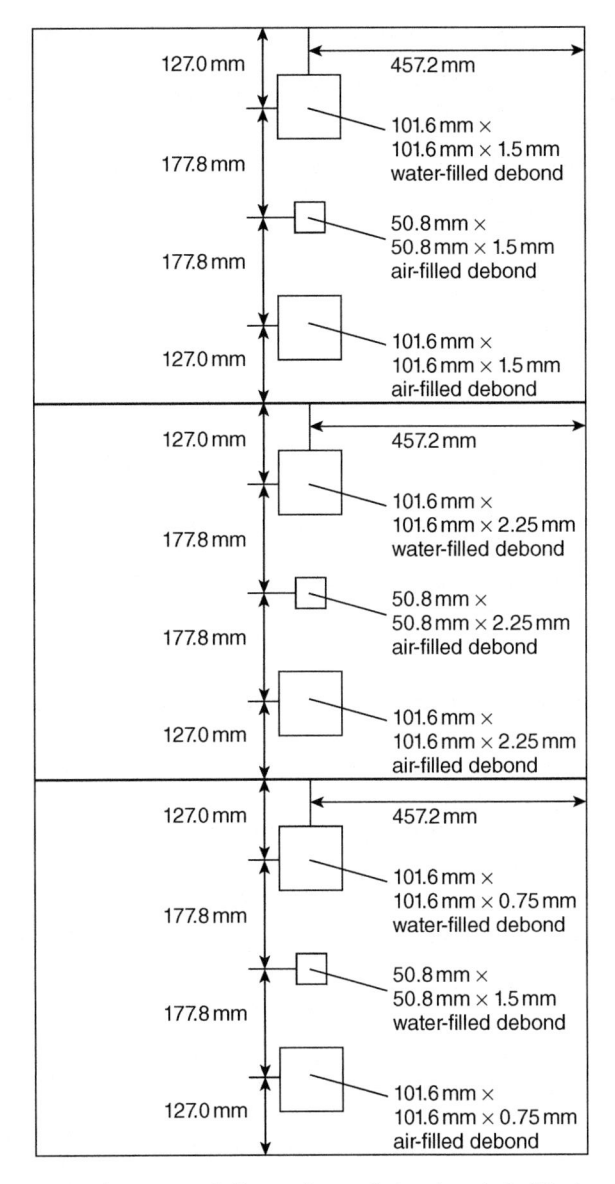

18.18 Layout and dimensions of simulated air-filled and water-filled debonds.

are clearly visible in this image but none of the air-filled debonds was detected. The features between the debonds are echoes produced by the flange–web junction in the underlying GFRP deck. The deck was also scanned along multiple longitudinal path and these scans were combined to form a three-dimensional image (C-scan) which is shown in Plate XII (between pages 296 and 297). This image shows the location of the water-filled debonds very clearly. While GPR antennas with 1–2 GHz frequency range has difficulty in detecting air-filled debonds, researchers have shown that microwaves at 10–24 GHz frequency range can detect air-filled debonds as well (e.g., Akuthota et al., 2004).

It should be noted that GPR offers superior penetrating capabilities compared to near-surface testing techniques such as infrared thermography. The GPR technique has shown potential in detecting defects in the web and bottom flange of the composite deck with scan taken from the top side which is more accessible (Hing and Halabe, 2010).

18.3.3 GPR testing for FRP wrapped columns

A study conducted by Halabe et al. (2010) showed that a GPR system with 1.5 GHz ground-coupled antenna was successful in detecting subsurface water-filled debonds in 152.4 mm × 304.8 mm (6″ × 12″) concrete cylinders wrapped with GFRP fabrics. Figure 18.19 shows the laboratory set-up for GPR testing of the FRP wrapped cylinders. Figure 18.20 shows the GPR scans from specimens with subsurface water-filled debonds of various sizes placed between the GFRP fabric and the underlying concrete cylinder. For cylinders wrapped with CFRP fabric, the debonds did not show up very well in the GPR scans, and infrared thermography was shown to be a more effective NDE technique (Halabe et al., 2010). Also, infrared thermography

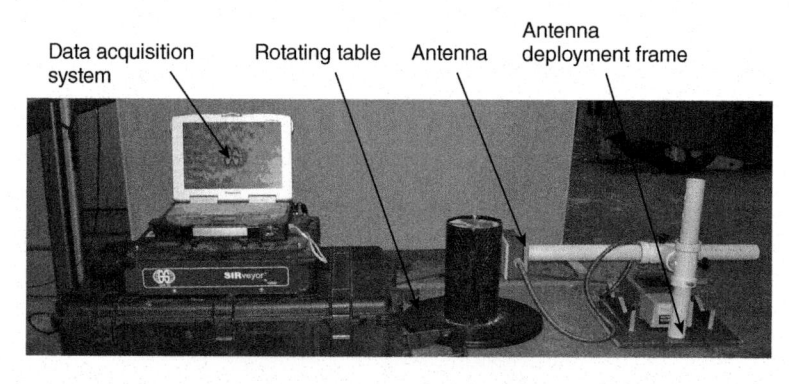

18.19 Laboratory set-up for GPR testing of FRP wrapped cylinders.

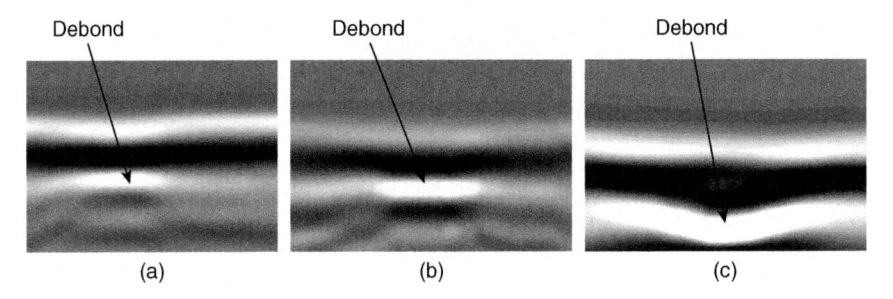

18.20 GPR image of GFRP wrapped cylinders with water-filled debonds of sizes (a) 35.6 mm × 35.6 mm, (b) 50.8 mm × 50.8 mm and (c) 76.2 mm × 76.2 mm.

was able to detect both air-filled and water-filled debonds as explained earlier in this chapter.

18.4 Digital tap testing

Tap testing or sounding technique is commonly used to detect subsurface debonds between FRP laminates or fabrics and the underlying structural component. Areas with air-filled subsurface debonds often produce a hollow sound when tapped with objects such as wooden stick, metallic rod, or FRP rebar. However, this kind of testing is very subjective and depends on the hearing ability of the testing personnel. A significant improvement in the area of tap testing has been achieved through the use of a digital tap hammer (Fig. 18.21). Such a device uses an instrumented hammer to record the echo signal received after tapping. This signal can be converted into a reading using parameters such as peak amplitude, area under the Fourier transform or power spectral density of the received echo signal. In the case of debonded areas, the echo from the debond results in a larger reading while defect-free areas produce smaller readings since the elastic waves propagate right through the test object and a small echo from the back wall is received. In other words, the more rigid or stiff the area, the smaller is the reading.

Advanced tap testing devices often record the contact surface echo as well as the slightly delayed echo coming from the subsurface debonds or voids. This enables the normalization of the debond echo by dividing it with the corresponding parameter for the surface echo. Such normalization makes the displayed tap test reading relatively immune to the tapping force. In other words, different users would achieve more or less similar normalized reading from a given area and the readings are repeatable within a narrow range. However, it is important to first establish a baseline reading

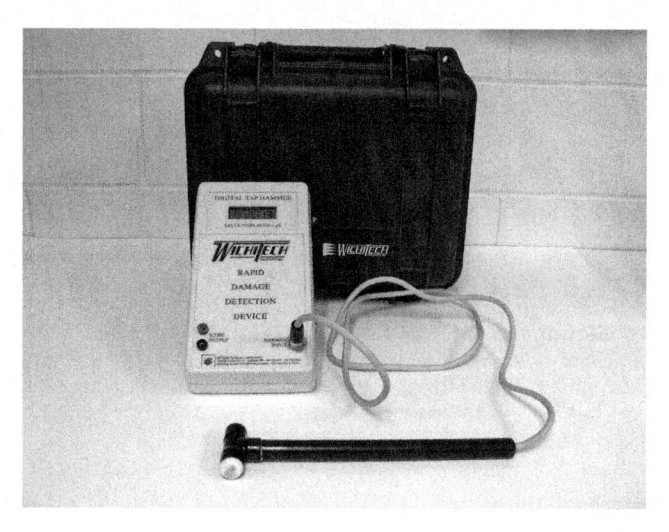

18.21 Digital tap hammer.

from a defect-free area using average of several tap tests before testing other areas to detect possible subsurface debonds or voids. Any area which results in a reading that is at least 10% higher than the baseline would have a good probability of a subsurface debond or void. Also, the higher the reading, the higher is the probability that a subsurface debond or void exists. Table 18.1 illustrates the readings from defective and defect-free regions in case of a GFRP bridge deck specimen and concrete cylinders wrapped with GFRP fabrics. It can be seen from Table 18.1 that the digital tap testing results for any given case are fairly consistent and repeatable. Also, the data from the third specimen shows that the air-filled subsurface debond leads to much higher readings compared to the water-filled subsurface debond.

The above results show that tap testing is a very easy-to-use and effective NDE technique for detecting subsurface debonds. However, it is a slow technique and cannot cover large areas quickly. If coverage of large areas is needed, then infrared thermography and GPR offer more efficient alternatives. Infrared thermography is very effective for detecting near-surface defects while GPR can provide significantly higher penetration depths.

18.5 Issues and challenges in using non-destructive evaluation (NDE) techniques

Unlike the aerospace industry, the civil infrastructure community invests only a small percentage of their budget on NDE related work, and most of the budget is designated for new construction or repair work. Agencies such

Table 18.1 Tap testing results obtained from laboratory specimens

Specimen type	Type of test area	Tap test readings from successive tests	Average tap test reading	Percentage change compared to defect-free area (%)
Deck specimen BD1a with subsurface debonds between wearing surface and GFRP deck	Defect-free	1134, 1124, 1160, 1178, 1107, 1142, 1177, 1162, 1149, 1141, 1144, 1150	1147	–
	Above 76 mm × 76 mm air-filled debond	1513, 1539, 1491, 1436, 1589	1514	31.9
	Above 51 mm × 51 mm air-filled debond	1441, 1584, 1455, 1546, 1518	1509	31.5
Concrete cylinder wrapped with one layer of GFRP fabric	Defect-free	1089, 1090, 1104, 1088, 1101, 1105, 1098, 1109, 1120, 1121, 1123, 1101	1104	–
	Above 51 mm × 51 mm air-filled debond	1796, 1821, 1803, 1805, 1813, 1778, 1792, 1839, 1774, 1781, 1850, 1797	1804	63.4
Concrete cylinder wrapped with two layers of GFRP fabric	Defect-free	1225, 1209, 1154, 1255, 1187, 1129, 1225, 1223, 1228, 1132, 1161, 1206	1195	–
	Above 25 mm × 25 mm water-filled debond	1447, 1390, 1420, 1390, 1361, 1415, 1413, 1433, 1405, 1457, 1396, 1407	1411	18.1
	Above 12.7 mm × 12.7 mm air-filled debond	2152, 2180, 2079, 2190, 2456, 2760, 2587, 2328, 2917, 3024, 2339, 2451	2455	105.5

as the Federal Highway Administration and State Departments of Transportation are continually looking for highly cost-effective NDE solutions. This requires the use of low-cost NDE equipment which is fairly easy for a field inspector or highway personnel to use. For in-service testing, portability of the NDE equipment is very crucial. Also, development of user-friendly data analysis software that can be used in the field to quickly identify problem areas during field inspection is important. Future development of automated data processing algorithms for various NDE techniques can greatly help to enhance the data analysis procedure and possibly minimize human errors in data interpretation. If such algorithms are incorporated into user-friendly software, it can help in increasing the use of NDE by field engineers and testing personnel. Also, it is important to understand the challenges faced during field testing. For example, structural components under the bridge are often difficult to access and require lane closures for NDE testing. Development of NDE systems with accessories adapted for rapid testing of structural members under the bridges can greatly facilitate the increased use of NDE. At the same time, such equipment and associated data processing software need to be available at a reasonable cost for increased use of NDE for condition assessment of civil infrastructure.

18.6 Future trends

Significant research and development work in the area of NDE of composites for civil infrastructure has taken place in the last ten years. This includes development of more robust and sensitive NDE systems, as well as easy-to-use field testing equipment. A significant number of researchers are devoting their time towards development of quantitative data analysis algorithms. For example, in the infrared thermography area, algorithms have been developed to predict width and depth of subsurface debonds by analyzing a sequence of infrared images acquired during the cooling cycle, that is, immediately after removal of heat. These algorithms make use of parameters such as break time, contrast analysis, derivative images, and determination of inflection points in the surface temperature profile over the debonds (Halabe and Dutta, 2010; Maldague, 2000, 2001; Maldague and Moore 2001b; Plotnikov, 1999; Ringermacher et al., 2007; Shepard et al., 1999, 2003, 2007; Starnes et al., 2003; Shepard, 2006, 2007; Vavilov, 2007). Computer-controlled short-duration flash lamps are often used in conjunction with more robust cameras (with capture rates of up to 500 frames per second) and advanced data acquisition and processing methodologies such as pulsed thermography, lock-in thermography, and vibrothermography (Spring et al., 2011). However, owing to the very high cost and bulky nature of these systems, they have been predominantly used only in the laboratory setting for aerospace and automotive applications, especially for thin composites. It is conceivable that in the near future these advanced algorithms and robust NDE hardware will be incorporated into small low-cost handheld units so that they can be conveniently used for inspection of civil structures in the field environment. In addition, it is expected that better heating systems will become available for application of uniform heating to field structural components.

In the area of GPR applications, algorithms are being developed to combine two-dimensional GPR cross-sectional scans into three-dimensional images for better rendering of the structural components' subsurface features. The three-dimensional GPR imaging techniques have been extended from flat objects such as composite bridge decks to cylindrical objects such as columns. Further development in this area includes the ability to create circular cross-sectional scans and the capability to slice a three-dimensional cylindrical image into vertical and horizontal slices at various distances and depths for easy viewing of the subsurface defects or position of inclusions such as steel rebars (Halabe and Pyakurel, 2010; Pyakurel, 2009; http://www.gpr-survey.com/). Currently, these algorithms require significant amount of intermediate steps with human intervention to produce three-dimensional images. In the future, it is expected that these algorithms will be packaged into more user-friendly software with automated data processing features and incorporated into portable field testing equipment so that their use will become widespread within the civil structures community.

Other NDE techniques such as fiber optics-based strain measurement, ultrasonics, acoustography, laser shearography, Raman spectroscopy, electromagnetic and eddy current, microwaves, and millimeter wave techniques are being developed and utilized for aerospace composite applications (e.g., Akuthota et al., 2004; Fallahpour et al., 2011; Kharkovsky et al., 2008; Smith, 2007; Wang et al., 2011). It is conceivable that these NDE techniques, which are also being tested in the laboratory or manufacturing setting for composite structural components (e.g., Akuthota et al., 2004; Kharkovsky et al., 2008; Washer and Blum, 2011; Wang et al., 2011), will be adapted for field application to composite components in civil structures in the near future. The data acquisition hardware for some of these NDE techniques could also be made more robust and adapted for permanent installation on field structures and the acquired data could be transmitted wirelessly to a central data processing location. This approach would become feasible for large field structures with the development of automated data processing algorithms that can integrate the results from different NDE techniques to reliably predict the condition of the structure as a whole and help in making timely repair and maintenance decisions so that a long service life for the structure can be achieved.

18.7 References

ACI 440.2R-08 (2008), Guide for the Design and Construction of Externally Bonded FRP Systems for Strengthening Concrete Structures, Report by Committee 440, American Concrete Institute, Farmington Hills, MI 48331.

Akuthota, B., Hughes, D., Zoughi, R., Myers, J., and Nanni, A. (2004), 'Near-Field Microwave Detection of Disbond in Carbon Fiber Reinforced Polymer Composites Used for Strengthening Cement-Based Structures and Disbond Repair Verification,' *Journal of Materials in Civil Engineering, ASCE*, **16**(6), 540–546.

ASTM D 4788 – 03 (2007). 'Standard Test Method for Detecting Delaminations in Bridge Decks Using Infrared Thermography,' *Annual Book of ASTM Standards*, Vol. 04.03, American Society of Testing and Materials (ASTM), West Conshohocken, Pennsylvania, USA.

Clark, M. R., McCann, D. M., and Forde, M. C. (2003), 'Application of Infrared Thermography to the Non-Destructive Testing of Concrete and Masonry Bridges,' *NDT&E International*, **36**, 265–275.

Dutta, S. S. (2010), 'Nondestructive and Destructive Evaluation of FRP Composite Wrapped Concrete Cylinders with Embedded Debonds,' Ph.D. Dissertation, West Virginia University, Department of Civil and Environmental Engineering, Morgantown, WV.

Fallahpour, M., Ghasr, M. T., Case, J. T., and Zoughi, R. (2011), 'A Wiener Filter-based Synthetic Aperture Radar Algorithm for Microwave Imaging of Targets in Layered Media,' *Materials Evaluation*, **69**(10), 1227–1237.

Halabe, U. B., and Dutta, S. S. (2010), 'Quantitative Characterization of Debond Size in FRP Wrapped Concrete Cylindrical Columns Using Infrared Thermography,' *Proceedings of the Fourth Japan-US Symposium on Emerging NDE Capabilities for a Safer World*, conference jointly conducted by the Japanese Society for Non-Destructive Inspection (JSNDI) and the American Society for Nondestructive Testing (ASNT), Maui Island, Hawaii, U.S.A., June, 64–68.

Halabe, U. B., and Perisetty, N. K. (2011), 'Composite Materials Use for Railroad Infrastructure – Rehabilitation of Timber Railroad Bridges Using Glass Fiber Reinforced Polymer Composite: Nondestructive Testing and Evaluation (Task 3),' USDOT-FRA Agreement No. FR-RRD-0010-10-01-00, Final Report, submitted to US Department of Transportation – Federal Railroad Administration (USDOT-FRA), Washington, DC, pp. 1–26.

Halabe, U. B., and Pyakurel, S. (2010), '2D and 3D GPR Imaging of Cylindrical Wooden Logs and Concrete Columns,' *Proceedings of the Fourth Japan-US Symposium on Emerging NDE Capabilities for a Safer World*, conference jointly conducted by the Japanese Society for Non-Destructive Inspection (JSNDI) and the American Society for Nondestructive Testing (ASNT), Maui Island, Hawaii, U.S.A., June, 81–87.

Halabe, U. B., Sotoodehnia, A., Maser, K. R., and Kausel, E. A. (1993), 'Modeling the Electromagnetic Properties of Concrete,' *American Concrete Institute (ACI) Materials Journal*, **90**(6), 552–563.

Halabe, U. B., Maser, K. R., and Kausel, E. A. (1995), 'Condition Assessment of Reinforced Concrete Structures Using Electromagnetic Waves,' *American Concrete Institute (ACI) Materials Journal*, **92**(5), 511–523.

Halabe, U. B., Steele, W. E., GangaRao, H. V. S., and Klinkhachorn, P. (2003a), 'NDE of FRP Wrapped Timber Bridge Components Using Infrared Thermography,' *Review of Progress in Quantitative Nondestructive Evaluation*, Vol. 22 (AIP CP657), ed. D. O. Thompson and D. E. Chimenti, American Institute of Physics, Melville, NY, 1172–1177.

Halabe, U. B., Vasudevan, A., and GangaRao, H. V. S. (2003b), 'Field Testing of the La Chein Composite Bridge Deck Using Infrared Thermography,' Final Report No. CFC-03-112, submitted by Constructed Facilities Center, West Virginia University, to West Virginia Division of Highways (WVDOH), District 9, Lewisburg, WV, October 2003, pp. 1–17.

Halabe, U. B., Vasudevan, A., and GangaRao, H. V. S. (2005), 'Field Testing and Evaluation of FRP Wrapped Timber Railroad Bridge Components Using Infrared Thermography,' *CD-ROM Proceedings of the Third US-Japan Symposium on Advancing Applications and Capabilities in NDE*, American Society for Nondestructive Testing and the Japanese Society for Nondestructive Inspection, Maui, Hawaii, June, 342–348.

Halabe, U. B., Roy, M., Klinkhachorn, P., and GangaRao, H. V. S. (2006), 'Detection of Air and Water-Filled Subsurface Defects in GFRP Composite Bridge Decks Using Infrared Thermography,' *Review of Progress in Quantitative Nondestructive Evaluation*, Vol. 25 (AIP CP820), ed. D. O. Thompson and D. E. Chimenti, American Institute of Physics, Melville, NY, 1632–1639.

Halabe, U. B., Vasudevan, A., Klinkhachorn, P., and GangaRao, H. V. S. (2007), 'Detection of Subsurface Defects in Fiber Reinforced Polymer Composite Bridge Decks Using Digital Infrared Thermography,' *Nondestructive Testing and Evaluation*, **22**(2–3), 155–175.

Halabe, U. B., Dutta, S. S., and GangaRao, H. V. S. (2010), 'Infrared Thermographic and Radar Testing of Polymer-Wrapped Composites,' *Materials Evaluation*, **68**(4), 447–451.

Hing, C. L. C., and Halabe, U. B. (2010), 'Nondestructive Testing of GFRP Bridge Decks using Ground Penetrating Radar and Infrared Thermography,' *Journal of Bridge Engineering, ASCE*, **15**(4), 391–398.

Kaplan, H. (1999). *Practical Applications of Infrared Thermal Sensing and Imaging Equipment*, SPIE Press, Bellingham, Washington.

Kharkovsky, S., Ryley, A. C., Stephen, V., and Zoughi, R. (2008), 'Dual-Polarized Near-Field Microwave Reflectometer for Noninvasive Inspection of Carbon Fiber Reinforced Polymer-strengthened Structures,' *IEEE Transactions on Instrumentation and Measurement*, **57**(1), 168–175.

Liang, R., and GangaRao, H. V. (2004), 'Applications of Fiber Reinforced Polymer Composites,' *Proceedings of Polymer Composites III Conference on Transportation Infrastructure, Defense and Novel Applications of Composites*, Robert C. Creese and Hota Gangarao, eds., Des*tech* Publications Inc., Lancaster, PA, 173–187.

Maldague, X. P. V. (1993), *Nondestructive Evaluation of Materials by Infrared Thermography*, Springer-Verlag, London.

Maldague, X. (2000), 'Applications of Infrared Thermography in Nondestructive Evaluation,' *Trends in Optical Nondestructive Testing* (invited chapter), ed. P. Rastogi, 591–609 (paper also available at http://www.gel.ulaval.ca/~maldagx/r_1123.pdf).

Maldague, X. P. V. (2001), *Theory and Practice of Infrared Technology for Nondestructive Testing*, John Wiley & and Sons, New York.

Maldague, X. P. V. and Moore P. O. (2001), *Nondestructive Testing Handbook – Infrared and Thermal Testing*, American Society for Nondestructive Testing, Columbus, OH.

Plotnikov, Y. A. (1999), 'Modeling of the Multiparameter Inverse Task of Transient Thermography,' *Review of Progress in Quantitative Nondestructive Evaluation*, Vol. 18, ed. D. O. Thompson and D. E. Chimenti, Kluwer Academic/Plenum Publishers, New York, 873–880.

Pyakurel, S. (2009), '2D and 3D GPR Imaging of Wood and Fiber Reinforced Polymer Composites,' Ph.D. Dissertation, Department of Civil and Environmental Engineering, West Virginia University, Morgantown, WV.

Ringermacher, H. I., Knight, B., Li, J., Plotnikov, Y. A., Aksel, G., Howard, D. R., and Thompson, J. L. (2007), 'Quantitative Evaluation of Discrete Failure Events in Composites Using Infrared Imaging and Acoustic Emission,' *Nondestructive Testing and Evaluation*, **22**(2–3), 93–99.

Shepard, S. M. (2006), 'Understanding Flash Thermography,' *Materials Evaluation*, **64**(5), 460–464.

Shepard, S. M. (2007), 'Thermography of Composites,' *Materials Evaluation*, **65**(7), 690–696.

Shepard, S. M., Rubadeux, B. A., and T. Ahmed (1999), 'Automated Thermographic Defect Recognition and Measurement,' *Nondestructive Characterization of Materials IX*, AIP Conference Proceedings 497, ed. R. E. Green, Jr., American Institute of Physics, Melville, New York, 373–378.

Shepard, S. M., Lhota, J. R., Rubadeux, B. A., Wang, D., and Ahmed, T. (2003), 'Reconstruction and Enhancement of Active Thermographic Image Sequences,' *Optical Engineering*, **42**, 1337–1342.

Shepard, S. M., Lhota, J. R., and Ahmed, T. (2007), 'Flash Thermography Contrast Model Based on IR Camera Noise Characteristics,' *Nondestructive Testing and Evaluation*, **22**(2–3), 113–126.

Smith, R. A. (2007), 'Advanced NDT of Composites in the United Kingdom,' *Materials Evaluation*, **65**(7), 697–710.

Spring, R., Huff, R., and Schwoegler, M. (2011), 'Infrared Thermography: A Versatile Nondestructive Testing Technique,' *Materials Evaluation*, **69**(8), 934–942.

Starnes, M. A., Carino, N. J., and Kausel, E. A. (2003), 'Preliminary Thermography Studies for Quality Control of Concrete Structures Strengthened with Fiber-reinforced Polymer Composites,' *Journal of Materials in Civil Engineering, ASCE*, **15**(3), 266–273.

Vasudevan, A. (2004), 'Application of Digital Infrared Thermography for NDE of Composite Bridge Components,' M.S. Thesis, Department of Civil and Environmental Engineering, West Virginia University, Morgantown, WV.

Vavilov, V. P. (2007), 'Pulsed Thermal NDT of Materials: Back to the Basics,' *Nondestructive Testing and Evaluation*, **22**(2–3), 177–197.

Wang, C., Faidi, W., Tralshawala, N., Vadde, A., and May, A. (2011), 'Online Monitoring of Composite Joining Process,' *Materials Evaluation*, **69**(12), 1393–1398.

Washer, G., and Blum, F. (2011), 'Developing Raman Spectroscopy for the Nondestructive Testing of Carbon Fiber Composites,' *Materials Evaluation*, **69**(10), 1219–1226.

19
Non-destructive evaluation (NDE) of composites: application of thermography for defect detection in rehabilitated structures

A. SHIRAZI, Simpson Gumpertz & Heger, USA and
V. M. KARBHARI, University of Texas at Arlington, USA

DOI: 10.1533/9780857093554.4.515

Abstract: Infrared (IR) thermography provides the ability to obtain real-time inspection capabilities in the field which when coupled with data interpretation provides a means of rapid assessment of the integrity and future serviceability of a rehabilitated structure. The chapter provides details on a program that enables the detection of defects and the establishment of a standardized protocol for quantifying progression of these defects in FRP rehabilitated concrete structures. The use of IR thermal imaging, using a means of progressive testing and a mathematical model from a time history of the thermal responses is shown to provide valuable information pertaining to the life-span of FRP rehabilitated systems.

Key words: non-destructive evaluation, NDE, infrared thermography, defect, debond, rehabilitation, external bonding, fracture.

19.1 Introduction

The rapid aging of civil infrastructure worldwide has created a need for the development and implementation of efficient methods of rehabilitation. Among the methods that have gained acceptance over the last decade is the use of externally bonded fiber-reinforced composites. This method has been shown to be a rapid and structurally efficient means of strengthening deficient concrete components in the civil infrastructure. While the efficacy of a number of rehabilitation techniques, including this one, has been demonstrated both through large-scale laboratory tests and use in the field, there is a continuing need for monitoring and inspection of strengthened structural components both to ensure public safety and also to build a database on the long-term in-field performance of these structures. Current methodologies for inspection of highway infrastructure are generically based on time intervals using subjective condition ratings. These have been reported to result in significant variability in assessment, and even in-depth inspections do not consistently give accurate results [1]. A number of non-destructive evaluation (NDE) techniques being evaluated for bridge inspection by the Federal Highway Administration (FHWA) are described by Washer [2] and a review of techniques has recently been provided by Kaiser and coworkers [3,4]. The need for advanced techniques for field NDE is perhaps even more pronounced with the use of newer rehabilitation techniques, such as the use of externally bonded fiber-reinforced polymer (FRP) composites, since the efficacy of the rehabilitation depends on the integrity of both the FRP material and the bond to the substrate.

A successful rehabilitation process requires the two major steps of load transfer and load distribution to take place effectively. Any abnormality can reduce the effectiveness of the technique. Defects can be classified based on the stage of formation and/or level of criticality. Kaiser and Karbhari [5,6] have conducted research in regard to the classification of defects in terms of type, size and origin. Defects are developed at three main stages (i.e. (a) use of improper materials such as outdated resins or damaged fabrics; (b) errors during fabrication and application, and (c) defects arising during the life of the rehabilitated structures), with each stage having many subcategories. Effects of environmental exposure, accidental impact and normal aging (including wear and tear) can both deteriorate materials and increase the size and/or effect of pre-existing defects. Since defects and their progression over time are likely to have a substantial effect on both quality and long-term integrity of the rehabilitation measures it is essential to develop methods to not only enable their detection using means of non-destructive testing (NDT) in the field but also for the criticality assessment of defects. Without the development and implement of such an approach both inspection for purposes of quality control and service-life determination will not be possible.

The detection of defects is an important part of the process through which safety and reliability of components and systems can be assured. The main challenge in NDE of composites is that despite the numerous failure mechanisms the composite is intrinsically elastic up to failure and therefore unlike steel does not provide a warning substantially prior to failure. Hence the method of NDE used must be able to cover the entire process by which engineers can:

- determine the existence of a defect;
- accurately pinpoint its location within the part;
- estimate its extent; and
- monitor its progression over time.

All of these tasks need to be performed without changing the integrity or serviceability of the structure.

With the exception of some visual inspection-based techniques, most NDE methods have a common characteristic in that they involve the application of a specific excitation to the structure and the measurement of the response of the structure to that specific excitation. A comprehensive review of NDE methods applicable to FRP rehabilitated concrete is presented by Kaiser and Karbhari [3] and is hence not repeated herein. However, it is important to emphasize that despite the wide range of available NDE methods, ranging from use of acoustic means and ultrasonic, to dye penetrants, radiography, thermography, eddy currents, microwaves and ground penetrating radar, optical means, conventional and advanced strain measurement techniques, modal analysis and load testing, only a small number can be used effectively in the field. Further, of these, only a much smaller subset has the potential to be used effectively for the four aspects of NDE mentioned earlier. To date only visual inspection and acoustic impact testing (coin-tap testing) have been used, albeit on a qualitative, rather than quantitative, basis.

Infrared (IR) thermography, used extensively in the aerospace arena, however, provides immense field potential in terms of being able to obtain real-time inspection capabilities and data interpretation which could provide value through rapid assessment of the integrity and future serviceability of a rehabilitated structure and has the added advantage of not requiring contact or a coupling medium (other than air) between the equipment and the structure being inspected [7]. Recent studies have demonstrated its effectiveness in identifying weak areas and defects at the bond line [8,9]. While it is evident from previous research [6–13] that IR thermography has the potential to be used for both condition assessment as well as health monitoring of FRP rehabilitated infrastructure, there is a need for substantial research related to the area of FRP composite–concrete bond as well as the identification and quantification of defects. Furthermore, for efficient use as a method of NDE, as well as monitoring of FRP rehabilitated structures, there needs to be a well-classified data-set of defect signatures, and effects of these defects.

19.2 Principles of infrared (IR) thermography

IR thermography is a technique that uses an IR imaging and measurement camera to see and measure IR energy being emitted from an object. The technique can be as simple as observing an object with an IR camera while heating it with a lamp or heat gun. Variations in the IR radiation are sensed by an IR camera and converted to a visible image which maps the region inspected. In aircraft manufacturing environments, however, far more sophisticated pulsed thermography systems use a very short, uniform pulse of light to heat the surface of the sample. Recent advances in pulsed thermography incorporate both improved analytical methods and hardware, resulting in highly accurate quantitative results that clearly distinguish subsurface defects and provide accurate estimates of part integrity.

The principle of IR thermography is to input heat energy on the object being inspected and measure the resulting discontinuity in IR flow. The difference in heat measured makes this an extremely valuable diagnostic tool. Thermography systems are used throughout the world to solve a wide range of NDE challenges, and equipment developed ranges from portable hand-held systems to fully automated in-place quality assurance systems. The method has been used to detect aspects such as delaminations, voids, corrosion, paint adhesion and thickness, skin to core debonds, porosity, wall thickness, impact damage, contamination, changes in thermal diffusivity, spot weld quality, efficiency of adhesive bonding and cracks. The ability to cover large areas rapidly without significant support equipment and without disruption of use of the structure as well as the ability to be used even in areas with restricted access make this method superior to other competitive methods for rapid inspection of concrete structures rehabilitated with composites.

It is noted that IR light lies between the visible and microwave portions of the electromagnetic spectrum. Near infrared light (~0.76–1.4 micron) is closest in wavelength to visible light. Far infrared light (3.0–1000 micron) is closer to the microwave region of the electromagnetic spectrum and middle infrared light (1.4–3.0 micron) is between these two extremes. Since the primary source of IR radiation is heat, or thermal radiation, any object which has an amount of energy in the form of heat emits IR rays. Even objects that we think of as being very cold, such as an ice cube, emit IR spectra. When an object is not hot enough to radiate visible light, it will emit most of its energy in the IR region. If a defect exists, the heat flow is affected by the presence of the defect and causes a change in surface temperature and consequently a different IR emission at the location of the defect. The IR emission is then measured by an IR camera and after data manipulation, the image can be shown in the form of a contour map representing regions with identical IR discharge, i.e. isothermal regions. Defects are thus detectable if thermographic images are taken after appropriate times of incident radiation, processed appropriately, and interpreted correctly. Reaching an appropriate level of quality requires the achievement of all of the above in an orderly fashion.

19.3 Using infrared (IR) thermography in practice: application to a bridge deck assembly

In order to ensure that the configuration and condition used in this study were similar to those that would be seen in the field, all tests were conducted using a model, three-girder, two-bay bridge specimen which was being used for a proof-test for the study of FRP rehabilitation of concrete. Thus, tests were conducted on a specimen of scale and configuration similar to that found in the field. Although all tests were conducted in a controlled laboratory setting, the placement and access available to the IR thermography camera and operator were also similar to that available in the field with positioning of the camera being controlled only by manual means or through a temporary wooden scaffold.

The three-girder, two-bay bridge deck specimen was designed and fabricated in accordance with the California Department of Transportation Bridge Design Specifications (BDS). The center-to-center spacing between longitudinal girders was 1680 mm (5.5′) and the assembly consisted of a slab with 152 mm (6″) depth, and girders having a total depth, including the slab flange, of 559 mm (22″) and a width of 203 mm (8″). The deck assembly had a total longitudinal span of 3600 mm (12′4″) including overhangs of 607 mm (24″) on either side of the edge girders. Details related to structural assessment are given in Ghosh and Karbhari [9] and are hence not repeated herein except to provide details pertinent to the IR thermography conducted as part of this research. The bridge-deck specimen was monotonically loaded under increasing load and cycled at predetermined levels to check for stability until a load equivalent to 75% of transverse steel yield at which point the slabs underwent cracking and deterioration warranting rehabilitation. Carbon fiber-reinforced composites in the form of prefabricated strips of 101.6 mm (4.0″) wide repeated every 381 mm (15″) center to center in both longitudinal and transverse direction were externally bonded to the deck soffits for purposes of rehabilitation.

After completion of the rehabilitation, the specimen was loaded under monotonically increasing loads with cycling at periodic intervals. The cyclic loading consisted of a reduction of load to levels of 11 kN (24 kips) and 0 kN (0 kips) followed by an increase to the next higher load level. Thermographic images were taken at both the 0 kN and 11 kN levels to enable assessment of the rehabilitated specimen and to detect the gaps and cracks that might close as the load was released. Thermographic images were then used for this research through an orderly study in order to investigate and quantify the presence and growth of defects and to enable the assessment of thresholds of criticality.

Active thermography was used for this experiment. In active thermography, excitation is induced artificially. In this research the excitation was provided by a pair of xenon flash lights (the apparatus is shown in Fig. 19.1). The cross-sectional dimension of the heat box on this particular instrument is 320 mm by 240 mm and therefore a number of shots were necessary to cover the area of interest within the experimental specimen. A scheme was set for the locations at which the camera was placed to help avoid taking unnecessary shots and consequently ensure lower-temperature elevation of the background while ensuring good overall coverage. The first step was to adjust parameters on the instrument to ensure the best performance. Some

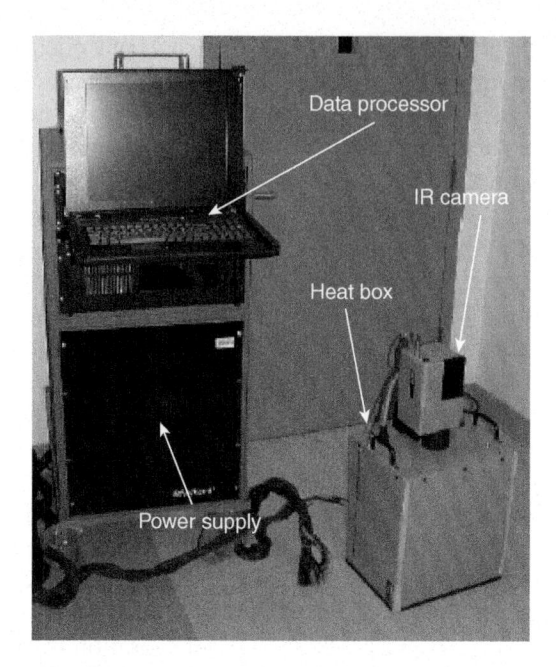

520   Non-destructive evaluation (NDE) of polymer matrix composites

19.1 IR thermography apparatus.

of these variables such as capturing frequency and camera focal point make permanent changes on the output image, and hence a great amount of care has to be taken through adjustment of these parameters. Other sets of parameters such as gate number and gate length change only the display of required data and can be readjusted even after the testing was performed

Capturing, or sampling, frequency is defined as number of images taken per second by the IR camera. This must be adjusted with respect to thermal conductivity and the maximum depth of the object that needs to be monitored. Since a defect is in continuous thermal interaction with its surrounding, an over-sampled or under-sampled job can easily result in missing important indicators. In this experiment, since only one layer of composite was used, a sampling frequency of 60 Hz (60 frames per second) was chosen based on a series of calibration tests to ensure that no important data was missed during the thermal interaction.

Duration time is the time period during which image capturing occurs. Adjusting this variable allows the heat to penetrate to a desired depth of the object being tested. This variable, like the sampling frequency, must be adjusted with respect to a desired monitoring depth and thermal conductivity of the object being tested. The duration time used in this research was set at 10 seconds.

A test done at 60 Hz sampling frequency with a 10-second duration time yields 600 images. The use of a large number of images decreases the possibility of missing a defect partially or completely but increases the level of effort to ensure the catching of effects due to very gradual and minor changes in thermal intensities from one image to another. The gate is a secondary time frame nested inside the main duration time. All of the images in one particular gate are superimposed and shown as one single image. The user can customize the length and the starting and ending points for each individual gate. For the purposes of the current research 20 gates were used. Images were taken in nine loading stages, of which four were at a low load of 11 kN (24 kips) to enable the detection of crack openings that might be obscured at lower loads. Tables 19.1 and 19.2 present a list of images taken at different load steps where the letters T and L denote transverse and longitudinal shots respectively, as designated in Fig. 19.2 and as an aggregate of images in Fig. 19.3.

Table 19.1 List of longitudinal images

Load steps	Ok	130-24	150-0	150-24	170-0	170-24	190-0	190-24	210-0
L01	×	×	×	×	×	×	×	×	×
L02	×	×	×	×		×	×	×	×
L03	×	×	×	×		×	×	×	×
L04	×	×	×	×	×	×	×	×	×
L05	×	×	×	×	×	×	×	×	×
L06	×	×	×	×		×	×	×	×
L07	×		×	×					×
L08	×		×	×					×
L09	×		×	×					×
L10	×		×	×					×
L11	×		×	×					×
L12	×		×	×	×				×
L13	×	×	×	×		×	×	×	×
L14	×	×	×	×		×	×	×	×
L15	×	×	×	×		×	×	×	×
L16	×	×	×	×		×	×	×	×
L17	×	×	×	×		×	×	×	×
L18	×	×	×	×		×	×	×	×

Table 19.2 List of transverse images

Load steps									
	Ok	130-24	150-0	150-24	170-0	170-24	190-0	190-24	210-0
T0I	x	x	x	x		x	x	x	x
T02	x	x	x	x	x	x	x	x	x
T03	x	x	x	x	x	x	x	x	x
T04	x	x	x	x	x	x	x	x	x
T05	x	x	x	x		x	x	x	x
T06	x	x	x	x		x	x	x	x
T07	x	x	x	x	x	x	x	x	x
T08	x	x	x	x	x	x	x	x	x
T09	x	x	x	x	x	x	x	x	x
T10	x	x	x	x		x	x	x	x
T11	x	x	x	x		x	x	x	x
T12	x	x	x	x	x	x	x	x	x
T13	x		x	x					x
T14	x		x	x					x
T15	x		x	x					x
T16	x		x	x					x
T17	x		x	x					x
T18	x		x	x					x
T19	x	x	x	x	x	x	x	x	x
T20	x	x	x	x	x	x	x	x	x
T21	x	x	x	x	x	x	x	x	x
T22	x	x	x	x	x	x	x	x	x
T23	x	x	x	x	x	x	x	x	x
T24	x	x	x	x	x	x	x	x	x
T25	x	x	x	x		x	x	x	x
T26	x	x	x	x		x	x	x	x
T27	x	x	x	x	x	x	x	x	x
T28	x	x	x	x		x	x	x	x
T29	x	x	x	x		x	x	x	x
T30	x	x	x	x		x	x	x	x

19.4 Data collection methodology

Since a large number of IR images have to be aggregated, differences in thermal intensities and backgrounds had to be considered through normalization with respect to a control undamaged region of each image. Two approaches were considered and employed to reconstruct the assembly out of individual images. In the first approach, normalized images were assessed

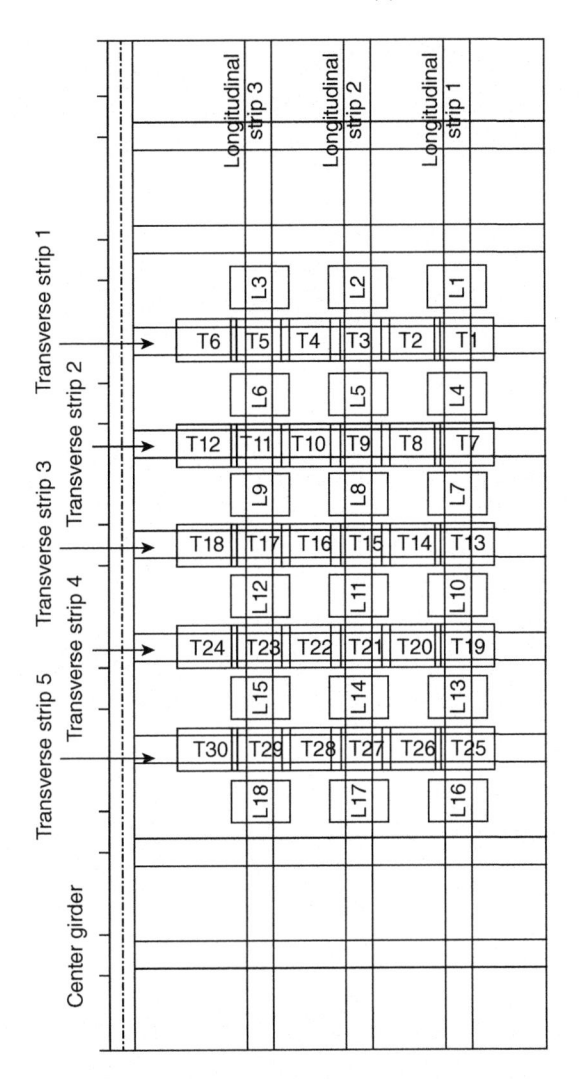

19.2 Schematic showing strips and areas of images.

using AutoCAD to find the overlap lengths between adjacent images. After recording the overlap lengths, individual images were examined using Matlab in a global context. The advantage of this approach is that it sets no limitation for the size of the test specimen or the number of images within the bound of computational ability of the selected computer system. In the second approach, reconstruction of the assembly was completed on its appropriate location on the global geometric image using AutoCAD. This single result was saved and transferred into a Matlab environment for further analysis. The advantage of this approach was that it was rapid and

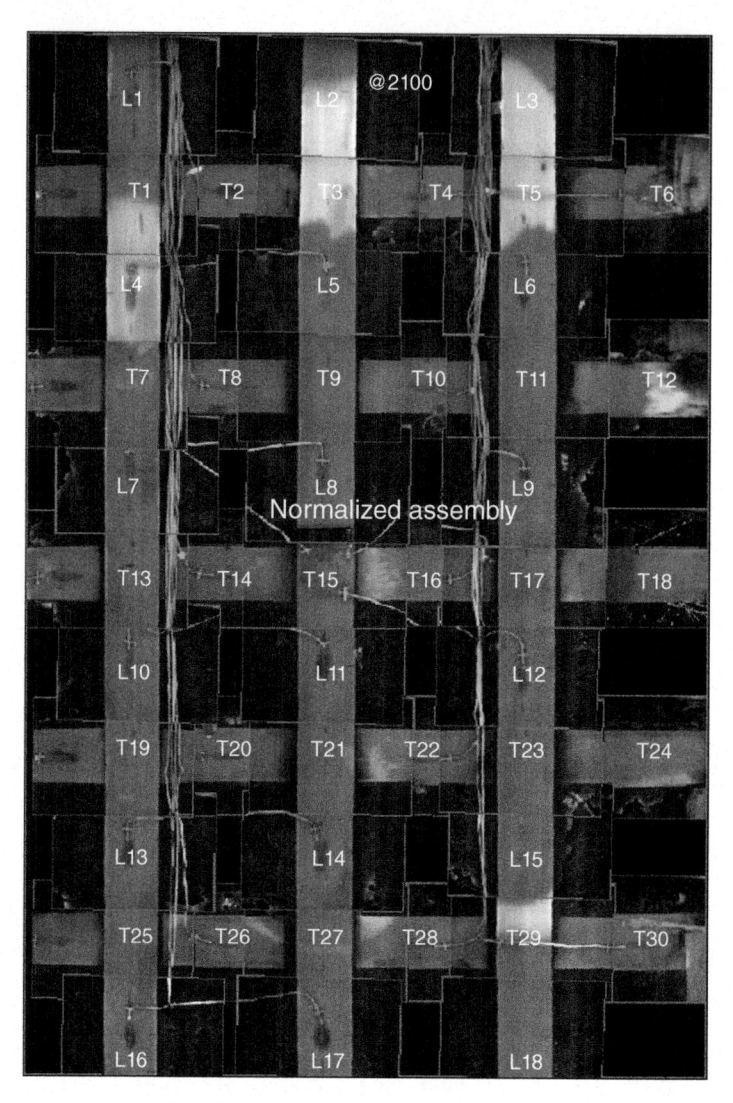

524 Non-destructive evaluation (NDE) of polymer matrix composites

19.3 Aggregation of images.

led to a more precise image of the global configuration. The disadvantage, however, was that the maximum size of the image that be analyzed was restricted by visual limitation of the screen.

In order to provide a frame of reference to allow for the retrieval of the actual extent of defects, as well as their relative position from neighboring edges or defects, a pixel–mm conversion factor was determined. Knowing that images were taken at 1:1 zoom ratio and the actual dimension of the heat box was 32 cm by 24 cm, a relation of 1 mm = 1 pixel was obtained.

Data from thermographic inspection was saved in the form of multidimensional matrices having 240 rows by 320 columns with the third dimension, k, representing the total number of images taken each time a test was run In the current case k is 600 (capturing frequency times capturing duration); however, since the changes in thermal intensity are extremely small in adjacent images the use of all 600 images is not practical. Rather than looking at each image one by one, an averaged image resulting from the superposition of a number of images at a selected time frame was used. Although the gate number and duration are adjustable parameters, a default setting of 20 gates was used in this research. The 600 images were thus reduced to 20 that were taken from three heat transfer stages as shown in Fig. 19.4.

A defect and its surroundings are exposed to a specific amount of energy, causing both to start raising temperature at different rates depending on their heat capacities, C, the amount of energy required to change the temperature of a unit mass of an object by one degree Celsius. The defect and its surrounding will then reach a transient equilibrium at one or two points of time which are called nodes. The two objects cool down at rates which are functions of their thermal diffusivities (stage B in Fig. 19.4). The cooling process continues until the change in slope is negligible (<5%) (stage C in Fig. 19.4). Images taken at stage A, i.e. data from the first few gates, are not suitable for analysis because the temperatures are still increasing. Images at stage C are also not suitable because defects cannot be distinguished from the bulk due to establishment of thermal equilibrium. Therefore, data from stage B, where the temperature difference between defects and their surrounding is maximum, is used.

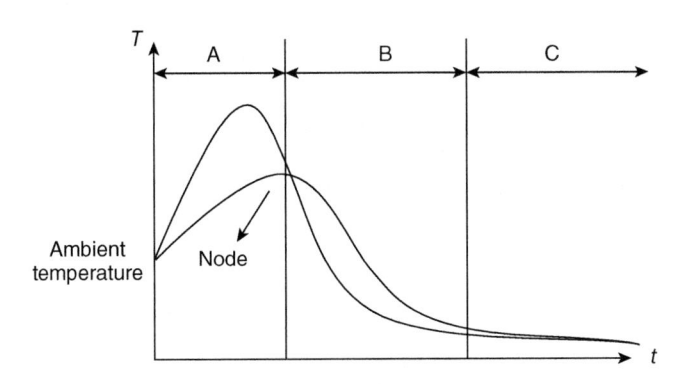

19.4 Heat transfer profile of two materials with different thermal conductivity in thermal interaction with each other. Stage A is when the two materials are absorbing heat; stages B and C show diffusion and equilibrium stages respectively.

In order to select the optimum gate number (or time window) at which defects would distinguishable, all 20 images from 20 gates of a shot with obvious defects were assessed in terms of thermal intensities along a single line of reference. In this case, the centerline of a strip was chosen. The variation of thermal intensity as a function of length was then plotted at all 20 gates using a single coordinate system and a simple Matlab code. Based on the response from gates 11 and 12 showing the sharpest differentiation at the boundary between the bonded and disbonded regions, gate 12 was selected as the basis for further analysis. It is noted that at higher gate numbers the material approaches thermal equilibrium which is eventually attained at gate 19, at which point IR camera is unable to differentiate between the intact material and a defect.

Normalization of raw images to a common baseline was necessary due to the occurrence of one or a combination of the following conditions: (a) heat from the flash used for region affects the image taken subsequently in an adjacent region due both to overlap and to thermal conductivity of the carbon fibers, (b) change in environmental conditions can cause an overall change from image to image, and (c) a defect can appear to have different severity in two different images if the background temperature changes from one image to another. In order to normalize images obtained from gate 12, markers representing the apparent undamaged area of each image were selected and used as the basis for each image. AutoCAD was used to enable the aggregation of the normalized images. AutoCAD was chosen because it allowed for multiple drawings to be made and be stacked on the top of each other for purposes of comparison. This aspect allowed for layering of images to smooth out the effects of overlapping photographs. The aggregated image was taken from the AutoCAD environment and saved as a single bitmap file which was then transferred into a Matlab environment for further analysis. A completed image is shown in Fig. 19.5 with regions of interest (ROI) also mapped for local quantification.

19.5 Assessing results

The process of IR testing was initiated through the visual inspection of the IR images to detect any abnormalities such as unexpected changes in brightness of the images. These were then classified in terms of type of defects, such as delamination, debonding, damaged edge, excessive adhesive, etc. or based on geometrical configuration and location (near edge, through-width, etc.). The challenge, however, is that it is difficult to catch minor changes and differences in brightness in the IR images.

Not every change in thermal intensity indicates the existence of a defect. The image of a structure in service might include many ancillary objects such as wires, clamps, surface coatings, etc., which can significantly affect

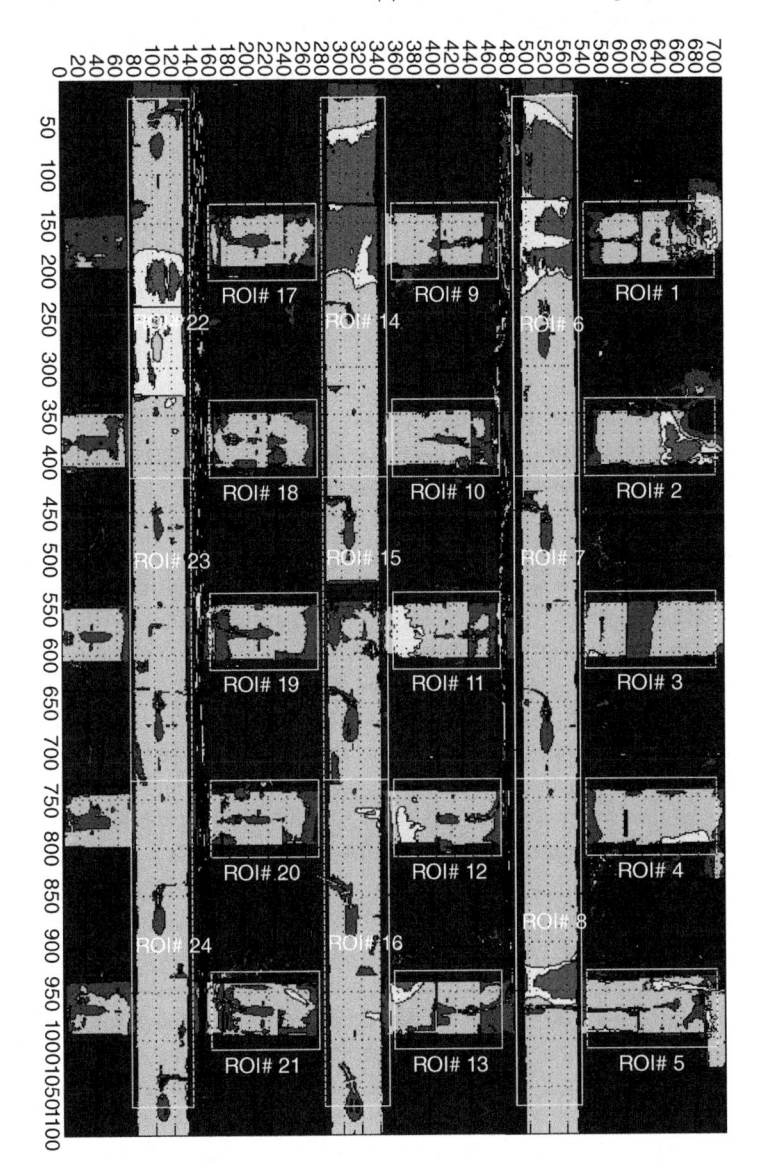

19.5 Typical completed plot showing defect extent and ROIs.

the apparent thermal conductivities of the composites and the neighboring areas. It is necessary to carefully review the resulting image contour plot to assess levels of thermal intensity fluctuation which could represent an actual defect, and in order to differentiate the regions with false indicators from those with actual defect signatures. Figure 19.5 shows the assembly of images representing the damage state as seen by IR-image at the 210k-0k load case (which is the indicative of the unloaded state of the specimen immediately after a peak load of 210 kips). There are a total of five horizontal and three vertical strips, with all five horizontal strips being overlaid by vertical strips at intersections. It should be noted that the dark areas at each side of the vertical strips at the intersections represent the accumulation of excessive adhesive in those areas. It is critical that this not be viewed as a defect and that one is aware of such false indicators. There are also some other oval-shaped dark marks at the center of vertical strips, indicating the locations with externally bonded strain gauges. The white coatings used to protect the strain gauges cause a variation in heat absorption and conductivity and hence affect the image of the underlying composite.

Three representations of image data were prepared at each load case. The first representation (Fig. 19.6(a)) shows the raw IR image after normalization without any further manipulation. This lay out helps to pinpoint the areas with possible anomalies. At this level, the possible defects can be selected only based on changes of relative brightness of the images, and hence it cannot be determined with a high degree of assurance whether or not an abnormality is really a defect. More information is required for a valid assessment about any of these areas.

The second representation (as shown in Fig. 19.6(b)) depicts a contour map plot and is used to separate the areas with differing thermal intensity and provides approximate location and extent of possible abnormalities. The purpose of graphing this plot is to meticulously mark locations with possible abnormalities with respect to the global reference system. At this preliminary stage, abnormalities cannot yet be validated as 'defects'; since investigation is required to determine an existing anomaly as a defect. This representation helps to differentiate between two adjacent areas with significantly different thermal intensities but cannot catch minor differences. It is important to know that the contour plot of the assembly is not an exact tool for quantification of defects. Further anomalies, which may be defects, are revealed using this representation. However, a wide range of data still falls within each band which could result in missing a defect. In addition, the image still does not show sufficient differentiation resulting in one possibly missing a defect. To get round these difficulties and ambiguities the threshold of failure was introduced.

The threshold of failure is defined as the minimum thermal intensity at which it is certain that debonding has taken place. The threshold of failure of this experiment was set at 10% above the baseline which turned out to be ~135 on the 256 gray scale. The numerical value of the failure threshold was obtained by comparison of the collected information of visually detectable defects at the 210k-0k load stage with the intensities of the same regions at the baseline intensity. A defects-only contour plot (Fig. 19.6(c)) is the final representation used in presenting defects with clear-cut boundar-

Application of thermography 529

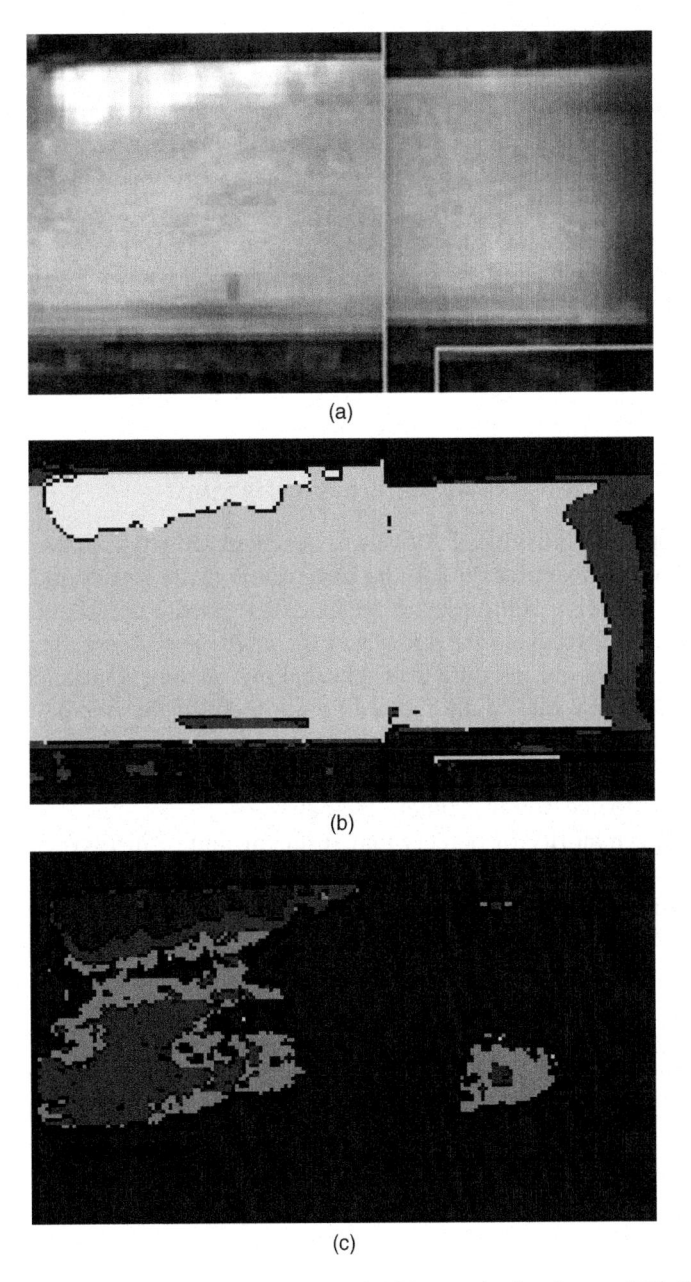

19.6 (a) Normalized IR image of ROI#4 at the load case 190k-24k, (b) contour plot of ROI#4 at the load case 190k-24k, (c) defects-only plot of ROI#4 at the load case 190k-24k.

ies and also eliminating insignificant effects that cause formation of distractive signals and false indicators. It is done by setting matrix elements smaller than the threshold value equal to zero in the assembly matrix. Another unique application of this layout is its use in monitoring of structures through autonomous means to pinpoint defects with a high level of accuracy. This can be done by projecting the layout over the surface of the structure being assessed such that changes can be easily identified. This facilitates the task of field monitoring significantly. Thus three representations can be prepared for each load level at which IR thermography assessments are conducted for the purpose of this research, and an example is shown in Fig. 19.7(a)–(c).

19.5.1 Global level quantitative assessment

The methodology introduced above was designed only to improve the reliability of qualitative assessment and does not provide a true quantitative measure. The global contour plots were used to locate defects when they were already fully developed. However, the challenge arises from the fact that the materials are in continuous thermal interaction, resulting in transfer of energy from one to another and attainment of thermal equilibrium faster, at which point the defect is completely hidden. Therefore, it is essential that the image to be viewed be taken at a transient point of time when it is possible to accurately estimate the extent of the defect.

By far the majority of defects that result in failure in FRP composite rehabilitated structures initiate along the edges of the strips (edge-defects) and develop slowly without visually observable growth until they result in complete failure by forming through-width debonds. Hence, it is crucial to detect the edge-defects at an early stage of development. A prominent cause of the debonding phenomena is from the difference in strain experienced by the two adherents under load (strain compatibility). This difference is significantly higher at the boundary of FRP composite strips. The edge-defects cannot be easily detected for two general reasons, namely:

- The direction of heat flow through a material that is in contact with another material at a lower temperature is always outwards; therefore, the edges of a strip of FRP composite are always initially warmer than its interior and consequently it may appear as a false defect in the IR image.
- Upon the activation of any FRP system a series of micro-cracks may appear in the adhesive that has been squeezed out between the strip and the concrete. These, in the normal progression, do not have any effect on the system.

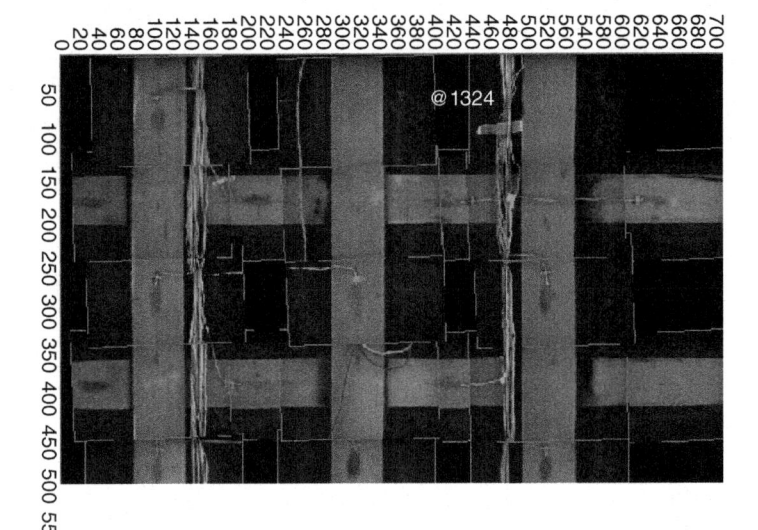

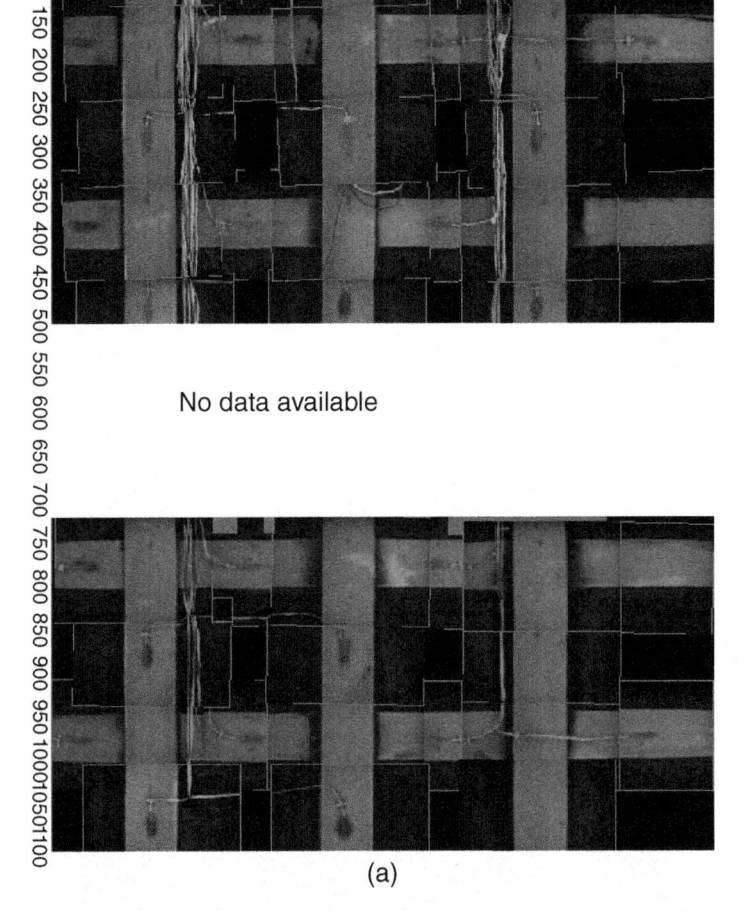

Application of thermography 531

(a)

19.7 (a) Normalized IR image of the structure at the 130k-24k load level (regions marked as 'no data available' do not have sufficient number of images for aggregation). (b) Contour plot of the structure at the 130k-24k load level. (c) Defects-only plot of the structure at the 130k-24k load level.

A global assessment can be made by plotting thermal intensities as functions of the length of the entire strip along three separate slices on the left, right and middle of each strip. Figures 19.8(a)–(c) show examples of this indicating specific types of defects identified by level of intensity. The graphs show the location, extent and the load stages at which each defect started

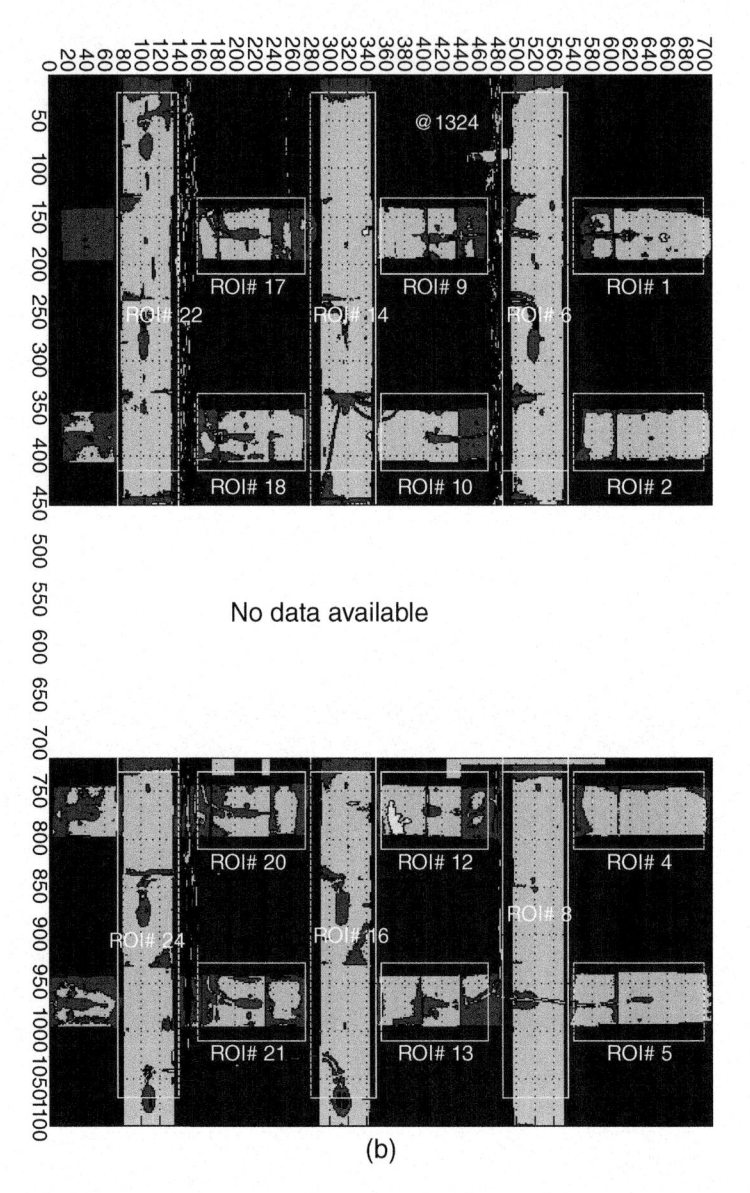

532 Non-destructive evaluation (NDE) of polymer matrix composites

19.7 Continued

to form and ultimately the defects that resulted in complete failure of the structure. The location in Fig. 19.8(a) shows that a small edge debond formed at load stage 150k-0k and grew as load increased to 170k-0k and 190k-0k but at the load stage at 210k-0k it dropped in intensity because the sustaining load was removed and consequently the debonding gap was

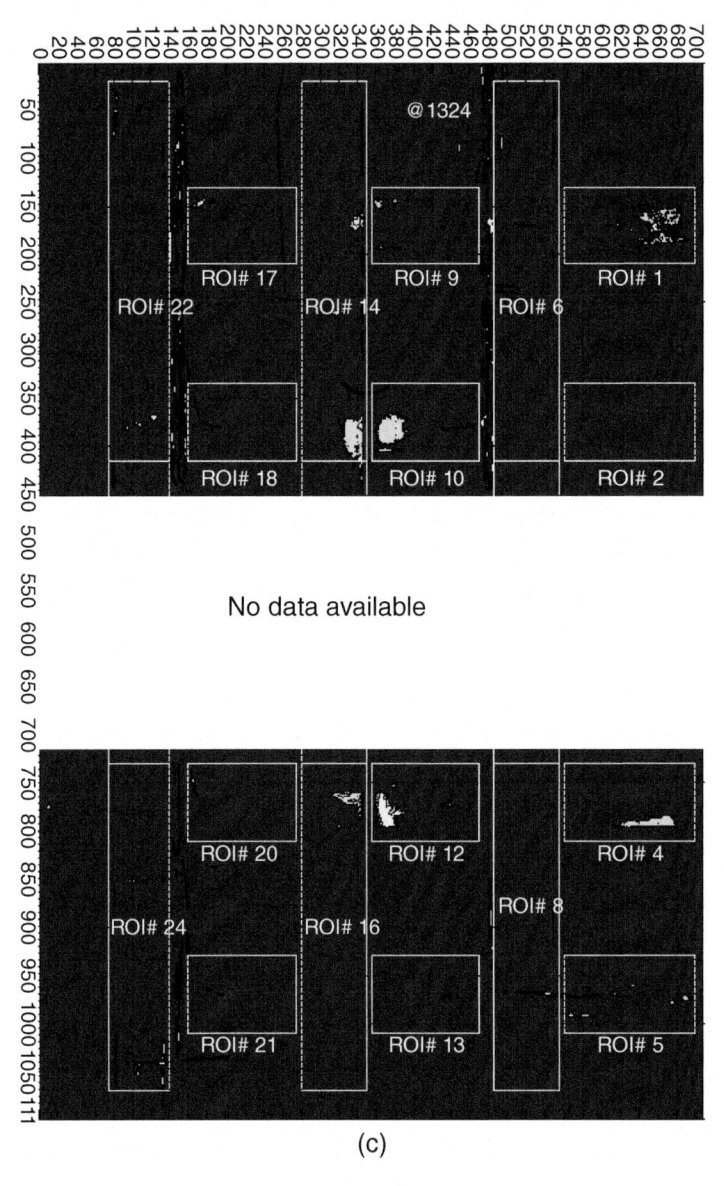

19.7 Continued

closed. Two locations, however, show that two defects formed at two different locations and load stages grew with load increase and merged to form a full through-width defect that resulted in complete failure of that region of the rehabilitation system. Local spikes that are extremely sharp are indicative of fiber rupture.

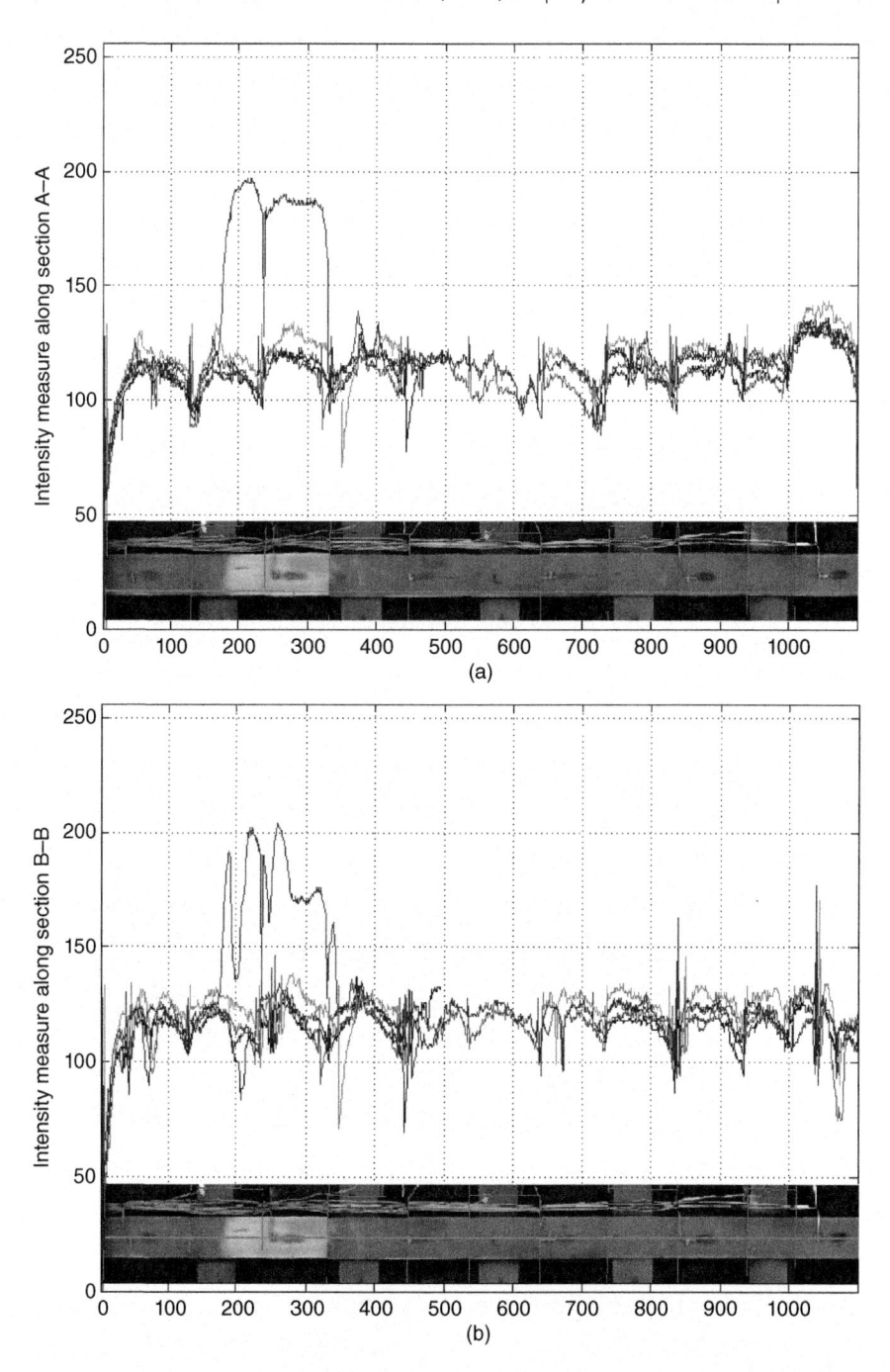

19.8 (a) Sliced strip analysis of section A–A; the two defects initiated independently at different loadings but merged at 210 kips where they formed a full through-width debond. (b) Sliced strip analysis of section B–B. (c) Sliced strip analysis of section H-H. Local spikes are representative of fiber rupture.

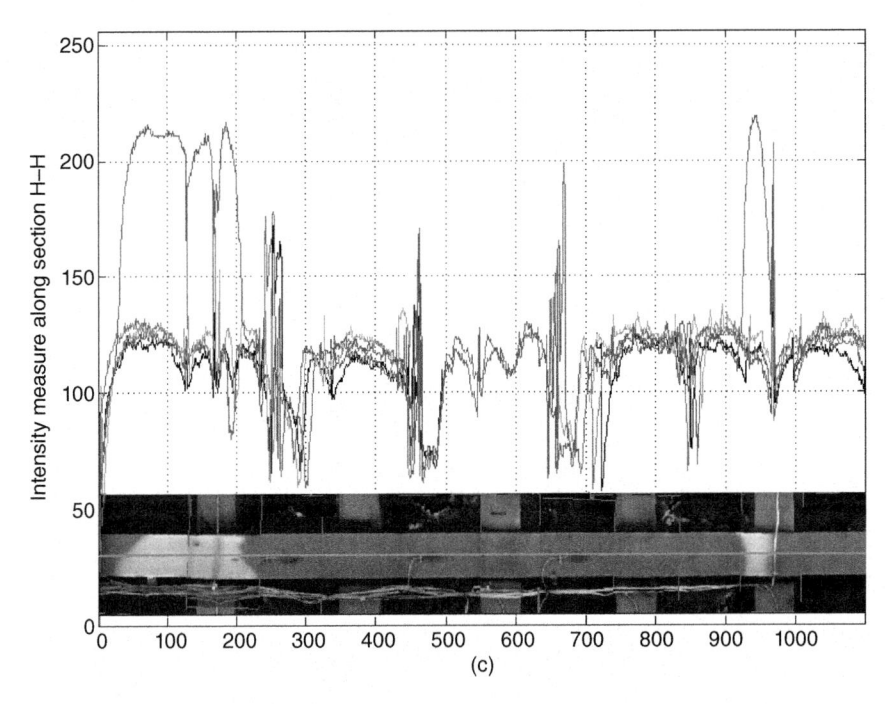

19.8 Continued

19.5.2 Assessment of defects in a localized area

Once anomalies are globally identified and tagged, they must be assessed for criticality and the rate of extension. This assessment should be based on a full understanding of the nature of defects and the stage of the structure's life at which they occur. In the context of the current research, the term 'defect' includes anomalies introduced during manufacturing/processing, placement operations and, damage incurred during the service life of the component. A critical aspect in the establishment of a validated NDE method is the classification and documentation of defect types such that a database can be developed for future reference. This section provides examples of defect types found during the course of the current research with the ROI being related to the overall image shown in Fig. 19.3. Only examples are shown in this section, rather than a full evaluation of the rehabilitated system. In order to provide a comprehensive view of assessment the three stages of assessment as shown previously in Fig. 19.8 are used in each case.

Figure 19.9 shows areas in ROI 1 that are indicative of voids in the adhesive between the FRP strip and the concrete. These are indicative of defects

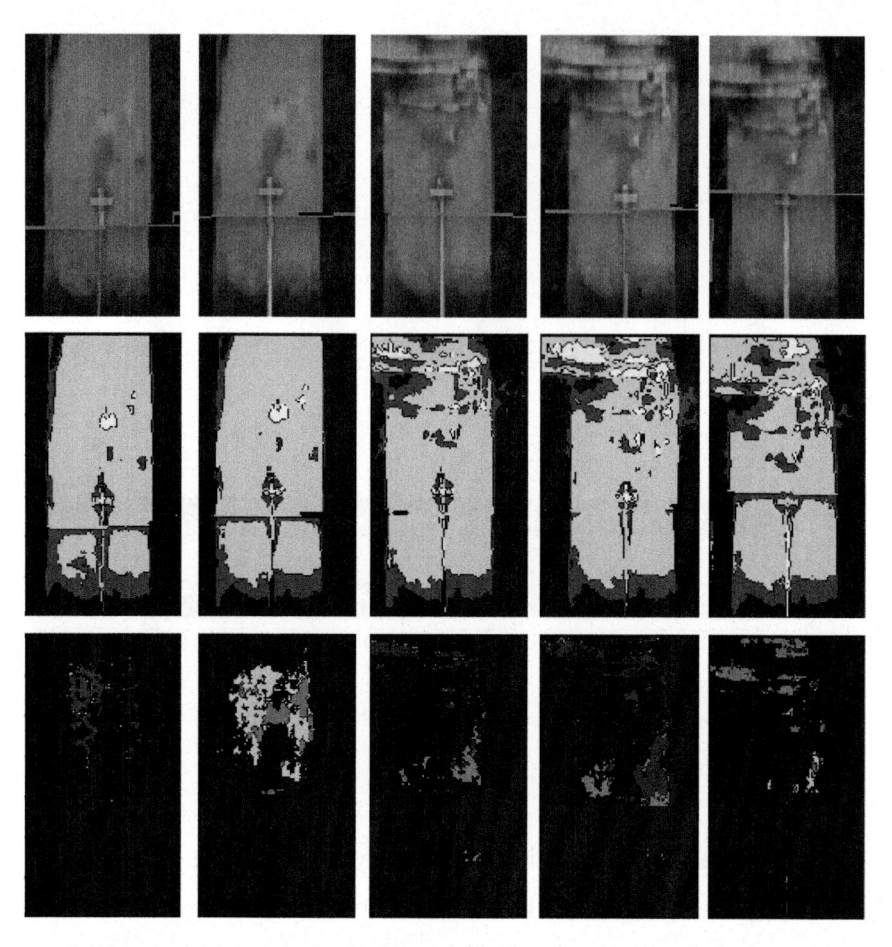

19.9 Region of interest #1 showing a pre-existing defect formed by insufficient adhesive at concrete–composite interface.

caused during initial application of the material. Figure 19.10 clearly illustrates the effectiveness of the method in finding hard-to-see anomalies. It is noted that a thorough investigation of the initial IR image did not reveal any indication of defects. However, through continued examination locally at the defects-only plots of ROI#6, two edge defects at the top and bottom of the region were observed. As seen in Fig. 19.10 these two defects enlarge through the stages of progressive loading until the defect at the top results in failure through debonding. The difference between 0k load and 24k sustaining load causes the defect at the bottom to appear less severe than the defect at the top. Figure 19.11 shows that three defects were initially seen at 130k-24k load stage. The fourth emerged later at the170k-24k load stage

19.10 Region of interest #6.

on the left side of the strip at about the time when the other defects merged thereby increasing stresses on a local region. Eventually a full through-width defect emerged at the 210k-0k stage by growth and joining of all four individual defects resulting in a debond that covered nearly 50% (by area) of the strip in ROI#14. Figure 19.12 shows a region wherein no indication

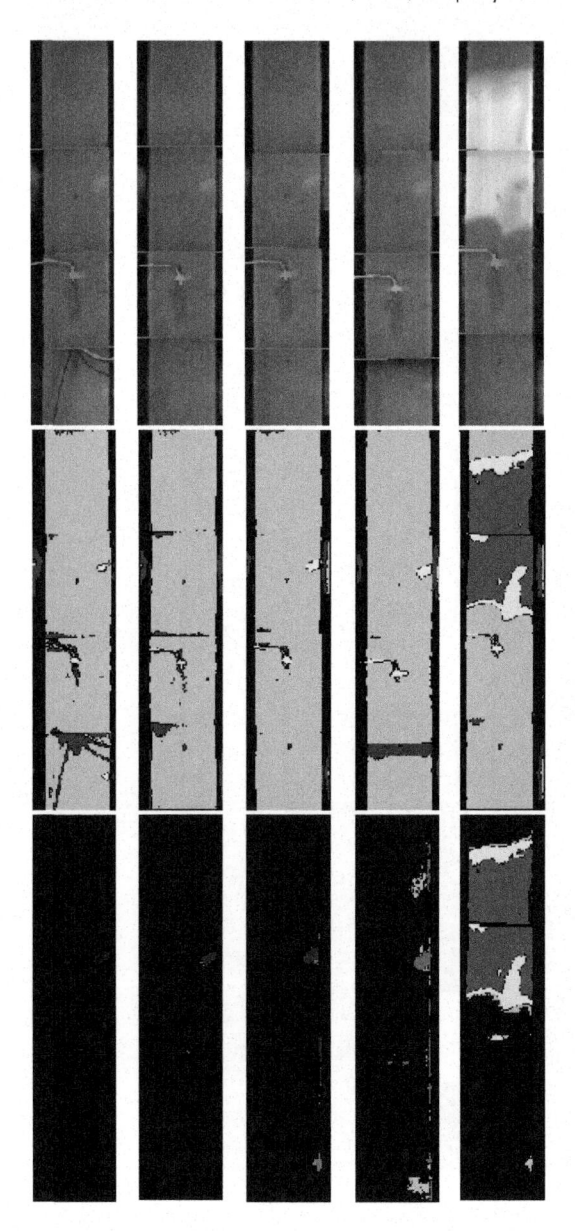

19.11 Region of interest #14.

of defects was observed on the images until the last stage wherein the inclined bright region is indicative of local fiber rupture. It is noted that this type of defect can be easily identified in Fig. 19.12 through a very localized spike in intensity.

19.12 Region of interest #21.

19.6 Conclusion

This chapter shows the development of a procedure to detect defects and the establishment of a standardized protocol for quantifying progression of these defects in FRP rehabilitated concrete structures. At the initial stage, optimal images were selected by plotting thermal intensity as a function of length along pre-selected cross-sections for images taken at the incrementally increasing times (gate numbers). Use was made of the sharpest differentiation at the boundary between a bonded and disbonded region (sharpest change in slope of the graph) to pinpoint the optimum image for further analysis. Then, in order to establish a uniform baseline (thermal intensity for undamaged areas), the selected images were normalized. The normalized images are aggregated to form an assembly that simulates the actual test structure, on the basis of which further analysis can be conducted using contour plots for both qualitative and quantitative analysis. As a means of quantitative assessment, numerical measures at which defects are initiated are used as thresholds, which also allows for the monitoring of their progression through increasing load steps. This procedure also enables a global–local focus so as to select regions of interest for in-depth analysis. Using the above-mentioned procedure a set of orthogonally intersecting strips used to rehabilitate a deck slab were investigated. The methodology not only enables the identification of anomalies, but also allows for distinction between actual defects and aspects caused on transient heating and scatter from neighboring objects. The analysis procedure clearly differentiates between common defect types and the progression can be easily followed through the global–local analysis.

It is noted that structural materials are intrinsically subjected to deterioration over time. The rate of degradation can be significantly affected by a number of environmental as well as serviceability conditions. The use of IR thermal imaging coupled with progressive testing and a mathematical model from a time history of the thermal responses can provide valuable information pertaining to the lifespan of FRP rehabilitated systems.

19.7 References

1. Phares, B.M., Rolander, D.D., Graybeal, B.A. and Washer, G.A. (2001), 'Reliability of Visual Bridge Inspection', *Public Roads Magazine*, Vol. 64, No. 5.
2. Washer, G.A. (2003), 'Improving Bridge Inspections', *Public Roads Magazine*, Vol. 67, No. 3.
3. Kaiser, H. and Karbhari, V.M. (2004), 'Non-Destructive Testing Techniques for FRP Rehabilitated Concrete: I – A Critical Review,' *International Journal of Materials and Product Technology*, Vol. 21[5], pp. 349–384.
4. Kaiser, H., Karbhai, V.M. and Sikorsky, C. (2004), 'Non-Destructive Testing Techniques for FRP Rehabilitated Concrete: II – Assessment,' *International Journal of Materials and Product Technology*, Vol. 21[5], pp. 385–401.
5. Kaiser, H. and Karbhari, V.M. (2003), 'Identification of Potential Defects in the Rehabilitation of Concrete Structures with FRP Composites', *International Journal of Materials and Products Technology*, Vol. 19[6], pp. 498–520.
6. Kaiser, H. and Karbhari, V.M. (2003), 'A Fracture Mechanics of Approach to Determination of the Defects in FRP Strengthening of Concrete', *Proceeding of the 48th International SAMPE Symposium*, Long Beach, CA, May 11–15, 2003, Book 2, pp. 1566–1580.
7. Shepard, S.M., Rubadeux, B.A. and Ahmed, T. (1999), 'Automated Thermographic Defect Recognition and Measurement', CP 497, *Nondestructive Characterization of Materials IX*, American Institute of Physics, pp. 373–378.
8. Valluzzi, M.R., Grinzato, E., Pellegrino, C. and Modena, C. (2009), 'IR Thermography for Interface Analysis of FRP Laminates Externally Bonded to RC Beams,' *Materials and Structures*, Vol. 45, pp. 25–34.
9. Ghosh, K.K. and Karbhari, V.M. (2012), 'Modal Testing as a Means of Quantitative Monitoring of Damage Progression in a Model FRP Rehabilitated Bridge Deck System,' *Structure and Infrastructure Engineering*, Vol. 8[3], pp. 227–250.

10. Starnes, M.A., Carino, N.J. and Kausel, E.A. (2003), 'Preliminary Thermography Studies For Quality control of Concrete Structures Strengthened With FRP Composites,' *ASCE Journal of Materials in Civil Engineering*, Vol. 15[3], pp. 266–273.
11. Brown, J.R. and Hamilton, H.R., III (2004), 'NDE of Fiber-Reinforced Polymer Composites Bonded to Concrete Using IR Thermography,' SPIE 5405, *Thermosense XXVI*, pp. 414–424.
12. Ghosh, K.K. and Karbhari, V.M. (2006), 'A Critical Review of Infrared Thermography as a Method for Non-Destructive Evaluation of FRP Rehabilitated Structures', *International Journal of Materials and Product Technology*, Vol. 25[4], pp. 241–266.
13. Bouvier, C.G. (1995), 'Investigating Variables in Thermographic Composite Inspections', *Materials Evaluation*, Vol. 53[5], May, pp.544–551.

20
Non-destructive evaluation (NDE) of composites: using shearography to detect bond defects

F. TAILLADE, M. QUIERTANT and K. BENZARTI,
Institut Français des Sciences et Technologies des Transports, de
l'Aménagement et des Réseaux (IFSTTAR), France,
C. AUBAGNAC, CETE de Lyon, France and
E. MOSER, Dantec Dynamics, Germany

DOI: 10.1533/9780857093554.4.542

Abstract: A shearographic method is proposed that (i) allows bonding defects (delamination or adhesive disbonds) between concrete structures and externally bonded fiber-reinforced polymer (FRP) reinforcements to be detected, (ii) assesses the width of these defects, and (iii) evaluates the quality of the adhesive bond in the case of partial delamination, damage or poor properties of the polymer adhesive. The performance of this method is first demonstrated on laboratory specimens containing calibrated defects, i.e. non-adherent polytetrafluoroethylene discs placed at the concrete-to-FRP interface. Then, a case study illustrates the interest of such method for on-site routine inspections of FRP strengthened structures.

Key words: non-destructive testing (NDT), shearography, disbonds, concrete, carbon epoxy, fiber-reinforced polymer (FRP).

20.1 Introduction

In civil engineering, strengthening or retrofitting of reinforced concrete (RC) structures by externally bonded fiber-reinforced polymer (FRP) systems is now a commonly accepted and widespread technique (Hollaway, 2010; Quiertant, 2011). However, the use of bonding techniques always implies following rigorous installation procedures (FIB, 2001; ACI, 2008; AFGC, 2011) and operators have to be trained in accordance with these procedures to ensure both durability and long-term performance of FRP reinforcements. The presence of bonding defects may indeed affect the structural performance and durability of the strengthening systems significantly. It is thus desirable that defects should be detected, located and evaluated in order to estimate if injection or replacement is needed. In these conditions, conformance checking of the bonded overlays through *in situ* non-destructive evaluation (NDE) techniques offers a major interest. The quality-control program should involve a set of adequate inspections and tests.

Visual inspection and acoustic sounding (hammer tapping) are commonly used to detect delaminations (disbonds) (Fig. 20.1). However, these current practices are unable to provide relevant information about the depth (in the case of multilayered FRP systems) and width of debonded areas and they are not capable of evaluating the level of adhesion between the FRP and the substrate (partial delamination, damage or poor mechanical properties of the polymer adhesive). Adherence properties of FRP systems installed on concrete substrates can be evaluated by conducting in-place pull-off adhesion tests on witness panels specifically bonded on test zones (Fig. 20.2). Consequently, different authors have developed non-destructive methods to assess the quality of the FRP/concrete adhesive bond, based on microwave (Akuthota *et al.*, 2004), acousto-ultrasonic (Ekenel and Myers, 2007), impact-echo (Maerz and Galeck, 2008), shearography (Hung, 2001; Taillade *et al.*, 2006; 2011a), infrared thermography (Galietti *et al.*, 2007; Valluzzi *et al.*, 2009) or a coupling of these two latter techniques (Lai *et al.*, 2009; Taillade *et al.*, 2011b).

This chapter is devoted to the shearographic method associated with a partial vacuum excitation set-up and applied to the detection of adhesion defects (delaminations/adhesive disbonds) in FRP strengthened structures and to the characterization of the depth and width of such defects as well. In a first part, the principle of shearography is recalled. A second part is devoted to a laboratory study, which aims at demonstrating that the proposed method provides relevant information on both the size of the bonding defects and the quality of the adhesive bond in the analyzed zone. Finally,

20.1 Inspection with acoustic sounding (hammer tapping).

20.2 In-place pull-off method.

the feasibility of the shearographic method for on-site inspection of carbon FRP (CFRP) strengthened infrastructures is discussed in the framework of a case study. Such a simple technology enables real-time NDE in the field with a high efficiency.

20.2 Shearography

Shearography is a speckle interferometric technique providing full-field and quasi-real-time quantitative images of the surface displacements of a loaded structure (Hung, 2001). The basic principle of the method is described in several references (Leendertz and Butters, 1973; Hung, 1974; 1989; Hung and Liang, 1979). This technique can be applied to the detection of debonded areas in a structure composed of a concrete substrate, one layer of polymer adhesive (most of time, an epoxy formulation is used), and one layer of FRP. The evaluation of the adhesion quality (in the case of partial or imperfect bonding) was proposed by Taillade *et al.* (2006; 2011a).

The principle of an interferometer with a video split, also called shearography, is to create the interference of two waves that have been submitted to nearly the same random fluctuations in optical path during their trajectories from the studied object to the charge-coupled device (CCD) matrix camera (Fig. 20.3). The optical phase is measured in the reference state and after a solicitation of the object. The optical phase difference depends on the deformation of the object.

In the case of plane waves, for directions of illumination and observation that are normal to the plate, the phase difference is expressed at the first order by:

$$\Delta\varphi \approx \frac{4\pi}{\lambda}\left(\frac{\partial w}{\partial x}\right)\delta_x \qquad [20.1]$$

where w is the amplitude of the displacements normal to the object surface, λ is the illumination laser wavelength of the order 0.5 μm and δ_x the shear distance. By performing an uncertainty budget (Taillade, 2006), it was found that the phase difference equivalent to noise is roughly $2\pi/50$ (noise including calibration procedure). Without any particular precaution, displacement difference can thus be mapped with a 5 nm uncertainty.

The excitation method consists in applying a partial vacuum (or depressure) ΔP to the surface of the sample by means of a suction cup. The depressure ΔP is the most current practice in non-destructive testing (NDT) by shearography (Clarady and Summers, 1993; Deaton and Rogowshi, 1993) to diagnose aircraft structures (Newman, 1991; Bobo, 1991) or cryogenic tanks of rockets (Burleigh et al., 1993). It makes it possible to accurately detect various kinds of defects (Hung et al., 2009) and especially delaminations in composite materials and disbonds in metal structures. The difference in pressure between the blade of air inside the defect and the surface subjected to the stress creates a 'bump'-shaped deformation into the defect. The depressure required to obtain a measurable deformation by shearography can be very low (only a few pascal). It depends primarily on the mechanical characteristics of the surface material (elastic properties of FRP,

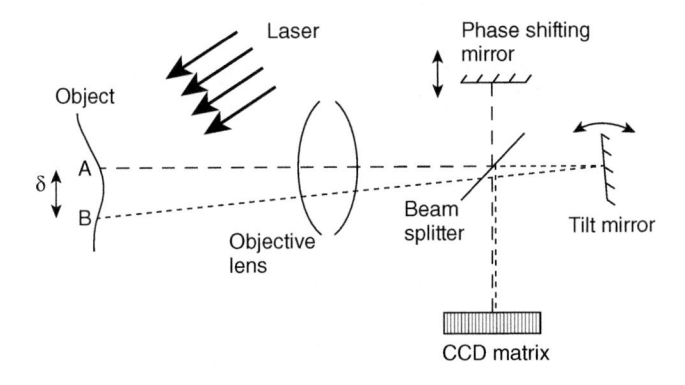

20.3 Principle of a shearographic interferometer set-up.

polymer adhesive and concrete) and on the width-to-depth ratio of the defect.

In order to predict the lowest depressure level which has to be applied and evaluate the resulting field of deformation, finite element (FE) simulations can be carried out for plotting abacuses usable by the operators. From the experimental point of view, mechanical loading can be achieved by applying a low depressure on the analysed surface using a suction cup.

20.3 The role of shearography in detecting defects

20.3.1 FE simulations

In a first step, it is proposed to calculate the mechanical static response of a layered sample (CFRP/polymer adhesive/concrete) with imperfect bonding and subjected to an external vacuum. Loading is applied to the upper side of the bonded composite and it is assumed that the depression is uniform over the entire surface of application. For the sake of simplicity, an orthotropic behavior is assumed for the FRP laminate whereas concrete and polymer adhesive are considered as isotropic materials. Defects are idealized as circular disbonds of finite diameter, and characterized by a lack of adhesive (Fig. 20.4). The problem has a symmetry of rotation around the z axis, the geometry of the finite element model can be limited to a study 2D axi-symmetric. A commercially available FE software, COMSOL, was used to numerically solve the equilibrium equation:

$$\text{div}\sigma = 0 \qquad [20.2]$$

where σ is the stress tensor given by Hooke's law $\sigma = C\varepsilon$ with C the rigidity tensor and ε the strain tensor.

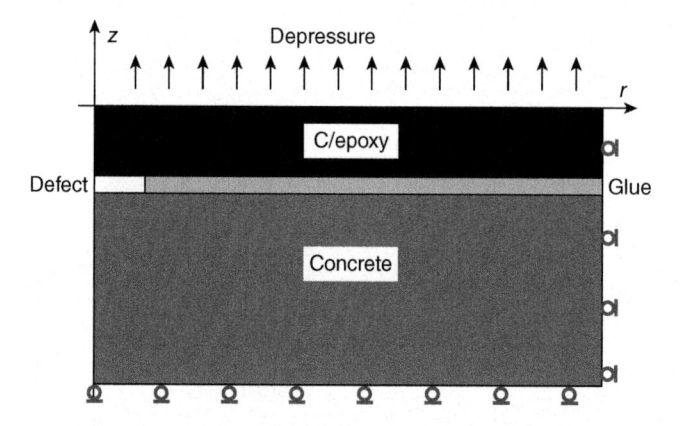

20.4 Geometry of the problem in the case of a depressure loading.

Table 20.1 Material properties considered for the finite element simulations

Concrete	Epoxy adhesive	Composite (T300/914) characterized by the rigidity tensor
$E = 33$ GPa $\upsilon = 0.21$	$E = 5.2$ GPa $\upsilon = 0.38$	$C_{(GPa)} = \begin{bmatrix} 143.8 & 6.2 & 6.2 & 0 & 0 & 0 \\ 6.2 & 13.3 & 6.5 & 0 & 0 & 0 \\ 6.2 & 6.5 & 13.3 & 0 & 0 & 0 \\ 0 & 0 & 0 & 3.6 & 0 & 0 \\ 0 & 0 & 0 & 0 & 3.6 & 0 \\ 0 & 0 & 0 & 0 & 0 & 5.7 \end{bmatrix}$

The thickness of the FRP plate, glue layer and concrete substrate were respectively 1, 0.2 and 20 mm. Various diameters have been considered for the bonding defects in the range 10–40 mm. Common values have been chosen for the mechanical properties of the various materials, as shown in Table 20.1. The material elasticity matrix C is expressed in a coordinate system that coincides with the principal directions of the FRP.

As an illustration of FE simulations, Plate XIIIa (between pages 296 and 297) and b show respectively the displacement field (w) and the derivative displacement curve ($\partial w/\partial x$) for a defect diameter equal to 40 mm and an applied depressure of 100 Pa. To visualize a defect, the phase difference $\Delta\varphi$ must be higher than the measurement uncertainty $u(\Delta\varphi)$, corresponding to the derivative displacement uncertainty of the order of 5×10^{-7} (see Plate XII: undetected defect zone), considering $u(\Delta\varphi) = 2\pi/50$, $\lambda \approx 0.5$ µm and $\delta_x = 10$ mm.

Simulations made it possible to evaluate the depressure ΔP requested, with respect to the defect diameter, to produce a phase difference $\Delta\varphi = 2\pi$ corresponding to one fringe pattern variation and a derivative displacement of the order of $\partial w/\partial x = 25 \times 10^{-6}$. Computed values of ΔP will be presented and compared to experimental data in the next section.

20.3.2 Experimental results and comparison with FE simulations

An experimental approach was conducted on strengthened concrete specimens containing calibrated defects:

- Two concrete slabs (300 × 300 mm²) were manufactured from a standard concrete formulation and then strengthened by externally bonded carbon fibre sheets. A commercial epoxy polymer was used to both impregnate the carbon fabrics and bond the FRP laminate to the concrete surface. Bonding defects were simulated by locally replacing the epoxy adhesive by polytetrafluoroethylene (PTFE) discs (0.5 mm thick) at the concrete/composite interface, as shown in Fig. 20.5.

- A first specimen, denoted S1, contained four discs of different diameters ranging from 10 to 40 mm. This range of defect diameters was chosen due to the fact that a delamination of surface area 6.5 cm^2 is considered as the threshold above which repair should be undertaken (Maerz and Galeck, 2008). The sample S1 was intended to demonstrate the sensitivity of the shearographic method to the size of the bonding defects.
- The second specimen S2 was designed with four PTFE discs of identical diameters (Φ = 40 mm) drilled with holes (in number and size represented by the percentage of remaining disc mass) in order to simulate different qualities of the adhesive bond. In this case, the objective was to evaluate the ability of the shearography technique to assess the level of the adhesive bond.

Shearographic evaluations were carried out using a home-made set-up which allowed sample deformations through a suction cup (consisting of a 180 × 180 × 70 mm^3 Plexiglas® chamber, with a 20 × mm wall thickness) to be visualized, as shown in Fig. 20.6. Figure 20.7 presents results for specimen S1 and shows the phase differences measured by shearography through the suction cup. To impose an optical phase difference of the order of 2π, the applied depressure was found to increase when the defect diameter decreased. The maximum depressure was around 190 hPa, which was not detrimental to the structure.

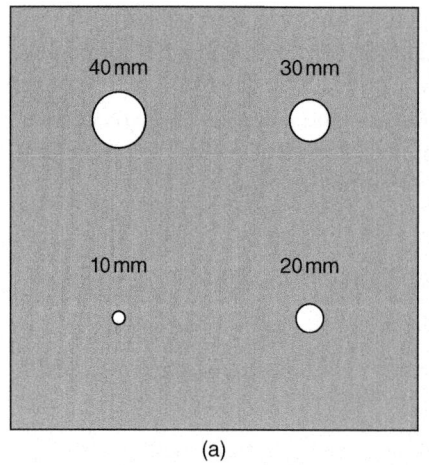

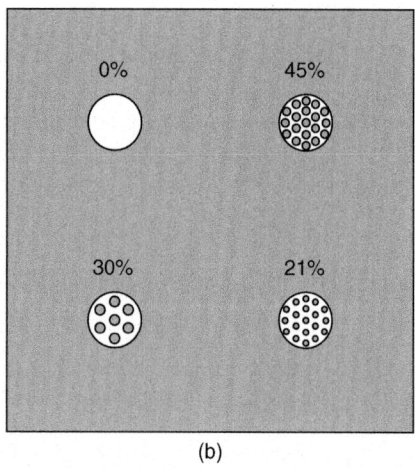

20.5 Concrete slabs strengthened with bonded FRP overlays containing calibrated defects (made of PTFE discs of variable size and shape): (a) sample S1 and (b) sample S2.

Using shearography to detect bond defects 549

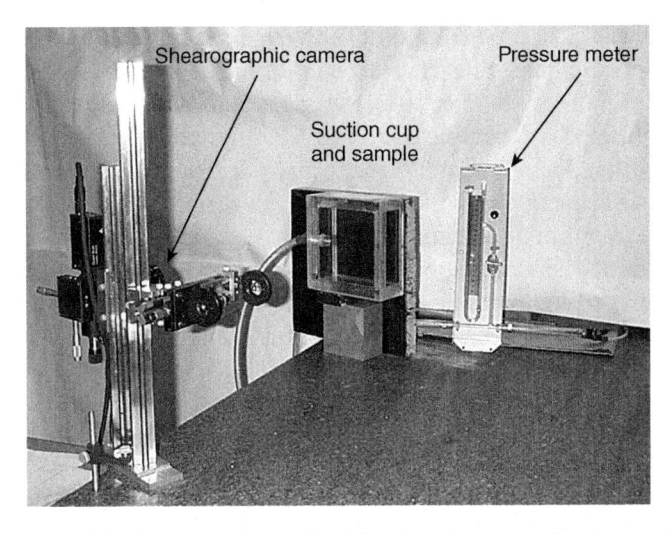

20.6 Experimental set-up used for the shearography evaluation.

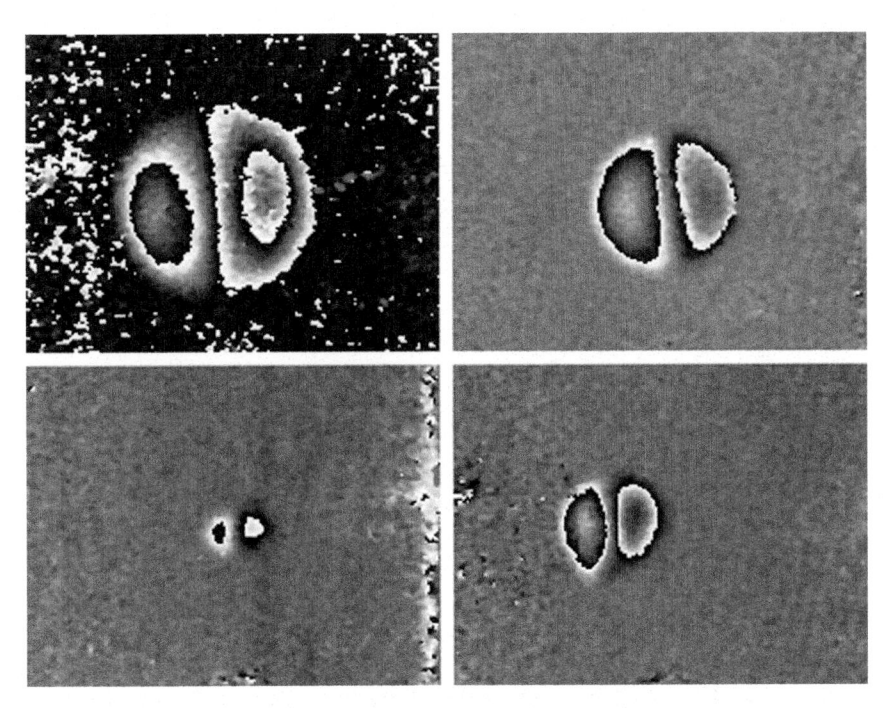

20.7 Visualization of strain fields on specimen S1 for various sizes of defects and different levels of partial vacuum.

Figure 20.8 depicts the evolution of the applied depressure (required to obtain an optical phase difference around 2π or a derivative displacement contrast of the order to 25×10^{-6}) as a function of the defect diameter. Experimental data are compared to the FE simulations previously introduced in Section 20.3.1. Taking the derivative displacement measurements and depressure uncertainties into account, experimental and theoretical results were found in good agreement.

As regards the shearographic evaluation of specimen S2, Figure 20.9 shows the phase differences that were measured for the defects with various qualities of the adhesive bond. Experimental results show that it is necessary to impose a higher depressure in order to measure a difference in optical phase of the order of 2π when the percentage of holes decreases on the PTFE disc (i.e., when the disbond is less severe or the quality of adhesion increases). Such a trend was validated by FE simulations. In the end, this laboratory study has demonstrated that shearography is very sensitive to both the diameter of the bonding defects and the quality of the adhesive bond in the defect region. Figure 20.10 shows the evolution of the required depressure as a function of the adhesive percentage. If realistic mechanical properties of the materials were computed into a model, the proposed non-destructive test would enable to assess the actual adhesive properties of the polymer adhesive in the defect zone.

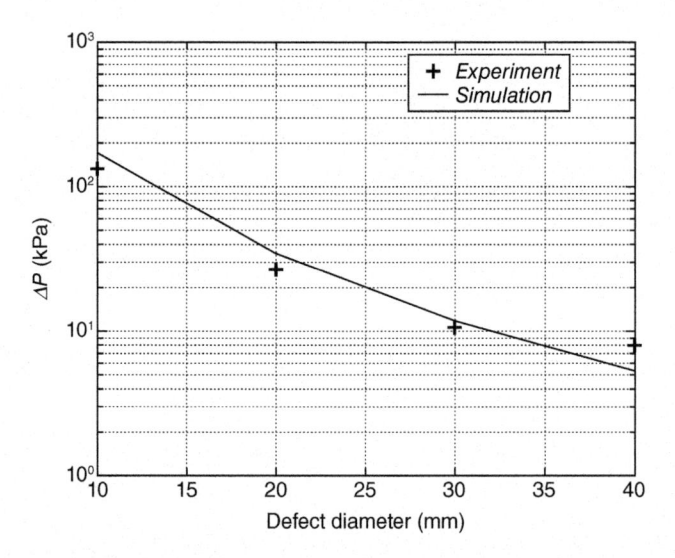

20.8 Applied depressure vs. defect diameter, for a derivative displacement $\partial w/\partial x = 25 \times 10^{-6}$; Experimental data and results from the FE simulations.

Using shearography to detect bond defects 551

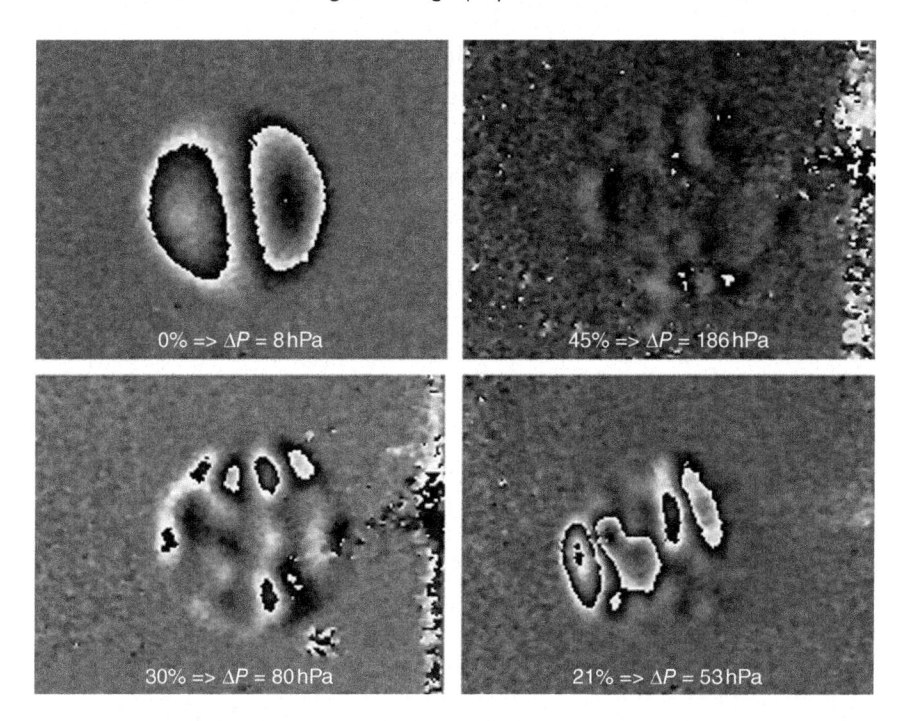

20.9 Visualization of strain fields on specimen S2 with a defect diameter of 40 mm, and for different qualities of the adhesive bond (expressed as the ratio between surface of the bonded areas and the surface of a plain PTFE disc).

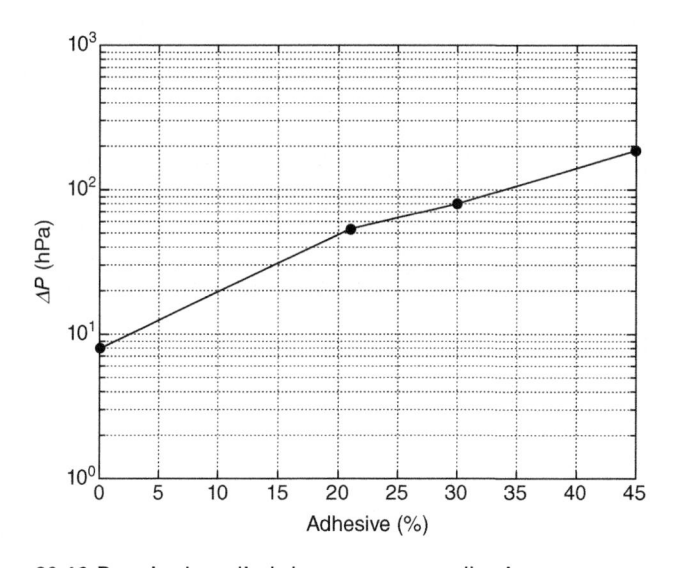

20.10 Required applied depressure vs. adhesive percentage.

© Woodhead Publishing Limited, 2013

20.4 Field inspection of a fiber-reinforced polymer (FPR)-strengthened bridge: a case study

In this part, the feasibility of the shearography associated to partial vacuum excitation for routine inspection of FRP strengthened concrete structures is illustrated through a case study conducted on an existing RC structure. The field test presented in this section only focuses on the detection of bonding defects and does not provides quantitative measurements.

20.4.1 Description of the bridge and repair works

The bridge under study is located near Besançon in France, over the Doubs river. It was built in the 1960s. The bridge consists of three distinct and independent sections, i.e. two access spans and a main central structure. The latter is divided into three spans, respectively 29, 54 and 29 m long, and composed of two box-girders made of prestressed concrete. A visual inspection conducted in the 1990s revealed extensive transverse cracking of lower slabs of box-girders at mid-span. Such deterioration was mainly attributed to an inadequate consideration of thermal gradients in the initial design and to a lack of the inter-element continuity of longitudinal prestressing in the lower slabs. In order to prevent brittle failure at mid-span, it was decided to repair the cracked box-girders by bonding carbon fiber sheets according to the wet lay-up process (onsite impregnation). A recalculation of the structure was performed in order to optimize the repair design with respect to the shear stress distribution. Finally, composite reinforcements were installed at the outer side of the web of girders as shown in Fig. 20.11.

20.11 View of the bridge under consideration; FRP repaired zones correspond to the white parts on the girders.

20.4.2 Shearographic inspection of the FRP repairs

In situ validation of the shearographic method was carried-out with a portable Q-810 Laser Shearography System (Dantec Dynamics), which is a fully integrated NDT system based on laser shearography and commonly used in the automotive and aeronautic industries. This system consists of a vacuum hood with integrated shear optics and a laser diode illumination array, which are both hermetically sealed to protect against dust and debris. The hood has an interchangeable flexible seal adapter and adjustable feet to ensure a solid contact to the test object and tight pressure seal. The vacuum hood of the Q-810 provides lock-on and pressure excitation down to 150 hPa and an optional heat source can be fitted to provide additional heat excitation (not used here). The vacuum hood is connected to a base unit (housing PC, control electronics and vacuum pump) by means of a 10 m long umbilical cable. The complete system is controlled by a specific integrated software (ISTRA 4D). Such a simple set-up offers a fully portable real-time assessment system. The main difficulty of the inspection was the accessibility to the FRP strengthened zones, which was resolved by using a truck mounted lift-platform (Fig. 20.12).

Two examples of localized defects with diameters around 10 mm are presented on Fig. 20.13, with applied depressure of 40 and 80 hPa, respectively. Figure 20.13a shows a disbond area, whereas Fig. 20.13b shows very clearly the overlap of two FRP overlays. These examples demonstrate the feasibility of the method for onsite evaluation. However, further work is needed in order to establish guidelines for the routine inspections of FRP strengthened bridges.

20.12 Inspection operations using the portable Q-810 Laser Shearography System (Dantec Dynamics).

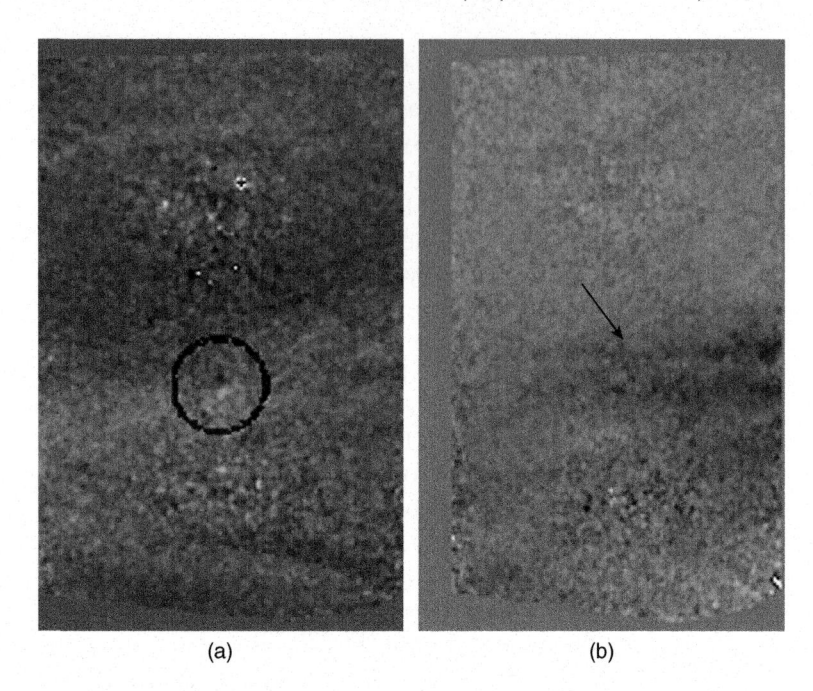

20.13 Examples of defects detected on the repaired infrastructure: (a) bonding defect, (b) overlap of two carbon fibre sheets.

20.5 Conclusion

In this chapter, the principle of shearography is reviewed and applied to the evaluation of the adhesive bond between concrete and external FRP reinforcements. A laboratory study first demonstrates the efficiency of the method to detect the bonding defects, as well as its sensibility to the size of the defect and the quality of the adhesive bond. In a final part, an example of field application illustrates the potential of the technique for onsite inspections.

20.6 References

ACI Committee 440.2R-08 (2008), *Guide for the Design and Construction of Externally Bonded FRP Systems for Strengthening Concrete Structures*, ACI, Michigan, USA.

AFGC (2011), *Réparation et renforcement des structures en béton au moyen des matériaux composites*, Technical report, Bulletin scientifique et technique de l'AFGC. in French.

Akuthota B, Hughes D, Zoughi R, Myers J, Nanni A. (2004). Near-field microwave detection of disbond in carbon fiber reinforced polymer composites used for strengthening cement-based structures and disbond repair verification, *J. Mater. Civil Eng.*, **16**(6), 540–546.

Bobo S. (1991), *Shearographic Inspection of a Boeing 737*. Technical report, Federal Aviation Administration.

Burleigh DD, Engel JE, Kuhns DR. (1993), Laser shearographic testing of foam insulating on cryogenic fuel tanks. *Rev. Prog. Quant. Nondestruct. Eval.*, **12**, 411–418.

Clarady JF, Summers M. (1993), Electronic holography and shearography NDE for inspection of materials and structures, *Rev. Prog. Quant. Nondestruct. Eval.*, **12**, 381–386.

Deaton JB, Rogowski RS. (1993), Electronic shearography: current capabilities, potential limitations and future possibilities for industrial nondestructive inspection, *Rev. Prog. Quant. Nondestruct. Eval.*, **12**, 395–402.

Ekenel M, Myers J. (2007), Nondestructive evaluation of RC structures strengthened with FRP laminates containing near-surface defects in the form of delaminations, *Sci. Eng. Compo. Mater.*, **14**(4), 299–315.

FIB (2001), *Externally bonded FRP reinforcement for RC structures*, Fib bulletin 14, Fib Task Group 9.3 Lausanne, Switzerland.

Galietti U, Luprano V, Nenna S, Spagnolo L, Tundo A. (2007), Non-destructive defect characterization of concrete structures reinforced by means of FRP, *Infrared Phys. Technol.*, **49**, 218–223.

Hollaway L. (2010), A review of the present and future utilization of FRP composites in the civil infrastructure with reference to their important in-service properties, *Construction Building Materi.*, **24**(12), 2419–2445.

Hung MYY. (2001), Shearography and applications in nondestructive evaluation of structures, *Proceedings of the International Conference on FRP Composites in Civil Engineering* (CICE 2001), 1723–1730.

Hung YY. (1974), A speckle-shearing interferometer – a tool for measuring derivatives of surface displacements, *Opt. Commun.*, **11**(2), 132–135.

Hung YYJ. (1989), Shearography: a novel and practical approach for nondestructive inspection, *Nondestruct. Eval.*, **8**(2), 55–67.

Hung YY, Liang CY. (1979), Image-shearing camera for direct measurement of surface strains, *Appl. Opt.*, **18**(7), 1046–1051.

Hung YY, Chen YS, Ng SP, Huang YH, Luk BL, Ip RWL. (2009), Review and comparison of shearography and active thermography for nondestructive evaluation, *Mater. Sci. Eng.* **64**(5–6), 73–112.

Lai WL, Kou SC, Poon CS, Tsang WF, Ng SP, Hung YY. (2009), Characterization of flaws embedded in externally bonded CFRP on concrete beams by infrared thermography and shearography, *J. Nondestructive Eval.*, **28**(1), 27–35.

Leendertz J, Butters J. (1973), An image-shearing speckle-pattern interferometer for measuring bending moments, *J. Phys. E : Sc. Inst.*, **6**(11), 1107–1110.

Maerz NH, Galecki G. (2008), *Preservation of Missouri transportation infrastructures: Validation of FRP composite technology*, Technical Report Volume 4 of 5 Non-Destructive Testing of FRP Materials and Installation, Gold Bridge, Prepared by Missouri S&T and Missouri Department of Transportation.

Newman JW. (1991), Shearographic inspection of aircraft structure, *Mater. Eval.*, **49**(9), 1106–1109.

Quiertant M. (2011), *Strengthening Concrete Structures by Externally Bonded Composite Materials*, ISTE-Wiley, Chapter 23 Organic Materials for Sustainable Construction, 503–525.

Taillade F. (2006), Metrological Analysis of Shearography, *Europ. Phys. J. – App. Phys.*, **35**, 145–148.

Taillade F, Quiertant M, Tourneur C. (2006), Nondestructive evaluation of FRP bonding by shearography, *Proceedings of the International Conference on FRP Composites in Civil Engineering* (CICE 2006), 327–330.

Taillade F, Quiertant M, Benzarti K, Aubagnac C. (2011a), Shearography and pulsed stimulated infrared thermography applied to a nondestructive evaluation of bonded FRP strengthening systems on concrete structures, *Construction Building Mater.*, **25**, 568–574.

Taillade F, Quiertant M, Benzarti K, Aubagnac C, Moser E. (2011b), Shearography applied to the non destructive evaluation of bonded interfaces between concrete and CFRP overlays, *Europ. J. Environ. Civil Eng.*, **15**(4), 545–556.

Valluzzi MR, Grinzato E, Pellegrino C, Modena C. (2009), IR thermography for interface analysis of FRP laminates externally bonded to RC beams, *Mater. Structures*, **42**(1), 25–34.

21
Non-destructive evaluation (NDE) of composites: use of acoustic emission (AE) techniques

W. CHOI, North Carolina A&T State University, USA and
H-D. YUN, Chungnam National University, Korea

DOI: 10.1533/9780857093554.4.557

Abstract: This chapter describes the application of acoustic emission (AE) techniques to monitor structural integrity of reinforced concrete (RC) beams strengthened in flexure with carbon fiber reinforced polymer (CFRP) sheets. The objective is to initiate the creation of a user-friendly health monitoring system for RC structures strengthened with CFRP sheets using AE techniques. The collected test results by AE were analyzed for four levels of damage based on initial crack, propagation, yielding of main bars, and fracture or rip-off of the CFRP sheets. The signal characteristics obtained by AE: event, amplitude versus frequency, and amplitude versus duration show clear differences in the different level of damage corresponding to the load.

Key words: acoustic emission (AE), structural integrity, carbon fiber reinforced polymer (CFRP) sheets, reinforced concrete (RC) beams.

21.1 Introduction

The use of carbon fiber-reinforced polymer (CFRP) materials in civil infrastructure for the repair and strengthening of reinforced concrete structures has become common practice. It allows an improvement in rigidity after the initial crack, as well as an enhancement of the loading and deflection capacity by maintaining the composite action with adequate bonding (Naaman *et al.*, 1999). However, the failure mode of the reinforced concrete (RC) beams strengthened with CFRP occurs in a brittle manner due to the rupture of the CFRP and/or the debonding of the concrete beams and CFRP sheets beyond the maximum loading capacity. In addition, it is hard to determine the crack in the beam and the debonding of the composite sheets. Accordingly, the microscopic damage process and a reliable monitoring technique of RC beams strengthened with CFRP sheets must be investigated fully to apply CFRP on damaged concrete structures.

Recently, several non-destructive methods have been applied to evaluate damage qualifications of structural members. Among them, the acoustic emission (AE) technique, a non-destructive method that is relatively easy to install and is capable of predicting the damage location, is used in this study. The AE technique is a method that analyzes the characteristics of elastic waves that are caused by microscopic damage in the concrete member. This technique has been used to assess microscopic damage that is internal to the concrete member and that is caused by various external loading conditions. The applicability of damage assessment of concrete has been previously reported (Quyang et al., 1991) and, thus far, the AE technique has been applied to determine the damage and failure behavior of plain concrete beams. However, this technique is currently being used to assess damage of fiber-reinforced concrete and RC members (Ohtsu et al., 2002). Investigations into the assessment of signal characteristics for RC members strengthened with CFRP sheets have only recently begun (Lee and Lee, 2002).

Lee and Lee (2002) fabricated four (100 mm × 100 mm × 440 mm) flexural beams: plain concrete, RC, plain concrete strengthened with CFRP sheets and RC strengthened with CFRP sheets. The AE technique applied to evaluating the characteristics of damage progress and the failure mechanism of the specimens. The two-dimensional AE source location was also performed to monitor initial crack occurrence, propagation, and the situation of defects in the concrete. Their test results showed that a large number of AE events were generated in every specimen except plain concrete reinforced with CFRP sheets when concrete was broken and for RC beam, the AE events occurred at the boundary between reinforcing bar and concrete before crack initiation. They concluded that the damage behavior and the microscopic fracture mechanism of the RC prism could be evaluated by using the AE parameters and crack initiation and propagation, and could be monitored by two-dimensional AE source location. Similarly, Henkel and Wood (1991) made a total of five RC beams strengthened with carbon fiber, glass fiber, aramid fiber composites and steel plate, and evaluated the AE signal characteristics of these specimens. The dependence of AE on the beam stiffness is investigated as well as a relation of the number of AE energy peaks to the number of visible cracks in the concrete. The Kaiser effect was studied in repeated loadings for all tested beams. The results suggested that the stiffness of the beam and repair materials had a pronounced effect on the AE signals and the Kaiser effect was observed in all the tests. They concluded that the AE testing method is a useful non-destructive technique for monitoring behavior in RC beams externally strengthened with thin bonded plates and an examination of cumulative AE hits and energy as a function of load could assist in analyzing the fracture behavior, previous loading history and composite stiffness of the beams.

Recently, Yun et al. (2010) applied the AE technique to assess the structural integrity and to predict the cracking location of a RC member strengthened with CFRP plate including installation imperfection. The relationship between an AE parameter and an applied load was also analyzed and AE characteristics on micro-cracking behavior of RC beam strengthened with CFRP plate were investigated. Their test results showed that crack source location using the AE technique was well in accordance with real crack distribution. The AE signal characteristics for the different levels of damage are evaluated so that the AE technique can be developed and used to determine the structural integrity for a RC member strengthened with CFRP sheets.

21.2 Testing acoustic techniques

21.2.1 Experiment

The specimen is 200 mm × 300 mm in cross-section with a span length of 2000 mm. The beams were reinforced with 4–13 mm (D13) and 3–13 mm (D13) diameter deformed bars at the tension and compression faces, respectively. They were provided with stirrups of 10 mm (D10) diameter with 100 mm, 80 mm spacing in the center and in the end, respectively. The detailed stirrup arrangement is shown in Fig. 21.1. Each specimen was identified according to a CFRP layer and control beam (CB). For example, S1-C1.0 represents the specimen strengthened with a single layer of a CFRP sheet. Also, each specimen was designed to prevent premature shear failure. The specified compressive strength at 28 days for concrete is 35 MPa. The compressive concrete strength at the day of testing was 37 MPa. The material properties for the reinforcement and CFRP sheets are 246 GPa and 4182 MPa for elastic modulus and tensile strength, respectively. The fabric type of CFRP sheets (SWHex-230C), with a 0.12 mm thickness, was used in this research.

All of the beams were tested under three-point loading. The tests were carried out using displacement control (1.0 mm/min) with a 1000 kN capacity servo-hydraulic UTM (universal testing machine). AE measurements were taken during the test using the Vallen AMSY4 instrument. The AE signals were captured using four broadband piezoelectric transducers (SE900, DECI model) attached to the specimen surface and pre-amplified

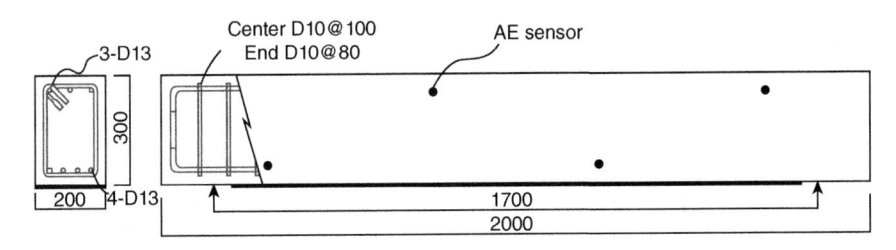

21.1 Configuration of the specimen (unit: mm).

(20 dB) before recording. In order to prevent a noise signal due to friction, a rubber sheet was placed between the beam and loading points. The threshold level was fixed at 40 dB to eliminate electrical and mechanical noises. The schematic test set-up is shown in Fig. 21.2.

Damage levels in the specimens are classified according to significant physical behaviors obtained from test results: (1) the initial cracking in the RC beam; (2) crack development; (3) the yielding of the reinforcement; and (4) crucial behavior such as CRFP debonding or rupture. The AE signal characteristics for each observed damage level shown in Fig. 21.3 are evaluated.

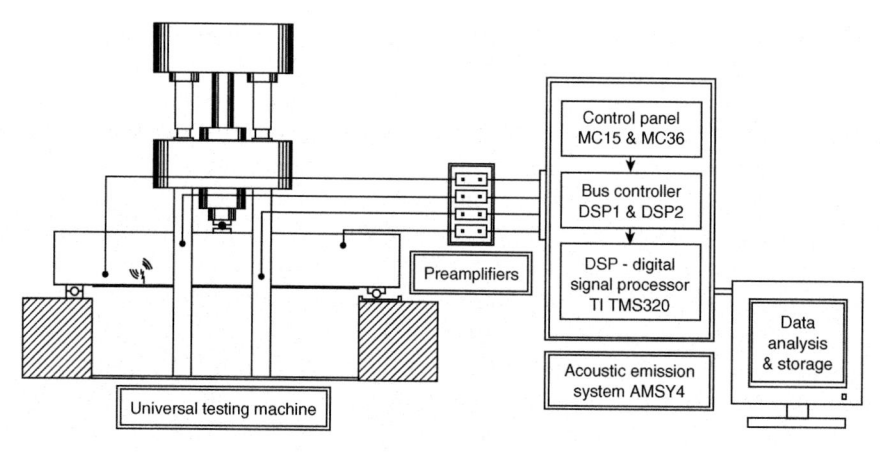

21.2 Schematic test-set up.

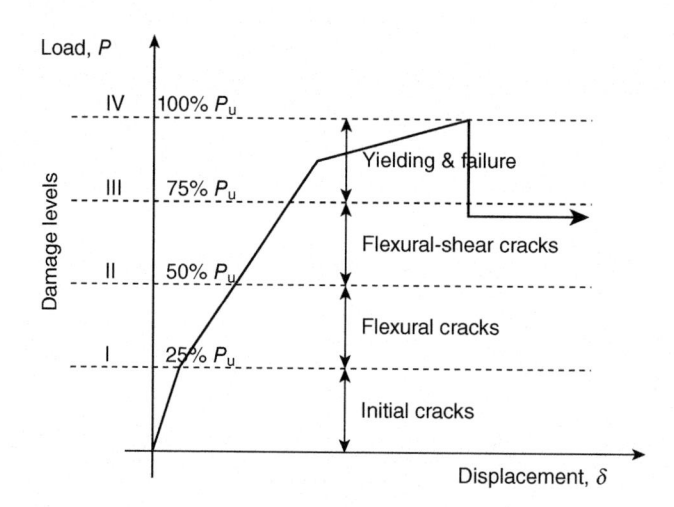

21.3 Definition of damage level of the beam.

21.2.2 AE event counts

Figure 21.4 indicates the typical characteristics of the relationship between the cumulative AE event counts and load-deflection of the specimen with and without strengthening. The event counts shown in this figure illustrate an interesting characteristic of the AE properties of RC beam with and without CFRP strengthening. This characteristic is that the rate of AE activity appears to increase from initial crack just prior to the yield load. This result is similar to a previous study on the cement-based materials by Landis (1999) in which the rate of AE activity increased just prior to the ultimate load. In Fig. 21.4, the rate of AE event counts jumps at

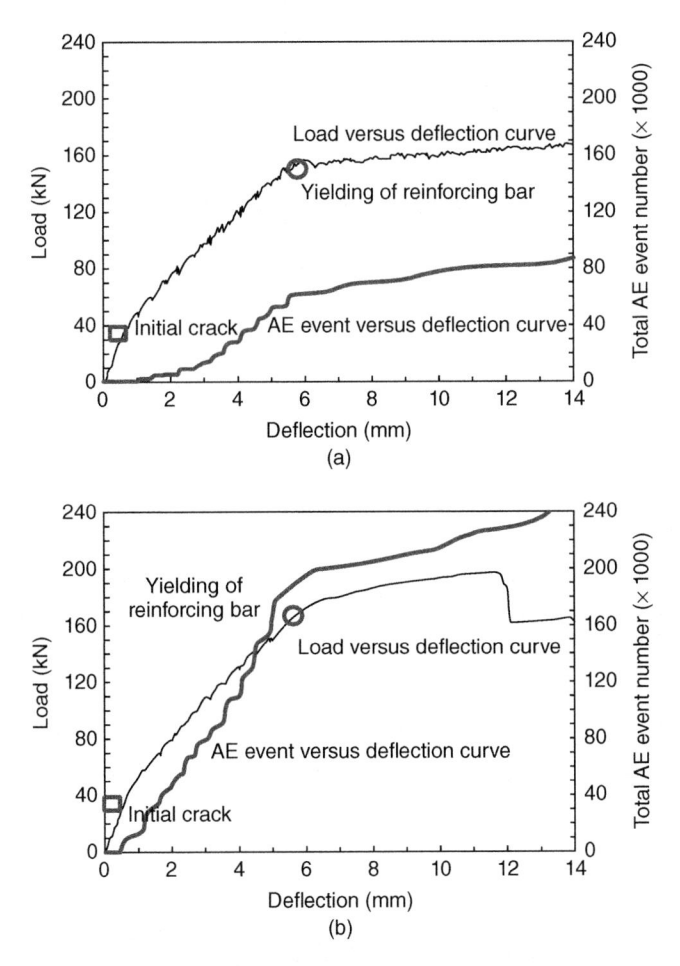

21.4 AE events versus load-deflection at midspan deflection: (a) CB specimen; (b) S1-C1.0 specimen.

approximately 60% and 68% of the ultimate load for the CB and the strengthened beams, respectively. Each beam begins to show a non-linear notable behavior at the point. Li and Shah (1994) indicate that this irregular event rate is due to the localization of critical micro-cracks.

For the CB, Fig. 21.4(a) shows limited occurrence of AE events before the initial cracking was observed as the load was increased. After the load reached the initial cracking load, the AE events increased and continued to increase along with the widely distributed cracking in the front face until the longitudinal tension steel yielded. When the longitudinal steel yielded, the rate of AE event counts rapidly decreased. The yielding of the steel reinforcement located at the flexural crack surface of the midspan results in the increment of the width of the existing crack instead of making a new crack occur. Compared with the occurrence of new cracks, increasing the width of existing cracks result in a decrease in the occurrence of elastic waves.

Figure 21.4(b) plots the load and the cumulative AE event counts versus midspan deflection of the beams, S1-C1.0 strengthened with CFRP sheet. The characteristics of the strengthened beam are similar to those of the RC beam, CB. However, a tremendous amount of AE activity, much greater than that of CB, was recorded because the elastic waves for the strengthened beam were additionally generated by the CFRP sheet and epoxy debonding at the interface between the CFRP sheets and the beam.

21.2.3 AE counts versus amplitude

The relationship between the count and the amplitude beyond 40 dB of the AE elastic waves that are due to the damages of the CB and S1-C1.0 beam is given in Fig. 21.5. This figure also includes a comparison for each damage level. There is a tendency for the amplitude to increase as the damage level increases. For damage level I, when the initial cracking is generated at the tension face, the AE elastic waves of the CB are in the amplitude range of 40 to 70 dB, and the count number is relatively small compared to those of the strengthened beams. On the other hand, the AE elastic waves of the strengthened beam, S1-C1.0, are in the amplitude range of 40 to 85 dB, because the characteristics of the signal are generated at the interface between the CFRP sheet and the epoxy. A large amplitude range of 40–100 dB was generated for the two specimens with installation imperfections due to the debonding in the interface between the concrete beam and the CFRP sheet at the beginning of loading. For damage level IV, the amplitudes of the AE signal of the CB are in the range of 65–85 dB. The amplitudes of elastic waves at the rupture of the CFRP sheets were determined to be in the 90–100 dB range by comparing the relationship of the count

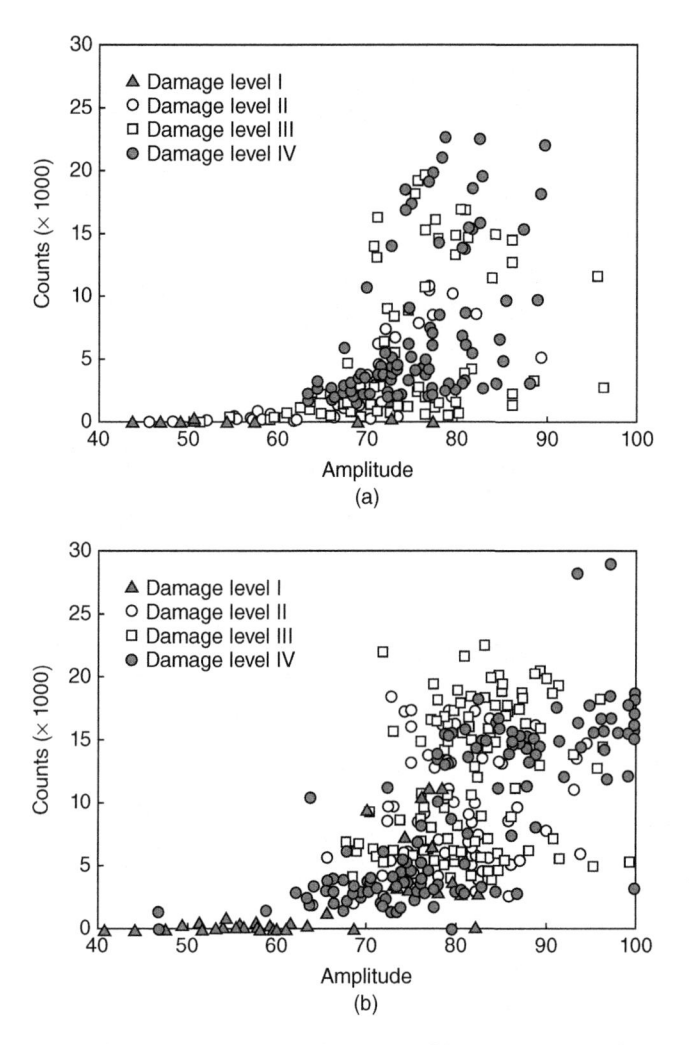

21.5 AE counts versus amplitude: (a) CB specimen; (b) S1.0–1.0 specimen.

and amplitude at damage level IV for the CB and S1-C1.0. For all beams, count increases with the evolution of damage up to ultimate load.

21.2.4 AE frequency versus amplitude

Figure 21.6 shows a comparison between the frequency and average amplitude for each damage level of tested beams. The AE signal with a similar frequency for S1-C1.0 was measured in damage levels II to IV. Based on

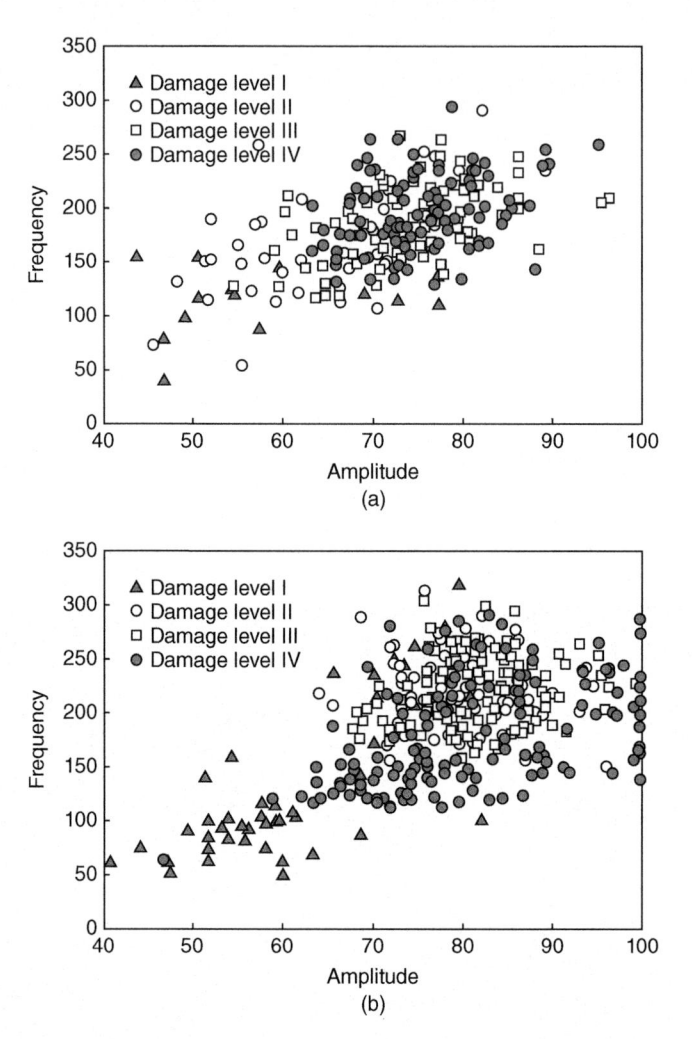

21.6 AE frequency versus amplitude: (a) CB specimen; (b) S1-C1.0 specimen.

the experimental results, the AE signals at the severe damage stage, such as the yielding of tension steel in the CB, were in the range of 125–275 kHz. As shown in Table 21.1, the amplitude and frequency in the CB tend to increase as the damage level number (i.e., I, II, III, IV) increases. However, there are AE signal characteristics that are high amplitude and low frequency. For the strengthened RC beams, when damage level I advances to damage level II, both the frequency and amplitude increase. On the other hand, no significant difference is evident between damage levels III and IV.

Table 21.1 Characteristics of AE parameters

Specimen	Damage level	Amplitude* (dB)	Frequency* (kHz)	Duration* (×1000 μs)	Event	% of total event	Count
CB	I	58	136	0.761	158	0.19	1611
	II	63	153	0.611	8935	10.98	117941
	III	69	190	1.959	27562	33.87	422391
	IV	77	158	1.329	44729	54.96	693552
S1-C1.0	I	69	201	0.795	11880	5.25	135479
	II	74	177	0.880	55905	24.71	782440
	III	76	189	1.512	82228	36.35	1235302
	IV	78	213	1.947	76201	33.69	1342630

*Frequency, amplitude and duration are average values at the each level of damage.

21.2.5 AE duration versus amplitude

For the CB and S1-C1.0 specimens, the plots of duration versus amplitude from AE signals are given in Fig. 21.7. In the case of the CB, the AE duration time, as shown in Fig. 21.7(a), was less than 800 μs until an initial crack occurred and then additional flexural cracks were generated on the surface of the specimen. When damage levels III and IV were achieved, the AE duration time was widely distributed between 1000–2500 μs. It is thought that this increment of AE duration time results from the characteristic of the AE signals due to the debonding between the reinforcement and the concrete. The AE duration time due to debonding between the reinforcement and concrete in damage levels III and IV was much longer than that due to crack initiation and propagation in damage levels I and II. In the case of the strengthened beam with CFRP, shown in Fig. 2.17(b) and Table 21.1, the AE duration time was in the range of 600 μs in damage level I. This duration time is due to the CFRP sheet's ability to restrain rapid cracking propagation at the beginning of loading. The wide range is a result of the debonding of the epoxy and damage to the CFRP. The duration time at the failure point in damage level IV ranged from 1000 to 3000 μs. In the case of S1-C1.0, the duration time from the initial crack and to failure ranged from 1000 to 3000 μs, again perhaps as a result from the CFRP sheet's ability to prevent crack propagation. Frequency and amplitude increase as the damage level number increases (i.e., from I to III, etc.). However, the frequency at damage level IV is low though the amplitude has increased.

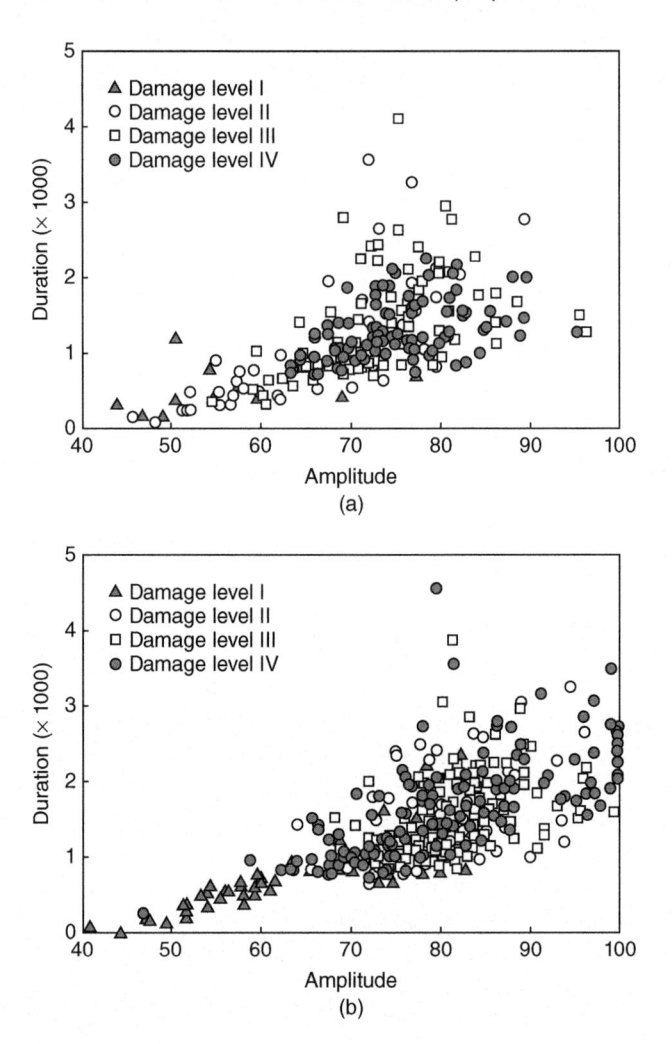

21.7 AE duration time versus amplitude: (a) CB specimen; (b) S1-C1.0 specimen.

21.3 Challenges in using acoustic emission

21.3.1 AE source locations

A two-dimensional AE source location technique was performed to monitor the crack initiation, propagation, and the location of damage in the beams. The AE source location could be determined by using the velocity of the longitudinal (P-wave) wave computed by the time differences among the first longitudinal wave arrival time of signals detected by an array of two AE sensors. In this study, the pencil breakage test was used to measure the velocity of the longitudinal wave. The longitudinal wave velocity of beams was 3674 m/s.

The source locations evaluated in the reinforced beam and the reinforced beam strengthened with a single-layer CFRP sheet (S1-C1.0) are shown at five different stages in Fig. 21.8 and Fig. 21.9, respectively. At damage level I, a few AE sources were generated at the center of the lower surface of the RC beam where the maximum bending moment applies. At damage level II, AE sources were continuously generated at the tension face of the RC beam where the crack was located as per the previous level. This continuous generation has caused the flexural tension crack at the tension face of the RC beam. At damage level III, the AE sources were generated at the tension face as well as the compression face where the crack was not detected visually. These AE signals were due to the local stress concentration of voids in the concrete at the compressive zone. At damage level IV, the AE signals were generated at the bonding face between the rebar and concrete in the tension face. Numerous AE sources were generated at the upper face of the RC beam, which resulted in the rapid propagation of the flexure and shear crack in the compression face due to the yielding of the main bars. After the maximum loads were achieved, the AE sources were measured at the compression zone due to the crushing of the concrete in the flexural compression zone.

Figure 21.9 shows two-dimensional AE source locations of RC beams strengthened with a layer of CFRP. At damage levels I and II, the cracks were not observed at the compressive zone of the beam, but numerous AE sources were generated at the compressive zone due to the local stress concentration of voids in the concrete. When the load was gradually increased, several AE sources were observed at the boundary between the steel reinforcement and the concrete (damage level III). At damage level IV, a few AE sources for S1-C1.0 were generated because the crack propagation was not observed due to the fact that the CFRP sheet restrained the propagation of the crack width at the tension face.

The AE sources derived from the crack between the CFRP sheet and the concrete as well as from debonding by the fracture of the epoxy were not detected because these cracks and damages were generated out of the range of the attached AE sensors. However, Fig. 21.9 shows that the locations of the surface cracks roughly correspond to the area covered by AE event locations. Therefore, the cracking area can be roughly evaluated by AE source location analysis.

21.3.2 Improved *b*-values

In order to apply AE techniques to the evaluation of damage and to the integrity of RC beams strengthened in flexure with CFRP sheets, it is

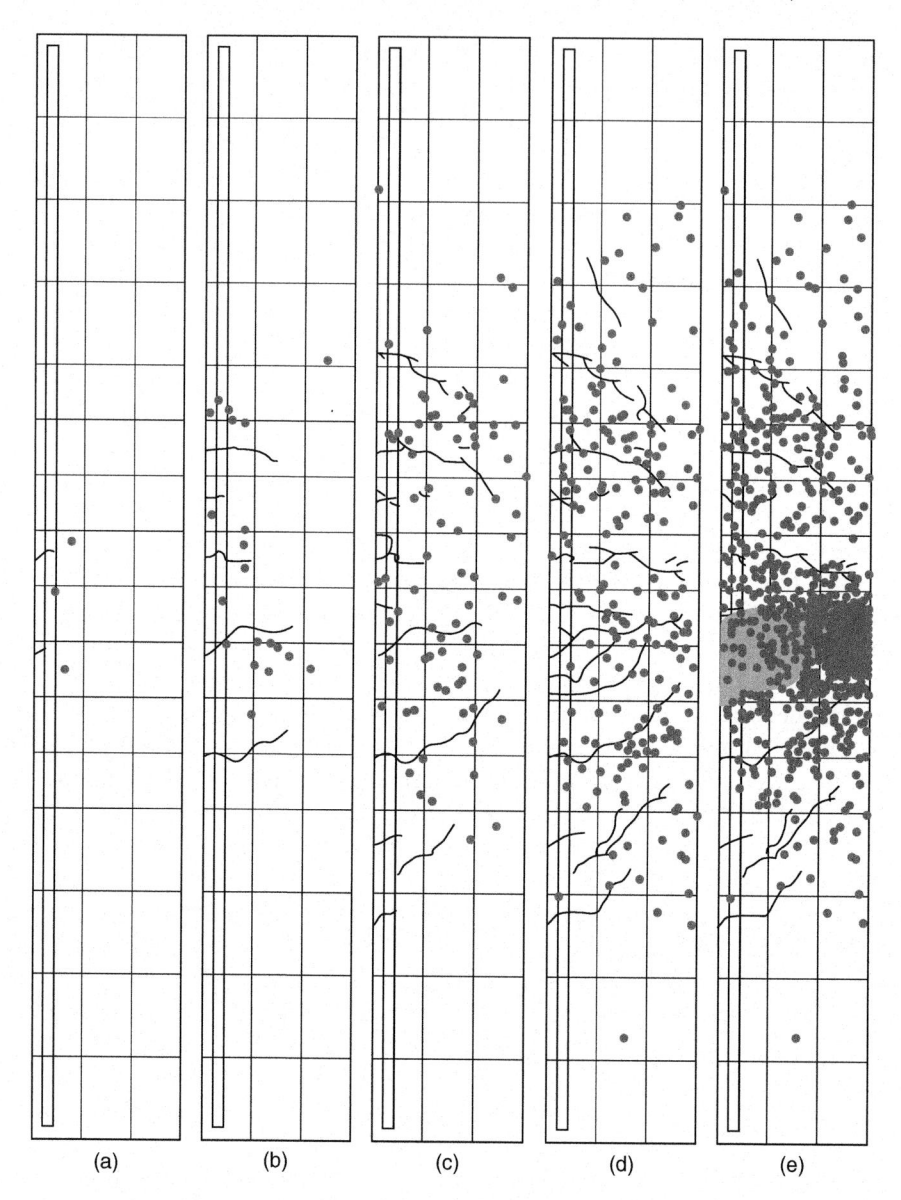

21.8 AE source locations for CB specimen: (a) damage level I; (b) damage level II; (c) damage level III; (d) damage level IV; (e) final failure (from damage level I to failure).

essential to study the characteristics of the AE parameters according to damage levels. The evolution of acoustic activity caused by micro-fractures within concrete is often quantified using the concise framework originated by Gutenberg and Richter in their analysis of earthquake magnitudes,

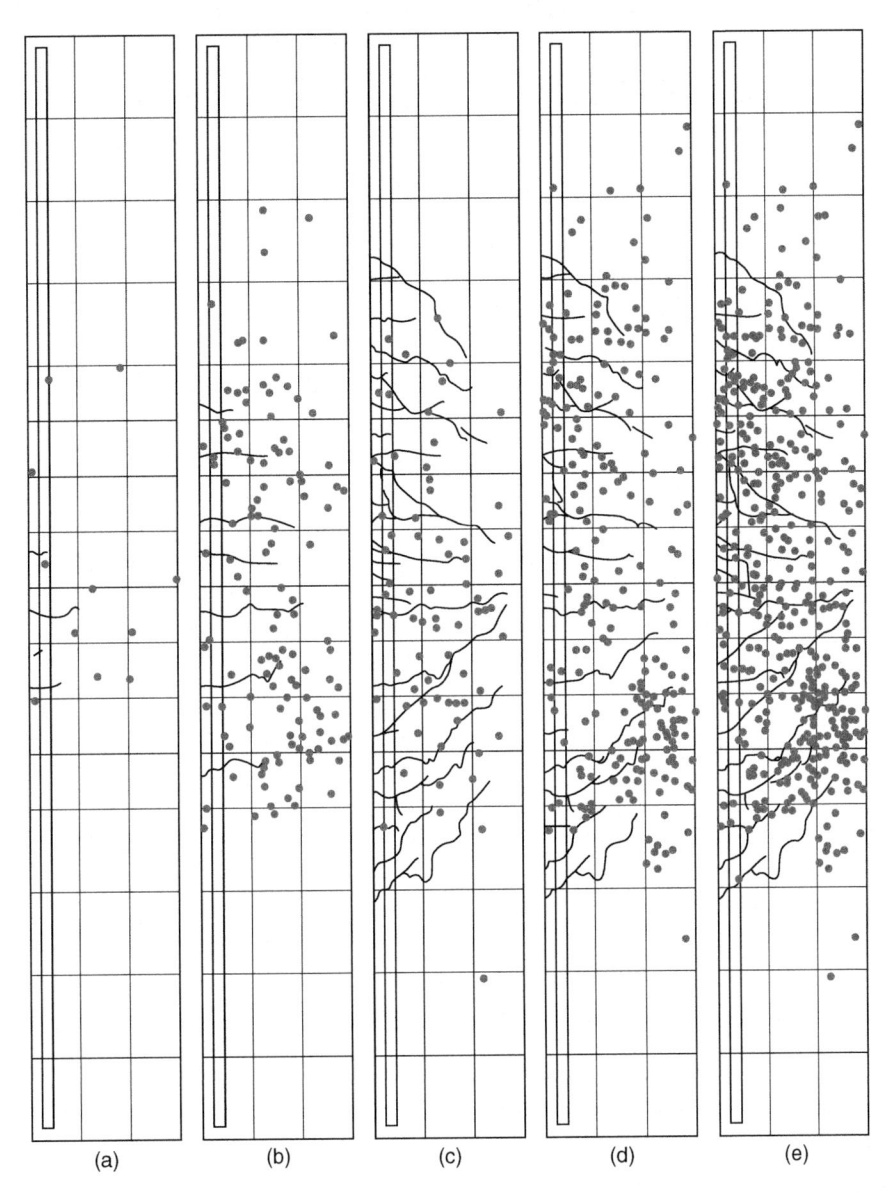

21.9 AE source locations for S1-C1.0 specimen: (a) damage level I; (b) damage level II; (c) damage level III; (d) damage level IV; (e) final failure (from damage level I to failure).

which is a reflection of the view that large-scale (i.e. geological) and small-scale (i.e. micro-fracture) acoustic events share a common origin in cascades of strain energy release events. In earthquake seismology, events of larger magnitude occur less frequently than events of smaller magnitude. This fact can be quantified in terms of a magnitude–frequency relationship, for which Gutenberg and Richter propose the empirical formula,

$$\log N(W) = a - bW \qquad [21.1]$$

where N(W) is the Richter magnitude of the events, the cumulative number of events having a magnitude greater than or equal to W, and a and b are the empirical constants. In AE data analysis, the coefficient b is known as the AE-b value (Mogi, 1962). The AE-b is given as the gradient of the linear descending branch of the cumulative frequency distribution. In the process where micro-fractures are more prevalent than macro-fractures, the b-value tends to increase, whereas in the process where macro-fractures occur more than micro-fractures, the b-value tends to decrease. The application of Eq. (21.1) to AE data is widespread in the study of micro-fractures in rock (Shiotani *et al.*, 2001) and brittle material (Brothers *et al.*, 2006).

The AE b-value corresponds to a negative gradient when the amplitude distribution is given by a straight line. However, the amplitude distribution does not express as one straight line. Thus, an amplitude range is needed that can be represented by a single straight line to obtain the AE b-value in the AE application. The improved b-value, proposed by Shiotani *et al.* (1994), is defined by utilizing statistical values of the amplitude distribution such as mean μ and standard deviation σ. The improved b-value (Ib) is proposed by

$$Ib = \frac{\log N(\mu - \alpha_1 \sigma) - \log N(\mu - \alpha_2 \sigma)}{(\alpha_1 + \alpha_2)\sigma} \qquad [21.2]$$

where α_1 is the coefficient related to the smaller amplitude and α_2 is the coefficient related to the fracture level. In this work, $\alpha_1 = 0$ and $\alpha_2 = 1$ were adopted. Eq. (21.2) can be applied directly to the analysis of AE data provided that the seismic b parameter is multiplied by 20 to account for the fact that the amplitude of AE events is recorded in decibels rather than logarithmic peak amplitude; this method value is referred to as the improved b-value. In experiments in this study that have been undertaken on RC beams strengthened with CFRP sheets, the improved b-value analysis is applied to investigate and evaluate the fracture process and damage level of RC beam strengthened with CFRP sheets.

In order to quantify the evolution of the damage process during flexural loading of RC beams strengthened with CFRP sheets, AE data were separated into four subpopulations representing each damage level (25, 50, 75, and 100 percentages of ultimate loads). The cumulative amplitude distribution of each subpopulation was then fitted to the improved b-value, Eq. (21.2). The results of the improved b-value calculations of CB and S1.0-C1.0 beams for each damage level are shown in Fig. 21.10. The improved b-value

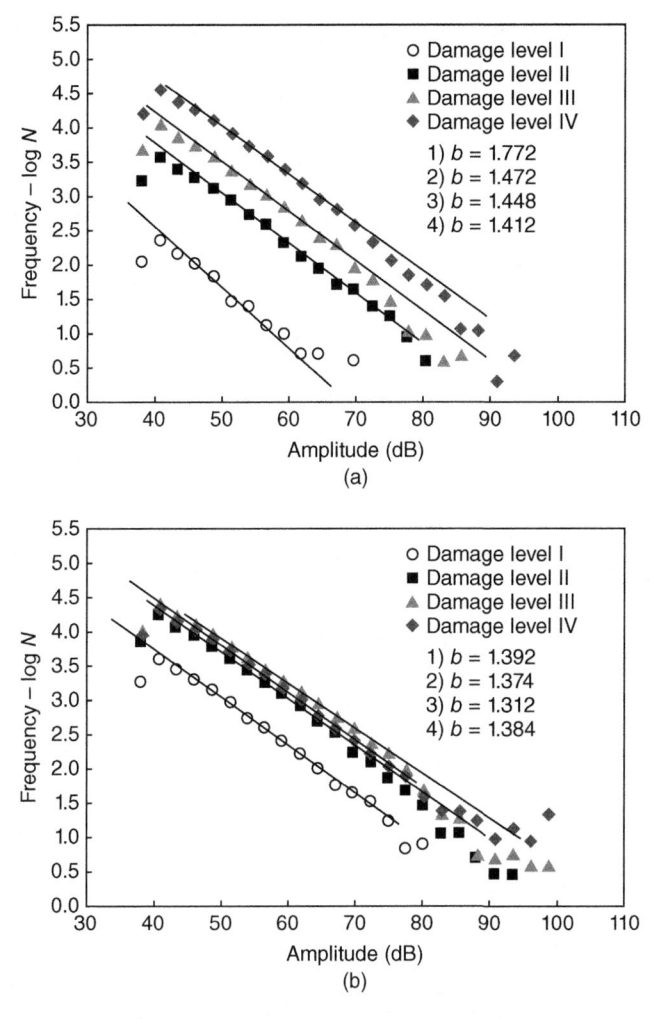

21.10 Amplitude distributions and improved *b* value with each damage level: (a) CB specimen; (b) S1.0-C1.0 specimen.

in the RC beam (the CB) falls in the range of 1.771–1.412, while the improved *b*-value in the S1.0-C1.0 strengthened with CFRP ranges from 1.392 to 1.304. The improved *b*-value quantifies the exponent of decay of the amplitude distribution with increasing amplitude; thus, large values of improved *b*-value reflect AE activity with few high-amplitude fractures, while smaller values of improved *b*-value reflect activity with comparatively more highly energetic fractures.

Based on the analysis of AE data, the failure of the reinforced member with the CFRP sheet is more brittle than that of the RC member. The fracture process for each specimen was observed by the damage accumulation due to the microcrack, which further led to damage localization. Figure 21.10 shows the variation in improved b-values for each level of damage. The improved b-value tends to decrease as the damage propagates. However, it is difficult to treat these findings as conclusive owing to the limited number of tested specimens. In order to make the improved b-value suitable for practical use, the accumulation and consolidation of comprehensive data under various conditions and a resultant database are needed.

21.4 Conclusion

This chapter provides the results from the limited tests for RC beams strengthened with CFRP. On the basis of the AE activities, the analysis of signal characteristics with regard to damage levels, and damage evolution of RC beams, the conclusions are presented below:

- Unique AE signal characteristics of the RC beam are evident depending upon whether the RC beam is strengthened with CFRP sheets. The high level of signal for the strengthened RC beam was measured on the damage of the interface between the CFRP and the epoxy in the beam that is a different characteristic compared with the AE signal caused by the generation and propagation of cracks.
- When the crack is generated and propagated (damage levels I and II), the characteristics of the average AE signal for RC beams strengthened with CFRP sheets, i.e., amplitude, frequency and duration time, are in the range of 52–74 dB, 83–201 kHz and 489–1099 µs, respectively. When significant crack development was observed and the main bars had yielded (damage levels III and IV), the characteristics of an average AE signal, i.e., amplitude, frequency and duration time, were in the range of 72–89 dB, 172–230 kHz and 891–2030 µs, respectively. The tendency is for the characteristic value of the AE signal to increase as the damage levels advance.
- This study results suggest that improved b-value analysis, on the peak amplitude distribution of AE elastic waves, would be a promising parameter, providing the structural integrity (damage degree) of RC beams strengthened with CFRP sheets quantitatively.

21.5 References

Brothers, A. H., D.W. Prine, and D.C. Dunand, 'Acoustic emissions analysis of damage in amorphous and crystalline metal foam,' *Intermetallics*, **14**, 2006, pp. 857–865. doi:10.1016/j.intermet.2006.01.029

Henkel, D.P. and J.D. Wood, 'Monitoring concrete reinforced with bonded surface plates by the acoustic emission method,' *NDT & E International*, **24**(5), 1991, pp. 259–264. doi:10.1016/0963–8695(91)90375-D

Landis, E. N., 'Micro-macro fracture relationships and acoustic emissions in concrete,' *Construction and Building Materials*, **13**, 1999, pp. 65–72.

Lee, J. and J. Lee, 'Nondestructive evaluation on damage of carbon fiber sheet reinforced concrete,' *Composite Structure*, **58**, 2002, pp. 139–147. doi:10.1016/S0263-8223(02)00029-6

Li Z. and S.P. Shah, 'Microcracking in concrete under uniaxial tension,' *ACI Material Journal*, **91**(4), 1994, pp. 372–381.

Mogi, K., 'Study shocks caused by the fracture of heterogeneous materials and its relations to earthquake phenomena,' *Bulletin of Earthquake Research Institute, University of Tokyo*, **40**, 1962, pp. 123–173

Naaman, A., S. Park, and M.D.M. Lopez, 'Repair and Strengthening of Reinforced Concrete Beams using CFRP Laminates; Volume 3: Behavior of Beams Strengthened for Bending,' Report No. UMCEE 98–21, the University of Michigan, Department of Civil and Environmental Engineering, Ann Arbor, April, 1999

Ohtsu, M., M. Uchida, T. Okamoto, and S. Yuyama, 'Damage assessment of reinforced concrete beams qualified by acoustic emission,' *ACI Structural Journal*, **99**(4), 2002, pp. 411–417.

Quyang, C., E. Landis, and S.P. Shah, 'Damage Assessment in Concrete Using Quantitative Acoustic Emission,' *Journal of Engineering Mechanics*, **117**(11), 1991, pp. 2681–2688. doi:10.1061/(ASCE)0733-9399(1991)117:11(2681)

Shiotani, T., K. Fujii, T. Aoki, and K. Amou, 'Evaluation of progressive failure using AE sources and improved *b*-value on slope model tests,' *Progress in Acoustic Emission*, **7**, 1994, pp. 529–534.

Shiotani, T., M. Ohtsu, and K. Ikeda, 'Detection and evaluation of AE waves due to rock deformation,' *Construction and Building Materials*, **15**, 2001, pp. 235–246.

Yun, H., W. Choi, and S. Seo. 'Acoustic emission activities and damage evaluation of reinforced concrete beams strengthened with CFRP sheets.' *NDT & E International*, **7** (Oct), 2010, pp. 615–628. doi:10.1016/j.ndteint.2010.06.006

22
Non-destructive evaluation (NDE) of composites: microwave techniques

M. Q. FENG, Columbia University, USA and G. ROQUETA and L. JOFRE, Universitat Politècnina de Catalunya (UPC), Spain

DOI: 10.1533/9780857093554.4.574

Abstract: Microwaves are electromagnetic (EM) waves with frequencies ranging from 0.3 to 300 GHz. In the last decades, microwaves have been considered as a potential tool for the assessment of materials or structures where visual inspection is not possible, owing to their capability to sense and penetrate light-opaque materials with a fair trade-off between penetration and resolution. This chapter is devoted to providing an insight into microwave non-destructive evaluation (NDE) techniques for inspection of a specific type of civil engineering structure, the reinforced concrete (RC) structures retrofitted, strengthened, or repaired with glass fiber-reinforced polymer (FRP) composites. The primary goal is to detect damage in FRP-wrapped concrete structures including debonds in the FRP – concrete interface and voids in concrete.

Key words: microwaves, microwave imaging, non-destructive evaluation (NDE), fiber-reinforced polymer (FRP), reinforced concrete (RC) structures, tomographic imaging, dielectric materials, material characterization.

22.1 Introduction

22.1.1 (FRP)-jacketed reinforced concrete (RC) structures

The jacketing technology using fiber-reinforced polymer (FRP) matrix composites has been applied for seismic retrofit, strengthening, and rehabilitation of reinforced concrete (RC) structures designed and constructed under older specifications. The enhanced structural performance of RC structures rehabilitated by FPP composite jackets has been well demonstrated. An increasingly large number of bridges and buildings have been retrofitted with such jackets in the United States and elsewhere.

However, debonding between the jacket and the structure caused by external agents, such as poor workmanship in installation and seismic damage, can considerably degrade the structural performance otherwise attainable by the FRP rehabilitation. As it was demonstrated by the results of an experiment performed by Haroun and Feng (1997), poor bonding conditions can significantly degrade the structural performance. In that

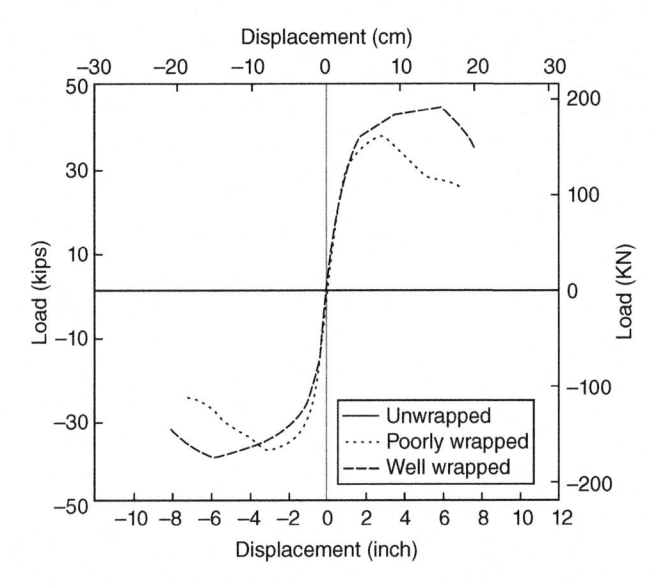

22.1 Cyclic loading tests for the determination of void and performance degradation of three concrete columns with different jacketing conditions: unwrapped, poorly wrapped and well wrapped. With permission from ASCE (Feng *et al.*, 2002).

study, three identical half-scale circular bridge columns were built. Two of the columns were wrapped with identical glass – FRP jackets but different bonding conditions: one was well wrapped while the other one was poorly wrapped with voids in the bonding interface. Force-displacement envelops resulting from the cyclic loading tests of the three columns (unwrapped, well wrapped, and poorly wrapped) are shown in Fig. 22.1. The unwrapped column had a ductility factor under two. The well-wrapped column performed excellently by increasing the column ductility factor up to six, while the poorly wrapped barely reached the ductility factor of three. This remains a significant concern as such debonds cannot be visually identified.

22.1.2 History review on non-destructive evaluation (NDE)

Manifold non-destructive evaluation (NDE) techniques have been studied for detecting damage in RC structures including cracks and delamination between concrete and rebar. Some of them appear promising but suffer from a number of disadvantages especially for testing volume materials.

- *Optical imaging* has found broad acceptance, but interacts only in a layer of a few microns at the surface of a material.

- *Ultrasonic testing* is a basic NDE technology that obtains volume information with spatial resolutions down to a submicron range. However, the technology suffers coupling problems when non-contacting operation is required. Also, it has been shown in Olson *et al.* (1993) that the results have low penetrability and are influenced by sizes and shapes of aggregates in concrete.
- *X–rays* can overcome this easily; their problem is the danger coming from the ionizing radiation leading to high and expensive requirements for safety measures. The same applies for *computer tomography based on radioactive materials*.
- *Nuclear magnetic resonance (NMR)* is restricted either to closed magnetic systems like in medicine or, for the case of open systems, to a limited depth of operation.
- *Ground penetrating radar (GPR)* is typically used in civil engineering for the detection of anomalies such as cracks, voids, reinforcement and other embedments. Radar has the ability to penetrate dielectric materials (e.g. concrete and FRP), but is reflected from conducting bodies (e.g. steel rebars) and interfaces. It has a certain polarization and is relatively small (Maser, 1989; Zoughi *et al.*, 1991). GPR has the ability to evaluate the properties of materials composed of a mixture of several constituents, and to collect data potentially at highway speeds. However, problems are (1) the large wavelengths (around 10 cm) used in common radar applications, which do not allow the identification of small cracks (around 1 mm), and (2) the time-consuming post-process of the measured radar images.

22.1.3 Microwave NDE

Microwave NDE methods have the potential to overcome all these disadvantages. Microwaves show a real volumetric interaction with the dielectric material properties for all non-metallic materials up to some decimeters in depth. They can work easily both in reflection and in transmission configurations. Electromagnetic (EM) waves can easily couple with a material through air. Microwave testing systems are safe to use owing to their very low power needs. They can be designed for a wide range of special applications and penetration depths.

All these positive features encourage the development of microwave scanners for material NDE testing. In this context, EM analysis of microwaves has been conducted for the NDE of different types of concrete problems such as the fibre density in steel fibre reinforced concrete (SFRC) (Roqueta *et al.*, 2010; Van Damme *et al.*, 2004), and corrosion of steel in RC structures (Roqueta *et al.*, 2012).

22.2 Electromagnetic (EM) properties of materials

Accurate assessment of civil infrastructure depends on reliable measurements of the EM properties of the elements forming the structures. Every material has a unique set of EM properties mostly affecting the way in which the material interacts with EM waves. The EM properties determine the speed, reflection, and the optimum illumination angle and are needed for numerical modeling of wave scattering for theoretical studies and for image reconstruction.

22.2.1 EM spectrum – microwaves

When analyzing EM properties, the EM spectrum deserves special attention. The wave frequency f is inversely proportional to the effective wavelength in the medium λ_e. As a rule of thumb, for EM visualization of a structure, one must place his choice of the frequency band in the region where $\lambda_e/2$ corresponds to the minimum desired discretization of the structure to be tested.

Figure 22.2 illustrates a well-established EM spectrum in wavelength and frequency with graphic examples illustrating the scale of the wavelength. Each segment of the spectrum has its own properties, such as resolution and penetration depth: both are a function of the wavelength. As the frequency increases, penetration depth decreases, but the resolution increases and so the technology cost, as illustrated in Fig. 22.3.

Microwaves (Pozar, 1998) are EM waves with wavelengths ranging from as long as one meter to as short as one millimeter, or equivalently, with frequencies between 0.3 and 300 GHz. Various molecular, atomic, and nuclear resonances of materials occur at microwave frequencies, creating a variety of unique applications. So far, microwaves have been shown to be a potential tool for the assessment of materials or tissues where visual inspection is not possible, thanks to their capability to sense and penetrate light-opaque materials with a fair trade-off between penetration and resolution. Owing to the radiation characteristics of microwaves, they interact in volumes up to some decimeters in depth. They allow flexible configurations of test systems and use very low EM power.

22.2.2 Dielectric materials and conductors

Every material is formed by molecules, which at a smaller scale, are formed by atoms. An atom of an element consists of a very small and compact nucleus that is surrounded by a number of negatively charged electrons revolving about the nucleus. The nucleus contains neutrons, which are neutral particles, and protons, which are positively charged particles. The

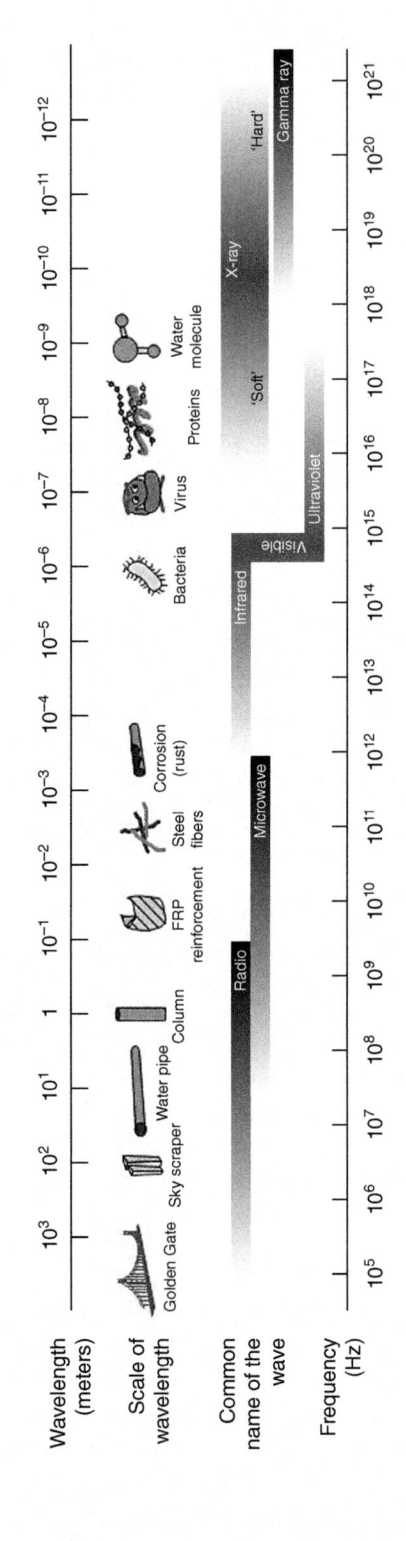

22.2 Electromagnetic spectrum with graphical examples.

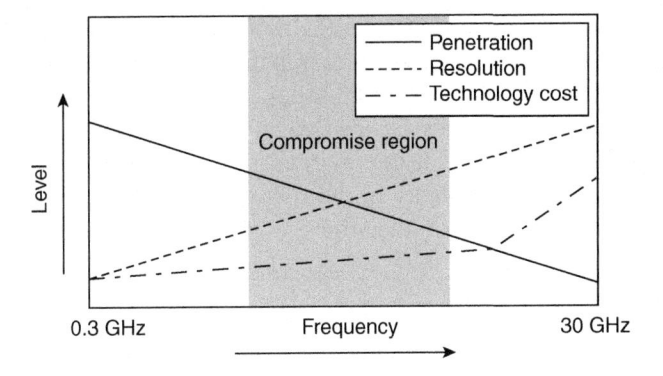

22.3 Relation between the resolution, penetrability, cost of the technology, and the frequency.

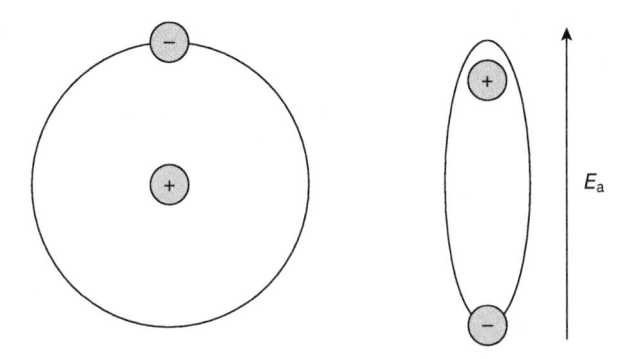

22.4 Typical atom in (left) the absence of an electric field, and (right) under an applied electric field © [1989] Wiley. Reprinted, with permission, from Balanis (1989).

negatively and positively charged particles are bound together by atomic forces (ultimately due to Coulomb forces) which keep the particles associated so as to form the material. Because of the mutual attraction between positively and negatively charged particles, every positively charged particle is very near to negatively charged particles. As a result, any small region of the material contains almost the same amount of positive and negative charge, so that the total charge in any small region of the material is ordinarily zero.

If a material is placed in an electric field due to external charged objects, the positively charged particles in the material experience electric forces along the direction of this field and the negatively charged particles experience electric forces opposite to the direction of this field (see Fig. 22.4). Thus the forces due to the applied electric field tend to separate the positively and negatively charged particles from each other. How much separation occurs depends crucially on the magnitude of the atomic forces which hold these charged particles together in the material. Thus the applied electric field produces only a very small separation of the charged particles if these are bound together by strong atomic forces. But the same field produces a larger separation of the charged particles if these are bound together by weaker atomic forces. In the following, two types of materials are distinguished, depending on the strength of the atomic forces.

Dielectrics

A dielectric material, or electric insulator, is a material whose negatively charged particles are strongly bound to some positively charged nearby particles, and so they are not free to travel (Balanis, 1989). When external electric fields are applied on dielectric materials, these charged particles then move only slightly away from their normal position in the material.

The movement of the charges against restraining molecular forces provides the material with the ability to store electric energy. The parameter used to represent the relative (compared to free-space) charge storage capability of a dielectric material is the dielectric constant or relative permittivity ε_r.

Conductors

By contrast, there are materials in which some of the charged atomic particles are bound so weakly by atomic forces that they are free to move throughout the entire material. Such materials are called 'electric conductors'.

Conductors are materials whose prominent characteristic is the motion of electric charges leading to the creation of a current flow. In a conductor material, when an external field is applied, outer shell (valence) electrons migrate very rapidly from one atom to another, thus creating a conduction current in the conductor.

The conductivity σ of a conductor is a parameter that characterizes the free-electron conductive properties. As temperature increases, electrons move faster and so the possibility of collision increases, which results in a decrease in the conductivity of the conductor. Materials with a very low value of conductivity are classified as dielectrics.

Complex permittivity

When a time-harmonic field $\mathbf{E}_0 e^{j\omega t}$ is applied to an atom, the relative permittivity of the medium is recognized to be complex ($\varepsilon_r = \varepsilon_r' - j\varepsilon_r''$). In turn, it can be related to other dielectric properties of the materials: the electric loss tangent $\tan\delta$ and the conductivity σ. That is,

$$\tan\delta = \frac{\sigma}{\omega\varepsilon_0\varepsilon_r'} = \frac{\varepsilon_r''}{\varepsilon_r'} \qquad [22.1]$$

where $\varepsilon_0 = 8.85\text{e}^{-12}$ F/m is the free space permittivity. The real part ε_r', known as the dielectric constant, describes the polarizability of the material under test. The imaginary part ε_r'' known as the dielectric loss factor, stands for dielectric losses according to the phase-shifted movement of polar molecules in an EM field. This parameter is always positive and usually much smaller than ε_r'. It is approximately proportional to the attenuation of a propagating wave.

In some materials, these parameters may present a certain variation with frequency. Such materials are referred to as dispersive materials. Thus, the specified dielectric properties should represent the respective relative permittivity, loss tangent, or conductivity at a given frequency range. Typical values of relative permittivity and loss tangent at microwave frequencies are listed in Table 22.1.

Table 22.1 Relative permittivity and loss tangents of typical dielectric materials

Material	T (°C)		0.3 GHz	3.0 GHz	30.0 GHz	Reference
Air		ε_r'	1.00	1.00	1.00	Balanis
		$\tan\delta$	0.0000	0.0000	0.0000	(1989)
Aluminum oxide	25	ε_r'	8.60	8.60	8.60	Von Hippel
		$\tan\delta$	<0.0001	<0.0001	<0.0001	(1995)
Concrete (dry)	25	ε_r'	7.00	4.00	1.70	Belrhiti
		$\tan\delta$	0.2800	0.0200	0.0300	et al. (2011)
Concrete (wet: w/c = 0.65)	25	ε_r'	45.00	30.00	–	Otto and Chew
		$\tan\delta$	2.8000	0.0300	–	(1991)
Nylon	25	ε_r'	3.06	3.02	3.02	Von Hippel
		$\tan\delta$	0.0140	0.0120	0.0107	(1995)
Polystyrene	25	ε_r'	2.54	2.54	2.54	Von Hippel
		$\tan\delta$	0.0023	0.0003	0.00053	(1995)
Soil (dry)	25	ε_r'	2.55	2.55	2.53	Von Hippel
		$\tan\delta$	0.0100	0.0062	0.0036	(1995)
Styrofoam	25	ε_r'	–	1.03	1.03	Von Hippel
		$\tan\delta$	–	0.0001	0.0001	(1995)
Teflon	25	ε_r'	2.10	2.10	2.08	Von Hippel
		$\tan\delta$	0.0001	0.0001	0.0006	(1995)
Water (distilled)	25	ε_r'	80.50	38.00	15.00	Von Hippel
		$\tan\delta$	0.3100	1.0300	0.4250	(1995)

*–6°C.

The main features of the dielectric spectrum of construction materials are temperature dependence (Awan and Shamin, 2007), moisture content dependence (Kharkovsky et al., 2002), frequency dispersion and other agents. All these features have to be taken into account when doing electromagnetic analysis of civil structures.

22.2.3 Wave propagation

The core of the presented microwave imaging technique is based on the analysis of microwaves sent toward and reflected back from a layered medium (Feng et al., 2002). It is well known that when a plane EM wave is launched from an illuminating device (typically an antenna) towards a layered medium, and encounters a dielectric interface, a fraction of the wave energy is reflected, while the rest is transmitted into the medium. The analysis of phase, frequency and amplitude differences between the transmitted and the received signals, provides information about the EM properties of the media, through which the signal is transmitted, reflected or scattered.

Let us consider the generic situation of Fig. 22.5, where a uniform plane wave is traveling across a medium of complex effective permittivity $\varepsilon_r' - j\varepsilon_r''$. The electric field is assumed to have an x component and the wave is travelling in the $+z$ direction. The electric field at any position can be expressed by

$$\mathbf{E}(z) = \mathbf{E}_0 e^{-\gamma z} \qquad [22.2]$$

In equation (22.2), γ is the complex propagation constant for the wave propagating through the material, and takes the form

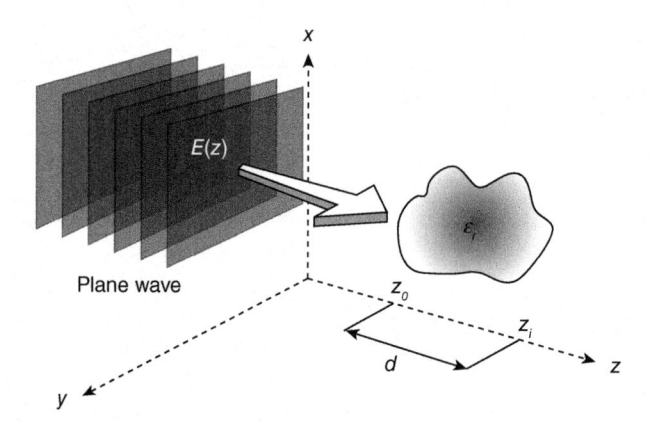

22.5 Plane wave with an x component travelling in the $+z$ direction across a lossy medium of complex effective permittivity $\varepsilon_{ri} = \varepsilon_{ri}' - j\varepsilon_{ri}''$.

$$\gamma = j\frac{\omega}{c}\sqrt{\varepsilon_r} = \alpha + j\beta \qquad [22.3]$$

Real α and imaginary β parts defining the propagation constant are the attenuation and phase constants, respectively. Based on the fact that the electric field has to satisfy the wave equation in lossy media, and considering equation (22.1), α and β can be related to the dielectric properties of the propagating medium through

$$\alpha = \frac{\omega}{c}\sqrt{\varepsilon_r'}\left\{\frac{1}{2}\left[\sqrt{1+(\tan\delta)^2}-1\right]\right\}^2 \text{ Np/m} \qquad [22.4]$$

$$\beta = \frac{\omega}{c}\sqrt{\varepsilon_r'}\left\{\frac{1}{2}\left[\sqrt{1+(\tan\delta)^2}+1\right]\right\}^2 \text{ rad/m} \qquad [22.5]$$

where c is the speed of light in free space. Notice that the velocity of propagation in a dielectric medium, v_p, is always smaller than c following the rule $v_p = c/\sqrt{\varepsilon_r'}$.

In terms of propagation, for a wideband incident signal, the propagation time delay between position $z = z_0$ and $z = z_i$ may be related to the phase constant as

$$\tau = d\frac{\delta\beta}{\delta\omega}, \quad d = z_i - z_0 \qquad [22.6]$$

The example in Fig. 22.6 shows how the knowledge of the medium attenuation (concrete walls) can help to predict the internet coverage inside a building. The propagation of microwaves ($f = 2.1$ GHz) inside a two-story building with a floor attenuation of 12.00 dB, partition wall attenuation of 3.40 dB and a load-bearing wall attenuation of 6.80 dB is predicted using the spectral iterative propagation technique (SIP) (Capdevila *et al.*, 2006). The source of microwaves (transmitter) is placed at the left side-close to the top floor. Results represented in Fig. 22.6 show the normalized power distribution along the first floor (a) and the top floor (b) (white is maximum power 0 dB m and dark is minimum power −80 dB m). Results of Fig. 22.6b show relatively high power levels in the top floor and close to the transmitter. Then as the propagation distance increases, and so the attenuation due to the walls does, the power decreases. Results of Fig. 22.6a show faster and stronger decrease of the power level due to the bigger distance between the transmitter and the first floor. In general, from the coverage prediction results it can be derived that there is a poor coverage in the first floor, which indicates the need of placing a repeater which enhances signal power in that area.

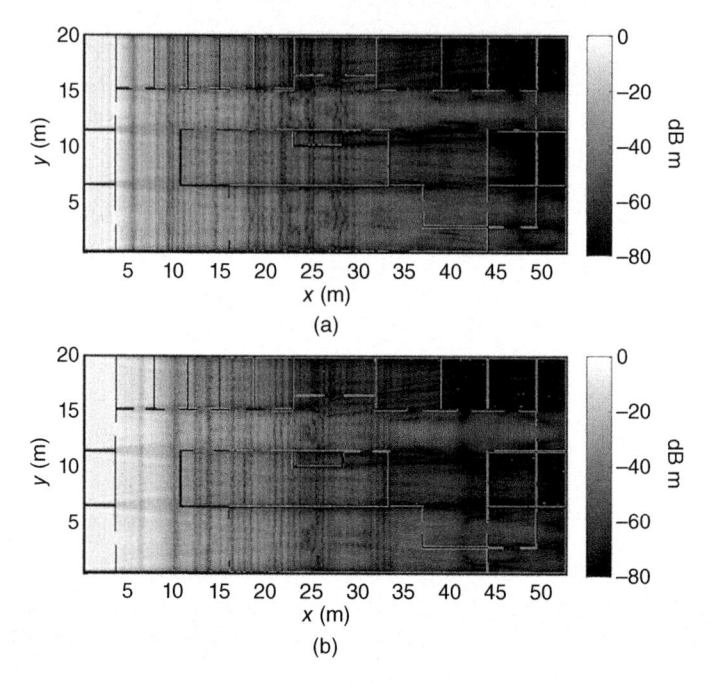

22.6 Prediction of radio system coverage (*f* = 2.1 GHz) inside a two-story building: (a) first floor, (b) top floor (Capdevila *et al.*, 2006).

22.2.4 Measurement techniques for material characterization

Many techniques have been developed for measuring the complex permittivity of materials and each technique is limited to specific frequencies, materials, and applications. Table 22.2 compares some of the mostly commonly used techniques.

The free space method and the open-ended coaxial probe are particularly attractive in practical applications such as characterization of civil engineering structures, due to its robustness in hostile environments while still being a simple microwave technique for material inspection. Also, measurements are performed without requiring physical contact between the structure under test and the antenna. In most instances there is no need for special structure preparation. Hence, these techniques offer fast bulk material testing suitable for rapid imaging of concrete structures.

By means of the free-space method, the reflection and transmission properties of the propagating waves can be related to the dielectric constant of the specimen. The convenience of using the reflection or the transmission

Table 22.2 Comparison between measurement techniques (MUT)

Measurement technique	Material under test (MUT)	Advantages	Disadvantages
Transmission/ reflection line (Torgovnikov, 1993)	Medium to high loss material embeddable in a coaxial line or waveguide	Determination of ε_r, μ_r	Band limited Machining of the MUT Measurement accuracy affected by air gaps Band limited
Open-ended coaxial probe (Rhim et al., 2004)	Liquids Biological tissues Semi-solids	No machining of the MUT Non-destructive measurements	Only reflection measurements (ε_r) Measurement accuracy affected by air gaps Band limited
Free space (Ghodgaonkar et al., 1989)	High temperature materials Large flat solids Gas Hot liquids	Determination of ε_r, μ_r Wideband Non-destructive measurements Measurements in hostile environments	Requires large and flat MUT Multiple reflections and edge diffractions on the MUT
Resonant cavity method (Komarov and Yakovlev, 2003)	Rod-shaped solid materials Liquids	Determination of ε_r, μ_r Very accurate Can measure very small MUT	Narrowband or single frequency. Need high-frequency resolution network analyzer

Source: Rohde and Schwas, (2006).

approach will be further explained in Section 22.4.1. Further description of this method can be found in Ghodgaonkar *et al.* (1989).

The open-ended coaxial probe method allows the dielectric constant of the specimen under test to be obtained from measurements of the reflection coefficient. In order to do so, three particular calibration standards are required (short circuit, open circuit, and a polar liquid). Further description of this method can be found in Rhim *et al.* (2004).

EM characterization of FRP composites and concrete

Glass and carbon fiber composites have found their applications in civil engineering structures, primarily for seismic retrofit, strengthening, and rehabilitation. Their dielectric properties are analyzed below, together with concrete.

Table 22.3 Measurement of the dielectric properties of FRP concrete structures components

Sample	Relative dielectric constant (ε_r)	Conductivity (S/M)
GFRP (epoxy)	3.84	0.18
Concrete	5.30	0.05

Source: Feng *et al.* (2000).

Free space method

The free space method is used for the dielectric characterization of these materials. A reflectometer is used to measure the dielectric constant and conductivity of carbon FRP (CFRP) and epoxy-glass FRP (GFRP). Flat material sample sheets of dimensions 25.5 cm × 21.5 cm, made of glass fibre-reinforced polymer, carbon fibre-reinforced polymer, adhesive epoxy, and concrete, respectively, are measured. The measurement results of the different materials are listed in Table 22.3.

From the high conductivity value shown in the table, CFRP composites behave more like a conductor rather than a dielectric material. As such, this kind of material will block the microwaves, making the microwave analysis ineffectual. In contrast, the dielectric properties of GFRP composites and concrete indicate that they are appropriate candidates for microwave NDE tests. The following studies will focus on structures made of these materials, such as concrete with GFRP jackets.

Open-ended coaxial probe

An open-ended coaxial probe method is used for the measurements of the dielectric properties of concrete and FRP in the range from 0.1 to 20 GHz in Rhim *et al.* (2004). The measurement set-up is shown in Fig. 22.7. Coaxial probe calibration was performed on air and 25 °C water with well-known properties prior to the measurement. The values of the dielectric constant and conductivity obtained from the measurements are plotted from Fig. 22.8 and Fig. 22.9 for concrete samples, and from Fig. 22.10 and Fig. 22.11 for GFRP and CFRP samples respectively. The measurement results of concrete agree well with the common trend observed in many well-known materials: dielectric constant decreases as frequency increases and conductivity or loss tangent increases as frequency increases. As shown in Fig. 22.8 and Fig. 22.9, the values of dielectric constant were almost constant after 8 GHz. The EM properties were significantly changed with water contents: dielectric constant and conductivity with more water contents have much

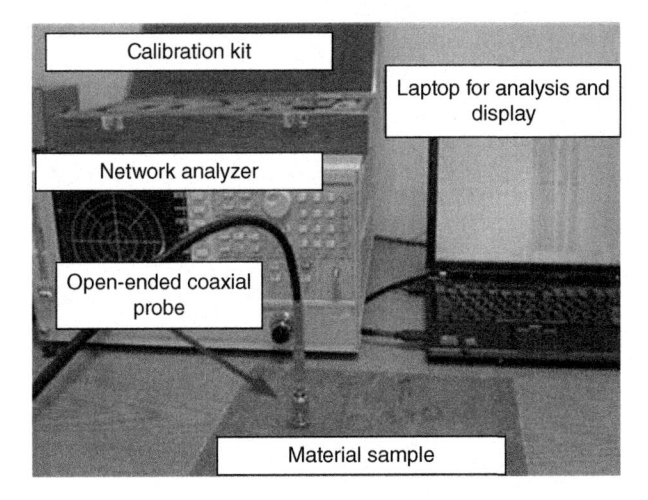

22.7 Photo of measurement set-up for the dielectric characterization of concrete and FRP composites. With permission from MCEER (Rhim *et al.*, 2004).

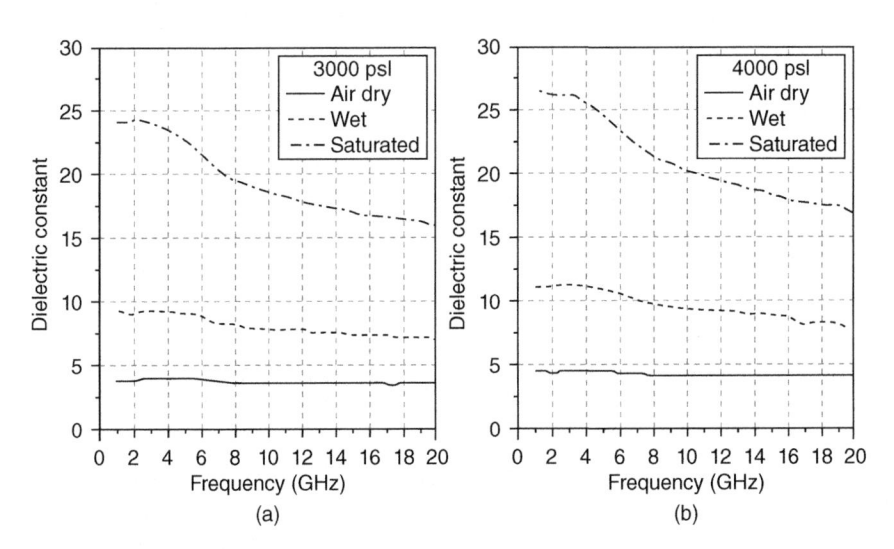

22.8 Dielectric constant of concrete samples. With permission from MCEER (Rhim *et al.*, 2004).

more large values than the ones with less water content. When the concrete was dry, both dielectric constant and conductivity were not affected by the strength. However, when the concrete contains more water, the dielectric constant and conductivity increased with the strength.

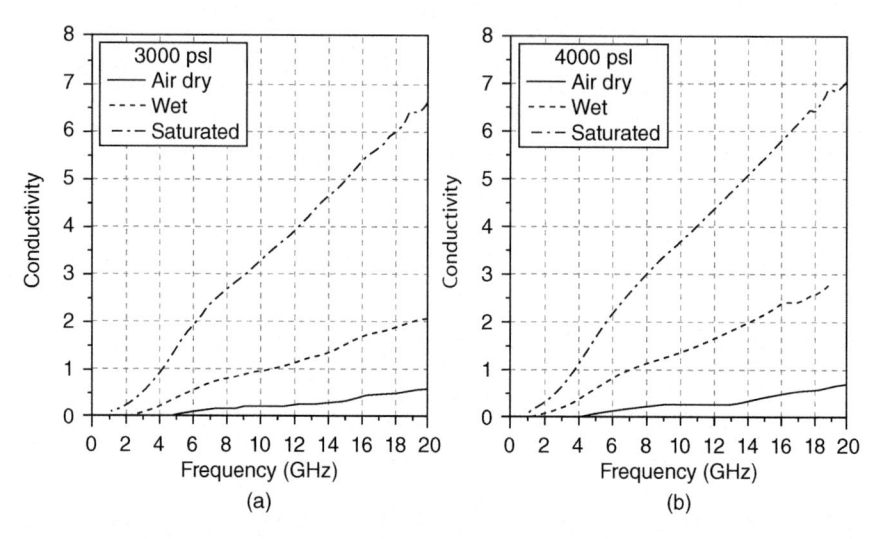

22.9 Conductivity of concrete samples. With permission from MCEER (Rhim *et al.*, 2004).

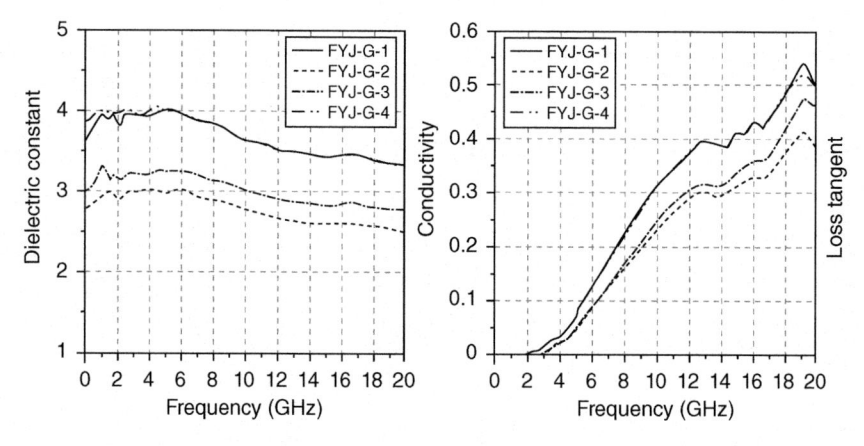

22.10 EM Properties of FRP sample (e-glass/epoxy). With permission from MCEER (Rhim *et al.*, 2004).

The measurement results of FRP show that the dielectric constant was almost constant over the measured frequency range but slightly decreased as the frequency increased, and was not much dependent upon the ages of the FRP samples, the number of layers, and the type of resins.

The difference between GFRP and CFRP samples was apparent. The conductivity and loss tangent increased as the frequency increased, which is in agreement with the trend observed in many, materials of well-known

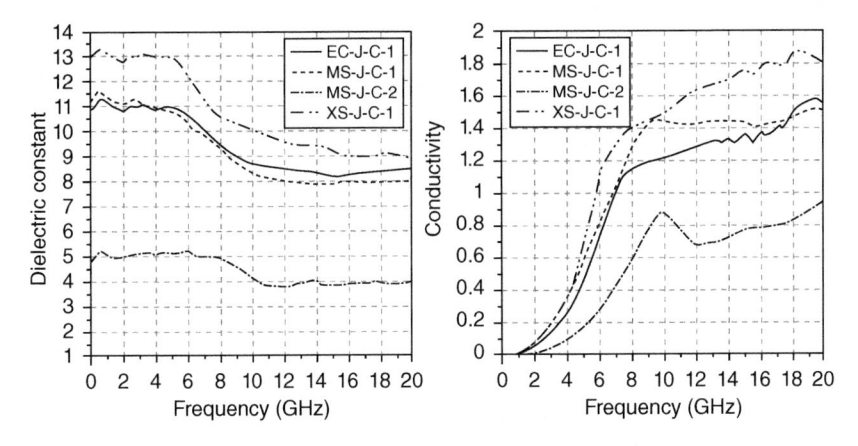

22.11 EM Properties of FRP sample (carbon/epoxy). With permission from MCEER (Rhim *et al.*, 2004).

properties. For the GFRP, the dielectric constant was observed in the range from 3 to 4 and the conductivity was 0.05 around 5 GHz. In Fig. 22.10, samples that have lower dielectric constant contained additional treatment such as cabosil outside and polyester stitches in diagonal form. In Fig. 22.11, the higher dielectric constant sample was a pultruded curing type and the other was a hand-layup curing type.

22.3 Sensing architectures

The choice of the sensing architecture affects the NDE capability of a microwave imaging system. The 'sensing architecture' broadly refers to (1) the type of sensing elements (typically antennas): radiation pattern, directivity, efficiency, and bandwidth, (2) the arrangement of the antennas: transmission or reflection, and (3) their spatial distribution: 2D array, 3D array, real array, or virtual array. There is a large choice of sensing architectures, depending on the morphology of the structure to be analyzed and the desired inspection depth (surface or subsurface inspection).

22.3.1 Transmission versus reflection conceptions

The antennas can be arranged in different configurations depending on the application requirements. The configuration based on the transmission principle (Fig. 22.12a), with its calibration simplicity, is more appropriate for laboratory measurements. The configuration based on a reflection principle (Fig. 22.12b), with a high sensitivity to interface reflections, is more

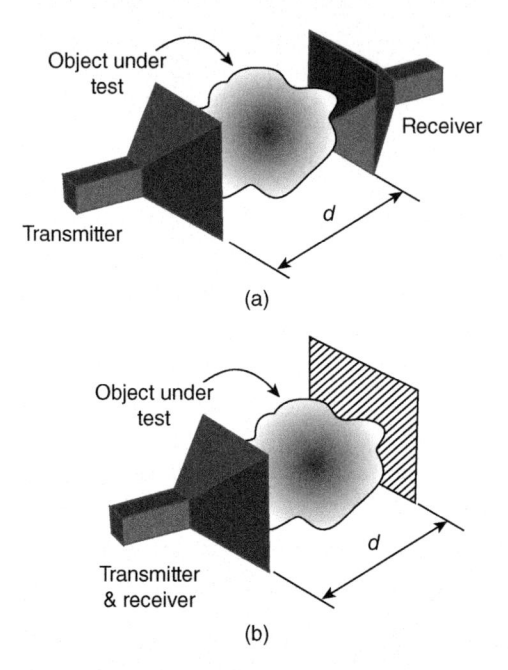

22.12 Generic measurement configurations: (a) transmission, (b) reflection.

appropriate for field applications where physical constraints prevent the transmission measurement.

In the *transmission configuration (Tx)*, the transmitting and receiving elements (in the example, horn antennas) are placed at different angles surrounding the structure under test. In the example in Fig. 22.12a, the transmitter is placed 180° apart from the receiver. The amplitude and phase of the wave propagated from the transmitter to the receiver will experience a variation that depends on the material properties, the thickness of the material, the illuminating frequency, and the alignment of the material with respect to the antennas. This variation will allow the characterization of the material under test. In a full imaging system, the transmission configuration forms a bistatic system. The added flexibility of using different antennas for transmitting and receiving permits the study of oblique transmission and reflection angles and improves the dynamic range of the measurements.

In the *reflection configuration (Rx)*, one antenna is used to transmit and receive the signal reflected from the material under test (Fig. 22.12b). The reflection coefficient is the ratio of reflected and transmitted waves. Reflection measurements allow integral measurement of permittivity across the depth of the material. The material properties can be measured by analyzing the magnitude and the phase of the reflection coefficient by using a contacting or a non-contacting antenna. An example of the contacting antenna for reflection measurements is an open-ended coaxial transmission line that is configured as a resonator and is pressed tightly against the object under test. A single broadband microwave horn antenna is an example of a non-contacting antenna, in which the antenna is located at a certain distance from the object under test. The major limitations of this configuration are the needs for a relatively large amount of material to achieve a sufficient sensitivity and for reducing the influence of interface reflections. In a full imaging system, the reflection configuration forms a monostatic system. When using this configuration, only normal incidence is analyzed, thus limiting the vision angle of the system. This kind of antenna arrangement is commonly used in surface imaging.

Figure 22.13a shows a monostatic array of antennas, and Fig. 22.13b a bistatic array of antennas. In the monostatic array, each antenna transmits and receives signal sequentially. In the bistatic array, one antenna acts as transmitter and the rest act as receivers (including the transmitting antenna), and this process is repeated sequentially with all the antennas.

22.3.2 2D and 3D imaging geometries

When analyzing the EM properties of a concrete structure (or any structure), one may be interested in obtaining as much information as possible. However, the price to pay is the high computational cost and complex measurement set-ups, as the image resolution improves.

The multidimensional feature of an EM visualization system is defined by the degrees of freedom of the elements forming the imaging system (the transmitters, the receivers, and the object itself). Depending on the nature of the structure under test (homogeneous or inhomogeneous), one may be interested in 2D or 3D imaging geometries.

For a z-homogeneous structure, let us first consider the 2D imaging system drawn on a cylindrical geometry, such as the one presented in Fig. 22.14a. In this system, N pairs of transmitters and receivers are arranged surrounding the object along the ϕ direction. Although the 2D configuration is often used in many imaging systems for its relatively low cost and high system feasibility (Broquetas *et al.*, 1991), 2D reconstructions are unable to distinguish features at different heights (in the z direction). In order to locate defects of the structure under test in z-inhomogeneous structures, a 3D configuration is needed. Consider the 3D imaging system of Fig. 22.14b, where $N \times N$ pairs of transmitters and receivers cover the object along the ϕ and θ directions. While providing detailed information with good resolution in both ϕ and θ directions, this kind of 3D reconstructions has high computational costs and requires a more complex experimental set-up (Guardiola *et al.*, 2011b).

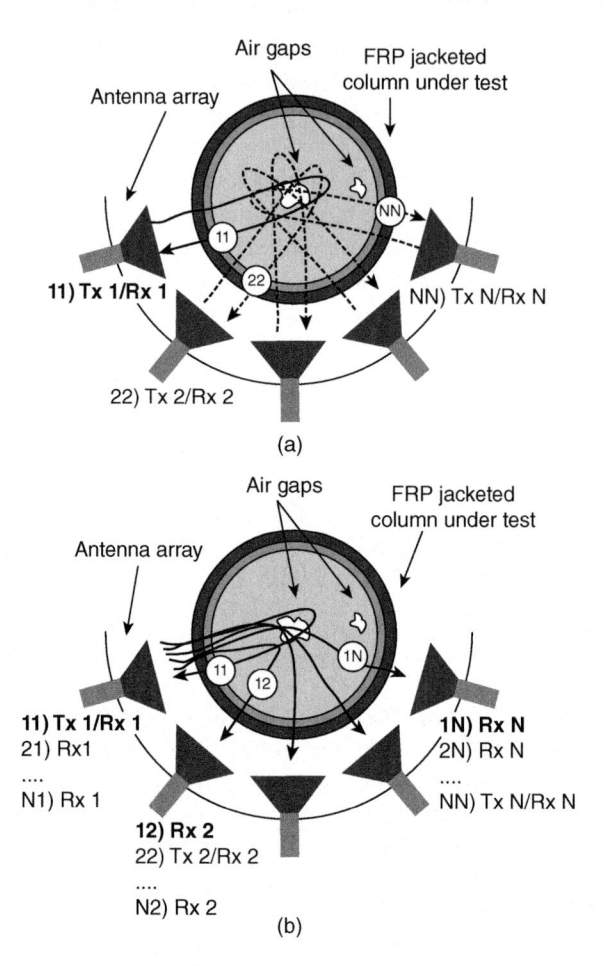

22.13 Scheme of (a) monostatic array and (b) bistatic array of antennas.

22.3.3 Real versus virtual array of antennas

The advantage of using several antennas is that each antenna illuminates the object from a different angle and receives the reflection from different viewpoints, thus retrieving more information of the object. However, the use of a large number of antennas (especially in 3D imaging geometries) can lead to highly complex set-ups, not only for the multiple antennas but also for the electrical circuitry designed to manipulate the signal transmission and reception.

Microwave techniques 593

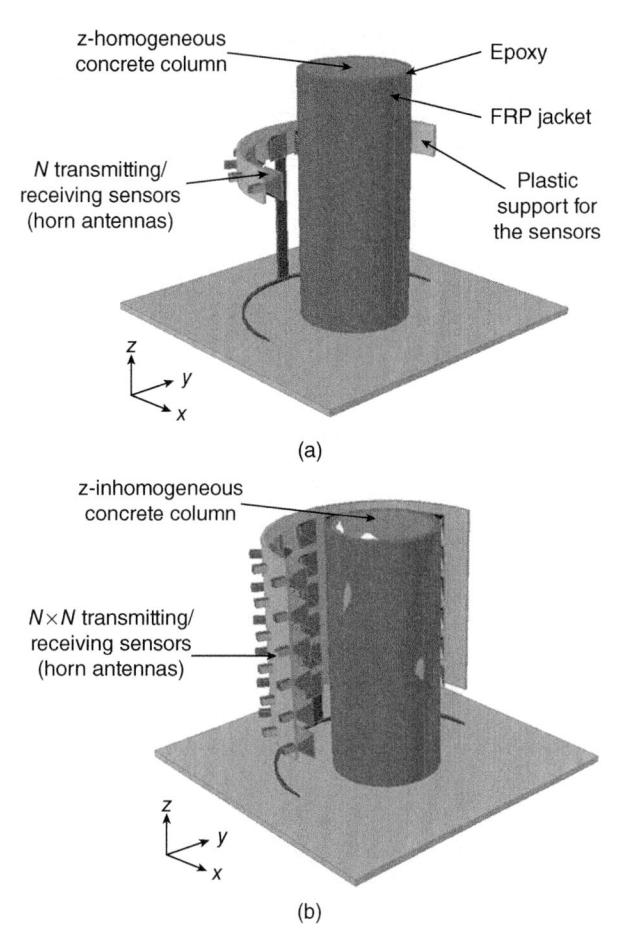

22.14 GFRP-jacketed concrete column placed in the centre of a cylindrical geometry system: (a) 2D imaging geometry, (b) 3D imaging geometry.

In an alternative imaging system, a set of N microwave antennas in a real array (Fig. 22.15a) could be replaced by a single microwave antenna moving through N positions surrounding a steady scene, transmitting and receiving the reflected signal at various points. This moving antenna is referred to as a virtual array (Fig. 22.15b). This kind of virtual array simplifies the measurement set-up in terms of the number of antennas, but requires the use of linear or rotary stages for the automatic movement of the antennas. In some cases, an equivalent set-up with a stationary antenna and a rotary target is used.

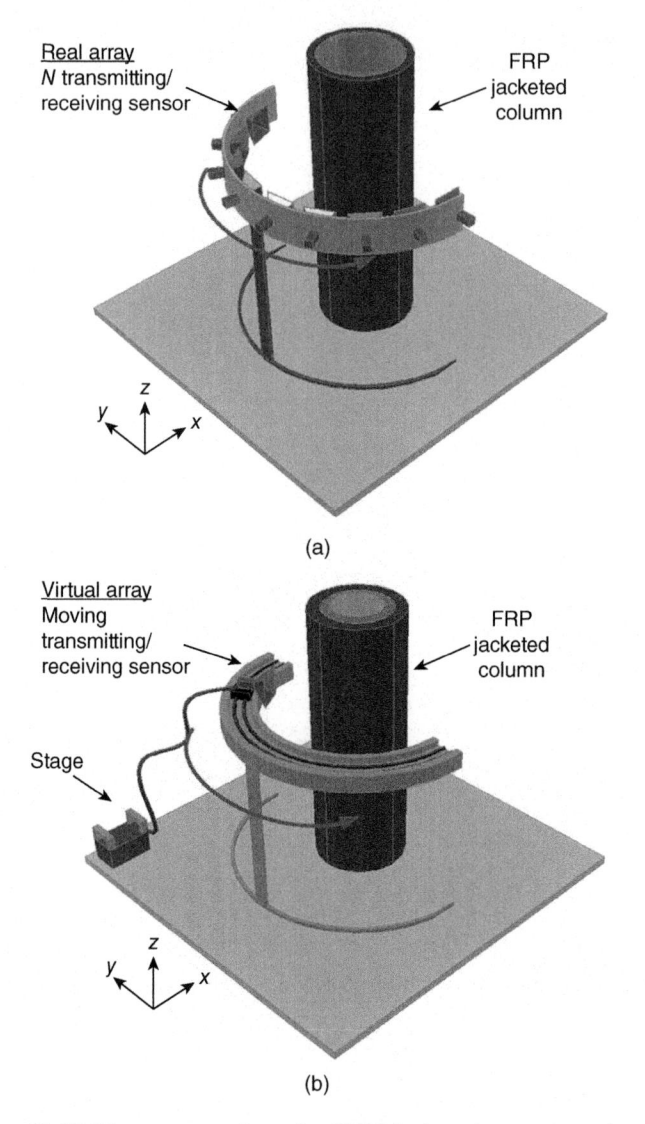

22.15 2D reconstruction of a CFRP-jacketed concrete column using (a) real array and (b) virtual array of antennas.

22.3.4 Surface and sub-surface imaging

The imaging technology can be categorized into surface and sub-surface imaging depending on the penetration capability. In surface imaging, any defect/damage in the structure is detected through analysis of the microwave reflection. This kind of scan allows the detection of the area and the location of such defect/damage. However, accurate information about the depth cannot easily be obtained. Instead, the correlation of the depth of the defect/damage and the measured signal has to be experimentally determined. To overcome this problem, sub-surface imaging, which relies on a numerical focusing procedure, allows the recovery of a 2D tomographic image of an object from its scattered field, so information under the surface can be retrieved.

22.3.5 Microwave antennas for NDE of civil engineering structures

A significant advantage of the microwave-based NDE technology lies in a broad variety of EM principles, based on which a large number of microwave antennas can be designed. Table 22.4 summarizes the antennas and their bases.

22.4 Microwave surface imaging of fiber-reinforced polymer reinforced concrete (FRP RC) structures

This technology is based on the reflections analysis of microwaves sent toward and reflected from layered FRP – adhesive – concrete medium: poor bonding conditions including voids and debonding generate air gaps which produce additional reflections of the microwaves. This study summarizes the work presented in Feng *et al.* (2002), where the authors developed a microwave imaging technology for detecting voids and debonding between the FRP jacket and an RC column.

22.4.1 Overview of the proposed technology

As shown in Fig. 22.16, a section of a RC column structure wrapped with a layer of FRP jacket is illuminated with the incoming wave. Assuming that the jacket is perfectly bonded to the structure without a void or debonding, the first reflection (#1) occurs at the surface of the jacket, while the second (#2) occurs at the interface between the jacket and the adhesive epoxy; the third (#3) at the interface between the adhesive epoxy and the concrete column surface and the fourth (#4) are the reflections coming from the reinforcement steel rebars. If an air gap produced by a void/delamination exists, there will be a fifth reflection (#5) from the air gap between the epoxy and the concrete. Therefore, imperfect bonding conditions can be, in principle, detected by analyzing these reflections in the time domain – related to equations (22.4) and (22.5) – and/or the frequency domain – related to equation (22.6). Fig. 22.17 shows the theoretical reflections expected from measurements on the FRP-jacketed RC structure.

Table 22.4 Brief description of microwave antennas for NDE of civil engineering structure

Antenna	Description	Picture	Frequency	Technical features
Horn	Hollow pipe of different cross-sections tapered to a larger opening.		Broadband (10:1 to 20:1)	Moderate directivity
Horn with dielectric lens	Horn in contact with a lens made of dielectric material aimed to focus the electric field towards the region of interest.		Broadband (10:1 to 20:1)	Free space measurements
Coaxial probe	Extension of a coaxial cable with a circular ground plane. Non-radiating element.		Broadband (9:1)	Good penetrability
Broadband monopole	Planar radiating structure with two metallic strips.		Broadband (4:1) Ex: UWB 3.0 to 10.0 GHz	Linear polarization
Ultra wideband slot	Planar radiating element. Radiation is produced by a hole in a metallic ground plane.		Broadband (3:1)	Mismatching to the medium for conventional horns (unless filled with appropriate matching medium)
RFID tags	Small probes that reflect part of the incident wave encoded with sensing information.	[RFID ALN-9529 Squiggle-SQ Inlay]	Single frequency 868.0 MHz/ 2.4 GHz (ISM in Europe)	Directional (no back radiation)

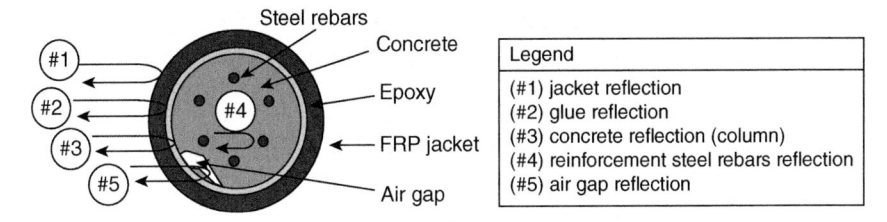

22.16 Horizontal cut of the reflection mechanism in a FRP-jacketed RC structure. With permission from ASCE (Feng *et al.*, 2002).

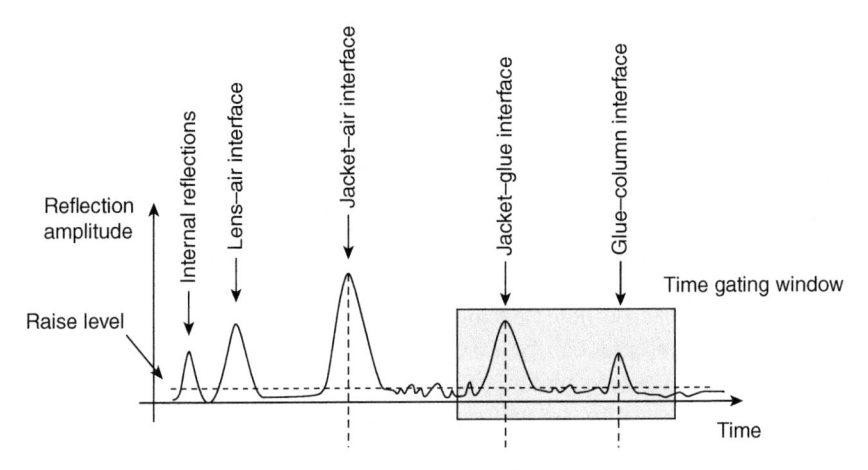

22.17 Time gating of reflection signal. With permission from SPIE (Feng *et al.*, 2000).

In order to remove spurious responses, the *time gating* technique is used. A specified frame of the time signal can be selected to remove unwanted reflections due to unavoidable obstacles different from the air voids, such as antenna internal reflections and antenna to lens reflection. The time gating is illustrated in Fig. 22.17 under the squared section.

22.4.2 The focusing beam technique

When illuminating with an EM plane wave, the differences between a perfect and a poor bonding conditions may not be detectable, since the reflections occurring at the surface of the jacketed column are much stronger than the reflections due to the air gaps. To overcome these difficulties, the use of a focused wave allows the signal into the bonding interface to be focused while defocusing the beam in other regions of no interest, as illustrated in Fig. 22.18. A focused wave can be obtained by using a dielectric

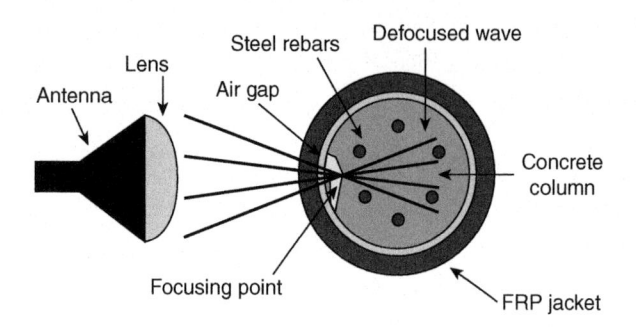

22.18 Horizontal cut of an electromagnetic wave focused on the bonding interface of a FRP-jacketed concrete column. With permission from ASCE (Feng *et al.*, 2002).

lens attached in front of an antenna that generates a plane wave (see 'Horn with dielectric lens' in Table 22.4).

When using focused waves, waves reflected from regions where the beam is defocused are weaker than those reflected from the focused region, and thus the difference between the perfect and poor bonding conditions in the focused region can be detected more effectively. Additionally, to improve sensitivity, the antenna can be placed in such a position that the angle of incidence of the waves makes most of the reflection from the jacketed column not to return to the measuring point.

Figure 22.19 shows the results, in both time and frequency domains, of the finite difference time domain (FDTD) simulation of the reflected microwaves in two GFRP-jacketed columns, with good and poor bonding conditions respectively. In the simulation, the jacketed columns are illuminated with focused microwaves forming a 45° angle with the jacket surface (referred to as 45° set-up). The reflections from the jacketed columns with and without the void are clearly distinguishable.

22.4.3 Experimental implementation

A series of experiments on GFRP-jacketed concrete columns have been conducted to investigate the effectiveness of the surface imaging using focused microwaves. The columns were wrapped with a three-layer glass FRP jacket (0.11 cm thick each). Columns with different bonding conditions were inspected. Poor bonding conditions were created by artificially introducing voids using Styrofoam strips ($\varepsilon_{r\_Styrofoam} \approx \varepsilon_{r_{air}} = 1.00$) between the FRP jackets and between the concrete columns.

Two types of measurement set-ups were tested. Set-up A consisted of a reflection measurement (S_{11}) using one circular lens on a horn antenna, and set-up B consisted of a transmission measurement (S_{21}) using two triangular

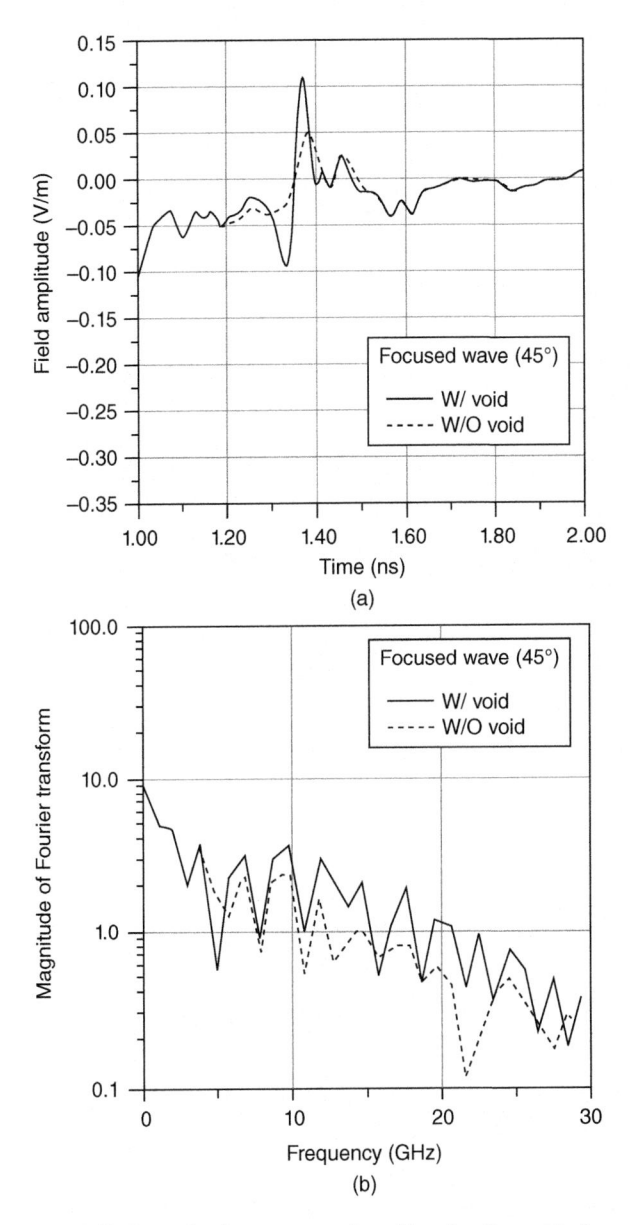

22.19 Reflected microwaves when illuminating with focused wave in 45° set-up: (a) time domain signal, (b) frequency domain signal. With permission from ASCE (Feng *et al.*, 2002).

lenses (one for transmitting and the other for receiving). The optimal angles between the two triangular lenses (40°) in set-up B, and between the wave direction and the surface of the jacketed column in set-up A (45°) were determined through simulations and experiments. Figure 22.20 shows a picture of the evaluation of one of the columns using set-up B with the two triangular lenses placed on a support that can slide up and down. The column was placed on a rotary stage and was rotated around its central axis,

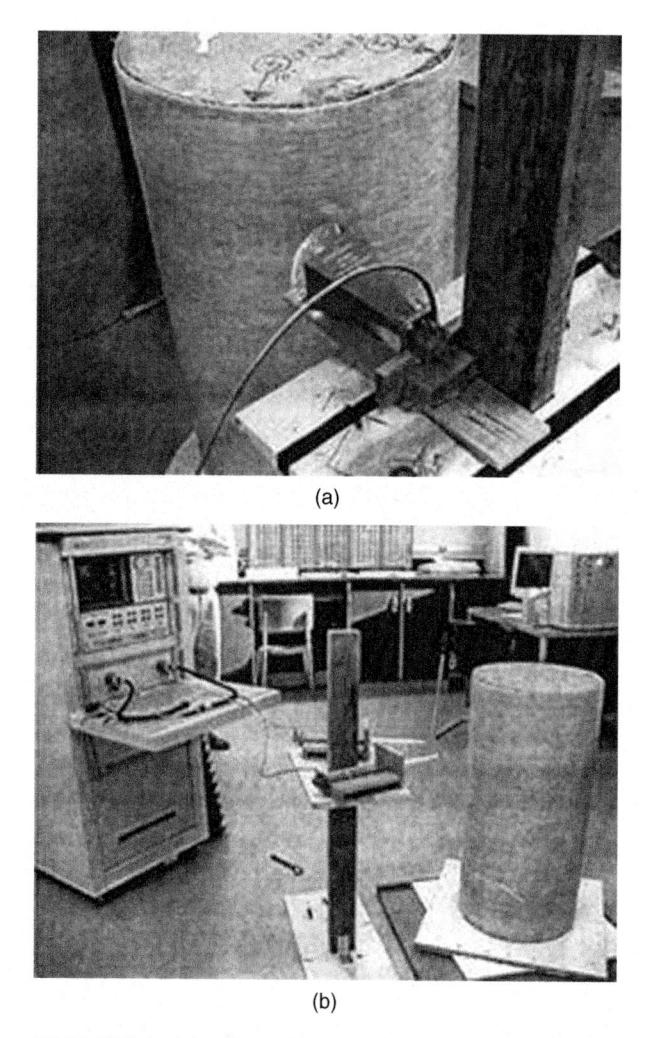

(a)

(b)

22.20 FRP-jacketed concrete column under testing. (a) Set-up A: reflection measurement (S_{11}) using a circular lens. (a) Set-up B: transmission measurement (S_{12}) using two triangular lenses. With permission from ASCE (Feng *et al.*, 2002).

so the entire surface of the jacketed column could be easily scanned. A network analyzer was used to send continuous sinusoidal EM waves sweeping from 8.2 GHz to 12.4 GHz and to receive the reflected wave. Then, the reflected wave was compared with a reference value representing a perfect bonding situation.

Scanned images of the concrete columns jacketed with the GFRP layers using set-ups A and B are respectively shown in Fig. 22.20. The plotted value is the sum of the square of the normalized response measured at each measuring point. The scanned area contained a void between the jacket and the column caused by a hole with a diameter of approximately 2 cm. Although the void couldnot be seen from the outside of the jacket, the scanned image from the S_{21} measurement using two triangular lenses (see set-up B in Fig. 22.21b) clearly shows the presence of an air void resulting from the hole in the concrete surface. The scanned image from S_{11} measurement using one circular lens (see set-up A in Fig. 22.21a) successfully detected the size and the shape of the hole, even with better agreement with the reality than the previous scan retrieved from set-up B. The voids and debonding areas at other locations were also successfully detected.

In Fig. 22.22, EM images of the concrete surface are obtained using the circular lens in reflection configuration (set-up A) at different frequencies ranging from 11.0 to 12.4 GHz. The image produced using the higher-frequency waves has a slightly better resolution.

In the second experiment (Feng et al., 2006), an evaluation test of the microwave NDE technology was performed in the field. The field evaluation took place on March 2, 2006, on an FRP-wrapped concrete girder of the Dang-Jeong Overcrossing located at Gun-Po, Korea. Eight years ago,

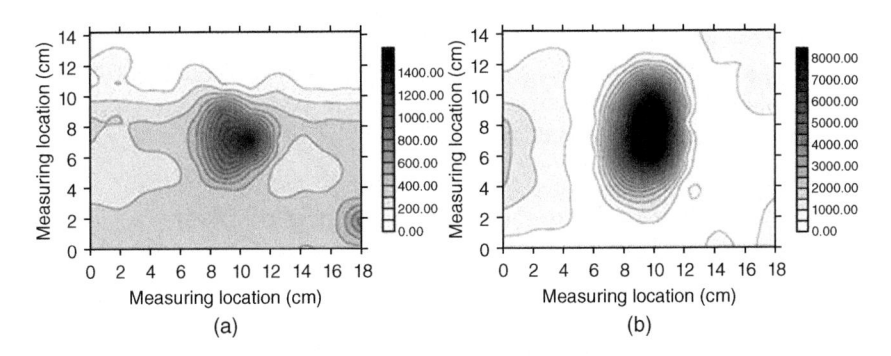

22.21 Scanned images of the surface of a FRP-jacketed concrete column without rebars and with a 2 cm diameter hole in its surface (Kim et al., 2001). (a) Set-up A: reflection measurement (S_{11}) using a circular lens. (b) Set-up B: transmission measurement (S_{21}) using two triangular lenses.

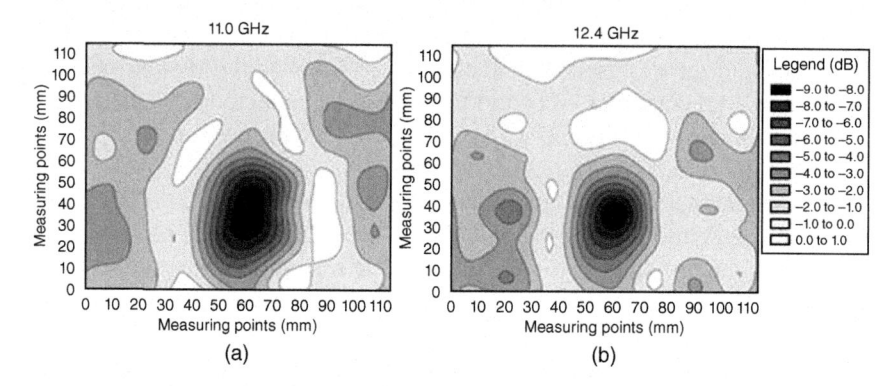

22.22 Scanned images of the surface of a GFRP-jacketed concrete column without rebars and with a 2 cm diameter hole in its surface: (a) reflection measurement (S_{11}) using two circular lenses at 11.0 GHz; (b) reflection measurement (S_{11}) using two circular lenses at 12.4 GHz. With permission from SPIE (Feng et al., 2000).

the bridge girder was retrofitted by Conclinic Co. Ltd. with two layers of E-glass FRP sheets and external post-tensioning as shown in Fig. 22.23. First, GAP-CAT-1100 (the device with light indicator) was used to pre-scan a large FRP-wrapped girder surface. Although majority of the area showed no sign of debonds (light areas), two debonding spots (dark areas) were identified. Then, more detailed inspection was performed by GAP-CAT-1200, which is capable of providing a real-time image of the scanned area. Fig. 22.23(b) shows the two real-time images revealing the details of the debonding areas. This field evaluation proved the ease of use and the effectiveness of the microwave NDE technology.

While this surface imaging technique is good at assessing the bonding condition in the interface between the column and the jacket, further analysis is required to determine the depth of such holes, or the presence of additional voids inside the structure. Thus, a sub-surface imaging technique is introduced in the next section.

22.5 Microwave sub-surface imaging of fiber-reinforced polymer reinforced concrete (FRP RC) structures

In the last three decades, microwaves have been exploited for the development of imaging algorithms able to produce readable images of the changes of some internal parameters of the scenario to be studied. Much effort has been put into developing robust tomographic approaches able to reconstruct harsh scenarios with inhomogeneous lossy media surrounding the

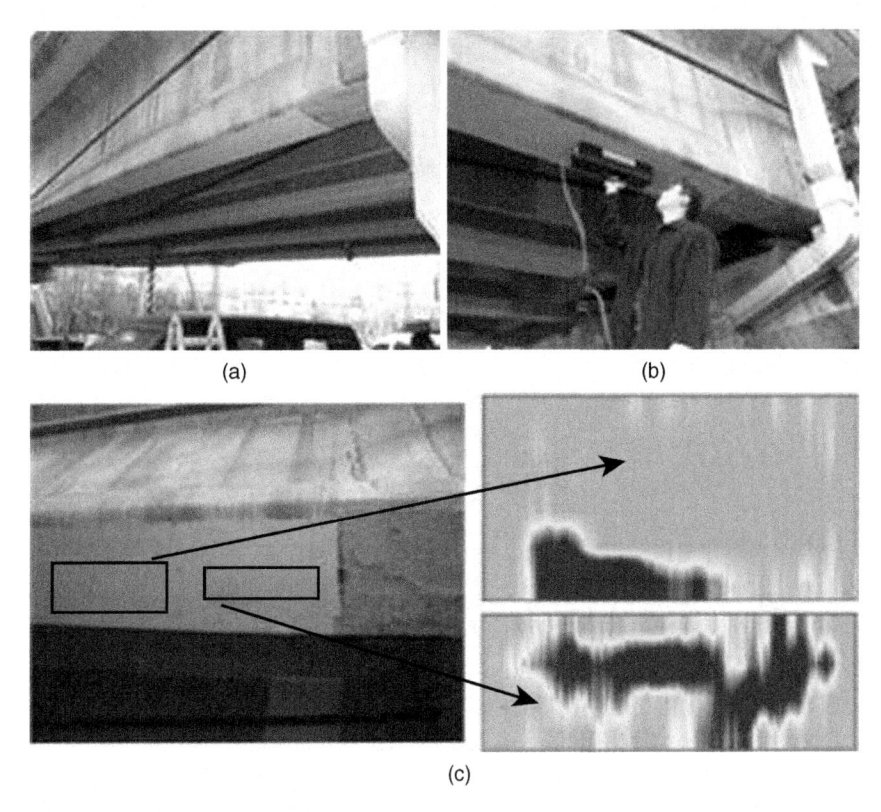

22.23 Field tests at Dang-Jeong overcrossing (Feng *et al.*, 2006): (a, b) inspection of FRP-wrapped concrete girders and; (c) scanned microwave images.

target, especially in the biomedical field (Guardiola *et al.*, 2009), and there is a growing trend to export these algorithms to other fields of applications such as civil engineering (Guardiola *et al.*, 2011a).

Tomographic techniques use different forms of back-projection algorithms and tend to be frequency domain. Wideband signals, such as those produced with ultra-wideband (UWB) systems and, in particular, the recent 3.1–10.6 GHz band, offer new possibilities to increase spatial resolution and material electrical parameter measurement accuracy.

22.5.1 UWB tomographic bi-focusing algorithm

The general idea for UWB tomographic imaging is based on the distribution of a given number of antennas (transmitters and/or receivers) on a certain region, surrounding as much as possible the object under test (herein

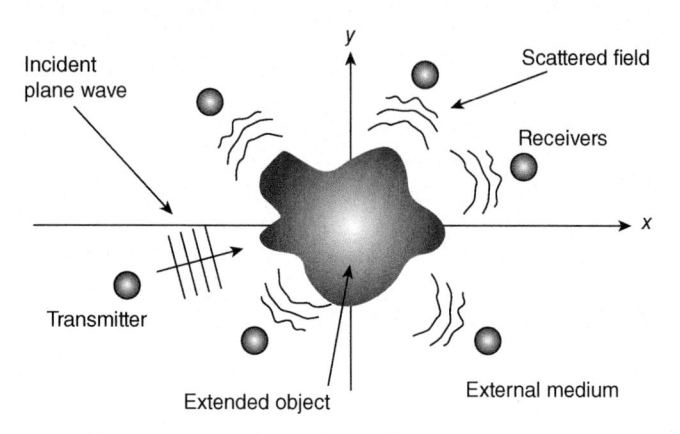

22.24 Generic scenario of a sensor network for imaging reconstruction of an extended object.

referred to as extended object), as shown in Fig. 22.24. When one of these antennas transmits a signal towards the area being investigated, this signal interacts with the obstacle and the result of this interaction is captured by the other antennas. The electrical field re-radiated by the obstacles is referred to as scattered field and gathers the effect that obstacles or inhomogeneities have on an incident wave. Hence, the interaction of the illuminating signal with the object under test provides spatial and electrical information of the object under investigation.

Scattered field analytical formulation

Considering that most of the FRP RC structures can be modeled as canonical cylindrical objects, a 2D circular imaging geometry is proposed. The presented formulation and images correspond to 2D geometry (as shown in Fig. 22.25) by supposing that the scenario is invariant in the z direction and the currents and fields are parallel to the z axis, allowing scalar formulation. This formulation can be easily extended to a 3D case using the appropriate Green function (Guardiola *et al.*, 2011a) which is a type of function used to solve inhomogeneous differential equations subject to specific initial conditions or boundary conditions. Further explanation about the Green's functions can be found in (Balanis, 1989).

The illumination of an object by the field $E_T(\vec{r}, f; \vec{r}_{t_i})$ generated by a transmitter located at position $\vec{r}_{t_i}$ at a frequency f induces an equivalent electric current distribution, $J_{eq}(\vec{r}, f; \vec{r}_{t_i})$, proportional to the electric contrast, $C(\vec{r})$ between the external medium of permittivity ε_{re} and the object internal permittivity ε_{ri}.

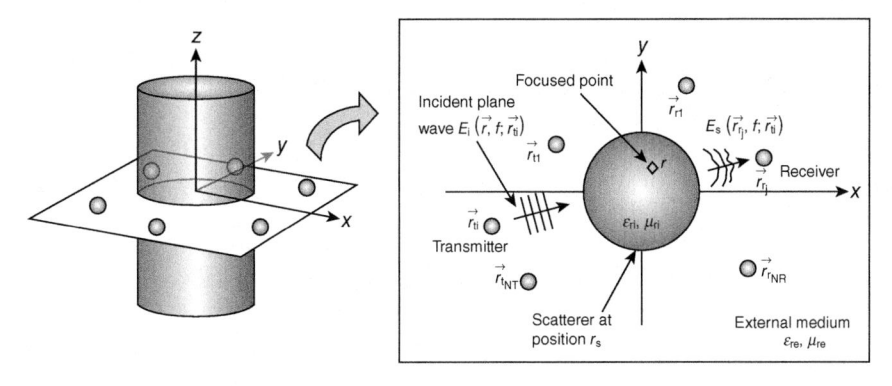

22.25 2D sensing geometry for the reconstruction of an object of permittivity ε_{ri} immersed in an external medium of permittivity ε_{re}.

$$C(\vec{r}) = 1 - \frac{\varepsilon_{ri}}{\varepsilon_{re}} \quad [22.7]$$

$$J_{eq}(\vec{r}, f; \vec{r}_{t_i}) = j\omega\varepsilon_{re}C(\vec{r})E_T(\vec{r}, f; \vec{r}_{t_i}) \quad [22.8]$$

Then the scattered field measured at a receiver positioned at $\vec{r}_{r_j}$ created by the equivalent current $J_{eq}(\vec{r}, f; \vec{r}_{t_i})$ induced on an imaginary scatter may be expressed as:

$$E_s(\vec{r}_{r_j}, f; \vec{r}_{t_i}) = j\omega\mu_0 \iint J_{eq}(\vec{r}, f; \vec{r}_{t_i}) G(|\vec{r}_{r_j} - \vec{r}|, f) dS \quad [22.9]$$

where $G(|\vec{r}_{r_j} - \vec{r}|, f) = \frac{1}{4j} H_0^{(2)}(k|\vec{r}_{r_j} - \vec{r}|)$ is the Green function for a 2D geometry. $H_0^{(2)}(k|\vec{r}_{r_j} - \vec{r}_s|)$ is the second kind zero-order Hankel function and $k = 2\pi f/\lambda$ is the wave number of the background reference medium.

The total electric field $E_T(\vec{r}_{r_j}, f; \vec{r}_{t_i})$ measured at the receiving position $\vec{r}_{r_j}$ can be related to the scattered field $E_s(\vec{r}_{r_j}, f; \vec{r}_{t_i})$ and to the incident field $E_i(\vec{r}_{r_j}, f; \vec{r}_{t_i})$ (incident electric field in absence of the object) through

$$E_T(\vec{r}_{r_j}, f; \vec{r}_{t_i}) = E_s(\vec{r}_{r_j}, f; \vec{r}_{t_i}) + E_i(\vec{r}_{r_j}, f; \vec{r}_{t_i}) \quad [22.10]$$

Under Born approximation, which supposes low scattering objects, the total electric field is equal to the incident field and the equivalent current can be replaced by the product of the contrast and the incident field. Then, the reconstructed contrast at the generic focusing point $\vec{r}_f$ can be expressed as

$$\tilde{C}(\vec{r}_{\mathrm{f}}) = A \sum_{i=1}^{N_{\mathrm{r}}} \sum_{j=1}^{N_{\mathrm{t}}} \frac{E_{\mathrm{s}}(\vec{r}_{\mathrm{r}_j}, f; \vec{r}_{\mathrm{t}_i})}{(2\pi f)^2} \frac{1}{G(|\vec{r}_{\mathrm{t}_i} - \vec{r}_{\mathrm{f}}|, f)} \frac{1}{G(|\vec{r}_{\mathrm{r}_j} - \vec{r}_{\mathrm{f}}|, f)} \qquad [22.11]$$

where A is a complex constant that models multiple system factors (illuminating field, gain of antennas, etc.) and can be removed by appropriate calibration.

Equation (22.11) represents the reconstruction algorithm in which every image point of the local electrical properties of the object is formed by means of the synthesis of two focused groups (bi-focusing) of antennas (transmitters and receivers). When evaluating this equation at a given object point, the received scattered fields resulting from all the transmitter/receiver combinations are numerically weighted by a focusing operator that restores the amplitude and phase changes suffered by a wave due to the scattering points in the object. By applying this bi-focusing operator to every single point of the image space grid, a replica of the extended object can be obtained.

UWB frequency combination

When highly inhomogeneous objects are inspected, multiple high-contrast scattering occurs resulting in frequency-dependent non-linear phenomena. A continuous frequency superposition is proposed in this formulation in order to smooth and reduce these effects. UWB signals are proposed as illuminating signals since they are able to provide enhanced image resolution and more clutter rejection when compared with mono-frequency reconstructions, and hence to obtain an image with fewer artifacts that could be interpreted as false positive detections. In this context, range resolution (inside the object, following the radius) responds to the expression given by:

$$\Delta r = \frac{c}{\sqrt{\varepsilon'_{ri}}} \frac{1}{B} \qquad [22.12]$$

where $B = f_{\max} - f_{\min}$ is the frequency bandwidth.

However, when combining reconstructions at different frequencies, the algorithm encounters frequency-dependent propagation effects that may corrupt the reconstruction by introducing undesired artifacts. Hence, the combination of such frequencies needs to be carefully chosen to ensure an improvement in the quality of the final image.

On one hand, the total contrast can be obtained as a coherent addition of the electrical contrast at every frequency. Then, the frequency averaged contrast (over a frequency range between $f_{\max}$ and $f_{\min}$) can be expressed as

$$\tilde{C}(\vec{r}_{\mathrm{f}}) = A \sum_{f_{\min}}^{f_{\max}} \sum_{i=1}^{N_{\mathrm{r}}} \sum_{j=1}^{N_{\mathrm{t}}} \frac{E_{\mathrm{s}}(\vec{r}_{\mathrm{r}_j}, f; \vec{r}_{\mathrm{t}_i})}{(2\pi f)^2} \frac{1}{G(|\vec{r}_{\mathrm{t}_i} - \vec{r}_{\mathrm{f}}|, f)} \frac{1}{G(|\vec{r}_{\mathrm{r}_j} - \vec{r}_{\mathrm{f}}|, f)} \qquad [22.13]$$

By performing this coherent combination, physical meaning of the images is preserved and therefore, dielectric information of the final image can be obtained.

When Born approximation is not fulfilled, frequency selective residual phase errors appear. Then, the coherent frequency summation may produce artefacts in the final image. To deal with these phase errors, an incoherent frequency combination is proposed in Guardiola *et al.* (2009). When using this incoherent approach, only the magnitude of the contrast profile at each frequency is combined and only the size of the objects is preserved.

Space and frequency sampling criteria

In order to obtain accurate spatial and electrical information of the electrically extended object using a network of antennas encircling a target, the minimum number of views and the frequency sampling must be accurately chosen.

On the one hand, the minimum number of views that guarantees an image resolution of $\lambda_{\mathrm{e}}/2$ (being λ_{e} the wavelength in the medium at the highest operating frequency inside the UWB interval) and free of angular aliasing, is equal to (Harrington, 1961):

$$N_\phi \geq 2 \frac{2\pi}{\lambda_{\mathrm{e}}} a = 2 \left(\frac{2\pi f_{\max} \sqrt{\varepsilon'_{\mathrm{re}}}}{c} \right) a \qquad [22.14]$$

where $\varepsilon_{\mathrm{re}}$ is the effective permittivity of the external medium, a is the object-encircling radius, and $c = 3 \times 10^8$ m/s is the speed of light in free space. So the maximum angular step can be written as

$$\Delta\phi = 2\pi / N_\phi = \lambda_{\mathrm{e}} / (2a) \qquad [22.15]$$

On the other hand, the frequency domain sampling Δf must ensure that the image is free of distortion and aliasing into the radial direction. To that extent,

$$\Delta f \leq \frac{c}{4a\sqrt{\varepsilon'_{\mathrm{re}}}} \qquad [22.16]$$

Then, given a certain frequency bandwidth of $B = f_{\max} - f_{\min}$ the number of frequency steps must be

$$N_{\mathrm{f}} = \frac{B}{\Delta f} = \frac{f_{\max} - f_{\min}}{\Delta f} \qquad [22.17]$$

22.5.2 Numerical simulations

The microwave sub-surface imaging formulation presented above is numerically tested in this section through multiple antenna configurations for its capability to detect a defect (air void) in a concrete column wrapped with GFRP-jacket. The geometry of the object under test is shown in Fig. 22.26. A concrete column with a diameter $d_{s\_1} = 25.00$ cm and relative permittivity $\varepsilon_{ri\_1} = 5.30$ is wrapped with an adhesive layer of epoxy of thickness $t_{s\_2} = 0.11$ cm and relative permittivity $\varepsilon_{ri\_2} = 3.10$, and a GFRP jacket of thickness $t_{s\_3} = 0.40$ cm and relative permittivity $\varepsilon_{ri\_3} = 3.80$. Concrete is assumed to be homogeneous and the effects from aggregate size are not considered in this theoretical model. A cylindrical air void of diameter $d_{o\_1} = 2.00$ cm and relative permittivity $\varepsilon_{o\_1} = 1.00$ is placed at position $\vec{r}_{o\_1} = (x_{o\_1} = 7.00$ cm, $y_{o\_1} = 7.00$ cm). Three steel rebars of diameter $d_{o\_2} = d_{o\_3} = d_{o\_4} = 1.2$ cm are placed at positions $\vec{r}_{o\_2} = (x_{o\_2} = -3.10$ cm, $y_{o\_2} = 5.10$ cm), $\vec{r}_{o\_3} = (x_{o\_3} = 3.10$, $y_{o\_3} = 5.40$) and $\vec{r}_{o\_4} = (x_{o4} = 6.20$ cm, $y_{o4} = 0.00$ cm) respectively. Steel rebars are considered to be perfect electric conductors (PEC). The antennas are placed in a matching medium with permittivity ε_{re} equal to concrete permittivity in order to minimize interface reflections. In order to apply the 2D formulation explained in the previous section, the scenario is considered to be homogeneous in the z-direction so the cylindrical geometry can be collapsed into the 2D circular geometry shown in Fig. 22.26.

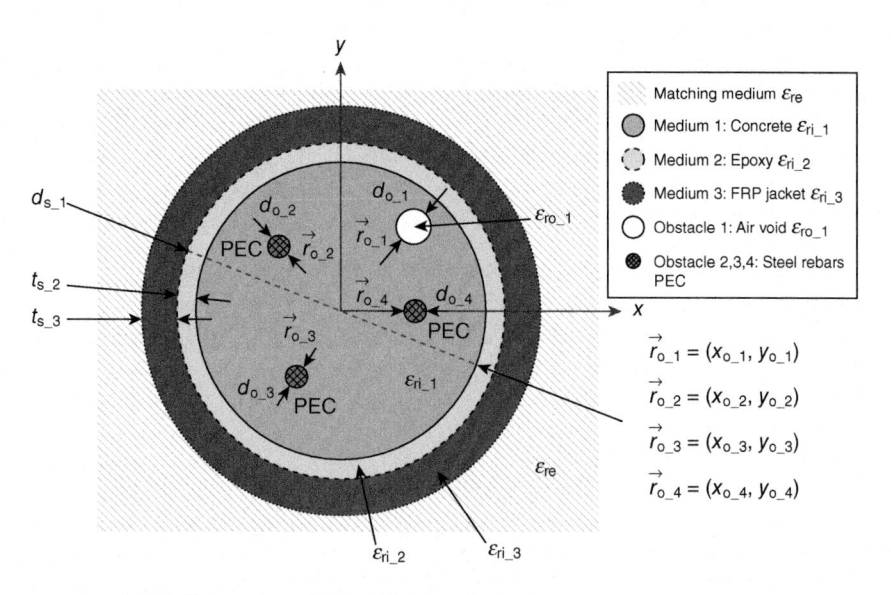

22.26 Geometry of the object under test.

The frequency bandwidth B for the simulations is chosen from 3.0 to 10.0 GHz. Concrete diameter d_{s_1} approximately corresponds to $20\lambda_e$ at $f_{max} = 10.0$ GHz. In terms of concrete column size, a more realistic scenario could be simulated at a lower frequency range, while keeping the electrical dimensions, at the expense of higher computational cost. Considering that the object encircling radius $a = \frac{d_{s\_1}}{2} + t_{s\_1} + t_{s\_2} = 13.00$ cm, the space and frequency sampling are chosen to be $N_\phi = 128$ and $N_f = 28$, according to equations (22.14) and (22.17) respectively.

Different antenna distributions (see Fig. 22.27) and UWB combinations are considered in order to test the resolution limits of the formulation. In the fully encircling geometry (Fig. 22.27a), the array of N_n transmitting elements is placed at the same position of the N_m receiving elements. In this way, a multiview set-up is considered. In the half encircling geometry (Fig. 22.27b), the transmitters are placed at one half of the scenario perimeter and the receivers are place at the other half. In the partially encircling geometry (Fig. 22.27c), the array of N_n transmitting elements is placed contiguous to an array of N_m receiving elements, with the first and the last antenna forming a 90° angle between their position vectors to the centre of the scenario. In this way, a semi-multiview set-up is considered. Four cases of analysis (A, B, C, and D) are described in Table 22.5. In all the cases, a lossless scenario is considered.

Without loss of generality in each case, an arrangement consisting of two arrays of antennas antennas, one for transmitting and the other for receiving signals, and a numerical focusing operator, is applied to the external signals, both in transmitting and receiving fields, in order to recover a 2D image from its scattered field. A measurement matrix of $n\, N_n \times N_m$ elements can be obtained as follows: for every selected transmitting element, the

Table 22.5 Description of the set-up used for the simulations in each case

Case		Geometry	f	N_ϕ N_m (Tx)	N_n (Rx)	N_f
A	1,2,3	Fully encircling	3 GHz, 5 GHz, 10 GHz	128	128	–
	4	Fully encircling	UWB: 3 GHz–10 GHz	128	128	28
B	1,2,3	Fully encircling	3 GHz, 5 GHz, 10 GHz	64	64	–
	4	Fully encircling	UWB: 3–10 GHz	64	64	28
C	1,2,3	Half encircling	3 GHz, 5 GHz, 10 GHz	32	32	–
	4	Half encircling	UWB: 3–10 GHz	32	32	28
D	1,2,3	Partially encircling	3 GHz, 5 GHz, 10 GHz	32	32	–
	4	Partially encircling	UWB: 3–10 GHz	32	32	28

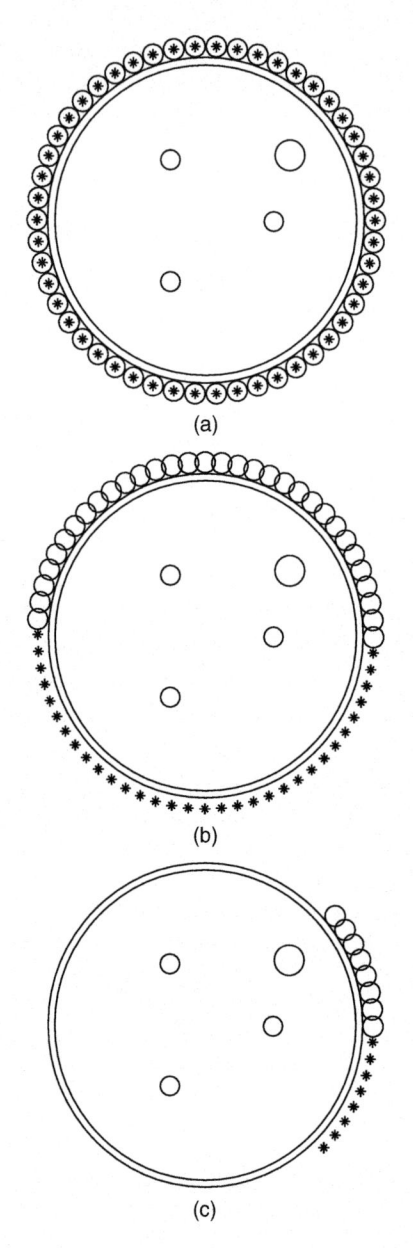

22.27 Simulation geometries where transmitting and receiving antennas are point-like sources represented by the symbol 'O' and '*' respectively: (a) fully encircling geometry, (b) half encircling geometry, and (c) partially encircling geometry.

Microwave techniques 611

receiving array is scanned obtaining a measurement column of N_m elements, then the procedure is repeated for the N_n elements of the transmitting array.

Simulations of the different scenarios have been conducted in order to verify the resolution capabilities of the system. The mono-frequency reconstruction at three different frequencies (f_{low} = 3.0 GHz, f_{medium} = 5.0 GHz and f_{high} = 10.0 GHz) is compared to the UWB combination in each case. The results are shown in Fig. 22.28.

In general terms, it can be observed that the UWB combination provides better results than mono-frequency reconstructions, which are full of interferences. The GFRP layer appears invisible at lower frequencies (f_{low}) and the targets are well located. However, low resolution at these frequencies falsifies the size of the targets so they appear bigger than they really are. Higher frequencies (f_{high}) provide better resolution capabilities; however,

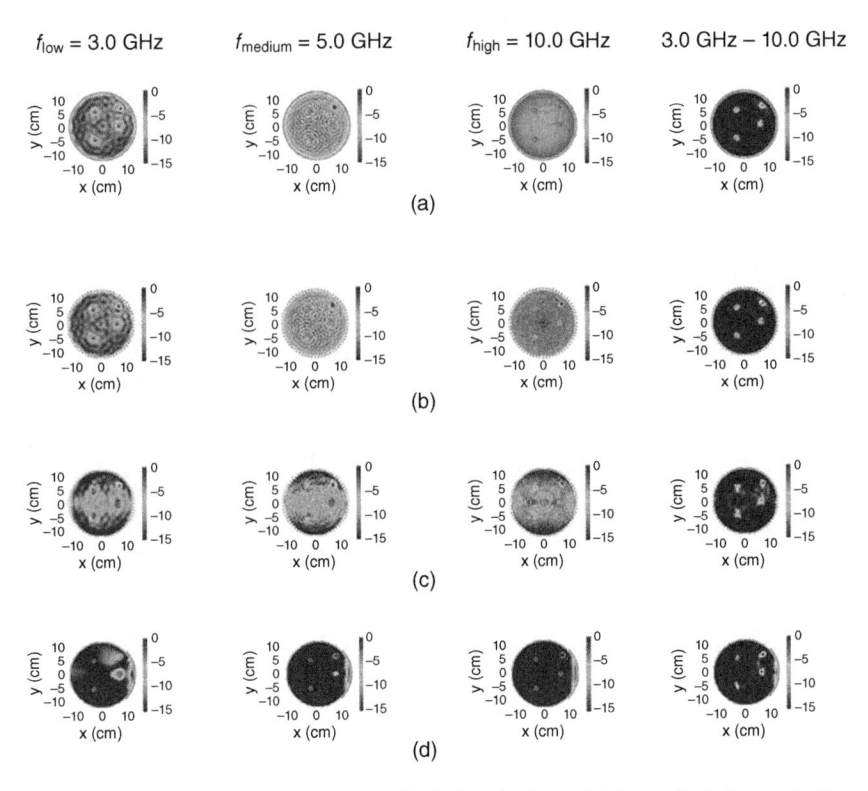

22.28 Results of the numerical simulations: (a) Case A, fully encircling geometry $N_n = N_m = 128$; (b) Case B, fully encircling geometry, $N_n = N_m = 64$; (c) Case C, half encircling geometry, $N_n = N_m = 32$; (d) Case D, partially encircling geometry, $N_n = N_m = 32$.

visibility of the FRP layer increases, which hinders the location of the targets.

The comparison between the two fully encircling geometries (Fig. 22.28a and b) demonstrates that the UWB combination allows a reduction of the number of viewpoints still providing good results. For example, results in Fig. 22.28b with 64 viewpoints are similar to results in Fig. 22.28b with 128 viewpoints. This clearly demonstrates the set-up simplification potential of the UWB combination.

Partially encircling geometries (Fig. 22.28c and d) have less resolution than the fully encircling geometry due to the reduced number of viewpoints. However, good performance of the reconstruction algorithm in Fig. 22.28d shows that most of the information is concentrated at the interactions between nearby antennas.

The results suggest the reconstruction algorithm works well in real conditions of extended damage areas. The system is able to achieve a resolution in the order of the wavelength in the dielectric medium ($\lambda_e \approx 1.00$ cm) thanks to the use of bi-focusing (focusing both the transmitting and receiving arrays) and the UWB combination. Additionally, the UWB combination allows a simplification both on the number the distribution of the antennas, still preserving the resolution capabilities of the system.

22.5.3 Experimental implementation

The effectiveness of the proposed sub-surface imaging technology using an antenna array was investigated through an experiment on two concrete blocks with a dimension of 30.00 cm × 30.00 cm × 30.00 cm. The first concrete block was used for calibration purposes. In the second concrete block, an artificial void using a rectangular Styrofoam bar of 5.00 cm × 2.00 cm × 2.00 cm was introduced during the pouring of concrete, with a distance of 3 cm from the face of concrete to the face of Styrofoam.

A continuous sinusoidal EM wave at 5.2 GHz was generated from a network analyzer and sent to test the specimen. Transmitting and receiving arrays of eight elements each acted alternately using a switch box and the calibrated transmission measurement of each transmitting and receiving array pair was assembled into a measurement matrix. The reconstructed image of a central cut is plotted in Fig. 22.29 by the amplitude of electric current distribution, J_{eq}.

Results demonstrate that the system was able to detect the air void inside the concrete block. The image of the square Styrofoam was successfully reconstructed in terms of size and location. The resolution, which is 2.50 cm at 5.2 GHz, could be improved by using a higher frequency, as demonstrated in numerical simulations of Section 22.5.

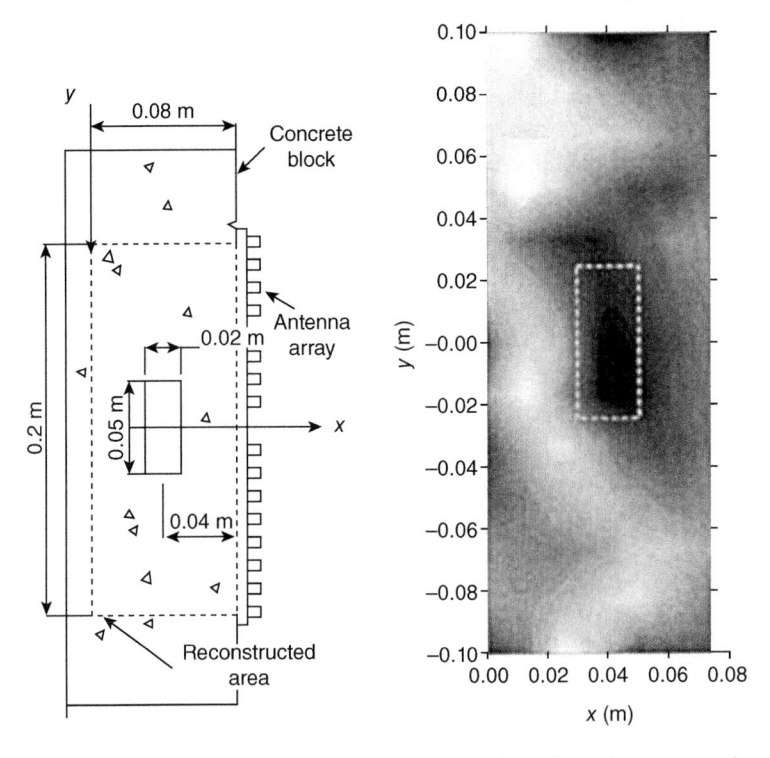

22.29 Experimental results of the microwave imaging of a rectangular void at (x = 4.00 cm, y = 0.00 cm) inside a concrete block. ≈ [2003] IEEE. Reprinted, with permission, from Kim *et al.* (2003).

22.6 Future trends

Improvements of the proposed imaging technology can be associated with several aspects. On the one hand, advanced illumination techniques such as ultra-wide band (UWB) combination can be used to improve the resolution of the reconstructed images. On the other hand, advanced processing techniques such as multiple-input-multiple-output (MIMO) processing can be used to improve the communication performance in the sense that it achieves higher spectral efficiency (more bits per second/Hz) and link diversity, thus reducing fading attenuation in the propagation. Finally advanced electronic systems such as radio frequency identification (RFID), micro-electro mechanical systems (MEMS) or CMOS may convert the current imaging technology into highly efficient, self-powered independent and automatic technology.

A future line of investigation is then related to the use of RFID tags to improve the resolution of the imaging system. As pointed out by Guardiola *et al.* (2011a), when embedding some of the sensors into the structure to be reconstructed, improved signal-to-noise ratio (SNRs) and simplified and cost-efficient imaging set-ups could be achieved while still providing readable images of the structure. This preliminary study reveals that the use of RFID tags embedded in FRP retrofitted concrete structures could be very useful to improve resolution in the images for the NDE inspection.

As a future step, the authors suggest the study of more complex structures such as real reinforced concrete structures jacketed using FRP, combined with the addition of RFID tags, to increase the sensitivity and resolution of the imaging system.

22.7 Sources of further information and advice

The interested reader can find more information about non-destructive testing in the book *Introduction to Nondestructive Testing: A Training Guide*, Second Edition, by Paul E. Mix, Ed. Wiley. Further information on antenna theory can be found in the book *Antenna Theory* by Balanis, Ed. Wiley.

22.8 References and further reading

Awan, E.H. and A. Shamin. 'Dielectric Properties of Novel Polyester Composite Material.' Journal of Natural Sciences and Mathematics (JNSMAC) PK ISSN 0022-2941; CODEN, 2007: Vol. 47, nos. 1&2, pp. 25–32.

Balanis, C.A. *Advanced Engineering Electromagnetics*. United States of America: John Wiley & Sons, New York, 1989.

Belrhiti, M.D., S. Bri, A. Nakheli, M. Haddad, and A. Mamouni. 'Complex permittivity measurement for dielectric materials at microwave frequencies using a rectangular waveguide.' *European Journal of Scientific Research*, 2011: Vol. 49, no. 2, pp. 234–248.

Born, M. and E. Wolf. *Principles of Optics*. New York: Pergamon, 1964.

Broquetas, A., J. Romeu, J.M. Rius, A. Elias-Fuste, A. Cardama, and L. Jofre. 'Cylindrical geometry: a further step in active microwave tomography.' *IEEE Transactions on Microwave Theory and Techniques*, 1991: Vol. 39, no. 5, pp. 836–844.

Capdevila, S., L. Jofre, A. Cardama, S. Blanch, J. Romeu, and G. Roqueta. 'Gap fillers coverage prediction at S frequency band.' XXI Simposium Nacional de la Unión Científica Internacional de Radio. Oviedo, Spain, 2006: 126–129.

Capdevila, S., L. Jofre, J.C. Bolomey, and J. Romeu. 'RFID Multiprobe Impedance-Based Sensors.' *IEEE Transactions on Instrumentation and Measurement*, 2010: Vol. 59, no. 12, pp. 3093–3101.

Fear, E.C., S.C. Hagness, P.M. Meaney, M. Okoniewski, and M.A. Stuchly. 'Enhancing breast tumor detection with near-field imaging.' *Microwave Magazine, IEEE*, 2002: vol. 3, no. 1, pp. 48–56.

Feng, M., F. De Flaviis, Y.J. Kim, and R. Diaz. 'Application of electromagnetic waves in damage detection of concrete structures.' Smart Structures and Materials 2000: Smart Systems for Bridges, Structures, and Highways. Newport Beach, CA, USA: SPIE, 2000. *Proc. SPIE 3988*, 2000: Vol. 118; doi:10.1117/12.383132.

Feng, Q.M., F. De Flaviis, and Y.J. Kim. 'Use of microwaves for damage detection of fiber reinforced polymer-wrapped concrete structures.' *Journal of Engineering Mechanics*, 2002: Vol. 128, no. 2, pp. 172–183.

Feng, M., Y.J. Kim, and K. Park. 'Real-time and hand-held microwave NDE technology.' *FRP International – The Official Newsletter of the International Institute for FRP in Construction*, 2006: Vol. 3, no. 2, pp. 2–5.

Ghodgaonkar, D.K., V.V. Varadan, and V.K. Varadan. 'A free-space method for measurement of dielectric constants and loss tangents at microwave frequencies.' *IEEE Transactions on Instrumentation and Measurement*, 1989: Vol. 38, pp. 789–793.

Guardiola, M., S. Capdevila, and L. Jofre. 'UWB bifosuing tomography for breast tumor detection.' *European Conference on Antennas and Propagation*. Berlin, 2009: pp. 1855–1859.

Guardiola, M., G. Roqueta, and L. Jofre. 'Semi-embedded RFID sensor networks for imaging applications.' 5th International Conference on Electromagnetic Near-field (NF) Characterization and Imaging (ICONIC 2011). Rouen (France), 2011a.

Guardiola, M., L. Jofre, S. Capdevila, S. Blanch, and J. Romeu. '3D UWB magnitude-combined tomographic imaging for biomedical applications. Algorithm validation.' *Radioengineering*, 2011b: Vol. 20, no. 2, pp. 366–372.

Haroun, M.A., and M.Q. Feng. 'Lap slice and shear enhancements in composite-jacketed bridge columns.' *Proceedings of the 3rd US–Japan Bridge Workshop*, Tsububa, Japan, 1997.

Harrington, R. *Time-harmonic electromagnetic fields*. New York: McGraw-Hill, 1961.

Johnson, E.C., J.P. Nokes, and G.F. Hawkins. 'NDE of composite seismic retrofits to bridges.' *Proceedings of the 9th International Symposium on Nondestructive Characterization of Materials*. Sydney, Australia, 1999.

Kharkovsky, S.N., M.F. Akay, U.C. Hasar, and D.A. Cengiz. 'Measurement and monitoring of microwave reflection and transmission properties of cement-based specimens.' *IEEE Transactions on Instrumentation and Measurement*, 2002: Vol. 51, no. 6, pp. 1210–1218.

Kim, Y.J., F. De Flaviis, L. Jofre, and M.Q. Feng. 'Microwave-Based NDE of FRP-Jacketed Concrete Structures.' *Proc. 46th Int. SAMPE Symp.*. Long Beach, 2001: Vol. 46.

Kim, Y.J., L. Jofre, F. De Flaviis, and M.Q. Feng. 'Microwave reflection tomographic array for damage detection of civil structures.' *IEEE Transactions on Antennas and Propagation*, 2003: Vol. 51, no. 11, pp. 3022–3032.

Klemm, M., I.J. Craddock, J.A. Leenderitz, A. Preece, and R. Benajmin. 'Radar-based breast cancer detection using a hemispherical antenna array. experimental results.' *IEEE Transactions on Antennas and Propagation*, 2009: Vol. 57, pp. 162–170.

Komarov, V.V., and V.V. Yakovlev. 'Modeling control over determination of dielectric properties by perturbation technique.' *Microwave Optice Technology Letters*, 2003: Vol. 39N(6), pp. 443–446.

Lister, M., J. Dovey, S. Giddings, I. Grant, and K. Kelly. *New Media: A Critical Introduction*. London: Routledge, 2003.

Maser, K.R. *New Technology for Bridge Deck Assessment*. Final Report, New England Transportation Consortium Phase I, Center for Transportation Studies, MIT, 1989.

Olson, L.D., F. Jalinoos, M.F. Aouad, and A.H. Balch. 'Acoustic tomography and reflection imaging for nondestructive evaluation of structural concrete.' NSF SBIR Final Report (Award No. 9260840), 1993.

Otto, G.P. and W.C. Chew. 'Improved calibration of a large open-ended coaxial probe for dielectric measurements.' *IEEE Transactions on Instrumentation and Measurement*, 1991: Vol. 40, no. 4, pp. 742–746.

Pozar, D.M. *Microwave Engineering*. New York, John Wiley & Sons, 1998.

Rhim, H.C., and O. Büyüköztürk. 'Wideband microwave imaging of concrete for nondestructive testing.' *Journal of Structural Engineering*, 2000: Vol. 126, no. 12, pp. 1451–1457.

Rhim, H.C., Y.J. Kim, M.Q. Feng, S.K. Woo, and Y.C. Song. 'Measurements of electromagnetic properties of concrete and fiber reinforced polymer for nondestructive testing.' *Proc. of the US-Korea Joint Workshop on Smart Structures Technologies*. Seoul, Korea, September 2–4, 2004.

Rohde & Schwarz. *Measurement of Dielectric Material Properties*. Application note, Rohde & Schwarz, 2006.

Roqueta, G., S. Irteza, J. Romeu, and L. Jofre. 'A novel compact UHF wideband antenna for near field electrical characterization of steel fiber reinforced concrete.' *3r. European Conference on Antennas and Propagation (EuCAP 2009)*. Berlin: IEEE Computer Society Publications, 2009. 2393–2397.

Roqueta, G., L. Jofre, J. Romeu, and S. Blanch. 'Broadband propagative microwave imaging of steel fiber reinforced concrete wall structures.' *IEEE Transactions on Instrumentation and Measurement*, 2010: Vol. 59, no. 12, pp. 3102–3111.

Roqueta, G., L. Jofre, and M. Feng. 'Analysis of the electromagnetic signature of reinforced concrete structures for nondestructive evaluation of corrosion.' *Instrumentation and Measurements Transactions on IEEE*, 2012: Vol. 61, no. 4, pp. 1090–1098.

Schantz, H. *The Art and Science of Ultra wide band Antennas*. Norwood, MA: Artech House, 2005.

Slaney, M., and A.C. Kak. 'Limitations of imaging with first-order diffraction tomography.' *IEEE Transactions on Microwave Theory Technology*, 1984: Vol. MTT-32, no. 8, pp. 860–874.

Torgovnikov, G.I. *Dielectric Properties of Wood and Wood Based Materials*. Berlin: Springer-Verlag, 1993.

Van Damme, S., A. Franchois, D. De Zutter, and L. Taerwe. 'Nondestructive determination of the steel fiber content in concrete slabs with an open-ended coaxial probe.' *IEEE Transactions on Geoscience and Remote Sensing*, 2004: Vol. 42, no. 11, pp. 2511–2521.

Von Hippel, A.R. *Dielectric Materials and Applications*. London: Artech House, 1995.

Zoughi, R., G.L. Cone, and P.S. Nowak. 'Microwave non-destructive detection of rebars in concrete slabs.' *Materials Evaluation*, 1991: Vol. 49, no. 11, pp. 1385–1388.

23
Non-destructive evaluation (NDE) of composites: using fiber optic sensors

Y. DONG, New Mexico Institute of Mining and Technology, USA

DOI: 10.1533/9780857093554.4.617

Abstract: Fiber-reinforced polymer (FRP) composites have been used to strengthen and retrofit deteriorated or deficient structures, especially concrete structures. The performance of the rehabilitated structure is highly influenced by the bonding between concrete and FRP, and the existence of potential defects at the concrete/FRP interface and in the FRP laminates. Research has shown that fiber optic sensing technologies provide more advantages over conventional non-destructive evaluation (NDE) methods in measuring strain distribution along the bond interface or in the FRP composite, and therefore to detect failure or monitor the performance of the rehabilitated structure. The chapter begins with a brief review of the recent development and advancement of fiber optic sensing technologies, including different types of fiber optic sensors (FOSs) and their applications to civil engineering structures. By embedding FOSs in the FRP composites or at the bond interface between concrete and FRP composites, the fiber optic sensing system provide a promising NDE tool for performance assessment of FRP rehabilitated structures. Field applications of FOSs to FRP strengthened or retrofitted bridges showed that smart structure systems with fiber optic sensors are reliable for real-time monitoring of FRP rehabilitated concrete structures.

Key words: non-destructive evaluation (NDE), fiber optic sensor (FOS), fiber-reinforced polymer (FRP) rehabilitation, reinforced concrete.

23.1 Introduction

Structures deteriorate over time due to material aging, environmental corrosion, long-term continuous use and overloading, lack of maintenance, or a combination of these effects. Usually built to last 50 years, the average bridges in the United States are near this service age. A recent study showed more than 26% of the bridges in the United States are either structurally deficient or functionally obsolete (ASCE 2009). To improve the performance of these structures, it is necessary on one hand to strengthen and retrofit the deteriorated structures, and on the other hand to upgrade the existing structures to meet the requirement of the current codes or service loads. Fiber-reinforced polymer or plastic (FRP) composites, as materials of high-strength and low-weight corrosion resistance, have found growing usage in strengthening and retrofit of existing deteriorated or deficient structures, especially concrete structures. The FRP composites used for rehabilitating applications fall into two major types. One type consists of pre-manufactured rigid strips or plates of carbon- or glass-reinforced thermosetting polymers which can be adhesively bonded to the surface of structure. The other type consists of layers of unidirectional sheets or woven/stitched fabrics of dry fibers that are saturated in the field with a thermosetting polymer which simultaneously bonds the FRP composite to the structure. For example, FRP composites can be used with bridge and building structures for flexural and shear strengthening of concrete beams and slabs, axial strengthening and ductility enhancement of concrete columns.

The performance of an FRP rehabilitated structure is highly influenced by its integrity which includes the composite and the adhesive layer in the case of bonded strip, and the composite and the saturating polymeric resin in the case of dry sheets and fabrics. The interface between the FRP and concrete substrate transfers loads from concrete to FRP composite. The polymer resin plays a critical role in the FRP composite in transferring load to the fibers and protecting the fibers. During manufacture and installation of the composite systems, it is common to see such defects as voids, inclusions, debonds, improper cure and delaminations (Kaiser and Karbhari 2003). These defects could also affect the performance of rehabilitated structures, and it is therefore necessary to identify and evaluate these defects not only during and immediately after installation of the FRP, but also throughout the service life of the rehabilitated structure.

Since the rehabilitated structures are partially or totally covered by non-transparent FRP composites, it is difficult to monitor the initiation and existence of underlying concrete failure or FRP debonding. From this perspective, non-destructive testing or evaluation (NDT/NDE) can fulfill the need by assessment of defects and damages and real-time monitoring of FRP property changes. Since strengthened structures are covered by the FRP plates, the mechanical properties of the concrete may not be measured or detected easily through conventional NDE methods, such as strain measurements using surface-mounted strain gauges or extensometers, radiography, thermography and acoustic emission methods, particularly in areas with micro-cracks and debonds underneath the externally bonded plate (Lau 2003). The conventional NDT methods usually require the location, or possible location, of damage to be known prior to the assessment. Many conventional damage detection techniques are not real time and do not allow for systematic comparison of the assessment results.

Lately, a great deal of research has been on development of fiber optic sensing technology and instrumentation, and great progresses have been made for the application of fiber optic sensors (FOSs) to civil infrastructures (Ansari 1997a, 1997b, 2007). For example, FOSs have been widely used in building and bridge structures to measure displacements, strains and temperature (Imai *et al*. 2003; Inaudi 2003; Katsuki 2003). An extra advantage of using FOSs in fibrous composite materials is the compatibility of these sensors with the manufacturing process of the material. They can be embedded directly into the composite during manufacturing. These globally non-invasive sensors can provide internal strain and temperature measurements from the moment processing starts until the final failure.

As is well known, fiber optic is free from electromagnetic interference, and is characterized by high-chemical and high-temperature resistance. The use of FOSs offers a very powerful tool to perform remote, on-line, *in situ* monitoring. Moreover, due to their ability to be multiplexed in a large number of independent channels, that fibers are readily embedded into the composite and the small size that makes them minimally intrusive in the host structures, FOSs provide useful tools for implementing integrated sensing networks within the material or structure.

This chapter briefly reviews the types of FOSs and the advancement of fiber optic sensing techniques, and their applications to FRP rehabilitated concrete structures for damage detection and performance assessment.

23.2 Fiber optic sensing technologies

Fiber optic sensing is based on the fact that light patterns and signals transmitted down an optical cable will change according to the condition of the cable. External perturbations such as strain, pressure or temperature variations will induce changes in the phase, intensity or wavelength of the light propagating through the optical fiber. A measuring device is able to convert the changes in one or more of the properties of the reflected light into numerical measurements of the change along the sensor length which are related to the external parameters. FOSs can measure temperature, strain, pressure, deformation, vibration, acceleration, etc. A detailed description of the basic principles of fiber optic sensing can be found in Dong and Ansari (2011).

23.2.1 Types of sensor

There are several methods to classify fiber optic sensors. Intensity, interferometric and spectrometric sensors are categorized by the transduction mechanisms of sensors. According to sensing range and application, FOSs are also classified as localized, multiplexed or distributed sensors.

The simplest FOSs are intensity sensors, or sometimes called amplitude sensors. By name, this type of sensor is based on light intensity losses that are associated with straining of optical fibers along any portion of their length (Dong and Ansari 2011). Intensity sensors are simple, but the measurements are only relative and variations in the intensity of the light source may lead to false readings unless a referencing system is used.

Spectrometric sensors relate wavelength change in the light to the measurement of interest. The Bragg grating sensor is the most common spectrometric sensor (Morey *et al.* 1989). In a fiber Bragg grating (FBG) sensor, an optical *grating* (a series of tiny reflectors) is placed along a length of the continuous fiber core and the grating spacing is proportional to the wavelength of light reflected when a light pulse is sent down the fiber. Mechanical or thermal strain induced at the location of the grating causes this grating spacing to change, and a shift in the wavelength of the reflected light occurs. Through a specialized optical technique, along with analysis and calibration of the FBG, the data obtained from the grating spacing can be converted to a measured strain value (Dong and Ansari 2011).

FBG sensors are sensitive to very small strains, suitable for strain and temperature sensing and have a number of advantages compared with conventional strain gauges. Although a single FBG sensor is intended to measure local 'point' strains, FBG sensors can be serially multiplexed for multi-point applications. FBG sensors are commercially available for both static and dynamic measurement. However, it should be noted that FBG sensors are sensitive to temperature and require thermal compensation during data collection.

Interferometric sensors require the interference of light from two identical single-mode fibers, one of which is used as reference arm and the other is the actual sensor (Dong and Ansari 2011). Usually a Mach Zender, Michelson or Fabry–Perot interferometer is used for this type of sensor (Ansari 1997b). The Fabry–Perot type sensor is actually a single fiber (Claus *et al.* 1993). In a Fabry–Perot type sensor, the fiber is manipulated with two parallel reflectors (mirrors) perpendicular to the axis of the fiber. The interference of the reflected signals between the two mirrors creates the interference pattern. The interference pattern generated at the output end of the phase sensors is directly related to the intensity of the applied strain field between the two reflectors. Like a FBG sensor, a Fabry–Perot sensor is only capable of providing localized measurements at the gap formed between the two mirrors. A Fabry–Perot sensor cannot be serially multiplexed, but Fabry–Perot sensors do have dynamic and static measurement capabilities.

The Brillouin scattering sensor is based on the change in light wave scattering at different frequencies due to displacements within an optical fiber (Dong and Ansari 2011). Most parts of a light will travel straight forward through a transparent media, while a small fraction is back-scattered. Different components of the back-scattered light can be identified. A Brillouin scattering sensor exploits the sensitivity of the Brillouin frequency shift for temperature and strain along an optical fiber at predefined gauge lengths that can be up to several meters in length. Brillouin scattering uses standard low-loss single-mode optical fiber offering the longest distance range with unrivalled performance and a compatibility with standard telecommunication components. In this technology, Brillouin scattering is usually optically stimulated, leading to the greatest intensity of the scattering mechanism and consequently an improved signal-to-noise ratio. The Brillouin frequency-based technique is unlike intensity-based techniques and is inherently more accurate and more stable in the long term, since intensity-based techniques suffer from a high sensitivity to drifts.

23.2.2 Distributed and long gauge sensing

The purpose of distributed or long gauge sensing is to determine locations and values of measurands along the entire length of the sensor. Most of the distributed sensing applications involve strain and temperature measurement at multi-points. A number of methodologies have emerged and have been employed in distributed sensing, including optical time domain reflectometry (Gu *et al.* 2000), Brillouin scattering (Brown *et al.* 1999), and white-light interferometric techniques (Zhao and Ansari 2001), etc.

FBG sensors are considered the most promising fiber optic sensing technology platform for distributed sensing (ISIS Canada 2004). FBG sensors permit continuous strain versus position measurement over the length of the Bragg grating (typically between 5 mm and 200 mm). A grating in a FBG sensor can be thought of as a series of smaller gratings. Each of these smaller gratings can be measured individually using a specialized optical procedure, and a spatial distribution of strains can hence be obtained. A series of these gratings, called sub-gratings, can be conceptualized one after the other along the optical fiber.

Brillouin scattering sensors measure static strain profiles using a single optical fiber. This means that these sensors can be used to measure the distribution of strains along their length. The gauge length of these sensors can vary from 15 cm to more than 1000 m (ISIS Canada 2004). The Brillouin scattering wavelength shift is dependent upon temperature and strain. To function as a strain sensor it must be configured to discriminate between strain and temperature in a manner similar to the approach taken for Bragg grating sensors. Use of this type of sensor requires extensive analysis of optical signals and data. Brillouin sensors were used to monitor the behavior of a steel pipe and an FRP-strengthened concrete column by Bao *et al.* (2005), and a bridge structure in Missouri, USA, by Bastianini *et al.* (2007).

Another distributed sensing technique is based on measurement of propagation time delays of light traveling in the fiber and the measurand-induced change in the transmission light. An optical time domain reflectometer

(OTDR) can be used for this purpose (Tateda and Horiguchi 1989; Dakin 1990). A pulsed light signal is transmitted into one end of the fiber, and light signals reflected from a number of partial reflectors along the fiber length are recovered from the same fiber end. By using this concept, it is possible to determine the location of the strain fields by the two-way propagation time delay and measurement of the reflected time signals (Ansari 1997b).

More recent development in distributed sensing involves an interferometric system (Zhao and Ansari 2001) which consists of two parts: the sensing interferometer module and the receiving unit. The sensing module comprises a number of individual single mode fibers of desired gauge lengths. The individual fibers are mechanically connected through ferrules and a portion of light is reflected when the light wave passes through them. The receiving unit consists of a Michaelson white light interferometer with a scanning translation stage, signal processing and the system control unit.

Distributed or long gauge FOSs directly measure the displacement, elongation or contraction of an object. Since the long gauge sensor is a flexible optical fiber, it can be used in many different configurations. For example, it can be wrapped around a column to measure circumferential contraction or expansion, or strung across a crack to monitor crack growth. The long gauge sensing system is highly versatile, and can be configured to gauge lengths in a wide range of 5 cm to 100 m.

Distributed sensors have typically very low sampling rates, and could therefore be used only for static applications. They are suitable for detecting long-term deterioration, such as corrosion-induced damage, by collecting readings due to static truck loading at different time intervals. They could also be used as damage identification tools following extreme events such as seismic excitations, by collecting readings before and after the event. Furthermore, distributed sensors are typically much more expensive.

23.2.3 Multiplexed sensing

Multiplexed sensors usually include a combination of individual sensors for measurement of perturbations over a large structure. The sensor network in multiplexed sensing usually is interrogated by a single sensor reading device (demodulation unit). Sensor multiplexing schemes are classified according to their physical geometry. When several sensors are distributed along one single optical fiber, they are called serial multiplexed, while sensors on separate fibers are called parallel multiplexed. A sensor network may include a combination of serial and parallel multiplexing with multiple fibers, each serving as host to more than one sensor (Tennyson 2001).

Multipoint detection of signals has been achieved through wavelength- and time-division multiplexing. The technique of wavelength division multiplexing includes using Bragg grating (Kersey and Morey 1993; Ansari 1997b). Takara (2001) developed a method in which pulses from individual sensors were separated in time and then guided into an individual return fiber for de-multiplexing and signal processing. Schemes were developed for multiplexing of the signals from both Michelson interferometric sensors (McGarrity *et al.* 1995; Pan *et al.* 1996) and Mach–Zehnder interferometric sensors (Brooks *et al.* 1985).

By serial multiplexing, FBG sensors along one single cable, which is in turn connected to the acquisition system, can realize multiple point measurement. However, this type of multiplexing is unreliable if the cable is damaged. In this case, all sensors readings will be lost. In addition, if strains due to environmental effects need to be accounted for, one option is to use two single-mode fibers in order to be able to compensate for temperature, a measurement fiber that is in contact with the structure and a reference fiber that is installed free on the structure. In this case, the field installation becomes even more complicated.

Fabry–Perot sensors can provide a very high sampling rate, up to 1000 Hz, which is necessary for collecting dynamic data under service loads. They are based on parallel multiplexing, which allows each sensor to be independently connected to the acquisition system. This type of multiplexing is more practical during the field installation process.

23.2.4 Summary

Of the sensor types discussed above, intensity-type sensors are simple to construct but their sensitivity is rather low. Interferometric sensors offer higher sensitivity but the required components are quite complicated. Most commercially available FOSs are FBG sensors and Fabry–Perot sensors. By using multiplexed or distributed sensing technologies, a single fiber can efficiently monitor structural performance at various locations.

All FOSs are small in size. Most commercially available FOSs are bondable or weldable, and can be easily embedded in most construction materials, including reinforcement bars and concrete, or bonded on the structural surface. The optical fibers do not degrade and corrode, and bond strongly to the matrix. They can provide a complete strain history from concrete curing, construction and dead load, to service loads, with creep and thermal changes included as well.

Another attractive feature of FOSs is that they serve as both the sensing element and the signal transmission medium, allowing the data acquisition and processing instrumentation to be located remotely from the project site. The signals are unperturbed by electromagnetic interference. Optical waves are suitable for long transmission distances of relatively weak signals. This is especially useful for remote monitoring of bridge structures.

23.3 Fiber optic sensors (FOSs) integrated with fiber-reinforced polymer (FRP) reinforcements

It is clear that FOSs, thanks to their compatibility with fibrous composite materials, are the best choice for internal strain monitoring of composite materials. In FRP or prestressed concrete structures, smart FRP bars or tendons are those embedded with FOSs in the process of pultrusion. The embedded sensors can monitor the internal strain and performance of the FRP bars. Innovative methods have been developed to embed short gauge Bragg grating and Fabry–Perot sensors and long gauge interferometric sensors into FRP rods during the pultrusion process (Kalamkarov *et al.* 2000). Zhou *et al.* (2003) studied the microstructure, mechanical, strain sensing and temperature-sensing properties of the FRP bars integrated with optical FBG used to reinforce concrete beams. Sim *et al.* (2005) reported the tensile strength and pull-out bonding characteristics of hybrid GFRP rod and FBG FOS. The strain acting on the hybrid FRP was measured by the embedded FBG sensor. From the results, it is concluded that a hybrid FRP rod can be successfully used for the purpose of reinforcement and smart monitoring in concrete structures.

Depending on the measurement requirements, the optical fiber can be terminated prior to or run through the anchorage system. In post-tensioning applications, tendon strains, anchorage slippage or prestressing losses are able to be monitored by the embedded FOSs in the FRP tendons (Ansari 2005). Kalamkarov *et al.*'s study (2000) showed the long-term performance and reliability of embedded FOSs after the pultruded smart FRP rods were exposed to mechanical tests at room temperature as well as under conditions of low and high temperatures, fatigue and short-term creep, under sustained loads in alkaline environments simulating conditions encountered in concrete structures.

In strengthening or retrofitting existing concrete structure, FOS can also be embedded into the FRP composite to monitor the internal deformation of the externally bonded FRP laminates. Jensen *et al.* (1992a, 1992b) studied the influence of orientation and numbers of embedded FOSs on the performance of host composite. The resin pockets caused by off-axis FOS embedment result in a high internal stress concentration. Experimental and analytical studies have showed the resin pockets have a negligible effect on the host material's performance under bending and tension (Shivakumar and Emmanwori 2004; Silva *et al.* 2005; Kousourakis *et al.* 2008). Such findings are further supported by Saton *et al.* (2001) and Jensen and Sirkis (1995). It is proven that small diameter optical fibers do not cause any significant reduction in strength of composites and standard 125 µm optical fibers produce a minimum perturbation of the host material when embedded parallel to the reinforcing fibers in laminates.

Another widely studied area is to use an FOS in detecting acoustic emissions (AE) from fiber fracture or FRP debonding. Giallorenzi et al. (1982) provides a detailed description pertaining to the principles and methodologies involved in fiber optic sensing of stress waves and acoustical signals. Various transduction mechanisms, including intensity, interferometric and spectrometric techniques have been utilized (Tran et al. 1992; Pierce et al. 1996; Shih 1998; Lim et al. 1999). Interferometric techniques have provided means for high-resolution detection of damage within different types of materials. For instance, Chen and Ansari (2000) developed a white-light AE sensor for detection of micro-cracks in concrete beams. In applications to composites, Pierce et al. (1996) used a fiber-optic Michelson interferometer for detection of lamb waves generated from the splitting of carbon fibers in carbon fiber-reinforced polymer (CFRP) composites. In another application, Liu et al. (1990) demonstrated detection of micro-air voids during manufacturing of Kevlar FRP composites by an interferometric sensor. A Michelson interferometer is employed for real-time interrogation of an AE event at two locations along the length of an optical fiber. The experimental program proved the system was able to detect the location of carbon and glass fiber fractures along the length of FRP rods (Sun et al. 2004).

23.4 Fiber optic sensors (FOSs) monitoring fiber-reinforced polymer (FRP) concrete interfacial bond behavior

When FRP composites are used for retrofit and rehabilitation of concrete structures, the improvement of strength and stiffness in a host structure depends on the bond between the host structure and the added FRP materials. When an FOS is used to monitor debonding or delamination at the interface of FRP and concrete, bare optical fiber can be directly adhered or embedded at the interface. In such an application (Zhao et al. 2007), long gauge fiber optic sensors were adhered to the CFRP fabric during the application of the epoxy (Fig. 23.1a). These sensors were supposed to detect peeling or debonding due to end zone stress concentrations. Besides monitoring of fabric deformations, unloading of any of the sensors would be indicative of interface debonding or peeling of concrete associated with a release in strain. Figure 23.1b demonstrates the peeling phenomenon as indicated by the reversal of the strain in sensor segment within 15 cm of the termination point of the FRP fabric. All the sensors were still operational following the completion of the destructive test.

A few other studies have showed that this technique is promising and can improve the knowledge of the strain state at a debonded region. Chan et al. (2000) examined the difference between the strain measurements from

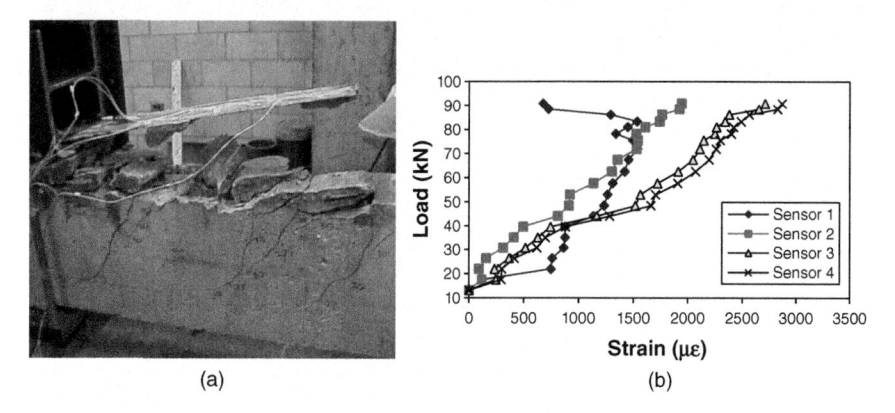

23.1 Fiber optic sensor embedded within the FRP laminate (Zhao *et al.* 2007).

multiplexed FBG sensors along the bond interface between concrete surface and FRP reinforcement and the strains from sensors externally bonded on the surface of the FRP reinforcement. Results showed that strain measured at the interface deviated noticeably from that at the surface. Embedded optical fibers allowed earlier detection of interfacial debonding between the concrete and composite than visible separation of the two components. Similar studies (Davis *et al.* 1996; Lau and Zhou 2000; Lau 2003) showed that the FBG sensors attached on the concrete surface and just in front of a crack-tip were highly susceptible to the high stress concentration at micro-cracks on the concrete surface and the occurrence of debond at the bond interface. This phenomenon could not be detected accurately through the use of surface-mounted strain gauges (Bonfiglioli and Pascale 2003). Zhu *et al.* (2003) used Mach–Zehnder interferometer sensors at the interface of FRP tube and inner concrete to measure the internal strain of the structure and the crack expanding of inner concrete. Imai and Feng (2009) used a distributed Brillouin-based FOS for detecting and monitoring FRP–concrete debonding. A model was proposed to take into consideration both the steady and the transient Brillouin interaction states, and showed that the stimulated Brillouin signal intensity distribution at the specific frequency, which corresponds to the maximum strain at the debonded region, is sensitive to the occurrence of debonding.

FOS embedded in FRP composites or at the interface between FRP and concrete can also be used to monitor the curing. Proper and full curing of the resins used to impregnate the fibers for externally bonded strengthening in the method of manual wet lay-up is very important. Antonucci *et al.* (2005)'s laboratory study showed the capability of embedding FOSs through the reinforcing fibers of the FRP laminates to measure the advancement of the cure reaction of the resin over time in conjunction with temperature measurements. Leng and Asundi (2002) embedded both extrinsic Fabry–Perot interferometer and FBG sensors in carbon/epoxy composite laminates to monitor the cure process simultaneously. The results show that both sensors can be used to monitor the strain development of composite laminates with and without damage during cure.

23.5 Field applications of fiber optic sensors (FOSs) to fiber-reinforced polymer (FRP) rehabilitated structures

As discussed above, the combination of advanced composites and fiber optic strain sensors is easy and offers important advantages to the traditional sensors. FOSs can be embedded in the composite material or at the interface between materials in the manufacturing or retrofitting processes. These sensors have been extensively evaluated for strain monitoring of engineering materials and structures in the laboratory and showed good response to thermal variations and both static and dynamic loading conditions. A few examples of field application of the these sensors are reviewed in this section.

The earlier application of FOS to FRP rehabilitated concrete structures in the USA included the historic Horsetail Fall Bridge in the state of Oregon (Laylor and Kachlakev 2000). During the strengthening of the bridge with FRP overwrap, FBG strain sensors were selected and installed for static and dynamic strain monitoring to evaluate the performance and long-term durability of FRP strengthened concrete beam. Another field application is the Fabry–Perot interferometric sensors used in a fiber optic network to monitor both static and dynamic load-induced strain of an aging reinforced concrete bridge that was upgraded with FRP wrap and rebar in the state of Missouri, USA (Watkins *et al.* 2007). Good agreement with design expectations, finite element computations, and co-located sensors was demonstrated. A health coefficient parameter was proposed as a single measure of load performance, but would require further development related to loading conditions and aging.

In development of innovative smart structures, ISIS Canada has been extensively employing FOSs for monitoring of bridges constructed of or rehabilitated with FRP composites (Mufti 2002, 2005). In one application, FBG sensors were used to monitor the strain along FRP tendons in pre-stressed concrete bridge girders on the Beddington Trail Bridge and Taylor Bridge. The sensors were able to monitor stress relaxation in the pre-tensioned tendons from the combined effects of destressing, concrete shrinkage and creep, the dead loading of the bridge deck, and the post-tensioning. It was verified that FOSs are durable and reliable for long-term monitoring (Mufti 2007). A dynamic truck test showed that these sensors were still operative six years later (Tennyson *et al.* 2001). In another application, FOSs, integrated with strain gauges and accelerometers, make up a structural health monitoring (SHM) system to Portage Creek Bridge (Mufti and Neale 2008). Long-gauge FOSs were attached to the circumference (outer layer) of FRP wrapped bridge piers. The SHM system monitors the structural performance under service conditions, diagnose faults, and quantify the risk of failure of the FRP-strengthened bridge. The data collected through high-speed Internet has proved the reliability of the instrumentation system. In reconstruction of the Joffre Bridge in Sherbrooke, Quebec, Fabry–Perot FOSs were bonded to the CFRP grids and steel girders to monitor the performance of the FRP bridge deck from the time the construction starts to several years after the completion of the construction (Benmokrane *et al.* 2004). Remote monitoring by collecting data of the FOSs through the Internet makes it possible to predict any potential degradation in any particular part of the bridge and provide the adequate maintenance in time.

The real-time monitoring of FRP reinforcement components with advanced optical sensing technology increased user confidence of the application of innovative composites in concrete structures.

23.6 Future trends

Various applications of FOS in civil engineering structures, such as monitoring of strain, displacement, vibration, cracks, corrosion and chloride ion concentration have been developed. Field tests on bridges have proved fiber optic sensing to be effective.

FOSs provide a multitude of capabilities for localized, multiplexed and distributed sensing. Fiber optic sensing technologies are especially compatible with concrete and FRP materials. FRP rehabilitated concrete bridges will greatly benefit from this technology if optical fibers are embedded into the concrete–composite interface to monitor strain distribution and indicate eventual long-term degradation. Embedded sensing allows for determination of the parameters in the future without re-installation, and invasive cutting procedures, that penetrate the concrete to expose rebar for installation.

FOSs can work in harsh natural environments, and have large sensing scope, joining with low transmission loss and anti-electromagnetic interference, so they are highly advantageous when performing SHM of civil engineering structures. However, because the study of FOS in civil engineering structures is relatively recent, their long-term sensing ability under field experimental conditions due to aging has to be investigated further. They are fragile in some configurations, and the damage is difficult to repair when embedded. The optical connection parts, which connect the embedded optical fiber with the outer data recording system, are also weak elements of the FOS system. Fiber optic sensing systems must be investigated in the field environment to develop practical protocols and to establish confidence in long-term performance. A current area of research is field validation of sensing techniques using in-service structures.

Another trend is the combination of smart composites and remote monitoring. The technology of smart composites, in which the optical fiber sensors are integrated into the composites to form a single part compound, has developed. Smart composites can be used as reinforcement as well as sensors for repairing and strengthening damaged concrete structures. The composite can be formed as a patch to bond on the concrete surface to improve the tension properties of the concrete structure. The sensors are integrated into the composite, which results in reducing the risk of damage of the fiber, and they can be manufactured in house without affecting the *in situ* environment.

Fiber optic sensing is applicable for remote monitoring the structural performance and aging where a segment of the fiber is used as a sensor gauge and a long length of the same or another fiber convey the sensed information to a remote station. On-site inspections will only be necessary when this remote sensing system indicates potential problems have developed such as excessive loads, structural failure and environmental degradation.

23.7 References

Ansari, F 1997a, 'Theory and applications of integrated fiber optic sensors in structures,' *ASCE-EMD, SP, Intelligent Civil Engineering Materials & Structures*, 2–28.

Ansari, F 1997b, 'State-of-the-art in the applications of fiber optic sensors to cementitious composites,' *Cement Concrete Composites*, **19**(1), 3–19.

Ansari, F 2005, 'Fiber optic health monitoring of civil structures using long gage and acoustic sensors,' *Smart Mater Structures*, **14**(3), 111–117.

Ansari, F 2007, 'Practical implementation of optical fiber sensors in civil structural health monitoring,' *J Intelligent Mater Systems Structures*, **18**(8), 879–889.

Antonucci, V, Giordano, M and Prota, A 2005, 'Fiber optics technique for quality control and monitoring of FRP installations,' *7th International Symposium on Fiber Reinforced Polymer Reinforcement for Reinforced Concrete Structures (FRPRCS-7)*, Kansas City, Missouri, November 6–10, 195–208.

ASCE 2009, *Report Card for America's Infrastructure*, http://www.infrastructurereportcard.org/fact-sheet/bridges [January 15, 2012].

Bao, X, Ravet, F and Zou, L 2005, 'Distributed Brillouin sensor based on Brillouin scattering for structural health monitoring,' *Fiber Optic Sens Syst*, **20**(6), 9–11.

Bastianini, F, Matta, F, Rizzo, A, Galati, N and Nanni, A 2007, 'Overview of recent bridge monitoring applications using distributed Brillouin fiber optic sensors,' *J Nondestruct Test*, **12**(9), 269–276.

Benmokrane, B, El-Salakawy, E, El-Ragaby, A, Desgagné, G and Lackey, T 2004, 'Design, construction and monitoring of four innovative concrete bridge decks using non-corrosive FRP composite bars,' presentation at the Annual Conference of the Transportation Association of Canada, Quebec City, Que, www.tac-atc.ca/English/pdf/conf2004/benmok.pdf [January 15, 2012].

Bonfiglioli, B and Pascale, G 2003, 'Internal strain measurements in concrete elements by fiber optic sensors,' *ASCE J Mater Civil Engg*, **15**(2), 125–133.

Brooks, JL, Wentworth, RH, Youngquist, RC, Tur, M, Kim, BY and Shaw, HJ 1985, 'Coherence multiplexing of fiber-optic interferometric sensors,' *J Lightwave Technol*, **LT-3**, 1062–1072.

Brown, AW, DeMerchant, MD, Bao, X and Bremner, TW 1999, 'Spatial resolution enhancement of a Brillouin-distributed sensor using a novel signal processing method,' *J Lightwave Technol*, **17**, 1179–1183.

Chan, PKC, Jin, W, Lau, KT, Zhou, LM and Demokan, MS 2000, 'Multi-point strain measurement of composite-bonded concrete materials with a RF-band FMCW multiplexed FBG sensor array,' *Sensors and Actuators A: Phiscal*, **87**(1–2), 19–25.

Chen, Z and Ansari, F 2000, 'Fiber optic acoustic emission crack sensor for large structures,' *J Structure Control*, **7**(1), 119–129.

Claus, RO, Gunther, MF, Wang, AB, Murphy, KA and Sun, D 1993, 'Extrinsic Fabry–Perot sensor for structural evaluation,' in F Ansari, *Applications of Fiber Optic Sensors in Engineering Mechanics*, ASCE-EMD Spect. Pub., ASCE, New York, 60–70.

Dakin, JP 1990, 'Multiplexed and distributed optical fiber sensor systems,' in JP Dakin, *The Distributed Fiber Optic Sensing Handbook*, IFS Publications, London, 3–20.

Davis, MA, Bellemore, DG, Putnam, MA, Kersey, AD, Slattery, KT, Corona, K and Schowengerdt, M 1996, 'High-strain monitoring in composite-wrapped concrete cylinders using embedded fiber Bragg grating arrays,' *Proc. SPIE 2721*, 149–154.

Dong, Y and Ansari, F 2011, 'NDE/NDT of rehabilitated structures,' in LS Lee and V Karbhari, Eds, *Service Life Estimation of Civil Engineering Structures: Reinforced Concrete and FRP Rehabilitation*, Woodhead Publishing, Cambridge, Chapter 18.

Giallorenzi, TG, Bucaro, JA, Dandridge, Sigel Jr. AGH, Cole, JH, Rashleigh, SC and Priest, RG 1982, 'Optical fiber sensor technology,' *IEEE J Quantum Electron*, **QE-18**, 626–664.

Gu, X, Chen, Z and Ansari, F 2000, 'Embedded fiber optic crack sensor for reinforced concrete structures,' *ACI Struct J*, **97**, 468–476.

Imai, M and Feng, M 2009, 'Detection and monitoring of FRP–concrete debonding using distributed fiber optic strain sensor,' in HF Wu, AA Diaz, PJ Shull, DW Vogel, Eds, *Nondestructive Characterization for Composite Materials, Aerospace Engineering, Civil Infrastructure, and Homeland Security 2009*, Proc. of SPIE Vol. 7294.

Imai, M, Sako, Y, Miura, S, Yamamoto, Y, Ong, SSL and Hotate, K 2003, 'Dynamic health monitoring for a building model using a BOCDA based fiber optic distributed sensor,' *Proceedings of the First International Conference on Structural Health Monitoring and Intelligent Infrastructure*, Tokyo, Japan, 241–246.

Inaudi, D 2003, 'State of the art in fiber optic sensing technology and EU structural health monitoring projects,' *Proceedings of the First International Conference on Structural Health Monitoring and Intelligent Infrastructure*, Tokyo, Japan, 191–198.

ISIS Canada, 2004, Educational Module 5: an introduction to structural health monitoring, http://www.isiscanada.com/education/Students/ISIS%20EC%20Module%205%20-%20Notes%20(Student).pdf [January 15, 2012].

Jensen, DW and Sirkis, JS 1995, 'Integrity of composite structures with embedded optical fibers,' in E Udd, Eds, *Fiber Optic Smart Structures*, Wiley, New York, 109–129.

Jensen, DW, Pascual, J and August, JA 1992a, 'Performance of graphite/bismaleimide laminates with embedded optical fibers. Part I: uniaxial tension,' *Smart Mater Struct*, **1**, 24–30.

Jensen, DW, Pasaual, J and August, JA 1992b, 'Performance of graphite/bismaleimide laminates with embedded optical fibers. Part II: uniaxial compression,' *Smart Mater Struct*, **1**, 31–35.

Kaiser, H and Karbhari, VM 2003, 'Identification of potential defects in the rehabilitation of concrete structures with FRP composites,' *Int J Mater Product Technol*, **19**(6), 498–520.

Kalamkarov, AL, MacDonald, DO, Fitzgerald, SB and Georgiades, AV 2000, 'Reliability assessment of pultruded FRP reinforcements with embedded fiber optic sensors,' *Composite Structures*, **50**(1), 69–78.

Katsuki, K 2003, 'The experimental research on the crack monitoring of the concrete structures using optical fiber sensor,' *Proceedings of the First International Conference on Structural Health Monitoring and Intelligent Infrastructure*, Tokyo, Japan, 277–284.

Kersey, AD and Morey, WW 1993, 'Multiplexed Bragg grating fiber-laser strain sensor system with modelocked interrogation,' *Electronic Lett*, **29**, 112.

Kousourakis, A, Bannister, MK and Mouritz, AP 2008, 'Tensile and compressive properties of polymer laminates containing internal sensor cavities,' *Composites A*, **39**, 1394–1403.

Lau, KT 2003, 'Fibre-optic sensors and smart composites for concrete applications,' *Mag Concrete Res*, **55**(1), 19–34.

Lau, KT and Zhou, LM 2000, 'Using fibre-optic Bragg grating sensor for strain measurement in composite-strengthened concrete structure,' *Proceedings of the Seventh International Conference on Composite Engineering*, Denver, CO, 501–502.

Laylor, HM and Kachlakev, DI 2000, http://onlinepubs.trb.org/onlinepubs/trnews/rpo/rpo.trn208.pdf [January 15, 2012].

Leng, JS and Asundi, A 2002, 'Real-time cure monitoring of smart composite materials using extrinsic Fabry–Perot interferometer and fiber Bragg grating sensors,' *Smart Mater Struct*, **11**, 249–255.

Lim, TK, Zhou, Y, Lin, Y, Yip, YM and Lam, YL 1999, 'Fiber optic acoustic hydrophone with double Mach–Zehnder interferometers for optical path length compensation,' *Opt Commun*, **159**(4–6), 15, 301–308.

Liu, K, Ferguson, MS and Measures, RM 1990, 'Fiber optic interferometric sensor for the detection of acoustic emission within composite materials,' *Opt Lett*, **15**(22), 1255–1257.

McGarrity, C, Chu, BCB and Jackson, DA 1995, 'Multiplexing of Michelson interferometer sensors in a matrix array topology,' *Appl Optics*, **34**(7), 1262–1268.

Morey, WW, Meltz, G and Glenn, DH 1989, 'Fiber optic Bragg grating sensors,' *Proc SPIE Fiber Optic and Laser Sensors*, **1169**, 98.

Mufti, A 2002, 'Structural health monitoring of innovative Canadian civil engineering structures,' *Steel Comp Struct*, **1**(1), 89–103.

Mufti, A 2005, 'Structural health monitoring of innovative bridge decks,' *Structure Infrastructure Eng: Maintenance, Management, Life-cycle Design Performance J*, **1**(2), 119–133.

Mufti, A 2007, 'Innovative, intelligent concrete structures using fiber reinforced polymer (FRP) materials,' *Proceedings of the International Workshop on Cement Based Materials and Civil Infrastructure (CBM-CI)*, Karachi, Pakistan, 10–11 December, 2007, 403–414.

Mufti, A and Neale, KW 2008, 'State-of-the-art of FRP and SHM applications in bridge structures in Canada,' *Compos Res J*, **2**(2), 60–69.

Pan, Y, Lankenau, E, Welzel, J, Birngruber, R and Engelhardt, R 1996, 'Optical coherence-gated imaging of biological tissues,' *J Select Topics Quantum Electron*, **2**, 1029–1034.

Pierce, SG, Philp, WR, Gachagan, A, McNab, A, Hayward, G and Culshaw, B 1996, 'Optical fibers as ultrasonic sensors,' *Appl Opt*, **35**(25), 5191–5197.

Saton, K, Fukuchi, K, Kurosawa, Y, Hongo, A and Takeda, N 2001, 'Polyimide-coate small-diameter optical fiber sensors for embedding in composite laminate structures,' SPIE, Newport Beach, CA, 285–294.

Shih, ST 1998, 'Wide-band polarization-insensitive fiber optic acoustic sensors,' *Opt Eng*, **37**(3), 968–976.

Shivakumar, K and Emmanwori, L 2004, 'Mechanics of failure of composite laminates with an embedded fiber optic sensor,' *J Comp Mater*, **38**, 669–680.

Silva, JMA, Devezas, TC, Silva, AP and Ferreira, JAM 2005, 'Mechanical characterization of composites with embedded optical fibers,' *J Comp Mater*, **39**, 1261–1281.

Sim, J, Moon, D, Hongseob, OH, Park, C and Park, S 2005, 'Hybrid FRP rod for reinforcement and smart-monitoring in concrete structures,' in JF Chen and JG Teng, Eds, *Proceedings of the International Symposium on Bond Behaviour of FRP in Structures* (BBFS 2005), 393–400.

Sun, C, Liang, Y and Ansari, F 2004, 'Serially multiplexed dual-point fiber-optic acoustic emission sensor,' *J Lightwave Technol*, **22**(2), 487–493.

Takara, H 2001, 'High-speed optical time-division-multiplexed signal generation,' *Optical Quantum Electron*, **32**, 795–810.

Tateda, M and Horiguchi, T 1989, 'Advances in optical time domain refectormetry,' *IEEE J Lightwave Technol*, **7**(8), 1217–1223.

Tennyson, R 2001, *Installation, Use and Repair of Fiber Optic Sensors*, ISIS Canada Design Manual No. 1.

Tennyson, RC, Mufti, A, Rizkalla, S, Tadros, G and Benmokrane, B 2001, 'Structural health monitoring of innovative bridges in Canada with fibre optic sensors,' *J Smart Mater Structures*, **10**(3), 560–572.

Tran, TA, Miller III, WV, Murphy, KA, Vengsarkar, AM and Claus, RO 1992, 'Stabilized extrinsic fiber-optic Fizeau sensor for surface acoustic wave detection,' *J Lightwave Technol*, **10**, 1499–1505.

Watkins, SE, Fonda, JW and Nanni, A 2007, 'Assessment of an instrumented reinforced-concrete bridge with fiber-reinforced-polymer strengthening,' *Optical Eng*, **46**(5) 051010.

Zhao, Y and Ansari, F 2001, 'Quasi-distributed fiber-optic strain sensor: principle and experiment,' *Appli Optics*, **40**(19): 3176–3181.

Zhao, M, Dong, Y, Zhao, Y and Ansari, F 2007, 'Monitoring bond in FRP retrofitted concrete structures,' *J Intelligent Mat Systems Structures*, **18**(8), 753–890.

Zhou, Z, Ou, JP and Wang, B 2003, 'Smart FRP-OFGB bars and their application in reinforced concrete beams,' *Proceedings of the First International Conference on Structural Health Monitoring and Intelligent Structure*, 13–15, Nov. 2003, Tokyo, Japan, 861–866.

Zhu, Q, Fang, R, Cao, Z and Zeng, W 2003, 'Measurement of internal strain of FRP concrete structure by fiber optic sensor,' in GX Shen, S Cha, FP Chiang and CR Mercer, *Optical Technology and Image Processing for Fluids and Solids Diagnostics*, Proceedings of the SPIE, Vol. 5058, 538–542.

24
Non-destructive evaluation (NDE) of Composites: infrared (IR) thermography of wind turbine blades

N. P. AVDELIDIS, Université Laval, Canada and
T-H. GAN, Brunel University, UK

DOI: 10.1533/9780857093554.4.634

Abstract: This chapter presents the technique of infrared thermography (IRT) in the inspection and assessment of glass-reinforced polymer (GRP) wind turbine blades. It presents and explains both the passive and active thermography approaches (pulsed, step heating, lock-in, etc.) when inspecting composite–GRP wind turbine blades, outlining the advantages and limitations of the technique(s). Furthermore, it reviews some of the most interesting signal processing techniques that can be used for the active thermography inspection and analysis of composite materials (differential absolute contrast – DAC, thermographic signal reconstruction – TSR, principal component thermography – PCT, pulsed phase thermography – PPT). Finally, it concludes the use of IRT in inspecting and evaluating wind turbine blades.

Key words: thermography, blades, glass-reinforced polymer (GRP), composites, signal processing.

24.1 Introduction

Wind turbine blade manufacture is the fastest growing market for composites. The market in blade manufacture is dominated by fibre reinforced epoxy-based resin composites. Commonly glass fibres, carbon and Kevlar fibres are increasingly used in larger blades. Polyester and vinyl ester plastics are also used for smaller blades. An appreciation of the size of the market can be gained from the fact that 44 000 blades were manufactured in 2007 [1]. In service, turbine blades can suffer a variety of degradation mechanisms – disbonding, delamination, fibre cracking, fatigue, ultraviolet (UV) degradation, etc.

Non-destructive evaluation (NDE) methods are employed in damage analysis and repair design of composite structures in three ways:

- damage location;
- damage evaluation, i.e. type, size, shape and internal position; and
- post-repair quality assurance.

The first and most important activity is to identify the damage. Assessment of the damage is initially achieved by visual inspection. This localizes the damaged area, and is then followed by a more sensitive NDE method, which maps the extent of any internal damage. Detailed NDE is particularly important when dealing with composites, because the majority of the damage is usually within the composite structure. A number of NDE techniques are currently available to identify damage in composite structures:

- visual, including optical magnification;
- dye penetrant testing;
- acoustic methods such as the tap test;
- ultrasonic testing (UT) methods, e.g. A-scan, C-scan and phased array;
- material property changes, especially stiffness
- thermography and thermal imaging techniques, especially active approaches.

The appropriate application of NDE techniques is of major importance to ensure the proper assessment of the blades.

24.2 Wind turbines

In the EU, wind energy provides over the 4.2% of total electric energy, and in 2008 the growth of 15% was higher than all the other energy sources. Thanks to wind energy in the EU over 12 000 new jobs/year have been created in the last 5 years and is forecast that by 2020 the number of workers in this field will reach 330 000. For the coming wind turbines, the development of better and really comprehensive condition monitoring systems is already a main issue and in demand. There are several reasons for this: wind turbines are increasing in size, large wind farms are placed offshore, and there is an increase in the size of proposed wind farms. Wind-farm owners require improved cost-efficient techniques for the surveillance of the health of turbines. These factors add to the increasing need of having better systems to monitor the condition of the turbines and predict when failure is likely to happen. The EU draws attention to the growing importance of offshore wind farms and to the fact that more R&D activities are required to solve problems linked to offshore farms (reliability, maintenance). These problems of reliability, etc., also apply to onshore farms.

In addition the trend is that as more turbines are built, more accidents occur. Numbers of recorded accidents reflect this. A summary of wind turbine accident data to 30 September 2010, in the EU [2] shows an average of 16 accidents per year from 1995 to 1999 inclusive; 48 accidents per year from 2000 to 2004, and 106 accidents per year from 2005 to 2010. Since 2006, 52 accidents regarding human injury have been documented: 80% involved wind industry or maintenance workers, and 20% involved members of the public or workers not directly dependent on the wind industry (e.g. transport workers). By far the biggest number of incidents found (220 incidents) was due to blade failure. Blade failure can arise from a number of possible sources, and results in either whole blades or pieces of blade being thrown from the turbine.

The European Wind Energy Association (EWEA) warns that fatigue loading on wind turbines is more severe than that on helicopters, aircraft wings or bridges. When the turbine turns to face the wind, the rotating blades act like a gyroscope. As the turbine pivots, gyroscopic precession tries to twist the turbine into a forward or backward somersault. This cyclic twisting causes fatigue and may crack the blade roots. Lately, the inspection methods have become very costly and labour consuming, and at the same time unreliable, by surveying a few square centimetres of blade at any one time. In addition, the inspection of the turbine blades requires exposed work at dangerous heights. Falls from height are the major cause of fatalities in the wind power generation industry. Inspection and maintenance costs for an average wind power installation can be estimated on a lifetime (30% of initial investment) or annually (40% of operating costs) basis [2].

The crucial need for developing dedicated damage detection and analysis tools in the wind energy community has been recognized for some time; however, research and development for dedicated systems for wind turbine blades is lagging behind R&D for other wind turbine component parts. The European Wind Energy Technology Platform's Working Group 'Wind Power Systems' included 'reliable monitoring systems for early failure detection and accurate interpretation of signals' as a key research priority both for onshore and offshore applications.

In this framework, thermography is an inspection technique that it is cost effective and reliable for wind turbine blade assessment, as it is able to:

- detect the presence of defects;
- quantitatively determine the extent and rate of damage;
- provide crucial information for assessing stability and serviceability;
- provide all above in real time;
- save costs in relation to existing inspection methods (some countries are already cutting subsidies), as it is an area investigation technique make it a prompt assessment tool.

24.3 Infrared thermography (IRT)

Passive and active thermography approaches (pulsed, step-heating, lock-in, etc.) will be explained and discussed when inspecting composite materials and particularly glass-reinforced polymer (GRP) wind turbine blades. Advantages and limitations will be presented. Infrared monitoring (IRM) is a technique capable of locating various forms of damage on composite materials and structures (i.e. fibre breakage, matrix cracking, debonding and delamination). It is an appropriate monitoring method that can locate and evaluate such defects at an early stage and thus before they become critical for the structure, and so cause component failure.

This approach uses various sensor types that operate in the infrared region that is located between the visible and microwave regions of the electromagnetic spectrum. Because heated objects radiate energy in the infrared, it is often referred to as the heat region of the spectrum. All objects radiate some energy in the infrared, even objects at room temperature and frozen objects such as ice. The higher the temperature of an object, the higher the spectral radiant energy, or emittance, at all wavelengths and the shorter the predominant or peak wavelength of the emissions.

Although IRM can detect damage on composites, it cannot really distinguish or even identify the form or extent of damage. Nonetheless, image analysis tools could be used (i.e. PPT, pulsed phase thermography, TSR, thermal signal reconstruction, etc.) in order to categorize each type of defect-damage. Work has already been done to detect fatigue damage in the skin of sandwich composite materials. Nonetheless, no work has been done on identifying the skin-from-core debonding or core damages caused by fatigue loading.

Infrared measurement and investigation is a recognized technique used in the inspection of different materials and/or structures [3]. During active infrared thermographic inspections, there are various approaches and thermal excitation sources that can be used; transient thermography is one of these. Transient thermography is one of the most effective non-destructive testing (NDT) techniques in the assessment of composite materials, as it is a high-speed, portable, non-contact and large area inspection technique. In this technique the surface under investigation is pulse heated (time period of heating varying from milliseconds to seconds) by one or more powerful optical lamps and the resulting thermal transient at the surface is monitored using an infrared camera. Nonetheless, the duration of the heating pulse depends on the thermal and physical properties of the materials, as well as its thickness. So, the heat flow into the sample is altered in the presence of a subsurface defect or feature, creating temperature contrast at the surface which as a result can be detected – recorded by an infrared system. Thus, transient thermography is a prompt technique that has the ability to provide quantitative information about hidden defects – features in a material. Analysing the transient temperature in the time domain typically attains this [4].

24.3.1 Infrared thermography (IRT) for condition monitoring of composite wind turbine blades

Up to now infrared thermography (IRT) has been intensively used as an NDT technique at manufacturing and assembly stage of wind turbine blades. The adhesive joints are critical points in the blade structure. That is why they are inspected with particular care. Infrared scanners are used to examine the blade throughout its length, measuring exactly the same points each time. The scanner is able to see through the laminate and check the adhesive joint. It records temperature differences in the adhesive, possibly identifying flaws, and takes a series of pictures. If there are any doubts, a point can be highlighted and later analysed using electronic image processing. If flaws are found, they can almost always be repaired immediately.

IRT has the potential for providing full-field non-contacting techniques for the inspection of wind turbine blades. The sensitivity of the thermal imaging has been shown to be suitable for non-destructive examination during fatigue testing; furthermore, it is thought that for blades in situ, the wind loading conditions may be sufficient to create effects detectable by thermal imaging [5].

Transient thermal NDT requires an external source of energy to induce a temperature difference between defective and non-defective areas in the specimen under examination. As depicted in Fig. 24.1, a wide variety of energy sources are available and can be divided into optical, if the energy is delivered to the surface by means of optical devices such as photographic

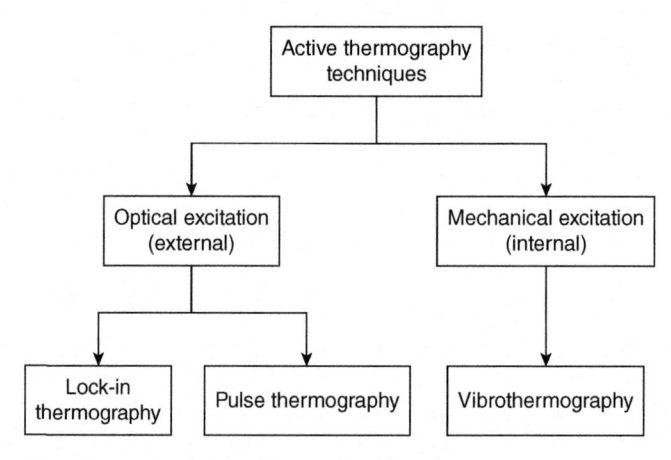

24.1 Active thermography NDT techniques.

flashes (for heat pulsed stimulation) or halogen lamps (for periodic heating), or mechanical, if the energy is injected into the specimen by means of mechanical oscillations, e.g. with a sonic or ultrasonic transducer. Optical excitation stimulates the defects externally, i.e. the energy is delivered to the surface of the specimen where the light is transformed into heat. Thermal waves propagate by conduction through the specimen until they reach a discontinuity that acts as a resistance reflecting the thermal waves back to the surface. Mechanical excitation, on the other hand, heats up the defects internally, i.e. mechanical oscillations injected in the specimen travel in all directions dissipating their energy at the discontinuities in the form of heat, which travels to the surface by conduction.

There are three classical active thermography techniques based on these two excitation modes, lock-in thermography (LT) and pulse thermography (PT), which are optical techniques applied externally, and vibrothermography (VT), which uses sonic or ultrasonic waves (pulsed or amplitude modulated) to excite surface or internal features. The experimental and theoretical aspects are different for each of these techniques and so are the typical applications.

PT is one of the most investigated thermography techniques and also one of the fastest. However, there are several variables involved in the technique such as the surface reflections, surface condition and roughness, optical arrangement, quality of light source, etc., which create problems in practice. It is also to be noted that the data acquired is complex and requires extensive analysis. LT requires periodic excitation of the sample and can be used to improve the signal-to-noise ratio of the output. For deeper defects, greater acquisition times are required when using LT. VT is a useful technique since ultrasound waves which travel through the specimen and facilitate the detection of defects are independent of their orientation [6]. The combined and integrated use of the above-mentioned thermal NDT approaches would provide reliable results in relation to the assessment of composites in aircraft transport applications [4].

24.4 Signal processing techniques

This section will present a review of some of the most interesting signal processing techniques that can be used for active thermography inspection and analysis of composite materials:

- differential absolute contrast (DAC);
- thermographic signal reconstruction (TSR);
- principal component thermography (PCT);
- pulsed phase thermography (PPT).

24.4.1 DAC

Thermal contrast is a basic operation that despite its simplicity is at the origin of many PT algorithms. Various thermal contrast definitions exist, but they all share the need for specifying a sound area S_a, i.e. a non-defective region. For instance, the *absolute* thermal contrast $T(t)$ is defined as:

$$\Delta T(t) = T_d(t) - T_{S_a}(t) \qquad [24.1]$$

with $T_d(t)$ the temperature of a pixel or the average value of a group of pixels on a defective area at time t, and $T_{Sa}(t)$ the temperature at time t for the S_a. No defect can be detected at a particular t if $T(t) = 0$. However, in practice raw data is contaminated with noise and other signal degradations and a threshold of detectability ($T(t) > 0$) is generally established.

The main drawback of classical thermal contrast is establishing S_a, especially if an automated analysis is intended. Even when S_a definition is straightforward, considerable variations in the results are observed when the location of S_a is changed.

In the DAC method, instead of looking for a non-defective area, an *ideal* S_a temperature at time t is computed locally assuming that on the first few images (at least one image at time t' in particular, see below) this local point behaves as if there is no visible defect. The first step is to define t' as a given time value between the instant when the pulse has been launched, and the precise moment when the first defective spot appears in the thermogram sequence, i.e. when there is enough contrast for the defect to be detected. At t', there is no indication of the existence of a defect yet. Therefore, the local temperature for S_a is exactly the same as for a defective area:

$$T_{S_a}(t') = T_d(t') = \frac{Q}{e\sqrt{\pi t'}} \qquad [24.2]$$

By substituting Eq. (24.2) into the absolute contrast definition, Eq. (24.1), it follows that:

$$\Delta T_{dac} = T_d(t) - \sqrt{\frac{t'}{t}} \cdot T(t') \qquad [24.3]$$

Actual measurements diverge from the solution provided by Eq. (24.3) for later times and as the plate thickness increases with respect to the non-semi-infinite case. Nevertheless, the DAC technique has been proven to be very efficient as it reduces the artefacts caused by non-uniform heating and surface geometry even for the case of anisotropic materials at early stages. Originally, proper selection of t' required an interactive procedure, for which a graphical user interface was developed, although automation is also possible. Furthermore, a modified DAC technique has been proposed. It is based on a finite plate model and the thermal quadrupoles theory that includes the plate thickness L explicitly in the solution to extend the validity of the DAC algorithm to later times. The Laplace inverse transform is used to obtain a solution of the form:

$$\Delta T_{\text{dac,mod}} = T_{\text{d}}(t) - \frac{\ell^{-1}\left[\coth\sqrt{\frac{pL^2}{\alpha}}\right]_t}{\ell^{-1}\left[\coth\sqrt{\frac{pL^2}{\alpha}}\right]_{t'}} \cdot T(t') \qquad [24.4]$$

where p is the Laplace variable.

Finally, it has also been proposed to use the DAC algorithm in combination with pulsed phase thermography data to eliminate the need for manual definition of a reference area and provide an automated tool for the determination of the subsurface defects depth.

24.4.2 TSR

TSR is an attractive technique that allows increasing spatial and temporal resolution of a sequence, while reducing the amount of data to be manipulated. TSR is based on the assumption that temperature profiles for non-defective pixels should follow the decay curve:

$$\ln(\Delta T) = \ln\left(\frac{Q}{e}\right) - \frac{1}{2}\ln(\pi t) \qquad [24.5]$$

As stated before, to fit the thermographic data, Shepard proposed to expand the logarithmic form into a series by using an m-degree polynomial of the form:

$$\ln(\Delta T) = a_0 + a_1 \ln(t) + a_2 \ln^2(t) + \ldots + a_p \ln^m(t) \qquad [24.6]$$

The thermal profiles corresponding to non-defective areas in the sample will approximately follow a linear decay, while a defective area will diverge from this linear behaviour. Typically, m is set to 4 or 5 to avoid 'ringing' and assure a good correspondence between acquired data and fitted values. At the end, the entire raw thermogram sequence is reduced to $m + 1$ coefficient images (one per polynomial coefficient) from which *synthetic* thermograms can be reconstructed.

Synthetic data processing brings interesting advantages such as: significant noise reduction, possibility for analytical computations and data compression (from N to $m + 1$ images). As well, analytical processing provides the possibility of estimating the *actual* temperature for a time between acquisitions from the polynomial coefficients. Furthermore, the calculation of first and second time derivatives from the synthetic coefficients is straightforward. The first time derivative indicates the rate of cooling while the second time derivative refers to the rate of change in the rate of cooling. Therefore, time derivatives are more sensitive to temperature changes than raw thermal images. There are no purpose using derivatives of higher order; since, besides the lack of a physical interpretation, no further defect contrast improvement can be observed. Finally, TSR synthetic data can be used in combination with other algorithms to perform quantitative analysis as described at the end of the next section.

24.4.3 PPT

In PPT, data is transformed from the time domain to the frequency spectra using the 1D discrete Fourier transform (DFT):

$$F_n = \Delta t \sum_{k=0}^{N-1} T(k\Delta t) \exp^{(-j2\pi nk/N)} = \text{Re}_n + \text{Im}_n \qquad [24.7]$$

where j is the imaginary number ($j^2 = -1$), n designates the frequency increment ($n = 0, 1, \ldots N$), t is the sampling interval, and Re and Im are the real and the imaginary parts of the transform, respectively.

Real and imaginary parts of the complex transform are used to estimate the amplitude A, and the phase:

$$A_n = \sqrt{\text{Re}_n^2 + \text{Im}_n^2} \qquad [24.8]$$

$$\phi_n = \tan^{-1}\left(\frac{\text{Im}_n}{\text{Re}_n}\right) \qquad [24.9]$$

Although very useful, Eq. (24.7) is slow. Fortunately, the fast Fourier transform (FFT) algorithm is available to be implemented or can be found (integrally or simplified) in common software packages. DFT can be applied to *any* waveform, hence it can be used with lock-in and VT data.

The phase, Eq. (24.9), is of particular interest in NDE given that it is less affected than raw thermal data by environmental reflections, emissivity variations, non-uniform heating, and surface geometry and orientation. These phase characteristics are very attractive not only for qualitative inspections but also for quantitative characterization of materials. For instance, a depth inversion technique using the phase from PPT has been proposed. The technique relies on the thermal diffusion length equation, in a manner similar to LT. The depth of a defect can be calculated from a relationship of the form:

$$z = C_1 \cdot \mu = C_1 \sqrt{\frac{\alpha}{\pi \cdot f_b}} \qquad [24.10]$$

where C_1 is an empirical constant and f_b [Hz] is the blind frequency, defined as the limiting frequency at which a defect located at a particular depth presents enough (phase or amplitude) contrast to be detected in the frequency spectra.

It has been observed that $C_1 = 1$ when using amplitude data, whilst reported values when working with the phase are in the range $1.5 < C_1 < 2$, with $C_1 = 1.82$ typically adopted. The phase is of more interest in NDT than the amplitude, since it provides deeper probing capabilities. A reason for this is the ratio involved in Eq. (24.9), which cancels part of the artefacts related to non-uniform heating, environmental reflections, emissivity variations and surface geometry variations. PPT results agree with these numbers for homogeneous materials: $C_1 \sim 1.72$ for steel, and $C_1 \sim 2.0$ for Plexiglas®, and for anisotropic materials: $C_1 \sim 1.73$ for carbon fibre-reinforced polymer (CFRP). The inversion problem in PPT is reduced to the estimation of f_b from the phase. The phase contrast has been proposed to determine f_b, but automatic determination without sound area definition is also possible.

As any other thermographic techniques, PPT is not without drawbacks. The noise content in phase data is considerable, especially at high frequencies. This causes a problem for the determination of the blind frequency. A de-noising step is therefore often required. The combination of PPT and TSR has proven to be very effective for this matter, reducing noise and allowing the depth retrieval of defects. Another difficulty is that, given the time-frequency duality of the Fourier transform, a special care must be given to the selection of the sampling and truncation parameters prior to the application of the PPT. These two parameters depend on the thermal properties of the material and on the depth of the defect, which are often precisely the subject of the investigation in hand. An interactive procedure has been proposed for this matter.

As a final note, the FFT is typically used to extract amplitude and phase information in PPT. Nevertheless, it is also possible to use different transformation algorithms such as wavelet transforms. The latter uses *wavelets* as the basis function instead of sinusoids and has the additional advantages of preserving the temporal information after the transformation. Wavelets are periodic waves of short duration that allow better reproduction of a transient signal and usage of different scales or resolutions. These advantages of the wavelet transform are currently under investigation.

24.4.4 PCT

As explained above, the Fourier transform provides a valuable tool to convert the signal from the *temperature–time* space to a *phase–frequency* space but it does so through the use of sinusoidal *basis functions*, which may not be the best choice for representing transient signals (as the temperature profiles typically found in pulsed thermography). Singular value decomposition (SVD) is an alternative tool to extract spatial and temporal data from a matrix in a compact or simplified manner. Instead of relying on a basis function, SVD is an eigenvector-based transform that forms an *orthonormal* space. Assuming data are represented as a $M \times N$ matrix **A** ($M > N$), then the SVD allows writing:

$$\mathbf{A} = \mathbf{URV}^T \qquad [24.11]$$

with **R** being a diagonal $N \times N$ matrix (with singular values of **A** present in the diagonal), **U** is a $M \times N$ matrix, $\mathbf{V}^T$ is the transpose of an $N \times N$ matrix (characteristic time).

The columns of **U** represent a set of orthogonal statistical modes known as empirical orthogonal functions (EOF) that describes spatial variations of data. On the other hand, the principal components (PCs), which represent time variations, are arranged row-wise in matrix $\mathbf{V}^T$. These characteristics of the SVD approach are valuable for PT applications. The four techniques just described are intended to process PT data.

24.5 Quality assurance and structural evaluation of glass fibre reinforced polymer (GFRP) wind turbine blades

Pulsed and pulsed phase thermography were used for the evaluation of GFRP wind turbine blades. Representative thermal images of the investigated samples were obtained during the transient phase of the thermographic inspections and are presented. Although the best possible results concerning the size of a detected defect were attained at particularly short transient times, the highest contrast images were acquired at relatively longer times.

Since the thermal conductivity of all these tested samples is relatively low, the samples were tested using a reasonably low maximum frame rate. In the trials, it was also made clear that PT and/or PPT could be limited in some cases of defect detection (i.e. impact damage detection, near-surface defects, relatively large defects – greater in diameter than depth).

Transient PT is a common method for detecting defects in composites. The flash lamp(s) can give a uniform heat to the surface at a relative short pulse (a few milliseconds). It is thus a prompt investigation thermography approach (i.e. much faster than other thermography approaches such as VT, optical LT, etc.).

Plate XIV (between pages 296 and 297) shows the first time derivative images after processing on the thermography results, as well as the thermal profiles of the investigated defects where deeper defects present higher temperatures. At low frequencies, Plate XVa and b (between pages 296 and 297), the high phase values (red) at the left part correspond to the plastic and the low phase values at the right (blue) correspond to the composite. The plastic forms a bump that can be seen in Plate XVc, left and right correspond to the composite material. In Plate XVd phase values are now high for the right side corresponding to the foam material, and low for composite at the left side. Figure 24.3 shows an investigated by thermography section of a wind turbine blade, whilst Fig. 24.2 presents the thermogram and the thermal contrast plot of the detected hole inside the blade.

Thermography is currently performed in NDT facilities, requiring the blade to be dismounted and transported. Furthermore, depth penetration results and thick complex geometry results are limited. Nonetheless, there have been intensive efforts in the EC (FP7 Funded Research Projects) in order to develop *in situ* robotic application of NDT tools for wind turbine blades inspection and assessment. Of course, thermography is one of the most promising tools in this development, aiming also to determine the exact dimensions (depth and size) of the detected defects on the inspected blades.

24.6 Infrared thermography (IRT) standards

Even though IRT is a well-developed technology, its applications to NDT and NDE are merely nowadays starting to approach maturity. There have been many efforts to have IRT recognised as the sixth major NDT, next to those already clearly established (eddy current, liquid penetrant, magnetic particle, radiography and ultrasonic). Such efforts have been thrown back because of the shortage of technical standards for the technique. However, there are highly regarded efforts made in the last few years by the international technical group of thermal infrared sensing (currently with 120 members) that meets every spring in the USA, prior to the start of the Thermosense Conference, a major conference dealing with IRT that it is organised by SPIE (International Society for Optical Engineering), as well as, from the QIRT working group that meets every two years during the course of the European Conference on Quantitative Infrared Thermography (QIRT). Various NDT institutes or research organizations or societies, such as the British Institute of NDT, the Hellenic Society of NDT and Canada's Research Chair in Multipolar Infrared Vision, as well as, major NDT conferences, such as the Advances in Signal Processing for Non Destructive Evaluation of Materials (IWASPNDE) international workshop and the Hellenic Society of NDT international conference, have also been contributing to such efforts for the establishment of IRT worldwide.

Nonetheless, there are numerous standards concerned with IRT and its applications.

24.7 Conclusion

The main advantage of IRT over destructive testing techniques is that large areas can be scanned fast and with no need to be destroyed during testing. This results in major savings in time, people, work and machinery. In

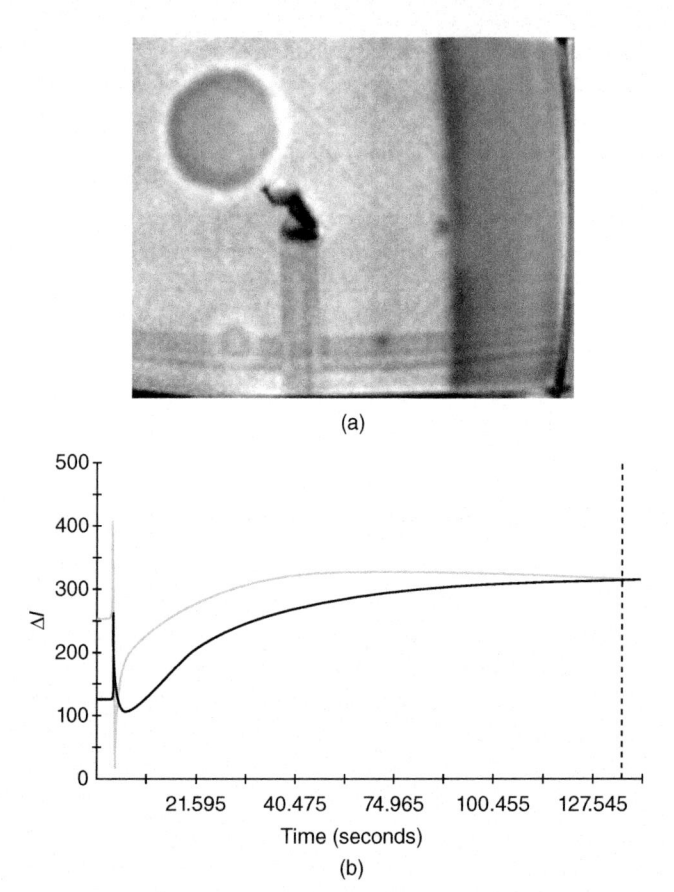

646 Non-destructive evaluation (NDE) of polymer matrix composites

24.2 At 30 531 seconds after flash heating – 2nd time derivative image at a frame rate of 2 Hz and contrast curve between open holes and sound area of investigated turbine blade.

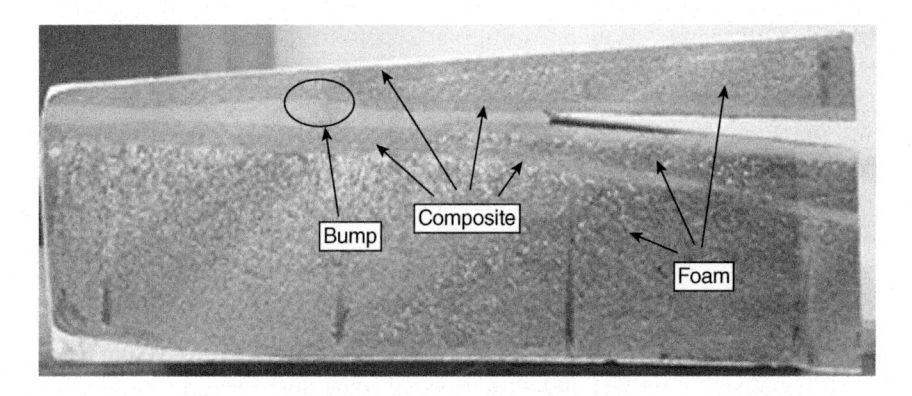

24.3 Wind turbine sample inspected using active thermography.

addition, there are advantages of infrared thermography over the classical non-destructive techniques. The IRT device is risk-free, as it does not emit any radiation. It only records the infrared radiation emitted from the material that is under assessment, hence no physical contact is required. Moreover, IRT is an area investigating technique, whereas most of the other non-destructive methods are either point or line testing methods. Furthermore, infrared thermographic testing may be performed during both day and night.

Thermography, because it uses infrared technology it is not possible to penetrate in extended depths (only a few millimetres). That of course is one of the main limitations of the technique. Furthermore, environmental conditions also play an important role on outdoor infrared thermographic surveys utilizing the passive approach (i.e. cloud cover, solar radiation, wind speed).

As a result of innovative technology, nowadays, highly developed infrared cameras are being produced and thus new applications of IRT are emerging continuously. This tendency is certainly encouraging and nonetheless groundbreaking. New applications are expected in both thermographic approaches (passive and active). In the active approach though, due to new heating sources (i.e. microwave heating, lateral heating), things are more promising.

24.8 Acknowledgements

The authors acknowledge Prof. Xavier Maldague and Dr Clemente Ibarra-Castanedo from Université Laval in Quebec for their help. Finally the authors would like to dedicate this chapter to the memory of Alexander P. Avdelidis, who left us so suddenly and without having the time to be involved with thermography.

24.9 References

1. *Prioritising wind energy research – strategic research agenda of the wind energy sector*, The European Wind Energy Association.
2. T. Moan, 'Recent research and development relating to platform requalification', *Trans. ASME*, **122**, p. 20, 2002.
3. X. Maldague, 'Applications of infrared thermography in nondestructive evaluation' in: *Trends in optical nondestructive testing*, eds. P.K. Rastogi, D. Inaudi (Elsevier Science, 2000).
4. N.P. Avdelidis, D.P. Almond, A. Dobbinson, B.C. Hawtin, C. Ibarra-Castanedo, X. Maldague, 'Invited Review Paper: Aircraft composites assessment by means of transient thermal NDT', *J. Progress in Aerospace Sciences*, **40**(3), pp. 143–162, 2004.
5. G.M. Smith, B.R. Clayton, A.G. Dutton, A.D. Irving, 'Infrared thermography for condition monitoring of composite wind turbine blades: feasibility studies using

cyclic loading tests', *Wind Energy Conversion 1993, Proceedings of the 15th British Wind Energy Association Conference*, pp. 365–371, 1993.
6. C. Ibarra-Castanedo, J.M. Piau, S. Guilbert, N.P. Avdelidis, M. Genest, A. Bendada, X.P.V. Maldague, 'Comparative study of active thermography techniques for the nondestructive evaluation of honeycomb structures', *Res. Nondestructive Eval.*, **20**(1), pp. 1–31, 2010.

25
Non-destructive evaluation (NDE) of composites for marine structures: detecting flaws using infrared thermography (IRT)

A. SURATKAR, A. Y. SAJJADI and K. MITRA,
Florida Institute of Technology, USA

DOI: 10.1533197808570935544.649

Abstract: This chapter discusses the detection of flaws in marine structures using infrared (IR) thermography. The equipment, measurement devices, different samples and methods used are described and the effectiveness of infrared thermography in detection is assessed.

Key words: non-destructive testing (NDT), thermography, composites, boat hull defects, COMSOL multiphysics software.

25.1 Introduction

Non-destructive testing (NDT) is the analysis of the properties of a material, structure, or component without any physical damage to the system. The detection of defects in material is an important part of quality control for its safety of system because flaws and cracks weaken the performance of a structure or material (Cartz, 1996; Raj et al., 2002). Many methods are prevalent for detection of defects in materials. Some of them are NDT, non-destructive evaluation (NDE), non-destructive characterization, and non-destructive inspection (NDI) (Mix, 2005). NDT is used for a large number of industrial applications. Many industries use NDT as a part of their manufacturing procedure to improve the quality of the product. In the manufacturing industry, NDT is used to detect cracks, tears, imperfect welds and junctions, tomography, surface contamination effects, and inclusions (Berger, 1977). NDT is effectively used in the military and fire departments. In medicine, non-invasive testing is used for such purposes as detecting bone cracks, tumor imaging, and embryo scanning. NDT also has a wide range of applications in geology, forensic studies, aerial temperatures and weather surveys, sculpture examinations, and ceramic fragments. NDT involves many methods like ultrasonic, radiographic, magnetic-particle, liquid penetrant, remote visual inspection (RVI), eddy-current testing, and low coherence interferometry, among others. These methods are time-saving and cost-reduction techniques for inspecting change in material structure and different anomalies.

There are different types of non-destructive techniques that are applied in many industries and research fields. As per the application, different NDT methods implemented for composite material subsurface defect detection are radiography, eddy current testing, ultrasonic testing method, acoustic emission technique, and thermography testing method. In this chapter, thermography testing is described for defect detection in composite materials.

25.2 Infrared thermography (IRT)

When a hot material is placed in an open atmosphere, the body cools. In this process, the energy from the hot object is converted from one form to another. This transfer of energy from the hot body source to the outer atmosphere is called thermal energy in transition (see Plate XVI between pages 296 and 297). Conduction, convection, and radiation are three modes of heat transfer. Radiative heat flow is measured by infrared thermography (IRT). The wavelength range of radiative heat transfer falls in the infrared ray's spectrum. Generally, the infrared spectrum wavelength range varies from 0.75 to 100 µm; however, practical measurements cannot go below 20 µm. Thermography and non-contact temperature measurement is based on measurement of thermal infrared radiation (Vavilov et al., 2008). Existence, or radiosity, is thermal infrared radiation emitted from a surface. The modern infrared detectors are precise for measuring thermal energy radiating from a surface. IRT is used to measure a wide temperature range of cool surfaces (well below 0 °C) to hot surfaces up to 700 °C (Minkina and Dudzik et al., 2009).

Since 1990, the infrared thermal imaging technique has been used in many applications such as non-destructive defect detection for automobile engines (Kaplan, 1993), marine ships (Vavilov and Taylor, 1982; Gerlach, 2006), aircraft structure wings analysis (Avdelidis et al., 2004, 2006), biomedical thermal imaging (Barone et al., 2006; Diakides and Bronzino 2007), military night vision (Servais and Gerlach, 2005), and firefighting (Therianult, 2001; Barndo and Chang, 2003). The infrared thermal detection method has been used to identify thermo-physical properties and to detect the coating thickness and hidden structures of objects (Pitarresi et al., 2009; Sheppard et al., 2009).

Generally subsurface defects are found in composite boat material after a period of operation for 15 to 20 years. These defects occur because of manufacturing defects, filling materials on structures to create a smooth profile of a boat, and change of climatic pressure on material. The mostly frequently observed ones are void defects and delamination, which occur in composite material because of climatic pressure change. These defects make a structure weak. If these defects are not detected during their primary stage, then they will be the root cause for the structural damage of the boat. There are some more defects like carbon fiber, extra layers of soft materials, etc., which occur mainly due to design requirements and manufacturing defects.

The thermography technique is used for the detection of defects in a boat hull or wall. Experiments are performed on a testing panel, provided by Holyes Electronics LTD. This testing panel is made up of resin layers, which simulate a boat wall. For experimental purposes, different defects such as carbon fiber, dry void, acid, delamination, and dry mat are included in the testing panel. The active thermography approach is considered to detect defects by an infrared camera. A thermal camera is used to measure the temperature decay with time when heated for a finite time. Parametric studies are performed on different samples. The boat hull model is simulated using finite element analysis (COMSOL). The boat hull is modeled with identical geometrical dimensions to those used in thermography experiments. The various defect temperature decay studies are performed for obtaining the theoretical values. Experimental measurements are finally validated using finite element simulation values. Therefore, NDT plays a major role in conducting experiments to detect defects in structures or bodies without causing any damage to the body.

25.3 Case study: non-destructive evaluation (NDE) of defects in a boat hull

This section describes the equipment and measurement devices, as well as the different samples and methods used, for NDI detection of defects in a boat hull in which a thermal camera and heat gun are the primary items of equipment. Next, a brief description of various samples and simulation material used for experiments is given. A resin-bonded boat hull with a different defect simulation panel is used for the thermography experiment. Finally, the experimental methodology along with data analysis procedures for the experiment are discussed.

25.3.1 Simulated test panel 1

The simulated test panel used in this experiment contained different defects generally found in boat hull material (Fig. 25.1). Two testing panels were obtained from Holyes Electronics Inc. The testing panel consisted of simulations of seven different anomalies: acid, dry void, delamination, carbon fiber, West 401, dry mat, and Kevlar. Every defect was distinguished by the location and the thermal properties of the material. These defects generally occurred in the hull or boat wall, and hence were selected for analysis in this research.

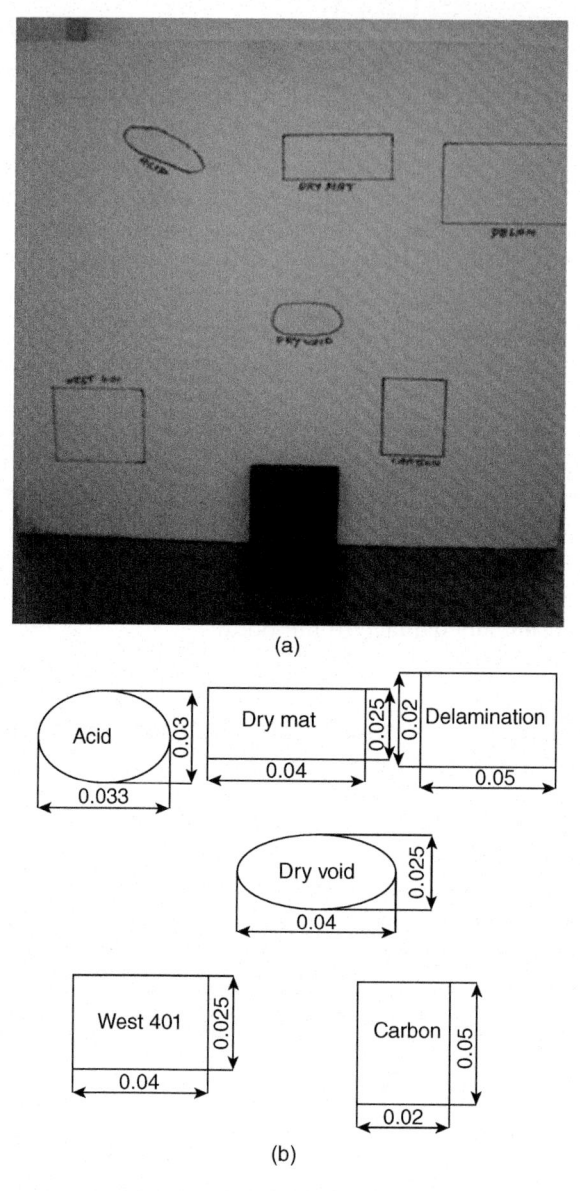

652 Non-destructive evaluation (NDE) of polymer matrix composites

25.1 (a) Defect simulation panel 1 and (b) defect simulation panel 1 with dimensions (in meters).

A brief description of different anomalies in the test panel simulating a boat wall or hull material is provided below:

- Dry void – A manufacturing defect where air is trapped between layers of fiberglass so that there is glass fiber but no resin to bond it.

- Acid – An area caused by osmosis where the ingress of water has reacted with any of various substances that are hygroscopic, left during construction.
- Carbon fiber – An area of the hull where carbon fiber has been used instead of glass fiber for reasons of added strength or repair.
- Delamination – An area of the hull where one layer of fiberglass has separated from the adjacent layer, possibly accompanied by an air gap. This may be caused by damage due to build-up of pressure at an osmosis site or an overstress condition as caused by an impact or collision with another vessel, or improper positioning of straps or supports.
- West 401 – This is one of many filler materials used to restore surface defects and repair the boat's shape.
- Dry mat – A layer of fiberglass which is not properly filled with resin during manufacturing, causing the entrapment of air and consequent weakness of structure.
- Kevlar – Kevlar is a kind of material used to strengthen the boat. If Kevlar is located at the machining location when assembling the boat walls, it is difficult to machine at that position.

25.3.2 Simulated test panel 2

Another testing panel was provided by Hoyles Electronics Inc. for further parametric studies. This testing panel or prototype also consisted of joints of boat hull or wall material. The internal structure of the boat is developed by joining the boat walls by bolting two separate walls. In this testing panel, simulations of real defects that have been observed in real boat walls have been included. This panel consistsed of the same defects as test panel 1, but the locations of defects were at different layers (Fig. 25.2).

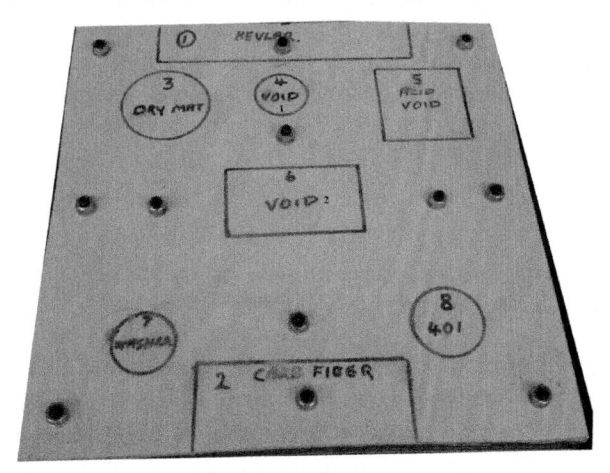

25.2 Defect simulation test panel 2.

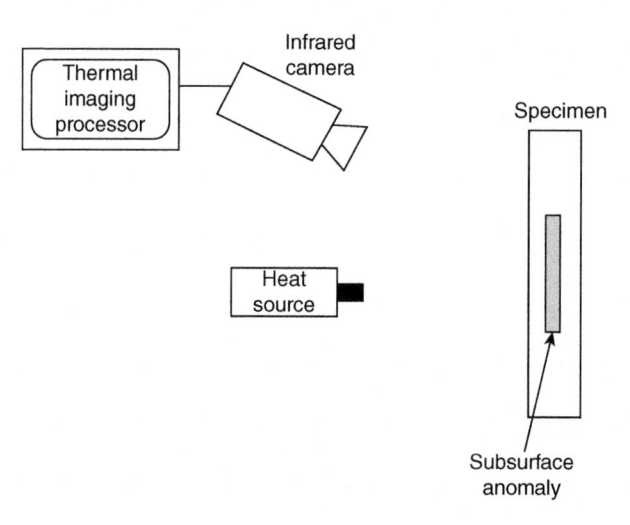

25.3 Schematic of experimental set-up.

25.3.3 Experimental set-up

The schematic of experimental set-up is shown in Fig. 25.3. The infrared camera (IR Flexcam Pro, Infrared Solutions) provided measurement of a wide range of thermal images. This camera recorded the radial surface temperature of objects. The distance between the thermal camera and testing panel was maintained. Hence, the focused area on the testing panel for every defect was the same. Computerized data acquisition was used to record thermal images, and the data obtained was processed with National Instruments IMAQ Vision Builder Image processing software. A heat gun (Master-Mite Heat gun – 10008) was used as a continuous heat source to heat the defects. The distance between the heating element and the test panel for each experiment was kept constant. A gap of approximately 2 cm was maintained between the heat source and the test panel while heating the test panel. The heat gun dissipated heat over a 10×10 cm^2 area on the testing panel.

25.4 Assessing the effectiveness of infrared thermography (IRT)

In this research, a non-destructive thermography method was used for defect detection in a boat hull or wall. Experiments were performed on a boat wall prototype testing panel which included different defects that might be found on a boat hull or wall. The primary parameter considered while using the thermography technique was particular time periods for heating the testing panel. Experiments were performed for different heating periods in order to find the optimum time period and efficient heating time to recognize all defects. These results were compared with computerized numerical analysis obtained using COMSOL.

25.4.1 Results of non-destructive detection of defects in composite boat hulls

The results obtained from experiments and numerical simulations for non-destructive detection of defects in boat hulls and walls are presented and discussed here. The goal of this thermography imaging technique was to measure the temperature decay for different defects and no-defect cases. This detection method has been developed for the boat maintenance industry; hence, minimal time consumption and higher accuracy of results in identification of defects in boat walls or hulls were necessary. In practical applications, there are many defects other than the ones studied in this research. In order to validate the results, some of the defects are compared with simulation studies. Hence, these defect thermal response results are compared with simulations of COMSOL Multiphysics finite element analysis (FEA) software. For further research to study new defects, simulation studies would be sufficient. Hence, this method gives different material properties and helps identify defects using the thermal response of different materials.

Time history analysis

All the defects considered in this research were heated for different time periods: 15 seconds, 30 seconds, 45 seconds, and 1 minute. Since initially the thermal camera was set at a particular level, it was necessary to check whether the temperature range of heated defects was in the range of thermal camera. The thermal camera was set at particular temperature level so that it could record maximum thermal changes. For the thermography decay tests performed, the thermal camera was set at level 73, thus covering a temperature range from 30.3 °C to 73.7 °C.

This range level was selected based on the maximum temperature observed during the time history analysis. A time history plot shows the temperature range of all defects after heating for different time periods. Figure 25.4 shows all defects in the range of the set thermal camera level. Hence, level 73 of the camera is the optimal level of experiment.

Characterization of different defects in test panel 1

In order to characterize the different defects by using a non-destructive thermography technique, a defect-embedded prototype testing panel was

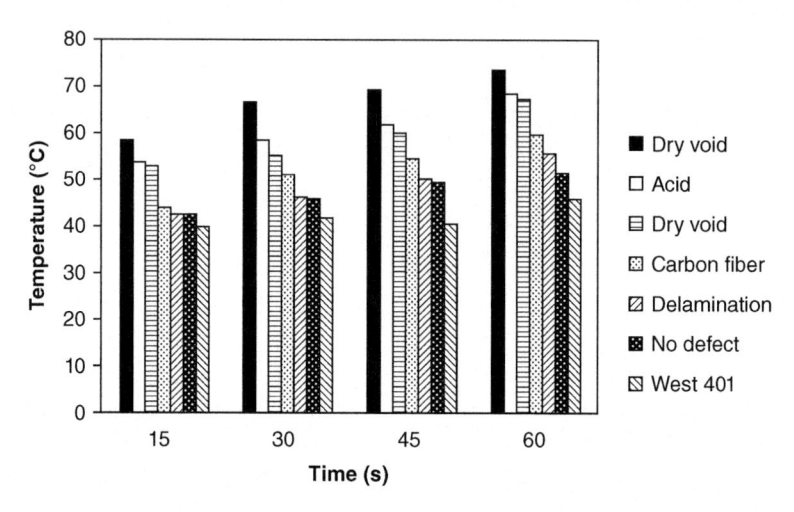

25.4 Time history plot for dry void.

used. All the defects considered in this experiment were heated for different time periods with the thermal camera set at level 73. The thermal camera was used to record the thermal intensity decay data with time after samples had been heated for a finite time with a heat gun.

When the testing panel had been heated for 15 seconds, a significant temperature rise was not observed for all defects. Also, thermal decay plots of some of the defects overlapped after a particular time interval. Hence, it is difficult to determine thermal decay of the defect. However, when the defect is heated for 1 minute, the thermal camera produced saturated images for some defects such as dry void and acid. Therefore, data obtained after heating the defects for 1 minute is not desirable. Hence, for the cases considered in this research, it was observed that optimal heating times for defect detection are 30 and 45 seconds, depending upon the type of defect.

Figure 25.5 shows the normalized temperature decay in different defects after heating each defect uniformly and separately for 30 seconds. In this figure, temperature is normalized with respect to the peak temperature (340 K) of defect, which is dry void for the defects considered in this experiment. This figure shows that dry void attains the highest peak temperature value when heated for 30 seconds followed by acid. Dry void is a defect of a trapped air vacuum, which is less dense than other defects; hence, a higher temperature occurs for dry void than for other defects. However, West 401 attains lowest temperature decay when heated for 30 seconds because West 401 is filler material (glass, wool, etc.) and has the lowest heat absorptivity. It is also observed from Fig. 25.6(b) that carbon fiber, delamination, dry mat, and no-defect cases are difficult to distinguish.

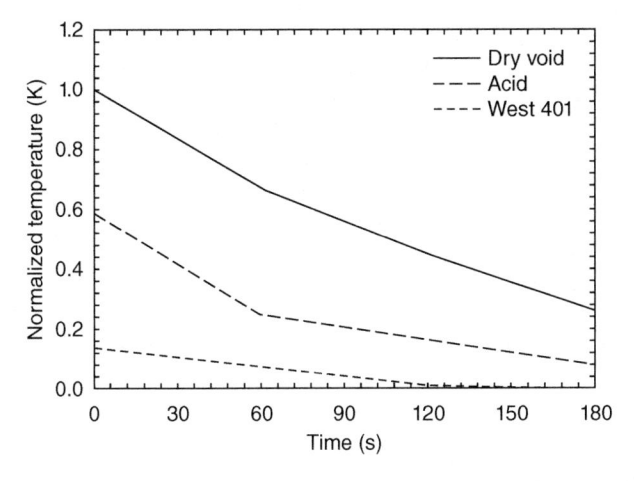

25.5 Temperature decay for dry void, acid and West 401 heated for 30 seconds.

Sample thermal images showing intensity decay as a function of time are shown in Figs 25.7(a) and (b) when a carbon fiber defect is heated for 30 seconds and 45 seconds respectively. Thermal intensity images show a more prominent thermal decay when heated for 45 seconds compared with heating for 30 seconds. Since dry void, acid, and West 401 are clearly distinguishable for 30-second heating cycles, remaining defects (i.e. carbon fiber, delamination, and dry mat) and no-defect cases were heated for a time period of 45 seconds; the corresponding temperature decay is shown in Fig. 25.8.

Figure 25.8 shows that dry mat defect attains a maximum temperature when heated for 45 seconds, followed by the temperature decay of carbon fiber and delamination defects. The no-defect case results in a minimal temperature rise. Thus, the defects that did not give conclusive results for 30 seconds of heating produced a satisfactory temperature decay when heated for 45 seconds.

Characterization of different defects in test panel 2

In order to study more defects and the effect of defect location (depth), another testing panel was provided by Hoyles Electronics Inc. This testing panel or prototype also consisted of joints of boat hull or wall material and a Kevlar defect together with the defects previously mentioned in panel 1. The defect location in this panel was much deeper than in the first test panel, thus providing a more realistic prototype with similar practical defects to those observed in a boat wall. This test panel consisted of two

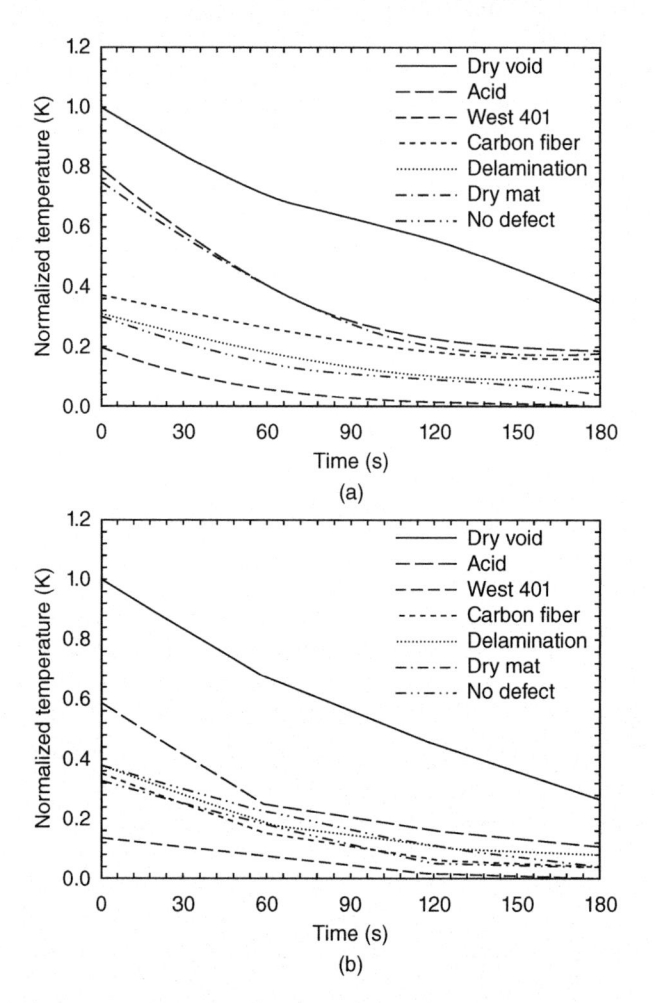

658 Non-destructive evaluation (NDE) of polymer matrix composites

25.6 Normalized temperature decay for different defects heated for (a) 15 seconds, (b) 30 seconds, (c) 45 seconds, (d) 1 minute.

types of dry void defect. These defects were located in different layers of the testing panel. Time history analysis and defect characterization tests were performed on this testing panel similar to those performed on the first panel. This test panel was also suitable to examine for 30 and 45 seconds to identify different defects.

Figure 25.9 shows the normalized temperature decay in different defects after heating each defect uniformly and separately for 30 seconds. In this figure, temperature is normalized with respect to the peak temperature (340 K) of defect, which is dry void for the defects considered in this experiment. Dry void 1 and dry void 2 were located at different layers in testing

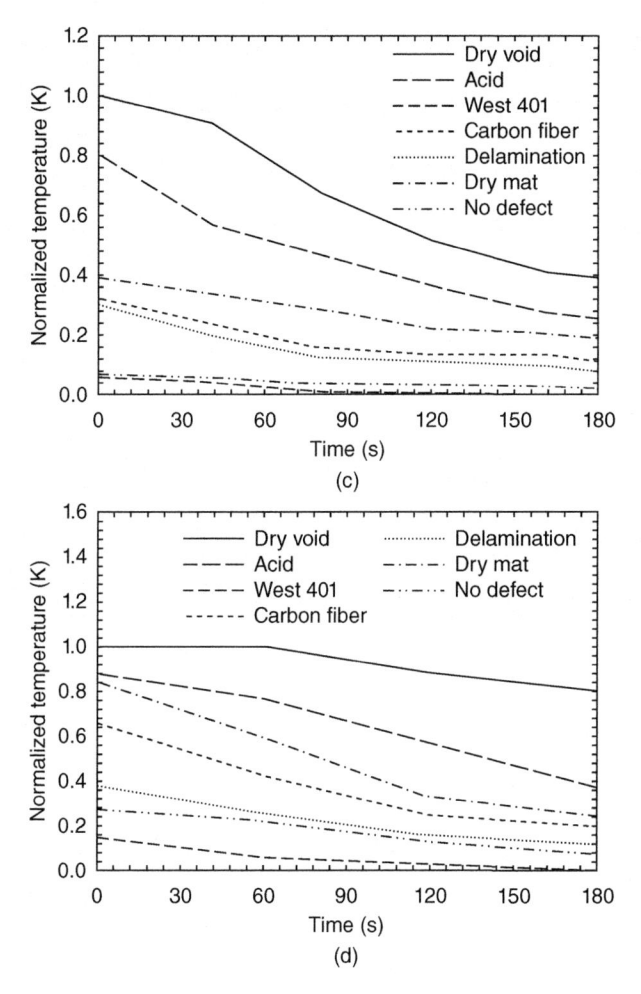

25.6 Continued

panel. Dry void 2 was deeper than dry void 1, and it also had residual resin fibers, which simulated dry void forming condition, whereas dry void 1 simulated a completely formed void without any material in the boat wall. Therefore, the density of the defect dry void 1 was less than that of dry void 2. Moreover, the heat response depended on the location (depth) of the dry void. Hence, this figure shows different thermal responses for two dry voids after 30 seconds of heating the defects. The thermal response of acid is followed by that of dry void.

The defects where significant temperature decay was not observed when heated for 30 seconds were heated for 45 seconds (Fig. 25.10). Satisfactory results were obtained when the defects were heated for 45 seconds. Carbon

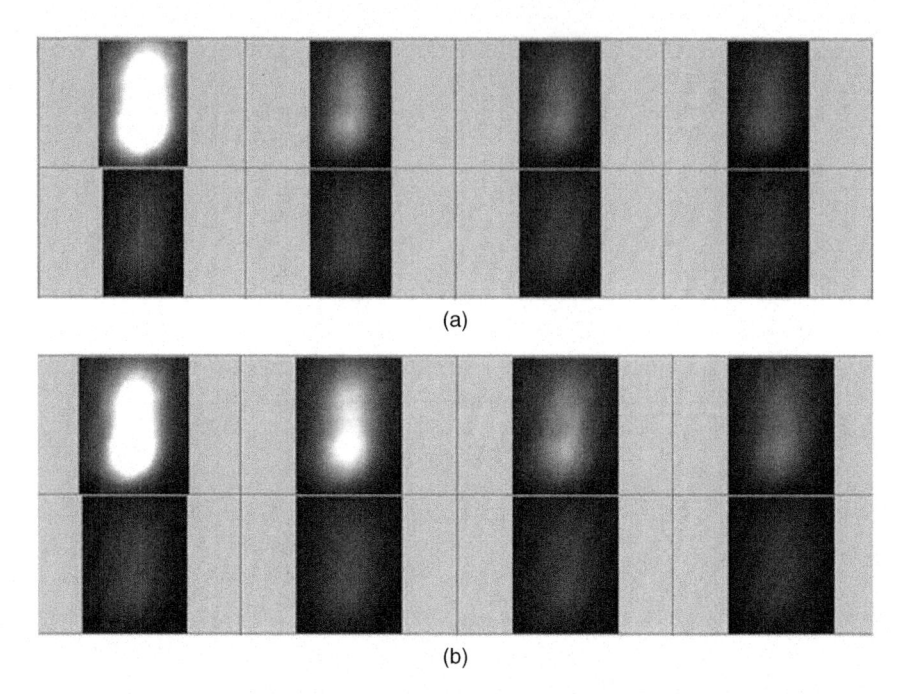

25.7 Thermal camera images of intensity decay after (a) 30 seconds and (b) 45 seconds heating for carbon fiber.

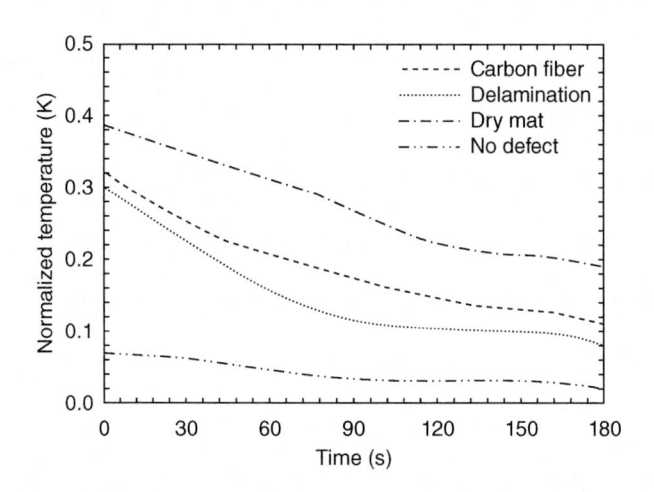

25.8 Temperature decay for different defects heated for 45 seconds.

fiber was first detected, followed by dry mat, washer, and no-defect cases. Figure 25.11 shows the comparison study of temperature decay patterns of all the dry void defects from both the panels. Dry void from panel 1, which is closer to the surface, was detected first. Hence, it absorbed heat energy

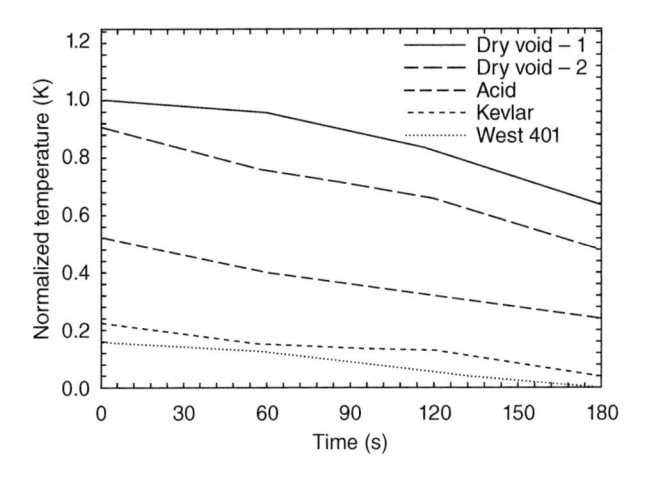

25.9 Temperature decay for different defect heated for 30 seconds – sample 2.

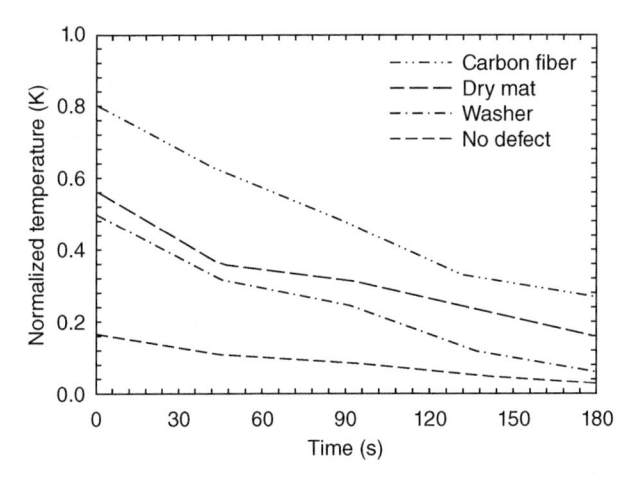

25.10 Temperature decay for different defects heated for 45 seconds – sample 2.

faster than the other defects. Hence, the temperature obtained from panel 1 dry void is higher than the others. This is followed by dry void 1 from panel 2. Dry void 1 is deeper than the dry void defect from panel 1. Dry void 2 from panel 2, which was the deepest located dry void defect of the three, was detected last.

Numerical simulation

COMSOL multiphysics simulation software was used to illustrate temperature decay in defects by considering FEA. For the computational

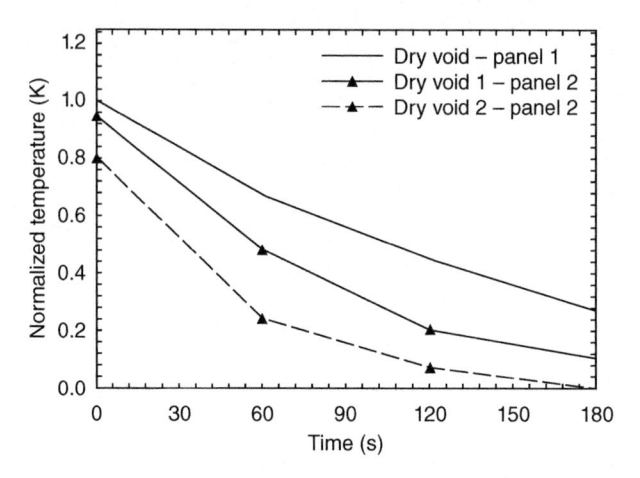

662   Non-destructive evaluation (NDE) of polymer matrix composites

25.11 Comparison of temperature decay for different dry void defects heated for 30 seconds.

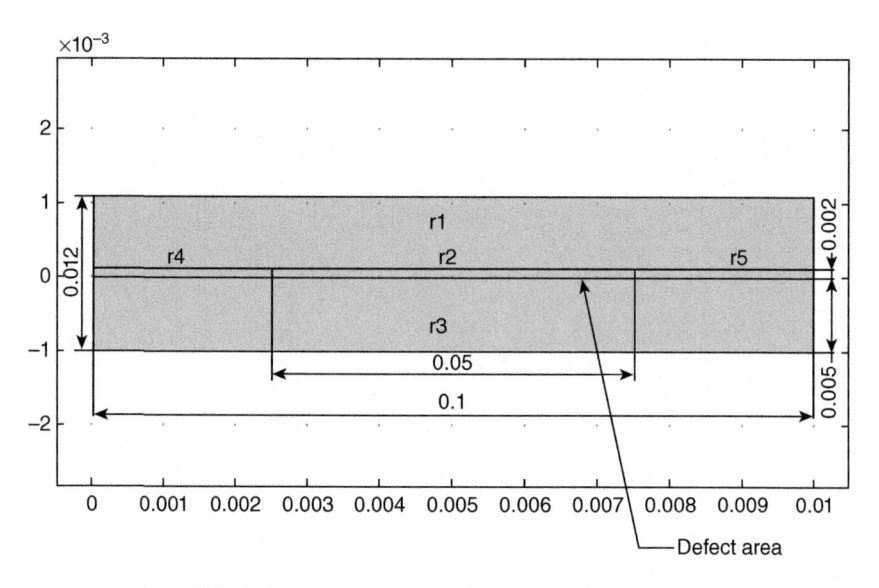

25.12 COMSOL model of defect simulation (in meters).

simulation, two defects, carbon fiber and delamination, were considered. Figure 25.12 shows the modeled defect in COMSOL software for the simulation of results. Areas r1, r3, r4, and r5 were developed in the COMSOL model feature. These areas were represented as normal boat wall resin material, and for COMSOL simulations 'resin-bonded glass fiber' was used from the material list. However, r2 was represented as a defect area. This defect area was placed in such way that it simulated a subsurface defect.

Table 25.1 Thermophysical properties of defects

	Dry void	Carbon fiber	Delamination	West 401	Resin-bonded fiberglass
Density kg/m^3	1.005	2.0	1.057	25	1.35
Thermal conductivity W/m K	0.0257	0.06	0.0272	0.04	0.541
Heat capacity J/kg K	1005.0	1850.0	717.8	840	1130

The thermophysical properties of carbon fiber and delamination as used in the simulation are given in Table 25.1.

A 2D thermal heat diffusion equation in the Cartesian coordinate system was applied to analyze the thermal response in composite boat hull material:

$$\frac{\partial T}{\partial t} = \alpha \frac{\partial^2 T}{\partial x^2} + \alpha \frac{\partial^2 T}{\partial y^2} + \frac{\dot{q}}{\rho c_p} \quad [25.1]$$

Here, ρ represents the density of material, C_p is the heat capacity, K is thermal conductivity, and α is the thermal diffusivity.

The subsurface defects were represented as inclusions of material with thermal properties different from the boat hull properties. An initial heated temperature equal to 350 K was considered. The ambient temperature conditions were taken as 300 K with surface boundary conditions as normal air cooling heat flux 40 W/m^2. With these boundary conditions, COMSOL software was used to compute the temperature decay for the case of the boat hull having delamination and carbon defects.

Plate XVII (between pages 296 and 297) shows the thermal response obtained using numerical simulation for the case of delamination defect. The color ranges display temperature variation across the defect area. Dark red and blue represent the highest and lowest temperatures in the delamination defect model respectively. The goal of simulation is to validate the temperature decay in defects as shown in Fig. 25.6 and 25.7. Figure 25.13 shows a comparison between experimental measurements and a numerical simulation for temperature decay in carbon fiber and delamination in the r2 region (i.e. defect area). Error bars in Fig. 25.13 represent the deviation of the thermal response during experiment sets. The standard deviation and precision index for three complete runs are calculated at each time point.

For each time point, uncertainty values, which are a product of the precision index and standard deviation, are plotted (those plotted in Fig. 25.13

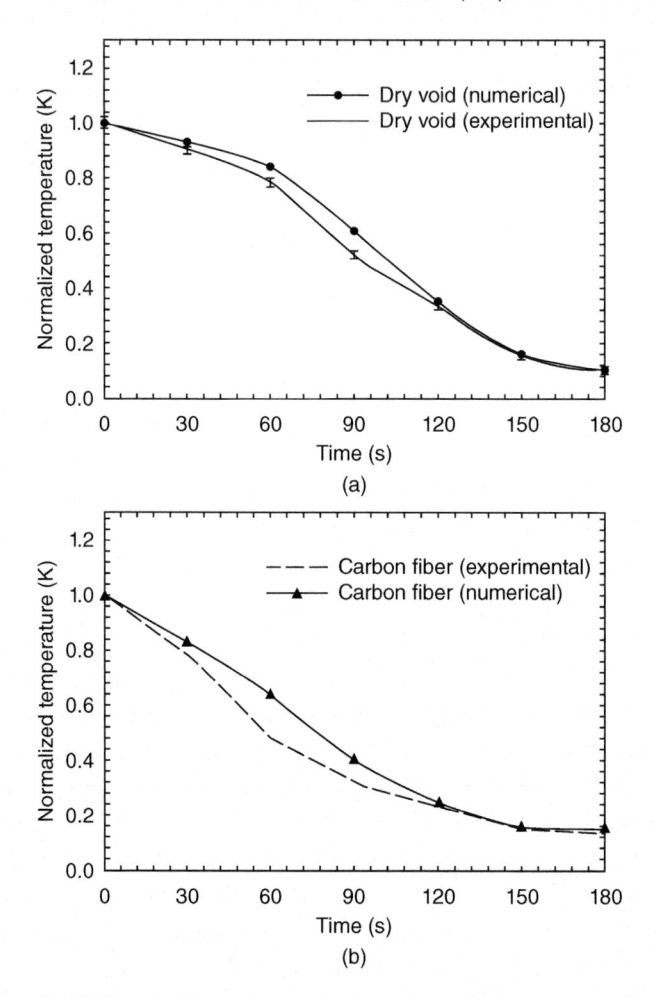

25.13 Comparison of temperature decay between experimental data and numerical simulation in (a) dry void, (b) carbon fiber, (c) delamination, and (d) West 401 defects.

for carbon and delamination defects for other tests were similar but were omitted from the figure for clarity). This error can be due to uncertainties in the thermal camera angle and variations in the surrounding atmosphere. Moreover, there is a slight difference in numerical and experimental results because the thermophysical properties of the actual defect may be different from the values used in numerical simulation. Thus NDT is helpful in detecting material defects. Results show that thermographic testing yields the best and fastest results for big structures such as a boat wall or hull.

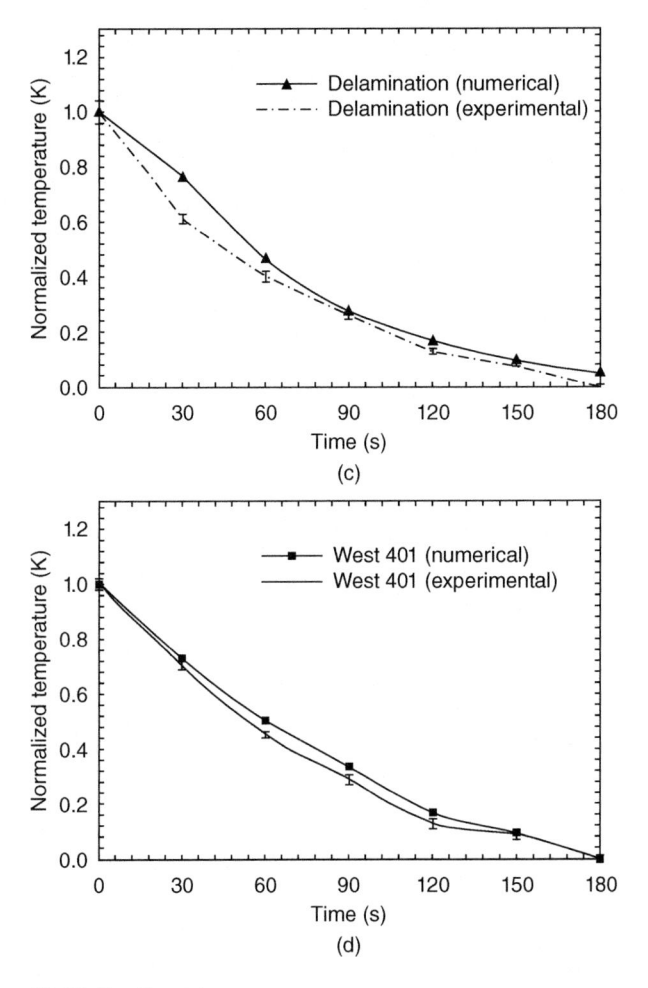

25.13 Continued

25.5 Conclusion

In this research, non-destructive testing methods have been used to identify and characterize different subsurface defects. The thermography technique was used to characterize the different types of subsurface defects on a boat hull by its thermal response. Thermographic techniques were mostly applied to detect defects in large structures. The thermography non-destructive technique was used to detect various types of defects in a boat hull or wall by analyzing the temperature decay with time. The defects considered in the experimental analysis were dry void, acid, West 401, dry mat, delamination, and carbon. The heating times of dry void, acid, and West 401 defects were shorter, and hence, easily detected, whereas identification of carbon, delamination, and dry mat defects took longer. A variety of defects were differentiated by using this method, and quick results could be obtained by it. The results were validated using parametric experimental studies. However, for real-time applications, simulation studies were performed using COMSOL Multiphysics. The defects used for simulation studies were delamination and carbon. The challenging task for numerical validation with experimental measurements is to obtain accurate thermophysical properties of various defects. Parametric study needs to be performed to obtain thermophysical properties for defects not available in the literature. The person performing the experiment needs to take the precaution of setting the temperature level of the camera to avoid saturation of thermal images.

The main advantages of the thermography technique are its speed and accuracy in detecting subsurface defects in composite materials. Moreover, it provides us with information regarding the shape and size of the defect, thereby making it easy to cure these defects in composite boat wall material. The instruments used in detecting these defects are handy so it is easy to work in the field. However, there are some limitations of this technique. During thermography there is reflection from other surfaces. Also, multiple defects can hinder accurate temperature measurements. Care must be taken to ensure that the thermal camera used does not get saturated during heating, otherwise it will not give accurate results.

25.6 References

Avdelidis, N. P., Moropoulou, A. & Almond, D. P., 2004. Passive and active thermal non-destructive imaging of materials. *Electro-Optical Imaging of Materials*, pp. 126–140.

Avdelidis, N. P., Almond D. P., Marioli-Riga Z. P., Dobbinson A. & Hawtin B. C., 2006. Pulse thermography: philosophy qualitative and quantitative analysis on aircraft materials and application. *Thermal NDT Journal, Insight*, Volume 48, pp. 289–289.

Barndo, A. & Chang, C., 2003. Fire fighter–Robot interaction during a hazardous materials incident exercise. *11th International Conference on Advanced Robotics*, Volume 2, pp. 658–663.

Barone, S., Paoli, A. & Razionale, A. V., 2006. A biomedical application combining vision and thermal 3D imaging, XVIII Congraso International de Ingenieria Grafica Siteges.

Berger, H., 1977. *Nondestructive testing standard*, ASTM International.

Cartz, L., 1996. *Nondestructive Testing – Radiography, Ultrasonic, Liquid Penetrant, Magnetic Particle, Eddy Current*, ASM International.

Diakides, N. A. & Bronzino, J. D., 2007. *Medical Infrared Imaging*, 10 ed, CRC Press.

Gerlach, N., 2006. *Comparison of Thermal Imaging Systems Used in Thermography as a Nondestructive Testing Method for Composite Parts*, European Federation for Non-destructive Testing.

Kaplan, H., 1993. *Practical Applications of Infrared Thermal Sensing and Imaging Equipment.* 3 ed., SPIE.

Minkina, W. & Dudzik, S., 2009. *Infrared Thermography: Errors and Uncertainties*, John Wiley and Sons.

Mix, P. E., 2005. *Introduction to Nondestructive Testing*, John Wiley and Sons Inc.

Pitarresi, G., Licari, A. & Pasta, A., 2009. *Thermal NDT of glass reinforced composited for noval applications by means of a linear infrared scanner.* Messina, 16th International Conference of Ship and Shipping Research.

Raj, B., Jayakumar, T. & Tavasimuthu, M., 2002. *Practical Non-destructive Testing* 2nd edn, Woodhead Publishing Limited.

Servais, P. & Gerlach, N., 2005. Development of a NDT method using thermography for composite inspection on aircraft using military thermal imager. Quebec City, Canada, Fifth Workshop Advances in Signal Processing for Non Destructive Evaluation of Materials.

Sheppard, P. J., Phillips, H. J. & Cooper, I., 2009. The practical use of NDE methods for the assessment of damaged marine composite structures. Edinburgh, UK, 17th International conference on composite material (ICMM 17).

Therianult, A. J., 2001. *Thermal Imaging Camera: Developing a Firefighter Proficient Evaluation,* Fire Service Finance.

Vavilov, V. P. & Taylor, R., 1982. *Theoretical and Practical Aspects of the Thermal NDT of Bonded Structures,* 5 ed., Academic Press.

Vavilov, V., Marinetti, S. & Nesteruk, D., 2008. *Accuracy Issues in Modeling Thermal NDT Problem,* SPIE International Society for Optical Engineering.

Index

ABAQUS/EXPLICIT, 247
acid, 653
acoustic emission (AE), 4, 12–30, 230, 625
  challenges, 566–72
    improved b-values, 567–72
  fundamentals, 13–23
    burst and continuous emission, 22
    detection, 20–3
    extensional and flexural wave modes, 20
    sources in PMC, 14–15
    types of noise, 15–16
    wave basics, 16–20
    wave propagation at and across the interface of two medium, 17
    waveform and measurable parameters, 22
  future trends, 28–30
  non-destructive evaluation (NDE) of composites, 557–72
    testing acoustic techniques, 559–66
  source locations, 566–7
    CB specimen, 568
    S1-C1.0 specimen, 569
  technique comparisons, 27–8
  testing, 23–7
    data analysis, 25–7
    EMI signal identified from waveform data analysis, 26
    specimen loading test setup, 24
    system setup and calibration, 23–5
    test conduction, 25
    test plan, 23

acoustic microscopy, 424, 426–7, 434
  case study of damage analysis and discussion, 431–40, 435, 438–9
  effects of stacking sequence, stacking parameter βT2, 435, 438–9
  relative impact resistance on various impact conditions, 439
  NDE of aerospace composites, 423–46
    damage analysis using scanned image microscopy, 424–31
    future trends, 440–4
acousto-ultrasonic (AU) technique, 404–5
acoustography, 511
Active Fibre Composite (AFC), 452
active IR thermography, 233
active non-destructive evaluation, 310, 312–13
active thermography, 519–20
adhesive bond, 195
  aerospace industry, 221–2
    advantages, 221–2
    evolution, 221
    prerequisite, 222
  challenges of NDT, 233–5
    bond performance assessment, 234
    extended methods, 234–5
  non-destructive testing (NDT), 222–7
    defects in a bonded honeycomb/skin construction, 225
    defects in bonded joints, 223
    defects presentation, 223–4

Index

quality assessment of adhesive bonded parts, 227
quality assessment prior to bonding processes, 224, 226
quality assessment within bonding processes, 226–7
testing, 202–3
adhesively bonded composites, 220–36
  adhesive bonding in aerospace industry, 221–2
  challenges in non-destructive testing (NDT) of adhesive bonds, 233–5
  methods of NDT, 227–33
  NDT in testing adhesive bonds, 222–7
aerospace composites
  experimental demonstration, 359–63
    case study, 359
    log–log decay of the temperature response, 360
    plot of temperature profile, 363
    time variation in the temperature profile, 361
    traces of log decay and second derivative on the laminate, 362
  heat propagation in dynamic thermography, 315–20
  infrared thermography, 309–32
    overview, 309–15
  inspection, 398–402
    cross-sectional view of woven carbon composite containing composite, 399
    damage morphology of impacted honeycomb sandwich, 401
    disbond failure between facesheet and honeycomb core, 402
  NDE of detecting impact damage, 367–93
    different composite materials, 384–92
    future trends, 392–3
    infrared thermography effectiveness, 369–72
    on-line monitoring, 372–84
  NDE of flaw characterisation, 335–64
    delaminations and planar inclusions, 346–54
    fundamentals of heat diffusion, 337–46
    future trends, 363–4
    image damage, 354–6
    porosity, 356–9
  NDE using acoustic microscopy, 423–46
    acoustic microscopy for damage analysis, 431–40
    future trends, 440–4
    scanned image microscopy for damage analysis, 424–31
  NDE using guided wave ultrasonics, 449–68
    GW SHM systems for composite structures, 455–66
    SHM transducer systems, 451–5
  NDE using ultrasonic techniques, 397–420
    NDT instruments, 419–20
    ultrasonic inspection methods, 402–6
    ultrasonic inspection of sandwich structures, 415–19
    ultrasonic inspection of solid laminates, 406–15
  non-destructive evaluation (NDE), 220–36
    adhesive bonding in aerospace industry, 221–2
    challenges in NDT of adhesive bonds, 233–5
    NDT in testing adhesive bonds, 222–7
    NDT methods, 227–33
  thermography, 321–32
    case studies of applications, 321–9
    other applications, 329–31
    vs other NDE techniques, 331–2
aerospace industry
  adhesive bonding, 221–2
    advantages, 221–2
    evolution, 221
    prerequisite, 222

Index

Agilent E1441, 254
Agilent E1437A, 254
Agilent E3242A, 254
air-coupled ultrasonic test, 405–6
  sandwich structures, 417
aluminium, 201–7
amplitude sensors *see* intensity sensors
anisotropy, 404–5
ASTM D 4788 - 03 (2007), 490
attenuation, 16
AutoCAD, 523, 526
autoclave curing, 175–6
  ultrasound measuring assembly, 176
automotive engineering, 160–78

B-scan *see* longitudinal GPR scan
b-values
  improvement, 567–72
    amplitude distributions and improved *b*-values with each damage level, 571
barely visible impact damage (BVID), 321, 354, 400
belt pulleys, 160–3
  correlation between sound velocity and belt pulley strength, 163
  mould for injection compression moulding, 161
  phenolic moulding compound, 160
  sound velocity vs process time and injection moulding process, 162
boat hull
  NDE of defects, 651–4
    experimental set-up, 654
    schematic diagram of defect simulation panel, 652
    simulated test panel 1, 651–3
    simulated test panel 2, 653
  results of non-destructive detection of defects, 655–65
Born approximation, 605
Bragg grating sensor, 620
Bridge Design Specifications (BDS), 519
Brillouin scattering sensor, 620–1
broadband piezoelectric transducers, 559

capturing frequency, 520
carbon fibre, 653
  bonded joints, 211–16
    dielectric permittivity and loss changes of CFRP-epoxy laminate, 214
    TDR traces as a function of exposure time and changes for a CFRP-epoxy laminate, 215
    water absorption against square root of time for CFRP-epoxy bonded joints, 213
  eddy current testing (ECT), 37–40
    electric and dielectric behaviour, 38
    frequency properties of dielectric materials, 39
  high frequency eddy current (HF EC), 43–8
    method, 43–5
    testing non-woven laminated materials, 45–8
    testing raw multiaxial materials without resin, 45
carbon fibre-reinforced (CFR) composites
  ultrasonic 'C' scanning, 292–5
carbon fibre-reinforced polymer (CFRP), 385–8
  CFRP-F1, 386
  CFRP-U_{tk} phase images, 386
  high frequency eddy current (HF EC), 43–8
    method, 43–5
    testing non-woven laminated materials, 45–8
    testing raw multiaxial materials without resin, 45
  phase contrast distribution, 387
  phase image taken at 0.5 Hz of specimen type after impact at 2.8 J, 385
charge-coupled device (CCD), 545
Charpy pendulum, 371
civil structures
  non-destructive evaluation (NDE) of composites, 483–511

Index

digital tap testing, 507–8
future trends, 510–11
ground penetrating radar (GPR), 501–7
infrared thermography, 484–500, 509
issues and challenges, 508–10
comparative vacuum monitoring (CVM) systems, 454–5
operating principle of CVM sensor, 454
complex permittivity, 580–2
relative permittivity and loss tangents, 581
composite assembly
bonding integrity evaluation, 107–9
adhesive-bonded composite shearographic inspection, 109
composite-composite joints
ageing, 209–16
adhesive thickness effect on dielectric measurements, 211
carbon fibre bonded joints, 211–16
water uptake effect on the adherent dielectric properties, 209–11
composites
NDE for civil structures, 483–511
digital tap testing, 507–8
future trends, 510–11
ground penetrating radar (GPR), 501–7
infrared thermography, 484–500, 509
issues and challenges, 508–10
NDE using shearography to detect bonded defects, 542–54
field inspection of FPR strengthened bridge, 552–4
shearography, 544–6
shearography role in detecting defects, 546–51
non-destructive evaluation (NDE), 12–30, 33–54, 56–79, 84–110, 116–33, 136–78, 185–217
AE fundamentals, 13–23
AE technique comparisons, 27–8

AE testing, 23–7
analytical methods and data processing, 48–54
automotive engineering, 160–78
curing degree and mechanical properties monitoring, 143–5
curing monitoring, 154–60
dielectric testing in bonded joints ageing assessment, 200–16
dielectric testing in bound integrity checking, 193–200
dielectric testing in cure monitoring, 188–93
digital shearography practical application, 98–101
digital shearography principles, 85–97
digital shearography to rest composites, 101–9
eddy current image of dry 3 axial carbon fibre non-woven fabric, 34
eddy current testing (ECT), 35–43
future trends, 28–30, 76–9
high-frequency dielectric measurement, 127–32
high-frequency eddy current imaging, 43–8
low-frequency dielectric cure monitoring, 120–3
low-frequency dielectric measurement of partially conductive and insulating composites, 118–20
low-frequency dielectric measurement of water ingress, 124–7
monitoring methods, 139–42
online process monitoring using ultrasound, 145–54
practical application of shearography, 62–70
shearography, 70–4
shearography vs other techniques, 75–6
theoretical principles of shearography, 57–62

Index

thermosets type, 137–9
  various types of epoxy resin polymerisation, 186
non-destructive evaluation (NDE) and acoustic emission (AE) techniques usage, 557–72
  challenges, 566–72
  testing acoustic techniques, 559–66
thermography for defect detection in rehabilitated structures, 515–40
  assessing results, 526–39
  data collection methodology, 522–6
  infrared thermography application to bridge deck assembly, 518–22
  infrared thermography principles, 517–18
computed tomography, 232
COMSOL, 546, 651, 655, 661–3
  model of defect simulation, 662
conductors, 580
coupon monitoring, 187
crack
  detection on a composite turbine blade using pulsed thermal stressing, 108
  micro-cracks in a glass fibre-reinforced plastic plate under thermal stress, 107
critical sized flaw, 6

damage analysis
  acoustic microscopy, 431–40
    gelatin impact damage, 435
    overview, 432–3
  experimental procedure, 433–4
    acoustic microscopy, 434
    gas-gun system and gelatin projectile, 434
    specimens, 433
  sample preparation, 424–5
    fabric reinforcement specification, 425
    properties of fibre and matrix, 425
    specimens stacking sequence, 425
  scanned image microscopy, 424–31
    acoustic microscopy, 426–7
    delamination image analysis, 428–30
    impact testing, 425–6
    sample preparation, 424–5
damage index, 242–3
data collection methodology, 522–6
  heat transfer profile of materials with different thermal conductivity, 525
  plot showing defect extent and ROIs, 527
debonding
  assessment, 256–8
    experimental results, 257
  detection and assessment in sandwich beams, 252–9
  detection in laminated composites, 101–2
    fringe patterns, 102
  severity, 258–9
    debonding extent and reflection coefficient, 260
    extent and delay in ToF in sandwich composite beams, 259
    magnitude of debonding-reflected signals, 260
defects, 280, 516
  assessment of localised area, 535–9
    region showing enlarging defect, 537
    region showing pre-existing defect, 536
    region showing three defects, 538
    region with no indication of defect, 539
  NDE using thermography for defects in rehabilitated structures, 515–40
    application to bridge deck assembly, 518–22
    assessing results, 526–39
    data collection methodology, 522–6
    principles of infrared thermography, 517–18
delamination threshold load (DTL), 368
delaminations, 280, 653

characterisation of depth, 347–51
  case descriptions and results, 348
  material thermal properties, 348
  modelled axisymmetric structure, 348
  second derivative of the surface response, 349
depth profiling for lateral size, 353–4
  plots the depth profile, 354
image analysis, 428–30
  acoustic images at 30 MHz, Plate VII
  delamination areas calculated from acoustic images, 428
  SEM images, 429, 430
lateral sizing, 351–3
  surface temperature profiles, 351
NDE using mechanical impedance, ultrasonic and IRT techniques, 279–305
  carbon fibre-reinforced (CFR) composites, 292–5
  comparison of techniques, 301–5
  disbonding in aluminium honeycomb structures, 281–92
  infrared thermography, 295–301
  non-destructive evaluation (NDE), 346–54
dethermalisation factor, 357
diagnostic signal generation, 8
dielectric loss, 190
dielectric permittivity, 199
dielectric techniques
  high-frequency dielectric measurement, 127–32
  low-frequency dielectric cure monitoring, 120–3
  low-frequency dielectric measurement of partially conductive and insulating composites, 118–20
  low-frequency dielectric measurement of water ingress, 124–7
  partially or non-conducting composite materials, 116–33

dielectric test, 4
  adhesive bonds testing in composites, 185–217
  bond integrity checking, 193–200
    high-frequency dielectric measurement system, 194
    joint structure visualisation, 194–200
    reflection traces for time domain reflectivity measurements, 196–7
    Smith's chart for adhesive bonded structure, 198
    theoretical simulation and experiment for reflection coefficient, 201
  bonded joints ageing assessment, 200–16
    composite–composite joints, 209–16
    metal–composite joints, 207–9
    metal–metal joints, 201–7
  cure monitoring, 188–93
    dielectric permittivity and loss, 189
    time temperature transformation, 192
    variation of conductivity and dipole relaxation location, 191
    various types of epoxy resin polymerisation, 186
dielectrics, 580
  materials and conductors, 577, 579–82
differential absolute contrast (DAC), 640–1
differential scanning calorimetry (DSC)
  heat flow vs temperature for epoxy resin, 141
diffraction, 16
diffractive optical element (DOE), 59
diffusion length, 339–40
digital shearography, 84–110
  practical application, 98–101
    flaw detection, 98–9
    holographic and shearographic inspection results, 98
    holographic results measurement, 99

stressing methods, 100–1
  vs ultrasound, 99–100
principles, 85–97
  delaminations in a composite honeycomb sample, 97
  derivation of optical path length change, 91
  fringe anomaly, 87
  fringe enhancement and multiplication technique, 96–7
  fringe formation, 90
  fringe interpretation, 91–3
  fringe patterns depicting the deflection derivatives, 87
  fringe phase determination, 93–6
  schematic diagram, 86
  shearing device, 88–90
  typical speckle pattern, 86
  wrapped and unwrapped phase distribution, 95
rest composites, 101–9
  bonding integrity evaluation, 107–9
  crack detection, 107
  debond detection in laminated composites, 101–2
  NDT of pressure vessels, 105–7
  pneumatic tire inspection, 102–3
  skin-core debonds in honeycomb materials, 103–5
digital tap testing, 507–8
  digital tap hammer, 508
DISPERSE, 461–2
dispersion, 16
distributed sensing, 621–2
distribution density function, 243–5
  damage position, S_0 mode captured by sensor and after Hilbert transform, 245
  representation of damage presence, 244
doubly refractive crystal, 88–9
  Wollaston prism, 88
dry mat, 653
dry void, 652
DuraAct, 452
duration time, 520

dye-penetrant testing, 4
dynamic mechanical analysis (DMA), 227
  storage and loss modulus vs temperature for epoxy resin, 142
dynamic scanning calorimetry (DSC), 227
dynamic thermography
  factors limiting inspections, 318–19
  heat propagation, 315–20
    transient thermal behaviour of a point, 316
  material interfaces, delaminations, disbonds and substructures, 316–17
dynamics of impact load (DIL), 368

eddy current, 4, 33–54, 511
  analytical methods and data processing, 48–54
    depth resolution in consolidated components, 53–4
    phase independent filter types image processing, 51
    textural analysis, 51–3
    2D fast Fourier transformation (2D FFT), 49–50
  dry 3 axial carbon fibre non-woven fabric, 34
  high-frequency imaging of carbon fibre and CFRP composites, 43–8
  testing principles and technology, 35–43
    carbon fibre materials, 37–40
    complex impedance plane, 37
    device technology for high eddy current imaging, 40–3
    probe and specimen configuration, 35
    standard penetration depth and density, 36
EddyCus, 40–1, 54
effective permittivity, 130
eigenvectors, 346
electric impedance, 457

electrical magnetic interference (EMI), 15
electromagnetic acoustic transducers (EMAT), 403, 412
electromagnetic current, 511
electromagnetic (EM) properties, 577–89
  dielectric materials and conductors, 577, 579–82
    atom in the absence of and under an applied electric field, 579
  EM spectrum, 577
    graphical example, 578
    resolution, penetrability, technology cost and frequency, 579
  measurement techniques for material characterisation, 584–9
    characterisation of FRP composites and concrete, 585
    comparison of techniques, 585
    dielectric properties, 586
    free space method, 586
    open-ended coaxial probe, 586–9
  wave propagation, 582–4
    plane wave across a medium of complex permittivity, 582
    prediction of radio system coverage, 584
electron probe imaging, 4
electronic speckle pattern interferometry (ESPI), 56
EN ISO 9000:2000, 222
Epoch 4, 420
epoxy resins, 187
event identification, 8
event interpretation, 8
excitation methods, 545
explicit finite element (eFE), 462
extensional wave, 19

Fabry–Perot type sensor, 620, 623, 627
failure threshold, 528–9
far-field approach, 129–32
  plane wave propagation through a multilayered laminate, 129
  Teflon films in a rubber-Teflon sandwich structure, 132

far infrared light, 518
Faraday's law, 36
fast Fourier transform (FFT), 408, 442–3
  basic principles, 50
  HF EC of 3 axial non-woven fabric, 50
fibre optic Bragg grating systems, 452–4
  fundamental operating principle, 453
fibre optic sensor (FOS)
  field applications to FRP rehabilitated structures, 627–8
  integration of fibre-reinforced polymer (FRP) reinforcements, 624–5
  monitoring FRP-concrete interfacial bond behaviour, 625–7
    sensor embedded within the FRP laminate, 626
  non-destructive evaluation (NDE), 617–29
  technologies, 619–23
    distributed and long gauge sensing, 621–2
    multiplexed sensing, 622–3
    types of sensor, 619–21
fibre optics-based strain measurement, 511
fibre-reinforced polymer (FPR) strengthened bridge
  description of bridge and repair works, 552
    FRP repaired zones, 552
  field inspection, 552–4
  shearographic inspection of FRP repairs, 553–4
    bonding defect and overlap of two carbon fibre sheets, 554
    inspection operation using portable Q-810 Laser Shearography System, 553
fibre-reinforced polymer (FRP), 617–18
  NDE using fibre optic sensor, 617–29
    field applications, 627–8
    future trends, 628–9

integration of FRP reinforcements, 624–5
monitoring FRP-concrete interfacial bond behaviour, 625–7
technologies, 619–23
fibre-reinforced polymer reinforced concrete (FRP RC)
  jacketing technology, 574–5
  cyclic loading tests, 575
finite difference (FD) formula, 463
finite element (FE) simulations, 546–7
  defect diameter equal to 40 mm and applied depressure, Plate XIII
  first time derivatives from a fifth degree polynomial, Plate XIV
  geometry of problem in case of depressure loading, 546
  material properties, 547
  3D GPR image of water-filled debonds, Plate XII
finite element method (FEM), 432, 441
  sandwich structure, 246–8
    mesh and geometry of honeycomb unit cell, 247
    properties of CF/EP sandwich panel, 248
flexural wave, 19
Focal Plane Array (FPA), 370
focusing beam technique, 597–8
  electromagnetic wave focused on bonding interface of FRP concrete column, 598
  time and frequency domain signal, 599
Fokker bond tester (FBT), 228, 229–30
Fourier space, 466
free space method, 586
frequency analysis, 466
full width at half maximum (FWHM), 352–3, 355

GAP-CAT-1100, 602
GAP-CAT-1200, 602
gelatin impact damage, 435
  damage profile on acoustic images, 435
  C-scan images, 436
  split damage, 437
  relation between damage area, 435
  projected damage area vs impact energy, 437
gelation, 193
glass fibre-reinforced polymer (GFRP), 377, 388–92
  comparison between images of specimen GFRP-G, 391
  infrared thermography testing for bridge decks, 487–8
  phase images comparison of GFRP-F, 390
  phase images of GFRP-F, 389
global level quantitative assessment, 530–5
  sliced strip analysis, 534–5
Green function, 604
ground penetrating radar (GPR), 501–7, 576
  GPR testing for FRP wrapped columns, 506–7
    GPR image of FRP wrapped cylinders with water-filled debonds, 507
    laboratory set-up of FRP wrapped cylinders, 506
  GPR testing for GFRP bridge decks, 504–6
    layout and dimensions of simulated air-filled and water-filled debonds, 505
    longitudinal GPR scan, Plate XI
    simulated debonds between wearing surface and underlying composite deck, 504
    three-dimensional image (C-scan), 506
  testing basics, 501–4
    components of GPR system along with air-launched radar antenna, 502
    high- frequency ground-coupled radar antennas, 503

guided waves, 239
  NDE of aerospace composites and SHM, 449–68
    GW SHM systems for composite structures, 455–66
    SHM transducer systems, 451–5
  NDE of debonding in sandwich panels, 238–74
    detection and assessment of debonding, 252–9
    future trends, 271–4
    numerical simulation of wave propagation, 246–52
    processing of wave signals, 240–6
    time reversal for debonding detection, 259, 261–71
Gutenberg and Richter formula, 570

Hanning function, 248
headlights, 164
  BMC produced by injection moulding process, 164
  compression mould with incorporated ultrasound sensors, 165
  sound velocity vs process time and curing behaviour of two BMC, 165
heat diffusion, 337–46
  estimate for slab thickness, 342–6
    derivative as a function of time, 343
    derivative of response curves, 344
    response to impulse heating, 342
  finite slab: impulse excitation, 340–1
    temperature response, 341
  half space: impulse excitation, 339–40
  harmonic excitation, 337–9
heat transfer mechanisms, 379–80, 382–4
high-frequency dielectric measurement, 127–32
high frequency eddy current (HF EC), 35
  device technology, 40–3
    EddyCus scanner, 41
    EddyCus scanner technical specification, 42
    imaging system setup for laboratory use, 40
    scalable 16 sensor demonstrator line array, 43
    sensor mounting kit and anisotropic single sensor, 42
  imaging of carbon fibre and CFRP composites, 43–8
    CFRP samples, 47
    data visualisation, 44
    delaminating of CFRP, 48
    method, 43–5
    Nyquist plot visualisation of complex impedance, 44
    RCF samples, 46
    testing non-woven laminated materials, 45–8
    testing raw multiaxial materials without resin, 45
Hilbert transform, 242
honeycomb sandwich structures, 415–16
Hooke's law, 546
Hsu–Nielsen source, 24–5

image construction, 246
image damage, 354–6
  thermographs from a flash inspection, 356
image shear, 59
IMAQ Vision Builder Image, 654
impact damage, 379–80, 382–4
  NDE of aerospace composites, 367–93
    different composite materials, 384–92
    future trends, 392–3
    infrared thermography effectiveness, 369–72
    on-line monitoring, 372–84
  impact testing, 425–6
    air-gun type of impact apparatus, 426
impedance, 16
inclusion, 317
infrared camera, 486–7
infrared monitoring (IRM), 637
infrared thermography, 484–500, 650–1
  application to bridge deck assembly, 518–22

aggregation of images, 524
  IR thermography apparatus, 520
  list of longitudinal images, 521
  list of transverse images, 522
  schematic showing strips and areas of images, 523
assessing effectiveness, 654–65
  characterisation of different defects in test panel 1, 654–7
  characterisation of different defects in test panel 2, 657–61
  comparison of temperature decay for dry void defects, 662
  normalised temperature decay for different defects, 658–9
  numerical simulation, 661–5
  temperature decay for different defects, 660, 661
  temperature decay for dry void, acid, and West 401, 657
  temperature decay of experimental data vs numerical simulation, 664–5
  thermal camera images of intensity decay, 660
  thermal response of delamination defect, Plate XVII
  thermophysical properties, 663
  time history analysis, 654, 656
assessing results, 526–39
  assessment of defects in localised area, 535–9
  global level quantitative assessment, 530–5
  normalised IR image, contour plot, and defects only plot of ROI#4, 529
  normalised IR image, contour plot, and defects only plot of structure, 531–3
boat hull defects, 651–4
delamination defects, 295–301
  detectability of defect in carbon fibre reinforced composite specimen, 301
  lock-in thermography, 297–300
  pulse thermography, 295–6
  different applications, Plate XVI
  effectiveness, 369–72
  materials, 369–70
  lock-in thermography test set up, 371
  set-up for impact tests, 371
NDE of aerospace composites, 309–32
  heat propagation in dynamic thermography, 315–20
  overview, 309–15
  thermography in aerospace composites, 321–32
NDE of marine structure composites, 649–66
NDE of wind turbine blades, 634–47
  infrared thermography (IRT), 636–9
  IRT standards, 645
  quality assurance and structural evaluation of GFRP wind turbine blades, 644–6
  signal processing techniques, 639–44
  wind turbines, 635–6
principles, 517–18
testing basics, 484–7
  experimental set-up using digital infrared camera and close-up view, 487
  infrared camera with SD card that stores radiometric JPEG images, 488
  various active heating and cooling sources, 489
testing for FRP wrapped columns, 497–500, 509
  images of CFRP cylinders with air-filled debonds of different sizes, 499
  images of CFRP cylinders with water-filled debonds of different sizes, 499
  images of GFRP cylinders with water-filled debonds of different sizes, 499

images of GFRP wrapped cylinders with air-filled debonds of different sizes, 498
photograph and infrared images showing good bond, Plate X
photographs and infrared images showing debond, Plate IX
tap testing results obtained from laboratory specimens, 509
testing for glass fibre-reinforced polymer (GFRP) bridge decks, 487–8, 490–7
digital photographs and infrared images of various debonded areas, 495
front and cross-sectional view with and without wearing surface overlay, 493
front and cross-sectional views of wearing surface, 490
infrared image of specimen BD1, 491
infrared image showing loose gravel on the surface, 497
photograph of guard rail area and infrared image, 496
schematic view of debond locations and corresponding infrared image, Plate VIII
surface temperature time-curves for air-filled debonds, 492
infrared thermography (IRT), 335
intensity sensors, 619–20
interferometric sensors, 620
inverse problem, 198
IR Flexcam Pro, 654
ISTRA 4D, 553

Kevlar, 653
kissing bond, 234

Lamb waves, 239, 455
wave propagation velocity, 254–6
pulse-echo signal and Hilbert transform, 255
skin panel vs sandwich beams, 256
laser shearography, 511

laser speckle, 57–8
coherent addition, 58–9
formation of objective and subjective speckle, 58
law of mixtures, 357
leak test, 228
line scanning thermography (LST), 314–15, 321–2
parameter optimisation, 319–20
sample generated images, 320
set-up, 315
liquid epoxy
curing monitoring, 157–60
sound velocity and thermocouple resin temperature vs process time, 159
sound velocity vs process time at different temperatures, 158, 159
lock-in thermography (LT), 297–300, 304, 314, 510
actual vs lock-in thermographic size, 298
images of phase differences, Plate II
phase differences, 299
phase differences vs testing frequency, 300
schematic diagram, 298
specimen image, Plate IV
thermograms and phase profile plots, Plate III
longitudinal GPR scan, 504
lorry cabins, 164–9
insertion of the sensors into the mould from the back, 168
RTM mould upper part showing sensors position, 167
RTM mould with partial filling and possible sensor positions, 166
RTM part produced by Fritzmeier GmbH, 167
sound velocity vs process time and flow front detection by signal onset, 168
sound velocity vs process time for the complete RTM process, 169
Love wave, 455

low-frequency dielectric measurement
  cure monitoring, 120–3
    dielectric loss variation, 123
    dielectric permittivity and loss variation for cure of thermoset resin, 122
    Wheatstone bridge configuration, 121
  partially conductive and insulating composites, 118–20
    interdigitated electrode, 119
    voltage and field distribution for an electrode, 120
  water ingress, 124–7
    composite high-pressure trunking, 124
    dielectric permittivity and loss variation for GRP composites, 125
    dielectric permittivity variation, 126

marine structures
  NDE using infrared themography, 649–66
    assessing effectiveness, 654–65
    boat hull, 651–4
    infrared thermography, 650–1
Matlab, 523–4
mechanical impedance, 281–92, 304
  dimensions of specimen, 282
  experimental set-up, 281
  results, 284–91
    experimental vs theoretical resonant frequency, 285
    frequency variations, 290
    irregularly shaped defects, 286–8
    resonant frequency against distance, 287, 288
    resonant frequency change across irregularly shaped defects, 289–91
    resonant frequency plot, 289
    resonant frequency vs defect size, 285
    resonant frequency vs depth, 284
    resonant frequency vs size, 284–6
    scanning directions for different defects, 286
    schematic illustration of an elliptical defect, 288
    schematic outline of defect, 290
  test procedure, 283–4
  test specimens, 282–3
mechanical noise, 15
metal–composite joints
  ageing, 207–9
    dielectric permittivity and loss for a CFRP–epoxy–aluminium joint, 208
    TDR traces for the CFRP–epoxy–aluminium joint, 210
metal–metal joints
  aluminium ageing, 201–7
    adhesive bond testing, 202–3
    joint testing with poor interfaces, 203–7
    variation of TDR traces as a function of time of exposure to water, 204, 206
    variation with ageing time of the surface resistivity, 206
    variation with time and frequency of the dielectric permittivity, 202, 205
MIA 3000, 283–4
  frequency spectra, 283
Michelson shearing interferometer, 89–90
  pase-shifter
    digital shearography phase-shift capability, 96
    schematic diagram, 89
microwaves, 511
  electromagnetic properties of materials, 577–89
  NDE, 576
  sensing architectures, 589–95
  sub-surface imaging of FRP RC structures, 602–13
    experimental implementation, 612–13
    experimental results of microwave imaging, 613

numerical simulations, 608–12
UWB tomographic bi-focusing algorithm, 603–7
surface imaging of FRP RC structures, 595, 597–602
experimental implementation, 598, 600–2
field tests at Dang–Jeong overcrossing, 603
focusing beam technique, 597–8
FRP-jacketed concrete column under testing, 600
horizontal cut of the reflection mechanism, 597
images of the FRP-jacketed concrete column surface, 601
images of the GFRP-jacketed concrete column surface, 602
overview, 595
time gating of the reflection signal, 597
techniques, 4
millimeter wave techniques, 511
mobile automated scanner, 420
multiple channel shearography, 64–6
dual-beam shearography system, 65
locations of illumination positions, 66
multiple-input-multiple-output (MIMO), 613
multiplexed sensors, 622–3

near-field approach, 127–8
near infrared light, 518
noise equivalent temperature difference (NETD), 319
Nomex, 248, 370, 416
non-destructive evaluation (NDE), 3–11, 309
acoustic emission (AE) techniques, 557–72
challenges, 566–72
testing acoustic techniques, 559–66
acoustic microscopy for aerospace composites, 423–46
case study of damage analysis using acoustic microscopy, 431–40

case study of damage analysis using scanned image microscopy, 424–31
future trends, 440–4
aerospace composites, 220–36
adhesive bonding in aerospace industry, 221–2
challenges in NDT of adhesive bonds, 233–5
NDT in testing adhesive bonds, 222–7
NDT methods, 227–33
civil structure composites, 483–511
digital tap testing, 507–8
future trends, 510–11
ground penetrating radar (GPR), 501–7
infrared thermography, 484–500, 509
issues and challenges, 508–10
composites, 12–30, 33–54, 56–79, 84–110, 116–33
automotive engineering, 160–78
curing degree and mechanical properties monitoring, 143–5
curing monitoring, 154–60
dielectric testing in bonded joints ageing assessment, 200–16
dielectric testing in bound integrity checking, 193–200
dielectric testing in cure monitoring, 188–93
digital shearography practical application, 98–101
digital shearography principles, 85–97
digital shearography to rest composites, 101–9
eddy current image of dry 3 axial carbon fibre non-woven fabric, 34
eddy current testing (ECT), 35–43
future trends, 28–30, 76–9
monitoring methods, 139–42
online process monitoring using ultrasound, 145–54

practical application of
   shearography, 62–70
shearography, 70–4
shearography vs other techniques,
   75–6
theoretical principles of
   shearography, 57–62
thermosets type, 137–9
damage of tolerance concept, 6
debonding in sandwich panels using
   guided waves, 238–74
  detection and assessment of
     debonding, 252–9
  future trends, 271–4
  numerical simulation of wave
     propagation, 246–52
  processing of wave signals, 240–6
  time reversal for debonding
     detection, 259, 261–71
detecting impact damage on
   aerospace composites, 367–93
  future trends, 392–3
  infrared thermography
     effectiveness, 369–72
  on-line monitoring, 372–84
different composite materials, 384–92
  CFRP materials, 385–8
  GFRP materials, 388–92
  sandwich structures, 392
fibre optic sensor, 617–29
  field applications, 627–8
  future trends, 628–9
  integration of fibre-reinforced
     polymer (FRP) reinforcements,
     624–5
  monitoring FRP-concrete
     interfacial bond behaviour,
     625–7
  technologies, 619–23
flaw characterisation aerospace
   composites, 335–64
  delaminations and planar
     inclusions, 346–54
  experimental demonstration,
     359–63
  fundamentals of heat diffusion,
     337–46

future trends, 363–4
image damage, 354–6
porosity, 356–9
frequency spectrum, 5
future trends, 10–11
history, 575–6
infrared thermography for aerospace
   composites, 309–32
  heat propagation in dynamic
     thermography, 315–20
  overview, 309–15
  thermography in aerospace
     composites, 321–32
infrared thermography for marine
   structure composites, 649–66
  assessing effectiveness, 654–65
  boat hull, 651–4
  infrared thermography, 650–1
infrared thermography for wind
   turbine blades, 634–47
  infrared thermography (IRT),
     636–9
  IRT standards, 645
  quality assurance and structural
     evaluation of GFRP wind
     turbine blades, 644–6
  signal processing techniques,
     639–44
  wind turbines, 635–6
mechanical impedance, ultrasonic
   and IRT techniques for
   detection of delamination,
   279–305
  carbon fibre-reinforced (CFR)
     composites, 292–5
  comparison of techniques, 301–5
  disbonding in aluminium
     honeycomb structures, 281–92
  infrared thermography, 295–301
microwave techniques for
   composites, 574–614
  electromagnetic properties of
     materials, 577–89
  future trends, 613–14
  imaging of FRP RC structures,
     595, 597–602
  overview, 574–6

684   Index

sensing architectures, 589–95
sub-surface imaging of FRP RC
    structures, 602–13
shearography usage to detect
    bonded defects, 542–54
  field inspection of FPR
    strengthened bridge, 552–4
  in-place pull-off method, 544
  inspection with acoustic sounding
    (hammer tapping), 543
  shearography, 544–6
  shearography role in detecting
    defects, 546–51
structural health monitoring of
    aerospace using guided wave
    ultrasonics, 449–68
  GW SHM systems for composite
    structures, 455–66
  SHM transducer systems, 451–5
thermography for defect detection in
    rehabilitated structures, 515–40
  assessing results, 526–39
  data collection methodology,
    522–6
  infrared thermography application
    to bridge deck assembly, 518–22
  infrared thermography principles,
    517–18
ultrasonic techniques for aerospace
    composites, 397–420
  inspection, 398–402
  NDT instruments, 419–20
  ultrasonic inspection methods,
    402–6
  ultrasonic inspection of sandwich
    structures, 415–19
  ultrasonic inspection of solid
    laminates, 406–15
non-destructive inspection (NDI),
    335–7
non-destructive testing (NDT), 637,
    649–50
  adhesive bonds, 222–7
    defects in a bonded honeycomb/
      skin construction, 225
    defects in bonded joints, 223
    defects presentation, 223–4

quality assessment of adhesive
    bonded parts, 227
  quality assessment prior to
    bonding processes, 224, 226
  quality assessment within bonding
    processes, 226–7
adhesive bonds, challenges, 233–5
  bond performance assessment, 234
  extended methods, 234–5
pressure vessels, 105–7
  delamination in a honeycomb
    panel of an aircraft, 106
  glass fibre-reinforced plastic tube
    shearographic inspection, 106
normal incident longitudinal waves,
    402–3, 409
nuclear magnetic resonance (NMR),
    576
numerical simulation, 347
  sub-surface imaging of FRP RC
    structures, 608–12
    geometry of object, 608
    results of simulation, 611
    set-up used for simulation, 609
    simulation geometries, 610

on-line monitoring, 372–84
  damage evolution and heat transfer
    mechanisms, 379–80, 382–4
    thermal images taken at 96 Hz at
      specimen GFRP-G impacted at
      12 J, 382
    variation of warm area with time,
      384
  qualitative analysis, 372–4
    CFRP-F1 impacted at 1.2 J, 373
    GFRP-A impacted at 12.3 J, 375
    GFRP-B impacted at 12.3 J, 376
    GFRP-F1 impacted at 2.7 J, 374
  quantitative analysis, 374–9
    maxima and minima ΔT values,
      378
    maxima and minima ΔT values
      variation with impact energy,
      379
    plots of temperature difference
      against time of GFRP-C, 376

thermal images taken after impact, 380
thermal images taken at 96 Hz at specimen GFRP-F impacted at 12J, 381
online process monitoring
  ultrasound, 145–54
    complete computer controlled measuring system, 153
    compression mould for laboratory testing, 151–4
    controlled ultrasound system US-plus system, 155
    crosslinking testing, 151
    longitudinal waves propagation between sensors, 146
    measured parameters, 147–8
    measuring set-up for hidden ultrasonic sensors, 148
    opened measuring cell, 152
    output signal for different numbers of excitation pulses, 150
    signal detection, 150–1
    signal evaluation, 151
    sound velocity and attenuation of the ultrasonic signal with increasing temperature, 152
    transmitter excitation, 148–50
    ultrasonic measuring arrangement, 146
    ultrasonic sensor, 147
    ultrasonic transmitter, 147
    US-plus system, 154
    working principles and use of transducers, 146–7
open-ended coaxial probe, 586–9
  conductivity of concrete sample, 588
  dielectric constant of concrete sample, 587
  EM properties of FRP sample (carbon/epoxy), 589
  EM properties of FRP sample (e-glass/epoxy), 588
  measurement set-up for dielectric characterisation, 587
optical imaging, 575

optical lock-in thermography (OLT), 370
optical time domain reflectometer (OTDR), 621–2
optimal transducer networks, 464–5
  optimisation for T-shaped plate by global detection and false alarm rate, 466
optimal transducer placement (OTP), 465
out-of-plane sensitive shearography, 62–4
  conventional shearography arrangement, 63

passive non-destructive evaluation, 310, 312
penetration depth
  resolution in consolidated components, 53–4
  characterisation, 53
phase independent filter
  image processing
    EC image on triaxial non-woven fabrics, 51
phase-shifter
  Michelson shearing interferometer digital shearography phase-shift capability, 96
  Wollaston prism shearing device, 94–6
  digital shearography with phase-shift capability, 95
phenolic resins
  curing testing, 155–7
    four termination times the degree of curing (α-DSC) and the HDT, 156
    sound velocity vs process time for compression moulding, 156
    stress–strain curve and the flexural strength, 157
piezoceramics, 451–2
piezoelectric (PZT), 240, 249, 253
piezoelectric transducer systems, 451–2, 460
  SMART Layer and Active Fibre Composites, 452

PiezoSys EPA-104, 254
Plexiglas, 548
ply waviness, 414–15, 415
pneumatic tire
 inspection, 102–3
  fringe patterns revealing debonds using partial vacuum stressing, 103
polyester resins
 curing monitoring, 157–60
polymer matrix composites (PMC), 14–15, 29–30
 acoustic emission (AE) sources, 14–15
porosity, 356–9, 398–9
 ultrasonic NDE, 406–8
pressure vessels
 hydrogen storage, 169–75
  CFK-hydrogen cylinders, 170
  change of sound velocity and DMA storage modulus of prepreg, 171
  change of sound velocity at linear heating and cooling of prepreg, 174
  change of sound velocity by prepreg crosslinking, 174
  mini-compression gadget with two heat regulated plates, 170
  reaction rate vs process time in prepreg crosslinking, 175
  sound velocity of fully cured epoxy resin, 173
  sound velocity of glass fibre prepreg material, 176
  sound velocity vs temperature at linear heating and cooling of prepreg, 172
  storage modulus of fully cured epoxy prepreg, 173
principal component thermography (PCT), 643–4
probabilistic optimisation techniques, 465
pulse thermography (PT), 295–6, 304, 314, 510
 stages, 296

temperature profile image from a thermal camera, Plate I
pulsed phase thermography (PPT), 642–3
PZT-5A, 441

Q-810 Laser Shearography System, 553
quasi-longitudinal wave (QL), 404
quasi-shear vertical wave (QSV), 404

radiative heating, 337
radio frequency identification (RFID), 613–14
radiosity, 650
Raman spectrosopy, 511
Rayleigh wave, 455
 transducer, 415
Redux, 185–6
reflection, 16
 coefficient, 130
 configuration, 590–1
reflectometer, 586
refraction, 16
regions of interest (ROI), 526
rehabilitated structures
 NDE using thermography for defect detection, 515–40
  assessing results, 526–39
  data collection methodology, 522–6
  infrared thermography application to bridge deck assembly, 518–22
  infrared thermography principles, 517–18
rejectable attenuation, 407
resin transfer moulding (RTM), 136, 137
rheology, 139–41
 crosslinking reaction of epoxy resin, 140
 process temperature influence on epoxy resin, 141
RiteLok SL65, 253

sandwich beam, 249–50, 253–4
 excitation frequency of 20 kHz and 150 kHz, 249

sandwich carbon fibre/epoxy (CF/EP)
  detection and assessment of debonding, 252–9
    experimental set-up, 254
    results, 254–9
    sandwich beams, 253–4
  geometry of sandwich beam, 253
  NDE using guided waves, 238–74
  numerical simulation of wave propagation, 246–52
    FEM, 246–8
    results for wave propagation and interaction with damage, 249–52
  processing of wave signals, 240–6
    damage image construction, 242–6
    Hilbert transform, 242
    magnitude of A_0 and S_0 modes, 241
    schematic diagram of experimental set-up, 241
    selection of wave modes and locating of debonding, 240–2
  specimen configurations, 254
  time reversal for debonding detection, 259, 261–71
sandwich plate, 250–2
  excitation of Lamb wave at 150 kHz, 250
  experimental and simulation wave signal at 10 kHz, 252
  wave signal at 10 kHz, 251
sandwich structures, 392, 415–19
  composite sandwich inspection with perforated facesheet
    air coupled ultrasonic scan image printed on transparency, 418
    field inspection of honeycomb sandwich on aircraft, 418–19
    transverse tree-strike damages on Black Hawk rotor blade, 419
  phase image of sandwich panel taken at 0.5 Hz, 393
scanning acoustic microscope (SAM), 424, 426
  mechanical scanning acoustic reflection microscope, 427
scanning electron microscopy (SEM), 424

semi-analytical finite element (SAFE), 462
sensing architectures, 589–95
  2D and 3D imaging, 591
    GFRP-jacketed concrete column, 593
  microwave antennas for NDE of civil engineering structures, 595
    antennas for NDE of civil engineering structure, 596
  real vs virtual array of antennas, 592–4
    2D reconstruction of a CFRP-jacketed concrete column, 594
  surface and sub-surface imaging, 594–5
  transmission vs reflection, 589–91
    generic measurement configurations, 590
    monostatic and bistatic array of antennas, 592
sensing technology, 8
shear horizontal wave (SH), 404, 409
shearography, 4, 56–79, 232–3, 544–6
  experimental results and comparison with FE simulations, 547–51
    applied depressure vs defect diameter for derivative displacement, 550
    concrete slabs with calibrated defects, 548
    experimental set-up used for evaluation, 549
    required applied depressure vs adhesive percentage, 551
    visualisation of strain fields on specimen S1, 549
    visualisation of strain fields on specimen S2, 551
  future trends, 76–9
    principle of the Mach–Zehnder shearing interferometer, 78
  NDE of bonded defects, 542–54
    field inspection of FPR strengthened bridge, 552–4
  non-destructive evaluation (NDE) of composite materials, 70–4
    loading techniques, 72–4

practical application, 62–70
principle of shearographic interferometer set-up, 545
role in detecting defects, 546–51
FE simulations, 546–7
theoretical principles, 57–62
coherent addition of speckle patterns, 58–9
fringe pattern interpretation, 59–62
interferometric speckle patterns correlation, 60
laser speckle, 57–8
optical paths from the point source S to two surface points P and Q, 61
vs other techniques, 75–6
thermography and ultrasound, 77
signal processing, 8, 465–6
pattern recognition scheme using neural networks, 467
wavelet coefficients C of 80 kHz Lamb waves, 467
signal transmission, 8
single-pole-double-throw (SPDT) switch, 426
skin-core debonds
honeycomb materials, 103–5
debond in a filament-winded composite pressure vessel, 105
portable digital shearographic NDT and a honeycomb panel, 104
SMART Layer, 452
solid laminates, 406–15
detection and mapping of layup and stacking errors, 408–13
angular distribution plot of ply interface in CFRP laminate, 411
azimuthal scan of transmitted shear wave amplitude for 24-ply laminate, 412
ultrasonic C-scan and angular distribution plot, 410
detection of ply waviness in laminates, 414–15

air-coupled pitch-catch set-up and ultrasonic pitch-catch scan, 416
impact damage ultrasonic imaging, 413–14
morphology of microcracks and delaminations, 414
ultrasonic NDE of porosity, 406–8
ultrasonic attenuation, 407
Sonix Software, 434
spectral iterative propagation technique (SIP), 583
spectrometric sensors, 620
static loading, 72–3
non-destructive testing of a composite panel using shearography, 73
steady-state limit, 357
Stefan-Boltzmann law, 486
step heating (SH), 314
Stonely wave, 455
stress-coating testing, 4
structural health monitoring (SHM), 3–11, 14, 450
damage of tolerance concept, 6
future trends, 10–11
GW systems for composite structures, 455–66
optimal transducer networks, 464–5
signal processing, 465–6
NDE of aerospace composites using guided wave ultrasonics, 449–68
SHM transducer systems, 451–5
simulating system performance, 461–4
guided wave propagation and scattering, 461–3
measured vs calculated impedance spectra, 464
scatter directivity patterns from explicit finite element simulations, 462
transducer-structure interaction, 463–4
transducer performance and integration, 456–61

electrical admittance vs frequency plots to measure quality of PZT sensors, 457
transducer systems, 451–5
  comparative vacuum monitoring systems, 454–5
  fibre optic Bragg grating systems, 452–4
  piezoelectric transducer systems, 451–2
surface resistance testing, 226

tap testing, 4, 229
temporal phase stepping, 67–9
  procedure followed for shearography, 68
testing acoustic techniques, 559–66
  AE counts vs amplitude, 562–3
    CB specimen and S1.0-1.0 specimen, 563
  AE duration vs amplitude, 565–6
    CB specimen and S1-C1.0 specimen, 566
  AE event counts, 561–2
    vs load-deflection at midspan deflection, 561
  AE frequency vs amplitude, 563–5
    CB specimen and S1-C1.0 specimen, 564
    characteristics of AE parameters, 565
  experiment, 559–60
    definition of damage level of beam, 560
    schematic test set-up, 560
    specimen configuration, 559
textural analysis, 51–3
  fibre orientation measurements, 53
  processed image results, 52
thermal contrast, 318
thermal diffusivity, 317, 338, 345, 358
thermal effusivity, 338, 350
thermal energy in transition, 650
thermal mismatch, 318
thermally induced noise, 15
thermograms, 485

thermographic signal reconstruction (TSR), 344, 641–2
thermography, 4, 310–11, 336
  case studies in aerospace applications, 321–9
    composite propellers for light aircraft, 328–9
    composite with pulled tabs, flat bottom holes and Grafoil, 326
    flat bottom holes of different sizes, depths and substructural elements, 324
    honeycomb sandwich structure composite, 324, 326–8
    interaction of heat front with embedded defects, 327
    LST thermal of a composite structure, 323
    test coupon with flat bottom holes, pillow and Grafoil inserts, 325
    thermal response of test coupons, 327
    thickness variations, embedded defects and substructures, 322–4
    UT and LST scans of honeycomb sandwich panels, Plate V
  fundamentals, 311–13
  other applications, 329–31
    LST technique for delamination detection, 330
    LST thermal images of composite of wind turbine blades, 330
  types of dynamic or active techniques, 313–15
    general classification of techniques, 313
  vs other NDE techniques, 331–2
    LST-scans of the four hardbacks, Plate VI
thermosets, 137–9
  crosslinked network structure of epoxy bisphenol A diglycidyl ether and diamine hardener, 138
  epoxy bisphenol A diglycidyl ether and diamine hardener, 138
3DIMPACT, 383

through-transmission ultrasonic (TTU) test, 403, 415–16
time domain reflection (TDR), 203
time domain spectral element (TDSE) analysis, 463, 464
time gated ultrasonic C-scan, 408–9
time-of-flight (ToF), 256, 258
time reversal
   debonding detection in sandwich panels, 259, 261–71
      generation and acquisition of signals, 262
      wave signals, 261–3
   experimental investigation, 263–4
      experimental set-up and sensor network, 265
      set-up, 264
   results, 264–71
      damage constructed with two actuator-sensor paths, 270
      damage detection using A_0 and S_0 modes, 266–70
      damage identification in CF/EP sandwich panels, 268
      debonding detection using S_0 modes, 270–1
      estimation of debonding, 272
      reconstructed and original signals of S_0 mode, 266
      reconstructed signals of the S_0 mode, 271
      sensor network paths, 269
      time reversal of A_0 modes, 267
      tuning of frequency and number of cycles, 264, 266
transducers, 146–7
   integration, 458–61
      direct and cut-out methods for embedding monolithic PZT sensor, 459
      micrographs of an integrated sensor in glass fibre epoxy, 459
transient thermography, 637
transmission configuration, 590
2D fast Fourier transformation (2D FFT), 49–50

UH Pulse 100, 426
ultra-wideband (UWB) tomographic imaging, 603–7
   scattered field analytical formulation, 604–6
      2D sensing geometry for reconstruction of an object of permittivity, 605
      sensor network for imaging reconstruction, 604
   space and frequency sampling criteria, 607
   UWB frequency combination, 606–7
Ultran NDC700, 434
ultrasonic 'C' scanning
   carbon fibre-reinforced (CFR) composites, 292–5
   results, 292–5
      actual defect sizes vs ultrasonic C scan size, 294
      cross section of delamination defect in honeycomb specimen, 295
      images of plastic insert defects in carbon fibre-reinforced specimens, 294
   test specimen, 292
      composite with implanted delamination defects, 293
ultrasonic inspection, 4
ultrasonic non-destructive testing (NDT) instruments, 419–20
ultrasonic sensors, 440–4
   overview, 440–1
   results, 442–4
      waveform simulations and FFT for generation parallel to panel delamination, 445
      waveform simulations and FFT for piezoelectric generating source, 445
      waveform simulations for two orthogonal orientations, 443
      waveform simulations of Fourier transform for two orthogonal orientations, 444

technical approach, 441–2
  FEM mesh for source placed perpendicular and parallel to plate, 442
ultrasonic techniques
  methods, 230–1, 402–6
    air-coupled ultrasonic testing, 405–6
    effects of anisotropy and acousto-ultrasonics, 404–5
    interaction of shear waves and fibre directions, 403–4
    normal incident longitudinal waves, 402–3
    water-coupled ultrasonic testing, 405
  NDE for aerospace composites, 397–420
    inspection, 398–402
    NDT instruments, 419–20
  sandwich structures, 415–19
    air-coupled ultrasonic inspection of sandwiches, 417
    composite sandwich inspection with perforated facesheet, 417–18
    field inspection of honeycomb sandwich on aircraft, 418–19
    TTU scan, 415–16
  solid laminates, 406–15
    detection and mapping of layup and stacking errors, 408–13
    detection of ply waviness in laminates, 414–15
    impact damage, 413–14
    ultrasonic NDE of porosity, 406–8
ultrasonic testing, 576
ultrasonics, 511
ultrasound
  automotive engineering, 160–78
  curing degree and mechanical properties monitoring, 143–5
    broken carbon composite bicycle fork, 144
    glass transition temperature vs degree of curing for epoxy resin, 143
    impact strength vs process time from UF-moulding compound, 145
    PF-part (Bakelite) in an incomplete state of crosslinking, 144
  curing monitoring, 154–60
    liquid epoxy and polyester resins, 157–60
    phenolic resins, 155–7
  monitoring composites curing, 136–78
  monitoring methods for composites, 139–42
  online process monitoring, 145–54
    complete computer controlled measuring system, 153
    compression mould for laboratory testing, 151–4
    controlled ultrasound system US-plus system, 155
    crosslinking testing, 151
    longitudinal waves propagation between sensors, 146
    measured parameters, 147–8
    measuring set-up for hidden ultrasonic sensors, 148
    opened measuring cell, 152
    output signal for different numbers of excitation pulses, 150
    signal detection, 150–1
    signal evaluation, 151
    sound velocity and attenuation of the ultrasonic signal with increasing temperature, 152
    transmitter excitation, 148–50
    ultrasonic measuring arrangement, 146
    ultrasonic sensor, 147
    ultrasonic transmitter, 147
    US-plus system, 154
    working principles and use of transducers, 146–7
  thermosets type, 137–9
  vs shearography, 99–100
    C-scan ultrasound, 100
universal testing machine (UTM), 559

692 Index

Vallen AMSY4 instrument, 559
vibration, 73–5
    typical time-averaged shearography fringe pattern, 75
vibrothermography, 314, 510
virtual array, 593
visual inspection, 4, 226, 228
vitrification, 193
vulcanisation
    tyre rubber, 177–8
        torque from curemeter and sound velocity, 177, 178

water break test (WBT), 226
water-coupled ultrasonic testing, 405
water uptake
    effect on the adherent dielectric properties, 209–11
        dielectric TDR of different carbon fibre lay-up plate joint, 211
        variation of the mass change and resistivity, 212
wave propagation, 17–19
wavelet analysis, 466
weak adhesive bond, 234
West 401, 653
wind turbine blades, 635–6
    infrared thermography (IRT), 636–9
        active thermography NDT techniques, 638
        condition monitoring, 638–9
        IRT standards, 645
        NDE using infrared themography, 634–47
        quality assurance and structural evaluation of GFRP wind turbine blades, 644–6
        40s heating with 5 lamps and grey phasegram, Plate XV
        signal processing techniques, 639–44
            DAC, 640–1
            PCT, 643–4
            PPT, 642–3
            TSR, 641–2
wiring, 460–1
Wollaston prism shearing device
    phase-shifter, 94–6
        digital shearography phase-shift capability, 95
Woodpecker, 229
wrapped phase map, 69–70
    noise reduction, 70
    process of phase unwrapping, 71

X-ray radiography, 4, 232
X-rays, 576

Printed and bound by CPI Group (UK) Ltd, Croydon, CR0 4YY
01/11/2025
01989662-0004